HANDBOOK OF WEED MANAGEMENT SYSTEMS

BOOKS IN SOILS, PLANTS, AND THE ENVIRONMENT

Soil Biochemistry, Volume 1, edited by A. D. McLaren and G. H. Peterson
Soil Biochemistry, Volume 2, edited by A. D. McLaren and J. Skujiņš
Soil Biochemistry, Volume 3, edited by E. A. Paul and A. D. McLaren
Soil Biochemistry, Volume 4, edited by E. A. Paul and A. D. McLaren
Soil Biochemistry, Volume 5, edited by E. A. Paul and J. N. Ladd
Soil Biochemistry, Volume 6, edited by Jean-Marc Bollag and G. Stotzky
Soil Biochemistry, Volume 7, edited by G. Stotzky and Jean-Marc Bollag
Soil Biochemistry, Volume 8, edited by Jean-Marc Bollag and G. Stotzky

Organic Chemicals in the Soil Environment, Volumes 1 and 2, edited by C. A. I. Goring and J. W. Hamaker
Humic Substances in the Environment, M. Schnitzer and S. U. Khan
Microbial Life in the Soil: An Introduction, T. Hattori
Principles of Soil Chemistry, Kim H. Tan
Soil Analysis: Instrumental Techniques and Related Procedures, edited by Keith A. Smith
Soil Reclamation Processes: Microbiological Analyses and Applications, edited by Robert L. Tate III and Donald A. Klein
Symbiotic Nitrogen Fixation Technology, edited by Gerald H. Elkan
Soil-Water Interactions: Mechanisms and Applications, Shingo Iwata and Toshio Tabuchi with Benno P. Warkentin
Soil Analysis: Modern Instrumental Techniques, Second Edition, edited by Keith A. Smith
Soil Analysis: Physical Methods, edited by Keith A. Smith and Chris E. Mullins
Growth and Mineral Nutrition of Field Crops, N. K. Fageria, V. C. Baligar, and Charles Allan Jones
Semiarid Lands and Deserts: Soil Resource and Reclamation, edited by J. Skujiņš
Plant Roots: The Hidden Half, edited by Yoav Waisel, Amram Eshel, and Uzi Kafkafi
Plant Biochemical Regulators, edited by Harold W. Gausman
Maximizing Crop Yields, N. K. Fageria
Transgenic Plants: Fundamentals and Applications, edited by Andrew Hiatt
Soil Microbial Ecology: Applications in Agricultural and Environmental Management, edited by F. Blaine Metting, Jr.
Principles of Soil Chemistry: Second Edition, Kim H. Tan

Water Flow in Soils, edited by Tsuyoshi Miyazaki
Handbook of Plant and Crop Stress, edited by Mohammad Pessarakli
Genetic Improvement of Field Crops, edited by Gustavo A. Slafer
Agricultural Field Experiments: Design and Analysis, Roger G. Petersen
Environmental Soil Science, Kim H. Tan
Mechanisms of Plant Growth and Improved Productivity: Modern Approaches, edited by Amarjit S. Basra
Selenium in the Environment, edited by W. T. Frankenberger, Jr., and Sally Benson
Plant–Environment Interactions, edited by Robert E. Wilkinson
Handbook of Plant and Crop Physiology, edited by Mohammad Pessarakli
Handbook of Phytoalexin Metabolism and Action, edited by M. Daniel and R. P. Purkayastha
Soil–Water Interactions: Mechanisms and Applications, Second Edition, Revised and Expanded, Shingo Iwata, Toshio Tabuchi, and Benno P. Warkentin
Stored-Grain Ecosystems, edited by Digvir S. Jayas, Noel D. G. White, and William E. Muir
Agrochemicals from Natural Products, edited by C. R. A. Godfrey
Seed Development and Germination, edited by Jaime Kigel and Gad Galili
Nitrogen Fertilization in the Environment, edited by Peter Edward Bacon
Phytohormones in Soils: Microbial Production and Function, W. T. Frankenberger, Jr., and Muhammad Arshad
Handbook of Weed Management Systems, edited by Albert E. Smith

Additional Volumes in Preparation

Soil Erosion, Conservation, and Rehabilitation, edited by Menachem Agassi

Photoassimilate Distribution of Plants and Crops: Source–Sink Relationships, edited by Eli Zamski and Arthur A. Schaffer

Soil Sampling, Preparation, and Analysis, Kim H. Tan

HANDBOOK OF WEED MANAGEMENT SYSTEMS

EDITED BY
ALBERT E. SMITH
University of Georgia
Griffin, Georgia

Marcel Dekker, Inc.   New York • Basel • Hong Kong

FIRST INDIAN REPRINT, 2015

Library of Congress Cataloging-in-Publication Data

Handbook of weed management systems / edited by Albert E. Smith.
p. cm. — (Books in soils, plants, and the environment)
Includes index.
ISBN 0-8247-9547-4 (hardcover : alk. paper)
1. Weeds—Control—United States—Handbooks, manuals, etc.
2. Herbicides—United States—Handbooks, manuals, etc. 3. Weeds-
-Control—Handbooks, manuals, etc. 4. Herbicides—Handbooks,
manuals, etc. I. Smith, Albert E. (Albert Ernest)
II. Series
SB612.A2H36 1995
632'.58—dc20 95-22582
CIP

The publisher offers discounts on this book when ordered in bulk quantities. For more information, write to Special Sales/Professional Marketing at the address below.

This book is printed on acid-free paper.

Marcel Dekker, Inc.
270 Madison Avenue, New York, New York 10016

Printed and bound in India by Bhavish Graphics.

FOR SALE IN SOUTH ASIA ONLY

Preface

The purpose of this book is to develop a basic structure of knowledge required for the development of weed management strategies for major crops grown in the United States. Although a number of excellent reference books have been published on the various aspects of weed science, there is a need for a handbook illustrating weed management systems. This handbook is intended to be a guide for the development of weed management systems for important crops without following the classic format of a handbook that includes numerous tables. This book is written by respected authors who are knowledgeable in their respective disciplines. It is designed to be of use to extension specialists, crop advisors, practitioners of crop management, and scientists worldwide. Additionally, this book is an excellent reference for undergraduate and graduate courses on crop management and weed science.

The first nine chapters include information basic to understanding the dynamics involved in developing the best weed management systems for crops. The final seven chapters illustrate weed management systems for major crops. It may appear that these final seven chapters will become outdated as soon as new herbicides are registered and government regulations are imposed. However, the basic concepts for these management systems will not change for many decades, and this information can serve as a structure for the development of new management systems.

It is not the intent of the authors, the represented organizations, or Marcel Dekker, Inc., to espouse the use of any herbicide that is not approved for a use, in any country, by a controlling agency. Some of the chemicals included in the discussions may not be registered for that use. Furthermore, mention of trademarks or proprietary products is for illustrative purposes only and does not constitute an endorsement or warranty, implied or otherwise, by the authors, represented organizations, or Marcel Dekker, Inc. It is advised that the instructions on the labels of all chemicals must be followed.

Albert E. Smith

Contents

Contributors

Prasanta C. Bhowmik, Ph.D. Professor, Department of Plant and Soil Sciences, University of Massachusetts, Amherst, Massachusetts

S. Wayne Bingham, Ph.D. Professor, Department of Plant Pathology, Physiology and Weed Science, Virginia Polytechnic Institute and State University, Blacksburg, Virginia

Rodney W. Bovey, Ph.D.* Research Agronomist, Agricultural Research Service, U.S. Department of Agriculture, Texas A&M University, College Station, Texas

David C. Bridges, Ph.D. Associate Professor, Department of Crop and Soil Sciences, University of Georgia, Griffin, Georgia

Orvin C. Burnside, Ph.D. Professor, Department of Agronomy and Plant Genetics, University of Minnesota, St. Paul, Minnesota

John Cardina, Ph.D. Research Scientist, Department of Horticulture and Crop Science, The Ohio State University, Wooster, Ohio

---

**Current affiliation*: Adjunct Professor, Department of Rangeland Ecology and Management, Texas A&M University, College Station, Texas

William J. Chism, Ph.D. Associate Research Scientist, Research and Development, Rhône-Poulenc Ag Company, Research Triangle Park, North Carolina

William W. Donald, Ph.D. Research Agronomist, Cropping Systems and Water Quality Research Unit (CSWQRU), Agricultural Research Service, U.S. Department of Agriculture, University of Missouri, Columbia, Missouri

E. Ford Eastin, Ph.D. Professor and Research Leader, Department of Crop and Soil Sciences, University of Georgia, Tifton, Georgia

Susan Erhardt-Zabik, Ph.D. Senior Research Chemist, North American Environmental Chemistry Laboratory, Residue Method Development, DowElanco, Indianapolis, Indiana

Warwick L. Felton, M.Sc., H.D.A. Special Research Agronomist, New South Wales Agriculture Center for Crop Improvement, Tamworth, New South Wales, Australia

S. Kent Harrison, Ph.D. Associate Professor, Department of Horticulture and Crop Science, The Ohio State University, Columbus, Ohio

Patrick L. Havens, Ph.D. Senior Research Engineer, North American Environmental Chemistry Laboratory, Ecosystem Fate and Assessment Group, DowElanco, Indianapolis, Indiana

David L. Jordan, Ph.D. Assistant Professor, Northeast Research Station, Louisiana State University Agricultural Center, St. Joseph, Louisiana

Mark M. Loux, Ph.D. Associate Professor, Department of Horticulture and Crop Science, The Ohio State University, Columbus, Ohio

Lionel D. Martin, B.Sc., Ph.D. Senior Lecturer, Muresk Institute of Agriculture, Curtin University of Technology, Northam, Western Australia, Australia

Ken Mc Nabb, Ph.D. Associate Professor, School of Forestry, Auburn University, Auburn, Alabama

Robert J. Mitchell, Ph.D. Associate Scientist, Joseph W. Jones Ecological Research Center, Newton, Georgia

H. Erdal Ozkan, Ph.D. Professor, Department of Agricultural Engineering, The Ohio State University, Columbus, Ohio

Gerald K. Sims, Ph.D. Microbiologist, Agricultural Research Service, U.S. Department of Agriculture, University of Illinois, Urbana, Illinois

Reid J. Smeda, Ph.D. Plant Physiologist, Southern Weed Science Laboratory, Agricultural Research Service, U.S. Department of Agriculture, Stoneville, Mississippi

Albert E. Smith, Ph.D. Professor, Department of Crop and Soil Sciences, University of Georgia, Griffin, Georgia

David B. South, Ph.D. Professor, Southern Forest Nursery Management Cooperative, School of Forestry, Auburn University, Auburn, Alabama

Robert H. Walker, Ph.D. Professor, Department of Agronomy and Soils, Auburn University, Auburn, Alabama

Leslie A. Weston, Ph.D. Associate Professor, Department of Horticulture and Landscape Architecture, University of Kentucky, Lexington, Kentucky

Gail A. Wicks, Ph.D. Professor, Department of Agronomy, University of Nebraska, North Platte, Nebraska

John W. Wilcut, Ph.D. Associate Professor, Crop Science Department, North Carolina State University, Raleigh, North Carolina

Alan C. York, Ph.D. Professor, Crop Science Department, North Carolina State University, Raleigh, North Carolina

Robert L. Zimdahl, B.Sc., M.Sc., Ph.D. Professor, Department of Plant Pathology and Weed Science, Colorado State University, Fort Collins, Colorado

HANDBOOK OF WEED MANAGEMENT SYSTEMS

1

Introduction

Robert L. Zimdahl
Colorado State University, Fort Collins, Colorado

I. CHARACTERISTICS OF WEEDS

There it is, a plant you don't want in your field, lawn, garden, pond, or some other site where you have decided it shouldn't grow. Why? What's the matter with it? It is growing quite well, probably better than what you planted and want to grow there. However, you don't want it because you regard it as a weed and the decision is yours. A weed is a plant growing where it is not desired (Buchholtz, 1967). The definition is clear. It is also clear, on examination, that the burden and responsibility for defining what plants are weeds rest with people who decide when a plant, in a particular place and time, is not desired. Human attitudes play a fundamental role in determining what plants are weeds. Weeds share the characteristic of being unwanted by some people.

Not everyone agrees that a particular plant is a weed.The plants someone doesn't want growing in a front lawn may be valued or ignored by a neighbor, who doesn't regard them as undesirable or weedy. Most cattle ranchers know that leafy spurge is a weed. It is an aggressive perennial plant that reproduces by seed and vegetatively and can dominate rangeland and pastures. Cattle don't eat it. The sheep rancher, on the other hand, may know it is a weed but doesn't regard it with the same disdain because sheep readily eat leafy spurge.

Do these plants we label as weeds have some characteristics in common other than a human attitude toward them? There are many weedy plants. The Weed Science Society of America recognizes no fewer than 2076 plants as weeds and the list is growing (*Composite List*, 1989). They share some traits, but not all weeds have all the traits of weediness. Among the shared traits is the fact that weeds often grow better than what was planted or what is desired and they are difficult to control. Weeds are successful and often succeed precisely where we don't want them. They interfere with human activities.

It is not just location—a plant growing where it is not desired—that defines what a weed is; it is something it does. It interferes with our activities, whether growing a crop, having a weed-free flower bed or lawn, or having a pond free of aquatic weeds.

To define a weed we need to understand the characteristics that allow them to be where we don't want them and to interfere with our activities. We might be able to tolerate a plant that is present but doesn't interfere with anything.

All weeds share some characteristics that make them plants capable of growing where they are not desired and that make controlling and managing them difficult. Not all weeds have identical characteristics but all share some weedy traits. The characteristics are usually divided into those related to 1) weed physiology, growth, and competitiveness; 2) reproduction; or 3) cultural practices (see Chapter 2 for a complete discussion).

II. HISTORY OF WEED MANAGEMENT

Weed management, a relatively new concept in weed science, has not yet achieved the status of having an official definition. Integrated pest management, developed for insect control, is older; it is

> a pest control strategy based on determination of an economic threshold that indicates when a pest population is approaching the level at which control measures are necessary to prevent a decline in net returns—that is—when the predicted value of the impending crop damage exceeds the cost of controlling the pest. (*Composite List*, 1989)

Weed control, on the other hand, has been defined as "the process of reducing weed growth and/or infestations to an acceptable level" (*Herbicide Handbook*, 1989). Control rather than management is what weed science has been about almost since its inception. The history of weed science is not the history of weed management; it is the history of weed control. This is not to say or imply that weed science is diminished because of its history. It is a recognition of reality. The reality is perhaps best exemplified in a statement by the eminent California extension weed specialist W. A. Harvey. Writing in the introductory chapter to *Principles of Weed Control in California*, Harvey said, "All (in weed science) before 2,4-D became ancient history" (Harvey, 1985).

Any new science is built upon preceding science. Indeed, weed scientists can now see so far because they stand on the shoulders of giants.* However, weed science was not created with a philosophical foundation based on the concepts that will surely be included in the eventual definition of weed management. Weed science was created with a pragmatic, problem-solving orientation. Weed scientists proudly and justifiably claim to be problem solvers and have solved many problems. Harvey (1985) suggests the process of solving those problems helped weed control evolve into weed science. Another way to think about the evolution of weed science is that it has moved from "how" questions to "why" questions.

Available historical accounts of weed science are accurate and are, almost exclusively, histories of chemical weed control (Crafts, 1960; Smith and Secoy, 1975 and 1976; Timmons, 1970; Upchurch, 1969; Zimdahl, 1983). Upchurch (1969) said that "one approach to studying the beginnings of weed control is to examine the appearance of various herbicides." He is correct, but his approach is the one taken by all writers. Weed science began with herbicides, the technical ability to control weeds, not with a concept of the need to manage crops and their associated weeds.

Many writers cite the early work of Bolley (1908) (a plant pathologist) in North Dakota and the nearly concurrent work of the Frenchmen Bonnet, Martin, and Duclos and the German Schultz (Crafts and Robbins, 1962). Each of these men used solutions of copper salts for selective weed control in small grains; later iron sulfate and sulfuric acid were used.

Succeeding work in Europe observed the selective herbicidal effects of metallic salt solutions or acids in cereal crops. Names encountered frequently include Rabate in France (1911), Morettini in Italy (1915), and Korsmo in Norway (1932). More recent historical accounts almost always cite the synthesis of 2,4-D by Pokorny (1941), the discovery of its growth-regulating properties

---

*This saying is usually attributed to Sir Isaac Newton, who, in a 1675 letter to Robert Hooke, said, "If I have seen further (than you and Descartes) it is by standing upon the shoulders of Giants." It is also traceable to earlier sources, e.g., Lucan (A.D. 39–65), Coleridge, "The Friend" (1828), or R. Burton.

(Zimmerman and Hitchcock, 1942), and the first field reports of its herbicidal activity (Hamner and Tukey, 1944a,b; Marth and Mitchell, 1946). At the same time, work was under way in England on the herbicidal properties of MCPA (Slade et al. 1945). A short but complete history of the development of the growth regulator herbicides has been written (Kirby, 1980). The first paper on a nonphenoxy acid derivative useful as a herbicide was by Bucha and Todd (1951), who reported the herbicidal properties of monuron, the first of many phenylurea herbicides. The date of the first U.S. patent for several other herbicides of current or historical interest is shown in Table 1. The tabulation does not attempt to list all herbicides ever patented or even most of those patented after 1970. It shows that almost no important herbicides were patented prior to 1940 and most were patented after 1944–1945, with a great surge of development in the 1950s and 1960s, a time coincident with the development of weed science.

Weed science textbooks almost universally note that weeds have been with us since settled agriculture began. Smith and Secoy (1976) state that the harmful effects of weeds were known to early historic humans and cite, as evidence, the hoes and grubbing implements that have been found. Similar tools are still used for weeding and other tasks in many parts of the world and the conclusion is logical. They (Smith and Secoy, 1976) also found that nearly all books from the time of Theophrastus (372?–287? B.C.) to the modern period have mentioned weeds and their detrimental effects. Familiar examples of humans' battle with weeds are easy to find. The thistles and thorns of Genesis (3:17–18) and the parables of the sower (Matt. 13:17–18) and the tares (Matt. 13:25–30) are among the earliest references.

The importance of weeds is immediately obvious in many third world countries. There, a farmer is a person with a hoe, and weeding is part of the burden of agriculture. It is what farmers do. Weed control, if achieved, is an incidental part of production. Weeding is something the world's developing country farmers must always do; it is not a task to be completed. If control technology has changed for most of the world's farmers in developing countries, it has progressed from primitive hand tools to animal powered implements. Weed science, that enabled planned weed control and the thought of annual completion of the task, really began after World War II with the discovery of the herbicidal properties of the phenoxy acid herbicides.

There were herbicides before 2,4-D but none as cheap, as effective, or as selective. Crafts (1960) traces the history of herbicide development from the early workers mentioned to the 1920s discovery of the field bindweed control potential of a dilute solution of sodium arsenite and its apparent translocation in plants. In Colorado, carbon bisulfide was applied as a soil fumigant to control *Phylloxera*, the root-borne disease of grapes. Carbon bisulfide had been used

Table 1 United States Patent Date of Some Important Herbicides

Common name	Patent date	Common name	Patent date
Pre 1940		*1960s*	
Copper sulfate	1905	Linuron	1960
Borax	1920	Fluometuron	1960
Sodium chlorate	1927	Dichlobenil	1960
Pentachlorophenol	1936	Chlorbromuron	1961
		Dicamba	1961
1940s		Pyrazon	1962
DNBP	1940	Prometryn	1962
2,4-D	1942	Bromoxynil	1963
2,4,5-T	1944	Bromacil	1963
Propham	1945	Picloram	1963
Ammonium sulfamate	1945	Diphenamid	1964
MCPA	1945	Ametryn	1964
TCA	1946	Propachlor	1965
Naptalam	1949	Cycloate	1965
		Terbacil	1965
1950s		Benefin	1966
Chlorpropham	1950	Isopropalin	1966
Monuron	1951	Oryzalin	1968
2,3,6-TBA	1952	Oxadiazon	1968
Dalapon	1953	Trifluralin	1968
Propanil	1953	Alachlor	1969
Diuron	1954	Metribuzin	1969
Amitrol	1954		
2,4,5-TP	1954	*1970s*	
MCPB	1955	Phenmedipham	1970
DSMA and MSMA	1956	Cyanazine	1970
Simazine	1956	Profluralin	1970
Amitrol-T	1956	Norflurazon	1971
CDAA	1957	Dinitramine	1971
Mecoprop	1957	Dalapon (Mg salt)	1972
4(2,4-DB)	1957	Bifenox	1972
Cacodylic acid	1958	Difenzoquat	1972
Chloramben	1958	Bentazon	1973
Barban	1958	Triclopyr	1973
Paraquat and diquat	1958	Chlorsulfuron	1978
Acrolein	1959		
EPTC	1959	*1980s*	
Dipropetryn	1959	Sethoxydim	1981
Butylate	1959	Tebuthiuron	1983
Triallate and diallate	1959	Sulfometuron	1983
Atrazine	1959	Imazapyr	1989
Propazine	1959	Imazethapyr	1989
Prometon	1959	Imazaquin	1989

first as an insecticide, in 1854, in France. The Colorado researchers noted that this chemical killed field bindweed, and it was subsequently developed as a soil fumigant for weed control. It is said that over 300,000 acres was treated in 1 year in Idaho.

Crafts (1960) outlined the early work on sodium chlorate, sodium arsenite, and sodium dinitrocresylate. These accounts confirm that most of the history of weed control is coincident with herbicide development and is understood as the history of chemical control. Other methods were, and still are, widely used, and the need for control has long been widely acknowledged. The introduction of the ability to control weeds selectively provided the stimulus for development of what we now know as weed science.

Timmons (1970) noted that state weed laws were first enacted between 1721 and 1766, but they were not weed laws as we presently know them. The early laws were designed to control plant diseases spread by weeds that often serve as alternate hosts. There were only a few Extension Service publications related to weeds or their control between 1860 and 1900 (Timmons, 1970). Few other publications about weeds were added until the 1950s. People who worked on weeds were minor actors in departments of agronomy, botany, horticulture, and plant physiology until the mid-1950s (Timmons, 1970). The journal *Weeds* (*Weed Science* since 1968) was not founded until 1951 and the Weed Science Society of America was founded in 1956. The origin and evolution of the North Central Weed Control Conference (Andersen, 1991) and the Western Society of Weed Science (Appleby, 1993) are new contributions to the story of the development of weed science societies. The primary European weed journal, *Weed Research*, was not begun until 1964.

Weed science shares with entomology the historical tradition that the pragmatic, applied sides of both were created in the United States. Although the history of weed science is not as old as that of entomology, they share the fact that an applied orientation has been highly effective. Weed scientists have met the challenge of solving many difficult weed control problems in a wide range of cropping situations. The scientific study of weeds may be comparatively young, but it is resplendent with the achievements of many dedicated workers. Domination of the science by pragmatism has yielded a nearly unparalleled record, within the community of agricultural scientists, of successful problem solving.

Thus, it may be concluded that weed science is young, has been dominated by pragmatists and problem solvers, and that its development has been closely tied to the rise of chemical weed control with herbicides. During the last decade, within and outside the community of weed scientists, there has been increasing attention to the development of systems of weed management stimulated, in part, by the problems created by herbicides. Weed scientists are beginning to deemphasize complete control of weeds and to conceive of weed management as a problem in applied ecology. Herbicides are recognized as only one technology useful in weed management programs. There has been increasing recognition of the need to treat weeds as part of the farming system rather than a problem to be solved independently of the system. Weed scientists have not forgotten the need to control, but it has been subsumed under the concept of management.

For many years, weed scientists have debated the importance of achieving 100% weed control. It is not, per se, a bad idea. The need for 100% control is based on the logical, and easily verified, assumption that if weeds are allowed to go to seed or develop vegetative reproductive structures future weed problems are assured. Solving (controlling) this year's weed problems reduces the need for equally strenuous measures in future years. Therefore, the closer one can come to eliminating production of weed seeds and growth of perennial reproductive parts, the better. Control of only 50% of the weeds in a field is uniformly regarded as unacceptable by weed scientists and growers. Good control is 85% or greater. It was, and in some minds remains, a sound, logical approach to minimizing the grower's weed control problems.

There is another school of thought that seems to find its foundation in the idea that weeds will always be with us. It is also a sound, logical proposition. Most weed scientists would testify to the omnipresence of weeds in agriculture, forestry, aquatic environments, or any other environment humans alter to suit their perceived needs or wants. As surely as night follows day, weeds will be there when the environment is modified by humans. If this is true, and it seems to be, then why not learn to live with them? Why not manage weeds, presumably to achieve an acceptable (manageable) population level, rather than direct maximum effort to control a high percentage or all of them? Control is straightforward. The goal is obvious—eliminate as large a number or percentage of the population as possible, without allowing the cost of purchase and application of control measures to exceed the benefit of increased yield from reduced weed competition.

Management is more subtle and harder to design and achieve. The goal is not completely clear. To manage is not to control or eradicate. Growers may have to live with some weeds. How many? Which ones are easiest to live with? We all know that some neighbors are easier to tolerate than others. How do we decide, before they move in, and what do we do about them after they have arrived if they aren't good neighbors? We really can't have zoning laws for weeds. We have to manage our relationships. Weed management becomes a type of social problem and those have always been, and probably always will be, more difficult than strictly scientific problems that can be solved by straightforward application of technology. Weed scientists may not have to know more to manage weeds than to control them, but they will have to know different things, presently unknown things.

III. THE SYSTEMS CONCEPT FOR WEED MANAGEMENT

Weed scientists and many other people know what weed control is. We do not yet know what weed management is. We know it will include the dictionary sense of the term: "taking charge of and directing" the growth of weeds. It will also include handling carefully, a rarer definition of management that includes the concept of husbandry, i.e., managing economically, conserving.

Weed management has been defined as "an environmentally sound system of farming using all available knowledge and tools to produce crops free of economically damaging, competitive vegetation" (Fischer et al., 1985). This definition differs from the definition of integrated pest management in that it lacks the specificity of an economic threshold and advocates no yield loss (i.e., no economic damage). It could be interpreted as bordering on advocacy of a high level of weed control because it mentions crops free of economically damaging, competitive vegetation.

Fryer (1985) defined weed management as the "rational deployment of all available technology to provide systematic management of weed problems in all situations." Unfortunately there is not universal agreement on what is rational or systematic.

Weed management will be a systematic approach to minimize weed impacts and optimize land use, and it will be an approach that combines prevention and control (Aldrich, 1984). Weed management is going to emphasize minimizing the effects of, but not eliminating, weed populations. Apparently, we are going to learn to live with weeds. They will become part of the community, albeit a part most folks wish would have settled somewhere else. The central objective of an effective weed management program will be to manipulate the crop-weed relationship so that growth of the crop is favored over that of the weeds.

Weed scientists would like to incorporate weed population thresholds in the definition and practice of weed management but don't because they are not known. An economic threshold is the weed density at which control measures should be employed to prevent the weed population from causing economic damage. A damage threshold is the population level at which the crop begins to suffer losses in yield, but the losses are not high enough to justify control measures. Regression analysis of experiments designed to measure weed-crop competition have shown that losses can be measured with extremely low weed populations, and common sense tells us that the cost of controlling a small population is well below the expected return. Weed scientists lack the research data to establish the economic or damage thresholds for the many possible weed-crop combinations. There are some noteworthy research efforts in this arena, especially those of Harold Coble, North Carolina State University; E. E. Schweizer, U.S. Department of Agriculture at Colorado State University; and Steve Radosevich, Oregon State University.

When the basis for weed management is fully developed it will surely include the following components:

1. Incorporation of ecological principles
2. Full utilization of plant interference and crop-weed competition
3. Definition and incorporation of economic and damage thresholds
4. Full integration of several weed control techniques including selective use of herbicides
5. Supervised weed management, probably by a professional pest manager employed to develop a program for each farm

The ecological principles essential to successful weed management will capitalize on natural weed mortality factors including interspecific and intraspecific competition, plant density, and genetic manipulation. The latter may develop weed populations that are less competitive or even complementary to the crop. Genetic manipulation also may develop populations that are more susceptible to standard control techniques such as tillage or use of herbicides. Successful weed management will incorporate precise timing of cultural practices such as tillage to maximize its benefit and careful selection of the rate and application time for herbicides.

The developing concept of weed management may one day trace its history to a seminal paper by Warren Shaw (1982), who played a key role as a visionary thinker, proponent of action, and planner for weed science during his 35-year career with the U.S. Department of Agriculture. For many years, he was the principal spokesman for weed science as a member of the National Program staff of the Agricultural Research Service of the USDA in Washington, D.C. Shaw's paper was one of nine presented at a symposium on integrated weed management systems held during the 1981 annual meeting of the Weed Science Society of America (Shaw, 1982). Others have published their concepts of weed management (Baldwin and Santleman, 1980; Buchanan, 1979; Fryer and Matsunaka, 1977; Harwood and Bantilan, 1974; Knake, 1978). However, because of Shaw's position, the respect he had earned, and national journal exposure, his paper may properly be regarded as a beginning.

Shaw (1982) foresaw the need for what he called integrated weed management systems. He knew that "most of the scientific disciplines that include the study of pests claim, with some justification, to have developed the "integrated pest management (IPM) approach." Shaw (1982) noted that senior leaders in "disciplines that study pests say that pest management is not new; we've been managing pests for years." The title of Fryer's (1985) article implies clearly that weed management is not a new concept. Shaw (1982) would have regarded this as an accurate interpretation but saw that we were entering a new era where the methods had to fit together in a system. He said that when weed scientists used the term integrated weed management systems (IWMSs) they meant "a directed agroecosystem approach for the management and control of weed and other pest populations at threshold levels that prevent economic damage in the current and future years." Here we see the inclusion of an agroecosystem approach, management *and* control, and threshold population levels. Shaw's (1982) approach used all available technology and he foresaw the need for new technology. The concepts of management and control were mixed, but he explicitly advocated knowledge of threshold levels before control measures were used. IWMSs were to integrate "effective, dependable, and workable weed and crop management practices into crop and livestock production systems that can be used economically by producers." IWMSs were not to replace selective, safe, and efficient herbicides. Herbicides were and always would be part of IPM and IWMSs. The objectives of IWMSs were to reduce 1) losses caused by weeds, 2) costs of control, 3) energy and labor requirements for weed control, and 4) tillage and consequent soil erosion caused by water and wind to assure an adequate supply of high-quality food, feed, and fiber, safeguarded against poisonous weed seeds and contaminants. Their goals were to improve environmental quality, to reduce hazards to man, and to maximize producer profits. This is a large order for any system, but it was and remains an appropriate agenda for action on IWMSs.

In 1982, when Shaw's paper was published, weed scientists didn't really subscribe to working in the system known as agroecology. Few people worked on and understood what agroecology was all about. Weed scientists, like their colleagues in other agricultural disciplines, were working within the prevailing system of agricultural practice, but they were studying it according to the dictates of Cartesian logic. René Descartes (1596–1650) the French philosopher and mathematician, taught that the thing to do with complexity was to disassemble it into its component parts and study them, each in its turn. His second rule of reasoning dictated that one was to divide each difficulty into as many parts as possible and necessary to solve a problem. This became the basis for the reductive science and reductive reasoning that characterizes the Western intellectual tradition (Checkland, 1981). Weed management demands integration and combination, not just division into component parts. The agricultural system has to be examined in pieces and as a whole. The great problem-solving ability of weed science had been derived from reductive analysis, and it was not to be abandoned easily. Weed management and the development of weed management systems demand that all parts of the agricultural system—crops, livestock, machinery, irrigation, fertilization, economics, and, perhaps of greatest importance, people and the environment—be considered.

Weed scientists have been prisoners of a frame of reference dominated by weed control. They are being asked to and are leaving the bounds of that frame of reference and moving to a holistic view of their discipline. Checkland (1981) pointed out that concepts such as weed management are frames of reference but are larger and more acceptable because they are holistic. A system to study anything resembles an epistemology, a way of thinking, and it can emphasize the whole or the parts (Checkland, 1981). A holistic concept is explicitly non-Cartesian. Holistic weed management approaches combine scientific study of weeds emphasizing why weeds behave as they do with weed technology and its emphasis on what can be done to solve weed problems. Weed management systems will use what is known *and* reveal what still must be learned. Systems thinking will define what it is we still need to know, how we should proceed to acquire the knowledge, and finally how the knowledge should be applied.

Weed management is developing as part of an agricultural system that will be governed by biology and ecology as well as chemistry and technology. Its eventual form is unknown, and weed scientists are now developing total weed management systems for various crops. However, the questions or research areas are clear. The first question may be properly placed in the category of megaquestions: questions that cannot be answered by a simple experiment or even a series of experiments. Such questions determine what experiments will be done, and Soule et al. (1990) have formulated these questions well. In relation to weed management, they ask whether rather than looking for ways to kill weeds without exerting selection pressure for resistance, we might do better to turn the question around and ask, How can we successfully do agriculture without having to kill weeds? Soule et al. (1990) do not disparage weed control. They change the scope of research by changing the frame of reference from control to management.

A megaquestion is interesting and challenging but really doesn't reveal what to do. What must be known are the subsidiary questions that will govern the development of weed management.

First among the important questions is one the major professor tries to help the graduate student ask. It is a primary question when one develops a research project and one that demands careful thought: What is the objective of the study?

Agriculture is a form of engineering (Weiner, 1990) designed to produce food and fiber. In the world's developed countries agriculture has been a system characterized by maximization of short-term yield. Weed scientists have contributed to achieving this, frequently unstated, objective. The shift to a weed management focus provides a chance to ask whether that is the correct objective for the future. Weiner (1990) defined three choices:

1. Maximum short-term yield
2. Maximum sustainable yield
3. Maximum yield stability (minimum risk)

Maximum yield stability characterizes third world agriculture, a system dominated by low, reasonably stable (barring environmental disasters such as floods or drought) long-term production. Weiner (1990) and Jackson (1984) suggest that the first choice "requires high input costs, high environmental costs, and high nutrient and capital fluxes." Both authors advocate low input, low environmental cost, and low nutrient and capital flux agricultural systems. It is apparent that weed scientists are moving toward the second choice. The choice and clarity of the objective of any weed management strategy should be clear and determining that objective is a crucial decision.

Answering the megaquestion and defining the objective of any research program are important steps, but when both tasks are completed one still may not know what to do. Aldrich (1984) and Altieri and Liebman (1988) provide guidance on appropriate objectives for weed management programs. Their ideas are similar but not identical. The dissimilarity is undoubtedly due to their different experiential origins. Aldrich was a weed scientist for many years and served as the editor of *Weed Science*. Altieri is an entomologist with the University of California and a leader in the emerging field of agroecology. Liebman is working to develop low input, locally adapted, economically feasible, productive cropping systems at the University of Maine.

The major difference in their positions is that Aldrich (1984) argues that if we expand our knowledge in several areas it will be possible to develop "fully effective" weed management systems. Altieri and Liebman (1988) state that it is impossible "to know everything about a weed species in order to develop an effective weed management system." They argue that our inability to know everything does not preclude knowing enough to develop "practical weed suppression in an agroecosystem." Both acknowledge the need and the developing ability to manage weeds effectively. Aldrich (1984) projects more faith in our ability to be fully effective, whereas Altieri and Liebman (1988) argue that nature is always going to be a step ahead of us. Ecological wisdom sides with the latter view, but pragmatic weed scientists, skilled at problem solving, will continue to develop systems that work.

Aldrich (1984) begins his recommendations with a plea for more precise knowledge about weed-crop competition. We do not know, except in the most general terms, what plants compete for and when competition is most severe. He wonders why "are crop yields sometimes reduced when weeds are present for only 2 or 3 weeks after crop emergence." Aldrich emphasizes the need to know precisely what is being competed for at particular points in the development of the crop and the weed(s). Altieri and Liebman (1988) agree with Aldrich (1984) and point out that competition, in any cropping system, depends on the crop and weed(s) species: their relative proportion, time of emergence, spatial distribution, relative growth rates, and environmental constraints. They also note that reduced crop growth is not always due to competition because of allelopathy, an acknowledged biological reality that has not been consciously exploited for weed control. Weed scientists readily agree that competition is dependent on the factors they mention. They also know that much research needs to be done to use these factors in weed management.

The critical weed-free period (Zimdahl, 1988) is known approximately for a few crop-weed situations but not for all. Altieri and Liebman (1988) suggest that critical periods should be determined for individual crop-weed associations and that they may change with environmental conditions. They suggest that there are "thresholds of weed tolerance that may have to be adjusted when criteria besides crop loss are used to measure the economic and ecological effects of weeds." They do not say, but I assume, that other criteria might include reducing soil erosion, changing insect or disease relationships, reduction of purchased inputs, or achievement of long-term system sustainability.

Altieri and Liebman (1988) state that "crop monocultures rarely use all moisture, nutrients, and light available." This creates ecological niches where weeds can succeed. Weed management systems must recognize this and be designed to prevent weed growth. If we grant the initial premise that there are more of the primary elements for which plants compete than they need, can weed management systems be designed to reduce the surplus and thus the opportunities for weed success? This combined preventive and curative approach is acknowledged as the best strategy in human medicine but is not widely practiced in agriculture.

The second major research area advocated by Altieri and Liebman (1988) is to study the influence of the cropping pattern. They argue that the cropping pattern has a powerful effect on reducing weed densities by preemption of resources. They also argue that weed control practices cause shifts in the composition, density, and spatial distribution of weed species. The concern about herbicide resistance is sufficient evidence to validate their point. In addition weed scientists are well aware of shifts in weed populations that have been caused by the success of herbicides. Replacement of annual broadleaf weeds by wild oats after extensive (and intensive) use of phenoxy acid herbicides and the subsequent succession of other annual grass weeds after wild oats were controlled further illustrate the point. They advocate research on rotation of crops with different life cycles and cultural practices as a means of combating and preventing this problem.

Aldrich (1984) advocates research on ways to predict future weed problems. He correctly notes that any production system will develop a unique set of weeds and that prediction of future levels of infestation from soil seed bank size, survival rate, and replenishment rate must be part of a successful system. At this point the approach of the two reports diverges. Aldrich wants this information to predict thresholds and implement the right plan to "deal with the weed problem before it becomes severe." Altieri and Liebman (1988) are closer to a preventive strategy, while Aldrich leans toward managed, albeit reduced control. Both are reasonable approaches to weed management.

Aldrich (1984) advocates much more information on dormancy of seed and perennial reproductive structures. Altieri and Liebman (1988) say the long-term success of any weed control program depends on understanding the mechanisms that regulate population size. Dormancy is a major regulator of seed banks, but populations consist of the live plants that are regulated by a "combination of factors (predation, parasitism, environmental stress, herbicides, interference, etc.)," and a particular factor may be effective during several stages of the life cycle. These authors are appealing for the same type of research, but the latter request is broader, with its emphasis on studies of population biology.

Both presentations make another point in different ways. Aldrich (1984) asks when control is necessary. His question concerns thresholds and could be framed as the farmer's question, Should I control? Altieri and Liebman (1988) ask the same question differently. They begin by stating that the crop-weed balance in any agroecosystem is determined by physical, biological, and cultural management factors *and* (emphasis added) that this balance influences crop and weed yield. They point out that in mixed cultures, complex cropping patterns (mixed cropping) aid weed competition, whereas in intensive, monocultural cropping systems that characterize developed country agriculture "the weed community is a direct product of the crop and its management. If the crop-weed balance is determined by a multiplicity of factors, the appropriate research question becomes how these are balanced to minimize weed impacts. The farmer's question remains as before, Should I control?, with the implied secondary question, At what weed population? Both authors affirm the need to control in any management process. Altieri and Liebman (1988) advocate recognition and use of the cropping system as a control technique in the management program. Their advocacy of preventive techniques extends to their final recommendation for evaluation of what they call direct and indirect control. Direct control is physical or chemical removal. Indirect control includes techniques designed to give the crop an advantage such as adjusting planting date, crop plant population, or rotational sequence.

Aldrich (1984) concludes his research recommendations with a plea for ways to evaluate the effectiveness of long-term programs. He includes questions such as, What is it worth to have a competitive crop variety developed by the plant breeders and What is it worth to have an allelopathic variety developed? His emphasis on valuation of techniques or strategies demands economic analysis, and thus cooperation with economists will be essential for full development of weed management strategies. This thinking is consistent with a holistic approach to agricultural problems. The task will not be just to solve weed problems but rather to develop techniques that can be fully integrated into a farm management scheme that optimizes production and maximizes profits.

Each of these approaches is consistent with 10 steps cited by Altieri and Liebman (1988) but probably attributable to Ennis et al. (1975). Ten key words describe the recommended process and they are a good guide for a weed management program.

1. *Monitor* seed and vegetative populations.
2. *Identify* problem weed species and their density.
3. *Study* the farmer's present methods.
4. *Assess* the dominant weed species and their interactions.
5. *Predict* weed populations and population shifts.
6. *Decide* whether control should be done.
7. *Choose* the control technology compatible with the system.
8. *Consider* the farmer's needs and resources.
9. *Integrate* weed and other crop protection measures.
10. *Evaluate* long-term environmental, social, and economic impacts.

Even a casual reading of these steps will lead to the conclusion that weed management is a holistic approach to weed problems *and* that weed research will become more multidisciplinary than it has been. Weed scientists cannot solve these problems alone, nor can anyone else, and weed scientists should not try to act alone.

Weiner (1990) argues that plant population biology has not contributed much to agriculture. He says this is partially because it has not been given much opportunity. In retrospect, it is an intuitively natural liaison that has not been promoted by either group. Agriculturalists have not looked to those who study population biology to solve their problems. Population biologists and ecologists have emphasized study of "natural systems," and if they have thought about agriculture they have regarded it as inherently unnatural. The link has not been promoted by funding agencies, where programs for research in agricultural ecology do not exist (Weiner, 1990). Liaison between population biology and weed science may be fundamental to development of appropriate weed management systems. Agriculturalists may have to initiate the projects and establish the linkages to benefit from those things the ecologists know and can do for agriculture. Ecologists can assist weed scientists in determining (Altieri and Liebman, 1988; William, 1981; Norris, 1982) 1) the factors governing weed numbers, 2) when weeds are most vulnerable to control, 3) weed response to control practices and cropping patterns, and 4) the relationships among weeds, crops, and other species.

These cooperative activities are consistent with the recommendations of Aldrich (1984) and Altieri and Liebman (1988) described. They are also consistent with the beliefs and actions of many weed scientists who are developing a philosophy of weed science that includes an array of strategies and the necessary fundamental research base to implement them.

It is the purpose of this book to set forth the current thinking on weed management techniques and strategies for a wide array of crops and cropping systems. It is the author's hope that the book will serve as a valuable information source for field practitioners of weed science and as a reference source for researchers. The authors hope the book will become a basic reference text in the development of weed management systems that will replace the now-dominant weed control technology.

The task will be difficult because of the complexity of biological systems. Weed scientists will have to change the way they think about weeds as weed control is replaced by weed management. They will have to cooperate and actively seek the knowledge other disciplines can bring to bear on their task. Some cherished dogmas will have to be abandoned. Surely they will have to learn to live with weeds in crops and the best fields may not be weed-free. The emerging challenge will be met. Difficult tasks have not deterred weed scientists in the past.

APPENDIX

List of Plants in Chapter

Common name	Scientific name	BAYER code
Wild oats	*Avena fatua* L.	AVEFA
Wild oats	*Avena* spp.	
Field bindweed	*Convolvulus arvensis* L.	CONAR
Leafy spurge	*Euphorbia esula* L.	EPHES

REFERENCES

Alternative Agriculture (1989). National Academy Press, Washington, D.C.

Andersen, Robert N. (1991). *The North Central Weed Control Conference: Origin and Evolution*, North Central Weed Science Society, Champaign, Illinois.

Aldrich, R. J. (1984). *Weed-Crop Ecology: Principles in Weed Management*. Breton Publishers, North Scituate, Massachusetts.

Altieri, M. A. and Liebman, M. (1988). Weed Management: Ecological Guidelines, in: *Weed Management in Agroecosystems: Ecological Approaches* (M. A. Altieri and M. Liebman, eds.), CRC Press, Inc., Boca Raton, Florida, pp. 331–337.

Appleby, Arnold. P. (1993). *The Western Society of Weed Science 1938–1992*. Western Society of Weed Science, Newark, California.

Baldwin, F. L. and Santleman, P. W. (1980). Weed science in integrated pest management, *Bioscience 30*: 675–678.

Bolley, H. L. (1908). Weeds and methods of eradication and weed control by means of chemical sprays, *N. Dakota Agric. Coll. Expt. Stn. Bull.* No. 80. p. 511–574.

Bucha, H. C. and Todd, C. W. (1951). 3-(p-chlorophenyl)-1,-1-dimethylurea: a new herbicide, *Science 114*: 493–494.

Buchanan, G. A. (1979). Role of weed management in cotton production, *Proc. Beltwide Cotton Prod. Res. Conf.* p. 246–248.

Buchholtz, K. P. (1967). Report of the terminology committee of the Weed Science Society of America, *Weeds 15*: 388–389.

Checkland, P. (1981). *Systems Thinking, Systems Practice*, John Wiley & Sons, New York, p. 46.

Crafts, A. S. (1960). Weed control research: past, present, and future, *Weeds 8*: 535–540.

Crafts, A. S. and Robbins W. W. (1962). Weed control: a textbook and manual, 3rd edition, McGraw-Hill Book Co., Inc., New York, p. vii–viii and 173.

Composite List of Weeds (1989). Weed Science Society of America. Champaign, Illinois.

Ennis, W. B., Jr., Dowler, W. M., and Klassen, W. (1975). Crop protection to increase food supplies, *Science 188*: 593–598.

Fischer, B. B., Yeary, E. A., and Marcroft, J. E. (1985). Vegetation management systems, in *Principles of Weed Control in California*. Thomson Publishers, Fresno, California, pp. 213–228.

Fryer, J. D. (1985). Recent research on weed management: new light on an old practice, in *Recent Advances in Weed Research* (W. W. Fletcher ed.), The Gresham Press, Old Working, Surrey, United Kingdom.

Fryer, J. D. and Matsunaka, S. (1977). *Integrated Control of Weeds*. University of Tokyo Press, Tokyo.

Hamner, C. L. and Tukey, H. B. (1944-a). Selective herbicidal action of midsummer and fall applications of 2,4-dichlorophenoxyacetic acid, *Bot. Gaz. 106*: 232–245.

Hamner, C. L. and Tukey, H. B. (1944-b). The herbicidal action of 2,4-dichlorophenoxyacetic acid and 2,4,5-trichlorophenoxyacetic acid on bindweed, *Science 100*: 154–155.

Harwood, W. A. (1985). *Principles of Weed Control in California*, Thomson Publishers, Fresno, California, p. 4.

Harwood, R. R. and Bantilan, R. T. (1974). Integrated weed management. I. Key factors affecting crop-weed balance. *Philippine Weed Sci. Bull. 1*: 14.

Herbicide Handbook (1989). 6th Edition, Weed Science Society of America, Champaign, Illinois, p. 267.

Jackson, W. (1984). Toward a unifying concept of an ecological agriculture, in *Agricultural Ecosystems* (R. Lowrance, B. R. Stinner, and G. J. House, eds.), John Wiley & Sons, New York, pp. 209–221.

Kirby, C. (1980). *The Hormone Weedkillers: A Short History of Their Discovery and Development*, British Crop Protection Council Pub., London.

Knake, E. L. (1978). The weed science phase of pest management, Department of Agronomy of the University of Illinois, Urbana.

Korsmo, E. (1932). Undersokelser 1916–1923. *Over ugressets skadevirkninger og dets bekjempelse I. Akerbrucket.* Johnson and Neilsens Boktrykkeri, Oslo, Norway.

Marth, P. C. and Mitchell, J. W. (1946). 2,4-dichlorophenoxyacetic acid as a differential herbicide, *Bot. Gaz. 106*: 224–232.

Morettini, A. (1915). L'impegio dell'acido solforico per combattere le erbe infeste nel frumento, *Staz. Sper. Agr. Ital. 48*: 693–716.

Norris, R. F. (1982). Interactions between weeds and other pests in the agro-ecosystem, in *Biometerology in Integrated Pest Management* (J. L. Hatfield and I. J. Thomason eds.), Academic Press, New York, pp. 343–406.

Pokorny, R. (1941). Some chlorophenoxyacetic acids, *J. Am. Chem Soc. 63*: 1768.

Rabate, E. (1911). Destruction des revenelles par l'acide sulfurique, *J. d'Agr. Prat. (n.s. 21)75*: 497–509.

Shaw, W. (1982). Integrated weed management systems, *Weed Sci. 30 (Suppl. 2)*: 2–12.

Slade, R. E., Templeman, W. G., and Sexton, W. A. (1945). Plant growth substances as selective weed killers: differential effect of plant-growth substances on plant species. *Nature (Lond.) 155*: 497–498.

Smith, A. E. and Secoy, D. M. (1975). Forerunners of pesticides in classical Greece and Rome, *J. Agric. and Food Chem.* 23: 1050–1055.

Smith, A. E. and Secoy, D. M. (1976). Early chemical control of weeds in Europe, *Weed Sci., 24*: 594–597.

Soule, J., D. Caree' and Jackson, W. (1990). Ecological impact of modern agriculture, in *Agroecology* (C. R. Carroll, J. H. Vandermeer, and P, M. Rosset, eds.), McGraw-Hill Publishing Co., New York.

Timmons, F. L. (1970). A history of weed control in the United States and Canada, *Weed Sci. 18*: 294–306.

Upchurch, R. P. (1969). The evolution of weed control as a science, *Indian J. of Weed Science 1*: 77–83.

Weiner, J. (1990). Plant population ecology in agriculture, in *Agroecology* (C. R. Carroll, J. H. Vandermeer, and P. M. Rosset, eds.), McGraw-Hill Publishing Co., New York, pp. 235–262.

William, R. D. (1981). Complementary interactions between weeds, weed control practices and pests in horticultural cropping systems, *Hortsci. 16* (4-spec. insert): 10–15.

Zimdahl, R. L. (1983). Weed science: a brief historical perspective, *Weeds Today 14(1)*: 10–11.

Zimdahl, R. L. (1988). The concept and application of the critical weed-free period, in *Weed Management in Agroecosystems: Ecological Approaches* (M. A. Altieri and M. Liebman, eds.), CRC Press, Boca Raton, Florida, pp. 145–155.

Zimmerman, P. W. and Hitchcock, A. E. (1942). Substituted phenoxy and benzoic acid growth substances and the relation of structure of physiological activity, *Contribs. Boyce Thompson Inst. 12*: 321–343.

2

Ecology of Weeds

David C. Bridges
University of Georgia, Griffin, Georgia

I. HISTORY OF WEEDS

For the purposes of weed management it seems most appropriate to consider the biology and ecology of plants, both crops and weeds, as they interact in an agricultural ecosystem. From a purely ecological perspective weeds may be regarded as colonizers, plants that are adept at capitalizing on disturbance. Bunting (1960) eloquently stated, "Weeds are pioneers of secondary succession of which the weedy arable field is a special case." However, all weeds are not colonizers. Recently, efforts to compile lists of characteristics or attributes of weeds have failed to distinguish weeds as a subset. Rather, they have further defined or delineated ruderals and/or colonizers. Requisite to a working definition of the term *weed* is the human element or criterion of being unwanted. Given that element, the list of ideal characteristics of weeds proposed by Baker (1974) represents features that are commonly observed in species that are universally regarded as weeds.

Weeds find most disdain in production agriculture. Of the world's 250,000 plant species the world's population relies on 12 to 15 species for 90% of our food supply while approximately 250 species occur as weeds (Holm et al., 1977). Holm categorized 76 of these species as the "world's worst weeds." Approximately 70% of the 250 species that are regarded as weeds throughout the world belong to 12 families. It is not surprising that so many weeds come from these families. The Gramineae, or grasses, include many of the world's major food crops: rice, corn, wheat, sorghum, millet, oats, barley, etc. The Compositae do not include many food crops, but with approximately 20,000 species universally it is the second largest plant family and is represented in virtually all parts of the world (Smith, 1977). Similarly, of the 76 species constituting the worst weeds, the Euphorbiaceae (spurge family), Leguminosae (pea family), and Solanaceae (nightshade family) are also represented. Several crops of worldwide importance belong to these families.

Weeds share many characteristics with crops. Among these common characteristics is the fact that many crops and weeds rely upon human intervention for establishment. Mankind and his activities contribute greatly to the dispersal of seed and propagules to new sites and introduce weeds to exotic habitats. Harlan and de Wet (1965) noted also that weeds are species that thrive in habitats disturbed by man. They regarded weeds and domesticated species, or crops, as so similar in their requirement for intervention that they concluded, "Since both (weeds and domesticates) are adapted to the same habitat, practices that tend to favor domesticates also tend to favor weeds."

Many practitioners and researchers of weed management believe that volunteer crop plants and escaped domesticated plants represent a special category of weeds. Certainly volunteer plants of one species occurring where another species is being cultured meet the criterion of being unwanted. Should these volunteers be considered "weeds"? They certainly are of economic concern! For example, many acres of soybean are treated annually in the midwestern United States for the control of volunteer corn. Likewise, hairy vetch, an important soil-improving legume, are difficult-to-manage plants in wheat throughout much of the southeastern United States. Are they to be considered weeds?

The same question may be asked about exotic plants that are introduced as desirable species but subsequently are deemed of less or no value or become established in unintended habitats. Examples include kudzu and showy crotalaria. Kudzu is native to Japan and China and was introduced to the United States in the late 1800s as an ornamental. In the early 1900s it gained widespread use in the southeastern United States as a forage crop and for soil conservation purposes. However, today it is regarded by many as having little or no redeeming value. Likewise, showy crotalaria was introduced as a soil-improving legume, but currently it is not recommended for this purpose because the seed, which commonly contaminate harvested grain, are toxic to livestock and poultry. Are these species weeds? Harlan and de Wet rhetorically state, "One man's weed is often another man's crop." However, Aldrich (1984) suggests that volunteers and escapes should be referred to precisely as what they are, plants representing the respective group of plants to which they belong rather than weeds.

So, is the concept of unwantedness a requisite part of the definition of a weed? Following Harlan (1975), "It is true that weeds are often unwanted, but that is not what makes them weeds." This statement illustrates the relationship between man and weeds. Therefore it is critical that one understand why the plant is unwanted. What is the effect of the weed that makes it undesirable in the eyes of man? Weeds impact the life of man by affecting his health, his crops, or his domesticated animals or simply by being aesthetically displeasing. Weeds reduce crop yield by competing for resources, by allelopathic effects, and by reducing food, feed, and fiber quality. Therefore a useful working definition of a weed must recognize the ecology and biology of the plant as well as the impact on man. A definition that recognizes all of these elements was proposed by Navas (1991), who defined a weed as "a plant that forms populations that are able to enter habitats cultivated, markedly disturbed or occupied by man, and potentially depress or displace the resident plant populations which are deliberately cultivated or are of ecological and/or aesthetic interest."

Regardless of the nomenclature one ascribes to these plants it is readily apparent that weeds and crops share a sufficient number of characteristics to ask, What is the origin of the plants that we commonly call weeds? Do weeds and crops share common origins and progenitors? Apparently, they do! Some crops are believed to have evolved from weeds (Anderson, 1971; Harlan and de Wet, 1965; Harlan, 1975). Alternatively, some weeds apparently have evolved from crops. It is important to recognize that the evolutionary process for weeds is generally the same as for all other groups of plants. Harlan and de Wet indicated that the evolutionary sequence is likely to be from wild plants that are adapted to naturally disturbed habitats to a crop-weed complex. They further reported that many of the plants that we refer to as weeds did not exist in their present form during the preagricultural era. The distribution of crops and weeds, relative to the families to which they belong, indicates that while some weeds, especially those referred to as "crop mimics," may have evolved parallel to the evolution of the crops with which they are most often associated, many other weeds evolved independently. It is not incidental that many weeds come from families that contribute many crop plants, whereas other families contribute disproportionate numbers of either weeds or crops, respectively (Table 1).

Table 1 Families Contributing Crops and Weeds

I.	Many major crops; many important weeds Gramineae, Leguminosae
II.	A few crops; many important weeds Compositae, Euphorbiaceae, Labiatae, Convolvulaceae, Chenopodiaceae, Cruciferae, Polygonaceae, Umbelliferae, Amaranthaceae
III.	Many minor crops; some weeds Rosaceae, Cucurbitaceae
IV.	A few crops; some weeds Solanaceae, Malvaceae, Palmaceae, Dioscoreaceae, Rubiaceae
V.	No important crops; some weeds Cyperaceae, Ranunculaceae, Cactaceae, Caryophyllaceae, Portulacaceae, Onagraceae, Asclepiadaceae, Plantaginaceae, and others

Source: Harlan and de Wet, 1965.

A knowledge of where, how, and when weeds evolved is important as weed management is explored. Less important is this knowledge if one focuses solely on weed control. It is apparent that weeds and crops are closely related, even beyond the fact that some people vigorously cling to the notion that weeds are simply undesirable plants in controlled, unnatural habitats composed of desirable species. That weeds and crops often share the requirement of human intervention for establishment and perpetuation in a given habitat is fundamental to understanding the difficulty often faced in managing weeds. Harlan and de Wet concluded that weeds have come from two main sources: (1) wild species that have long been adapted to sites of natural disturbance and (2) new species or varieties that have evolved since the inception of agriculture. The same claim could easily be made for many of our modern-day crops.

II. CLASSIFICATION

Weeds are often classified to facilitate communication. Classification may be based on taxonomic nomenclature, morphology, origin, adaptation, or life cycle, to name a few characteristics. Classifying schemes often reflect important biological and/or ecological features that are important to weed management. Several schemes will be considered along with their implications for weed management.

A. Life Cycles

One of the more common methods of classifying weeds is to divide them on the basis of the length of their life. The season of growth and timing of reproductive activity are also considered in classifying weeds.

Annuals

An annual is a plant that completes its life cycle from emergence to physiologically mature seed in 1 year or less. Annuals are often subdivided into summer annuals and winter annuals. Summer annuals are plants that emerge in the spring or early summer, grow during the summer, and produce seed by the autumn. Winter annuals emerge in the autumn, grow vegetatively during the winter, and produce seed by spring.

How do annual weeds respond to weed management? Since weeds are often considered colonizers, or ruderals, it is not surprising that many weeds are annual plants. Annuals typically grow quickly and reproduce even under adverse growing conditions. Seeds of many annual weeds will germinate, often discontinuously, over a broad range of environmental conditions and emerge under the conditions of disturbance that are associated with agriculture. Despite these adaptive features that tend to favor perpetuation many consider annual weeds to be the group of weeds most easily managed. Annuals respond well to cultural, mechanical, and chemical management. Cultivation, mowing, hoeing, and handpulling are typically effective for the control of annual weeds, especially if applied during the early vegetative phase of growth. Likewise, annual weeds respond well to chemical management. Since these plants reproduce by seed they can often be controlled with herbicides, especially soil-applied herbicides. They also can be controlled with postemergence-applied herbicides. However, because of their often rapid growth during the vegetative phase, chemical control efforts must be undertaken in a timely fashion to be effective. In general, older and larger annual weeds are more difficult to control with herbicides than are younger and smaller weeds. Cultural control of annual weeds is an integral component of weed management. Because annuals rely exclusively on seed for perpetuation of the species within a habitat, it is particularly important to minimize seed production within the habitat and to prevent seed introduction into the habitat. Regardless of the control tactic, efficacy in controlling annual weeds decreases with size and age.

Biennials

Biennials are plants that live for more than 1 year, but less than 2 years. They germinate, emerge, and grow vegetatively, forming a rosette during the first year. After a cold period they typically undergo floral initiation, flower, and produce seed during the second year of the life cycle. Wild carrot and common mullein are examples of biennial species. The terminal event for biennial species is flowering. If flowering is prevented during the second year some biennials may continue to live and flower in a subsequent year. However, they are not regarded as perennials because once they flower, their life cycle is complete. Relative to annuals and perennials few weed species are biennials.

Tillage or other forms of disturbance often interrupt the life cycle of these plants. Therefore tillage and cultural practices are effective methods of managing biennial weeds. However, regardless of the control tactic, biennial weeds are best controlled when they are in the seedling or rosette stage.

Perennials

Perennials are plants in which the vegetative structures live for more than 2 years, typically by renewed growth from the same root system. Another distinctive feature is the mechanism(s) by which the species perpetuates itself in a habitat. Perennials are often capable of vegetative perpetuation, which according to Kigel and Koller (1985) offers means of spread and propagation without recourse to any reproductive process that involves flowering. Simple herbaceous perennials normally reproduce by seed and do not naturally spread vegetatively. However, if a portion of the plant is damaged or removed, regeneration may occur. Simple herbaceous perennials survive unfavorable seasons and resume growth via dormant roots, rhizomes (specialized underground stem tissue), bulbs, tubers, or crowns. They may or may not produce seed during the year of establishment. Creeping herbaceous perennials in addition to producing seed may also spread via a variety of organs: roots, stems, stolons, rhizomes, or tubers. Woody perennials, including shrubs and trees, undergo secondary thickening which is expressed as incremental growth. These plants often flower only after reaching maturity, which may occur shortly after establishment or which may take many years. Woody perennials often reproduce by seed, but they may also regenerate from buds and spread by means of root sprouts.

Perennial weeds are among the most difficult to manage, because of the multitude of mechanisms by which they are propagated, especially since many perennial weeds spread and regenerate via specialized vegetative structures. Timing of control efforts depends largely on the control strategy or strategies. Perennial weeds, like annual and biennials, that establish from seed are easiest to control in the seedling stage. However, control is often much more difficult if a perennial plant propagates principally by vegetative means or has passed beyond the seedling stage. In fact, integrated strategies that capitalize on the collective efficacy of mechanical, cultural, and chemical tactics generally are the most effective for managing perennial weeds. The interaction(s) of the aforementioned control tactics relative to management of perennial weeds is well illustrated in a recent treatise by Donald (1990). Canada thistle is a herbaceous perennial that spreads by the formation of adventitious shoots from adventitious root buds. Donald notes that cultural, mechanical, and chemical tactics all contribute to the management of Canada thistle (Figure 1). Historically Canada thistle control strategies have relied on reducing shoot growth in

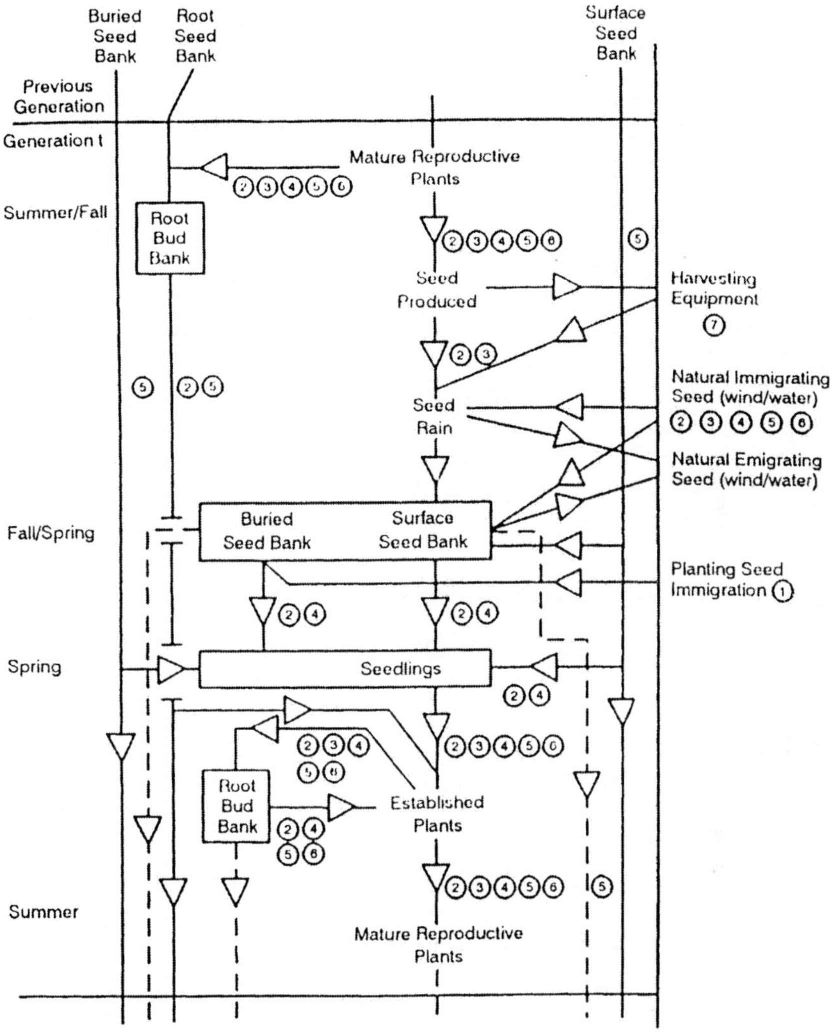

Figure 1 Flow chart outline of the life cycle of Canada thistle over two generations. The life cycle stages that can be manipulated by man are indicated by *1* planting weed-free crop seed, *2* tillage, *3* mowing or grazing, *4* herbicide application, *5* biological control, *6* planting of competitive crops, and *7* cleaning of combines. (From Donald, 1990.)

hopes of depleting root numbers and root energy reserves. Repeated tillage reduces the persistence of roots, especially where shoot growth is minimized either by mowing or by herbicide use. Donald reported that the time of spraying (herbicides) during the Canada thistle life cycle may contribute to the variability in its control. The difficulty in managing Canada thistle is largely due to the anatomical, physiological, and propagative complexities that are associated with this perennial weed.

Such difficulty is the rule with perennial weeds, not the exception. Similar difficulty has been met when attempting to manage weeds such as purple nutsedge, silverleaf nightshade, horsenettle, field bindweed, and johnsongrass. Therefore, consideration of the life cycle of a weed is critically important to the selection and timing of weed management (Figure 2). Concerning classifying weeds according to their life cycle, it is noteworthy that individual plants of a given species may not necessarily behave consistently with their specified life cycle. Life cycle is greatly influenced by environment and expression is often very plastic. Species, and especially individual plants within a population, may exhibit life cycle characteristics inconsistent with their normally ascribed life cycle. For example, plants belonging to species normally regarded as annuals may function as perennials, and vice versa, depending on the environmental condition to which they are exposed.

B. Growth Form and Habitat

Weeds are often classified on the basis of their growth form. They may be divided into herbaceous species and woody species, a division which often reflects an aspect of their life cycle. Herbaceous species may be subdivided into vines, those with prostrate growth, and those with upright habitat. Woody species may be subdivided into trees, shrubs, and woody vines.

Monocots and Dicots

Another typical classification scheme is based on the number of cotyledons (seed leaves) present. The seed-producing plants (Spermatophytes) comprise two classes, Angiosperms and Gymnosperms. Most weeds are Angiosperms. The Angiosperms are further subdivided into two subclasses: Monocotyledoneae and Dicotyledoneae. The Monocotyledoneae, often referred to as the monocots, include the grasses (family Gramineae), sedges (family Cyperaceae), lillies (family Liliaceae), and cattails (family Typhaceae). Colloquially, weeds that are monocots are referred to as the "grassy" or "grasslike" weeds because of their leaf shape and growth form. However, the terminology "grassy" or "grasslike" is often misleading, thus causing confusion when discussing issues such as herbicide selectivity. Therefore, care must be taken when using this terminology. The monocots can be distinguished from the dicots by the characteristics listed in Table 2. The Dicotyledoneae, often referred to as the dicots, include many of the weeds that are often referred to as "broadleaf weeds" or simply

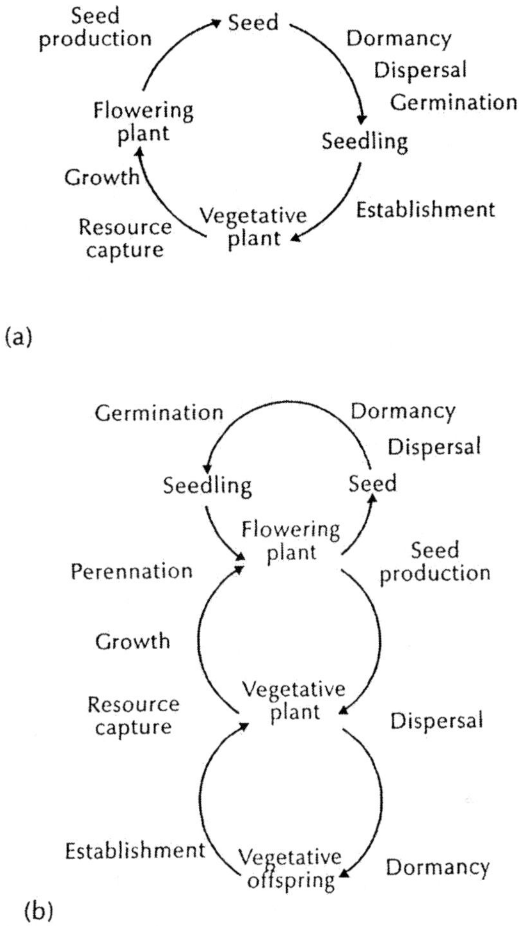

Figure 2 Schemes illustrating the life cycle of an annual flowering plant (a) and a perennial plant producing both seed and vegetative progeny (b). (From Radosevich and Holt, 1984.)

"broad-leaves." Again, this terminology must be used with great care since all plants that have broad leaves are not dicots.

Classifying weeds as monocots versus dicots, or grassy weeds versus broad-leaves, is well entrenched in the plant science literature. This classification scheme has become an integral part of much literature on the technologies of weed control and weed management, including many herbicide labels. How-

Table 2 Distinguishing Characteristics of Monocots and Dicots

Characteristic	Dicots	Monocots
Flower parts	Usually in fours or fives	Usually in threes
Cotyledons	Two	One
Leaf venation	Usually netlike	Usually parallel
Primary vascular bundles in stem	In a ring	Scattered
True secondary growth with vascular cambium	Commonly present	Absent

Source: Raven et al., 1976.

ever, the pervasive use of this nomenclature does not free it of ambiguity nor improve its inaccuracy.

Habitat Distinctions

Another classification scheme often used to separate weeds plays on their typical habitat or other special characteristics. Trees and shrubs are often referred to as "woody" weeds. Likewise, these woody, perennial species when found in pasture, rangeland, or noncropland sites may be referred to as "brush." In fact, brush and brush management are so important to range productivity that brush management has evolved into a highly specialized area of weed management (Scifres, 1980) (refer to Chapter 13).

Aquatic weeds represent another class of weeds that is recognized on the basis of habitat in which they occur. Aquatic weeds may be either annual or perennial and are often subdivided into smaller groups based upon their growth habit such as floating, emergent, or submersed (Radosevich and Holt, 1984). Sometimes aquatic weeds are referred to as either "floaters," "bankers," or "sinkers," depending on their location or position of rooting in the aquatic habitat.

Epiphytic weeds are another group recognized by their relationship to the habitat; they include such plants as mosses, ferns, lichens, and orchids. Epiphytes are organisms that grow nonparasitically on another organism or on some nonliving structure, deriving moisture and nutrients from the air (Raven et al., 1976). Orchids growing in coffee trees are often regarded as weeds because of the damage that results from moisture accumulation. Another example of an epiphytic weed is Spanish moss, which commonly grows in deciduous trees in the southeastern United States.

Yet another group of weeds recognized through special characteristics are parasitic weeds. Parasitic weeds are those plants that live on a plant host and derive nutrients or energy from that host. Parasitic plants can be either obligate or facultative parasites. However, many parasitic plants contain chlorophyll and are therefore capable of semiautotrophism, whereas others depend completely on the host for photosynthate. Parasitic plants occur in approximately 10 families, but the majority of economically important parasitic plants belong to four families: Cuscutaceae (the dodders), Loranthaceae (the mistletoes), Orobanchaceae (the broomrapes), and Scrophulariaceae (the witchweeds) (King, 1966).

III. WEED ADAPTATION

What features and attributes do weeds possess that result in their remarkable success in their environments? Are they "super" plants? For the most part, weeds are simply well adapted to the environments in which they are found. Weeds that commonly occur in arable agricultural lands are often species that benefit from the repeated disturbance that is characteristic of most fields. They are well adapted to, and in fact depend on, the disturbances and modifications that accompany much agricultural production: tillage, irrigation, and fertilization. Weeds compete with crops for resources such as light, water, and nutrients. They are generally regarded as competitive species. However, from an ecological perspective many weeds are ruderals rather than competitors. They are far more akin to the r-strategist than the k-strategist.

Following Grime (1979), many weeds are ruderals or competitive ruderals. Ruderals thrive under conditions of disturbance, which ensure their successful establishment. Patterson (1985) noted that most agricultural ecosystems are not situations in which "competitors" thrive but rather ecosystems relatively free of stresses. A priority of most modern-day agricultural systems is to minimize plant stresses such as drought, low soil fertility, and invasion by pests. It is not surprising that weeds, which are equally as well adapted to perturbation and man's intervention as are crops, respond to management much as crops do. In fact, the same adaptations that maximize crop growth, development, and productivity likely do the same for weeds.

Even though weeds are adaptively like any crops, certain features that are associated with their success in invading, becoming established, and persisting in new environments are identifiable.

A. Features Related to Establishment

Many weeds are ruderals. Germination of these species is favored by soil disturbance, soil moisture, and possibly high soil nitrogen levels compared to undisturbed, nonagricultural sites. One characteristic of ruderals is that total reproductive effort is relatively high. Large numbers of seed are produced but with little investment in carbon per seed. Fecundity is high, but the chance of successful establishment from an individual seed is relatively low. Many weed species exhibit discontinuous germination, meaning that all seed within the seed bank do not germinate simultaneously. Rather various mechanisms of dormancy and safe-site protection ensure that seed germinate over extended periods, and only when environmental conditions favor successful establishment. For example, dimorphism results in two seed within the common cocklebur fruit that have different germination requirements, thus ensuring that they do not germinate coincidentally. The production of large numbers of seed also ensures perpetuation of the species where repeated tillage operations tend to deplete the seed bank. These adaptive strategies confer survival under the conditions of disturbance common to many agricultural sites.

Seed dispersal is another adaptive feature that confers survival potential in most arable sites. Secondary dispersal features such as the potential for wind, water, animal, or other mechanical seed transport increase the chances of establishment in most ecosystems. Species that rain seed only beneath the producing plant are much less likely to encroach to new sites. Therefore, secondary dispersal mechanisms may contribute to weediness of a species.

The conditions required for seed germination and seedling emergence may have a great impact on the potential of a species to become established as a weed in a given site. The ability to germinate over an extended range of soil temperatures or water potentials is a clear advantage to establishment and perpetuation. Weeds are generally more heterogeneous populations than the crops with which they are associated. Most crop species (and their cultivars and/or varieties) are the product of repeated selection processes, whether purposely by breeders or not. One of the prime objectives of modern-day plant breeding is selection for cultivar stability, or homogeneity. While weeds also undergo constant selection pressure because of the constant intervention of man in most agricultural systems these species often have not been exposed to the selection pressure that crops have. The result is a more heterogeneous population, often comprising more than one biotype at a given location. The "universal" weed genotype has not been identified, but certainly the heterogeneity often expressed within weed populations does confer survival advantage.

B. Features Related to Growth and Development

Once weeds have emerged, what characteristics favor their survival and competition with crops? It is a widely accepted ecological principle that the ability to capture (occupy) space, especially early, is strongly correlated with competitiveness, resource capture, and growth. Therefore, plants that emerge and have high relative growth rates during the early vegetative phases of growth are often good competitors. Rapid early growth requires efficient resource capture. These plants develop extensive root systems within days to weeks of emergence. They also exhibit high rates of leaf expansion and stem elongation. In order to accomplish this rapid early growth most species rely on extensive and readily convertible energy reserves in the seed and relatively high net assimilation rates. Perennial species such as johnsongrass and field bindweed rely on energy reserves stored in the rhizomes and roots, respectively. Likewise, yellow and purple nutsedge derive energy from their tubers, enabling them to establish themselves quickly after emergence.

Plasticity within the species is an adaptive feature that enables plants to tolerate widely divergent environmental conditions. Acclimation is the primary mechanism by which weeds exhibit plasticity. When exposed to drought, plants may undergo physiological, anatomical, and/or biochemical changes that confer survivability. When experiencing drought stress, plants may undergo osmotic adjustment or may exhibit heliotropic responses such as modifying leaf angle and therefore minimizing thermal loading. Likewise, plants subjected to shade may acclimate by altered specific leaf area, leaf area per unit dry weight. By becoming thinner leaves reduce maintenance cost and therefore maintain photosynthetic efficiency. These are but a few of the plastic responses that may occur in weeds that render them better adapted to the environment to which they are exposed.

C. Weed Interference

Plant interactions can be positive or negative. The combined effect of all negative plant interactions is called interference. Interference between crops and weeds includes competition, allelopathy, and parasitism.

Weed Competition

Competition is a process that occurs when the combined resource demands of plants within a given area exceed the available supply. As previously stated weeds compete with crops for resources such as light, nutrients, and water. Competition between weeds and crops is expressed by altered growth and development of both species. Interspecific competition occurs when two or more species coexist in time and space and simultaneously demand a limited resource. Intraspecific competition occurs when two or more plants of the same species coexist in time and space and simultaneously demand a limited resource. Competitiveness of various weeds with crops has been the focus of much research aimed at establishing economic and/or action thresholds for weeds with various crops. Crop yield declines nonlinearly to successive increases in weed density. Yield losses at weed densities insufficient to result in intraspecific competition by the weed are generally additive with successive increases in weed population. However, as density increases, loss per weed diminishes (Figure 3). The economic threshold is the weed density at which the value of loss due to weed competition exceeds the cost of control. Action thresholds may include consideration of other factors such as the effect of allowing seed rain, which will potentially affect weed management in succeeding crops.

With respect to time, or period of competition, weeds that emerge before or at the time of crop emergence generally reduce crop yield more than weeds that emerge after crop emergence. Likewise, the longer weeds are permitted to compete with a crop the greater the crop yield loss will be. The temporal relationships of weed competition with crops are illustrated by the concepts of weed-free requirement and weed competition period (Figure 4). The weed-free requirement is the minimum period following crop emergence that the crop must

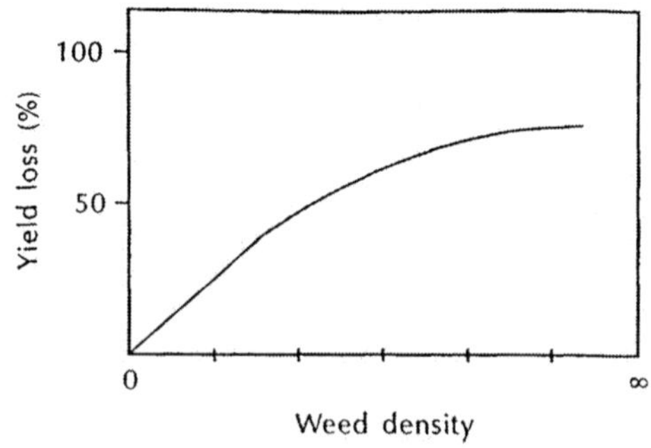

Figure 3 Crop yield loss as a function of density of a single weed species.

be maintained free of weeds in order to prevent crop yield loss. Conversely, the weed competition period is the maximum amount of time that weeds can be allowed to compete with the crop and not result in yield loss. Often interactions between density and these two periods occur. However, these two intervals define the period during which weed management is most critical relative to crop yield.

Allelopathy

Allelopathy is a type of negative interference that occurs when one plant produces and releases chemicals into the environment (air or soil) which are deleterious to the growth and development of another plant. Plants may be self-allelopathic (autotoxic) or they may be allelopathic to other species. The effects and potential effects of allelopathy have been widely established and are known to play an important role in some ecosystems. Some practical applications of allelopathy have occurred. However, allelopathic effects often are not under-

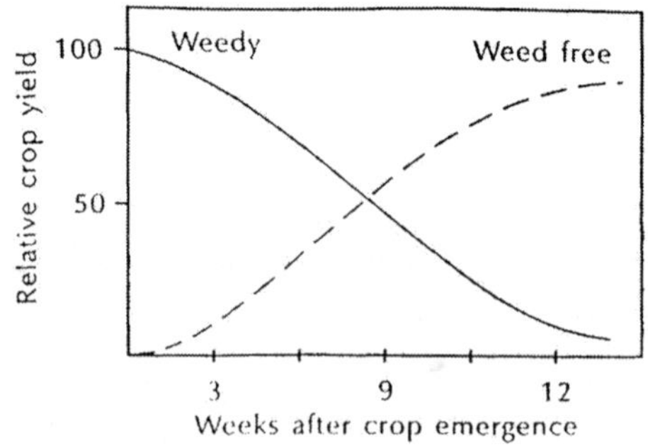

Figure 4 The influence of weed-free period and weed competition period on crop yield and critical period of weed control.

stood or are misunderstood. Therefore, functional application of the process has been limited.

Parasitism

Parasitism is a process by which one plant lives on a plant host and derives nutrients or energy from that host via a living linkage, typically through a haustoria. Parasitic plants can have great impact on the productivity of the host plant. The host plant–parasitic plant relationship is often very complex and may involve chemical regulators that control germination and/or attachment to the host.

D. Weed Population Dynamics

Weed populations at a given site are typically very dynamic. Population density often changes as a result of selection pressures such as herbicide use, tillage, cultivation, grazing, and cropping sequence changes. The intensive use of herbicides over the past 30 years has introduced a selection pressure that has altered the species composition of arable sites probably more than any other single production activity. For a given herbicide use pattern the species that are not controlled by the herbicide(s) are selected for and the species that are sensitive to the herbicide(s) are selected against. Likewise, changes in rotational and tillage practices may greatly impact species composition and population density. For example, continuous no tillage or minimum tillage may favor perennial species over annual species. The persistence of annual species is generally favored by tillage that continually restores the seedbed to a condition favorable for germination and establishment as opposed to sites that are not tilled annually. In contrast, no tillage may favor perennial species that are perpetuated by underground storage organs such as rhizomes, tubers, corms, and/or root stocks.

APPENDIX

List of Crops in Chapter

Common Name	Botanical name	BAYER code
Barley	*Hordeum* spp.	
Coffee trees	*Coffea arabica* L.	
Corn	*Zea mays* L.	ZEAMX
Millet	*Setaria, Panicum, Pennisetum* spp.	
Oats	*Avena sativa* L.	
Rice	*Oryza sativa* L.	
Sorghum	*Sorghum bicolor* (L.) Moench	
Soybean	*Glycine max* (L.) Merr.	
Wheat	*Triticum aestivum* L.	

REFERENCES

Aldrich, R. J. (1984) *Weed-Crop Ecology: Principles in Weed Management*, Breton Publishers, North Scituate, Massachusetts, p. 6.

Anderson. E. (1971). *Plants, Man, and Life*, University of California Press, Berkeley.

Baker, H. G. (1974). The evolution of weeds, in *Annual Review of Ecology and Systematics* (R. F. Johnson, ed.), Annual Reviews, Palo Alto, California, pp. 1–24.

Bunting, A. H. (1960). Some reflections on the ecology of weeds, in *The Biology of Weeds* (J. L. Harper, ed.), Blackwell Scientific Publications, Oxford, pp. 11–26.

Donald, W. W. (1990). Management and control of Canada thistle, *Reviews of Weed Science*, (C. L. Foy, ed.), Weed Science Society of America, Champaign, Illinois, pp. 193–249.

Grime, J. P. (1979). *Plant Strategies and Vegetation Processes*, John Wiley & Sons, Inc., New York.

Harlan, J. R. (1975). *Crops and Man*, American Society of Agronomy, Madison, Wisconsin, pp. 85–104.

Harlan, J. R. and de Wet, J. M. J. (1965). Some thoughts about weeds, *Econ. Bot. 19*: 16–24.

Holm, L. G., Plucknett, D. L., Pancho, J. V., and J. P. Herberger (1977). *The World's Worst Weeds: Distribution and Biology*, University Press of Hawaii, Honolulu, pp. 3–4.

Holzner, W. (1982). Concepts, categories and characteristics of weeds, *Biology and Ecology of Weeds* (W. Holzner and M. Numata, ed.), Dr. W. Junk Publishers, The Hague, Boston, London, pp. 3–20.

Kigel, J. and Koller, D. (1985). Asexual reproduction of weeds, in *Weed Physiology: Reproduction and Ecophysiology* (S. O. Duke, ed.), CRC Press, Inc., Boca Raton, Florida, p. 68.

King, J. J. (1966). *Weeds of the World: Biology and Control*, Interscience, New York, pp. 1–48.

Navas, M. L. (1991). Using plant population biology in weed research: a strategy to improve weed management, *Weed Res. 31*: 171–179.

Patterson, D. T. (1985). Comparative ecophysiology of weeds and crops, in *Weed Physiology: Reproduction and Ecophysiology* (S. O. Duke, ed.), CRC Press, Inc., Boca Raton, Florida, p. 68.

Radosevich, S. R., and Holt, J. S. (1984). *Weed Ecology: Implications for Vegetation Management*, John Wiley & Sons, Inc., New York, p. 7.

Raven, P. H., Evert, R. F., and Curtis, H. (1976). *Biology of Plants*, 2nd edition, Worth Pub. Inc., New York.

Scifres, C. J. (1980). Brush management in perspective, in *Brush Management: Principles and Practices for Texas and the Southwest*, Texas A&M University Press, College Station and London, pp. 3–22.

Smith, J. P., Jr. (1977). *Vascular Plant Families*. Mad River Press, Inc., Eureka, California, p. 320.

3

Preventive Weed Management

Robert H. Walker
Auburn University, Auburn, Alabama

I. INTRODUCTION

The most basic of all weed control methods is prevention, that is, those measures taken to forestall the introduction and spread of weeds. Such measures include production of weed-free seeds, cleaning of weed-infested seed, and proper quarantine and regulation of contaminated crop seed. In this modern day, planting of contaminated crop seed continues to be a major source of spread of noxious weeds. In this way, weeds are spread from continent to continent, region to region, and farm to farm. Dissemination of weeds by this means is largely preventable; only common sense and diligence are needed to ensure success (Day, 1972).

Other preventable means of dissemination include transport of seeds and plant parts from farm to farm on planting, tilling, harvesting, and processing machinery. Weeds are disseminated by livestock by adhering to their hair or feathers or passing through their digestive tracts. The seeds and vegetative parts of weeds are often transported in fill earth, potting soils, and manures. Weed seeds are often carried into new land as contaminants of irrigation and drainage waters. Infested hay, grain, straw, and other agricultural products constitute a major means of spreading weeds; this is particularly true for animal feed transported from farm to farm, often over long distances (Day, 1972).

The overall concept of prevention has been overshadowed during the last two decades by the development and use of modern herbicides and effective mechanical weed-control implements. These tools have led to the mistaken impression that effective control always can be obtained after weeds have invaded an area. This impression may hold true in some instances, but it will eventually lead to added cost and increased potential for environmental problems (Ross and Lembi, 1985).

Ross and Lembi (1985) believe promotion of the pest management concept of an economic threshold has been another reason for deemphasis of preventive weed management. This concept holds that a control practice should not be employed until the point at which damage to the crop equals the cost of control. The concept for economic thresholds was developed for highly mobile and somewhat predictable insect populations whose numbers can fluctuate greatly during the season or between years. Theoretically and in practice, relatively high populations of insects can be present without causing sufficient damage to justify treatment.

The economic-threshold concept is probably species-specific for weeds. For example, if a newly introduced weed requires a control practice other than those currently being used in a crop management system, or if it has the potential of causing more damage than those weeds already present, the economic threshold may be reached when the first plant produces seed or other propagule. Although the damage caused by a single plant may not have much of an impact on crop yields, the potential for loss from the progeny of that plant plus the cost of control in future years make it imperative to try to eliminate the plant before it produces seed or other propagules, not after its population has reached an economic threshold level (Ross and Lembi, 1985).

The success of preventive weed management programs varies with the weed species, its means of dissemination, and the amount of effort applied. Preventative weed management programs are most successfully applied in those situations in which man himself is the vector. Preventative weed programs usually require community action through the enactment and enforcement of appropriate laws and regulations (Day, 1972).

II. FARM-MANAGEMENT PRACTICES

A. Clean Crop Seed

Most states have enacted laws that standardize the maximum number of weed seeds allowable in various crop seeds in commercial trade. If producers buy and plant certified or purer seeds, the chance of introducing weed problems into a field is small, but it can still occur. However, there are producers who save their seed for planting next year's crop. Soybean and cereal-grain seed are those most commonly saved on farms. The potential is high for this seed to contribute to increased weed problems (Walker and Buchanan, 1982; Dastgheib, 1989).

Jenson (1962) reported on a drill-box survey in Utah in which 1232 samples of seed grain were collected from grain drills in the field at planting time. After analyzing each sample for the presence of weed seeds, he found that over half of the samples contained wild oat seed, the most prevalent. The worst sample contained seed of five noxious weed species, with a total of 167,600 noxious weed seeds per 45 kilograms of crop seed. This farmer was planting an average of 50 noxious weed seeds on each square meter of his field. Tracing the source of the surveyed grain seed showed that the farmers' own seed contained the most weed seeds. Grain seed that had been cleaned, tested, and tagged had fewer weed seeds. Ninety-nine percent of the uncleaned grain seed and 44% of the cleaned seed contained weed seeds.

In addition to wild oat, small grain seeds are often contaminated with seeds of vetches, wild radish, and ryegrass. Common weed seed contaminants in soybean seeds include balloonvine, annual morningglories, ragweeds, and common cocklebur.

B. Smother Crops

Smother crops are highly competitive for light, moisture, and nutrients with the weeds infesting an area. Crop plants used for this purpose are grains (barleys, millets, rye, sorghum); legumes (alfalfa, clovers, cowpea, vetches); grasses (wheatgrasses, sorghum/sudan hybrids, sudangrass); and others (common buckwheat, common sunflower). Any highly competitive crop (rapid growth rate, high dry matter production) that is well adapted to an area may be suitable for use as a smother crop. However, care must be taken not to use a crop that may, in itself, develop into a problem weed. If in doubt, destroy the crop before it has an opportunity to produce seed. Smother crops may be grown in rotation with less competitive crops (Anderson, 1977).

C. Crop Rotation

Crop rotation is practiced as a means of weed management to prevent or reduce the buildup of high populations of certain weeds common to a particular crop. Any given crop, cultivated or not cultivated, is plagued by certain weed species that thrive under the same cultural conditions provided for the crop and possess similar growth habits. The particular weed species involved may vary from area to area but, in general, are characteristic to the crop (Walker and Buchanan, 1982).

When the same cultural practices are followed year after year on the same land, crop-associated weeds tend to multiply rapidly and successfully compete with the crop, reducing yields, quality, and economic return (Walker and Buchanan, 1982).

The choice of the kind of crop plants to include in a crop rotation sequence and their sequential planting, when good management is a prime factor, should give preference to those crop plants whose growth and cultural characteristics are in sharp contrast to those of the previous crop and the predominant problem weeds (Walker and Buchanan, 1982). A good rule of thumb is to rotate grass crops with broadleaf crops. The purpose of this rotation is that grass weeds are more common in grass crops and broadleaf weeds are more common in broadleaf crops. This is even more evident when selective herbicides are used for weed control in the respective grass and broadleaf crops.

Leighty (1938) pointed out that crop rotation's first and foremost purpose was to maintain the fertility of the soil. He recognized its importance from the standpoint of weed management as well as its advantages in control of plant disease and erosion. Other advantages for rotating crops include the following: 1) legumes increase the nitrogen content of the soil, 2) labor can be better distributed as a result of crop diversification, and 3) crop yields increase.

The use of rotations began to decline in the late 1940s with the introduction of synthetic-nitrogen sources as a replacement for legume nitrogen. The abandonment of crop rotation without suitable herbicides led to some intense weed problems as reported by Slife and Wax (1976). For example, 2,4-D was used extensively for broadleaf weed control in corn, and, in less than 10 years, annual-grass weeds had emerged as the dominant weed problem. New herbicides soon became available, however, to alleviate this problem.

Several investigators (Buchanan et al., 1975; Frans, 1969; Weber et al., 1974) have reported shifts in weed populations attributable, in part, to the continuous and extensive use of chemicals for weed control in crops. Although conclusive evidence for such effects exists, the effects are not necessarily detrimental, but rather have added another dimension to the crop-weed relationship. Crop rotation now implies rotating both crops and herbicides. Fortunately, rotating both the crop and the herbicide has proved more successful in influencing weed populations. However, the influences of herbicides should not be overemphasized because the crop(s) affects the associated weed spectrum. Buchanan et al. (1975), in a 3-year study, reported that higher populations of common cocklebur were associated more with soybeans than with corn regardless of whether weed-control methods were mechanical or chemical.

A classic experiment, initiated in 1965 at Urbana, Illinois, illustrates the value of rotating crops and herbicides for increasing crop production and influencing weed management. Treatments included 1) continuous corn, 2) continuous soybean, 3) continuous wheat, 4) corn-corn-soybean, and e) corn-soybean-wheat. Herbicide treatments included 1) the same major herbicide each year, 2) varied herbicides from year to year, and 3) cultivation only. After 9 years, the following conclusions were drawn:

1. A single herbicide treatment in combination with cultivation produced 25% more corn than did cultivation alone.
2. Corn yields were 8% higher when corn was grown in a corn-corn-soybean or a corn-soybean-wheat rotation than in continuous corn.
3. Soybean treated with herbicides plus cultivation produced 27% more than when they were only cultivated.
4. Soybean grown in a corn-corn-soybean or a corn-corn-wheat rotation yielded 14% more than in continuous soybean.
5. Perennial weed species invaded the monoculture systems but not the multiculture systems (Aldrich et al., 1975; ARS National Research Program, 1976).

The previous discussion focused on the influence of crop rotation as it directly affects weed management. The indirect effects associated with better insect, disease, and nematode control must, however, be considered concomitantly, because they can often be a more limiting factor than weeds. Crops free of these limiting factors grow better and thus compete more effectively with weeds.

Zimdahl (1981) conducted a survey among 41 weed scientists in 31 continental states to determine their opinions on the extent of use of the six principal nonchemical methods for weed control in ten agronomic crops. The survey included a question on the extent to which crop rotation was employed specifically for weed control, and he reported that the answers were confusing.

For almost every crop, weed scientists estimated that 0% to 100% of the growers employed crop rotation for weed control. It is obvious that some weed control can be accomplished by crop rotation and that it is employed to control certain problem weeds. However, it was not equally obvious that weed control is the primary reason for most crop rotations. Economic and other cultural considerations such as avoidance of pesticide residues, disease infestations, or soil problems (e.g., salt toxicity) are equally plausible reasons for rotating crops while accomplishing some weed control. Because of these factors and the unknown extent to which they interact, it was not possible for him to derive an accurate estimate of the use of crop rotation for weed control even though weed control can be accomplished when crops are rotated (Zimdahl, 1981).

III. CLEAN FARM EQUIPMENT

Harvesting, mowing, hay baling, tilling, and earth moving equipment all have potential to disseminate seeds and/or vegetative reproductive structures of weeds. The combine harvester is considered an almost perfect device for the dispersal of weed seeds. Research has shown that less than 10% of the weed seeds unshed prior to harvest were discharged by the combine harvester. Of the total seeds shed, 4.5% were removed by straw baling, 4% remained on the soil surface, and the vast majority (91.5%) were removed in the grain (Froud-Williams, 1987). Once combines are contaminated with weed seeds they can disseminate them over long distances. For example, in the central Great Plains much of the small grain is harvested by custom operators, so weed seeds can be moved many miles as combines are transported from state to state. Jointed goatgrass has been spread from Kansas to Nebraska as combines have moved northward (Wilson, 1987).

Removal of straw by baling has been implicated in the dissemination of weed seeds, notably wild oat. As many as 19,360 wild oat seeds from a single straw bale have been recorded. However, on average, losses in straw accounted for only 2% of seed shed; the majority (59%) contaminated the grain or was shed onto the soil (39%) (Wilson, 1987). It is not uncommon for a bale of straw/hay or a portion of one to remain with the baler when transported to another field or farm. Proper disposal of contaminated straw/hay and cleaning of the baler are needed to enhance preventive weed management.

Moldboard plows, disk harrows, power tillers, cultivators, etc., are all capable of collecting vegetative reproductive structures of weeds. Many of the most troublesome species have been and continue to be disseminated with these implements. Common bermudagrass is a major weed in upland cotton fields of the South (Whitwell and Brown, 1983). This primarily results from tillage equipment that transports rhizomes and/or stolons from field borders into the field; then the bermudagrass is distributed over time by the mechanical cultivator. Very similar events occur with the introduction and/or spread of yellow and purple nutsedge. Other troublesome weed species distributed by tillage equipment include johnsongrass, field bindweed, Canada thistle, and quackgrass (Ahlgren, Klingman and Wolf, 1951; Ross and Lembi, 1985).

IV. MANAGEMENT OF ANIMAL FEEDSTUFFS AND MANURE

Livestock ingest many seeds when grazing pastures and consuming hay, silage, or grain products. Seeds that retain their viability after passage through the digestive tract may be desirable forage species but many are weeds (Blackshaw and Rode, 1991; Burton and Andrews, 1948). One study estimated that spreading sheep manure could contribute as many as 1.6 million weed seeds per hectare (Dastgheib, 1989). Although passage of weed seeds through animals is known to be an important means of dissemination, too little is done to manage manure and feedstuffs properly and thus improve preventive weed management. Also, research in this area is very limited.

Many mixed and ground feeds contain weed screenings separated from wheat or other grains prior to milling. These weed seeds are mixed with bran or other feeds with or without grinding. The grinding process usually destroys only the larger weed seeds; many of the smaller seeds escape unless the feed is very finely ground. Some feedstuffs may be heat treated to destroy viable weed seeds. Coarsely ground feeds have been found to contain as many as 9000 viable weed seeds per kilogram (Muenscher, 1955). Feed that contained 20% unground weed seeds fed to cattle had an average weed seed germination of 26% before and 4% after feeding (Blackshaw and Rode, 1991). The longer the seeds of several species of weeds remained in the digestive tract of dairy cows, the greater the reduction in germination. Germination was reduced 80% to 90% by passage through the animals (Atkeson et al., 1934; Burton and Andrews, 1948).

The passage of seeds of velvetleaf, field bindweed, and Pennsylvania smartweed through cattle, horses, sheep, hogs, and chickens was studied. The average percentage of viable seed recovered for the three species was 24% for hogs, 23% for cattle, 13% for horses, 11% for sheep, and 0.3% for chickens. Viability varied among the three species, ranging from 13% for Pennsylvania smartweed to 23% for field bindweed and velvetleaf (Harmon and Keim, 1934).

The effect of ensiling seeds on their subsequent germination and viability has been studied. Weed seeds removed from the midpoint of a 91-megagram silo after 3 months storage had unaltered seed germination. Some weed seeds actually have been shown to have a greater germination percentage after being soaked for 2 to 3 months in silage juice. Another study was conducted to determine the effect of ensiling and/or rumen digestion by cattle on the germination of several common grass and broadleaf weed species. Seed survival of grass species subjected to ensiling and/or rumen digestion tended to be less than that of broadleaf species. Downy brome, foxtail barley, and barnyardgrass were nonviable after either ensiling for 8 weeks or rumen digestion for 24 hours. Some green foxtail (17%) and wild oat (0% to 88%) seeds survived digestion in the rumen but were killed by the ensiling process. Varying percentages of seeds of kochia, redroot pigweed, common lambsquarters, wild buckwheat, round-leaved mallow, and field pennycress remained viable after ensiling (3% to 50%), rumen digestion (15% to 98%), and ensiling plus rumen digestion (2% to 19%) (Blackshaw and Rode, 1991).

To prevent further distribution and dissemination of weed seeds by livestock, measures should be implemented. For example, livestock grazing or consuming feedstuffs that are infested with weed seeds should be confined to areas and fed noncontaminated feedstuffs before being transferred to noninfested pastures or fields. Confinement time required to clear the digestive tract of livestock is not well documented. However, 10 to 12 days was required to clear the digestive tracts of dairy cows that were fed seeds of bahiagrass, common bermudagrass, carpetgrass, dallisgrass, johnsongrass, and lespedeza. But the seeds of all these species did not pass through the digestive tract with the same speed. The seeds of the small-seeded grasses passed through faster than those of the larger-seeded grasses (Burton and Andrews, 1948).

There is no doubt that manure contains viable weed seeds and if it is spread on pastures or fields, weeds will surely be introduced, distributed, and their density increased unless the manure has gone through proper fermentation. The effect of fermenting manure on the germination of weed seeds was studied in Iowa and Maryland. At both places horse and cow manures used separately and the two mixed in equal parts were used. At Maryland, 52 varieties of seeds were placed in the manure for 1 to 6 months. All seeds failed to germinate after the 1-month period. At Iowa, similar results were obtained in 1907 for the 2- to 6-month periods. For the 1-month period six species germinated. In 1908, 31 different weed seeds were left in horse manure for 5 weeks, the highest percentage germination of any weed seed being 1%. Most of them failed to germinate.

Other research determined the effect of different methods of storing chicken manure on the viability of certain weed seeds. Seeds of field bindweed, whitetop (hoary cress), and Russian knapweed were put in wire containers and placed in chicken manure, stored in different ways for various periods, and then germinated in soil and on blotter paper. Seeds were removed from the various piles and germinated at different intervals, the shortest being 10 days and the longest 4 months. In no case was the viability of field bindweed seed destroyed. However, a study conducted in Nebraska (Table 1) showed complete loss of viability of weed seeds at the end of 4 months' storage in cow manure (Harmon and Keim, 1934). The viability of whitetop and Russian knapweed seeds was destroyed after they had been in the moist, loose manure for 20 days or in the moist compacted manure for 1 month. In the unmoistened piles the viability of whitetop and Russian knapweed seeds was not completely destroyed at the end of 4 months (Stoker et al., 1934).

Proper fermentation requires that the manure be kept moist and the fermentation process continued for a minimum of 4 months. At this time, it would be wise to collect samples of the manure from the piles and conduct germina-

Table 1 The Effect of Length of Time in Cow Manure upon the Viability of Various Weed Seeds

	Period of storage				
Kinds of seeds	1 month	2 months	3 months	4 months	Viability before burial
Velvet weed	2.0	0.0	0.0	0.0	52.5
Bindweed	4.0	22.0	1.0	0.0	84.0
Sweet clover	22.0	4.0	0.0	0.0	68.0
Peppergrass	0.0	0.0	0.0	0.0	34.5
Smooth dock	0.0	0.0	0.0	0.0	86.0
Smartweed	0.0	0.0	0.0	0.0	0.5
Cocklebur	0.0	0.0	0.0	0.0	60.0
Puncture vine	0.0	0.0	0.0	0.0	52.0

Source: Harmon and Keim (1934).

tion tests. This can be done by spreading a thin layer of the manure (approximately 1.5 centimeters) over a clean potting mixture and allowing seeds to germinate for a few weeks under proper moisture and temperature conditions.

V. WEED-FREE IRRIGATION AND DRAINAGE WATERS

Producers who rely on surface water for irrigation can increase weed problems by applying water contaminated with weed seeds. For example, Kelley and Burns (1975) conducted a detailed study in Washington state and found surface-irrigation water to be a significant source of weed seeds on irrigated farmlands. A total of 137 different species were represented in weed seed in the irrigation lateral and 77 species in water from the Columbia River. They calculated that if the seeds were evenly distributed in the applied water during the season, the number of seeds disseminated would range from 1700 to 15,000 per hectare.

Aldrich (1984) showed the number of seeds of selected weeds found in water from the Columbia River and in an irrigation lateral during the irrigation season (Table 2).

Wilson (1980), in a more recent study, also showed irrigation water in western Nebraska to be a significant source of weed seed. During the 1978 growing season, irrigation water disseminated approximately 7900 seed per hectare in the sampled fields. Seventy-seven different plant species were found in the water with germination averaging 26%. Two to five times more weed seed were found in two irrigation canals than in the North Platte River. As

Table 2 Kind and Number of Weed Seeds in Water from the Columbia River and an Irrigation Lateral

	Seeds per 2.54 hectare centimeters of water		
Weed	Columbia River	Irrigation lateral	Percent germination[a]
Barnyardgrass	3.53	72.87	19
Cattail	0.12	8.87	34
Cutgrass, rice	0.69	8.99	9
Dandelion	0.12	1.24	65
Dock, curly	2.40	6.08	37
Dropseed, sand	0.79	6.89	66
Flixweed	1.01	22.66	52
Foxtail, green	0.69	7.56	2
Foxtail, yellow	0.12	4.97	52
Horseweed	3.76	15.20	80
Lambsquarters	21.72	307.24	5
Lettuce, prickly	8.18	16.83	57
Mustard, tumble	36.40	65.46	70
Pigweed, redroot	6.82	66.10	4
Quackgrass	0.00	10.97	80

[a]Average from all sources.
Source: Aldrich (1984).

water moved through the canals, the weed seed content of the water increased dramatically. The majority of the seed collected was found floating on the water surface.

These data show the potential of adding weed seeds to the soil from surface sources of irrigation water. This problem can be greatly reduced by controlling weeds along irrigation laterals and by removing weed seed from the irrigation water with appropriate screening devices.

VI. MANAGEMENT OF WEEDS HAVING WIND-BORNE SEEDS

Wind is a vehicle for dissemination of weed seeds that have suitable structures. Results of studies with tansy ragwort indicate that 60% of the shed seeds fell at the base of the plants, and only about 0.4% were beyond 4.6 meters. Regardless of wind direction, the majority of seeds was close to the source plant. Where movement occurred with the wind, seeds moved to about 35 meters and collected to a density of approximately 10 per square meter. If all survived to become a part of the viable seedbank, they would represent a significant addition. Beyond this distance, additions would be relatively insignificant (Aldrich, 1984). However, seeds that possess a light, feathery pappus (parachute-like structure) such as chicory, dandelion, Canada thistle, bull thistle, and cogongrass may blow for miles. However, in both cases these plants or plant parts are blown across the soil by the wind and seeds are disseminated in the process (Ahlgren et al., 1951). Preventing dissemination of weed seeds by the wind is a simple process, but a monumental task. The best defense is to keep weeds with wind-transported propagules away from cultivated fields. This is best done by keeping fence rows and waste areas clean, thus preventing seed production, thereby largely eliminating the source. Plants that do invade should be destroyed before they reproduce. It is possible that vegetative barriers that are sufficiently dense and tall could be used to retard dissemination of wind-borne weed seeds. Mechanical structures such as fences have been observed to retard wind dissemination of Russian thistle.

VII. HANDLING WEED-INFESTED CROPS

The underlying goal in producers' approach to weeds in the future must be the development of a long-term plan. This plan must have as its focus the production unit or possibly even the individual field. Individual fields will be more applicable for a monoculture cropping system. This plan must be founded on the principles of weed prevention, including a reduction in weed propagules, a reduction in emergence of weeds with crops, and a lessening of competition from those weeds that are present. Weed control should be relied upon to provide the final measure of crop protection rather than the only measure, as is now frequently the case (Aldrich, 1984). When producers rely heavily on chemicals and cultivation to control weeds, it is not unusual to find weed-infested crops, particularly when weather conditions have been less than optimal for the working of the herbicides and cultivation.

When handling weed-infested crops, producers should use those techniques and procedures that have been outlined for preventive weed control. The primary goal should be to keep the weed propagules confined to the area or field that they now infest. It is much easier to deal with a smaller area than larger and more numerous areas. An additional goal should be to prevent as much seed and/or other propagule production as possible. This may necessitate destroying certain areas in the field that are heavily infested before any propagules are produced. Crop production in such areas will be minimal anyway. Another consideration should be to fallow the field mechanically for at least one crop cycle. Bridges and Walker (1985) obtained significant reduction in sicklepod seed numbers in the soil reserve with summer mechanical fallow. The reduction was near 50% the first year and did not decline substantially in additional years. The success of mechanical fallow will depend to a degree on the species involved. Thus, germination characteristics of the species involved should be reviewed to determine susceptibility to the use of mechanical fallow.

VIII. LEGAL MEASURES FOR PREVENTATIVE WEED MANAGEMENT

A law in Scotland during 1212 to 1249 decreed severe penalties to tenants or bond servants who seeded weeds on the land or who did not destroy certain weeds. The enforcement was so effective that the weeds were practically eliminated; the area became famous for cereal grains free of weed seeds. "Gool riders" could no longer find weeds and collect enough fines to afford food and drink. These early regulatory practices were described in 1814 by Sinclair, who suggested the need for their renewal (Timmons, 1970). These types of regulatory practices do not exist today but laws do exist, some of which are effective.

The Federal Seed Act, approved August 9, 1939, regulates interstate and foreign commerce in seeds. Its purpose is to protect purchasers from mislabeled crop seed and crop seed contaminated with weed seeds. All states have seed laws designed to regulate the quality of agricultural seeds for planting that are sold within their borders. These laws generally conform with the Federal Seed Act, although they differ from state to state as far as weed species are concerned (Anderson, 1977).

Weed laws are enacted to compel local officials and property owners to carry out weed sanitation measures on lands under their jurisdiction. Farms, roadsides, railroad tracks, waste areas, utility rights-of-way, and ditch banks are examples of such lands. Not all states have weed laws, but where they exist, the enabling legislation rests with the state legislature. The weed law in many states is similar to the model uniform state weed law that was drafted in 1963 by the U.S. Department of Agriculture (USDA), state departments of agriculture, and the Weed Science Society of America. The model weed law permits the establishment of weed control districts at the local level under the direction of a superintendent who is responsible for supervising control and eradication programs. The superintendent often is the chief executive officer of a county, city, or township district (county commissioner or township trustee). Funding for the program usually comes from local taxes levied for this purpose (Ross and Lembi, 1985).

Quarantines are another legislative tool that are used to mitigate the effects of weeds. A quarantine is defined as a restriction imposed by duly constituted authorities whereby the production, movement, or existence of plants, plant products, animals, animal products, any other article or material, or the normal activity of persons is brought under regulation, in order that the introduction or spread of a pest may be prevented or limited. If the pest has already been introduced and established, a quarantine is necessary so that it may be controlled or eradicated, thereby reducing or preventing losses that would otherwise occur through damage done by the pest (Sand, 1987).

Federal quarantines and regulations are in effect under provisions of Sections 8 and 9 of the Plant Quarantine Act of August 20, 1912, as amended in Section 106 of the Federal Plant Pest Act (7 U.S.C. 161, 162, 155, 150ee CFR 301.80-1 through 301.80-10) (Sand, 1987).

On January 3, 1975, the Federal Noxious Weed Act of 1974 (Public Law 93-629) was signed into law. It directs the USDA to protect American agriculture from designated noxious weeds of foreign origin. Under this authorization and direction from Congress, the USDA, through the Animal and Plant Health Inspection Service (APHIS), has a stated obligation to protect American agriculture from such species as witchweed, which is under quarantine/eradication in North and South Carolina. Also, these states have statutory authority contained in the North Carolina Plant Pest Law and the South Carolina Pest Act of 1912. North and South Carolina have right-of-entry laws to support the necessary actions of the eradication program (Sand, 1987).

APPENDIX

Weed and Crop List Cross-Reference

Common name	Botanical name	BAYER code
Wild oat	*Avena fatua* L.	AVEFA
Common cocklebur	*Xanthium strumarium* L.	XANST
Jointed goatgrass	*Aegilops cylindrica* Host.	AEGCY
Common bermudagrass	*Cynodon dactylon* (L.) Pers.	CYNDA
Yellow nutsedge	*Cyperus esculentus* L.	CYPES
Purple nutsedge	*Cyperus rotundus* L.	CYPRO
Johnsongrass	*Sorghum halepense* (L.) Pers.	SORHA
Field bindweed	*Convolvulus arvensis* L.	CONAR
Canada thistle	*Cirsium arvense* (L.) Scop.	CIRAR
Quackgrass	*Agropyron repens* (L.) P. Beauv.	AGRRE
Velvetleaf	*Abutilon theophrasti* Medik.	ABUTH
Pennsylvania smartweed	*Polygonum pensylvanicum* L.	POLPY
Downy brome	*Bromus tectorum* L.	BROTE
Foxtail barley	*Hordeum jubatum* L.	HORJU
Barnyardgrass	*Echinochloa crus-galli* (L.) P. Beauv.	ECHCG
Green foxtail	*Setaria viridis* (L.) P. Beauv.	SETVI
Kochia	*Kochia scoparia* (L.) Schrad.	KCHSC

(*continued*)

Appendix Continued

Common name	Botanical name	BAYER code
Redroot pigweed	*Amaranthus retroflexus* L.	AMARE
Common lambsquarters	*Chenopodium album* L.	CHEAL
Wild buckwheat	*Polygonum convolvulus* L.	POLCO
Round-leaved mallow	*Malva pusilla* Sm.	MALPU
Field pennycress	*Thlaspi arvense* L.	THLAR
Whitetop	*Cardaria draba* (L.) Desv.	CADDR
Russian knapweed	*Centaurea repens* L.	CENRE
Tansy ragwort	*Senecio jacobaea* L.	SENJA
Chicory	*Cichorium intybus* L.	CICIN
Dandelion	*Taraxacum officinale* Weber in Wiggers	TAROF
Bull thistle	*Cirsium vulgare* (Savi) Tenore	CIRVU
Russian thistle	*Salsola iberica* Sennen & Pau	SASKR
Witchgrass	*Panicum capillare* L.	PANCA
Sicklepod	*Cassia obtusifolia* L.	CASOB
Witchweed	*Striga* spp.	
Wild radish	*Raphanus raphanistrum* L.	RAPRA
Balloonvine	*Cardiospermun halicacabum* L.	CRIHA
Cogongrass	*Imperata cylindrica* (L.) Beauv.	IMPCY
Annual morningglories	*Ipomoea* spp.	
Ragweeds	*Ambrosia* spp.	
Soybean	*Glycine max* (L.). Merr.	GLXMA
Vetches	*Vicia* spp.	
Ryegrasses	*Lolium* spp.	
Barleys	*Hordeum* spp.	
Millets	*Setaria, Panicum, Pennisetum* spp.	
Rye	*Secale cereale* L.	SECCE
Sorghum/Sudan hybrids	*Sorghum bicolor* (L.) Moench	
Alfalfas	*Medicago* spp.	
Clovers	*Trifolium* spp.	
Cowpea	*Vigna sinensis* (L.) Endl.	
Wheatgrasses	*Agropyron* spp.	
Sudangrass	*Sorghum sudanense* (Piper) Stapf.	
Buckwheat, common	*Fagopyrum esculentum* Moench.	FAGES
Sunflower, common	*Helianthus annuus* L.	HELAN
Corn	*Zea mays* L.	ZEAMX
Wheat, soft	*Triticum aestivum* L.	TRZAX
Cotton, upland	*Gossypium hirsutum* L.	GOSHI
Bahiagrass	*Paspalum notatum* Fluegge.	PASNO
Carpetgrass, narrowleaf	*Axonopus affinis* Chase	AXOAF
Dallisgrass	*Paspalum dilatatum* Poir	PASDI
Lespedezas	*Lespedeza* spp.	

REFERENCES

Ahlgren, G. H., Klingman, G. C., and Wolf, D. E. (1951). *Principles of Weed Control*, John Wiley and Sons, Inc., New York, pp. 16–17.

Aldrich, R. J. (1984). *Weed-Crop Ecology, Principles in Weed Management*, Brewton Publishers, North Scituate, Massachutte, pp. 52-54, 373–398, 421–422.

Aldrich, S. R., Scott, W. O., and Leng, E. R. (1975). *Modern Corn Production*, A and L Publishers, Champaign, Illinois, pp. 223–237.

Anderson, W. P. (1977). *Weed Science Principles*, West Publishing Co., St. Paul, Minnesota, pp. 63–65.

ARS National Research Program. (1976). Weed control. NRP No. 20280. U.S. Department of Agriculture, Agricultural Research Service, Washington, D.C., pp. 80–83.

Atkeson, F. W., Hulbert, H. W., and Warren, T. R. (1934). Effect of bovine digestion and of manure storage on the viability of weed seeds, *J. Am. Soc. Agron. 26*: 390–397.

Blackshaw, R. E. and Rode, L. M. (1991). Effect of ensiling and rumen digestion by cattle, *Weed Sci., 39*: 104–108.

Bridges, D. C. and Walker, R. H. (1985). Influence of weed management and cropping systems on sicklepod seed in the soil, *Weed Sci. 33*: 800–804.

Buchanan, G. A., Hoveland, C. S., Brown, V. L., and Wade, R. H. (1975). "Weed Population Shifts Influenced by Crop Rotations and Weed Control Programs," Proceedings of the Southern Weed Science Society, Memphis, Tennessee, pp. 60–71.

Burton, G. W., and Andrews, J. S. (1948). Recovery and viability of seeds of certain southern grasses and lespedeza passed through the bovine digestive tract, *J. Agric. Res. 76*: 95–103.

Dastgheib, F. (1989). Relative importance of crop seed, manure, and irrigation water as sources of weed infestation, *Weed Res. 29*: 113–116.

Day, B. E. (1972). Nonchemical weed control, *Pest Control: Strategies for the Future*, National Academy of Sciences, Washington, D.C., pp. 330–338.

Frans, R. E. (1969). "Changing Ecology of Weeds in Cotton Fields," Proceedings of the Beltwide Cotton Production-Mechanization Conferences, New Orleans, Louisiana, pp. 24–30.

Froud-Williams, R. J. (1987). Changes in weed flora with different tillage and agronomic management systems, in *Weed Management in Agroecosystems: Ecological Approaches* (M. A. Altieri and M. Liebman, eds.), CRC Press, Boca Raton, Florida, pp. 217–218.

Harmon, G. W. and Keim, F. D. (1934). The percentage and viability of weed seeds recovered in the feces of farm animals and their longevity when buried in manure, *J. Am. Soc. Agron. 26*: 762–767.

Jensen, L. A. (1962). "Are we planting weeds?" Proceedings of the Western Weed Control Conference, Las Vegas, Nevada, p. 52.

Kelley, A. D. and Burns, V. F. (1975). Dissemination of weed seeds by irrigation water, *Weed Sci. 23*: 486–493.

Leighty, C. E. (1938). Crop rotation, in *U.S Department of Agriculture Yearbook of Agriculture, Soils and Men*, U.S. Government Printing Office, Washington, D. C., pp. 406–430.

Muenscher, W. C. (1955). *Weeds*. Cornell University Press, Ithaca, New York, pp. 16–18.

Ross, M. A., Lembi, C. A. (1985). *Applied Weed Science*, Burgess Publishing Co., Minneapolis, pp. 26–29.

Sand, P. F. (1987). The American witchweed quarantine and eradication program, in *Parasitic Weeds in Agriculture, Vol. 1., Striga*, (L. J. Musselman, ed.), CRC Press, Boca Raton, Florida, pp. 207–223.

Slife, F. W. and Wax, L. M. (1976). Weed and herbicide management, in *World Soybean Research* (L. D. Hill, ed.), Interstate Printers and Publishers, Inc., Danville, Illinois, pp. 397–403.

Stoker, G. L., Tingy, D. C., and Evans, R. J. (1934). The effect of different methods of storing chicken manure on the viability of certain weed seeds, *J. Am. Soc. Agron. 26*: 600–609.

Timmons, F. L. (1970). A history of weed control in the United States and Canada, *Weed Sci. 18*: 294–307.

Walker, R. H. and Buchanan, G. A. (1982). Crop manipulation in integrated weed management systems, *Weed Sci. 30(Suppl.)*: 17–24.

Weber, J. B., Best, J. A., and Witt, W. W. (1974). Herbicide residues and weed species shifts on modified-soil field plots, *Weed Sci. 22*: 427–433.

Whitwell, T., and Brown, S. M. (1983). "Report of 1982 Cotton Weed Loss Committee," Proceedings of the Beltwide Cotton Production-Mechanization Conferences, San Antonio, Texas, pp. 259–264.

Wilson, R. G. (1987). Biology of weed seeds in the soil, in *Weed Management in Agroecosystems: Ecological Approaches,* (M. A. Altieri and M. Liebman, eds.), CRC Press, Boca Raton, Florida, p. 34.

Wilson, R. G., Jr. (1980). Dissemination of weed seeds by surface irrigation water in western Nebraska, *Weed Sci. 28*: 87–92.

Zimdahl, R. L. (1981). Extent of mechanical, cultural, and other nonchemical methods of weed control, *Handbook of Pest Management in Agriculture,* Vol. II. (D. Pimentel, ed.), CRC Press, Inc., Boca Raton, Florida, pp. 79–83.

4

Mechanical Weed Management

Gail A. Wicks
University of Nebraska, North Platte, Nebraska

Orvin C. Burnside
University of Minnesota, St. Paul, Minnesota

Warwick L. Felton
New South Wales Agricultural Center for Crop Improvement, Tamworth, New South Wales, Australia

I. INTRODUCTION TO MECHANICAL WEED MANAGEMENT

Mechanical weed management includes hand and machine tillage, mulching, flooding, draining, heating, cutting, pulling, and dragging. Mechanical methods are the oldest methods of managing weeds. They can be done with unskilled labor and generally are economical, nonpolluting, without residual hazards, and relatively safe to the operator. Mechanical approaches are the traditional and predominant methods of weed management in developing countries. Machine tillage is still a common weed management method in developing countries, but chemical weed management methods predominate. The primary reason for tillage in crop protection worldwide is weed control, and this need varies among countries and crop production systems, depending mainly upon the extent of herbicides and other weed management methods.

A. History of Mechanical Weed Management

Initial attempts to remove competing species in crop production mechanically probably occurred in woodlands with little understory vegetation rather than in grasslands with numerous, competitive, rhizomatous perennial plants. This same farming method is being mimicked today in "slash and burn" agriculture, where land is cleared, burned, and farmed until excessive competition from rhizomatous grasses ensues, when the land is abandoned and reverts to forest. Early agriculturists began with vegetative propagation and progressed to planting seeds. They also changed from the culture of perennial to more annual species. Agriculturists first farmed grasslands along eroded riverbanks and floodplains where few competing plant species existed. Some cultivation combined with hand removal of invading plant species was practiced. In many areas of the world these weeds are fed to livestock. Hand hoeing has continued to be an effective weed management method that is still used on over half of the food produced worldwide from plants today.

Animal and machine powered tillage allowed our ancestors to farm the prairie and forest soils of the world. During the latter part of the nineteenth century various tillage and planting implements (Table 1) were patented for use in the United States (Rosenberg, 1979). Tillage was the predominant weed management method in the United States and Australia until the middle of the twentieth century, when herbicides found an increasing role in reducing manpower and horsepower requirements in crop production. Tillage practices in agriculture vary from crop to crop the world over, but in all cases their basic

Table 1 First United States Patents Issued for Implements Used in Preparing Seedbeds, Planting Crops, and Cultivating Crops

Implement	US patent issued
Plow, cast iron	1797
Plow, gang	1884
Harrow, disk	1867
Harrow, spring-tooth	1869
Harrow, disk	1877
Cultivator, straddle row	1879
Cultivator, parallel riding	1884
Planter, check row	1883
Seeder, grain drill	1884

Source: Rosenberg (1979).

objectives are to destroy existing vegetation and produce suitable soil conditions for crop establishment and growth, permit proper management of plant residues, and control weeds (Hunt, 1904).

The day to day cultivation of soil is both an art and a science. Often tillage is used to undo past soil management mistakes. Tillage may be used to stimulate weed germination, manage weeds, level the ground, prepare a seedbed, incorporate various materials, break up compacted soil layers, manage soil temperatures, aerate the soil, control pests, aid root development, alter soil structure, manage soil erosion, manage soil water, alter soil relief, affect crop yields, harvest root crops; occasionally it is done because of custom or habit or to keep someone busy (Burnside, 1985). Nevertheless tillage can be detrimental in many situations. For example, it was a major contributor to the dust bowls of the 1930s and 1950s in the Great Plains. The dust bowl in the 1930s was brought about by lack of vegetation due to tillage, overgrazing, high temperature, and drought. Then high winds moved the fragile soils. Hardy Webster Campbell, a homesteader in Brown County, South Dakota, was referred to as the father of dryland farming (Berg, 1983). He advocated summer fallow, surface packing, seeding grain in widely spaced rows, mulching with dust through frequent cultivation, and deep plowing, These practices began in the 1890s and were widely adapted in the Great Plains.

Some problems with tillage are that it incorporates crop residue, exposes soil to erosion, destroys soil structure, wastes soil water, compacts soil, stimulates weed growth, enforces weed seed dormancy, damages crop roots, speeds up organic matter breakdown, consumes energy and labor, increases soil temperature, affects many production practices, increases crop production costs, and destroys fauna and flora (Burnside, 1985). However, the advantages of tillage generally outweigh its limitations.

B. Types of Mechanical Weed Management

Mechanical methods are still among the major techniques of weed management worldwide. Even farmers on the most advanced technological farms still use a number of mechanical weed management methods. Thus, we need to understand the various methods of mechanical weed management so that they can be used more effectively and economically.

Manual Weed Removal

Hand pulling is one of the oldest weed management methods, and it is most effective against annual and biennial weeds when the entire root system can be removed from the soil. It is still effectively used on large farms to eliminate weed escapes rather than allowing such weeds to produce seeds and proliferate. Many high school students earn their spending money roguing weeds in soybean fields in the midwest (A. R. Martin, University of Nebraska, Lincoln, personal communication, 1992). Present day farmers still rogue a weed escape such as shattercane out of corn, field bean, soybean, sugar beet, and sorghum fields. Also, many farmers choose to rogue common sunflower, velvetleaf, marsh elder, and other weed escapes in fields.

Spudding is often used in lawns to remove tap-rooted weeds such as various thistles and dandelion physically. Spudding is an immediate and selective method of removing scattered lawn weeds rather than allowing them to proliferate or resorting to herbicide treatment (Pammel and King, 1926). A spud is a 5 cm chisel with a handle. The spade is another tool that works well on scattered biennial weeds such as common burdock or musk thistle.

Hoeing is the selective uprooting or cutting of unwanted plants or weed escapes. It is most effective against annual and some biennial species, but established perennials may even succumb after repeated hoeings. Hand hoeing and horse hoeing were common at the turn of the century. Hand hoeing is faster and easier than hand pulling or spudding and thus is more widely used in developing countries, in gardens, and for high value crops such as sugar beets. Both long handled and short handled hoes are used. The short handled hoes, for example, are used carefully to remove weeds near sugar beets and this requires bending the back. Hoe blades have various widths and weights, depending upon the job to be performed. Hoes must be kept sharp for maximum effectiveness and ease of use.

Close cultivation (1.5 to 3 cm from the row) and hand hoeing by migrant workers were the standard weed management practices for sugar beet until the development of cycloate in the 1960s. As weed management systems improved, the need for hand labor decreased. Also increasing costs of hand labor forced farmers to use more herbicides. Schweizer (1980) stated,

> The cost of weeding sugar beets in the United States varies considerably from one farm to another, and maximum costs are closely associated with the indigenous weed population on the farm. In 1970, the federal wage scale for removing weeds from sugar beets once with a hoe was $41/ha; in 1978, this cost was escalated to $69/ha. A second weeding cost $26/ha in 1970 and $45/ha in 1978. Since labor for weeding sugar beets has become more expensive and scarce each year, continued profitable production of this crop will depend on mechanization of all phases of production.

During the 1980s farmers had to use herbicides on their sugar beets to control most weeds because the migrant workers would not hoe weedy fields.

Smothering with Nonliving or Living Mulch

Mulches are effective in weed management because they exclude light, provide a physical barrier to emergence, and affect soil temperature and moisture. Common nonliving mulches are plastic sheets, straw, grass clippings, chips of bark, sawdust, and various other plant parts. Plastic sheets are usually black to absorb more heat from the sun and thus readily destroy weed growth beneath the plastic. One of the recent advances is the use of permeable black plastic for newly planted trees in shelterbelts (Sunbelt Weed Barrier or Water Conservation Mulch, DeWitt Corp., Rt 3, Box 338, Sikeston, MO 63801). It allows water to pass through the plastic, reduces evaporation, and acts as a barrier to weed emergence.

Smother crops were used as an aid in managing specific weed species in the 1920s (Pammel and King, 1926). Quackgrass could be controlled by first plowing 5 to 8 cm deep as soon as the small grain crop was removed. Then the common harrow was used to expose the quackgrass roots, and disk when quackgrass began to grow. In the spring the farmer plowed 15 to 18 cm deep, dragged, and planted a dense stand of sorghum, Sudangrass, millet, buckwheat, or rye. After harvest of the smother crop the farmer plowed 15 to 18 cm deep and harrowed in the fall.

Smother and cover crops are receiving more attention because of their efficiency in reducing soil erosion and inhibiting weed growth. Moore et al. (1991) stated:

> Today's increased awareness of the potential environmental impacts of conventional agricultural practices has promoted the concept of ecologically sound crop production. Soil erosion and an increased use of herbicides are two of the main concerns associated with traditional field crop production. The use of cover crops for mulch has potential to reduce the risks of environmental contamination due to erosion and excessive herbicide use.

They found that early season weed biomass could be reduced 70% to 93% with cover crop mulches. Short lived cover or smother crops would protect the soil but not require their mechanical or chemical destruction before planting the crop. This approach to weed management would be similar to planting into a "stale seedbed" (Muzik, 1970).

Water Management

Aquatic weeds can be managed by water drainage and terrestrial weeds by flooding. Mechanical methods of aquatic weed management include cutting, chaining, crushing, and dredging, but these methods are expensive and only moderately effective.

Hard to control perennial weeds have been eradicated from the fields rotated to rice or fish production. Successful control of many perennial weeds occurs after several weeks to months of submergence. Flooding for 2 to 4 weeks during the growing season can prevent johnsongrass from establishing from rhizomes (McWhorter, 1972).

Comes et al. (1978) found that one-third of 82 species tested produced some seedlings after storage in water for 60 months. Annual grasses were the most susceptible weed seed to become nonviable after water storage. When stored in water for longer than 12 months, 22% of the annual grass species had some seeds that germinated, compared to 75% for the annual broadleaf weeds.

Weed seeds are disseminated by irrigation water. In Nebraska 77 plant species were found in water taken from irrigation canals (Wilson, 1980). During the 1978 growing season irrigation water disseminated 48,400 seeds/ha. The only way to prevent this influx of weed seeds into fields is to prevent weed seed production along irrigation canals and use screens to intercept the weed seeds.

Burning or Heat

Vegetation in the interface between grasslands and woodlands fluctuates as a result of the occurrence of fires. Fire is most destructive to woody species, and thus controlled burns are still being used for rangeland and pasture weed management and the destruction of understory woody plants in pine forests. However, some undesirable shrubs, such as beaked hazel, require two burns to control 80% of the seedlings and 60% of the small clones (Tappeiner, 1979). More burns are required for larger clones. Prescribed burning is effective for controlling eastern redcedar in the Great Plains states (Stritzke and Bidwell, 1990).

Flame cultivators and weed burners run on petroleum fuels and have been used for selective and nonselective weed management. Selective flaming is used to kill cambial tissue in young weeds (not burn the weeds) in crops such as castor bean, cotton, corn, sorghum, fruits, and vegetables (Peacock et al., 1965). Flaming has decreased in popularity during the past two decades because of economical and environmental concerns associated with this method of weed management.

Solarization of soil has been accomplished by covering soil with sheets of plastic. Heating of wet soil surface by sunlight to temperatures greater than 65°C kills nondormant weed seeds and greatly reduces weed seedling populations (Egley, 1983). This process may require several weeks to destroy seed viability of some weed species. Soil sterilization by heat is widely used in greenhouses, but generally the method has proved uneconomical under field conditions. However, there is considerable interest in developing economical methods of destroying the weed seed bank in soil because this would be the ultimate method of managing weeds. One possible method is use of a Fresnel lens, which is a curved-linear lens made of acrylic plastic (Johnson et al., 1990). Parallel grooves refract solar radiation at an angle needed to converge radiant energy to a point focus. The temperature in a 1 by 150 cm area averaged

309°C, which destroyed seed viability of the weed species subjected to a 20-sec exposure on the soil surface. Destroying seed viability at deeper soil depths would require greater exposure times, but destruction at increasing soil depths would not be feasible.

Cutting or Mowing

Cutting or mowing are weed management methods used during the earliest forms of agriculture and are still widely practiced today. Manual cutting implements, including the machete, scythe, sickle, hoe, and axe, are used on small infestations or areas inaccessible to power mowers. There are various power mowers using an oscillating sickle, pivoting blade, or rotary blade. The latest in weed cutters is a trimmer/brush cutter or nylon-line trimmer (John Deere & Co., Moline, IL). It is operated by a small two-cycle engine to provide power to cut weeds, grass, and brush with a nylon line or a blade. Machine mowing of vegetation is common in residential areas, pastures, rangelands, and woodlands and along roadways; reel mowers are used on lawns and golf courses. Mowing prevents or delays seed production of many weed species, is most effective against tall growing plants, and will deplete the root reserves of perennial weeds after repeated mowings. Nevertheless, many weed species, such as crabgrass in lawns or barnyardgrass in seedling alfalfa, adapt to mowing by developing a prostrate growth habit. However, mowing does remove undesirable plant growth while preventing bare ground and excessive soil erosion.

Mechanical shearing has been effective on small eastern redcedar. The shears are operated hydraulically. Ranchers are using this method to control eastern redcedar that are infesting range and pastures.

Mowing has been a common management practice for controlling weeds in shelterbelts and orchards. Perennial grasses are planted between tree rows, while mulches or herbicides are used to control weeds beneath the trees. However, when grass is controlled only by mowing, trees generally receive less nitrogen, are stunted, and yield less fruit (Lord and Vlach, 1973).

Mechanical Weed Pullers

Several implements have been designed to pull tall growing weeds out of short growing crops. Generally, this involves two rubber tires rotating together that pull some weeds out by their roots while pulling the tops off others. The machines work best in removing the roots of weeds in moist soil, providing one is able to access the fields when the soil is moist. However, weed pullers have been largely replaced by rope wick applicators using nonselective herbicides (Dale, 1981). Rope wick applicators control tall growing weeds much more effectively and economically than weed pullers and have become the method of choice for controlling tall growing weeds in crops. However, short-statured shatter cane biotypes have increased, thus reducing the effectiveness of rope wick applicators (Burnside et al., 1980). Short-statured shatter cane biotypes are still a problem in sorghum seed production fields and make shatter cane identification more difficult (P. T. Nordquist, University of Nebraska, North Platte, personal communication, 1992).

Mechanical Tillage

Tillage with implements powered by animals or motors is effective and economical for weed management and is practiced worldwide. The plow dates back over 2000 years in Europe. In the United States Charles Newbold obtained a patent for the first cast iron plow in 1797 (Rosenberg, 1979). Farmers were reluctant to buy the plow because they believed that the cast iron injured the fertility of the soil and promoted weed growth. This implement was the forerunner of the moldboard plow that allowed our ancestors to farm the grasslands of the world. However, the moldboard plow has decreased in popularity in North America and Australia in the past several decades because of increasing concern over soil erosion. Agriculturists also are questioning what they consider to be excessive or "exercise" tillage. These concerns have led to increased research and implementation of reduced tillage systems of crop production. Reduced and no-tillage systems of growing crops have resulted in more judicious usage of tillage and are expected to continue to increase in popularity (Burnside, 1985; Sprague and Triplett, 1986).

Hunt (1904) suggested that killing of weeds is the most important purpose of tillage. This is done by burial, cutting, desiccation, or depletion of food reserves of weeds. Also, tillage stimulates bud and seed growth of weeds, which facilitates their subsequent destruction. Even though the major reason for tillage is weed management, on some clay or heavy bottomland soils, tillage benefits crop growth beyond weed control by loosening and aerating compacted soil (Sprague and Triplett, 1986). Tillage is also important to smooth out ruts left by harvesting equipment during wet periods.

Tillage must be timely or valuable soil water for the next crop is lost. Campbell (1907) suggested that the ground be disk harrowed as soon as the crop has been harvested. In the winter wheat–summer fallow area of the Great Plains Zook and Weakley (1944) found that land left untilled until the latter part of June contained 60% less water in the fall than land kept free of weeds by tillage for the entire fallow period. The major loss of soil water is through plants (Pratley and Rowell, 1987).

Tillage implements, pulled by tractors with increasing horsepower, have shown a steady growth in size and soil movement following increasing farm size and decreasing number of farm workers. This trend toward bigger equipment and greater soil disturbance is slowing or reversing because of problems with soil compaction and erosion as well as economical, environmental, and ethical considerations (Sprague and Triplett, 1986).

There is a wide range of tillage implements designed for residue management, seedbed preparation, and weed management. It is important in planning weed management programs that various tillage schemes be considered in sequence, for example, moldboard plowing followed by disking, harrowing, and/or cultivation. The addition of different crops also changes the potential for different species of weeds to survive. Ball and Miller (1990) found that different weed species increased or decreased, depending upon which tillage and cropping system were used. The environment influences the tillage and cropping sequences. The various tillage implements, using ASAE Standards (1993) definitions, will be discussed under primary, secondary, and tertiary tillage methods.

Primary Tillage. Primary tillage equipment is used initially to break and loosen the soil to depths of up to 75 cm with such tools as the ripper, moldboard plow, chisel plow, tandem-disk harrow, sweep plow, paraplow, or rotary tiller (Wiese, 1985). A list of tillage and planting equipment and the amount of residue remaining after field operations is found in Table 2. Plant residue can be incorporated most effectively by implements that invert the surface soil and bury the residue such as the moldboard and disk plows. Moldboard plows are still commonly used for this purpose in Great Britain, in Europe, and on heavy bottomland soils in the United States. The moldboard plow breaks up and loosens the soil and buries plant residue, weeds, and their propagules. If weeds are not buried they may recover. Optimum depth of plowing is about 18 cm. After plowing, the soil surface is rough, readily takes in water, and is more resistant to wind erosion on most soils. Subsequent tillage operations will further pulverize the soil and increase the risk of soil loss by erosion.

Disk plowing and tandem-disk harrowing are used more commonly for incorporating plant residue in Australia and the Great Plains of the United States. These implements are more effective in controlling vining-type weeds, such as prostrate knotweed, than are chisel or sweep plows or similar equipment.

In much of the "Corn Belt" in the United States and in Australia the chisel plow has become the most popular primary tillage tool. Chisel plows have the advantage of being able to operate effectively in a wide range of soil types and conditions, they help to break up compacted soil layers, and a wide range of chisel points are available to meet different requirements of soil penetration, residue flow, and weed management. The optimum tillage depth varies with soil and may be 15 to 30 cm. Greater depths are necessary when hard pans are present.

The sweep plow is a primary tillage implement that does not incorporate as much plant residue as a moldboard plow, disk plow, or tandem-disk harrow, and this is important for erosion control in semiarid regions. It is sometimes

Table 2 Residue Cover and Effectiveness of Weed Management After Various Tillage Operations in Corn, Small Grains, and Soybean Production Systems

Machine or operation	Percentage of initial residue left after field operations		Weed management effectiveness
	Corn or small grain	Soybean	
Overwinter weathering	80–95	70–80	Suppresses weeds
Moldboard plow	0–10	0–5	Buries weeds and weed seed
Paraplow	80–90	66–75	Destroys weeds
Subsoiler	70–90	60–70	Kills weeds in disturbed area
Chisel plows with			
Sweeps	65–86	35–55	Cuts off weed roots
Straight chisel points	55–80	30–50	May remove or cover some weeds
Twisted points	40–60	15–35	Covers more weeds
Coulter chisel plows with			
Sweeps	60–80	30–60	Cuts off weed roots
Straight chisel points	59–70	25–45	May remove or cover some weeds
Twisted points	35–55	10–30	Covers more weeds
Disk chisel plows with			
Sweeps	55–75	25–45	Cuts off weed roots, poor control of weeds
Straight chisel points	45–65	20–40	Not as good as sweeps
Twisted points	30–50	10–25	Not as good as sweeps
Disk plows:			
5 to 8 cm deep	65–75	50–60	Inverts most weeds but may not kill them
15 to 18 cm deep	25–35	15–25	Covers weeds with soil
Disks harrows			
Offset light duty	45–55	30–40	Inverts most weeds but may not kill them
Offset heavy duty	35–45	25–35	Covers weeds with soil
Tandem (as a secondary operation)	40–60	35–45	Controls most weeds
Tandem after harvest, before other tillage	80–90	50–60	Some weeds escape
Field cultivators as primary tillage operation			
Duckfoot points	–	30–55	Weeds escape before shanks
Sweeps or shovels 15–30 cm	–	50–70	Some weeds escape
Sweeps 30–51 cm	–	55–75	Some weeds escape
Field cultivators as secondary operation			
Duckfoot points	60–80	50–70	Controls small weeds
Sweeps or shovels 15–30 cm	75–85	60–75	Kills small weeds
Sweeps 30–51 cm	80–90	65–80	Kills small weeds

(continued)

Table 2 Continued

Machine or operation	Percentage of initial residue left after field operations		Weed management effectiveness
	Corn or small grain	Soybean	
Finishing tools			
Soil finisher	45–65	30–50	Kills small weeds
Seedbed conditioner	75–95	50–70	Kills small weeds
Culti-mulcher	70–90	60–70	Kills small weeds
Harrows	70–90	65–85	Kills small weeds
Drills			
Hoe openers	50–80	40–60	Covers small weeds, provide ridges; weeds may desiccate in ridge
Disk openers	80–90	60–80	No effect on weeds
No-till coulters	75–85	70–80	No effect on weeds
Cross slot openers	90–95	90–95	No effect on weeds
Planters			
Runner planters	85–95	80–90	No effect on weeds
Double disk opener planters	80–90	70–80	Little effect on weeds
Sweeps or double row cleaning disks	60–80	40–60	Removes small weeds and weed seeds from row
Ridge-till planter	60–70	30–50	Removes weeds from the ridge
No-till planters with			
Offset double disk openers	90–95	85–95	Removes small weeds and seeds from row
Smooth coulter	90–95	85–95	No effect on weeds
Ripple coulter	85–90	80–90	No effect on weeds
Fluted coulter	80–85	70–80	May kill some small weeds
Two or three fluted coulters	75–85	65–75	Will kill small weeds
Anhydrous applicator	75–85	45–70	Little effect on weeds
Knife-type fertilizer applicator	60–80	40–60	No effect on weeds
Cultivators			
Rotary hoe			Kills weeds in white stage
Spike tooth harrow			Kills weeds in white stage
Rolling cultivator			Will kill seedling weeds
Sweeps			Will kill weeds that are 3 to 6 cm tall
After harvest[a]	75–95	65–90	

Source: Adapted from Conservation Technology Impact Center.
[a]Begin calculations with residue remaining after harvest.

equipped with sweeps of various sizes that range from 46 to 210 cm in width. The wide sweep plow has decreased in popularity in Australia and some areas of the United States because of reduced utility in clay or rocky soils, and it can produce an undesirable tillage pan and does not kill weeds under wet conditions and heavy residues.

The rotary tiller finds more limited use because it pulverizes the soil, destroys soil structure, and is expensive to operate. The most common use for the rotary tiller nowadays is to till soil in a narrow strip in front of a planter unit. Fertilizer and herbicides can also be band applied and incorporated into the soil during planting with the rotary tiller; this has been referred to as the "weed and feed" method of crop management. Depth of incorporation needs to be monitored so it is not too deep for optimum herbicide efficacy.

Secondary Tillage. Secondary tillage is used for weed management, herbicide incorporation, erosion management, and seedbed preparation. Destroying weeds is one of the main reasons for preparing a seedbed for crops. Best results are obtained when sufficient time has elapsed between tillage operations to allow weed seeds to germinate so they can be destroyed by the next tillage operation. The more weed seedlings that are killed with frequent tillage operations before planting, the less weed interference there is in the succeeding crop. Secondary tillage implements are used to work the soil up to 15 cm deep; they include the field cultivator, tandem-disk harrow, blade plow, harrow, rodweeder, seedbed conditioner, roller harrow, and various combinations of these tillage implements (Wiese, 1985). Generally, ground driven cultivators such as mulch treaders need to be operated at 8 km/hr and rotary hoes should be pulled at 13 km/hr to control weeds effectively (Gunsolus, 1990). Each implement leaves varying amounts of plant residue on the soil surface (Table 2) and causes varying amounts of soil disturbance depending on the depth and speed of operation, stubble type and age, soil type, and water content. The field cultivator and the tandem-disk harrow are the most popular implements for secondary tillage. Both are fast, wide-pass operations that do not require large tractors, can incorporate herbicides, and leave a level, weed-free seedbed. Rodweeders maintain plant residue on the surface, but there is a restricted range of soil conditions necessary for satisfactory operation. They are not suited to areas where there are stones or stumps present. Seedbed conditioners are used mostly in the southern United States to reduce soil clods, incorporate herbicides, and level the field. The roller harrow has limited use, but it effectively reduces clods to produce smaller soil aggregates and a firm seedbed for small seed crops such as perennial grasses and legumes.

Tertiary Tillage. Corn seeding rate needs to be increased 5% to 12% to compensate for stand reduction cost by rotary hoeing and row cultivation (Doll and Mulder, 1991). Tertiary tillage begins after planting and includes the cultivation of row crops. Most cultivator tools work best in the absence of plant residue. The rotary hoe, finger weeder, round-wire-teeth harrow, and spike-tooth harrow are most effective on weeds in the "white stage" that have germinated but not yet emerged. Spike-tooth harrowing should be done before the corn has emerged, with a second harrowing, with teeth set at a slant, occurring between corn emergence and the three to four leaf stage. Round-wire teeth harrowing should be done across the rows. Round-wire teeth harrowing can be done until corn is 46 cm tall. Intrarow tillage implements used after crop emergence should be employed after the early morning turgidity has left the plants; they are then less susceptible to injury (Rathers and Harrison, 1951). In corn a single rotary hoeing at 12 to 14 days after planting when weeds have germinated but not emerged followed by one cultivation reduced weed biomass 75% (Moomaw and Robinson, 1973a). In sorghum a single rotary hoeing at 12 days after planting plus one cultivation reduced weed biomass 100% but only 50% when rotary hoeing was delayed until 22 days by rain (Moomaw and Robinson, 1973b). A 36% grain sorghum yield loss resulted when rainfall occurred shortly after planting compared to 20% when weeds emerged after sorghum had become well established (Wiese et al., 1964). Schweizer et al. (1992) found that weed populations emerging 3 weeks later than corn were reduced 98% with intrarow cultivation but only 75% when weeds emerged with the corn. The rotary hoe was more effective than the round-wire teeth harrow in reducing weed populations. These implements are less effective on weeds that germinate from deeper than 5 cm, on no-till fields, or on fields with more than 30% plant residue cover (Gunsolus, 1990).

Prior to herbicides, careful interrow crop cultivation was used to manage weeds in row crops. For example, corn land to be planted was disked early to encourage weed seed germination before moldboard plowing. Corn was surface planted about May 15. One or two rotary hoeings were done soon after the corn emerged, and 3 or 4 days later the corn was cultivated. Cultivators were equipped with vertical fins which moved soil laterally into the corn row. Cultivator shovels next to the row were set 4.5 cm from the row. The first cultivation was done at a speed that would not move soil over the corn. Two weeks later the corn was cultivated again, and a third cultivation occurred when corn was about 50 cm tall. This was called a hilling operation and soil was moved into the corn row to cover as many weeds as possible (Furrer and Burnside, 1962). In years with excessive rain, often cultivation was not timely and weeds were poorly controlled.

Row crop cultivation is an important weed management tool for row crops in fields that are conducive to cultivation. Corn grain yields were greater with three than with two cultivations and four cultivations were unnecessary. Grain sorghum grain yields increased with two cultivations, but three cultivations were not needed. In the absence of weeds there was a slight advantage of one cultivation and slight decrease in grain sorghum yields with three cultivations (Table 3) (Burnside and Wicks, 1965).

Table 3 Effect of Cultivation on Weed Control and Sorghum Yields in Nebraska

Number of cultivations	Check		Atrazine at 2.2 kg/ha
	Untreated	Handweeded	
	Sorghum yields, kg/ha		
0	1760	4820	4700
1	3570	5020	4890
2	4770	5080	5080
3	4770	4890	4890
Avg	3720 b	4950 a	4890 a
	Weed biomass, kg/ha		
0	2520	11	490
1	750	11	67
2	120	2	45
3	100	0	78
Avg	870 a	6 c	170 b

Source: Burnside and Wicks (1965).
Numbers followed by the same letter within rows are not significantly different at the 0.05 level.

Wilson (1948) stated,

> Theoretically from a weed control view point, the cultivated crops should reduce the annual weeds to a point where they will be less of a problem during the remainder of the rotation. Actually, this goal is not attained in many cases. The numerous annual weeds often grow in the corn after the last cultivation and produce an abundance of seed to become a problem in succeeding crops. Much may be done to reduce these late weeds by extra tillage at the start of the crop season to encourage germination and by eliminating the small plants with a harrow or weeder. Some extra hand cultivation after the cultivated crop is laid by will reduce the weed population.

In the 1990s weeds that emerge after the last cultivation of corn still cause concern for farmers. Examples of weeds that germinate in midsummer and have been a concern of farmers are large crabgrass, bristly foxtail, and shatter cane. Fields that have received hail are especially vulnerable to late season weeds. These weeds may not reduce grain yields but produce enough to cause weed management problems in future years. Perennial weeds, such as Canada thistle, hemp dogbane, and johnsongrass, also can be a late season problem in corn.

During the 1950s and early 1960s weed control in grain sorghum depended upon three or four tillage operations to prepare a seedbed and encourage weed seed germination. In the central Great Plains grain sorghum was planted from

June 1 to June 20. Interrow crop cultivation occurred after grain sorghum emergence. Thus the introduction of herbicides allowed grain sorghum to be planted earlier and benefited yield because of the longer growing season.

Interrow crop cultivators include rolling, vertical, and horizontal shovels; combination disk and vertical and horizontal shovels; and power driven rotary cultivators. Most cultivators are effective on weeds up to 15 cm tall but are not effective on intrarow weeds unless sufficient soil can be moved into the row to cover the weeds. The first cultivation should be 5 to 7 days after the white root stage of weed seedlings, repeated in 7 days (Erbach et al., 1970). Hooded shields, rolling shields, and plant shields are available to prevent soil from covering the crop. There are numerous options available for sweeps. Selection depends upon soil type, soil condition, weed growth, crop, crop stage, soil water, and amount of plant residue on the soil surface. Interrow cultivation is economical, nonselective, and rapid. Various tractor guidance systems have facilitated setting cultivator shovels closer to the row, but most farmers do not have a guidance system or choose not to set the shovels very close to the rows. They would rather band apply a herbicide, thus reducing the risk of crop root damage from the cultivator. Cultivation of corn should begin after the five-leaf stage and depth should be shallow enough (about 4 to 7 cm deep) to minimize injury to root crops. Interrow cultivation is limited by soil moisture, and it is often ineffective against intrarow weeds. Another cultivation, as soon as soil dries out, should follow cultivation done when the soil is too wet.

Interrow crop cultivation timeliness is essential to obtain adequate weed control without reducing grain yields. Interrow crop cultivation generally should begin 2 to 3 weeks after crop emergence, but weather may prevent timely cultivation. Delvo (1971) used weather records over a 10-year period to determine the effect that timely cultivation had on grain yields. He found that crop yield (corn, grain sorghum, and winter wheat) was reduced by 23% when cultivation alone was compared to preemergence herbicide application plus cultivation.

Row crop cultivation of no-till fields where herbicides failed in the 1970s was difficult because of cultivator design. Subsequently, cultivator frames were strengthened and first-generation cultivators met with greater success. In Nebraska, interrow crop cultivation of no-till grain sorghum planted into undisturbed winter wheat stubble improved weed control and grain yields (Table 4) (Wicks and Burnside, 1981). One cultivation was sufficient at Lincoln for adequate weed control but not at North Platte because of greater summer annual grass density.

Row crop cultivators are available now that operate efficiently in heavy residues. However, large green weeds may interfere with trash flow and trash and soil can build up quickly and clog cultivators.

Table 4 Effect of Cultivation on Sorghum and Weeds Grown on No-Till Winter Wheat Stubble Ground That Had Been Treated with Atrazine plus 2,4-D at 2.2 plus 1.1 kg/ha plus 9 L/ha Crop Oil After Winter Wheat Harvest

	Lincoln			North Platte		
No. of cultivations	Untreated	Handweeded	Atrazine 2.7 kg/ha	Untreated	Handweeded	Atrazine 1.8 kg/ha
			Weeds biomass, kg/ha			
0	690 a	0 b	56 b	2860 a	45 d	1440 b
1	120 b	0 b	0 b	1690 b	22 d	500 c
			Sorghum grain, kg/ha			
0	3280 c	4830 a	4520 ab	720 d	3070 a	2080 b
1	4260 b	4690 ab	4450 ab	1490 c	3010 a	2920 a

Source: Wicks and Burnside (1981).
Numbers followed by the same letter along cultivation and within columns are not significantly different at the 0.05 level.

Planters and Drills for Surface Tillage. Indians initially planted four corn kernels per hill spaced 122 cm apart. Therefore, our first corn planters placed corn seed in hills within rows. This arrangement was thought to allow corn to emerge better; fertilizing was easier and controlling weeds was more efficient. After the development of cultivators, hill planting probably still was thought to have an advantage over corn drilling. There was less chance of covering a clump of corn than single plants with soil. Corn was always planted into a seedbed prepared by moldboard plowing as cultivation was easier and there were less weed problems.

As farmers moved away from prepared seedbeds there were a number of planter units introduced that also did some secondary tillage. Fluted or rippled coulters are needed to cut through plant residue and generally a seed firming wheel is needed to improve seed-soil contact. There are now many narrow-row drills that will seed through heavy plant residue. Strip-tillage planters generally move plant residue from intrarow to interrow to facilitate soil warming and prevent crop seed inhibition from plant residue. Weed seeds on the soil surface are also swept between rows and planting usually places the new crop row over the previous row (Wicks and Somerhalder, 1971). Row cultivators for fields with high plant residue use disks to move soil from the row and back again to cover small weeds. A sweep operates between rows to control weeds without incorporating the plant residue. Advantages of sweep strip-tillage planters are lower planting and tillage costs, soil erosion, labor requirements, and weed management needs while maintaining crop yields.

Seed Conditioning

Grain that is harvested from the field is rarely free of weed seeds. Combines remove many weed seeds by air and sieves during harvesting, but these weed seeds are deposited back onto the fields. Efficiency of removing weed seeds from crop seeds depends upon the size of crop seeds relative to the various weed seeds, equipment, and adjustment of equipment. Admixtures of weed seeds in crop seeds must be separated on the basis of some physical or chemical difference. Such differences include seed size, weight, shape, surface area, specific gravity, stickiness, pubescence, texture, color, or electrical properties (Burnside, 1970b). Cleaning also reduces inert material that adds to seed bulk or, if moist, may cause heating in storage. Inert material can inhibit seed drop during planting so that the crop may not be as competitive with weeds because of misses in the row.

Initial seed conditioning operations should be designed to reduce bulk; facilitate mechanical seed handling, especially for obtaining a uniform seed flow; or reduce seed drying time. Final seed conditioning can then be performed with a number of different machines such as an air-screen cleaner, specific gravity separator, indent disk and indent cylinder separators, pneumatic and aspirator separators, velvet roller, spiral or inclined draper, vibrator, timothy bumper mill, horizontal disk separator, electronic separator, magnetic separator, buckhorn plantain machine, and color separator (Burnside, 1970b). Even with seed cleaning a small amount of weed seeds may remain with the crop seeds. Cross-contamination may also occur during the cleaning process. Farmers take their harvested grain for conditioning and may bring home other farmers' weeds. This has occurred with admixtures of jointed goatgrass, rye, and triticale in winter wheat (C. R. Fenster, Professor Emeritis, University of Nebraska, Gering, personal communication, 1992). Winter wheat seed that was sent to a seed conditioner for removal of small kernels returned with triticale. This seed had 17 triticale plants/ha in one field and 19 triticale plants/ha in another. Winter wheat not cleaned had no triticale (G. A. Wicks, personal observation, 1992). Triticale had never been grown on the farm. Farmers should have their seed cleaned at approved seed conditioners. Gravity and length graders can remove jointed goatgrass but may not remove all the triticale and rye.

Planting crop seeds containing weed seeds has been the most common method of spreading weeds for centuries. Recent drill box studies have shown a high number of different weed seeds being sown back into the winter wheat fields (Stanelle, 1986). In Nebraska, only 68% of drill box samples of winter wheat were conditioned (Prentice, 1991). Downy brome (18%) and Japanese brome (17%) were among the 29 weeds species identified, and these are serious competitors with winter wheat. Admixtures should not occur with large seeded crops such as corn or soybean, but sometimes they do.

II. EVALUATION OF MECHANICAL WEED MANAGEMENT METHODS

Farmers and equipment manufacturers have developed many implements that aid in managing weeds and preparing seedbeds for crops. These developments have occurred over centuries, and there are constant changes being made by farmers in tillage equipment. Evaluation of efficiency is a continuous process. There are several advantages and disadvantages of tillage management that farmers should consider when planning weed management programs.

A. Advantages

Tillage can have positive influences on weed management and crop production practices. There are situations where it would be advantageous to manage weeds with mechanical rather than herbicidal methods. For example, it may be too windy to spray but not to till the ground, or the weeds may have grown beyond the stage labeled for use of chemical control. Rain is threatening, there is enough time to till the field, but not to use herbicides that require 6 to 8 hours to be rain-fast. In this situation the proper tillage tool must be selected to bury weeds rather than to cut off the roots below the ground. Sometimes a combination of tillage and herbicides may be advantageous for weed management without losing crop residue necessary for erosion control.

Weed Seed Burial

Moldboard plowing buries weed seeds at depths from which many germinated seedlings will not emerge. Therefore, it is an excellent method of controlling weeds whose seeds have little dormancy such as downy brome (Wicks et al., 1971). However, moldboard plowing does not control weeds that have seed dormancy and this occurs in many of our weeds. It initially buries weed seeds, enforces dormancy, and initiates the seed decay process, but when these fields are moldboard plowed again these weed seeds are returned to a germinable zone in the soil (Burnside et al., 1977). Those weed seeds that have not lost viability then germinate when conditions are favorable. Field cultivation every 10 to 14 days would control seedlings and be helpful in encouraging additional seeds to germinate. Shattercane seeds that are buried will undergo accelerated decay but some remain dormant for many years (Burnside et al., 1977). Roberts (1970) and Roberts and Feast (1972) showed that weed seeds dissipated more rapidly from cultivated than untilled soil.

Weeds are very resilient opportunists, and regardless of the tillage system, herbicide usage, or cropping system, some weed species will adapt and survive. Kapusta (1979) found that specific weed species and populations will vary greatly from year to year and in relation to seedbed preparation after conventional planted crops on seedbeds prepared by moldboard plowing, tandem-disk harrowing, and harrowing. Fall panicum, giant foxtail, and ivyleaf morningglory were the most difficult to control. Horseweed was only present on the no-till plots. This plasticity of weeds makes specific weed management recommendations difficult.

Perennial Weed Management

Tillage has been successfully used for managing perennial weeds for generations. The basic principle is to till at 2 week intervals to prevent weeds from producing top growth and seeds and thus reduce food reserve in their roots. Field bindweed can be eliminated after 3 or 4 years of intensive tillage. Research in Kansas (Russ and Anderson, 1960) showed that intensive tillage at 2 to 3 week intervals plus an application of 2,4-D during the growing season for 3 years was the most effective means of controlling field bindweed. Stands of leafy spurge and Russian knapweed were reduced 90% and 80%, respectively, after 4 years with treatments involving moldboard plowing during late May then field cultivating at 2-week intervals until mid-August, planting competitive crops, and timely applications of 2,4-D (Derscheid et al., 1963a, 1963b). Thereafter annual applications of 2,4-D in the spring were necessary to prevent reinfestations. Derscheid et al. (1961) rated Canada thistle and perennial sow thistle easier to control than field bindweed and field bindweed easier to control than leafy spurge and Russian knapweed.

Moldboard plowing incorporates perennial weeds, and subsequent surface tillage operations further suppress regrowth. Perennial grasses such as quackgrass should be treated differently than Canada thistle or field bindweed. With Canada thistle and field bindweed, tandem-disk harrow the ground deep enough so that subsequent tillage tools can operate efficiently. Then every 10 to 14 days use field cultivators equipped with 25- to 46-cm-wide sweeps or shovels. These should operate parallel to the soil surface at a depth of 10 to 15 cm. They need to be sharp and overlap each other so roots are cut off by the shovels. Worn shovels or sweeps should be resurfaced or replaced.

Maximilian sunflower, a perennial, was controlled during the 1920s by shallow plowing and then dragging to expose the roots to the sun. Corn was planted and cultivated; the following year small grain was planted, and then clover (Pammel and King, 1926).

Tillage operations for quackgrass begin with the moldboard plow and follow up with tillage tools that pull the rhizomes from the soil and leave them on the soil surface to desiccate. Late fall plowing aids control since roots and rhizomes are exposed to freezing and desiccation over winter.

It is more difficult to manage perennial weeds in row crops than in small grains and fallow areas. Fields should be moldboard plowed before planting row crops. Row crop cultivation plus herbicides effectively suppress weed growth after the crops are planted. At least two row crop cultivations are necessary. Additional postemergence herbicides can be applied in row crops with a high clearance sprayer. These weed management procedures have been very successful against Canada thistle growing in corn.

Stimulate Weed Seed Germination

Weed seeds near the soil surface are likely to germinate with adequate soil water, because they have more suitable temperature, light, and oxygen conditions. Repeated cultivation exposes more seeds to conditions favorable for germination (Chepil, 1946). Therefore, frequent shallow tillage during peak germination periods should greatly reduce seed bank populations and also control the weed seedlings. In arid environments, cultivating the soil will hasten soil water loss, which may inhibit weed seed germination.

Suppression of weed germination with tillage at night is being investigated in Europe and the United States. Light triggers germination of some weed seed. Photocontrol of weeds with nighttime tillage is encouraging. Tillage at night has reduced weed emergence by 50% (Hartman and Nezadal, 1990). The moldboard plow was more efficient in reducing weed emergence than the tandem-disk harrow when used in the daytime. But when it was used at night weed emergence was similar for moldboard plowing and tandem-disk harrowing (D. A. Mortensen, Associate Professor, Department of Agronomy, University of Nebraska, Lincoln, personal communication, 1993). Weed emergence was significantly less when tandem-disk harrowing was done at night than in the daytime. One could regulate weed emergence depending upon whether and when weed emergence was to be suppressed or encouraged. Much more research is needed on this fascinating approach to managing weed emergence.

Plant Weed Seeds

Weed seeds that fall on the soil surface usually have to be covered with soil in order to germinate. However, if the surface soil does not dry out, seeds of some species may germinate even when they are on the soil surface (e.g., green foxtail and jointed goatgrass in no-till fields). In field crops winter annual weeds germinate close to the soil surface; this explains why horseweed is more abundant in no-till than tilled soils.

In the winter wheat–fallow rotation, sweep tillage after wheat harvest helps plant winter annual weed seeds such as downy brome, jointed goatgrass, and volunteer wheat (Fenster and Wicks, 1982). It is important to have these seeds germinate in the fall so they can be killed with additional tillage. Delaying tillage until the following spring may mean that some winter annual weeds will not germinate until late summer or fall. If emergence is delayed until September then these weed seedlings will establish themselves in the newly planted winter wheat field.

In row crops tillage tools may plant seed at various depths depending upon the implement used. Yenish et al. (1992) found over 60% of weed seeds in the surface 1 cm of soil with no-tillage, 30% of weed seeds in the same depth with chisel plowing, and in equal distribution of weed seeds in the 0- to 19-cm depth with moldboard plowing. The chisel plow and moldboard plow were disked twice in the spring.

Ball and Miller (1990) found more weed seeds located near the soil surface after chisel plowing than moldboard plowing in different cropping sequences. Moldboard plowing followed by the tandem-disk harrow uniformly distributes weed seeds vertically in the soil profile. They found that the chisel plowing was favorable for increasing hairy nightshade and stinkgrass in continuous pinto bean; redroot pigweed, stinkgrass, and common lambsquarters in sugar beet for 2 years followed by corn for 1 year; and green foxtail in continuous corn.

Prepare a Seedbed

An important function of tillage is to prepare a seedbed for crops. Timely tillage will control small weeds that have emerged and eliminate those that have germinated but not yet emerged. Delaying weed management too long may mean that weeds have desiccated the soil and drills will not place crop seeds into an environment favorable for germination.

For shallow planted crops such as sugar beet and alfalfa, seedbeds need to be firm, smooth, and free of clods. This may require several tillage operations. Corn is a deep planted crop (4 to 8 cm) so surface roughness has less influence on obtaining satisfactory corn stands.

Corn grain yields have been increased on a silty clay loam where the field was chisel plowed and disked once as compared to rotary tillage or no-tillage plus atrazine because of better water infiltration (Hakimi et al., 1973). Row crop cultivation in the absence of weeds after crop emergence is sometimes beneficial on finer textured soils because it loosens and aerates the soil (Sprague and Triplett, 1986).

Reduce Herbicide Use

Tillage can reduce herbicide use in many situations. Some soil applied herbicides should be soil incorporated and a lower rate can be used because less herbicide is then lost as a result of volatility and photodecomposition. Herbicides can be applied in a band versus broadcast with row crops, thus reducing the amount applied per hectare. Cultivation is then done to control weeds between the rows. Effective seedbed and early crop growth tillage may allow the farmer to switch from preemergence to postemergence herbicides. Then the farmer should scout the fields to determine whether a postemergence herbicide is needed.

In grain sorghum, application of atrazine at 1.1 kg/ha plus row crop cultivation was equal to application of atrazine at 2.2 kg/ha without cultivation in weed management and grain yield (Burnside et al., 1964). Recent research has shown that herbicide rates could be reduced in soybean fields when used in combination with tillage in soybean and corn fields (Devlin and Maddux, 1991; Doll et al., 1991). The amount of herbicide used in corn and soybean is greatly decreased with many postemergence herbicides as rates are in grams per hectare (g/ha) versus kilograms per hectare (kg/ha) as in the case of most preplanting or preemergence herbicides.

Interrow weeds can be transplanted into the row when cultivating if the soil is wet or it rains soon after cultivation. This is avoided by cultivating when the surface soil is dry and the cultivator disks are adjusted to move soil away from the row during the first cultivation. Then a second cultivation is needed to move soil back into the row to cover weed seedlings in the row after the crop has attained greater height. Moving soil away from the row also works well for 20- to 30-cm-tall interrow weeds when herbicides are banded to maintain the intrarow weed-free and weather has prevented timely cultivation.

Aid in Controlling Insects and Diseases

Tillage can be used to bury plant residue to reduce certain insect and disease populations. Timely late summer and fall tillage also aids in control of certain pests. One of the classic cases of insect and disease control with tillage and/or herbicides is with wheat streak mosaic (Wiese, 1987). Volunteer wheat and summer annual grasses are hosts to the leaf curl mite. Destroying weeds for a 10- to 14-day period before winter wheat planting eliminates the host plants, the leaf curl mite dies, and thus this vector is not able to transmit the virus to winter wheat.

Yellow leaf spot severity in wheat is decreased if crop residue is destroyed with tillage before the winter wheat is seeded (Summerell and Burgess, 1989). Crop residues under a no-till situation could be burned if delayed until near planting to provide maximum erosion control during the fallow period (Rees, 1987).

Damage from cutworms can be reduced if weeds are controlled. Late fall plowing exposed cutworms and buried their food (Hunt, 1904). Fields need to be kept free of weed growth in the spring, and planting can be delayed to prevent cutworm damage.

Aid in Managing Herbicide Resistant Weeds

A concern facing corn farmers with triazine-resistant weeds is the control of these weeds without increasing production costs and destroying crop residue. Herbicide-resistant weeds are increasing around the world as there are 57 weed species with biotypes resistant to herbicides (LeBaron, 1991). Weeds have become resistant to herbicide families such as carbamates, substituted ureas, phenoxys, dinitroanilines, sulfonylureas, and triazines (Holt and LeBaron, 1990). The development of herbicide-resistant weeds is favored by using the same herbicide year after year, using persistent herbicides, using herbicides that kill a high percentage of target weeds, using herbicides with a single site of action, and using herbicides for target plants that have little seed dormancy. Lack of tillage during seedbed preparation, no cultivation after corn and sorghum planting, or failure to use tillage after small grain harvest also contribute to development of herbicide-resistant weeds. Possible solutions include changing the crop rotation, changing to banding preemergence herbicides in corn, using herbicide mixtures or postemergence herbicides, cultivating twice, and/or adopting ridge-tillage. In small grains, control weeds after harvest with tillage plus herbicides.

Economics

Weersink et al. (1992) compared the production costs of chisel plow, ridge-till, and no-till with a moldboard plow system for growing corn and soybean. They found that total farm costs per hectare were higher for moldboard plowing and chisel plow tillage systems compared to those of no-till and ridge-till systems because of their larger machinery complement. Herbicide costs were higher for the ridge-till and no-till systems. These systems also reduced labor costs up to 61% annually compared to those of the moldboard plow system. It is expensive to grow weeds as they are costly to control by tillage and/or herbicides. Weeds also reduce yield and quality of crops. Weeds cause an estimated annual loss in the United States of $7.5 billion and in Canada of $909 million (Chandler et al., 1984).

Tillage or herbicide-tillage combinations are generally more economical than a weed management program that relies exclusively on herbicides or tillage (Delvo, 1971). However, increased cost of machinery, repairs, and fuel may change some weed management programs in the future. Farmers will use the most economical methods of weed management.

Lal et al., (1990) suggest that substantial energy can be saved by the following:

-Using no-till planting where possible
-Using chisel plowing or disking instead of moldboard plowing
-Avoiding primary tillage deeper than 20 cm
-Reducing the number of secondary tillage operations
-Avoiding secondary tillage deeper than 10 cm
-Combining various operations, e.g., pulling a spring-tooth harrow behind a disk or applying herbicides along with secondary tillage
-Traveling in the highest gear practical and throttling down
-Maintaining proper drive-wheel slippage
-Ensuring that tillage equipment is properly adjusted
-Matching implement size to tractor size

In addition weeds must wilt within 30 minutes after tillage or a repeat operation will be needed. Therefore, do not till when ground is wet and cold and avoid nightime cultivation. If cultivation is too deep, only root pruning occurs and weeds recover. Use the right herbicide for weeds present and environmental conditions.

Farmers with triazine-resistant weeds realize that control measures to replace atrazine will increase costs. Rotation of crops and herbicides may not always be economically feasible. Government programs often prevent changing to a crop for which triazine-resistant weeds are easily controlled. Crop production change also may mean the purchase of new planting and tillage equipment. Production systems involving several crop, herbicide, and tillage practices should be considered.

Furrow Openers

In semiarid areas, row crop planters are often equipped with furrow openers. These attachments remove the dry surface soil, clods, and plant residue over the crop row, enhancing soil-seed contact, soil moisture, and subsequent crop emergence. Furrow openers also remove germinating weeds and sufficient weed seeds from over the row to aid in reducing the weed population in the row (Duley and Matthews, 1947). Weed seed germination between the rows may be delayed because of lack of soil water for germination. Then the crop can emerge ahead of the weeds and thus allow a height difference that can improve the efficiency of subsequent cultivation. Wicks and Somerhalder (1971) found that 81% of the weed seeds were removed from 0- to 7.6-cm depth over the row in ridge-till corn during the planting operation.

Improve Water Infiltration

Tillage generally improves water infiltration by reducing rainfall runoff. Cultivation in row crops often improves water infiltration on clay pan soils, and it has increased corn grain yields when compared to no-tillage plus herbicides after planting (Whitaker et al., 1973).

A tillage tool (contour basin listing) that not only controls weeds but reduces water runoff was developed in the southern Great Plains (Steward and Musick, 1982). The dike cultivator forms small dams between crop rows that trap water and ensures greater water infiltration. It is used on about 800,000 ha of cropland in the Great Plains (Papendick and Campbell, 1988).

A similar type of practice (Duley and Matthews, 1947) in which the implement left shallow pockets on the soil surface was not widely accepted in the 1940s because when the pockets filled with water or silted in, their effectiveness was reduced. Often water erosion was serious with moderate to heavy rain. Dams or pockets on sloping land would break or overflow near the top of the hill, causing serious erosion down slope. In addition weed management was a greater problem with this tillage system.

Incorporate Plant Residue That Is Toxic to Crop Seeds

Crop residues that are selectively toxic are used as a weed management tool in conservation tillage and no-till systems. The most commonly used crops are rye, oat, wheat, sorghum, and certain winter legumes (Worsham, 1989). Sometimes crop residue inhibits germination and emergence of the subsequent crop. An example is winter wheat planted into soybean residues. The winter wheat will emerge faster and more uniform stands will be obtained if the soybean ground is lightly tilled to incorporate the residue before planting (G. W. Hergert, 1992, Cropping systems for stretching limited water supplies in the central Great Plains, unpublished data, University of Nebraska, North Platte). Faster, earlier growth of winter wheat also would aid in managing weeds by shading the ground sooner. Large seeded legumes, corn, transplanted vegetables, and tobacco show good tolerance to allelochemicals from crop residues (Worsham, 1989).

Prevent Weed Seed Production

Timely tillage may prevent seed production if performed before weeds have pollinated. Plowing soon after small grain harvest was advocated during the 1920s to prevent weed seed production (Pammel and King, 1926). Weeds may survive the tillage operation and continue to grow, so additional tillage is needed to kill the escapes. Controlling green and yellow foxtail before they set seed in wheat stubble after harvest is important in the long-term management of these weed species. However, using tillage to manage scattered weed escapes may not justify the tillage expense because of the small change in the weed seedbank.

B. Disadvantages

Tillage exposes the soil to wind and water erosion. This is the price we pay for crop production because the competing natural vegetation must be managed in order to produce a profitable crop. Thus the challenge to agriculturists is to maximize crop production while minimizing adverse effects on our soil, natural resources, and environment.

Increased Erosion

The development of the steel moldboard plow and other implements that leave the land bare of vegetation increases exposure of soils to erosion. Water erosion increases on bare cultivated land free of vegetation, especially if it is sloping, compared to stubble land or soil protected by vegetation. Clean tillage was the major cause of wind erosion in the Great Plains (Duley and Mathews, 1947).

Tillage loosens the soil surface so that soil is more readily moved by wind or water when runoff events occur (Beatley, 1982). Occasionally the surface soil is even lost as a result of water or wind erosion to the depth of tillage.

Accelerated erosion takes place when the soil surface contains finer particles and little plant residue. Soil crusting also may follow heavy rains, causing quicker runoff, and crop seedlings may not be able to emerge. To prevent erosion during winter some farmers delay tillage until spring. Thousands of kilometers of terraces have been built with little thought of maintaining crop residue. Many of these terraces silt in quickly and become useless.

Tillage that leaves the soil surface smooth, free of clods and plant residue, subjects land to greater wind erosion because tillage destroys the growing plants and incorporates plant residue. Also at the soil surface tillage produces smaller aggregates which are more subject to wind erosion.

Increased Loss of Nitrogen and Organic Matter

Tillage hastens the loss of nitrogen and organic matter by stimulating mineralization and microbial decomposition of the organic matter. Eventually, tilled soils producing crops need to have fertilizer requirements replenished. On some tilled soils the water holding capacity may be lowered so leaching of nutrients and pesticides is increased. Also no-till fields retain more soil water and thus increase crop yield potential in semiarid regions, thus increasing the fertilizer requirement.

Reliability in Controlling Weeds

There are many tillage implements, which differ markedly in their ability to manage weeds. The moldboard plow, tandem-disk harrow, and power driven rotary tiller are the best tillage implements for managing weeds before the crop is planted. Depending upon the weed species, subsequent tillage is often needed to control remaining weeds. Size of weeds dictates the success of the various tillage tools. The rotary hoe and the spike-tooth harrow are most effective on seedling weeds that have not emerged. The spring-tooth harrow is capable of killing larger weeds and aids in managing quackgrass on plowed ground as the rhizomes are pulled from the ground and left on the soil surface to be exposed to desiccation.

There are various types of cultivators that vary in effectiveness for managing weeds. Shovel and sweep cultivators are most common and are probably most efficient (Sprague and Triplett, 1986). Rotary ground driven cultivators are effective in controlling seedling grass and broadleaf weeds, but summer annual grasses that have tillered are extremely difficult to kill since the tines do not remove the weed roots from the soil. These grass weeds continue to grow after use of the rolling cultivator. "Hillers" are "furrowers" or disks set to move soil within the crop row in order to cover up small weeds. The final cultivation in furrow irrigated corn moves soil from between the rows to within the row in order to provide a furrow for movement of irrigation water. This extra soil also can cover up small weeds within the crop row.

Weed management efficiency is improved when the crop is several centimeters taller than the weeds. This situation allows more soil to be moved within the row in order to cover the weeds. However, this is not feasible with soybean as the soil in the row reduces the seed set height so there is more seed loss during harvest. Partially covering peanut plants with soil increases the incidence of diseases caused by soil-borne fungi (Buchanan et al., 1982). If weeds are too tall, a band-applied herbicide may supplement cultivation. For example, a low rate of 2,4-D applied to small broadleaf weeds in corn or sorghum may not kill them but will slow their growth. Also weeds may be bent over after spraying; this improves the chance that soil moved into the row will cover these weeds. Applications of nicosulfuron or primisulfuron will control shatter cane in corn, but they do not consistently kill field or longspine sandbur. However, these herbicides will stunt the sandburs while allowing the corn to grow normally, creating a height differential. The sandburs can then be covered with soil during a subsequent cultivation.

In short growing crops such as sugar beet, intrarow weeds are difficult to control without herbicides or hand labor. Rotary hoeing at 2.8 km/hr when sugar beets were in the two-plus-leaf stage reduced weed biomass 7% as compared to that of the untreated check (Wicks and Anderson, 1969). In peanuts, the feasibility of cultivation alone may be limited as shallow cultivation does not control some weed species such as Texas panicum and pitted morningglory (Wilcut et al., 1987). Weed management with herbicides plus cultivation generally provides the most economical return to the producer.

In orchards, herbicides have definite advantages over tillage. Herbicides are economical, reduce erosion, enable the grower to establish a sod between trees, control weeds near the tree trunk more easily, facilitate better insect spray coverage, and do not disrupt the roots of trees.

Wet Ground Delays Timely Tillage

Tillage when the soil is relatively dry is most efficient in controlling weeds. Occasionally, when the soil is moist, weeds may even be transplanted within the crop row by cultivation. Weeds are most difficult to manage with tillage when rain occurs repeatedly after planting (Moomaw and Robinson, 1973b; Wiese et al., 1964). Rain aids rerooting of weeds after tillage, and when tillage is delayed by wet weather, weeds continue to grow undisturbed. By the time the soil is dry enough for cultivation, the weeds may be too large for effective management by cultivation.

Weed Propagules Are Spread by Tillage Equipment

Spreading of weed propagules is a concern of farmers as weed "patches" often enlarge when tillage tools pass through the weed infested areas. This is especially true with vines such as field bindweed or rhizomatous weeds such as quackgrass. Common purslane is a troublesome weed in the winter wheat-fallow areas of the Great Plains and is spread by plant parts attaching to tillage implements. Weeds are commonly spread by seed infested soil carried on tillage equipment. Anytime you transport surface soil you also spread weed seeds and other propagules. Vehicles and most farm implements (e.g., commercial combines) spread weed seeds (Burnside, 1970b). Weed seed such as puncture vine or longspine sandbur can attach to machinery tires or animals and be moved great distances before it is dislodged.

One of the worst cases of weeds being spread by tillage is rush skeletonweed in Australia. Rush skeletonweed was introduced in about 1917, and by the mid-1930s many wheat growers were going broke because of competition from rush skeletonweed and a depressed economy. Cultivation had actually encouraged the spread of this weed, which regenerated from root fragments (Pratley and Rowell, 1987).

Soil Compaction

Large wheeled tractors may add to compaction problems when farmers have the horsepower and traction to work the fields before soils have adequately dried. Implements such as the tandem-disk harrow can cause considerable compaction, especially when the soils are worked too wet. All primary and secondary tillage implements will cause varying amounts of soil compaction, depending upon soil conditions and size of the tractors (Sprague and Triplett, 1986).

Root Pruning

Most plants have a large portion of roots near the soil surface and are subject to root pruning during row crop cultivation. Cultivation done too closely to the row late in the season or too deep prunes the roots of row crops. Results of early research indicated that corn yields were reduced 7% to 13% when cultivation depth was increased from 4 or 5 cm to 10 cm (Hunt, 1904; Williams and Welton, 1915). To prevent this crop injury, interrow crop cultivation should be completed by the fifth or sixth week after corn emergence. "Shovels that do not penetrate deeper than 4 or 5 inches deep and do not come within 9 inches of the plant at their deepest point will do little if any permanent injury under Nebraska conditions. The possibility of serious injury increases with the severity of cultivation and under conditions which favor a very shallow, lateral growth of secondary root system" (Kiesselbach et al., 1935). A significant increase in corn lateral root development occurs after pruning by cultivation (Spencer, 1941). This may reduce the corn plant's ability to use water at greater depths in the soil profile. Pruning the roots may cause slight to severe reduction in corn yields. Sometimes it is necessary to cultivate deep in the case of compacted soil. If so, the first cultivation should be deep in order to reduce damage to crop roots. Sweeps operate at a shallower depth than cultivator shovels and are effective in killing small weeds. In a four state study, when cultivations were initiated at 1 or 2 weeks after emergence and cultivated weekly at zero, one, two, four, or six times, seed cotton yields were not affected by cultivation when weeds were eliminated (Snipes et al., 1992). Some soybean cultivars are more tolerant of cultivation injury than others (Russell et al., 1971). In potatoes each cultivation reduces potato yields in the absence of weeds about 2% (Nelson and Giles, 1989). No cotton yield loss was reported with root pruning by deep or repeated cultivation (Buchanan and Hiltbolt, 1977). Root pruning from cultivation may reduce the vigor of trees in shelterbelts and orchards.

Soil Water Loss

Tillage loosens the soil and decreases its overall organic matter content, allowing water to be lost by evaporation more readily from the plant root zone. This phenomenon is especially noticeable in fallow cropping systems in the semiarid Great Plains. Good and Smika (1978) showed that shallow tillage, that controlled all weeds during the growing season, resulted in the loss of 5 to 8 mm of water as compared to that of untilled plots where weeds were controlled with herbicides.

Weed Seed Germination Following Cultivation

Many weed seed are encouraged to germinate by shallow cultivation because the soil is loosened and aerated and weed seeds are exposed to light. Thus farmers generally observe a flush of weed seedlings following each cultivation. A single cultivation 4 to 6 weeks before planting wheat has encouraged germination of wild oats and annual ryegrass in Australia (Pratley and Rowell, 1987).

Tillage Tolerant Weeds

Some weed species are easier to manage with tillage than others. For example, annuals such as small redroot pigweed are easily controlled with tillage. Common purslane, a xeophyte, is extremely difficult to kill with tillage after the two- to four-leaf stage. Common purslane has the capability of rerooting after extended dry periods because it does not desiccate as rapidly as other weeds. Therefore, it is an increaser on summer fallow. Undercutting weeds too deep will cause a fibrous root development that makes weed management difficult with fallow operations. Half killed weeds are often difficult to kill. Some perennial weeds such as quackgrass develop their roots below the plow depth and thus tolerate plowing.

III. INTEGRATION OF MECHANICAL WITH OTHER WEED MANAGEMENT METHODS

A. Preventive

Preventing the introduction of a weed into a field is the most efficient weed management method. In weed management a gram of prevention is worth a kilogram of cure. Control of incipient infestations is necessary in good weed management programs. One of the most important methods of weed management involves identifying and mapping small weed infestations. Successful integration of mechanical and other weed management methods must consider their impact on erosion, water quality, and economics before implementation. For example, one would not want to destroy the crop residue before the rainy season or dry out the soil with excess tillage during the windy season prior to crop planting. These incipient weed infestations may be controlled by hand pulling, hoeing, spading, chopping, or mowing or with a herbicide. Weed control operations must be performed before seed development or weed seeds may have disseminated naturally or by man's intervention. Prevention also includes cleaning equipment of weed propagules and planting weed-free seed (for detailed discussion refer to Chapter 3).

B. Cultural

Many weeds have peak germination and emergence periods. The first summer annual weeds to emerge during the spring in winter wheat fields in the central Great Plains are kochia, waterpod, Russian thistle, and common lambsquarters, barnyardgrass, green foxtail, witchgrass, stinkgrass, and redroot pigweed peak emergence is in early June (Hoefer et al., 1981). Wilson et al. (1992) found that the emergence of common cocklebur, common lambsquarters, common sunflower, green foxtail, yellow foxtail, kochia, Pennsylvania smartweed, redroot pigweed, and Russian thistle was greatest in April and declined at later planting dates. Longspine sandbur, shattercane, and wild-proso millet emergence did not vary with planting date. Downy brome and jointed goatgrass peak emergence was April and September and lowest in June and July: a typical response for winter annual weeds. It should be possible to predict weed growth and determine whether the weed is in a susceptible stage of growth for control using mechanical methods or herbicides (Wilson et al., 1992).

Adjusting planting date and seedling rate can be useful in managing weeds when done in combination with tillage. Downy brome, Japanese brome, and jointed goatgrass are serious winter annual weeds in winter wheat. They germinate as soil temperatures cool and generally emerge with the winter wheat. Therefore, delaying planting of winter wheat is a method of managing these winter annual weeds. This works best when rainfall occurs in early September and rodweeding is performed about 6 days later.

Early planted corn requires more cultivation to control weeds (Hunt, 1904). Delayed planting of corn to control flushes of weeds with tillage may be practical, but planting can not be delayed beyond an optimum planting date. Forcella (1991) has developed weed seedling emergence models that are used in predicting timing of redroot pigweed emergence in soybean and indicating optimum timing for mechanical weed control. He was able to contain 85% control by

Table 5 Influence of Planting Date on Weed Control in Soybean When Field Was Disked in Late April and Rated on July 3

Weed species	Seedbed preparation and planting date: May 12	Seedbed preparation and planting date: June 2	Reduction
	Density/m^2		%
Common lambsquarters	13	2	85
Redroot pigweed	38	14	63
Giant foxtail	74	18	76
Soybean yield, kg/ha	470	1747	

Source: Gunsolus (1990).

delaying planting and cultivating to control emerged weeds. Delayed planting aids weed control in soybean (Table 5).

Delaying seedling of alfalfa until fall works well for controlling summer annual weeds but does not control winter annual weeds unless a herbicide is used. Alfalfa should be planted by mid-August in the Corn Belt in order to become established before winter. Therefore, winter annual weeds flourish with the alfalfa seedling. If there are mainly summer annual weeds, such as wild oat, shatter cane, redroot pigweed, and green foxtail, alfalfa should be planted in the fall. Delayed planting in the spring also allows time to kill winter annual broadleaf weeds with spring tillage. Delayed seeding with spring small grains in the central Great Plains does not work as crops are forced into grain fill during the hot part of the summer, and this condition significantly reduces grain yield. Delayed seeding for the control of wild oat in Canada is not practical as the yield loss from late seeded spring wheat is greater than the cost of the herbicide to control wild oat (Hunter, 1980). Some small grain crops are more competitive than others.

Selection of competitive crops is an important weed management method. Sow spring barley rather than spring wheat to reduce weed competition (Nalewaja, 1978). Barley is becoming more popular in northern New South Wales, Australia, because of weed competition, better frost tolerance and therefore earlier sowing, better performance where fertility is declining, and earlier harvest. Timeliness of sowing and harvest improves equipment and labor management. Tall growing crops with rapidly developing seedlings such as corn are more competitive than shorter, slow growing crops such as grain sorghum. Before preemergence herbicides for sorghum were available, the planting date was delayed and two or three tillage operations were used to kill summer annual weeds. Herbicides such as atrazine allowed planting dates to be moved up to an earlier date (Burnside and Wicks, 1965). This increased hybrid selection availability and greatly increased sorghum grain yield.

Flax is a poor competitor with weeds. In the days before herbicides flax was planted on ground that was broken out of sod. The virgin sod was relatively free of summer annual weeds. Sugar beet is also a poor weed competitor and initially weeds were controlled with moldboard plowing, close cultivation, and hand hoeing.

The plow was recognized as a weed management tool for corn in the 1920s in Iowa (Pammel and King, 1926). "One of the best and most effective methods of destroying the foxtail is by plowing the small grain field as soon as grain is removed. If this is not done a large amount of seed is produced. After this plowing in the fall the field should be disked and harrowed in the spring and then planted to corn. The corn should be cultivated as frequently as possible, at least four or five times." In Nebraska corn yielded 4% more with four cultivations compared to three (Table 6).

Early weed control in corn (before the 1950s) was accomplished by moldboard plowing, using a weed-free seedbed; planting the corn in checked rows; using spike-tooth harrowing, flexible-tine harrowing, or rotary hoeing; and practicing interrow cultivation. Weed control was the main consideration for corn cultivation (Kisselbach et al., 1928). They said, "The yields of grain increased materially as the number of cultivations was increased up to the number necessary for perfect weed control. The problem of weed control will vary, however, from year to year, depending on weather and soil conditions and the efficiency of the cultivations." Corn can be harrowed or rotary hoed until the three- or four-leaf stage. Check-row corn was planted with a planter that dropped seed in hills spaced 106 cm apart. The seed drop was triggered by "wire knots" spaced 106 cm apart on a wire laid the length of the field. After corn emergence, the check-rowed corn was cultivated in two directions at right angles to each other. Late emerging weeds after the last corn canopied were

Table 6 The Relation of Frequency of Cultivation to Corn Grain Yield from 1922 to 1932

Number of cultivations	Corn yield 1922–1927	Corn yield 1928–1932	Reduction 1922–1927	Reduction 1928–1932
	kg/ha		%	
0	445	—	81	—
1	1360	—	42	—
2	2110	1980	10	7
3	2250	2020	3	5
4	2330	2130	0	0

Source: Kisselbach et al. (1935).

controlled by one-horse cultivation or by lamb grazing of weeds in the cornfields. This also facilitated corn harvest, which was done manually. Ewes could not be used because they ate the corn ears. The practice of check-rowed corn disappeared because planting on contours increased, check rows slowed planting, better cultivators mounted on tractors became available, herbicides were introduced, and the practice reduced corn yield potential. Hill-dropped corn replaced check-row corn. Planting two to three seed per hill helped brace the corn from soil deposited during cultivation. Weed escapes were and still are controlled by hand hoeing in many fields. If weeds were to be a problem it was better to plant corn in hills rather than spacing seed evenly in the row (Williams and Welton, 1915).

In spring small grains, weeds are controlled by planting early so that the crop will shade the ground before summer annual weeds germinate. Moldboard plowing improves control of green foxtail and has resulted in higher spring wheat grain yields as compared to stubble mulch tillage (Bond et al., 1971). In the era before combines, grain binders were used to cut the crops; usually this was before much weed seeds shattered. The fields were shocked and the bundles were hauled to a central location for threshing. Threshing was a method of weed management as weed seeds in the grain bundles ended up in the straw pile rather than being spread back into the field as with combines (Burnside, 1970b).

Cultivation of small grains is not a common practice now, but it was used before the introduction of 2,4-D. As early as 1904, favorable results were reported when wheat was harrowed 1 to 2 weeks after seeding (Hunt, 1904). Campbell (1909) suggested that the best time to harrow was when the small grain was beginning to stool or when it was 8 to 10 cm tall. The flex-tine weeder caused less damage to the small grain than the common lever harrow.

Tall cultivars of small grain were the rule in the days before herbicides. These cultivars were selected by farmers over time for their weed competitiveness. After the advent of herbicides, plant breeders began to select for short cultivars that produced more grain and were less prone to lodging. This increased farmer dependency on herbicides as semidwarf cultivars are less competitive with weeds. Research is now showing that crop competitiveness to weeds is a feasible breeding objective (Rose et al., 1984).

Adjusting soil fertility is a method of weed management. Starter fertilizer helps corn seedlings grow faster, thus reaching a height that facilitates earlier cultivation. Adequate fertilizer increased leaf growth, which facilitated earlier shading of the soil surface and thus reduced weed seed germination. Increasing nitrogen rates applied to winter wheat decreased annual grass weed populations and weed yields (Valenti and Wicks, 1992). Application of too much nitrogen may allow the small grain crops to grow rank and produce more shade, but this situation tends to facilitate lodging and reduce grain yields.

Cover Crops

Rye is used to produce winter cover for protection of the soil from wind and water erosion. Rye also is an effective means for reducing herbicide inputs by suppressing weeds (Worsham, 1991). Perennial legumes are also used for soil cover and nitrogen production, but they are generally more expensive to establish. Rye and vetch on sandy soils increased the productivity of the crops and crop competition to weeds. In Australia, self-seeding annual legumes are used profitably as livestock graze the legumes for 1 year or more before wheat is planted. The success of this system depends on the timeliness and consistency of rainfall patterns. Cover crops may be killed with a herbicide, tillage, or a combination of treatments. Disadvantages of cover crops for some areas are that they may be expensive and difficult to establish, use considerable water, and may be difficult to kill. One has to be careful with plant selection as some plants such as kudzu may become serious weeds. Kudzu was introduced in the southern United States to aid in controlling soil erosion. Ryegrass, rye, and hairy vetch are other examples of plants that are difficult to control in some regions.

Smother crops are cover crops that are planted with the purpose of suppressing weed growth. Rye planted in the fall produces lots of top growth by corn planting time. Most weeds are controlled and rye must be controlled in a timely way to prevent corn yield losses. A recent innovative approach was planting white mustard, a short growing plant (10 cm), at 1000 to 2000 seeds/m^2 in band over the corn row. Weeds were suppressed for 4 to 8 weeks with no effect on corn development (DeHaan et al., 1990).

Mulch Till

Mulch till usually refers to using a tandem-disk harrow or chisel plow to replace the moldboard plow for primary tillage. Plant residue is left on or near the soil surface and it is the most common tillage practice for corn production. This method leaves many weed seeds near the soil surface. In corn, giant foxtail density in a chisel plow plus one disking prepared seedbed was seven times greater than in a field that was moldboard plowed and disked twice. However, velvetleaf density increased threefold with moldboard plow plus two disking versus chisel plow plus one disking seedbed preparation (Buhler and Daniels, 1988).

Moldboard Plow–Tandem Disk Harrow-Plant

The tillage moldboard plow–tandem disk harrow–plant tillage sequence was the standard conventional tillage practice in the Corn Belt until the 1980s. Moldboard plowing killed most growing weeds, buried weed seeds, and left the soil free of plant residue so that subsequent tillage operations were more effective in controlling late germinating weeds. The weed seed bank was reduced with moldboard plowing and summer annual grass weeds were reduced more than broadleaf weeds (Burnside et al., 1986).

The plow-plant system is used with spring small grains and sugar beet to bury weeds and plant immediately. This enables the crop seed to be firmly placed into moist soil for quicker and more uniform germination and to get ahead of germinating weeds.

Stale Seedbed

The present day stale seedbed techniques involve preparing a seedbed by tillage between crop harvest and planting corn, sorghum, or soybean the following year. This technique allows farmers to chisel or moldboard plow clay soils in the fall to prevent compacting wet soils in the spring. Once the seedbed is prepared, no further tillage occurs before planting. Weeds are controlled with herbicides after seedbed preparation. Bruff and Shaw (1992a, 1992b) demonstrated a well planned weed management program using the stale seedbed technique in soybean. Their initial tillage, after harvest in the fall, was followed by the final seedbed preparation, in which they incorporated herbicides with a tandem-disk harrow 1 to 2 months before planting soybean. They used nonselective herbicides to eliminate weeds at planting and postemergence herbicides to control late emerging weeds. Advantages of this system are that it used tillage to reduce or eliminate crop residues that could harbor diseases and insects, spread labor over a longer period, precluded delay of planting to till the ground, had less risk of compacting the soil when ground was wet, may save soil water, and offered more opportunities to control some weed species.

Ridge-Till

Ridge-till is a row crop production system using no-till followed by strip tillage. Fields are left untilled until planting and during the planting operation the surface soil within the row is moved to the interrows. About 80% of the weed seed is also moved to the interrow areas along with the surface soil (Wicks and Somerhalder, 1971). Cultivation controls weeds after planting and rebuilds the row ridges. It is important in semiarid regions that weeds be prevented from using soil water before planting the crop. Herbicides may not be needed before the crop is planted, depending upon the climatic conditions, weed species, weed density, and time of planting. Early planted corn may not require a herbicide to control emerged summer annual weeds. However, if winter annual weeds are present, a herbicide should be used before these weeds dry out the soil in the ridge or produce seed. Late planted crops such as soybean or sorghum need to have weeds controlled 1 to 3 weeks before planting. One cultivation and a ridging operation are needed to re-form the ridges after planting. Ridge-till is becoming more popular as equipment improves and farmers strive to keep more crop residue on the soil surface. Summer annual weeds are a greater problem in ridge-till than moldboard plowed fields (Forcella and Lindstrom, 1988; Wicks and Somerhalder, 1971). Rotations involving corn and soybean have fewer weed problems than continuous corn.

If corn is harvested for ensilage rather than grain, winter annual weeds will be more prevalent after harvest. In the Great Plains there were 7.80 versus 0.02 weeds/m^2 in the spring before ridge planting where corn growth the previous year was harvested for ensilage versus grain, respectively (G. A. Wicks, unpublished data, 1992). Shade, stalks, and leaves between the rows prevented weed emergence in later harvested corn for grain. Weeds in the ensilage area had to be killed by cultivation before the corn was ridge planted the following spring.

Crop Rotations

Weeds are easier to manage in crop rotations than in monocultures. Crop rotations provide increased opportunities for implementing different weed management strategies. Mechanical weed control should be used to help manage weeds, maintain or reduce crop residue, and prepare a seedbed. More efficient weed control is obtained when rotations include both summer and winter annual crops. An example is to change from continuous winter wheat to winter wheat–fallow rotation or winter wheat–sorghum–fallow. A third crop could be included by following grain sorghum with corn. If volunteer sorghum is a problem weed, corn can be planted after grain sorghum so nicosulfuron or primisulfuron can be used to control the volunteer sorghum. This type of rotation interchanging fall and spring planted crops greatly improves control of winter annual weeds such as downy brome, jointed goatgrass, and rye. A good rotation in the central Great Plains is winter wheat–sorghum or soybean-corn, depending on expected rainfall. Much of the Corn Belt follows corn with soybean because of better pest control and higher crop yields. Longer rotations may include alfalfa if moisture is adequate.

Wild oat is a serious weed in spring planted small grain areas. Pammel and King (1926) suggested that they can be controlled by cultivation, planting of clean seed, and good crop rotation. A good rotation to follow is oat followed by corn, then alsike clover and timothy. In New South Wales, Australia, rotating wheat, a winter crop, with grain sorghum, a summer crop, allows two winters to control wild oat with tillage or herbicides (Martin and Felton, 1993). Changing the crop rotation to a fall planted crop eliminates competition from wild oat in the winter annual crop. However, in central Oklahoma and northern Texas, wild oat has developed a winter annual life cycle that enables it to mimic winter wheat (Wiese, 1977). This emphasizes the plasticity of weeds and the importance of integrated control measures. Weed management methods are transient and there are no permanent solutions in weed management.

Narrow Rows

Small grains have traditionally been planted in narrow rows 15 to 36 cm apart. Disk drills with 18- to 25-cm spacings are popular, while hoe drills usually have 30- to 36-cm spacing. In the Pacific Northwest drills with 50-cm spacings will be used in dry years when necessary to reach moisture to germinate the winter wheat without covering the seed too deeply. Disk drills are most successful when surface soil water is adequate for germination or rainfall occurs during the seedling establishment period. If the soil is too dry for uniform crop emergence, weeds are not adequately suppressed by crop competition. Hoe drills are used in semiarid areas when there is sufficient soil water to a depth of 5 cm for germination. The points on the hoe drill transfer dry soil over the row to interrow areas and the press wheel firms the seed into moist soil. The drill destroys weed seedlings by removal or burial.

Row width for row crops such as corn was dictated originally by the width of a horse as horse drawn cultivators were used to cultivate corn. In 1904, the most common row width was 112 cm (Hunt, 1904). This allowed enough space between rows so that horses could pull a cultivator. Eventually a spacing of 106 cm became the most popular. Tractors and preemergence herbicides allowed farmers to explore reduced row widths. As weed management methods improved, further reductions in row widths were facilitated. Narrow rows shade the ground more quickly than wider spaced rows, thus improving weed control (Burnside and Colville, 1964; Felton, 1976). The most common corn row spacing in the 1990s is 76 cm. Soybean and sorghum are two row crops, where the practice of decreased row spacing below 76 cm has been on the increase.

Harvest Date

Timely harvest can be used efficiently as a weed management technique. Natural and mechanical losses of grain need to be prevented. Natural losses include those due to hail, rain, wind, disease, insects, weeds, birds, animals, shattering at maturity, and genetics. Some cultivars are very susceptible to shattering. Some early maturing winter wheat cultivars, such as 'Wichita', are susceptible to shattering. Field bean shatter easily at maturity. Partridge pea desiccate when ripe, causing the pod to open and throw the seed.

Mechanical losses include grain lost by the action of the sickle, by the reel, and by passage through the combine or spillage during unloading and transporting. Dodds (1967) measured 40 kg/ha loss when small grain was combine harvested at 35% moisture and a 84 kg/ha loss when it was harvested at 12% moisture. Natural loss could vary from 17 kg/ha to 100% of the potential yield. Overloading the straw walkers causes major grain losses when combining (Nyborg, 1964). Some custom combine operators overload their machines by traveling too fast and cut too high, leaving some grain heads in the field. Swathing and combining will complete small grain harvest 3 to 5 days sooner than straight combining (Dodds, 1967).

Green weeds contribute to harvesting losses with small grains mostly by inhibiting grain from falling through the straw walkers, chaffers, and sieves in the combine. Weedy small grain fields should be swathed to aid in drying out the weeds before combining. In corn, sorghum, and soybean, widely varying weed densities that had been desiccated by freezing did not reduce harvest efficiency (Burnside et al., 1969). The stripper header, a recent innovation, should aid weed management as only the grain goes through the combine. The crop is left standing so there is no windrow effect to cause tillage or planting problems in the succeeding crop.

Harvest dates can be adjusted by crop and cultivar selection, planting date, crop use, number of hectares to harvest, and size of harvest equipment. Delayed harvest increases the risk of more grain shattering, hence more weed problems (Burnside et al., 1969). During delays weeds also may grow and interfere more with combine efficiency. Timely harvest of crops offers farmers a better opportunity to remove weeds before they produce seeds.

C. Biological

Controlling weeds has also facilitated the control of certain diseases and insects. For example, wheat streak mosaic virus is spread by the leaf curl mite. This insect lives on volunteer winter wheat and other annual grasses. Killing these grass weeds with tillage or herbicides greatly reduces the risk of this virus spreading into adjacent newly emerged winter wheat fields. The destruction of host plants plus the late seeding of winter wheat controls the disease (Wiese, 1987). Farmers have also reported less damage from Russian wheat aphids by eliminating volunteer winter wheat and other host plants in the fallow strips. Controlling Russian thistle, plaintain, mustards, and Virginia pepperweed can reduce the number of leafhoppers carrying the curly-top virus into sugar beet and dry bean fields.

D. Chemical Plus Mechanical

Herbicides can be applied as broadcast, band, or spot treatments. Broadcast applications have been popular, but recent interest in band applications has increased because of environmental and economical concerns. For decades spot spraying of patches of perennial weeds has been an effective control program.

A combination of tillage and herbicides is an effective method in weed management (Table 7). Tillage is used to control weeds missed by broadcast and band herbicide applications. Initially the most efficient and dependable method of managing shattercane in corn was EPTC applied preplanting with soil incorporated plus two cultivations after corn emergence (Burnside, 1970a). Two cultivations were more effective than one, but a third cultivation was not required. After 25 years, this is still a viable treatment for the control of shattercane in corn. Doll and Mulder (1991) found that weed control, corn grain yield, and economic returns were generally higher in treatments combining chemical and mechanical weed management than in those using mechanical

Table 7 Influence of Cultivation and Cultivation plus Herbicides (herb) on Weed Biomass and Crop Yield for Corn, Sorghum, and Soybean

Number of cultivations*	Corn		Sorghum		Soybean	
	No herb	Herb	No herb	Herb	No herb	Herb
			Weed biomass, kg/ha			
0	2070 a	690 bc	3740 a	490 b	4100 a	2420 b
1	1150 b	240 cd	620 b	22 b	2000 bc	370 de
2	860 b	66 d	430 b	22 b	1200 cd	134 e
			Grain yields, kg/ha			
0	4160 b	5610 a	3710 b	4240 ab	1050 c	1480 b
1	5790 a	5770 a	4600 a	4350 a	1620 b	1890 a
2	5460 a	5310 a	4600 a	4580 a	1610 b	1860 a

Source: Burnside et al. (1969).
*Numbers followed by the same letter along cultivation for crops and within columns for weed biomass and grain yields are not significantly different at the 0.05 level.

weeding alone. In potatoes, the best weed management program consisted of a combination of pendimethalin plus one cultivation (Nelson and Giles, 1989). Tillage can take various forms such as the rotary hoe or shovel cultivation.

Mechanical injury to plains pricklypear plants before herbicide application has improved control. Use of a cultipacker before spraying silvex, 2,4,5-T, and 2,3,6-TBA (Thatcher et al., 1964) or the rotary hoe before spraying silvex or picloram improved control of plains pricklypear (Wicks et al., 1969). Cuticle injury facilitates increased herbicide penetration into pricklypear, plus it allows insects and/or diseases to enter the plants and help destroy them. Picloram is the only herbicide presently labeled for this treatment.

Herbicide booms have been mounted beneath sweep machines in order to place herbicides in a horizontal layer about 10 to 15 cm beneath the soil surface (Fenster et al., 1962; Solie et al., 1983). Applying herbicides by this technique allows for the use of volatile herbicides without the usual soil incorporation, reduces vapor and runoff losses, combines a tillage operation with a herbicide application, and leaves plant residue on the soil surface. Weed control was inadequate with only subsurface application so a second boom was mounted behind the sweep plow to place a portion of the herbicide on the soil surface. This procedure improved weed management and eliminated the need for contact herbicides. This method of herbicide application with sweep tillage was not adopted by farmers. Some of the reasons were that soil dried out too quickly, after winter wheat harvest in dry years they had difficulty operating the sweep plow, and preventing overlapping of herbicides was difficult.

Spray nozzles have been attached to the front of tandem-disk harrows so that spraying and incorporation can be accomplished with one pass. This remains a popular application method of applying volatile herbicides or where rainfall is not sufficient to incorporate a herbicide.

Spray nozzles have also been attached to the cultivator so that intrarow weeds can be sprayed during the cultivation operation. Some herbicides do not work as well with this application because they are adsorbed to dust particles and thus unavailable for plant uptake. In addition, translocation into the weeds is inadequate when weeds go into stress from root pruning.

Use of reduced herbicide rates in combination with tillage has become increasingly popular because of economical and environmental considerations. Reduced herbicide rates have been especially successful in soybean production (Delvin and Maddox, 1991).

Shifting weed management strategies from preemergence to postemergence herbicides in combinations with row crop cultivations requires judgment in deciding when to cultivate or spray to obtain maximum weed control. Newsom and Shaw (1991) showed that cultivation at any time between 3 days before to 7 days after a postemergence application of imazaquin improved sicklepod control over that of imazaquin with no cultivation. However, control was best when cultivation was accomplished 3 or 7 days after imazaquin was applied. When chlorimuron was used with cultivation, control of sicklepod was best when cultivation was delayed until 7 days after application. Poorest control was obtained with both herbicides when cultivation occurred at the time the postemergence herbicides were applied.

Sometimes combining chemical and mechanical weed management does not eliminate the weed problem. In small grain crops annual use of triallate or flamprop in four successive wheat crops and tillage between rows did not prevent massive buildup of wild oat seed (Martin and Felton, 1983).

IV. FARMING TRENDS AND FUTURE OF MECHANICAL WEED MANAGEMENT

Farm numbers in the United States (about 2 million) dropped 6.8% from 1982 to 1988 and average farm size thus increased from 178 to 187 ha (USDA, 1989). Currently 390 million ha of land is farmed in the United States, and a steadily increasing amount of erodible land is being retired into conservation reserve programs. Livestock production is being concentrated on fewer farms so that the bulk of our farm units have become cash grain or crop enterprises. Thus we are making less use of crop residue and this affects our mechanical weed control and farming methods. Efforts are being made to move the animal wastes to the field. This creates soil management and environmental problems as there may be too much manure in limited areas for incorporation into the soil. The manure also contains weed seeds that eventually have to be dealt with as these seeds germinate and infest the crops.

Mechanical weed management has been and will continue to be a major method of weed management because it is effective, economical, nonselective; provides immediate results; is not soil specific; mitigates perennial weeds; can be rapidly performed over large areas; leaves no harmful residues; and has longtime farmer acceptance. However, anytime soil is tilled it is exposed to erosion, and tillage requires considerable energy and time. Much poor weed management occurs with tillage because of improper selection, timing, and management of equipment. Thus, in the future, mechanical weed management will be used more judiciously than it has been in the past. Farmers in the United States must meet the conservation requirements set forth in the 1985 and 1990 U.S. Food Securities Acts by 1995. Highly erodible lands will require special management practices including 30% ground cover to remain in compliance as agricultural cropland.

Choices include crop residues, planting of grass in highly erodible fields, building of terraces, strip farming, and contour farming. A 27 state survey by the Soil and Water Conservation Society in 1992 indicated that more than half of the farmers were in compliance (*Omaha World Herald*, June 28, 1992, section M, page 1).

APPENDIX A

List of Herbicides in Chapter

Common name	Chemical name
Atrazine	6-chloro-*N*-ethyl-*N'*-(1-methylethyl)-1,3,5-triazine-2,4-diamine
Chlorimuron	2-[[[[(4-chloro-6-methoxy-2-pyrimidinyl)amino]carbonyl]amino] sulfonyl]benzoic acid
Cycloate	*S*-ethyl cyclohexylethylcarbamothioate
Dicamba	3,6-dichloro-2-methoxybenzoic acid
EPTC	*S*-ethyl dipropyl carbamothioate
Flamprop	*N*-benzoyl-*N*-(3-chloro-4-fluorophenyl-*DL*-alanine)
Imazaquin	2-[4,5-dihydro-4-methyl-4-(1-methylethyl)-5-oxo-1*H*-imidazol-2-yl]-3-quinolinecarboxylic acid
Nicosulfuron	2-[[[[(4,6-dimethoxy-2-pyrimidinyl)amino]carbonyl]amino]sulfonyl]-*N-N*-dimethyl-3-pyridinecarboxamide
Pendimethalin	*N*-(1-ethylpropyl)-3,4-dimethyl-2,6-dinitroenzenamine
Picloram	4-amino-3,5,6-trichloro-2-pyridinecarboxylic acid
Primisulfuron	2-[[[[[4,6-bis(difluoromethoxy)-2-pyrimidinyl]amino]carbonyl]amino] sulfonyl]benzoic acid
Silvex	2-(2,4,5-trichlorophenoxy)propionic acid
Triallate	*S*-(2,3,3-trichloro-2-propenyl) bis(1-methylethyl)carbamothioate
2,4-D	(2,4-dichlorophenoxy)acetic acid
2,4-5-T	(2,4,5-trichlorophenoxy)acetic acid
2,3,6-TBA	2,3,6-trichlorobenzoic acid

APPENDIX B

List of Crops and Weed Species in Chapter

Common name	Botanical name	BAYER code
Weeds		
Barnyardgrass	*Echinochloa crus-galli* (L.) Beauv.	ECHCG
Bindweed, field	*Convolvulus arvensis* L.	CONAR
Brome, downy	*Bromus tectorum* L.	BROTE
Brome, Japanese	*Bromus japonicus* Thunb. ex Murr.	BROJA
Burdock, common	*Arctium minus* (Hill) Bernh.	ARFMI
Cocklebur, common	*Xanthium strumarium* L.	XANST
Crabgrass, large	*Digitaria sanguinalis* (L.) Scop.	DIGSA
Dandelion	*Taraxacum officinale* Weber in Wiggers	TAROF
Dogbane, hemp	*Apocynum cannabinum* L.	APCCA
Foxtail, bristly	*Setaria verticillata* (L.) Beauv.	SETVE
Foxtail, giant	*Setaria faberi* Herrm.	SETFA
Foxtail, green	*Setaria viridis* (L.) Beauv.	SETVI
Foxtail, yellow	*Setaria glauca* (L.) Beauv.	SETLU
Goatgrass, jointed	*Aegilops cylindrica* Host	AEGCY
Hazel, beaked	*Corylus cornuta* Marsh.	CORCO
Horseweed	*Conyza canadensis* (L.) Cronq.	ERICA
Johnsongrass	*Sorghum halepense* (L.) Pers.	SORHA
Knapweed, Russian	*Centaurea repens* L.	CENRE
Knotweed, prostrate	*Polygonum aviculare* L.	POLAV
Kochia	*Kochia scoparia* (L.) Schrad.	KCHSC
Kudzu	*Pueraria lobata* (Willd.) Ohwi	PUELO
Lambsquarters, common	*Chenopodium album* L.	CHEAL
Marsh elder	*Iva xanthifolia* Nutt.	IVAXA
Millet, wild proso	*Panicum milicoceum* L.	PANMI
Morningglory, ivyleaf	*Ipomoea hederacea* (L.) Jacq.	IPOHE
Morningglory, pitted	*Ipomoea lacunosa* L.	IPOLA
Mustard, wild	*Sinapis arvensis* L.	SINAR
Nightshade, hairy	*Solanum sarrachoides* Sendtner	SOLSA
Oat, wild	*Avena fatua* L.	AVEFA
Panicum, fall	*Panicum dichotomiflorum* Michx.	PANDI
Panicum, Texas	*Panicum texanum* Buckl.	PANTE
Pepperweed, Virginia	*Lepodium virginicum* L.	LEPVI
Pigweed, redroot	*Amaranthus retroflexus* L.	AMARE
Plantain	*Plantago* spp.	
Pricklypear, plains	*Opuntia polyacantha* Haw.	OPUPO
Puncture vine	*Tribulus terrestris* L.	TRBTE
Purslane, common	*Portulaca oleracea* L.	POROL
Quackgrass	*Elytrigia repens* (L.) Nevski	AGRRE
Redcedar, eastern	*Juniperus virginiana* L.	IUPVI

Appendix B Continued

Common name	Botanical name	BAYER code
Ryegrass, annual	*Lolium rigidum* Gaud.	
Ryegrass, perennial	*Lolium perenne* L.	LOLPE
Sandbur, field	*Cenchrus incertus* M.A. Curtis	CCHIN
Sandbur, longspine	*Cenchrus longispinus* (Hack.) Fern.	CCHPA
Shattercane	*Sorghum bicolor* (L.) Moench	SORVU
Sicklepod	*Cassia obtusifolia* L.	CASOB
Skeletonweed, rush	*Chondrilla juncea* L.	CHOJU
Smartweed, Pennsylvania	*Polygonum pensylvanicum* L.	POLPY
Sow thistle, perennial	*Sonchus arvensis* L.	SONAS
Spurge, leafy	*Euphorbia esula* L.	EPHES
Stinkgrass	*Eragrostis cilianensis* (All.) E. Mosher	ERACN
Sunflower, common	*Helianthus annuus* L.	HELAN
Sunflower, Maximilian	*Helianthus maximilianni* Schrad.	HELMA
Thistle, Canada	*Cirsium arvense* (L.) Scop.	CIRAR
Thistle, musk	*Carduus nutans* L.	CRUNU
Thistle, Russian	*Salsola iberica* Sennen & Pau	SASKR
Velvetleaf	*Abutilon theophrasti* Medik.	ABUTH
Vetch, hairy	*Vicia villosa* Roth	VICVI
Waterpod	*Ellisia nyctelea* L.	ELSNY
Witchgrass	*Panicum capillare* L.	PANCA
Crops		
Alfalfa	*Medicago sativa* L.	MEDSA
Barley, common	*Hordeum vulgare* L.	
Beans, dry or field	*Phaseolus vulgaris* L.	
Beans, pinto	*Phaseolus vulgaris* L.	
Buckwheat	*Fagopyrum* spp.	
Castor bean	*Ricinus communis* L.	
Clover	*Trifolium* spp.	
Clover, alsike	*Trifolium hybridum* L.	
Corn	*Zea mays* L.	ZEAMX
Cotton	*Gossypium* spp.	
Flax	*Linum usitatissimum* L.	
Millet		
Mustard, white	*Brassica hirta* Moench	SINAL
Oat	*Avena sativa* L.	
Partridge pea	*Cassia fasciculata* Michx.	CASFA
Peanut	*Arachis hypogaea* L.	
Pine	*Pinus* spp.	
Potato	*Solanum tuberosum* L.	
Rice	*Oryza sativa* L.	
Rye	*Secale cereale* L.	
Ryegrass	*Lolium* spp.	

(*continued*)

Appendix B Continued

Common name	Botanical name	BAYER code
Sorghum	*Sorghum bicolor* (L.) Moench	
Soybean	*Glycine max* (L.) Merr.	
Sudangrass	*Sorghum bicolor* (L.) Moench formerly *S. sudanense* (Piper) Staph.	
Sugar beet	*Beta vulgaris* L.	
Timothy	*Phleum pratense* L.	
Tobacco	*Lycopersicon esculentum* Mill.	
Triticale	*X Triticosecale* Wittm.	
Vetch	*Vicia* spp.	
Wheat, spring	*Triticum aestivum* L.	
Wheat, winter	*Triticum aestivum* L.	

REFERENCES

ASAE Standards. (1993). Terminology and definitions for agriculture tillage implements, ASAE Standards, American Soc. of Agric. Eng., St Joseph, Michigan pp. 255–266.

Ball, D. A. and Miller, S. D. (1990). Weed seed populations response to tillage and herbicide use in three irrigated cropping sequences, *Weed Sci. 38*: 511.

Beatley, M. T. (1982). Soil erosion: its agriculture, environmental, and socioeconomic implications, Reports 92, 29. Council for Agricultural Science and Technology, Ames, Iowa.

Berg, F. M. (1983). *South Dakota Land of Shining Gold*, Flying Diamond Books, Hettinger, North Dakota, pp. 135–136.

Bond, J. J., Power, J. F., and Willis, W. O. (1971). Tillage and crop residue management during seedbed preparation for continuous spring wheat, *Agron. J. 63*: 789.

Bruff, S. A. and Shaw, D. R. (1992a). Early season herbicide applications for weed control in stale seedbed soybean (*Glycine max*), *Weed Technol. 6*: 36.

Bruff, A. S. and Shaw, D. R. (1992b). Tank-mix combinations for weed control in stale seedbed soybean (*Glycine max*), *Weed Technol. 6*: 45.

Buchanan, G. A. and Hiltbold, A. E. (1977). Response of cotton to cultivation, *Weed Sci. 25*: 132.

Buchanan, G. A., Murray, D. S., and Hauser, E. W. (1982). Weeds and their control in peanuts, Peanut Sci. and Tech. Am. Peanut Res. and Educ. Soc., (H. E. Pattee and C.T. Young, eds.), Yoahum, Texas, pp. 205–249.

Buhler, D. D. and Daniel, T. C. (1988). Influence of tillage systems on giant foxtail, *Setaria faberi*, and velvetleaf, *Abutilon theophrasti*, density and control in corn, *Zea mays, Weed Sci. 36*: 642.

Burnside, O. C. (1970a). Control of wild cane in corn, *Weed Sci. 18*: 272.

Burnside, O. C. (1970b). Progress and potential for nonherbicidal control of weeds through preventive weed control, FAO International Conference on Weed Control, (Holstein, J. T., Chairman), Davis, California, pp. 464–483.

Burnside, O. C. (1985). Weed control challenges with conservation tillage in the Great

Plains. Agricultural Chemicals of the Future, BARC Symposium 8, Rowman & Allanheld, Totowas (Hilton, J. L., ed.). pp. 199–209.

Burnside, O. C., Wicks, G. A., and Fenster, C. R. (1964). Influence of tillage, row spacing, and atrazine on sorghum and weed yields from nonirrigated sorghum across Nebraska, *Weeds 12*: 211.

Burnside, O. C., Wicks, G. A., and Fenster, C. R. (1977). Longevity of shattercane seed in the soil across Nebraska, *Weed Res. 17*: 139.

Burnside, O. C., Ross, W. M., and Martin, A. R. (1980). Occurrence and dissemination of short-statured shattercane, Res. Rep. North Cent. Weed Control Conf., *35*: 69.

Burnside, O. C. and Colville, W. L. (1964). Soybean and weed yields as affected by irrigation, row spacing, tillage, and amiben, *Weeds 12*: 109–112.

Burnside, O. C. and Wicks, G. A. (1965). Effect of herbicides and cultivation treatments on yield components of dryland sorghum in Nebraska, *Agron J. 57*: 21.

Burnside, O.C., Wicks, G. A., Warnes, D. D. Somerhalder, B. R., and Weeks, S. A. (1969). Effect on harvesting efficiency in corn, sorghum, and soybeans, *Weed Sci. 17*: 438.

Burnside, O. C., Wilson, R. G., Wicks, G. A., Roeth, F. W., and Moomaw, R. S. (1986). Weed seed decline and buildup in soils under various corn management systems across Nebraska, *Agron. J. 78*: 451.

Campbell, H. W. (1907). *Campbell's 1907 Soil Culture Manual,* The Woodruff-Collins Press, Lincoln, Nebraska, pp. 309.

Chandler, J. M., Hamill, A. S., and Thomas, A. J. (1984). Crop losses due to weeds in Canada and United States, Weed Sci. Soc. of Amer., Champaign, Illinois. pp. 22.

Chepil, W. S. (1946). Germination of weed seeds. I. Longevity, periodicity of germination and vitality of seed in cultivated soil, *Sci Agr. 26*: 307.

Comes, R. D., Bruns, V. F., and Kelley, A. D. (1978). Longevity of certain weed and crop seeds in fresh water, Weed Sci. 26: 336,

Dale, J. E. (1981). Control of johnsongrass (*Sorghum halepense*) and volunteer corn (*Zea mays*) in soybeans (*Glycine Max*), *Weed Sci. 29*: 708.

DeHaan, R. L., Wyse, D. L., Ehlke, N. J., and Putman, D. H. (1990). Development of a *Brassica* sp. smother plant for weed control in corn, *Proc. North Control Weed Sci. Soc. 45*: 27.

Devlin, D. L. and Maddux, L. D. (1991). Using reduced rates of postemergence herbicides in soybean, *Proc. N. Central Weed Sci. Soc. 46*: 97.

Delvo, H. W. (1971). Economic analysis of herbicide use restriction policies on Nebraska cash-grain farms, Ph.D. Thesis, Univ. of Nebr., Lincoln.

Derscheid, L. A., Nash, R. L., and Wicks, G. A. (1961). Thistle control with cultivation, cropping and chemicals, *Weeds 9*: 90.

Derscheid, L. A., Wicks, G. A., and Wallace, W. H. (1963a). Cropping, cultivation and herbicides to eliminate leafy spurge and prevent reinfestation, *Weeds 11*: 105.

Derscheid, L. A., Wallace, W. H., and Wrage, L. J. (1963b). Cropping, cultivation and herbicides to eliminate Russian knapweed and prevent reinfestation, *Weeds 11*: 237.

Dodds, M. E. (1967). "Wheat harvest-machines and methods," in Proceedings of the Canadian Centennial Wheat Symposium (K. F. Nielsen, ed.) Modern Press, Saskatoon, Saskatchewan, pp. 357–371.

Doll, J. D. and Mulder, T. A. (1991). Best management practices for corn weed control, *Proc. North Central Weed Sci. Soc. 46*: 13.

Doll, J. D., Proost, R., Viscocky, M., and Buhler, D. D. (1991). Using reduced herbicide rates in reduced tillage systems, *Proc. North Central Weed Sci. Soc. 46*: 48.

Duley, F. L. and Mathews, O. R. (1947). Ways to till the soil, in *Science in Farming. Yearbook of Agriculture 1943–1947*, (S. T. Gaines and H. E. Balaguer, eds.), U.S. Government Printing Office, Washington, D.C. pp. 518–526.

Egley, G. W. (1983). Weed seed and seedling reductions by solarization with transparent polyethylene sheets, *Weed Sci. 31*: 404.

Erbach, D. E., Lovely, W. G., and Wilkins, D. E. (1970). The effect of early shallow cultivation on weed control in corn production, *Res. Rep. N. Central Weed Control Conf. 25*: 95.

Felton, W. L. (1976). The influence of row spacing and plant population on the effect of weed competition in soybeans (*Glycine max*), *Aust. J. Ex. Agric. & An. Husb. 21*: 926.

Fenster, C. R., Hanway, D. G., and Burnside, O. C. (1962). Equipment for subsurface application of herbicides in fallow land, *Weeds 10*: 329.

Fenster, C. R. and Wicks, G. A. (1982). Fallow systems for winter wheat in western Nebraska, *Agron. J. 74*: 9.

Forcella, F. (1991). Weed seedling emergence models are practical tools for weed control, *WSSA Abstracts 31*: 41.

Forcella, F. and Lindstrom, M. J. (1988). Weed seed populations in ridge and conventional tillage, *Weed Sci. 36*: 500.

Furrer, J. G. and Burnside, O. C. (1962). Shattercane: once a crop, now a weed, *Nebraska Agric Exp. Sta. Quart. 8*: 10.

Good, L. G. and Smika, D. E. (1978). Chemical fallow for soil and water conservation in the Great Plains, *J. Soil & Water Conser. 33*: 89.

Gunsolus, J. L. (1990). Mechanical and cultural weed control in corn and soybeans, *American J. of Altern. Agric. 5*: 114.

Hakimi, A. H., Erami, A., and Ghorasky, S. R. (1973). Effect of different tillage methods on growth and yield of silage corn, *Agron. J. 65*: 509.

Hartman, K. M. and Nezadal, W. (1990). Photocontrol of weeds without herbicides, *Naturwissenschaften 77*: 158.

Hoefer, R. H., Wicks, G. A., and Burnside, O. C. (1981). Grain yields, soil water storage, and weed growth in a winter wheat–corn–fallow rotation, *Agron. J. 73*: 1066.

Holt, J. S. and LeBaron, H. M. (1990). Significance and distribution of herbicide resistance, *Weed Technol. 4*: 141.

Hunt, T. F. (1904). *The Cereals in America*, Orange Judd Co., New York, pp. 421.

Hunter, J. H. (1980). An economical assessment of delayed seeding for control of wild oats, *Proc. N. Cent. Weed Control Conf. 35*: 89.

Johnson, D. W., Krall, J. M., Delaney, R. H., and Thiel, D. S. (1990). Response of seed of 10 weed species to Fresnel-lens-concentrated solar radiation, *Weed Technol. 4*: 109.

Kapusta, G. (1979). Seedbed tillage and herbicide influence on soybean (*Glycine max.*) weed control and yield, *Weed Sci. 27*: 520.

Kiesselbach, T. A., Anderson, A., and Lynes, W. E. (1928). Tillage practices in relation to corn production, *Nebr. Ag. Exp. Sta. Bul. No. 232.*

Kiesselbach, T. A., Anderson, A., and Lyness, W. E. (1935). Cultural practices in corn production, *Nebr. Ag. Exp. Sta. Bul. No. 293.*

Lal, R., Eckert, D. J., Fausey, N. R., and Edwards, W. M. (1990). Conservation tillage in sustainable agriculture, in *Sustainable Agricultural Systems* (C. A. Edwards, R. Lal, P. Madden, R. H. Miller, and G. House, eds.), Soil and Water Conser. Soc., Ankeng, Iowa, pp. 203–225.

LeBaron, H. M. (1991). Herbicide resistant weeds continue to spread, *Resistant Pest Management* (M. Whalon, and R. Hollingworth, eds.), Pest Research Center, East Lansing, Michigan, pp. 36–37.

Lord, W. J. and Vlach, F. (1973). Response of peach trees to herbicides, mulch, mowing, and cultivation, *Weed Sci. 21*: 227.

Martin, R. J., and Felton, W. L. (1993). Effect of crop rotation, tillage practice, and herbicides on the population dynamics of wild oats in wheat. *Aust. J. Agric. 33*: 159.

McWhorter, C. G. (1972). Flooding for johnsongrass control, *Weed Sci. 20*: 238.

Moomaw, R. S. and Robinson, L. R. (1973a). Broadcast or banded atrazine plus propachlor with tillage variables in corn, *Weed Sci. 21*: 106.

Moomaw, R. S. and Robinson, L. R. (1973b). Broadcast or band atrazine plus propachlor with tillage variable in grain sorghum, *Agron. J. 65*: 274.

Moore, M. J., Swanton, C. J., and Gillespie, T. J. (1991). The effect of cover crops on soybean and weed development, *WSSA Abstracts 31*: 46.

Muzik, T. J. (1970). *Weed biology and control*, McGraw-Hill, Inc., New York, p. 65.

Nalewaja, J. D. (1978). "Weed control in cereals—now and in the future," Proceedings of the First Conference on the Council of Australian Weed Science Society, Melbourne, April 12–14, 1978, pp. 215–222.

Nelson, D. C. and Giles, J. F. (1989). Weed management in two potato (*Solanum tuberosum*) cultivars using tillage and pendimethalin, *Weed Sci. 37*: 228.

Newsom, L. J. and Shaw, D. R. (1991). Influence of cultivation timing on chemical control of sicklepod (*Cassia obtusifolia* L.) in soybean, *WSSA Abstracts 31*: 8.

Nyborg, E. O. (1964). A test procedure for determining combine capacity, *Can. Agric. Eng. 6*: 8.

Pammel, L. H. and King, C. M. (1926). Descriptive manual, *Iowa Ecological Survey, The Weed Flora of Iowa, Bul. No. 4.*, pp. 3–429.

Papendick, R. I. and Campbell, G. S. (1988). "Concepts and management strategies for water conservation in dryland farming," Challenges in Dryland Agriculture: A Global Perspective, Proceedings of the International Conference on Dryland Farming, Aug. 15–19, 1988, (P. W. Unger, W. R. Jordan, T. V. Sneed, and R. W. Jensen, eds.), Amarillo, Bushland, Texas, pp. 119–127.

Peacock, J. L., Burnside, O. C., Lavy, T. L., Hanway, D. G., and Kittock, D. L. (1965). Flaming, intertillage, and chemical weed control is castorbeans, *Weeds 13*: 290–292.

Pratley, J. E., and Rowell, D. L. (1987). From the first fleet—evolution of Australian farming, in *Tillage New Directions in Australian Agriculture*, (P. S. Cornish and J. E. Pratley, eds.), Inkata Press, Sydney, Australia, pp. 2–47.

Prentice, L. J. (1991). Hard red winter wheat (*Triticum aestivum* L.) seed quality relationship to yield, M.S. Thesis, Univ., of Nebr., Lincoln.

Rathers, H. C. and Harrison, C. M. (1951). *Field Crops*, McGraw-Hill Book Company, Inc., New York, p. 446.

Rees, R. G. (1987). Effects of tillage practices on foliar diseases, in *Tillage New Directions in Australian Agriculture*, (P. S. Cornish and J. E. Pratley, eds.), Inkata Press, Sydney, Australia, pp. 318–334.

Roberts, H. A. (1970). Viable weed seeds in cultivate soils, Re. Natl. Veg. Res. Stn., for 1969, pp. 25–38.

Roberts, H. A. and Feast, P. M. (1972). Fate of seeds of some annual weeds in different depths of cultivated and undisturbed soil, *Weed Res. 12*: 316.

Rose, S. J., Burnside, O. C., Specht, J. E., and Swisher, B. A. (1984). Competition and allelopathy between soybeans and weeds, *Agron. J., 76*: 523.

Rosenberg, S. H. (1979). *Rural America a Century Ago*, American Society of Agricultural Engineering, St. Joseph, Michigan, pp. 1–13.

Russ, O. G. and Anderson, L. E. (1960). Field bindweed control by combinations of cropping, cultivation, and 2,4-D, *Weeds 8*: 397.

Russell, W. J., Fehr, W. R., and Mitchell, R. L. (1971). Effect of row cultivation on growth and field of soybeans, *Agron J. 63*: 772.

Schweizer, E. E. (1980). Herbicides applied sequentially for economical control of annual weeds in sugarbeets, (*Beta vulgaris*), *Weed Sci. 28*: 152.

Schweizer, E. E., Lybecker, D. W., Westra, P. (1992). Alternative tillage systems to control weeds within the corn row, *WSSA Abstracts 32*: 14.

Snipes, C. E., Colvin, D. L., Patterson, M. G., and Crawford, S. H. (1992). Cotton (*Gossypium hirstrum*) yield response to cultivation timing and frequency, *Weed Technol. 6*: 31.

Solie, J. R., Wittmuss, H. E., and Burnside, O. C. (1983). Improving weed control with a subsurface jet injector system for herbicides, *Trans. ASAE 26*: 1022.

Spencer, J. T. (1941). The effect of root pruning and prevention of fruiting on the growth of roots and stalks of maize, *Agron. J. 33*: 481.

Sprague, M. A. and Triplett, G. B. (1986). No-tillage and surface-tillage agriculture, John Wiley & Sons, Inc., New York, 467.

Stanelle, J. (1986). Quality of 1984 Kansas wheat (*Triticum aestivum* L.) effect on quality and yield of grain, M.S. Thesis, Kans. State Univ., Manhattan.

Steward, B. A. and Musick, J. T. (1982). Conjunctive use of rainfall and irrigation in semiarid regions, *Adv. in Irrig. 1*: 1.

Stritzke, J. F. and Bidwell, T. G. (1990). Eastern redcedar and its control, in *OSU Extension Facts No. 2850*, Oklahoma State Univ., Stillwater.

Summerell, B. A. and Burgess, L. W. (1989). Factors influencing survival of *Pyrenophora tritici-repentis*: stubble management, *Mycol. Res. 93*: 38.

Tappeiner, J. D., II, (1979). Effect of fire and 2,4-D on the early stages of beaked hazel (*Corylus cornuta*) understories, *Weed Sci. 27*: 162.

Thatcher, A. P., Davis, G. W., and Alley, H. P. (1964). Chemical control of plains pricklypear in southeastern Wyoming, *J. Range Manage. 17*: 190.

USDA. (1989). *1987 Census of Agriculture, United States Summary and State Data,* U.S. Dept. of Commerce, Washington, D.C., Vol. 1, Part 51, AC87-A-51, p. 1.

Valenti, S. A. and Wicks, G. A. (1992). Influence of nitrogen rates and wheat (*Triticum aestivum*) cultivars on weed control, *Weed Sci. 40*: 115.

Weersink, A., Walker, M., Swanton, C., and Shaw, J. E. (1992). Costs of conventional and conservation tillage systems, *J. Soil and Water Cons. 47*: 328.

Whitaker, F. D., Heinemann, H. G., and Wischmeier, W. H. (1973). Chemical weed control affect runoff, erosion, and corn yields, *Soil and Water Cons. 28*: 174.

Wicks, G. A. and Anderson, F. N.l (1969). Weed control in sugarbeets with herbicides and cultivation, *Weed Sci. 17*: 456.

Wicks, G. A. and Burnside, O. C. (1981). Effect of cultivation and herbicides on ecofallow sorghum, *Proc. N. Centr. Weed Control Conf. 36*: 124.

Wicks, G. A., Burnside, O. C., and Fenster, C. R. (1971). Influence of soil type and depth of planting on downy brome seed, *Weed Sci. 19*: 82.

Wicks, G. A., Fenster, C. R., and Burnside, O. C. (1969). Selective control of plains prickly pear in rangeland with herbicides, *Weed Sci. 17*: 408.

Wicks, G. A. and Somerhalder, B. R. (1971). Effect of seedbed preparation for corn on distribution of weed seed, *Weed Sci. 19*: 666.

Wiese, A. F. (1977). Weed control in *Dryland Agriculture Agronomy Monograph No. 23*, (H. E. Dregne and W. O. Willis, eds.), American Society of Agronomy, Madison, Wisconsin, pp. 463–488.

Wiese, A. F. (1985). Weed control in limited-tillage systems in *Monograph No. 2,* Weed Science Society of America, Champaign, Illinois.

Wiese, A. F., Collier, J. W., Clark, L. E., and Havelka, U. D. (1964). Effect of weeds and cultural practices on sorghum yields, *Weeds 12*: 209.

Wiese, M. V. (1987). Diseases caused by viruses and viruslike agents, *Compendium of Wheat Diseases*, 2nd Edition, APS Press, St. Paul, Minnesota, pp. 80–85.

Wilcut, J. W., Wehtje, G. R., and Walker, R. H. (1987). Economics of weed control in peanuts (*Arachis bypogaea*) with herbicides and cultivations, *Weed Sci. 35*: 711.

Williams, C. G. and Welton, F. A. (1915). Corn experiments, *Ohio Sta. Bull. 282.*

Wilson, H. K. (1948). *Grain Crops*, McGraw-Hill Book Company, Inc., New York, p. 384.

Wilson, R. G. (1980). Dissemination of weed seeds by surface irrigation water in western Nebraska, *Weed Sci. 28*: 87.

Wilson, R. G., Jarvi, K. J., Seymour, R. C., Witkowski, J. F., Danielson, S. D., and Wright, R. F. (1992). Annual weed growth across Nebraska, *Univ. of Nebraska Res. Bull.*, 314-F, 53.

Worsham, A. D. (1989). Current and potential techniques using allelopathy as an aid in Weed Management, *Phytochemical Ecology, Allelochemicals, Mycotoxins,* and *Insect Pheromones* and *Allomones*, *Academia Sinica Monograph Series No. 9*, Academia Sinica, Taipei, ROC, pp. 275–291.

Worsham, W. D. (1991). Role of cover crops in weed management and water quality, Proceedings of Cover Crops for Clean Water (W. L. Hargrove, ed.), Jackson, Tennessee, pp. 141–145.

Yenish, J. P., Doll, J. D., and Buhler, D. D. (1992). Effect of tillage on vertical distribution and viability of weed seed in soil, *Weed Sci. 40*: 429.

Zook, L. L. and Weakly, H. E. (1944). Summer fallow in Nebraska, *Univ. Nebraska Exp. Sta. Bull. 362.*

5

Chemical Weed Management

S. Kent Harrison and Mark M. Loux
The Ohio State University, Columbus, Ohio

I. INTRODUCTION

The use of chemicals by man to control unwanted vegetation dates back several centuries, but modern chemical weed management essentially began with the discovery of the hormonelike herbicides in the 1940s. The first hormonelike herbicides developed were the substituted phenoxyacetic acids. The phenoxyacetic acid herbicides were discovered quite by accident as a result of research being conducted concurrently in the United States and England to develop synthetic organic compounds that mimicked the biological activity of auxins (Kirby, 1980). Development and use of 2,4-D, MCPA, and related hormonelike herbicides to control broadleaf weeds in cereal grains ushered in a new era of chemical weed management after World War II. It was during the 1950s and 1960s that the modern practice of using relatively low rates of synthetic herbicides for selective weed control in field crops was widely adopted in developed countries of the world.

The term *selective* is used to describe a herbicide that is toxic to some plants (e.g., broadleaf weeds) and much less toxic to others (e.g., grass crop plants) at a given dosage. Some plants may be more tolerant to a herbicide than others because of their ability to prevent herbicide entry or movement in the plant or to metabolize the herbicide rapidly to an inactive form after its absorption. Other plants may withstand exposure to some herbicides by having biochemical differences or alterations in the site normally attacked by the herbicide in susceptible plants.

Since the 1940s, several hundred herbicide products have been commercialized successfully in the United States by foreign and domestic agrochemical companies. Today there are over 130 compounds worldwide that are used as selective herbicides, and approximately 30 others that are used as nonselective herbicides (Zimdahl, 1993). In contrast, there are several hundred trademarked herbicide products commercially available since there may be more than one manufacturer of a particular herbicide and/or multiple formulations that all contain the same active ingredient. When one considers the fact that most herbicides are applied in combinations, or "tank-mixes" containing two or more herbicides, the list of possible herbicide combinations from which growers must choose to fit a weed management system becomes quite extensive.

The herbicide industry must synthesize and screen thousands of compounds in order to find a commercially viable new herbicide (Saggers, 1976). It has been estimated that in the near future, 70,000 to 80,000 compounds may have to be screened to find just one successful herbicide (Boger and Sandmann, 1989). In modern herbicide development, the discovery of novel organic compounds that exhibit phytotoxic properties often leads to the synthesis of related compounds, where chemists attempt to optimize herbicidal activity of the original compound by altering or modifying its chemical structure (Beyer et al., 1988). Between 1984 and 1988, 71% of the "new" herbicides introduced by the agrochemical industry were modifications or analogues of preexisting chemicals (Hopkins, 1989). During the same period, 170 novel compounds were synthesized; however, only 21% eventually reached the market as commercial products (Hopkins, 1989). Consequently, there are often several herbicides developed within a chemical "family" that are structurally related and have essentially the same mode of action in plants. However, it cannot be stated exclusively that all members of a chemical family have the same mode of action, as there are a few notable exceptions (Zimdahl, 1993). Herbicides within a chemical family also often vary in selectivity as a result of physicochemical differences that cause them to behave differently in the soil or plant system.

Herbicides are often grouped or classified according to their basic chemical structure and biochemical mode of action in plants. Triazines, dinitroanilines, sulfonylureas, and imidazolinones are examples of chemical families that each contain several commercial herbicides having the same mode of action in susceptible plants. As mentioned previously, selectivity differences among the members of a single family may largely determine their specific use as agricultural herbicides.

II. HERBICIDE SELECTIVITY

A selective herbicide is one that is much more toxic to some plants than others within the limits of 1) a specified dosage range, 2) the method of herbicide application, and 3) environmental conditions that follow application. Thus, being a relative term, *selectivity* indicates that a given herbicide is safer to some plants than others under a specific set of conditions, often including method of application. Herbicide selectivity is the basis for successful chemical weed management in crop production and is possible as a result of physical and/or biological factors that produce differential herbicide toxicity for crops and weeds.

Physical factors that impart selectivity by application method do so by causing a lethal dose of herbicide to be in physical contact with weeds but not crop plants. Biological factors that make some plants less susceptible to a herbicide than others include genetic resistance, differential absorption, differential translocation, and differential metabolism of the herbicide. Although selectivity of some herbicides among plants is largely attributable to a single mechanism, e.g., differential metabolism, often a combination of application method and biological tolerance is required to impart the greatest degree of selectivity over a range of herbicide dosages and environmental conditions. Herbicide selectivity is not absolute. Virtually all herbicides lose selectivity if applied improperly or under environmental conditions that adversely affect biological tolerance mechanisms.

A. Physical Placement

Herbicide selectivity by physical placement is discussed here in general terms and refers to any factor(s) that results in a spatial separation between sensitive crop tissues/absorption sites and a toxic dosage of herbicide. The desired weed and crop selectivity is achieved when a toxic concentration of herbicide is in direct contact with weeds but concentrations toxic to crop plants are prevented. Preemergence application of herbicides used for selective control of small-seeded grass and broadleaf weeds in large-seeded annual crops is one example of selectivity by placement. In this case, crop seed are planted and germinate below the zone of greatest herbicide concentration in which the smaller-seeded weeds are germinating. By a similar mechanism, some perennial crops avoid damage from soil-applied herbicides by having deep roots that avoid contact with toxic concentrations of herbicide. Directed postemergence sprayers, shielded sprayers, rope-wick applicators, and other specialized herbicide application equipment that directs herbicides onto weeds and away from crop plants are other examples of herbicide-application selectivity by placement.

A number of factors other than application method and application timing relative to crop or weed growth stage can ultimately influence physical placement of herbicides, including mechanical incorporation, irrigation, herbicide formulation, chemical properties of the active ingredient, and physicochemical properties of the soil. Restrictions on herbicide use regarding these factors are routinely included on the product label. Environmental variables interact with the physical factors affecting herbicide selectivity, and extremes in rainfall and temperature that affect herbicide movement or alter plants' physiological responses to a herbicide may reduce selectivity.

B. Plant Factors and Selectivity

Resistance and Tolerance

Although the term *resistance* may be used in a general way to describe the ability of some plants to withstand higher dosages of a herbicide than others, we choose to distinguish *resistance* from *tolerance* here in order to explain the mechanisms by which some plants withstand dosages of a herbicide that are normally lethal to other plants. A plant may be classified as being resistant to a herbicide if it grows and yields normally at the recommended agricultural dose of the herbicide (Gressel, 1985). When plants respond variably to a herbicide or are only partially affected by the normal herbicide dosage, they may be classified as tolerant or semitolerant. Resistance is associated here with the inability of a herbicide to bind or interact with its normal site of action, whereas tolerance is associated with plant anatomical or physiological mechanisms that prevent a lethal dosage of herbicide from reaching its target site.

Herbicide resistance in plants is usually conferred by one or two genes that result in either an absence or an alteration in the specific site of herbicide action. If the site of action for a herbicide lies within a single enzyme or other specific protein binding site, a difference in a single amino acid in the sequence of the targeted protein's structure can render a plant resistant (Hirschberg and McIntosh, 1983). Since the genetic alterations that confer resistance in plants to target-specific herbicides can be very minor, the natural genetic diversity that exists within many weed species enhances the probability that large populations contain at least some genotypes that are naturally herbicide-resistant (discussed later).

Herbicide tolerance may be conferred by many genes, resulting in additive levels of tolerance in some plant species (Gressel, 1985). Tolerance mechanisms include differential herbicide absorption, translocation, and metabolism. Herbicide selectivity between crop plants and weeds is usually greatest when crop plants are resistant to rather than tolerant of a herbicide, insofar as the two terms have been defined. However, some herbicide tolerance mechanisms are quite effective and may impart levels of tolerance that are comparable to those observed in resistant plants. For example, the ability of corn to tolerate high dosages of atrazine with no adverse effects is due to the natural abundance in corn tissues of an atrazine-detoxifying enzyme, gluthathione *S*-transferase.

Differential Absorption

Herbicides on plant foliage or in contact with subterranean plant organs must cross several natural barriers before ultimately reaching their biochemical site of action. The pathway of absorption and translocation is more tortuous for some herbicides than others and depends in part on morphological, anatomical, and physiological characteristics of the individual plant on which the herbicide is deposited. Perhaps the greatest variability in rates of herbicide absorption among plants occurs after foliar applications.

Genetically controlled and environmentally influenced morphological characteristics such as leaf angle, leaf arrangement, leaf area, and presence of hairs (pubescence) on the leaf surface first determine the maximum potential herbicide dosage from a herbicide spray that is intercepted and retained by a plant. Application variables affecting selectivity are explained in Chapter 6. Differential spray retention among plants may play a role in determining selectivity of some herbicides. For example, studies have shown that selective postemergence control of wild oat in flax with asulam is possible because flax retains less spray than wild oat (Hibbitt, 1969; Sharma et al., 1978). Addition of an adjuvant to the spray mixture can increase herbicide spray retention and thus herbicidal activity on some weeds, but this may be accompanied by a concomitant decrease in weed/crop selectivity. The role of differential spray retention as a factor in determining selectivity between weeds and crops is probably minor for most postemergence herbicides used in agriculture.

Herbicides deposited on a plant surface must first penetrate nonliving barriers before ultimately reaching their site of action in living plant tissue. The outermost layer covering all aerial plant parts is the cuticle, which represents the most significant barrier to foliar penetration by herbicides (Hess, 1985; Kirkwood, 1987). The cuticle is a complex layer consisting of waxes embedded in a matrix of polymeric cutins with an outermost coating of soluble waxes collectively called the epicuticular wax (Baker, 1980). The cuticle prevents excess water loss from transpiration and provides plants some protection against insects and pathogens. The impervious nature of the cuticle also makes it quite effective at preventing or impeding the entry of many foliar-applied chemicals.

Epicuticular waxes are hydrophobic (nonpolar and oil-soluble) and decrease leaf wettability by repelling water and other hydrophilic (polar and water-soluble) substances (Price, 1982). A simplified explanation of the solubility of one substance in another and its dependence on the relative polarities of each is given by the adage that "likes dissolve likes." This simply means that hydrophilic substances tend to be readily soluble in other hydrophilic substances, whereas they are insoluble in hydrophobic substances, and vice versa. This principle applies when discussing foliar penetration of the waxy cuticle by herbicides. Some herbicides, such as glyphosate, are highly polar and do not easily penetrate the nonpolar epicuticular wax layer. In such cases, use of an adjuvant is necessary to facilitate herbicide penetration. Conversely, herbicides of a more hydrophobic, nonpolar nature are more soluble in epicuticular waxes and thus tend to penetrate plant cuticles more readily than polar herbicides. Herbicides formulated as salts are relatively polar, whereas those formulated as esters are relatively nonpolar.

Cuticle thickness varies from plant to plant and even among leaves of the same plant. Cuticle composition and thickness depend on the age of the tissue and environmental conditions that exist during deposition of cuticular components, including the epicuticular wax layer. Leaves that develop under conditions of high light, low soil moisture, and low relative humidity tend to develop a thicker epicuticular wax layer than those which develop under low light, adequate soil moisture, and high relative humidity (Hunt and Baker, 1982). Subsequent environmental conditions can also affect cuticle thickness and permeability. Adequate soil moisture, maximum sunlight, and high relative humidity tend to keep cuticles hydrated and thus more conducive to penetration by herbicides. Drought or prolonged periods of hot, dry weather can result in an increase in epicuticular wax, cuticle dehydration, and plant stress, and foliar application of herbicides during such periods may result in a loss of effectiveness due to lack of herbicide absorption. Use of a proper surfactant or other adjuvant can aid foliar wetting and enhance cuticular penetration of herbicides under adverse conditions. Addition of one or more adjuvants to foliar herbicide sprays is often recommended even when environmental conditions are favorable; however, improper adjuvant selection or dosage may reduce selectivity between crop plants and weeds.

Roots, like leaves, have a cuticle through which herbicides must initially penetrate in the process of absorption. However, the cuticle on root surfaces generally lacks a waxy covering and is much thinner and more permeable to herbicides than the foliar cuticle (Hess, 1985). Absorption of water, mineral nutrients, and other dissolved substances by plants occurs primarily by root hairs, and herbicides in the soil solution readily move into roots by simple diffusion. Most herbicide absorption occurs in root hairs located in the region just behind the root apex (Bukovac, 1976). Soon after germination but before emergence, developing shoots of some plants may also serve as sites of herbicide absorption from soil and be a source of herbicide selectivity. For example, the coleoptile as well as the radicle of some grass seedlings are strong absorption sites for carbamothioate herbicides. As a result, those grasses are especially susceptible to soil-applied carbamothioates during and immediately after germination (Prendeville, 1968; Prendeville et al., 1968).

Differential Translocation

Translocation is the net internal movement of water and dissolved substances from one plant region to another. After penetration of the cuticle, most herbicides are translocated in the apoplast and/or symplast some distance before reaching their site of action. The apoplast is the nonliving, continuous network of cell walls, intercellular spaces, and xylem tissue that functions as the conduit for water and mineral nutrient transport from the roots to the shoot. The symplast is the continuous network of living cells, intercellular protoplasmic connections (plasmodesmata), and phloem tissue that transports sugars and other organic solutes from leaves and storage organs to other areas of the plant.

After absorption and before long-distance translocation in the xylem or phloem can occur, herbicides must first undergo short-distance translocation from the epidermis to the vascular tissue. In the foliar tissue, the herbicide moves to the xylem or phloem via normal apoplastic and/or symplastic routes. Herbicides absorbed by roots, however, may encounter an additional barrier if short-distance translocation occurs in the apoplast. Herbicides in the root apoplast that are traveling toward the xylem and phloem in the center of the root first encounter the endodermis, a layer of cells surrounding the vascular tissue. The endodermal cell walls are impregnated with suberin, creating a banded layer called the Casparian strip that is impermeable to herbicide penetration. Therefore, to reach the xylem or phloem tissue for long-distance translocation, the herbicide must penetrate the cell membrane and enter the symplast in order to bypass the Casparian strip. Having bypassed the Casparian strip, the herbicide may then enter the xylem or phloem for long-distance translocation (Hess, 1985).

Herbicides in the apoplast are able to penetrate the cell membrane and enter the symplast, and virtually all herbicides must penetrate the cell membrane to reach their biochemical site of action inside the cell. Therefore, no herbicide is totally confined to either the symplast or the apoplast (Devine, 1989; Devine and Vanden Born, 1991). In fact, some herbicides translocate extensively in both the apoplast and the symplast. However, long-distance translocation of many herbicides occurs predominantly in one system over the other. The primary route of herbicide translocation is dependent on physicochemical properties of the herbicide and conditions within the plant (Devine, 1989). Some herbicides, for example, may undergo ionization upon entry into the symplast, resulting in an anionic form of the herbicide that is unable to penetrate back across the cell membrane. The anionic herbicide is thus "trapped" in the cytoplasm and is largely confined to translocation in the symplast. Conversely, nonionic herbicides that move freely across cell membranes and into and out of living cells undergo translocation primarily in the apoplast (Devine and Vanden Born, 1991; Hess, 1985). Differential rates and extent of translocation sometimes play an important role in determining herbicide selectivity among plants.

The driving force behind apoplastic translocation is transpiration, so net movement of apoplastically translocated herbicides in plants is upward from the site of entry along the gradient from highest water potential (soil water) to lowest water potential (the atmosphere) via the transpiration stream. As a result, apoplastically translocated herbicides are generally soil-applied, where they are absorbed by roots and translocated upward via the water potential gradient to transpiring leaves. Some apoplastically translocated herbicides are effective against weeds when foliar-applied, but uniform spray coverage of foliar tissue and use of a surfactant are usually required for maximum effectiveness since little downward movement of the herbicide from the site of absorption can occur.

Differential apoplastic translocation after root uptake plays a role in determining selectivity of several herbicides, including linuron (Walker and Featherstone, 1973), fluometuron (Rogers and Funderburk, 1968), metribuzin (Smith and Wilkinson, 1974), simazine (Shone and Wood, 1972), bifenox (Leather and Foy, 1978), chloramben (Stoller, 1970), pyrazon (Stephenson and Ries, 1967), prometryn (Funderburk and Lawrence, 1963), and terbacil (Herholdt, 1968). In some instances, herbicides become compartmentalized after absorption and are immobilized or sequestered in roots or veins of tolerant species where damage is minimized. The mechanism by which some herbicides are immobilized in tissues of tolerant species is not clearly understood, but others are known to undergo partial metabolism (conjugation) which renders them immobile in the apoplast (Hess, 1985).

Translocation of herbicides in the symplast is bidirectional and direction of net movement depends on the location of areas of greatest assimilate supply and demand within the plant. The supply and demand for assimilate in plants are determined by the relationship between sites of net assimilate production ("sources") and sites of assimilate utilization ("sinks"), respectively. The direction of symplastic transport in green plants is from source to sink. Mature green leaves carrying on maximum rates of photosynthesis are primary sources, whereas meristematic regions or other tissues with high metabolic activity and rapid growth are sinks.

Source-sink relationships ultimately determine the rate, direction, and extent of phloem-mobile herbicide translocation and vary among weeds according to life cycle (Devine and Vanden Born, 1991). Maximum herbicidal activity on annual weeds with phloem-mobile herbicides generally is attained if plants are nonstressed and treated during a period of rapid growth. During this period, the root and shoot apical meristems are primary sinks into which phloem-mobile herbicides are readily translocated along with the flow of photosynthate from mature leaves, the primary sources.

Perennial weed source-sink relationships are more complex than that of the annuals, and response to phloem-mobile, symplastically translocated herbicides is highly dependent on stage of development. Since a key objective in perennial weed management is to destroy overwintering vegetative propagules (rhizomes, creeping roots, tubers, etc.) as well as the foliage, phloem-mobile herbicides should be applied during a period in which photosynthate is being accumulated in the propagules. As shoots of perennial weeds begin to emerge in the spring, the net source-to-sink carbohydrate movement is upward from the overwintering vegetative propagules to the developing shoot tissue. Foliar application of a phloem-mobile herbicide during rapid vegetative growth will result in little downward translocation and shoot regrowth from propagules is likely. Much better permanent control is obtained if perennial weeds are treated when the majority of the leaves are exporting photosynthate to replenish vegetative propagules. The shift in net carbohydrate flow from shoot tissue to underground propagules usually occurs as the plant slows or stops vegetative growth and begins to enter the reproductive stage. Successful perennial weed management with foliar-applied, phloem-mobile herbicides such as glyphosate, dicamba, and 2,4-D is based on application timing relative to source-sink relationships.

Differential translocation plays a role in determining the selectivity of some phloem-mobile herbicides, including 2,4-D (Ashton, 1958), dicamba (Quimby and Nalewaja, 1971), flamprop (Jeffcoat and Harries, 1975), glyphosate (Gottrup et al., 1976), benzoylprop (Hill et al., 1978), imazamethabenz (Shaner et al., 1983), and chlorsulfuron (Devine et al., 1990). As with differential apoplastic herbicide translocation, the mechanisms by which tolerant species compartmentalize phloem-mobile herbicides are poorly understood. In general, plants that exhibit tolerance to phloem-mobile herbicides by not translocating them must also be able to metabolize the absorbed herbicide into an inactive form in order to prevent delayed toxicity (Hess, 1985; Shimabukuro, 1985).

Differential Metabolism

The most common mechanism that contributes to herbicide selectivity is differential metabolism. A plant that is tolerant to a herbicide by this mechanism is capable of altering or degrading the chemical structure of a herbicide through enzymatic (and occasionally nonenzymatic) reactions that render the herbicide nontoxic. Most plant enzymes that metabolize herbicides are present in fixed amounts in the plant cell and have a relatively broad range of specificity that may allow a single species to metabolize and detoxify a number of different herbicides (Hatzios and Penner, 1982). The rate, in addition to the extent of herbicide metabolism, is often an important factor in determining selectivity.

Herbicides may be classified according to their ability to undergo biotransformations in plants as 1) stable, 2) metabolically deactivated, or 3) metabolically activated (Hatzios, 1991). Stable herbicides are those which do not undergo metabolic deactivation in plants. Herbicides that are metabolically deactivated may undergo oxidation, reduction, hydrolysis, and/or conjugation reactions that render them nontoxic in tolerant species. Metabolically activated herbicides, once they are absorbed by sensitive plants, undergo metabolic transformation that results in an enhancement of herbicide phytotoxicity rather than detoxification.

Glyphosate (Duke, 1988) and paraquat (Shimabukuro, 1985) are examples of stable herbicides that are not metabolized to any extent by most plants; this characteristic is not surprising since they are two of the most nonselective herbicides in existence. Examples of herbicides that are extremely selective by metabolic deactivation are the sulfonylureas (Beyer et al., 1988) and imidazolinones (Shaner and Robson, 1985; Shaner and Mallipudi, 1991), for which differences in sensitivity between tolerant and sensitive species can be several-thousand-fold. Tolerant species rapidly metabolize sulfonylureas and imidazolinones, whereas the rate of metabolic detoxication is much slower in susceptible plants. The classic example of a herbicide that undergoes metabolic activation is 2,4-DB. Plants susceptible to 2,4-DB such as common lambsquarters and redroot pigweed enzymatically convert the relatively nontoxic 2,4-DB to the phytotoxic 2,4-D acid via a normal cellular process called beta-oxidation (Synerholm and Zimmerman, 1947). Tolerance of several legumes to 2,4-DB is based on their rapid metabolic deactivation of 2,4-DB before lethal quantities of 2,4-D can accumulate (Smith, 1979).

Herbicide Antidotes

Selectivity for some herbicides may be attained by the use of chemicals that protect plants against toxic herbicidal action. Herbicide antidotes, also referred to as herbicide protectants or safeners, allow the use of certain herbicides for selective weed control in crops that otherwise would be susceptible to those herbicides. Chemical antidotes may be applied to crops as seed treatments or be included as components of the commercial herbicide formulation. The development and technology of herbicide antidotes are relatively new, and presently there are commercial antidotes which protect corn from carbamothioate herbicides and corn and grain sorghum from chloroacetamide herbicides. The primary mode of action of the herbicide antidotes studied thus far involves the stimulation of herbicide metabolism in the protected plants (Hatzios and Penner, 1982). Future antidotes may be developed that allow weed-crop selectivity for a wider range of herbicides than those for which antidotes are currently available.

Biotechnology and Future Developments in Selectivity

A greater understanding of herbicide detoxication enzymes in plants, herbicide structure-activity relationships, and gene transfer technology will have a major impact on selective chemical weed management in the future. One benefit of biotechnology may be the development of safer and more effective herbicides. Genes that encode resistant target sites or enzymes that detoxify specific herbicides can now be inserted into some crops and will likely play an important role in chemical weed management of the future. Crops for which research and development of herbicide-resistant varieties is being conducted include canola, tobacco, cotton, sweet potato, tomato, sugar beet, oilseed rape, alfalfa, corn, soybean, and flax (Dyer et al., 1993). However, the ecological impact of using herbicide-resistant crops is unknown, and development of resistant weed populations should be of great concern. Ideally, new herbicides and herbicide-resistant crops will not be relied solely upon for weed management; rather they will be utilized to complement existing tools in well-planned, long-term integrated weed management systems.

III. HERBICIDES AND ENZYME INHIBITION

The most recent major advance in herbicide chemistry has been the development of highly active herbicides that target specific plant enzymes. Some of the newer herbicide families exhibit a high degree of biological activity at relatively low rates of application and are low in mammalian acute toxicity. Since development of these selective, target-specific enzyme inhibitors constitutes the latest breakthrough in herbicide development, a more detailed discussion of their general mechanism of action is in order.

Enzymes are special proteins that perform the important biological function of catalyzing metabolic chemical reactions. The compounds on which biosynthetic enzymes act are called substrates, and enzymes have unique active sites that attract and/or interact with only very specific substrates that "fit" the active site (Figure 1). Enzymes are required for biosynthesis reactions that assemble complex molecules from simpler, intermediate compounds.

An enzyme often catalyzes only a single, yet critical biosynthesis reaction. Some classes of herbicides act as inhibitors of specific enzymes by blocking the enzyme's active site and preventing catalysis of the normal substrate molecules (Figure 1). This leads to a depletion of the critical product of the reaction, and the resulting deficiency leads to a disruption of normal plant metabolism. For example, sulfonylureas and imidazolinones are two chemical classes that contain herbicides which inhibit acetohydroxyacid synthase (AHAS, EC 4.1.3.18), a plant enzyme that catalyzes the formation of the branched-chain amino acids leucine, isoleucine, and valine. Amino acids are the building blocks of all

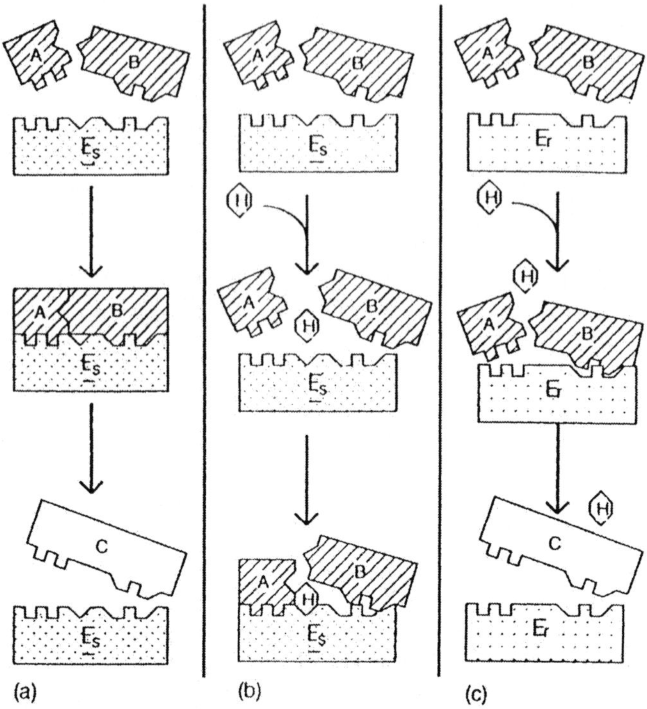

Figure 1 Simplified diagram representing (a) normal catalysis of substrates A and B by enzyme E_S to form product C; (b) binding and inhibition of the susceptible enzyme E_S by a herbicide, H, resulting in no formation of product C; (c) catalysis of substrates A and B by a herbicide-resistant enzyme E_r that lacks the herbicide binding site, resulting in normal production of product C despite the presence of herbicide H. E_S and E_r could also represent the susceptible and resistant forms, respectively, of the nonenzyme protein that is the binding site for herbicides which inhibit photosynthesis. (From Pfister et al., 1979.)

proteins including enzymes, so a depletion in any of the 20 amino acids required by plants quickly leads to a depletion in critical proteins necessary for normal metabolism, growth, and development.

The approach of developing herbicides that inhibit specific plant enzymes has been quite successful, resulting in highly selective compounds that are low in mammalian toxicity. A benefit of this strategy is that other organisms can be genetically resistant to a specific enzyme-inhibiting herbicide either by lacking the target enzyme (e.g., mammals do not have AHAS) or by having a different structural form of the enzyme for which the herbicide has low or no affinity (Stidham, 1991). However, the latter mechanism can also result in the occurrence and development of herbicide-resistant weed populations.

IV. HERBICIDE CLASSIFICATION AND MODE OF ACTION

Herbicides may be classified several ways, including by their method of application (soil vs. foliar-applied), general symptoms (contact vs. systemic), period of residual activity (persistent vs. nonpersistent), or selectivity (nonselective vs. selective, broadleaf vs. grass control). In order to identify and categorize herbicides that selectively control weed complexes or cause characteristic injury symptoms in plants, it is useful to classify herbicides according to both their general mode of action and their basic chemical structure. As a general rule, herbicides belonging to the same chemical family often produce similar symptoms on susceptible plants, although there are a few exceptions to this rule (Zimdahl, 1993). There also may be strong similarities in herbicidal symptoms on plants among herbicide families within a given mode of action class. Mode of action classes, chemical families belonging to each, and representative herbicides within each family are shown in Table 1.

The seven major mode of action categories shown in Table 1 are described in the following section. A description of each general mode of action category is followed by a brief discussion of the biochemical mechanism of action, if known, for each chemical family.

A. Amino Acid Synthesis Inhibitors

Amino acid synthesis inhibitors are compounds that inhibit specific plant enzymes involved in the synthesis of amino acids, which are the building blocks of all proteins. There are 20 amino acids essential for normal plant growth. Unlike animals, plants must internally synthesize each amino acid from biochemical intermediates. The inability of a plant to synthesize even a single amino acid causes profoundly adverse effects on plant growth and development. Inside the plant, amino acid synthesis inhibitors are readily translocated to areas of high metabolic activity (e.g., meristematic tissue). There they bind with the enzyme, and the resulting deficiency of critical amino acids leads in turn to a gradual depletion of proteins vital for normal plant growth and development. Visible injury symptoms on susceptible plants are rather slow to develop in comparison with the action of some other classes of herbicides and usually do not appear for a period of several days, although cessation of growth can be almost immediate.

Glyphosate and Glyphosate Analogues

Glyphosate and its analogues are rapidly inactivated in soil by biological and chemical reactions and thus are effective only when applied to plant foliage. Herbicides in this class are nonselective and are extremely effective in controlling a broad spectrum of annual and perennial weeds. Glyphosate, the primary member of this class of herbicides, specifically inhibits the enzyme 3-phospho-

Table 1 Selected Organic Herbicides Classified by General Mode of Action and Chemical Family

General mode of action	Chemical families	Representative herbicides	Application sites[a]
Amino acid synthesis inhibitors	Glyphosate and glyphosate analogues	Glyphosate, sulfosate	Foliar, aquatic
	Imidazolinones	Imazamethabenz, imazamox, imazapyr, imazaquin, imazethapyr	Foliar, soil
	Sulfonylureas	Bensulfuron, chlorimuron, chlorsulfuron, ethametsulfuron, halosulfuron, metsulfuron, nicosulfuron, primisulfuron, rimsulfuron, sulfometuron, thifensulfuron, tribenuron, triasulfuron, triflusulfuron	Foliar, soil
	Triazolopyrimidine sulfonamides	Flumetsulam, metosulam	Foliar, soil
Contacts-membrane disrupters	Bipyradiniums	Diquat, difenzoquat, paraquat	Foliar, aquatic
	Diphenylethers	Acifluorfen, bifenox, fomesafen, lactofen, oxyfluorfen	Foliar, soil
Hormonelike herbicides	Benzoic acids	Dicamba	Foliar
	Phenoxy acids	2,4-D, 2,4-DB, Dichlorprop, MCPA, MCPP	Foliar, aquatic
	2-Pyridinecarboxylic acids	Clopyralid, fluroxypyr, picloram, triclopyr	Foliar, soil
	Quinolinecarboxylic acids	Quinclorac	Foliar
Lipid synthesis inhibitors	Cyclohexenediones	Clethodim, cloproxydim, cycloxydim, sethoxydim	Foliar
	Aryloxyphenoxy-propanoates	Diclofop, fluazifop-P, fenoxaprop, quizalofop	Foliar

Photosynthesis inhibitors	Biscarbamates	Desmedipham, phenmedipham	Foliar
	Hydroxybenzonitriles	Bromoxynil, ioxynil	Foliar
	Triazines	Ametryn, atrazine, cyanazine, hexazinone, metribuzin, prometon, prometryn, propazine, simazine, terbutryn	Soil, foliar, aquatic
	Uracils	Bromacil, terbacil	Soil
	Ureas	Chloroxuron, diuron, fenuron TCA, fluometuron, linuron, methazole, siduron, tebuthiuron	Soil, foliar
	Unclassified	Bentazon	Foliar
		Ethofumesate	Soil, foliar
Pigment synthesis inhibitors	Aminotriazole	Amitrole	Foliar
	Isoxazolidinones	Clomazone	Soil
	Pyridazinones	Pyrazon, norflurazon	Soil, foliar
		(fluridone)[b]	Aquatic
Meristem mitotic inhibitors	Dinitroanilines	Benefin, dinitramine, ethalfluralin, fluchloralin, oryzalin, pendimethalin, trifluralin	Soil
	N-Phenylcarbamates	Asulam, barban, chlorpropham propham	Foliar, some soil use
	Carbamothioates	EPTC, butylate, cycloate, diallate, molinate, pebulate, triallate, vernolate	Soil
	Chloroacetamides	Acetochlor, alachlor, butachlor, dimethenamid metolachlor, propachlor	Soil
	Unclassified	Endothall	Soil, aquatic

[a]Applications listed for a chemical family may not be approved for every herbicide within that family. Herbicide label directions specify approved methods of application for individual herbicide products.
[b]Associated chemistry.
Sources: Herbicide Handbook (1989); Boger and Sandmann (1989); Ashton and Crafts (1981); Fedtke (1982).

5-enolpyruvylshikimate synthase (EPSP synthase, EC 2.5.1.19), which is involved in the synthesis of the aromatic amino acids phenylalanine, tryptophan, and tyrosine (Duke, 1988).

Glyphosate and related compounds are readily translocated via apoplast and symplast throughout plants to actively growing shoot tissue as well as to actively growing underground vegetative propagules of perennial species. As a result, they are effective in preventing vegetative reproduction and overwintering of perennial weeds if applied at the correct stage of plant growth. The optimum time of glyphosate application to creeping perennial weeds is when assimilates are being translocated downward into vegetative propagules for storage. Therefore, maximum effectiveness of glyphosate on many troublesome creeping perennials is achieved when the herbicide is applied to weeds reaching shoot maturity, usually anytime during or after the floral bud stage (for broadleaves) or boot stage (for grasses) of reproduction.

Glyphosate is popular as a preplant "burndown" herbicide in no-till crop or sod production because of its effectiveness on annual and perennial weeds, low mammalian toxicity, and lack of residual activity in soil. Toxicity symptoms of treated plants normally begin to appear 3 to 4 days after application and include stunting and chlorosis of newly developing leaf tissue in the shoot meristematic areas. Purple or reddish discoloration of older shoot tissue may also appear on perennial grasses in the early stages. Within 10 to 14 days treated plants turn totally necrotic, and vegetative propagules of perennial species shrivel, turn brown, and become desiccated.

Imidazolinones and Sulfonylureas

Imidazolinone and sulfonylurea herbicides, although chemically different, inhibit the same enzyme and generally produce similar phytotoxic symptoms in susceptible plants. Both classes exhibit activity as foliar- or soil-applied treatments. They are translocated in the apoplast and symplast to all actively growing plants parts, particularly the meristematic areas, and are most effective on small, rapidly growing annual weeds. Imidazolinones and sulfonylureas are effective at very low application rates (<0.15 kg/ha) and most are highly selective herbicides that control a broad spectrum of annual weeds. Because of its high degree of specificity for certain weeds and crops, each sulfonylurea or imidazolinone herbicide must be carefully matched with the crop and its target weed spectrum. The mechanism of selectivity for imidazolinones and sulfonylureas is differential metabolism or genetic resistance.

In susceptible plants imidazolinones and sulfonylureas inhibit acetohydroxyacid synthase (AHAS), an enzyme also referred to as acetolactate synthase (ALS) (Stidham, 1991). This key enzyme is necessary for the production of three branched chain amino acids: leucine, lysine, and isoleucine. Susceptible grass plants are stunted and develop interveinal chlorosis and/or purpling of the foliar tissue within 7 to 10 days after herbicide application. Newly emerging grass leaves may appear bleached and malformed with additional time. Lateral roots of susceptible plants may be pruned or poorly developed if herbicide is present in the soil, creating the so-called bottle-brush appearance of primary roots. Susceptible broadleaf plants are stunted and may show foliar chlorosis or purpling. Shoot apical meristems turn necrotic and die before older plant parts completely die. Youngest leaves may be small, chlorotic, and puckered, and mature leaves sometimes develop a dark reddish discoloration of the veins.

B. Contacts—Membrane Disrupters

Contact herbicides are applied postemergence, as they generally undergo limited translocation and exhibit little or no herbicidal activity in soil. After absorption of contact herbicides by foliar tissue of susceptible plants, secondary reactions generate toxic free radicals that disrupt cell membranes. Membrane disruption causes intracellular contents (water and cytoplasm) to leak out into the intercellular spaces and surrounding tissue. This phenomenon is initially manifested at the whole plant level by a "water-soaked" appearance and wilting of the foliage. As water from the spilled cellular contents evaporates, the foliage desiccates and turns necrotic, leading to an appearance similar to that observed several days after a killing frost. Off-target spray droplets containing a contact herbicide can produce necrotic spots on sensitive nontarget plants. Compared to those of herbicides with other modes of action, initial symptoms of contact herbicides develop rapidly, usually within hours after application. The rapid action and lack of soil activity make some contact herbicides ideal for nonselective "burndown" of vegetation prior to no-till crop planting. Other selective contacts are used to control annual weeds in tolerant annual and perennial crops.

Since contact herbicides have limited translocation in plant tissues, adequate spray coverage and a proper adjuvant are often required for maximum activity. For the same reason, contact herbicides generally have little direct effect on weed roots or the underground vegetative propagules of perennial weeds. Foliar activity of contact herbicides increases with sunlight, temperature, and humidity and is maximized on young plants that are nonstressed and growing actively.

Bipyridiniums

Bipyridinium herbicides are nonselective herbicides that are used as "burndown" herbicides or harvest-aid desiccants. Bipyridinium ions are positively charged molecules (cations) that bind strongly to negatively charged soil clay and organic matter; hence they are generally not available for root uptake and exhibit no residual activity at normal use rates in most agricultural soils. The bipyridinium herbicides penetrate plant cells and enter chloroplasts, the intracellular organelles that contain chlorophyll and carry on photosynthesis. The bipyridiniums intercept electrons as they flow through their normal pathway in the chlorophyll-mediated light reactions of photosynthesis (Calderbank and Slade, 1976). The extra electron that resides in the newly activated paraquat radical is transferred to molecular oxygen, creating superoxide radicals and the subsequent formation of other toxic free radicals which react spontaneously with plant cell membranes and other oxidizable cellular components.

The reliance of bipyridinium ion phytotoxic action on photosynthetic electron flow and oxygen, a photosynthesis by-product, explains why maximum herbicidal activity is observed under high sunlight and temperatures conducive to maximum rates of photosynthesis. Under such conditions, initial symptoms on susceptible plants may begin to appear within 1 hour after application. Extreme care should be used when handling paraquat or diquat, since if ingested they can also act as electron scavengers and free radical generators in mammalian systems, resulting in high acute mammalian toxicity.

Diphenyl Ethers

Diphenyl ether herbicides inhibit the enzyme protoporphyrinogen oxidase (PROTOX) in susceptible plants (Duke et al., 1991). This enzyme plays a key role in the synthesis of chlorophyll in green plants. When a diphenyl ether inhibits PROTOX, there is an abnormal buildup of chlorophyll precursors that are extremely light-sensitive. Upon exposure to light, the unstable precursors undergo photoreactions that result in formation of destructive free radicals that oxidize lipids in cell membranes and other oxidizable cell components. Thus, as in the bipyridinium herbicides, herbicidal activity is greatest under high light intensity and warm temperatures. However, the initial injury symptoms on susceptible plants usually takes several hours to develop. Tolerant crops are able to metabolize diphenyl ethers to a nontoxic form, although there is usually some desiccation of the mature foliage before the crop plants resume growth and completely recover.

C. Hormonelike Herbicides

Hormonelike herbicides are synthetic growth regulators that mimic the activity of natural plant hormones. They are phytotoxic to many plants because they are much more potent than natural plant hormones and cause secondary effects that inhibit plant growth. Hormonelike herbicides can exhibit activity after absorption by either roots or shoots, but some have much longer residual activity in soil than others. Hormonelike herbicides are highly systemic in susceptible plants and are translocated via the apoplast and symplast to meristematic regions and vegetative propagules of perennial weeds if applied at the proper stage of growth (see Differential Translocation, in Section II.B). They are toxic tc broadleaf plants primarily and many are used for selective control of broad leaf weeds in grass crops; however, they can injure grasses if applied during sensitive stages of growth, including early seedling development and reproduction.

The chemical families that comprise hormonelike herbicides, including phenoxy acids, benzoic acids, and 2-pyridinecarboxylic acids, are all believed to cause phototoxicity by interfering with nucleic acid metabolism (Fedtke, 1982). Specifically, some hormonelike herbicides are known to cause a hormonal imbalance in plant tissues that induces a rapid increase in biosynthesis of ribonucleic acid (RNA). Overproduction of RNA leads initially to a transient stimulation of protein synthesis, photosynthesis, ion uptake, and cell division. However, subsequent tissue swelling prevents translocation of photosynthate to vital areas, leading to aberrant growth of developing plant tissues and organs.

The earliest injury symptom of hormonelike herbicides in broadleaf plants usually occurs within a few hours and is characterized by epinasty, a downward twisting of the stem and leaf petioles. As other metabolic functions are affected, overall metabolism and normal cellular functions are disrupted. Callus tissue may be formed, along with malformations of new leaves such as leaf cupping, crinkling, or elongation with near-parallel veins. Hormonelike herbicides generally do not translocate well in most grasses and are metabolized over time. Grasses that are affected may develop tightly rolled leaves, stem brittleness, and/or malformation of vegetative or reproduction structures.

D. Lipid Synthesis Inhibitors

Several herbicides from different chemical families have been shown to inhibit lipid synthesis in plants, but in many instances it may be a secondary effect resulting from the primary herbicidal mechanism of action. The families included as lipid synthesis inhibitors in the classification scheme presented here are the cyclohexenediones and aryloxyphenoxy propanoates. These herbicides inhibit fatty acid biosynthesis as their primary mode of action. They are highly selective herbicides that are toxic to many annual and perennial grasses when applied at recommended dosages. In general, broadleaf crops and weeds are highly tolerant of or resistant to these herbicides, and some cereal and turf grass species also exhibit varying degrees of tolerance.

Aryloxyphenoxy propanoates and cyclohexenediones are foliar-applied, although some do have limited residual activity in soils for a short period (Buhler and Burnside, 1983). Translocation occurs in the apoplast and symplast with the herbicide tending to concentrate in the meristematic tissue and other areas of high metabolic activity, including underground vegetative propagules that are importing assimilates.

Fatty acid synthesis in susceptible plants is blocked when aryloxyphenoxy propanoates and cyclohexenediones inhibit acetyl-coenzyme A (acetyl-CoA) carboxylase (ACCase, EC 6.4.1.2), an enzyme responsible for the first committed step in fatty acid biosynthesis (Gronwald, 1991). Fatty acids are structural components of lipids, which play a crucial role in the structure and function of plant membranes. Injury symptoms observed on grasses begin to appear several days after application, although growth and leaf expansion cease almost immediately. Later symptoms include chlorosis followed by necrosis of newest leaf tissue emerging from the whorl. These emerging new leaves can be easily pulled from the rest of the plant after 5 to 7 days and will be necrotic at the base. Total plant necrosis usually occurs within 10 days to 2 weeks. Sublethal dosages in grasses can cause stunting, along with chlorosis and translucent "banding" of new leaves as they emerge from the whorl. Differential metabolism and/or insensitivity of ACCase (i.e., genetic resistance) appears to be the major mechanism of selectivity for the aryloxyphenoxy phenoxy propanoates and cyclohexenediones.

E. Photosynthesis Inhibitors

Photosynthesis inhibitors block the light reactions of photosynthesis whereby plants convert the energy from sunlight into the chemical forms required for plant metabolism. Most photosynthesis inhibitors are more toxic to broadleaves than grasses at selective dosages, but some herbicides in this class are also quite toxic to grasses. As shown in Table 1, some photosynthesis inhibitors are used primarily as foliar-applied treatments, whereas others are effective as both soil- and foliar-applied treatments.

Photosynthesis inhibitors are translocated mainly in the apoplast, so internal movement within the plant is upward and outward from the site of absorption. The exceptions are hydroxybenzonitriles and bentazon, which have limited symplastic translocation. Most photosynthesis inhibitors are soil-applied so that root uptake will ensure their translocation to all foliar plant parts along with water and minerals in the transpiration stream. Foliar applications of photosynthesis inhibitors can be effective but require uniform spray covering of the foliage and the inclusion of an adjuvant in the spray mixture to maximize herbicide-foliage contact and spray retention.

Inside the plant cell, photosynthesis inhibitors enter the chloroplast and bind to a specific protein in the photosynthetic apparatus (Pfister et al., 1981). Herbicide binding of the protein site blocks the flow of electrons normally initiated when chlorophyl absorbs solar energy. Consequently, light energy cannot be converted and "trapped" into chemical forms required by the plant for normal metabolism and growth. In addition to the effects of starvation for photosynthate, sunlight-induced secondary reactions can cause photooxidation of chlorophyll, resulting in contactlike herbicide symptoms and more rapid destruction of plant tissue than is caused by simple starvation. The most common symptom produced by photosynthesis inhibitors in susceptible plants is chlorosis (yellowing) of the leaf tissue. In grasses, symptoms first appear on the older leaves near the base of the plant. Leaf tips and margins first show chlorosis and eventually turn necrotic. In broadleaves, the oldest leaves are also affected first and show chlorosis of leaf margins and interveinal tissue, followed by necrosis. Selectivity mechanisms for photosynthesis inhibitors are differential metabolism and genetic resistance.

F. Carotenoid Synthesis Inhibitors

In contrast to the photosynthesis inhibitors, carotenoid synthesis inhibitors do not directly block the light reactions of photosynthesis but rather inhibit the production of carotenoid pigments. Herbicides that are known to inhibit biosynthesis of carotenoids are translocated in the apoplast and with a few exceptions are used primarily as soil-applied treatments. As with other apoplastically translocated herbicides, activity from foliar applications is maximized with thorough spray coverage and use of a spray adjuvant. The row-crop herbicides in this class vary widely in selectivity and are active on various annual grass and broadleaf weeds.

Carotenoids are auxillary pigments in green plants that have two main functions: 1) augmentation of the light reactions of photosynthesis by "harvesting" and transferring of sunlight energy to chlorophyll and 2) protection of chlorophyll and the photosynthetic apparatus from photooxidation. Herbicides in the pyridazinone family inhibit phytoene desaturase, an enzyme in the carotenoid synthesis pathway of plants (Sandmann et al., 1991). Aminotriazole and isoxazolidinones are believed to interfere with other related enzymes in the carotenoid biosynthesis pathway (Sandmann and Boger, 1989). The results of carotenoid biosynthesis inhibition are reduction in photosynthetic efficiency and photooxidation (i.e., bleaching) of chlorophyll upon exposure to sunlight. Susceptible plants develop bleached white and/or chlorotic foliage after absorption of the herbicide. Otherwise, susceptible plants appear normal for a short time with no apparent growth malformations or other morphological symptoms. Starvation and overall disruption of plant processes ensue, eventually resulting in tissue necrosis and plant death. Mechanisms governing selectivity of the carotenoid synthesis inhibitors are differential metabolism and possibly genetic resistance of the target enzyme.

G. Meristematic Mitotic Inhibitors

Meristems are regions of active cell division in plants from which new primary growth arises. Meristematic tissue occurs in plant shoot and root tips. Meristem mitotic inhibitors affect fundamental process(es) involved in mitosis, the stepwise process of cell division. As a result, they are soil-applied herbicides that are most effective when placed in direct contact with germinating seedlings.

Meristem mitotic inhibitors are often further subdivided into herbicides that are 1) primarily root meristem inhibitors (dinitroanilines) and 2) primarily shoot meristem inhibitors (acetamides, carbamothioates, and *N*-phenylcarbamates). Meristem mitotic inhibitors are absorbed by shoot and/or root tissue of germinating seedlings. Dinitroanilines show very limited or no translocation in susceptible plants, whereas other chemical families in this mode of action class translocate to other plant parts via the apoplast and/or symplast.

Dinitroaniline herbicides inhibit root meristematic activity in susceptible plants by disrupting the latter stages of mitosis. Cytological studies on meristematic cells have shown that dinitroanilines cause the disappearance of microtubules, fiberlike protein strands onto which newly replicated chromosomes must align and become attached before cell division can proceed (Hess, 1987; Vaughn and Lehnen, 1991). Microtubules are also required for cell wall development and elongation, so a second effect often observed in dinitroaniline-treated roots is abnormal enlargement (swelling) of root tips. This so-called club root symptom on root tips is often used in diagnosing dinitroaniline injury. Other general symptoms of susceptible plants include nonemergence of germinated seedlings or stunting of plants with lack of lateral roots or severely pruned roots in the herbicide-treated zone. Broadleaf plants may develop a swollen and cracked hypocotyl and callus tissue at the base of the stem. Emerged grass shoots are thick and stunted, often with a purple or reddish discoloration.

Herbicide families that are primarily shoot meristem inhibitors include the acetamides, carbamothioates, and *N*-phenylcarbamates. The specific biochemical site of action of these herbicides is unknown, but they are known to prevent mitotic entry of meristem cells and cause various cellular abnormalities in meristematic tissue. General symptoms include nonemergence of germinating seedlings and malformation and stunting of shoots. Affected grasses may leaf out underground or not unfurl properly, and injured broadleaf plants develop puckered leaves with leaf buds that do not open (carbamothioates) and/or a shortened midvein with a notched distal leaf margin (acetamides). Small-seeded grass and broadleaf weeds are most susceptible to meristem mitotic inhibitors in general, and selectivity in crops is at least partially due to differential seed size and herbicide placement relative to crop seeding depth.

V. HERBICIDE SELECTION AND WEED-CROP MANAGEMENT

A number of factors must be considered in determining the proper herbicides to integrate into a weed management system, including product and application costs, tillage system, crop rotation sequence, specific crop-weed associations, environmental conditions, and other general crop management practices. Historically, chemical weed management in crops other than cereal grains has relied heavily on herbicides with soil-residual activity that are applied before crop planting (preplant) or immediately after planting but before weed emergence (preemergence). In the past, a lack of effective, broad-spectrum postemergence herbicides and the prophylactic benefits of soil-applied herbicides limited the number of total weed management options available to growers. The recent advent of highly active, selective, and broad-spectrum postemergence herbicides has increased the total number of options available for chemical weed management in several major crops.

Combinations of soil-applied and/or postemergence herbicides will often provide the broadest spectrum of weed control and help prevent establishment of dominant weed species complexes. There is no single herbicide that will selectively control all weed species, so efficient weed management with chemicals requires that herbicide selection be based on the weed species present in the target area. Knowledge of prior weed infestations, in terms of both the species present and their relative abundances, permits accurate prescription of the most cost-effective chemical weed management strategy well in advance of crop planting and weed emergence. In terms of herbicide selection for various crop management systems, some general herbicide options and application methods available for various cropping systems are shown in Table 2. In crop production areas where weed species composition is diverse, an integrated approach to weed management may prove most effective and include cultural and mechanical weed management practices in addition to herbicide use (see Chapters 3 and 4).

Herbicides are often marketed in prepackaged mixtures containing more than one active ingredient. Premixed products are formulated to provide broader-spectrum weed control, prevent excessive persistence, or reduce the potential for crop injury. Similarly and for the same reasons just mentioned, many herbicides can be "tank-mixed" with other herbicides just prior to application or applied sequentially to the infested area in separate applications.

A. Soil-Applied Herbicides

In the production of annual crops, soil-applied herbicides may be used at various times including several weeks before crop planting (early preplant), just prior to crop planting (preplant or preplant incorporated), immediately after crop planting but before weed emergence (preemergence), or after crop emergence and directed onto soil or weeds while preventing contact with the crop (post-directed). Activity of soil-applied herbicides is influenced by soil texture, organic matter content, pH, water status, and tillage (see Chapter 8). Label rate recommendations for soil-applied herbicides are usually based on soil texture

Table 2 Chemical Methods Potentially Available for Weed Management in Various Crop Production Systems

	Annual crops				Perennial crops							
	Conventional till, wide rows	Conventional till, solid seeding	No-till, wide rows	No-till, solid seeding	Perennial crops at establishment, conventional-till	Perennial crops at establishment, no-till	Herbaceous, mowed weekly	Herbaceous, mowed two to five times	Herbaceous, not mowed	Herbaceous, limited tillage	Woody plants in rows	Woody plants, grown randomly
Preplant	R[a]	R	R	R	S[b]	R						
Preplant incorporated	R	R		R								
Preemergence	R	R	R	R	R	R	S		S	R	S	
Postemergence												
Broadcast selective	R	R	R	R	R	R	R	R	R		S	S
Foliar-directed	R		R							R	R	
Soil-directed	E		E							R	R	
Wick-applied	S	S	S	S	S	S		S	S	S	S	S
Spot treatment	R	R	R	R	R	R	R	R	R		R	R
Harvest-aid postemergence	R	R	R	R	R		S					
Long residual soil-applied herbicides								S		R		

[a]R = routinely available.
[b]S = sometimes available.
Source: Ross and Lembi (1985).

and organic matter content, and some carry restrictions for use within a specified soil pH range.

Herbicides applied to soil equilibrate between adsorption to soil colloids and free movement in the soil solution (and air-filled pore space, for some herbicides). The water solubility of a herbicide is not always a good indicator of its availability for plant uptake. Adsorption (binding) to colloids controls the concentration of herbicide in soil solution (Bailey and White, 1964; Helling, 1971), and only nonadsorbed herbicides are in solution and readily available for plant uptake. A soil with high adsorptive capacity, e.g., a fine-textured soil with large amounts of clay and organic matter, requires a higher dosage of herbicide to maintain a given level of activity on weeds than a coarse-textured soil low in organic matter. Selectivity of some soil-applied herbicides may be lost in sandy, coarse-textured soils low in organic matter, and label restrictions on use of such products with regard to soil type should be followed closely.

Soil pH may have an effect on the activity of some herbicides by influencing the extent to which they are adsorbed to soil colloids or the rate at which they are chemically degraded. For example, the triazine herbicides are generally more available for plant uptake in alkaline soils ($pH > 7$) than in acid soils ($pH < 7$) because of their greater adsorption to soil colloids under acidic conditions. Low soil pH-induced adsorption of imidazolinone herbicides to soil colloids may ultimately increase their persistence in soil and carryover potential to subsequent crops as the herbicide is later desorbed and becomes available for plant uptake (Loux and Reese, 1992; Che et al., 1992). In contrast, several sulfonylurea herbicides are degraded in soils primarily by acid-catalyzed hydrolysis. As a result, application of these sulfonylureas to alkaline soils ($pH > 7$) may result in delayed herbicide degradation, excessive persistence, and carryover injury to sensitive crops (Beyer et al., 1988).

Since soil-applied herbicides are absorbed by plants from the soil solution, adequate soil moisture is required for herbicide activity. Herbicide uptake may occur via roots, seeds, shoots, and vegetative propagules of perennial plants (Anderson, 1977). Soil moisture content below the permanent wilting point results in a loss of herbicidal activity because virtually all of the herbicide becomes adsorbed to soil and cannot move to plants by diffusion or mass flow in the soil solution (Moyer, 1987). As soil moisture content increases from the permanent wilting point to field capacity, there is usually a small initial increase in the herbicide concentration in soil solution followed by a further increase over time as the herbicide diffuses from the soil colloids back into the soil solution (Moyer, 1987). Thus under dry soil conditions, the herbicide is tightly adsorbed to soil colloids and insufficient herbicide may be available in the soil solution to provide acceptable weed control. Similarly, the amount of rainfall required to move a surface-applied herbicide by mass flow into the root zone for plant uptake is dependent on how strongly the herbicide is adsorbed to the soil.

Early Preplant Herbicides

Early preplant herbicides are applied to the soil weeks in advance of crop planting, so herbicides applied in this manner must have sufficient residual activity in soils to extend through the first few critical weeks after crop emergence. Examples of herbicides used successfully in this manner are atrazine and simazine for weed control in corn.

Advantages

- Early preplant application allows more time to receive rainfall necessary to move herbicide from the soil surface into the zone of weed seed germination.
- Early herbicide application reduces work load and time requirements later during crop planting operations.
- Early application prevents emergence of annual weeds and may eliminate the need for burndown herbicide in no-till systems.
- This application technique may reduce risk of carryover problems with persistent herbicides by providing a longer period for herbicide degradation to occur.

Disadvantages

- Herbicide selection is limited because many have insufficient soil residual activity to maintain weed control for the first critical few weeks after crop emergence.
- Preplant application may require the use of higher rates, split applications, or use of postemergence herbicides to augment the period of weed control provided.
- Early application of herbicides when soils are wet may cause soil compaction from application equipment.

Preplant Incorporated Herbicides

Preplant incorporated herbicides are physically mixed into the soil prior to crop planting. Incorporation is usually performed with some type of secondary tillage implement that distributes the herbicide evenly in the top 5 to 8 cm of soil. Soil incorporation of some herbicides, dinitroanilines for example, is necessary to prevent their loss through volatilization or photodecomposition. In addition, incorporation reduces dependence of some soil-applied herbicides on rainfall for weed control activity by directly placing the herbicide in the zone of weed seed germination.

Advantages

- Incorporation reduces dependence on rainfall to move herbicide into contact with weed seeds and roots.
- Herbicidal activity on some perennial weeds is greater than that with preemergence application.

Disadvantages
- Incorporation requires additional inputs of labor, time, and equipment.
- Incorporation may contribute to soil compaction, crusting, or erosion.
- Improper incorporation can result in loss of herbicidal activity or erratic herbicide performance.
- Incorporation is not possible in some conservation tillage production systems.

Preemergence Herbicides

Preemergence herbicides are applied to the soil prior to weed emergence. In annual crops, preemergence herbicides are usually applied immediately after the crop is seeded. Herbicides applied to the soil in perennial crops, turfgrass, pastures, or noncrop areas before weed seedlings emerge are also considered preemergence herbicides. Rainfall or sprinkler irrigation is required to move preemergence herbicides into the top few centimeters of soil for optimum activity.

Advantages
- Crop planting and herbicide application may be performed in a single field operation.
- Preemergence application followed by adequate rainfall may result in greater efficacy of some herbicides than mechanical incorporation.
- Preemergence herbicides can be used in any tillage system.

Disadvantages
- Activity of preemergence herbicides is dependent on adequate rainfall soon after application.
- Intense rainfall events may cause off-target movement of herbicides by runoff in surface drainage or, in coarse-textured soils, by excessive leaching.
- Some perennial and deep-germinating weeds are generally less affected by preemergence herbicides than by incorporated herbicides.

B. Foliar-Applied Herbicides

Where selective weed control is needed in a growing crop, some foliar-applied herbicides may be used in a broadcast spray to both the crop and weeds while others must be applied to weed foliage while avoiding the growing crop (directed postemergence or spot applications). Critical factors that determine effectiveness of foliar-applied herbicides are rate and timing of application, target weed species, use of spray adjuvants, application equipment, and environmental conditions.

The recommended application rates and timings specified on product label directions are based on weed species, weed and/or crop size, and climatic conditions. Annual weeds become more tolerant to a herbicide as they grow larger, so foliar-applied herbicides are usually most effective on small weed seedlings. Foliar penetration of herbicides is greatest in nonstressed plants under conditions of high temperature and relative humidity. A rain-free period following postemergence herbicide application is necessary to allow a toxic dosage of the herbicide to be absorbed by the target plants. The critical rain-free period necessary to maintain efficacy is specified on herbicide labels and varies widely, ranging from 1 to 8 hours.

Spray adjuvants are used to modify the physical characteristics of a spray solution or facilitate toxic action of a herbicide (Parr, 1982). Examples of adjuvants are surfactants, oil-surfactant blends, fertilizer solutions, spreaders, stickers, spray modifiers, antifoam agents, and drift control agents. Foliar spray adjuvants work by enhancing foliar wetting and facilitating herbicide penetration under less than optimal environmental conditions. In some cases, a mixture of one or more adjuvants (e.g., surfactant + liquid fertilizer solution) is required to optimize foliar activity on weeds while maintaining adequate weed-crop selectivity.

When foliar-applied, apoplastically translocated herbicides and those for which translocation is very limited (i.e., contacts) generally require more thorough spray coverage on target plants than systemic herbicides that are translocated in the symplast. Herbicides in the apoplast only translocate upward from the site of foliar contact, so underlying shoot tissue and other plant parts not directly contacted by those herbicides remain unaffected. Uniform spray coverage of foliar tissue is ensured by selection of the proper nozzles, spray volume, pressure, and adjuvant.

Postemergence Selective Herbicides

Postemergence selective herbicides are applied in broadcast sprays deposited on weed and crop foliage. Selectivity between weeds and crops may be based on genetic resistance of the crop or differential absorption, translocation, and/or metabolism between weeds and the crop. Postemergence selective herbicides are often used in combination with soil-applied herbicides in order to target problem weeds that emerge during the growing season.

Advantages

- Soil type does not directly affect herbicidal activity.
- Herbicide use decisions are based on knowledge of weed species and populations present at the time of application, a practice conducive to integrated weed management systems.
- Postemergence herbicides can be used in any tillage system.
- Application rates, timing, and location may make postemergence herbicides less likely to move off-target than soil-applied herbicides.

Disadvantages

- Application is restricted to a relatively brief period with respect to weed/crop growth stage in order to provide maximum efficacy and maintain crop safety.
- Efficacy may be reduced if environmental conditions cause weeds to be stressed at the time of application.
- Rainfall may prevent application of postemergence herbicides at the proper time with respect to weed/crop growth stage.
- Many postemergence herbicides have little residual activity in soil, so subsequent weed infestations may require additional control measures.

Directed Postemergence and Spot-Applied Herbicides

Directed postemergence and spot-applied herbicides are usually low in selectivity and require precise application to prevent crop injury. Directed postemergence herbicide application in row crops requires a sufficient height differential between crops and weeds so that spray can be directed onto weeds while avoiding crop foliage. Drop tubes that lower spray nozzles beneath the crop canopy or specialized nozzle tips may be used for directed applications. An advantage of directed postemergence herbicides is that selectivity is by placement. Weeds not easily controlled by other herbicides may be targeted with a more potent herbicide that is less selective if applied by other methods. A disadvantage is that specialized application methods are necessary and the risk of crop injury may be greater than with selective broadcast herbicides.

Spot herbicide application refers to any application method whereby the herbicide is applied intermittently and only as needed to specific target weeds. Recirculating sprayers, rope-wick applicators, trigger-controlled spray guns, and regular field spraying equipment that is used to apply herbicides only to isolated weed infestations are examples of spot application equipment. Advantages of spot application are that nonselective herbicides which are effective on a broad spectrum of weeds may be used, and herbicide costs are reduced since relatively little herbicide is used compared to broadcast applications. Two disadvantages of spot application are that fields require thorough scouting to locate target weeds accurately and it does not control other nontarget weeds present in the field.

C. Herbicide Selection and Tillage Systems

Annual weeds tend to be the predominant problem in annual crops grown under conventional tillage. A conversion from conventional tillage practices to minimum- or no-till crop production is often accompanied by a shift in weed species composition over time. Annual weed species also tend to predominate under reduced tillage, but there tends to be a shift from predominantly annual broadleaf to annual grass species. In addition, over time there is an increased occurrence of biennial and perennial weed species under reduced tillage (Triplett and Lytle, 1972; Froud-Williams et al., 1981; Loux and Berry, 1991). Creeping perennial weeds create particular weed management difficulties because they are not easily controlled by most selective herbicides. Some perennial weeds must be managed by a combination of tactics, including crop rotation, selective cultivation, handweeding, preplant tillage, and use of nonselective systemic herbicides applied preplant or postharvest.

VI. HERBICIDE RESISTANCE IN WEEDS

Differential herbicide absorption, translocation, and metabolism are some of the basic mechanisms of selectivity upon which modern herbicide use is based. As mentioned earlier, herbicide-resistant weeds may be defined as those which survive and grow normally at a usually effective agricultural herbicide dosage (Gressel, 1985). In an agronomic sense, the term *herbicide resistance* also implies a decreased response of a population of a weed species to a herbicide as a result of its repeated use (LeBaron and Gressel, 1982). In other words, the development and appearance of herbicide resistance mark a change in the genetic composition of a weed population in response to selection pressure, i.e., repeated use of the same herbicide. The mechanism of herbicide resistance in plants involves a genetically induced alteration in the biochemical site of action (Figure 1) (Arntzen et al., 1982). In contrast, some species just exhibit tolerance to a herbicide because of elevated levels of herbicide-detoxifying enzyme activity (Hatzios and Penner, 1982)

In weed populations throughout the world, large populations of single species can contain a relatively small number of less competitive biotypes that differ genetically from the majority of the population. Since the herbicide revolution began in the 1940s, research has shown that some weed biotypes are naturally resistant to herbicides applied at dosages normally lethal to the majority of the population (Warwick and Marriage, 1982). The first report of triazine-resistant weeds in the United States was published in 1970 (Ryan, 1970). Today at least 100 weed species have been reported to have biotypes resistant to herbicides, with more than half of those being resistant to the triazine herbicides (LeBaron and McFarland, 1990). In addition to the triazines, weed biotype resistance has been reported for 15 other herbicides or herbicide families (LeBaron and McFarland, 1990).

The development of resistant weed populations may be explained as follows. When a herbicide having a very site-specific mode of action (for example, a photosynthesis inhibitor) is applied to an area repeatedly over time, the portion of a weed population susceptible to that herbicide is gradually depleted. This creates an opportunity for naturally resistant weeds to become established, including resistant biotypes of species that are normally susceptible to the herbicide. If the same herbicide or others with an identical mode of action are applied on a year-to-year basis, there is constant selection pressure that favors resistant biotypes and no interruption of the resistant weeds' yearly reproductive cycle. As a result, the overall population of resistant biotypes increases rapidly. Characteristics of herbicides or herbicide families that contribute to the development of herbicide resistance in weeds are (LeBaron and McFarland, 1990) as follows:

1. Specific mode of action with a single target site
2. Extreme activity and effectiveness in killing a wide range of weed species
3. Long soil-residual activity and extended period of activity in soil
4. Frequent application for several growing seasons without rotating, alternating, or combining other types of herbicides

Growing concern has been generated by reports that some weed biotypes are showing resistance to the newer target-specific herbicides, including imidazolinones and sulfonylureas (Primiani et al., 1990; Mallory-Smith et al., 1990). This is of special concern since both classes of herbicides bind with the same enzyme (AHAS) in susceptible plants, and evidence is growing that some weed biotypes resistant to imidazolinones may also be resistant to sulfonylureas, a phenomenon known as cross-resistance. Perhaps of even greater concern is the fact that cross-resistance to herbicides with dissimilar modes of action has been reported (Powles and Howat, 1990).

Prevention is important to avoid the development of herbicide-resistant weed populations in agricultural land. Preventive measures are all designed to break the monotony of applying constant pressures that select for herbicide-resistant weeds. The objective of a resistance management strategy should be to reduce selection pressure so that resistant biotype populations do not develop and expand (Holt and LeBaron, 1990). Some management techniques that help prevent resistant populations from developing are listed below (LaBaron and McFarland, 1990; Gressel and Segal, 1990).

1. *Crop rotation:* Some weeds are more easily managed in certain crops than others, since different control options may be available. Crop rotation also helps disrupt weed life cycles and helps prevent any single weed species from becoming established and dominant.

2. *Herbicide rotation:* Herbicide rotation is generally practiced along with crop rotation and is defined here as the concurrent use of herbicides with dissimilar modes of action. For example, it could possibly speed the development of a resistant weed population if an AHAS-inhibiting herbicide were used for weed control in soybeans only to be followed the next year by application of a different AHAS-inhibiting herbicide in corn or wheat. If possible, herbicide

modes of action and other weed management strategies should also be rotated in a continuous cropping system.

3. *Timely postemergence practices:* In general, triazine-resistant weed populations appear to develop more rapidly in no-till cropping systems. Timely cultivation and/or postemergence herbicide applications are possible control options for conventional tillage. No-till growers have a number of postemergence herbicide options available for timely control of weed escapes. Control of escaped weeds is necessary to prevent reseeding and further development of resistant weed populations.

4. *If possible, avoid excessively persistent herbicides and limit herbicide rates to those that give cost-effective weed control:* Overapplication and excessive soil persistence of herbicides only intensify the selection pressure for resistant weeds.

It should be emphasized that any one of the measures discussed can provide temporary control of herbicide-resistant weed populations. However, long-term management strategies for herbicide-resistant weeds should involve an integration of all management strategies (cultural, chemical, mechanical, etc.) that act in unison to reduce herbicide selection pressure. For example, if weed cross-resistance to herbicides with dissimilar modes of action is encountered, then simple herbicide rotation or timely postemergence practices alone may not be adequate to control the weed even on a short-term basis (Powles and Howat, 1990). In such a scenario, the integration of crop rotation and other nonchemical weed management methods along with alternative chemical management strategies may provide the only successful means by which to manage the resistant weed population in the long term.

VII. HERBICIDE HAZARDS AND TOXICITY

In the United States, herbicides require federal approval by the Environmental Protection Agency (EPA) before being granted a label, that is, approval for commercial sale of the formulated product. Approval of a new herbicide by the EPA is based on the manufacturer's claims of the product's effectiveness, its toxicological effects on various organisms, and its potential environmental impact. Herbicide labels indicate specific safety precautions and requirements for handling the product. General guidelines to reduce risks associated with herbicide use are as follows:

Apply herbicides only to crops or other areas for which use has been approved and specified on the label.

Follow recommended application rates. Overapplication is illegal, may damage crops, or may cause deleterious off-target effects.

Apply herbicides only at times or growth stages specified on the label and observe all recommended posttreatment harvest or feeding intervals. Also closely observe intervals recommended between application and planting of follow-up crops (i.e., recrop restrictions).

Guard against drift injury from herbicide sprays or vapor to off-target plants.

Prevent unnecessary human and animal exposure to herbicide sprays and all treated areas.

When mixing or handling herbicides, wear goggles, rubber gloves, and other protective clothing recommended by the label.

Calibrate herbicide application equipment correctly before attempting to mix and apply a herbicide.

Consult the herbicide label for the proper method of container disposal and promptly return unused portions of herbicides, in their original containers, to a safe storage place.

Herbicide formulations and label requirements and recommendations often change from year to year, so *it is extremely important always to refer to the most recent product label* before applying any herbicide.

A herbicide manufacturer must determine whether a potential new herbicide poses an unreasonable toxic hazard to humans before proceeding with the expensive process of further testing and development. The toxicity of a herbicide is a measure of its inherent potency as a poison under controlled conditions. Toxicity is measured as the amount of herbicide required to cause harm or death to a test organism. Risk is actually more important than inherent toxicity for those handling and applying herbicides, since risk connotes the probability of encountering a poisonous dosage of herbicide. Risks are greatest for those who handle, mix, and apply herbicides, although bystander exposure and personnel reentry into treated areas must also be considered in this context. Thus, the actual risk associated with herbicide use depends on the combined effects of a herbicide's inherent toxicity and the likelihood of exposure to a toxic dosage (Ware, 1983). Obviously, the likelihood of exposure to a given amount of any pesticide, including herbicides, depends on safe handling and use.

Toxic hazards of pesticides vary greatly, and there are two types of risks associated with pesticide use: 1) acute poisoning, which results from exposure to a single toxic dose of the pesticide through handling, mixing, application, or entry into or near treated areas; and 2) chronic poisoning from repeated long-term exposure to low levels of the pesticide. Potential acute toxicity of a pesticide to the user is estimated by measuring its median lethal dosage in a population of test animals. Median lethal dosages, termed LD_{50} or LC_{50} values, are the single dosages or concentrations, respectively, required to kill 50% of a population of test animals. Thus a high LD_{50} or LC_{50} value indicates relatively low toxicity, whereas a low value indicates relatively high toxicity.

The LD_{50} is measured in terms of milligrams of herbicide per kilogram of body weight when administered to the test animals orally or dermally. The LC_{50} is determined to measure inhalation toxicity to mammals by respiratory exposure and is expressed as milligrams of herbicide per liter of air. The LC_{50} may also be expressed in units of milligrams per liter of water when used to measure toxicity to aquatic organisms. Genetically uniform populations of albino rats are most commonly used to determine acute toxicity of pesticides, although the EPA requires toxicological data on a number of additional organisms for registration.

The rat oral LD_{50} is commonly used to compare the relative toxicities of herbicides and estimate the amounts that are acutely toxic to humans. Fortunately, acute toxicity hazards of herbicides in general are relatively low compared to most other classes of pesticides. Herbicide labels must contain "signal words" in bold print which designate the herbicide as belonging to one of four toxicity categories established by the EPA (Table 3). Table 4 contains a list of the rat acute oral LD_{50} and rabbit acute dermal LD_{50} values ranked by toxicity for several commonly used herbicides.

VIII. HERBICIDE REGISTRATION

A. General Registration Requirements

United States federal law requires that herbicides and all other pesticides be registered with the administrator of the EPA before they can be distributed or sold. The registration of any pesticide essentially indicates that the EPA, upon extensive review and evaluation of research data, has reached the conclusion that substantial benefits may be derived from use of the pesticide and with little risk when used properly. The Office of Pesticide Programs is the EPA division that regulates pesticides. Specific regulations governing pesticide registration are given in Section 3 of the Federal Insecticide, Fungicide, and Rodenticide Act (FIFRA, revised in October 1988). The FIFRA Requirement of Registration states that no person in any state may sell or distribute to any person any pesticide that is not fully registered by the EPA. There are exemptions for use of unregistered pesticides under Section 5 (Experimental Use Permit) and Section 18 (Exemption of Federal and State Agencies), to be discussed later.

Under FIFRA, the EPA will grant registration for a herbicide only if the following criteria are met:

1. The herbicide's composition must be such as to warrant the proposed claims it set forth by the manufacturer.
2. Its label and other required documentation must comply with the requirements of FIFRA.

Table 3 EPA Toxicity Categories for Pesticide Labeling and Relative Toxicity Ratings

EPA toxicity category	Signal word[a]	Relative toxicity rating	Rat oral LD_{50} (mg/kg)	Probable lethal oral dose for humans
I	Danger-Poison (fatal)[b]	6 - Supertoxic	<5	<0.4 g
		5 - Extremely toxic	5–50	0.4–40 g
II	Warning (may be fatal)	4 - Very toxic	50–500	40–400 g
III	Caution	3 - Moderately toxic	500–5000	400–4000 g
IV	Caution	2 - Slightly toxic	5000–15,000	4–12 kg
		1 - Practically nontoxic	>15,000	>12 kg

[a]All labels must contain the statement "Keep Out of Reach of Children."
[b]Must be accompanied by skull and crossbones on the label.
Source: Adapted from EPA Pesticide Classification Procedures (1979); Gleason et al. (1976).

Table 4 Classification and Acute Toxicities of Herbicides Grouped by EPA Toxicity Categories

EPA toxicity category	Herbicidal active ingredient[a]	Rat acute oral LD_{50} (mg/kg)[b]	Rabbit acute dermal LD_{50} (mg/kg)[c]
I ($LD_{50} \leq 50$)	Acrolein	46	—
	Arsenic acid	48	—
	Endothall	38–51	m
II (LD_{50} = 50–500)	Bromoxynil	440	>3660
	Copper sulfate	470	>8000
	Cyanazine	334	>2000
	Diquat	230	>400
	Ioxynil	110	210 (rat)
	Paraquat	138	>480
	Tebuthiuron	400	>5000
III (LD_{50} = 500–5000)	Acetochlor	2148	3667
	Acifluorfen	1300	m
	Alachlor	930	13,300
	Ametryn	1110	>10,200
	Ammonium sulfamate (AMS)	3900	m
	Atrazine	1869–3080	9300
	Barban	1350	>20,000 (rat)
	Bensulide	770	3950
	Bentazon	1100	>2500 (rat)
	Borate	2330	m
	Butachlor	2000	>2000
	Butylate	4659	>4640
	Cacodylic acid	830	m
	Chloramben	3500	m
	Chloroxuron	3700	m
	Chlorpropham	3800	10,200
	Clethodim	1360	>5000
	Clomazone	2585	>2000
	Copper chelate	650–2420	m
	Cycloate	2000–3190	>4640
	Cycloxydim	3940	—
	2,4-D	300–1000	m
	2,4-DB	1960	m
	Diallate	1050	8700
	Dicamba	1707	>2000
	Dichlobenil	3160	1350
	Dichlorprop	800	1400 (mouse)
	Diclofop	557–580	640

Table 4 Continued

EPA toxicity category	Herbicidal active ingredient[a]	Rat acute oral LD_{50} (mg/kg)[b]	Rabbit acute dermal LD_{50} (mg/kg)[c]
	Diethatyl	2318	6834–10,250
	Difenzoquat	991	>2000
	Dimethenamid	1570	>2000
	Diphenamid	1373	m
	Dipropetryn	3900–4200	m
	Diuron	3400	m
	DSMA	1800	m
	EPTC	1652	10,000
	Fenac	1780	>2000
	Fenoxaprop	3310	<2000 (rat)
	Fenuron TCA	4000–5700	m
	Fluazifop-P	4096	>2420
	Fluchloralin	1550	m
	Fluometuron	1840	>3038
	Fluoroglycofen	1500	5000
	Glufosinate	2170	1400 (rat)
	Hexazinone	1690	>5278
	Linuron	1254	m
	MCPA	800	—
	Mecoprop	1060	900
	Mefluidide	>4000	>4000
	Metham	820	2000
	Methazole	2501	>12,500
	Metolachlor	2780	>10,000
	Metribuzin	1090	m
	Molinate	720	>4640
	MSMA	900	m
	Naptalam	1770	m
	Pebulate	921–1120	4640
	Prometon	2276	2000–2500
	Prometryn	3750	>10,200
	Propachlor	1800	>20,000
	Propanil	1384	m
	Pyrazon	3600	m
	Pyridate	4690	>2000
	Quinclorac	>2610	2000
	Quizalofop	1670	>5000
	Sethoxydim	2676–3125	m
	Terbutryn	2000	>10,200
	Thiobencarb	920–1903	2900 (rat)

(continued)

Table 4 Continued

EPA toxicity category	Herbicidal active ingredient[a]	Rat acute oral LD_{50} (mg/kg)[b]	Rabbit acute dermal LD_{50} (mg/kg)[c]
	Triallate	1100	8200
	Triclopyr	713	m
	Vernolate	1780	4640 (guinea pigs)
IV ($LD_{50} \geq 5000$)	Amitrole	>5000	>2000
	Asulam	>5000	m
	Benefin	>5000	>2000
	Bensulfuron	>5000	>2000
	Bifenox	>5000	>2000
	Borax	5560	—
	Boracil	5200	>5000
	Chlorimuron	>5000	>2000
	Chlorsulfuron	5545	>3400
	Clopyralid	>5000	>2000
	DCPA	>10,000	>10,000
	Desmedpiham	10,250	8700
	Ethalfluralin	>10,000	>2000
	Ethofumesate	6400	20,050
	Flumetsulam	>5000	>2000
	Fluridone	>10,000	>500
	Fluroxypyr	>5000	>2000
	Fomesafen	8160	>1000
	Fosamine	>5000	>1683
	Glyphosate	5600	>5000
	Imazamethabenz	>5000	>2000
	Imazapyr	>5000	>2000
	Imazaquin	>5000	>2000
	Imazethapyr	>5000	>2000
	Isopropalin	>5000	>2000
	Isoxaben	>10,000	>2000
	Lactofen	>5000	>2000
	Metsulfuron	>5000	>2000
	Napropamide	>5000	>4640
	Naptalam	8200	m
	Norflurazon	>8000	>20,000
	Oryzalin	>5000	>2000
	Oxadiazon	>5000	>2000 (rat)
	Oxyfluorfen	>5000	m
	Pendimethalin	>5000	>2000
	Phenmedipham	>8000	>4000 (rat)
	Picloram	8200	>4000
	Prodiamine	>5000	m
	Pronamide	8350	>3160

Table 4 Continued

EPA toxicity category	Herbicidal active ingredient[a]	Rat acute oral LD_{50} (mg/kg)[b]	Rabbit acute dermal LD_{50} (mg/kg)[c]
	Propazine	>5000	>10,200
	Propham	9000	10,200
	Siduron	>7500	m
	Simazine	>5000	>10,200
	Sodium chlorate	5000	—
	Sulfometuron	>5000	>2000
	Terbacil	5000–7500	>5000
	Thifensulfuron	>5000	>2000
	Triasulfuron	>5000	>2000
	Tribenuron	>5000	>2000

[a]Values shown are for technical grade herbicide. *Acute oral and dermal toxicities of formulated herbicide products may be lower or higher than the figures shown, depending on specific formulation components.*
[b]The values shown for acute oral toxicity are for male white albino rats. For several active ingredients, acute toxicity may differ somewhat in male versus female animals.
[c]Unless otherwise specified, dermal toxicity values are for albino rabbits. The "m" value for some herbicides indicates negligible to moderate skin irritation, but no lethality observed; "—" indicates data not available.
Source: *Herbicide Handbook* (1989); Worthing (1987).

3. It must perform its intended function without unreasonable adverse effects on the environment.
4. It must not generally cause unreasonable adverse environmental effects when used in accordance with widespread and commonly recognized practices.

In summary, the manufacturer is required to provide scientific evidence that a herbicide, when used as directed, will effectively control the weeds listed on the label; that it will not injure humans, crops, livestock, wildlife, or the environment; and that it will not result in illegal residues in food or feed. Residue tolerances are established by the Pesticide Tolerances Division of the EPA. Herbicide tolerances are based on the results of 2-year feeding studies that must be conducted on at least two species of mammals in order to establish the no-observable-effect level (NOEL) of residues.

In considering an application for herbicide registration, the EPA may waive data requirements for efficacy and register a herbicide without determining whether its label claims of efficacy are warranted. In such cases the manufacturer must decide whether the herbicide is effective enough to support its label claims. In any case, product performance in accordance with label claims is ultimately the responsibility of the manufacturer.

The initial step in applying for registration of a herbicide is the filing with the EPA a statement that includes the following:

1. The applicant's name and address and that of any other party whose name will appear on the proposed label
2. The name of the pesticide
3. A complete copy of the proposed pesticide label, a statement of all claims made for it, and directions for its use, carefully developed to provide clear instructions on proper use of the product
4. A full description of the tests conducted and results upon which the claims for the pesticide are based and/or citation of previously published or submitted data
5. The complete formula of the pesticide
6. A specification of whether the pesticide should be classified as a general use or restricted use pesticide

Once data in support of registration of a pesticide have been submitted to the EPA along with the rest of the complete application packet of the registrant, the EPA must notify the registrant within 45 days whether the application is complete or incomplete. Once a complete application is filed, the EPA will notify the registrant of its approval or disapproval within 90 days. Registration is approved if the EPA determines that the four essential criteria of FIFRA (outlined previously) have been sufficiently fulfilled.

The EPA data requirements for registration (item 4) include the results of a series of toxicological tests to determine acute, subchronic, and chronic toxicity to laboratory animals. The chronic toxicity tests are conducted to determine the herbicide's likelihood of causing tumors (oncogenicity) and birth defects (teratogenicity) and its effects on reproduction. Mutagenicity tests are also conducted to determine whether the herbicide affects chromosomal structure or induces gene mutations. A number of special tests are conducted to determine the herbicide's dermal penetration, general metabolism, and effects on domestic animals and other nontarget species. Toxicity threats to fish, wildlife, and endangered species are indicated under "Environmental Hazards" on the herbicide label.

Finally, the EPA requires that data be presented on various aspects of the environmental fate of the herbicide, including residues on crops and potential for off-target movement and groundwater contamination. If the herbicide is determined to be a possible hazard to groundwater, especially on coarse-textured soils overlying a shallow water table, the label must contain a groundwater advisory statement pertaining to the hazard. Examples of herbicides with labels that carry groundwater advisories are alachlor, atrazine, clopyralid, cyanazine, metolachlor, metribuzin, picloram, and simazine.

Development and registration of a single new herbicide require an investment of tens of millions of dollars and may require close to a decade from the time of initial herbicide discovery until product registration. An important consideration in this process is that a U.S. patent life is only 17 years, so a herbicide manufacturer may actually have less time to sell a product on the market exclusively than it takes to get the product registered from the time of its initial discovery. The manufacturer conducts most of the early research and development necessary for registration, and university and private researchers provide additional research and field testing in the latter stages of development.

B. Herbicide Use Classification: General and Restricted Use

The Federal Environmental Pesticide Control Act (FEPCA) was enacted in 1972 as an amendment to FIFRA and requires that all pesticides be classified as either general-use or restricted-use pesticides. A second component of FEPCA requires that anyone applying restricted-use pesticides be certified by the state in which he or she lives. A herbicide is classified as a general-use pesticide if the registration data provided by the manufacturer to the EPA indicate that it poses no unreasonable risk to the environment and is safe for use by the general public. The majority of herbicides available are classified as general-use, and no special training is required at the present time (1995) for the purchase or application of general-use herbicides.

Herbicides classified as restricted-use pesticides are those which may pose possible adverse health or environmental effects. They can only be sold to and applied by persons who have received special training and certification by EPA-sanctioned state agencies. Examples of restricted-use herbicides are alachlor, aminotriazole, atrazine, bromoxynil, cyanazine, diquat, paraquat, and picloram.

C. FIFRA Section 5: Experimental Use Permits

A herbicide manufacturer may apply to the EPA for an Experimental Use Permit (EUP) for a herbicide before it is granted full registration. The EPA may authorize any state to issue an EUP for an herbicide. The purpose of the EUP is to provide a mechanism by which the applicant can accumulate herbicide performance information under actual field/user conditions and provide supplemental data required for registration. The EUP is typically requested near the end of the registration process, usually 1 to 2 years before full registration is anticipated.

A crop treated with a herbicide under the EUP may be harvested or destroyed, depending on the amount of information (toxicology, residues, etc.) the EPA had on the product at the time the EUP was requested. If the EPA determines that use of the herbicide may result in any residues in food or feed, then it may establish a temporary tolerance level for residues before issuing the EUP, in which case the crop is allowed to be harvested. If tolerances are not established, then a "crop-destruct" EUP is specified and the crop must be destroyed after efficacy results are obtained.

The EUP is valid for a 1-year period or such other time as the EPA prescribes and allows the applicant to use or sell a limited quantity of the product under controlled conditions. Detailed records must be kept of the amounts of products sold, the locations of where the product is applied, and the performance results. In addition to the manufacturer, public or private agricultural research agencies or educational institutions may apply for an EUP for the purposes of experimentation with the product.

D. FIFRA Section 18: Exemption of Federal and State Agencies

The second situation in which a pesticide may be used before it is granted full registration is when the EPA determines that an existing emergency warrants immediate use of the pesticide. Section 18 of FIFRA gives the EPA the authority to exempt any federal or state agency from any provision of FIFRA if it determines that emergency conditions exist. With regard to the qualifications for exemption of a nonregistered pesticide, salient features of Section 18 are that the pest situations be "urgent" and "nonroutine," that no other registered pesticides which will effectively control the pest are available, and that no economically or environmentally feasible alternate practices that provide adequate control are available.

There are four types of emergency exemptions for unregistered pesticides in general: 1) Specific exemption: this authorizes use of an unregistered pesticide in an emergency to prevent a significant economic loss or a significant risk to endangered or threatened species, beneficial organisms, or the environment; 2) quarantine exemption: this may be authorized to control the introduction or spread of any pest not known to be widely prevalent or distributed in the United States; 3) public health exemption: this authorizes use of the pesticide to control a pest that poses a significant risk to human health; and 4) crisis exemption: this may be utilized in an emergency condition when there is an urgent, immediate need for the pesticide upon discovery of the emergency and time does not allow for the authorization of a specific, quarantine, or public health exemption.

E. FIFRA Section 24c: Authority of States

A state may regulate the sale or use of any federally registered pesticide in the state, but only to the extent that the regulation does not permit any sale or use prohibited by FIFRA. The 24c section of FIFRA provides a mechanism by which the designated state pesticide regulatory agency can register a new pesticide that contains already registered active ingredients or amend existing product labels for new uses determined to be special local needs. State experts first discover or determine special local needs that exist and request registration for those specific needs from the designated state agency. If in agreement with the applicant, the state agency can register the special local need for the product. The state agency must notify the federal EPA of the registration within 10 days. If the EPA does not cancel or question the registration within 90 days, the Section 24c state-registered use becomes a legal component of the pesticide label for that state only.

F. Reregistration and Special Review

Toxicology and environmental testing requirements established by the EPA have become more extensive in recent years, and many herbicides registered in the 1960s and early 1970s were granted registration before the more stringent requirements were established. In addition to requiring that pesticides be classified as either general- or restricted-use pesticides, the 1972 FEPCA amendment to FIFRA requires that older products be reregistered. The 1988 version of FIFRA requires pesticide manufacturers to fill in "data gaps" for older products by 1997 to determine whether they should be reregistered. Since the reregistration requirement was enacted, several products and certain uses for products have been voluntarily discontinued by manufacturers because the time and cost requirements for reregistration are too high to justify the costs of filling in the "data gaps" required for reregistration (Beall et al., 1991).

If the EPA determines that a registered pesticide exceeds established risk criteria, a process called special review may be initiated to gather scientific information and reassess the benefits and risks of the pesticide to determine whether registration restriction or cancellation is necessary. Risk criteria that may trigger a special review involve any one of the following: acute toxicity; chronic toxicity (oncogenicity, mutagenicity); other chronic effects (reproductive, spermatogenicity, testicular effects); significant reduction in wildlife, endangered species, or other nontarget species; or lack of emergency treatment or antidote.

A period of public response to a special review from the pesticide manufacturer and/or anyone wishing to comment on reasons for issuance of the special review, including states, grower or commodity groups, or private parties, is allowed for 45 days following the EPA's statement of its preliminary position on the potential risk of the pesticide. Once the scientific data are gathered and responses to special review are heard, extensive risk-benefit analyses are conducted and the EPA issues a final recommendation on the granting of reregistration.

APPENDIX A

Common and Scientific Names of Crops Mentioned in Chapter

Common name	Scientific name	Authority
Alfalfa	Medicago sativa L.	
Canola	Brassica napus	L.
Corn	Zea mays	L.
Cotton	Gossypium hirsutum	L.
Flax	Linum usitatissimum	L.
Oilseed rape	Brassica napus	L.
Soybean	Glycine max	(L.) Merr.
Sugar beet	Beta vulgaris	L.
Sweet potato	Ipomoea batatas	Lam.
Tobacco	Nicotiana tabacum	L.
Tomato	Lycopersicon esculentum	Mill.
Common lambsquarters	Chenopodium album	L.
Redroot pigweed	Amaranthus retroflexus	L.
Wild oat	Avena fatua	L.

APPENDIX B

Common and Chemical Names of Herbicides Mentioned in Chapter

Common name	Chemical name
Acetochlor	2-Chloro-N-(ethoxymethyl)-N-(2-ethyl-6-methylphenyl)acetamide
Acifluorfen	5-[2-Chloro-4-(trifluoromethyl)phenoxy]-2-nitrobenzoic acid
Acrolein	2-Propenal
Alachlor	2-Chloro-N-(2,6-diethylphenyl)-N-(methoxymethyl)acetamide
Ametryn	N-Ethyl-N-(1-methylethyl)-6-(methylthio)-1,3,5-triazine-2,4-diamine
Amitrole	1H-1,2,4-Triazol-3-amine
AMS	Ammonium sulfamate
Asulam	Methyl[(4-aminophenyl)sulfonyl]carbamate
Atrazine	6-Chloror-*N*-ethyl-*N'*-(1-methylethyl)-1,3,5-triazine-2,4-diamine
Barban	4-Chloro-2-butynyl 3-chlorophenylcarbamate
Benefin	N-Butyl-N-ethyl-2,6-dinitro-4-(trifluoromethyl)benzeneamine
Bensulfuron	2-[[[[(4,6-dimethoxy-2-pyrimidinyl)amino]carbonyl]amino]sulfonyl] methyl]benzoic acid
Bensulide	O,O-bis (1-Methylethyl) S-[2-[(phenylsulfonyl)amino]ethyl] phosphorodithioate
Bentazon	3-(1-Methylethyl)-(1H)-2,1,3-benzothiadiazin-4(3H)-one 2,2-dioxide
Bifenox	Methyl 5-(2,4-dichlorophenoxy)-2-nitrobenzoate
Bromacil	5-Bromo-6-methyl-3-(1-methylpropyl)-2,4(1H,3H)pyrimidinedione

Appendix B Continued

Common name	Chemical name
Bromoxynil	3,5-Dibromo-4-hydroxybenzonitrile
Butachlor	N-(Butoxymethyl)-2-chloro-N-(2,6-diethylphenyl)acetamide
Buthidazole	3-[5-(1,1-Dimethyletyhl)-1,3,4-thiadiazol-2-yl]-4-hydroxy-1-methyl-2-imidazolidinone
Butylate	S-Ethyl-bis(2-methylpropyl)carbamothioate
Cacodylic acid	Dimethyl arsinic acid
CDAA	2-Chloro-N,N-di-2-propenylacetamide
Chloramben	3-Amino-2,5-dichlorobenzoic acid
Chlorimuron	2-[[[[(4-Chloro-6-methoxy-2-pyrimidinyl)amino]carbonyl]amino] sulfonyl]benzoic acid
Chloroxuron	N′-[4-(4-Chlorophenoxy)phenyl]-N,N-dimethylurea
Chlorpropham	1-Methylethyl-3-chlorophenylcarbamate
Chlorosulfuron	2-Chloro-N-[[(4-methoxy-6-methyl-1,3,5-triazin-2-yl)amino] carbonyl]benzenesulfonamide
Clethodim	(E,E)-(±)-2-[1-[[(3-Chloro-2-propenyl)oxy]imino]propyl]-5-[2-(ethylthio)propyl]-3-hydroxy-2-cyclohexen-1-one
Clomazone	2-[(2-Chlorophenyl)methyl]-4,4-dimethyl-3-isoxazolidinone
Clopyralid	3,6-Dichloro-2-pyridinecarboxylic acid
Cyanazine	2-[[4-Chloro-6-(ethylamino)-1,3,5-triazin-2-yl]amino]-2-methyl-propanenitrile
Cycloate	*S*-Ethyl cyclohexylethylcarbamothioate
DCPA	Dimethyl 2,3,5,6-tetrachloro-1,4-benzenedicarboxylate
Desmedipham	Ethyl[3-[[(phenylamino)carbonyl]oxy]phenyl]carbamate
Diallate	*S*-(2,3-Dichloro-2-propenyl) bis(1-methylethyl)carbamothioate
Dicamba	3,6-Dichloro-2-methoxybenzoic acid
Dichlobenil	2,6-Dichlorobenzonitrile
Dichlorprop	(±)-2-(2,4-Dichlorophenoxy)propanoic acid
Diclofop	(±)-2-[4-Dichlorophenoxy)phenoxy]propanoic acid
Diethatyl	N-(Chloroacetyl)-N-(2,6-diethylphenyl)glycine
Difenzoquat	1,2-Dimethyl-3,5-diphenyl-1H-pyrazolium
Dinitramine	N^3,N^3-Diethyl-2,4-dinitro-6-(trifluoromethyl)-1,3-benzenediamine
Diphenamid	N,N-Dimethyl-α-phenyl-benzeneacetamide
Dipropetryn	6-(Ethylthio)-*N,N′*-bis(1-methylethyl)-1,3,5-triazine-2,4-diamine
Diquat	6,7-Dihydrodipyrido[1,2-α:2′,1′-c]pyrazinediium ion
Dithiopyr	*S,S*-Dimethyl 2-(difluoromethyl)-4-(2-methylpropyl)-6-(trifluoro-methyl)-3,5-pyridinedicarbothioate
Diuron	N′-(3,4-Dichlorophenyl)-N,N-dimethylurea
DSMA	Disodium salt of MAA
Endothall	7-Oxabicyclo[2,2,1]heptane-2,3-dicarboxylic acid
EPTC	*S*-Ethyl dipropyl carbamothioate
Ethalfluralin	N-Ethyl-N-(2-methyl-2-propenyl)-2,6-dinitro-4-(trifluoromethyl) benzeneamine

(continued)

Appendix B Continued

Common name	Chemical name
Ethofumesate	(±)-2-Ethoxy-2,3-dihydro-3,3-dimethyl-5-benzofuranyl methanesulfonate
Fenac	2,3,6-Trichlorobenzeneacetic acid
Fenoxaprop	(±)-2-[4-[(6-Chloro-2-benzoxazolyl)oxy]phenoxy]propanoic acid
Fenuron	N,N-Dimethyl-N'-phenylurea
Flamprop	N-Benzoyl-N-(3-chloro-4-fluorophenyl)-DL-alanine
Fluazifop-P	(R)-2-[4-[[5-(Trifluoromethyl)-2-pyridinyl]oxy]phenoxy]propanoic acid
Fluchloralin	N-(2-Chloroethyl)-2,6-dinitro-N-propyl-4-(trifluoromethyl) benzeneamine
Fluometuron	N,N-Dimethyl-N'-[3-(trifluoromethyl)phenyl]urea
Fluoroglycofen	Carboxymethyl 5-[2-chloro-4(trifluoromethyl)phenoxy]-2-nitro-benzoate
Fluridone	1-Methyl-3-phenyl-5-[3-(trifluoromethyl)phenyl]-4(1H)-pyridinone
Fluroxypyr	[(4-Amino-3,5-dichloro-6-fluoro-2-pyridinyl)oxy]acetic acid
Fomesafen	5-[2-Chloro-4-(trifluoromethyl)phenoxy]-N-(methylsulfonyl)-2-nitro-benzamide
Fosamine	Ethyl hydrogen (aminocarbonyl)phosphonate
Glufosinate	2-Amino-4-(hydroxymethylphosphinyl)butanoic acid
Glyphosate	N-(Phosphonomethyl)glycine
Hexaflurate	Potassium hexafluoroarsenate
Hexazinone	3-Cyclohexyl-6-(Dimethylamino)-1-methyl-1,3,5-triazine,2,4 (1H,3H)-dione
Imazamethabenz	(±)-2-[4,5-Dihydro-4-methyl-4-(1-methylethyl)-5-oxo-1H-imidazol-2-yl]-4(and 5)-methylbenzoic acid (3:2)
Imazapyr	(±)-2-[4,5-Dihydro-4-methyl-4-(1-methylethyl)-5-oxo-1H-imidazol-2-yl]-3-pyridinecarboxylic acid
Imazaquin	2-[4,5-Dihydro-4-methyl-4-(1-methylethyl)-5-oxo-1H-imidazol-2-yl]-3-quinolinecarboxylic acid
Imazethapyr	2-[4,5-Dihydro-4-methyl-4-(1-methylethyl)-5-oxo-1H-imidazol-2-yl]-5-ethyl-3-pyridinecarboxylic acid
Ioxynil	4-Hydroxy-3,5-diiodobenzonitrile
Isopropalin	4-(1-Methylethyl)-2,6-dinitro-N,N-dipropylbenzeneamine
Isouron	N'-[5-(1,1-Dimethylethyl)-3-isoxazolyl]-N,N-dimethylurea
Isoxaben	*N*-[3-(1-Ethyl-1-methylpropyl)-5-isoxazolyl]-2,6-dimethoxy-benzamide
Karbutilate	3-[[(Dimethylamino)carbonyl]amino]phenyl (1,1-dimethylethyl) carbamate
Lactofen	(±)-2-Ethoxy-1-methyl-2-oxoethyl 5-[2-chloro-4-(trifluoromethyl) phenoxy]-2-nitrobenzoate
Linuron	*N'*-(3,4-Dichlorophenyl)-*N*-methoxy-*N*-methylurea
MAA	Methylarsonic acid
MAMA	Monoammonium salt of MAA

Appendix B Continued

Common name	Chemical name
MCPA	(4-Chloro-2-methylphenoxy)acetic acid
MCPB	4-(4-Chloro-2-methylphenoxy)butanoic acid
Mecoprop	(±)-2-(4-Chloro-2-methylphenoxy)propanoic acid
Mefluidide	*N*-[2,4-Dimethyl-5-[[(trifluoromethyl)sulfonyl]amino]phenyl] acetamide
Metham	Methylcarbamodithioic acid
Methazole	2-(3,4-Dichlorophenyl)-4-methyl-1,2,4-oxadiazolidine-3,5-dione
Metosulam	*N*-(2,6-Dichloro-3-methylphenyl)-5,7-dimethoxy [1,2,4]triazolo [1,5-a]pyrimidine-2-sulfonamide
Metolachlor	2-Chloro-*N*-(2-ethyl-6-methylphenyl)-*N*-(2-methoxy-1-methylethyl) acetamide
Metribuzin	4-Amino-6-(1,1-dimethylethyl)-3-(methylthio)-1,2,4-triazin-5(4H)-one
Metsulfuron	2-[[[[(4-Methoxy-6-methyl-1,3,5-triazin-2-yl)amino]carbonyl] amino]sulfonyl]benzoic acid
Molinate	*S*-ethyl hexahydro-1*H*-azepine-1-carbothioate
Monuron	*N'*-(4-Chlorophenyl)-*N*,*N*-dimethylurea
MSMA	Monosodium salt of MAA
Napropamide	*N*,*N*-Diethyl-2-(1-naphthalenyloxy)propanamide
Naptalam	2-[(1-Naphthalenylamino)carbonyl]benzoic acid
Nicosulfuron	2-[[[[(4,6-Dimethoxy-2-pyrimidinyl)amino]carbonyl]amino] sulfonyl]-*N*,*N*-dimethyl-3-pyridinecarboxamide
Nitrofen	2,4-Dichloro-1-(4-nitrophenoxy)benzene
Norflurazon	4-Chloro-5-(methylamino)2-(3-(trifluoromethyl)phenyl)-3(2*H*)-pyridazinone
Oryzalin	4-(Dipropylamino)-3,5-dinitrobenzenesulfonamide
Oxadiazon	3-[2,4-Dichloro-5-(1-methylethoxy)-phenyl]-5-(1,1-dimethylethyl)-1,3,4-oxadiazol-2-(3*H*)-one
Oxyfluorfen	2-Chloro-1-(3-ethoxy-4-nitrophenoxy)-4-(trifluoromethyl)benzene
Paraquat	1,1′-Dimethyl-4,4′-bipyridinium ion
Pebulate	*S*-Propyl butylethylcarbamothioate
Pendimethalin	*N*-(1-Ethylpropyl)-3,4-dimethyl-2,6-dinitrobenzeneamine
Perfluidone	1,1,1-Trifluoro-*N*-[2-methyl-4-(phenylsulfonyl)phenyl] methanesulfonamide
Phenmedipham	3-[(Methoxycarbonyl)amino]phenyl 3-(methylphenyl)carbamate
Picloram	4-Amino-3,5,6-trichloro-2-pyridinecarboxylic acid
Primisulfuron	2-[[[[4,6-bis(Difluoromethoxy)-2-pyrimidinyl]amino]carbonyl] amino]sulfonyl]benzoic acid
Profluralin	*N*-(Cyclopropylmethyl)-2,6-dinitro-*N*-propyl-4-(trifluoromethyl) benzeneamine
Prometon	6-Methoxy-*N*,*N'*-bis(1-methylethyl)-1,3,5-triazine-2,4-diamine

(*continued*)

Appendix B Continued

Common name	Chemical name
Prometryn	*N,N'*-bis(1-Methylethyl)-6-(methylthio)-1,3,5-triazine-2,4-diamine
Pronamide	3,5-Dichloro (*N*-1,1-dimethyl-2-propynyl)benzamide
Propachlor	2-Chloro-*N*-(1-methylethyl)-*N*-phenylacetamide
Propanil	*N*-(3,4-Dichlorophenyl)propanamide
Propazine	6-Chloro-*N,N'*-bis(1-methylethyl)1,3,5-triazine-2,4-diamine
Propham	1-Methylethyl phenylcarbamate
Pyrazon	5-Amino-4-chloro-2-phenyl-3(2*H*)-pyridazinone
Quinclorac	3,7-Dichloro-8-quinolinecarboxylic acid
Quizalofop	(±)-2-[4-[(6-Chloro-2-quinoxalinyl)oxy]phenoxy]propanic acid
Sethoxydim	2-[1-(Ethoxyimino)butyl]-5-[2-(ethylthio)propyl]-3-hydroxy-2-cyclohexen-1-one
Siduron	*N*-(2-Methylcyclohexyl)-*N'*-phenylurea
Simazine	6-Chloro-*N,N'*-diethyl-1,3,5-triazine-2,4-diamine
Sulfometuron	2-[[[[(4,6-Dimethyl-2-pyrimidinyl)amino]carbonyl]amino]sulfonyl] benzoic acid
Tebuthiuron	*N*-[5-(1,1-Dimethylethyl)-1,3,4-thiadiazol-2-yl]-*N,N'*-dimethylurea
Terbacil	5-Chloro-3-(1,1-dimethylethyl)-6-methyl-2,4-(1*H*,3*H*)-pyrimidinedione
Terbutryn	*N*-(1,1-Dimethylethyl)-*N'*-ethyl-6-(methylthio)-1,3,5-triazine-2,4-diamine
Thifensulfuron	3-[[[[(4-Methoxy-6-methyl-1,3,5-triazine-2-yl)amino]carbonyl] amino]sulfonyl]-2-thiophenecarboxylic acid
Thiobencarb	*S*-[(4-Chlorophenyl)methyl]diethylcarbamothioate
Triallate	*S*-(2,3,3-Trichloro-2-propenyl) bis(1-methylethyl)carbamothioate
Triasulfuron	2-(2-Chloroethoxy)-*N*-[[(4-methoxy-6-methyl-1,3,5-triazin-2-yl) amino]carbonyl]benzenesulfonamide
Tribenuron	2-[[[[(4-Methoxy-6-methyl-1,3,5-triazin-2-yl)methylamino] carbonyl]amino]sulfonyl]benzoic acid
Triclopyr	[(3,5,6-Trichloro-2-pyridinyl)oxy]acetic acid
Trifluralin	2,6-Dinitro-*N,N*-dipropyl-4-(trifluoromethyl)benzeneamine
2,4-D	(2,4-Dichlorophenoxy)acetic acid
2,4-DB	4-(2,4-Dichlorophenoxy)butanoic acid
Vernolate	S-Propyl dipropylcarbamothioate
Xylachlor	2-Chloro-*N*-(2,3-dimethylphenyl)-*N*-(1-methylethyl)acetamide

REFERENCES

Anderson, W. P. (1977). *Weed Science Principles*, West Publishing Co., St. Paul.

Arntzen, C. J., Pfister, K., and Steinback, K. E. (1982). The mechanism of chloroplast triazine resistance: alterations in the site of herbicide action, in *Herbicide Resistance in Plants* (H. M. LeBaron and J. Gressel, eds.), John Wiley & Sons, Inc., New York, p. 185.

Ashton, F. M. (1958). Absorption and translocation of 2,4-D in sugar cane and bean plants, *Weeds 6*: 257.

Ashton, F. M. and Crafts, A. A. (1981). *Mode of Action of Herbicides, 2nd Edition*, John Wiley & Sons, New York.

Bailey, G. W. and White, J. L. (1964). Review of adsorption and desorption of organic pesticides by soil colloids with implications concerning pesticide bioactivity, *J. Agric. Food Chem. 12*: 324.

Baker, E. A. (1980). Effect of cuticular components on foliar penetration, *Pestic. Sci. 11*: 367.

Beall, G. A., Bruhn, C. M., Craigmill, A. L. and Winter, C. K. (1991). Pesticides and your food: how safe is safe? *California Agric. 45*: 4.

Beyer, E. M., Jr., Duffy, M. J., Hay, J. V., and Schlueter, D. D. (1988). Sulfonylureas, in *Herbicides: Chemistry, Degradation, and Mode of Action* (P. C. Kearney and D. D. Kaufman, eds.), Marcel Dekker, Inc., New York, p. 118.

Boger, P. and Sandmann, G. (1989). *Target Sites of Herbicide Action*, CRC Press, Inc., Boca Raton, Florida, p. 295.

Buhler, D. D. and Burnside, O. C. (1983). Soil activity of fluazifop, sethoxyim, and Dowco 453 {methyl 2-[4-((3-chloro-5′-(trifluoromethyl)-2-pyridinyl) oxy) phenoxy] propanoate}, *Abstr. Weed Sci. Soc. Am. 23*: 29.

Bukovac, M. J. (1976). Herbicide entry into plants, in *Herbicides: Physiology, Biochemistry, Ecology*, Vol. 1, 2nd Edition (L. J. Audus, ed.), Academic Press, New York, p. 335.

Calderbank, A. and Slade, P. (1976). Diquat and paraquat, in *Herbicides* (P. C. Kearney and D. D. Kaufman, eds.), Marcel Dekker, Inc., New York, p. 501.

Che, M. D., Loux, M. M., Traina, S. J., and Logan, T. J. (1992). Effect of pH on sorption and desorption of imazaquin and imazethapyr on clays and humic acid, *J. Environ. Qual. 21*: 698.

Devine, M. D. (1989). Phloem translocation of herbicides, *Rev. Weed Sci. 4*: 191–213.

Devine, M. D., Bestman, H. D., and Vanden Born, W. H. (1990). Physiological basis for the different phloem mobilities of chlorosulfuron and clopyralid, *Weed Sci. 38*: 305.

Devine, M. D., and Vanden Born, W. H. (1991). Absorption and transport in plants, in *Environmental Chemistry of Herbicides* (R. Grover and A. J. Cessna, eds.), CRC Press, Inc., Boca Raton, Florida, p. 119.

Duke, S. O. (1988). Glyphosate, in *Herbicides: Chemistry, Degradation, and Mode of Action* (P. C. Kearney and D. D. Kaufman, eds.), Marcel Dekker, Inc., New York, p. 1.

Duke, S. O., Lydon, J., Becerril, J. M., Sherman, T. D., Lehnen, L. P., Jr., and Matsumoto, M. (1991). Protoporphyrinogen oxidase-inhibiting herbicides, *Weed Sci. 39*: 465.

Dyer, W. E., Hess, D. F., Holt, J. S. and S. O. Duke (1993). Potential benefits and risks of herbicide-resistant crops produced by biotechnology, *Hort. Rev. 15*: 367.

EPA Pesticide Classification Procedures. (1979). *The U.S. Federal Register 40*: 28279.

Fedtke, C. (1982). Herbicides with auxin activity, in *Biochemistry and Physiology of Herbicide Action*, Springer-Verlag, Berlin, p. 159.

Froud-Williams, R. J., Chancellor, R. J., and Drennan, D. S. H. (1981). Potential changes in weed floras associated with reduced-cultivation systems for cereal production in temperate regions, *Weed Res. 21*: 99.

Funderburk, H. H. and Lawrence, J. M. (1963). Absorption and translocation of radioactive herbicides in submersed and emerged aquatic weeds, *Weed Res. 3*: 304.

Gleason, M. N., Gosselin, R. E., and Hodge, H. C. (1976). *Clinical Toxicology of Commercial Products*, 4th Edition, Williams & Wilkins Co., Baltimore, p. 6.

Gottrup, O., O'Sullivan, P. A., Schraa, R. J., and Vanden Born, W. H. (1976). Uptake, translocation, metabolism and selectivity of glyphosate in Canada thistle and leafy spurge, *Weed Res. 16*: 197.

Gressel, J. (1985). Herbicide tolerance and resistance: alteration of site of activity, in *Weed Physiology*, Vol. 2 (S. O. Duke, ed.), CRC Press, Inc., Boca Raton, Florida, p. 161.

Gressel, J. and Segal, L. A. (1990). Modelling the effectiveness of herbicide rotations and mixtures as strategies to delay or preclude resistance, *Weed Tech. 4*: 186.

Gronwald, J. (1991). Lipid biosynthesis inhibitors, *Weed Sci. 39*: 435.

Hatzios, K. K. (1991). Biotransformations of herbicides in higher plants, in *Environmental Chemistry of Herbicides* (R. Grover and A. J. Cessna, eds.), CRC Press, Inc., Boca Raton, Florida, p. 141.

Hatzios, K. K. and Penner D. (1982). *Metabolism of Herbicides in Higher Plants*, Burgess Publishing Co., Minneapolis.

Helling, C. S. (1971). Pesticide mobility in soils. III. Influence of soil properties, *Soil Sci. Soc. Am. Proc. 35*: 743.

Herholdt, J. A. (1968). The mode of action and metabolism of terbacil (3-tert-butyl-5-chloro-6-methyl-uracil) in bean (*Phaseolus vulgaris* L.) and citrus (*Citrus sinensis* L. Osbeck and *C. jambhiri* Lush), Ph.D. Thesis, Univ. Calif. Riverside.

Hess, F. D. (1985). Herbicide absorption and translocation and their relationship to plant tolerances and susceptibility, in *Weed Physiology*, Vol. 2 (S. O. Duke, ed.), CRC Press, Inc., Boca Raton, Florida, p. 191.

Hess, F. D. (1987). Herbicide effects on the cell cycle of meristematic plant cells, *Rev. Weed Sci. 3*: 203.

Hibbitt, C. J. (1969). Growth and spray retention of wild oat and flax in relation to herbicidal activity, *Weed Res., 9*: 95

Hill, B. D., Stobbe, E. H., and Jones, B. L. (1978). Hydrolysis of the herbicide benzoylprop ethyl by a wild oat esterase, *Weed Res. 18*: 149.

Hirschberg, J. and McIntosh, L. (1983). Molecular basis of herbicide resistance in *Amaranthus hybridus, Science 222*: 1346.

Holt, J. S. and LeBaron, H. M. (1990). Significance and distribution of herbicide resistance, *Weed Tech. 4*: 141.

Hopkins, W. L. (1989). A global evaluation of "new" herbicide activity, 1984-1988: its changing dynamics and a look at its future direction, Brighton Crop Protection Conference, *Weeds 1*: 231–236.

Hunt, G. M. and Baker, E. A. (1982). Developmental and environmental variations in plant epicuticular waxes: some effects on the penetration of naphthylacetic acid, in *The Plant Cuticle* (D. F. Cutler, K. L. Alvin and C. E. Price, eds.), Linnaeus Society Symposium Series No. 10, Academic Press, London.

Jeffcoat, B. and Harries, W. N. (1975). Selectivity and mode of action of flamprop-isopropyl, isopropyl (±)-2-[*N*-(3-chloro-4-fluorophenyl)benzamido] propionate, in the control of *Avena fatua* in barley, *Pestic. Sci. 6*: 283.

Kirby, C. (1980). *The Hormone Weed Killers*, British Crop Protection Council, Lavenham Press Ltd., Suffolk, U.K.

Kirkwood, R. C. (1987). Uptake and movement of herbicides from plant surfaces and the effects of formulation and environment upon them, in *Pesticides on Plant Surfaces* (H. J. Cottrell, ed.), John Wiley & Sons, New York, p. 2.

Leather, G. R. and Foy, C. L. (1978). Differential absorption and distribution as a basis for the selectivity of bifenox, *Weed Sci. 26*: 76.

LeBaron, H. M., and Gressel, J. (eds.). (1982). *Herbicide Resistance in Plants*, John Wiley & Sons, Inc., New York.

LeBaron, H. M. and McFarland, J. (1990). "Overview and prognosis of herbicide resistance in weeds and crops," Proceeding of the American Chemical Society Symposium on Fundamental and Practical Approaches to Combating Resistance, ACS Symposium Series, ACS Books, Washington, D.C.

Loux, M. M. and Berry, M. A. (1991). Use of a grower survey for estimating weed problems, *Weed Tech. 5*: 460.

Loux, M. M. and Reese, K. D. (1992). Effect of soil type and pH on imazethapyr persistence and carryover of mazaquin and imazethapyr to corn, *Weed Tech. 7*: 452.

Mallory-Smith, C. A., Thill, D. C., and Dial, M. J. (1990). Identification of sulfonylurea herbicide-resistant prickly lettuce (*Lactuca serriola*), *Weed Tech. 4*: 163.

Moyer, J. R. (1987). Effect of soil moisture on the efficacy and selectivity of soil-applied herbicides, *Rev. Weed Sci. 3*: 19.

Parr, J. F. (1982). Toxicology of adjuvants, *Adjuvants for Herbicides*, Weed Science Society of America, Champaign, Illinois, pp. 93–114.

Pfister, K., Radosevich, S. R., and Arntzen, C. J. (1979). Modification of herbicide binding to photosystem II (PS II) in two biotypes of *Senecio vulgaris L., Plant Physiol. 64*: 995.

Pfister, K., Steinback, D. E., Gardner, G., and Arntzen, C. J. (1981). Photoaffinity labeling of an herbicide receptor protein in chloroplast membranes, *Proc. Natl. Acad. Sci. U.S.A., 78*: 981.

Powles, S. B. and Howat, P. D. (1990). Herbicide-resistant weeds in Australia, *Weed Tech. 4*: 178.

Prendeville, G. N. (1968). Shoot zone uptake of soil-applied herbicides, *Weed Res. 8*: 106.

Prendeville, G. N., Oliver, L. R., and Schreiber, M. M. (1968). Species differences in site of shoot uptake and tolerance to EPTC, *Weed Sci. 16*: 538.

Price, C. E. (1982). A review of the factors influencing the penetration of pesticides through plant leaves, *The Plant Cuticle* (D. F. Cutler, K. L. Alvin and C. E. Price, eds.), Linnaeus Society Symposium Series No. 10, Academic Press, London, p. 237.

Primiani, M. M., Cotterman, J. C., and Saari, L. L. (1990). Resistance of kochia (*Kochia scoparia*) to sulfonylurea and imidazolinone herbicides, *Weed Tech. 4*: 169.

Quimby, P. C. and Nalewaja, J. D. (1971). Selectivity of dicamba in wheat and wild buckwheat, *Weed Sci. 19*: 598.

Rogers, R. L. and Funderburk, H. H. (1968). Physiological aspects of fluometuron in cotton and cucumber, *J. Agric. Food Chem. 16*: 434–440.

Ross, M. A. and Lembi, C. A. (1985). *Applied Weed Science*, Burgess Publishing Co., Minneapolis, p. 241.

Ryan, G. F. (1970). Resistance of common groundsel to simazine and atrazine, *Weed Sci. 18*: 614.

Saggers, D. T. (1976). The search for new herbicides, *Herbicides* (L. J. Audus, ed.), Vol. 2, Academic Press, London, p. 564.

Sandmann, G. and Boger, P. (1989). Inhibition of carotenoid synthesis by herbicides, *Target Sites of Herbicide Action* (P. Boger and G. Sandmann, eds.), CRC Press, Inc., Boca Raton, Florida, p. 25.

Sandmann, G., Schmidt, A., Linden, H. and Boger, P. (1991). Phytoene desaturase, the essential target for bleaching herbicides, *Weed Sci. 39*: 474.

Shaner, D. L. and Mallipudi, N. M. (1991). Mechanisms of selectivity of the imidazolinones, *The Imidazolinone Herbicides* (D. L. Shaner and S. L. O'Connor, eds.), CRC Press, Inc., Boca Raton, Florida, p. 91.

Shaner, D. L. and Robson, P. A. (1985). Absorption, translocation, and metabolism of AC 252,214 in soybean (*Glycine max*), common cocklebur (*Xanthium strumarium*) and velvetleaf (*Abutilon theophrasti*), *Weed Sci. 33*: 269.

Shaner, D. L., Simcox, P. D., Robson, P. A., Mangels, G., Reichert, B., Ciarlante, D. R., and Lox, M. (1983). "AC 222-293 translocation and metabolic selectivity," Proceedings of the British Plant Protection Conference, London, p. 333.

Sharma, M. P., Vanden Born, W. H., and McBeath, D. K. (1978). Spray retention, foliar penetration, translocation, and selectivity of asulam in wild oats and flax, *Weed Res., 18*: 169.

Shimabukuro, R. H. (1985). Detoxication of herbicides, *Weed Physiology*, Vol. 2 (S. O. Duke, ed.), CRC Press, Inc., Boca Raton, Florida, p. 215.

Shone, M. G. T. and Wood, A. V. (1976). Factors responsible for the tolerance of blackcurrants to simazine, *Weed Res. 12*: 337.

Smith, A. E. (1979). Metabolism of 2,4-DB by white clover (*Trifolium repens*) cell suspension cultures, *Weed Sci. 27*: 392.

Smith, A. E. and Wilkinson, R. E. (1974). Differential absorption, translocation, and metabolism of metribuzin (4-amino-6-*tert*-butyl-3-(methylthio)-*as*-triazine-5(4*H*)one) by soybean cultivars, *Physiol. Plant. 32*: 253.

Stephenson, G. R. and Ries, S. K. (1967). The movement and metabolism of pyrazon in tolerant and susceptible species, *Weed Res. 7*: 51.

Stidham, M. A. (1991). Herbicides that inhibit acetohydroxyacid synthase, *Weed Sci.: 39*: 428.

Stoller, E. W. (1970). Mechanism for the differential translocation of amiben in plants, *Plant Physiol. 46*: 732–737.

Synerholm, M. E. and Zimmerman, P. W. (1947). Preparation of a series of w-(2,4-D) aliphatic acids and some related compounds with a consideration of their biochemical role as plant growth regulators, *Contrib. Boyce Thompson Inst. 14*: 369.

Triplett, G. B. Jr. and Lytle, G. A. (1972). Control and ecology of weeds in continuous corn grown without tillage, *Weed Sci. 20*: 453.

Vaughn, K. C. and Lehnen, L. P., Jr. (1991). Mitotic disrupter herbicides, *Weed Sci. 39*: 450.

Walker A. and Featherstone, R. M. (1973). Absorption and translocation of atrazine and linuron by plants with implications concerning linuron selectivity, *J. Exp. Bot. 24*: 450–458.

Ware, G. W. (1983). Pesticides: Theory and Application, W. H. Freeman and Co., San Francisco, p. 179.

Warwick, S. I. and Marriage, P. B. (1982). Geographical variation in populations of *Chenopodium album* resistant and susceptible to atrazine. I. Between and within population variation in growth response to atrazine, *Can. J. Bot. 60*: 483.

Weed Science Society of America. (1989). *Herbicide Handbook of the Weed Science Society of America*, Champaign, Illinois.

Worthing, C. R., ed. (1987). *The Pesticide Manual: A World Compendium*, The British Crop Protection Council, Thornton Heath, U.K.

Zimdahl, R. L. (1993). *Fundamentals of Weed Science*, Academic Press, Inc., San Diego, p. 210, 222.

RELATED REFERENCES

Ashton, F. M. and Monaco, T. J. (1991). *Weed Science: Principles and Practices*, 3rd Edition, John Wiley & Sons, Inc., New York.

Cobb, A. (1992). *Herbicides and Plant Physiology*, Chapman & Hall, London.

Gunsolus, J. L. and Curran, W. S. (1991). *Herbicide Mode of Action and Injury Symptoms*, North Central Regional Extension Publication 377, Cooperative Extension Services, U.S. Department of Agriculture, Washington, D.C.

Hamm, J. G. (1982). *The Handbook of Pest Control*, Frederick Fell Publishers, Inc., New York.

Pesticides Guide: Registration, Classification, and Application. (J. J. Keller and Associates, Inc., eds.). (1982). J. J. Keller and Associates, Neenah, Wisconsin.

Stephens, R. J. (1982). *Theory and Practice of Weed Control*, Macmillan Press Ltd., London.

Weed Control in Limited-Tillage Systems (A. F. Wiese, ed.). (1985). Monograph Series, Number 2, Weed Science Society of America, Champaign, Illinois.

6

Herbicide Application Equipment

H. Erdal Ozkan
The Ohio State University, Columbus, Ohio

I. INTRODUCTION

Yield and crop quality depend on effective application of herbicides, insecticides, fungicides, fertilizers, growth regulators, etc. Inadequate quantity of chemical wastes time, money, and materials. An overdose adds to the cost, may damage the crop, and can cause environmental pollution. Proper selection, use, and maintenance of application equipment saves money and chemicals and preserves the environment.

There are many types of equipment for applying herbicide formulations. These include liquid and granular applicators and fumigation equipment. Within each group, there are various types of equipment, that depend on 1) the time of application (preemergence, postemergence, preplanting incorporated), 2) the growth stage of the target weed, and 3) the physical shape and size of vegetation dominant within the application site (crops, vegetables, trees, shrubs, grass etc.). This chapter includes information on some of the most common types of equipment used for application of herbicides; less common application equipment and techniques such as fumigation equipment, dusters, and dry granular application by aircraft, are not discussed. Detailed information on operation and selection of application equipment that is not covered can be found in a number of references (Akesson and Yates, 1979; Marer, 1988; Mathews, 1982; McWhorter and Gebhardt, 1987).

II. LIQUID APPLICATION EQUIPMENT

Since liquid formulations make up a great majority of the herbicides sold in the United States of America, there is more liquid application equipment in use than granular application equipment. Although there are many types and sizes of liquid application equipment, low-pressure boom sprayers are by far the most common type used for applying herbicides. Other types of liquid application equipment used for herbicide applications include hand-operated backpack sprayers, hand gun sprayers, air-assisted sprayers, and rope-wick applicators.

A. Low-Pressure Boom Sprayers

All types of liquid application equipment contain similar components such as a reservoir to store the spray liquid, a mechanism to move the liquid from the reservoir to the point of discharge, and various components to regulate the liquid flow. The following information relates to components of boom sprayers. However, much of this information is applicable to other types as well.

Nozzles

Although nozzles are among the least expensive components of a spraying system, they greatly influence sprayer performance. They not only meter the proper amount of liquid but atomize the spray liquid into droplets, and also control the variability of the spray over the entire width of the sprayer boom.

Nozzle flow rate depends on the size of the nozzle orifice and pressure. Flow rate increases as pressure increases. This increase, however, is not a linear relationship, the flow rate varies with the square root of pressure increase. Thus, the pressure must be increased four times to double the flow rate. For example, if 150 kPa produces 1 L per minute, 600 kPa would be required to produce 2 L per minute flow rate.

Manufacturers supply tables of flow rates for nozzles at various pressures. These tables are developed by measuring the flow rate of water through the nozzle. Flow rate of other liquids and chemical solutions may differ from that of water. Most herbicide and carrier mixtures are more dense or viscous than water and thus have lower flow rates. Therefore, sprayers must be calibrated for each type of liquid applied.

Nozzles are responsible for atomizing the spray liquid and producing droplets. Spray droplet diameters are measured in micrometers (μm). A micrometer is 1 millionth (10^{-6}) of a meter and is usually referred to as a *micron*. Volume median diameter (VMD) is often used to characterize the spray droplets produced by nozzles. For instance, a VMD of 200 microns indicates that one-half of the spray volume contains droplets smaller than 200 microns, while the other half contains droplets larger than 200 microns. Depending on the nozzle selected, VMDs of droplets produced may vary from just a few microns to over 1000 microns. As the size of the nozzle opening increases, droplet size generally increases. For instance, VMDs of XR8002, XR8004, XR8006, and XR8008 nozzles (Spraying Systems Co.) with flow rates of 0.8, 1.5, 2.3 and 3.0 L/min, respectively, are 146, 194, 231, and 305 microns, respectively.

As shown in Figure 1, popular fan-pattern nozzles have four parts: body, strainer, tip, and cap. The body and cap make a unit that holds a strainer and a tip. The nozzle body may be mounted on either a wet boom (material flows through the boom) or a dry boom (material flows through a hose attached to the boom).

Nozzle tips atomize liquid into distinctive spray patterns and droplets. Tip orifice size and type, and spray pressure determine the flow rate. Tips are available with a wide range of capacities and distribution patterns for specific spraying requirements.

Strainers (filters) are placed just ahead of nozzle tips to filter the liquid and prevent plugging. The filters in the nozzle tips are sized according to tip orifice. Each opening in the filter must be smaller than the size of the nozzle orifice. In North America, filters are usually classified by a mesh number which refers to the number of openings per square inch of the screen material. Strainers with high mesh numbers have smaller openings than strainers with low mesh numbers. Most filters have 50- to 200-mesh openings. In general, 100-mesh filters are recommended for most nozzles with a flow rate below 0.75 L/min, and 50-mesh strainers for nozzles with a flow rate between 0.75 and 4 L/min. If a satisfactory line strainer is used, no nozzle strainer is needed for nozzles with flow rates above 4 L/min. For spraying abrasive materials such as wettable powders, a slotted strainer with openings equivalent to 16- to 50-mesh filters should be used. Always check the manufacturer's catalogue to determine the strainer size recommended for a specific nozzle.

A quick shut-off at each nozzle prevents dripping while turning at ends of rows, filling the tank, or transporting the sprayer. Special strainers equipped with ball check valves are available for this purpose. Chemicals may corrode this type of valve. A diaphragm check valve, however, is useful in stopping drippage when the sprayer is not in use, and it is well protected from chemical corrosion. This check valve allows the nozzle tip to be changed without the spray material leaking from the boom. A ball-type check valve, on the left, and a diaphragm check valve are shown in Figure 2.

Spray nozzles are made from a variety of materials, including aluminum, brass, nylon, plastic, stainless steel, hardened stainless steel, and ceramic. Fan-

Figure 1 Parts of a nozzle.

they are relatively inexpensive. However, they wear rapidly when used to apply abrasive materials such as wettable powders. They also are susceptible to corrosion by some liquid fertilizers. Ceramic and stainless steel tips are the most resistant against wear and corrosion but are the most expensive.

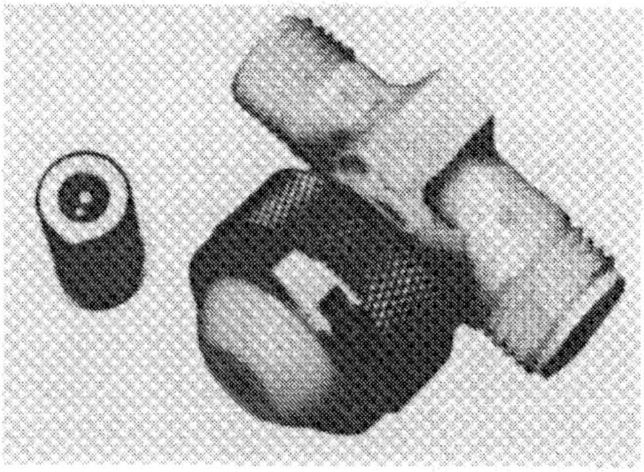

Figure 2 Ball-type (*left*) and diaphragm check valves.

Nylon or plastic tips are sometimes more resistant to wear than metal tips because molded tips have a smoother finish. The surface of metal tips (especially brass and aluminum tips) has microscopic grooves as a result of machining and drilling the orifice. This rough finish presumably causes turbulence and enhances the abrasive action of particles suspended in a spray liquid. A disadvantage of nylon or plastic nozzles is that the threads of some nozzle bodies and caps may be easily damaged by constant use, especially if they are overtightened. They are also subject to swelling when exposed to some solvents.

Ceramic spray tips are also very resistant to abrasion but are expensive and susceptible to breakage. Improved, shatter-resistant ceramic nozzles have been introduced recently. These ceramic nozzles are encased in plastic, which serves as a protective shield to prevent shattering or breakage.

Calibration clinics conducted in Ohio and some other states revealed that more than one-third of the sprayers surveyed were overapplying pesticides (Ozkan, 1987). The major reason for overapplication was worn nozzles. There has been considerable interest in determining wear rates of nozzles made from different materials. A summary of results of nozzle wear tests at several locations is included in a publication by Reed (1984). Reed reported that tests conducted at five different locations resulted in greatly different relative wear rates for brass and stainless steel flat fan tips. The relative wear rates ranged from no difference in wear rates between brass and stainless steel tips to stainless steel lasting 19 times longer than brass. The latest set of tests were conducted by Reichard et al. (1991) following the procedures outlined in ASAE Standard S471 (American Society of Agricultural Engineers, 1991). The tests included regular flat fan nozzles with initial flow rates of 0.8, 1.5, 2.3, and 3.0 L/min. Nozzle materials included brass, nylon, plastic, stainless steel, and hardened stainless steel. Table 1 shows the number of hours the nozzles were subjected to wear tests and the percentage flow rate increase at the end of the test period.

Figure 3 shows the wear rates of regular flat fan nozzles, made of different materials, with a flow rate of 2.3 L/min. The patterns of wear for nozzles with 0.8, 1.5, and 3.0 L/min nozzles were similar to that presented in Figure 3. Reichard et al. (1991) concluded the following:

1. For all nozzle capacities, the stainless steel tips had average use times 5.6 and 2.1 times longer than those of brass and nylon tips.
2. For all nozzles and materials tested, the flow rates from brass and hardened stainless steel tips increased the greatest and least amounts, respectively, with time of use.
3. The flow rates of some plastic tips decreased slightly with time of use before beginning to increase. At 10% increase in flow rate, some of the plastic tips had longer use times than stainless steel tips.
4. The percentage increase in flow rates for all brass, stainless steel, and nylon tips varied directly with the square root of time of use.
5. The time of use before 10% increase in flow rates increased rapidly with increased capacities of brass, stainless steel, nylon, and plastic tips.

Some applications require precise overlapping of adjacent nozzle patterns. When nozzles wear, they may no longer produce the spray pattern essential to uniform coverage. Such distortions in spray patterns could cause streaks from under- or overdosed areas. When tests conducted by Reichard et al. (1991) to determine the changes in flow rates were completed, Ozkan et al. (1992) used the same nozzles to determine the changes in spray distribution patterns of regular flat fan nozzles due to orifice wear. Test results showed that there was some difference between the spray deposit distributions of new and worn regular fan nozzles. Regardless of the size and material of the nozzles, the width of the spray pattern remained nearly constant, but the worn nozzles delivered greater volumes of liquid in the center of the pattern.

Nozzle wear may also affect droplet size, which is an important factor in achieving satisfactory pest control. Some research has shown that small droplets allow more efficient use of insecticides than larger droplets because of the small size of the target organism. However, small droplets are highly susceptible to spray drift (see Chapter 7 for more information on importance of droplet size in reducing spray drift). Not much information was available on the influence of nozzle wear on droplet size distributions. The most recent work in this area was conducted by Ozkan et al. (1992) using regular flat-fan nozzles. They concluded that, regardless of the type of material, generally there was little difference in droplet size distributions of new and worn nozzles with low flow rates. However, the difference was noticeable as the flow rate of nozzles increased.

Table 1 Duration of Wear Tests and Increase in Flow Rate of Nozzles at the End of the Test Period

Nozzle capacity L/min	Brass		Nylon		Plastic		Stainless steel		Hardened stainless steel	
	Test period (hr)	Increase in flow rate (%)	Test period (hr)	Increase in flow rate (%)	Test period (hr)	Increase in flow rate (%)	Test period (hr)	Increase in flow rate (%)	Test period (hr)	Increase in flow rate (%)
0.8	26	22.8	40	15.8	40	18.1	40	14.0	40	1.5
1.5	44	19.6	100	17.9	100	12.4	100	12.2	100	5.5
2.3	110	21.0	242	23.8	242	14.2	242	22.6	242	5.3
3.0	268	20.2	352	17.7	352	13.5	352	11.0	352	10.2

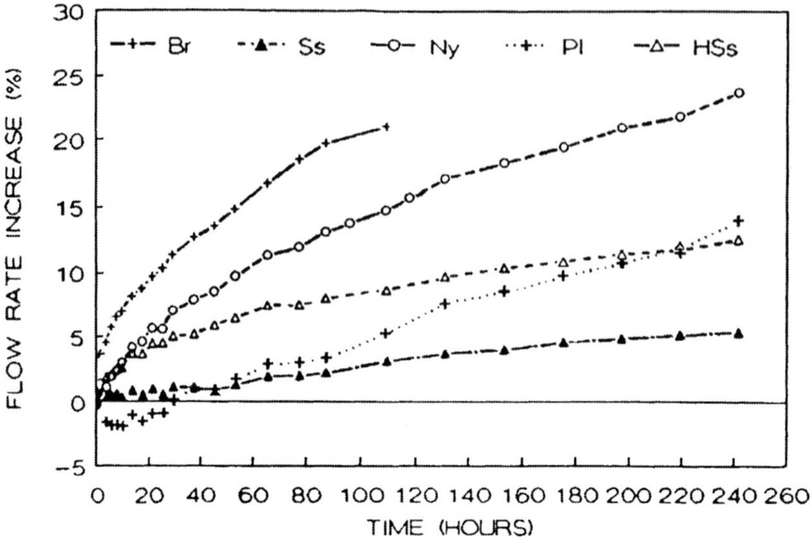

Figure 3 Regular flat-fan nozzle wear rates. Effects of tip material on wear rates of nozzles with nominal flow rates of about 2.3 L/min (0–6 gpm).

Select the proper nozzle for each type of spray application. There are many types of nozzles available to provide differences in throughput, spray angle, and pattern (Figure 4). Only the nozzles most commonly used on low-pressure sprayers are discussed in the following section.

Flat-fan nozzles are best used for broadcast application of preemergence herbicides and postemergence application of systemic herbicides. Spray droplets from a flat-fan nozzle form a fan shape as they leave the nozzle and deposit an elliptical pattern on the ground with more spray droplets in the center of the fan, tapering off at each end. This results in reduced volumes at both ends. To obtain uniform coverage across the spray swath, adjust boom height so that approximately 30% to 40% of the spray pattern on each side should overlap with the spray pattern from the adjacent nozzle (Figure 5a). For example, the proper overlap for a series of 80 degree flat fan nozzles spaced 50 cm apart on the boom is 15 to 20 cm. Flat-fan tips should be mounted on the boom directed downward, perpendicular to the ground. However, they should be turned 5 to 10 degrees away from the axis of the boom so that the spray pattern from one nozzle bypasses the spray patterns of adjacent nozzles.

Regular flat-fan nozzles are available with different fan angles. The shape of the nozzle orifice determines the spray angle. Although flat-fan nozzles with 65-, 73-, 80-, and 110-degree spray angles are available, nozzles with 80-de-

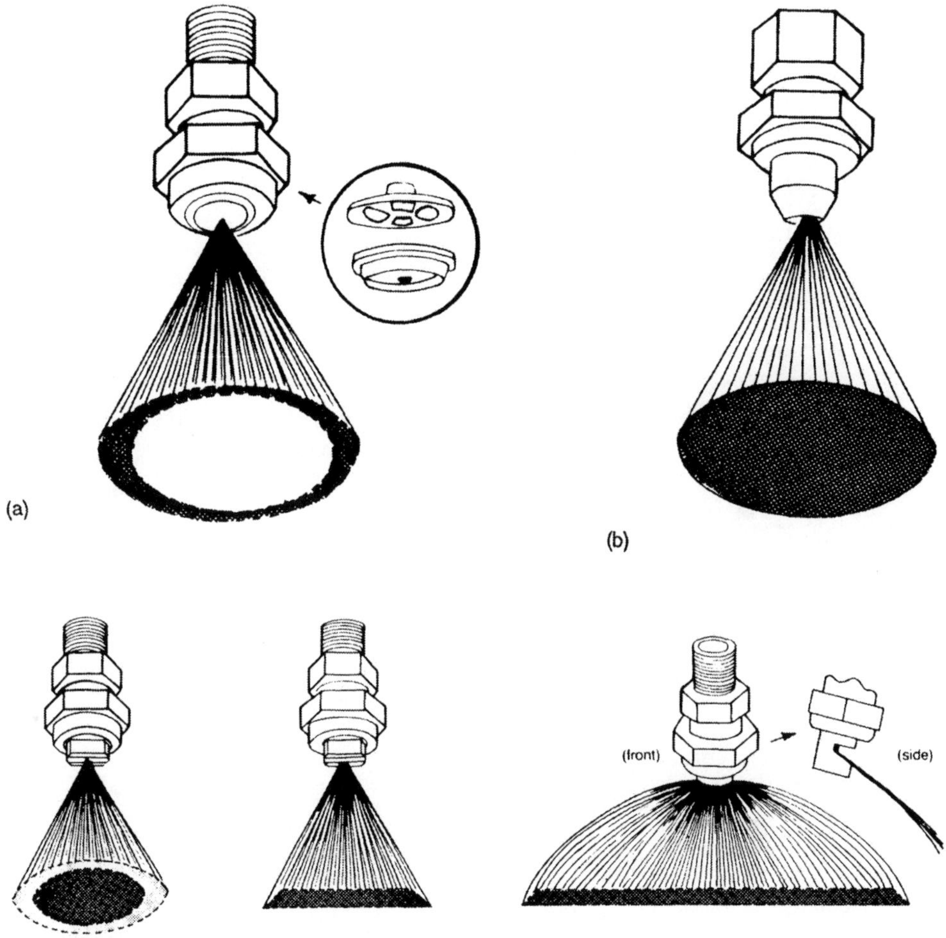

(c)

Figure 4 Common sprayer nozzles: (a) hollow cone nozzle; (b) solid cone nozzle; (c) (*from right to left*) regular flat-fan nozzle, even flat-fan nozzle, and flooding-fan nozzle.

gree spray angles are most common in North America while 110-degree tips are most popular in Europe. For a given nozzle spacing, the nozzles with wider spray pattern angles allow a lower boom height to be used. Additionally, the wider-angle tips produce somewhat smaller droplet sizes, which may or may not be desirable, depending on the type of chemical applied, target pest characteristics, and drift situation.

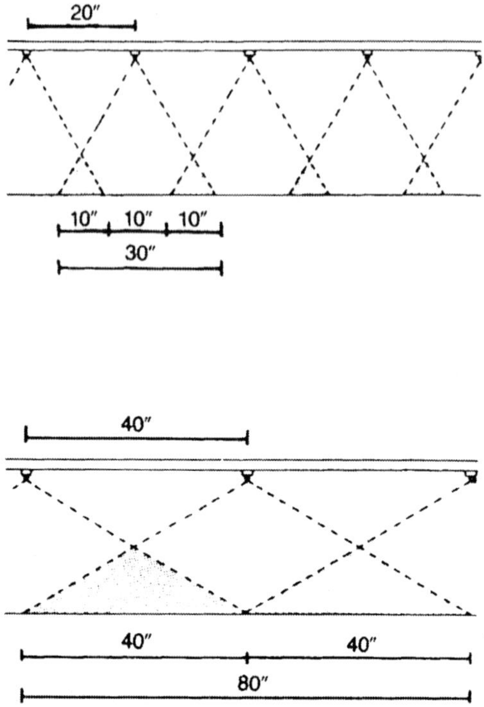

Figure 5 Proper spray overlap: (a) 50% overlap, regular flat-fan nozzles; (b) 100% overlap, flooding-fan nozzles.

Regular flat-fan nozzles are usually spaced 51 cm apart on the boom. For postemergence herbicides, nozzle height is measured from the top of the crop to the tip of the nozzle. To establish the best coverage, adjust boom height, which may also be referred to as nozzle height, for each spray angle and nozzle spacing. Proper nozzle heights for various spray angles of nozzles 51 and 76 cm apart are shown on Table 2. Spacings other than these require different boom heights.

When applying preemergence herbicides or postemergence-systemic herbicides, with flat-fan nozzles, keep the operating pressure between 100 and 200 kPa to reduce the number of drift-prone droplets. Drops produced at 300 kPa and higher are finer and more susceptible to drift. When spraying postemergence contact herbicides, however, use pressures of 300 to 400 kPa. Fine drops produced with high pressure cover plant surfaces better.

Low-pressure (LP), or extended range (XR) flat-fan nozzles are designed to provide better spray distribution over a range of spraying pressures from 100

Table 2 Proper Distances Between the Spray Target and Flat-Fan Nozzles with Various Spray Angles

Spray angle (degrees)	Nozzle height (cm) 51-cm Spacing	76-cm Spacing
65	56 to 61	84 to 89
73	51 to 56	69 to 74
80	43 to 48	61 to 66
110	25 to 30	33 to 38

to 400 kPa. The same nozzle can be used effectively for applying systemic herbicides at low pressure or contact herbicides at high pressure. Other than the improvement in pressure range, these tips have the same spray characteristics as those of regular flat-fan nozzle tips described earlier.

"Low drift" or "reduced drift" flat-fan nozzles are designed to reduce the number of drift-prone droplets. They are to be operated at the recommended pressure rates for conventional flat-fan nozzles. One nozzle company indicates that compared to a comparable size flat-fan nozzle, a reduced drift nozzle reduces the spray volume contained in droplets less than 200 μm by 65%. A removable preorifice is inserted just ahead of the nozzle tip cavity to reduce liquid velocity and to increase the VMD of the nozzle while reducing the number of small, drift-prone droplets.

Twin-orifice flat-fan nozzles have two orifices that direct one flat-spray pattern 30 degrees forward and the second flat-spray pattern 30 degrees to the rear. As a result of atomizing the same amount of liquid with two small orifices instead of one large orifice, the droplet size is smaller than an equivalent-capacity regular flat-fan nozzle tip. As a result, these nozzles are best for application of postemergence contact herbicides which require a thorough spray coverage.

Even flat-fan nozzles are best used for banding over the row of preemergence herbicides at 150 to 300 kPa, or for banding systemic postemergence herbicides at lower pressures (125–200 kPa). They are frequently used with planting equipment to apply herbicides in a narrow band where the seed is planted. Even-flat tips are similar to regular flat-fan tips except the ends of the spray pattern are not tapered, providing an even application across their entire widths, and require no overlapping of adjacent patterns. Coverage with even flat-fan nozzles is more uniform than that of the regular flat-fan nozzles. Generally, these nozzles have 40-, 80-, and 95-degree spray angles. The boom height is adjusted for the desired spray band width (Table 3).

Table 3 Proper Nozzle Height to Achieve Desired Bandwidths with Even-Flat-Fan Nozzles

Nozzle height (cm)		
80-degree Series	95-degree Series	Bandwidth (cm)
13	10	20
15	13	25
18	15	30
20	18	36

Flooding-fan nozzles work well for applying preemergence herbicides and herbicide-fertilizer mixtures. Flooding-fan nozzles deliver a wide (up to 130 degrees), flat spray with large droplets. The wide spray angle allows wider nozzle spacings and lower boom heights for broadcast spraying. Because of their wide spray angle, flooding nozzles are popular for attaching to a disk when incorporating herbicides. These nozzles operate at lower pressures (70–75 kPa) than flat-fan nozzles, resulting in less drift. Since this nozzle provides good performance at low pressures, it may be used with selected systemic herbicides. However, it is not suggested for application of contact herbicides because of its larger droplet size.

The flooding-fan nozzle has a higher application rate along the ends of the pattern than a flat-fan nozzle. Therefore, operating these nozzles with the 30% overlap, which is optimum for flat-fan nozzles, results in excessive application in the overlap region. For optimum spray coverage using flooding-fan nozzles, overlap adjacent spray patterns so that half of the width of the area covered with one nozzle is overlapped with the pattern from one adjacent nozzle, while the other half is overlapped by the other adjacent nozzle pattern. This is referred to as double overlap or 100% overlap (Figure 5b). Pressure affects the width of spray pattern more than with the regular flat-fan nozzle.

One manufacturer has introduced a slightly modified version of their regular flooding-fan nozzles. The new nozzle, called "Turbo FloodJet," has a pre-orifice just ahead of the main orifice to reduce liquid velocity reaching the main orifice. This in turn reduces the number of fine droplets and provides a much smoother spray pattern than that of the regular flooding-fan nozzles.

Mounting flooding-fan nozzles on the boom in various positions produces slightly different spray patterns. The position, however, is not critical as long as double coverage is obtained throughout the boom swath. However, the most uniform application usually results when the nozzle is mounted about 45 degrees above horizontal, in the direction away from the sprayer.

Wide-angle full-cone nozzles produce large droplets that are distributed uniformly in a full-cone pattern. They are widely used with incorporation equipment for spraying preplant incorporated herbicides. The optimum pressure range is from 100 to 175 kPa with a spray angle as high as 120 degrees. They should be spaced 1 m apart on the boom, as with flooding nozzles. To optimize broadcast uniformity, overlap spray patterns 30% to 50% on each edge with the nozzles angled back at a 30 to 45 degree angle from the straight down position. Suggested spray height at 45 degree angle for 1 m spacing is 35 to 45 cm.

Hollow-cone nozzles produce a spray pattern with the liquid concentrated on the outside of the conical pattern. The typical spray distribution is saddle-shaped with less liquid in the center of the distribution, tapering off rapidly at the edges. Therefore, these nozzles are not well suited for broadcast applications because proper overlap is difficult to achieve. Hollow-cone nozzles generally produce the smallest droplets of the commonly used nozzle types. Therefore, they are best suited for application of insecticides, fungicides, or growth regulators where penetration into the canopy and thorough coverage are essential. Hollow-cone nozzles can be used for band application of postemergence contact herbicides. To assure finely atomized spray droplets reach target areas, the nozzles are often directed from both the top and the sides of rows by using drop pipes and swivel connectors. Spray drift with hollow-cone nozzles is relatively high because of the number of fine droplets produced at the typical operating pressure of 300 kPa and over.

Other nozzles that produce a hollow-cone pattern include *whirl-chamber* and *RA Raindrop nozzles*. The whirl-chamber nozzle has a whirl-chamber above a conical outlet, and its spray angle goes up to 130 degrees. It is used mainly for herbicide application. The recommended pressure ranges from 135 to 150 kPa for minimum drift. RA Raindrop is a whirl-chamber hollow-cone nozzle with a special cap. It is used for incorporating preplant herbicides. When they are used for broadcast application, rotate Raindrop nozzles 30 to 45 degrees below horizontal to obtain the most uniform distribution.

Off-center nozzles emit a pattern of spray more to one side than to the other. These nozzle tips are commonly used at the end of a boom to increase the spray swath width. The spray pattern produced by these nozzles is very uneven; therefore, they should be used only for areas that are inaccessible to the spray boom and only when winds are very light to reduce spray drift.

Double outlet flat-fan nozzles are similar to regular flat-fan nozzles except that they have two orifices to obtain a wide spray angle of up to 150 degrees. They are commonly used on a drop pipe to direct herbicides to the ground under tall field crops.

Boomless or cluster nozzles are best suited for areas where the terrain or obstructions make it difficult or impossible to operate a boom sprayer or a cluster nozzle is best suited for these situations. The nozzle assembly mounts at the rear of the sprayer and can deliver a spray swath 10- to 20-m wide, depending on pressure and capacity. Spray distribution is not as uniform as with a boom sprayer.

Conventional hydraulic nozzles use liquid pressure to develop the droplets and a spray pattern. *Rotary nozzles* use centrifugal force to break the liquid into droplets. There are several types of rotary nozzles, some with horizontal spinning disks and some with vertical spinners, commercially available in Europe. However, the rotary nozzles with the horizontal spinning disks (Figure 6) are more commonly seen in North America. With horizontal disks, liquid enters at the bottom (or at the top for inverted-cup-type spinners) of a serrated spinning cup (or disk) and is propelled outward. This action forms liquid ligaments on the serrated surface of the cup. The ligaments leave the edge of the cup and are broken up on reaction with the air. The force to spin the cups comes from either a small electric motor mounted on each atomizer or from the tractor hydraulic power.

Droplets produced by this process, known as controlled droplet application (CDA), have a much narrower size range than those produced by hydraulic nozzles. Both very small and very large droplets can be eliminated by changing the rotational speed of the disk or the amount of liquid being fed into the spinner. The faster the disk turns, the smaller the droplets get. Because it is

Figure 6 Hydraulically driven rotary nozzle.

possible to eliminate extremely small and large sizes of droplets from the total droplet spectrum, a much lower volume of spray carrier (usually water) is required when using CDAs. This allows smaller tanks to be used and/or more area to be sprayed between tank refills. Typical application rates vary from 15 to 50 L/ha depending on the travel speed, spacing between units, and liquid flow rates into the spinner. The flow rates are generally controlled by changing the metering orifices and the pressure in the spray line. The units are generally spaced 1 m apart.

The primary advantages of CDAs are the uniformity of droplet sizes and the lower diluent volumes required as compared to hydraulic nozzles. In spite of these characteristics, CDAs have not been widely accepted in North America. Their high costs and the lack of reliable data about the effectiveness of pesticides at reduced rates of spray diluent are perhaps the two principal reasons for the lack of demand. In some experiments the weed control achieved by CDAs has been similar to or better than that obtained by conventional hydraulic nozzles, but in other cases the results were not satisfactory. Reluctance of chemical companies to accept responsibility for the failure of their products if the recommended rates are not used has further slowed the rate of acceptance of CDAs by the pesticide applicators. Research in various universities indicates that whereas reduction in drift can be achieved as a result of more uniform droplets, rotary nozzles have shown some susceptibility to pattern distortion from the wind. To eliminate pattern distortion and reduce drift, one manufacturer has developed a CDA with an integral fan that creates forced air flow directed downward (Figure 7). As the droplets are released from the edge of the serrated spinning cup, the droplets are entrained in the air flow and carried toward the target with a much higher velocity. The air flow also creates turbulent movement of the target plant, which improves coverage, droplet penetration into the canopy, and underside leaf coverage. Visual observation of water-sensitive cards that were placed inside the plant canopy, on the upper and under sides of leaves, and then sprayed with hydraulic and air-assisted CDAs indicates that the coverage in general is more uniform and efficient with the air-assisted CDA. Additional research is being conducted to evaluate the performance of controlled droplet applicators.

Once the proper type of hydraulic nozzle is determined for a specific application, the proper size can be selected as follows:

Step 1. Select the application rate in liters/minute (L/min). This is a management decision that the applicator will base on pesticide label recommendations, field conditions, and water supply. The application rate referred to here is the liters of spray mixture that includes the pesticide and the carrier (water, fertilizer, etc.).

Step 2. Select an appropriate ground speed in kilometers/hour (km/h).

Figure 7 Air-assisted controlled droplet atomizer.

Step 3. Determine the spray width per nozzle (W) in centimeters. For broadcast application,

W = nozzle spacing for band spraying,

W = bandwidth for directed spraying

$$W = \frac{\text{Row spacing (or bandwidth)}}{\text{Number of nozzles per row (or band)}}$$

Step 4. Determine the flow rate required (L/min) from each nozzle by using a nozzle manufacturer's catalogue or the following equation:

$$L/\text{min} = \frac{(L/\text{ha}) \times (\text{km}/\text{h}) \times W}{6000}$$

(6000 is a constant to convert liters per hectare, kilometers per hour, and centimeters to liters per minute.)

Step 5. Select a nozzle size from the manufacturer's catalogue that will give the L/min flow rate determined in step 4 when the nozzle is operated within the recommended pressure range. If a nozzle of this size is not available, change the travel speed in equation 1 and determine the new flow rate required.

Pumps

Pumps are needed to move spray solutions from the tank to the nozzles and to create enough hydraulic pressure at the nozzle to form droplets and transport them to the target. Selection of the pump for a particular application depends on the spray application rate desired, pesticide formulations, and the pressure required to apply them. Sprayers are often used for several operations which may require different flow rates. Select a pump with a capacity to handle the highest application rate. Pump capacity must be sufficient to satisfy the flow rate required at the nozzles and the agitation if the sprayer is not equipped with a mechanical agitator. Select a pump with a reserve capacity at least 20% greater than the requirement for the highest flow rate plus agitation. This will compensate for possible future reduction of the flow rate due to wear.

The type of chemical sprayed also affects pump selection. Some pumps handle abrasive materials like wettable powders with much less wear than others. Chemical reaction and corrosion affect some pump materials.

Another important consideration in pump selection is the pump's pressure rating. The pump should produce a desired operating pressure at the capacity required for specific spraying jobs. Some pumps are capable of delivering the desired flow rate at various operating pressures (Figure 8), whereas others cannot. Also, forcing a low-pressure pump to produce high pressure causes it to wear out rapidly.

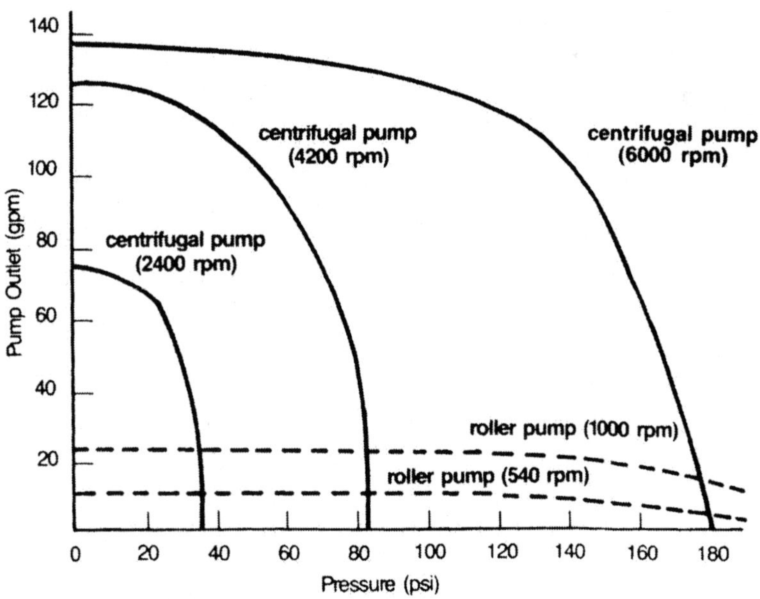

Figure 8 Pump performance curves.

Most tractor power take-off (PTO) shafts rotate at 540 and/or 1000 rpm. Some pumps require faster speeds than a standard farm tractor PTO can provide. When higher speeds are required, an auxiliary power source or a gear unit is required to increase the rotational speed of the pump to achieve optimum performance.

Most pumps are of the positive-displacement or non-positive-displacement type. A positive-displacement pump discharges a nearly constant volume of liquid with each stroke or revolution regardless of back pressure on the pump. When using these pumps a pressure-relief valve or pressure control device must be employed to bypass excess liquid, that is not used in the spray operation, back to the spray tank. A positive-displacement pump also creates a vacuum and, therefore, does not require filling the pump or placing it below the tank level to prime it. Piston, gear, and roller types of pumps are examples of positive-displacement pumps.

A non-positive-displacement pump, such as a centrifugal pump, does not require a bypass relief valve and does not prime itself by pumping air. In general, a non-positive pump has a longer life than a positive pump. Close fitting and reciprocating parts of a positive pump wear rapidly, especially with abrasive suspensions or wettable powders.

Characteristics of several pump types are outlined in Table 4. Of these types, roller, centrifugal, and diaphragm pumps are the most widely used on low-pressure sprayers. Following are brief descriptions of these three common pumps.

Roller pumps (Figure 9) are popular because of their low initial cost and compact size in relation to their capacity. They are easily repaired and suitable for a variety of pesticides. In addition, they are easy to mount, and they operate efficiently at standard tractor PTO speeds. Roller pumps are not suitable for application of nonabrasive formulations. Rollers held in a slotted rotor revolving in an eccentric (nonsymmetrical) case create the pumping action. A roller moves in and out of each slot radially and provides a seal against the wall

Table 4 Pumps for Sprayers and Common Operating Conditions

Pump type	Pressure ranges (kPa)	Operating speeds (rpm)	Flow rates (L/min)	Displacement type
Roller	70–2000	300–2000	40–200	Positive
Centrifugal	35–500	1000–4500	0–800	Nonpositive
Piston	150–7000	600–1800	20–250	Positive
Turbine	35–400	600–1200	40–300	Nonpositive
Diaphragm	500–5000	200–800	20–150	Semipositive

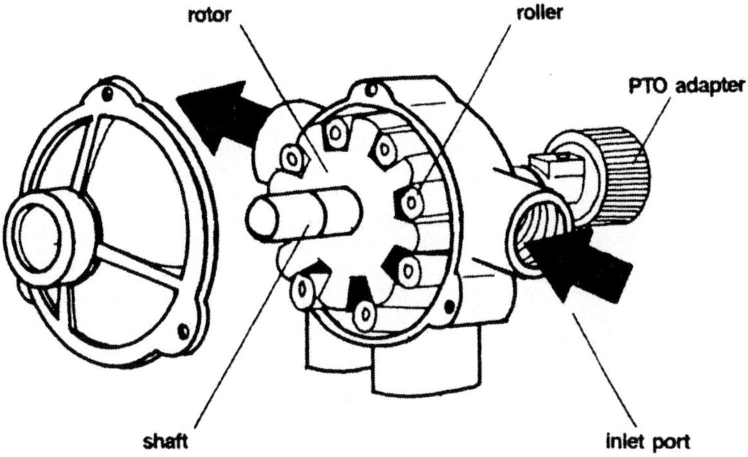

Figure 9 Typical roller pump (*cutaway*).

of the case by centrifugal force. As the rollers pass the inlet port, the space between rollers and housing becomes larger and draws liquid into the pump. The fluid remains in spaces between and under rollers as it moves toward the outlet port. As the rollers near the outlet port, the spaces become smaller as a result of the eccentric housing. Here the rollers are forced inward toward the center and the liquid is forced out the discharge port.

Roller pumps have a wide range of pressure and capacity ratings. Pump output varies from 49 to 200 L/min, depending on the length and diameter of the case, its eccentricity, and the speed of rotor rotation. The maximum operating pressures of roller pumps range from 70 to 2000 kPa. Although roller pumps are often classified as positive-displacement pumps, their output decreases substantially as operating pressure increases because the rollers tend to let more fluid "leak back" between rollers.

Roller pumps, constructed from different materials, are available. They are usually constructed with cast iron or corrosion-resistant housings and rotors; polypropylene, nylon, synthetic fluorine-containing resin (Teflon), or rubber rollers; and Viton, rubber, or leather seals. One disadvantage of roller pumps is the rapid wear when abrasive materials such as wettable powders are present in the spray liquid pumped. Abrasive materials such as wettable powders can cause rapid wear of pump housing, rotor slots, and rollers. The amount of wear can be reduced to a minimum by selecting the most durable types of roller, with proper maintenance and storage of the pump, and by keeping abrasive (sandlike) materials out of the sprayer. Compatibility between materials and chemicals to be applied must be considered when selecting the pump. Nylon and synthetic fluorine-containing resin (Teflon) rollers are highly resistant to most pesticides and fertilizers and are recommended for sprayers that are used to apply a variety of chemicals. Rubber rollers are preferable for water solutions and wettable powder slurries at pressures under 700 kPa. Solutions should be free of sand or scale, which are abrasive to the rollers. Polypropylene rollers are best when applying solutions with little or no lubricating quality.

Useful life of roller pumps can be extended by not leaving water in the pump during freezing weather and by eliminating excessive restriction in the suction line, which usually creates cavities in the pump. Tapered rollers are an indication of too much restriction in the inlet line.

Properly connecting a pump to the system is important. Figure 10 illustrates the recommended hookup for roller pumps. It is important to place the control valve in the agitation line. By controlling the bypass flow, it regulates spraying pressure. To adjust the system properly:

1. Close the control valve and open the boom valve (or spray gun).
2. Start the sprayer. Make sure flow is uniform from all spray nozzles, and adjust the relief valve until the pressure gauge shows about 69 to 103 kPa above the desired spraying pressure.
3. Slowly, open the control valve (agitation line) until spraying pressure is reduced to the desired value. If the pressure does not decrease to the desired value, even with the control valve wide open, replace the agitator nozzle with one that has a larger orifice.

If insufficient agitation results when spraying pressure is correct and the relief valve is closed, try a smaller agitation nozzle. This increases agitation and permits a wider open control valve for the same pressure.

Centrifugal pumps (Figure 11) are preferred when high volume of spray is required under low pressure. Centrifugal pumps have a wide range of applications. They are preferred especially for spraying abrasive materials because they do not have parts susceptible to wear and are easy to repair. An impeller with curved vanes rotates at high speed inside a disk-shaped casing. Liquid drawn inside the pump at its center is thrown by centrifugal force into a channel around the edge of the pump casing. The high-velocity spray liquid is forced out the outlet, generating pressure in the outlet line. The design of the pump, especially the diameter and speed of the rotating impeller, affects the efficiency in developing outlet pressure.

The pump impeller operates between 2000 and 4500 revolutions per minute (rpm) depending on the design. This requires using some kind of speed-up mechanism when operating centrifugal pumps from the tractor PTO shaft. The simplest and least expensive of these speed-up mechanisms are belts and pulleys. This method uses an industrial timing belt to speed up the pump impeller. Some pumps have a planetary gear system that is enclosed and mounted

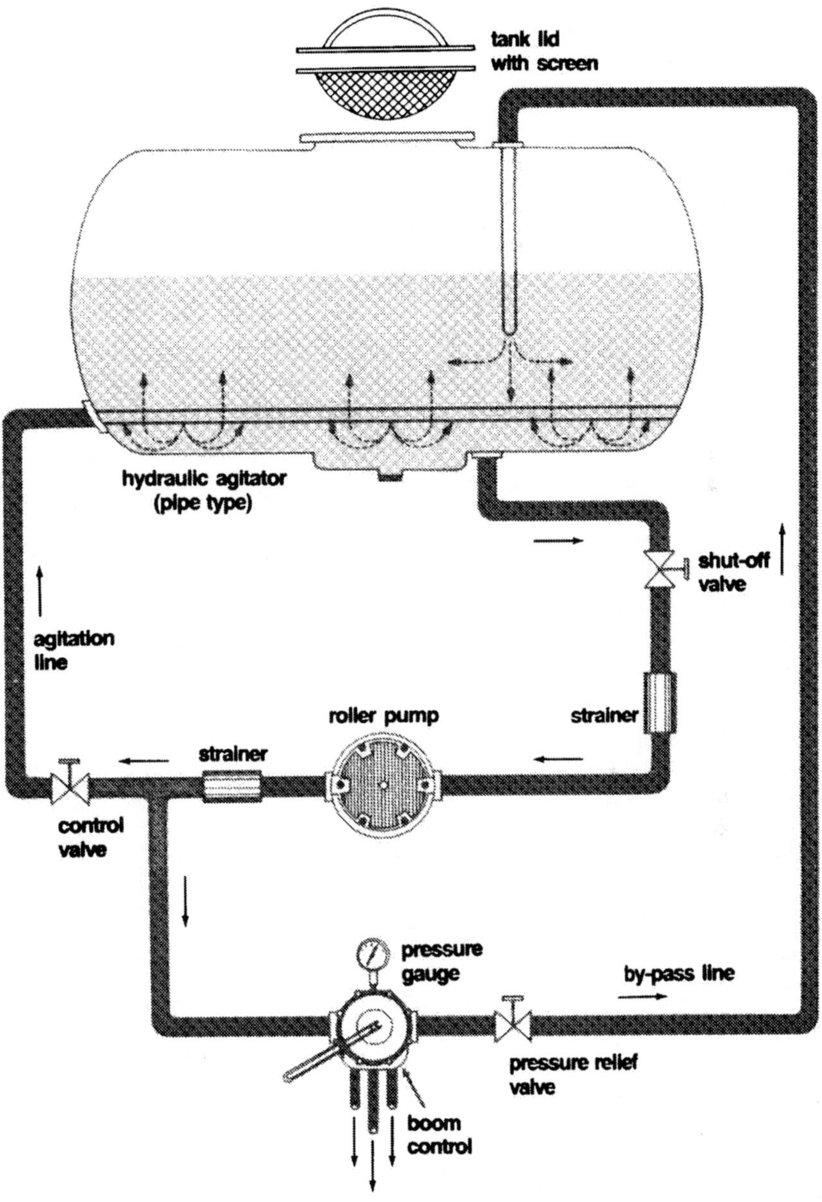

Figure 10 Spraying system with a roller pump.

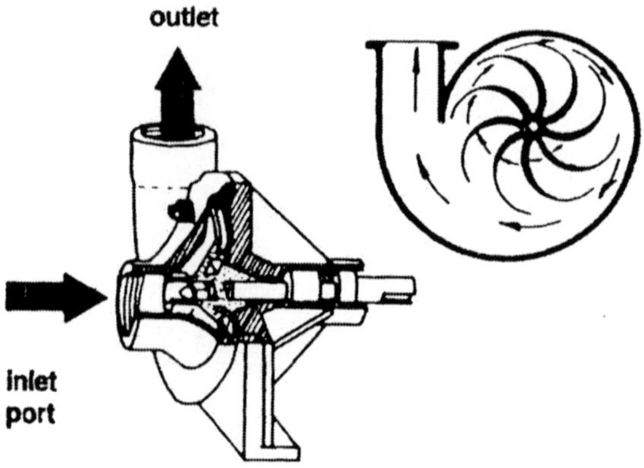

Figure 11 Typical centrifugal pump (*cutaway*).

directly on the PTO shaft. The pump may also be enclosed and mounted directly on the PTO shaft. The pump may also be driven by a direct-connected hydraulic motor that operates off the hydraulic system. The advantage of the hydraulic system is that the motor maintains a more uniform pump speed and output and frees the tractor PTO for other purposes.

Typical maximum working pressure of centrifugal pumps is 200 to 400 kPa. Some centrifugal pumps on air-blast orchard sprayers can develop pressures up to 1400 kPa. However, output volume drops off rapidly when outlet pressure exceeds 200 kPa. This characteristic of centrifugal pumps makes it possible to control pressure by a simple shut-off valve rather than a pressure regulating valve. Because centrifugal pumps are not positive-displacement pumps, they are not self-priming. Liquid must flow by gravity from the tank to the pump. Self-priming pumps discharge air from within the pump housing when they are started. The centrifugal pump cannot do this. To prime a centrifugal pump, mount it below the minimum liquid level of the tank. In addition, install a small vent tube from the pump housing to the supply tank. This positive vent line allows the pump to prime itself by "bleeding off" trapped air when the pump is not operating or upon starting. The small stream of liquid that returns to the tank during operation is negligible. Restricting flow of liquid to the inlet part of centrifugal pumps may result in loss of pressure and possible damage to the pump.

A partially clogged suction strainer, a collapsed suction line, or a suction line with insufficient capacity considerably restricts the liquid flow into the pump, resulting in loss of pressure control and possible damage to the pump. Therefore, a suction strainer need not be installed on the intake line of the centrifugal pump. The absence of a suction strainer generally does not affect the performance of centrifugal pumps because they are capable of handling small pieces of foreign material without damage. A strainer with large openings may be placed ahead of the pump if water is pumped from streams or ponds that contain large pieces of debris provided that the strainer can handle the large capacities of the pump with a minimum drop in pressure across the strainer, and the strainer is cleaned frequently. Generally, a small strainer beyond the control valve is adequate because much of the output is circulated back into the tank.

The centrifugal pumps have a second flow control valve in the system. As shown in Figure 12, one is in the agitation line; the other is in the line leading to the boom. This permits control of agitation flow independent of nozzle flow. Also, centrifugal pumps have no relief valves in the system.

Following are operational guidelines for using a spraying system with a centrifugal pump:

1. Prime the pump with all valves open.
2. Close the throttling valve and control valve at the agitation line; open the boom valve.
3. With the pump running, open the throttling valve until the pressure gauge indicates the desired pressure.
4. Open the agitation control valve until sufficient agitation is observed. Then, if spraying pressure drops, readjust the throttling valve to restore the desired pressure.
5. Make sure the flow is uniform from all nozzles.

Diaphragm pumps (Figure 13) are positive-displacement, are self-priming, and require less power to run than other pump types with similar flow and pressure ratings. They deliver low volume (20–150 L/min) at medium to high pressure (500–5000 kPa). A flexible diaphragm moves up and down in a chamber, creating suction and pressure. As the diaphragm is moved up, it creates a vacuum that opens the suction valve and draws liquid into the chamber. This action forces the diaphragm to move down, forcing the fluid out of the chamber through the discharge valve. The unique design of diaphragm pumps keeps all moving parts sealed in an oil bath and protected from corrosive and abrasive sprays. In addition to protecting moving parts, the oil bath balances the pressure on the diaphragms, increasing their useful life. One problem with these pumps is that some solvents may damage the diaphragm. Check the operator's manual to avoid such problems.

Selection of the proper pump size is important to achieve adequate agitation in the spray tank and sufficient pressure at the nozzle. At least 2% of the total tank volume should be returned to the tank for agitation. It is recommended that the pump capacity be 20% greater than the sum of these requirements.

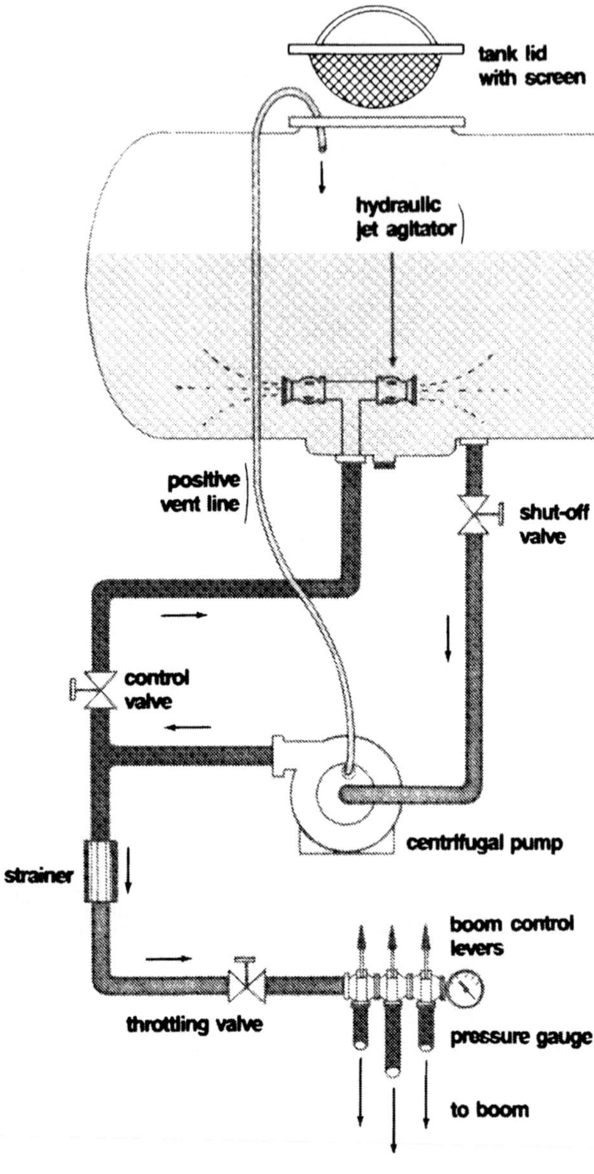

Figure 12 Spraying system with a centrifugal pump.

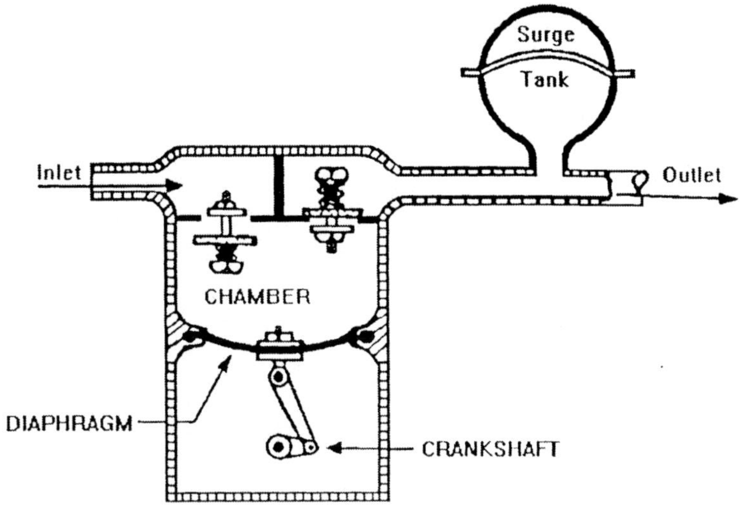

Figure 13 Diaphragm pump.

Pump capacity (L/min) = 1.2 × [(boom requirement [L/min] + agitation requirement [L/min]

Boom requirement (L/min) = No. of nozzles × flow rate per nozzle (L/min)

Agitation requirement (L/min) = 2% of total tank volume (L/min)

Tanks

The tanks should not be overlooked when selecting a sprayer. Choose sprayer tanks on the basis of their construction, shape, and durability, and not only their initial cost. A low-cost tank is seldom a good investment.

Points to consider when selecting a tank are 1) sufficient capacity, 2) markings indicating liquid level in the tank, 3) a sight gauge if the tank is not transparent, 3) a device at the bottom for adequate agitation, 4) shape suitable for easy mounting and effective agitation, 5) a large enough drain plug at the lowest point for rapid and complete drainage, 6) a nonabrasive surface characteristic for easy rinsing of pesticide residues and corrosion resistance, 7) a large opening for easy filling and cleaning, 8) a lid to prevent pesticides from spilling or splashing out, and 9) an 80 or 100 mesh screen covering the filler opening.

The chemicals that will be used and the cost will affect the choice of tank materials. Following are brief descriptions of common tank materials. Sprayer tanks are made from either metal (aluminum, galvanized steel, stainless steel), fiberglass, or thermoplastic material such as polypropylene.

Galvanized steel tanks were more common in the past than they are now. They are relatively inexpensive, and their modification and repair are fairly easy. A galvanized tank is subject to rust and corrosion. It is suitable for most pesticides, but check the product label for exceptional cases. For example, glyphosate reacts with metal to produce hydrogen gas, resulting in an explosion hazard (Marer, 1988). Often protective coatings do not prevent rust. Corrosion and rust plug strainers, screens, and nozzles and cause excessive wear and damage of pumps.

Stainless steel tanks are corrosion resistant to crop chemicals. Stainless steel is the most durable and long-lasting of all materials, but it is expensive. However, extensive use justifies the initial investment.

Polyethylene and polypropylene tanks are relatively inexpensive and available in various shapes and sizes. They are noncorrodible, resistant to deterioration from most pesticides, and compatible with most liquid fertilizers. Most commercially available plastic tanks are constructed from either high-density polyethylene or polypropylene. Although they are all called "poly," there are wide variations in the final chemical makeup of polyethylene tanks. Some poly tanks are much more resistant to deterioration by sunlight, some are more durable in cold weather, and some absorb much less chemical than others. For example, low-density polyethylene does not resist solvents well. If it is exposed to sunlight for a long period, and the weight of the liquid is not properly supported, polyethylene tanks can break as the sprayer bounces over obstructions or rough field surfaces. Some types of polyethylene tanks may be repaired with special equipment, whereas others must be replaced. Storing polyethylene tanks indoors where there is no direct sunlight is recommended.

Fiberglass tanks are used on all types of sprayers and as nurse tanks. Fiberglass is also a strong and durable material. Though it is possible that breakage and cracking can occur under impact, it is repairable. Fiberglass tanks can be used with most chemicals except some solvents. They are highly resistant to most organic solvents, have good to fair resistance to acids, but have only fair resistance to alkali materials. Deep scratches on surfaces of these tanks may allow some pesticides to be absorbed into the fiberglass material. This may cause contamination of future tank mixes. Scratched areas must be smoothed by recoating with resin.

Selection of the tank size is a compromise between having a large tank and reducing the time required for refilling at the expense of increased soil compaction, or having a small tank that requires less power and minimizes soil compaction but reduces the field efficiency of spraying. Even though small tanks require more frequent filling and consume more time in mixing chemicals, traveling, etc., remember that larger tanks may not be justified where different mixes of chemicals are applied to small fields.

Tractor sprayers may be either mounted or trailer-type. With mounted types, tanks and booms usually mount on a frame that hitches to the rear of a tractor (Figure 14). The added weight in the rear of a small tractor reduces load on front wheels, causing the tractor to slide when turning. Keep tanks as close to the tractor as possible to reduce the effect of tank weight. Saddle tanks, mounted to both sides of the tractor, eliminate this problem but may block the operator's view when spraying row crops. Saddle tanks are preferred when a tractor is spraying while pulling some other machine such as a row-crop planter or disk harrow. For one-pass incorporation of chemicals, tanks may be mounted on tillage equipment.

Sprayers with mounted tanks are maneuverable, particularly when spraying row crops. They do require more time in getting ready for spraying. This is particularly true of side mounted saddle tanks, which must be bolted to the tractor frame.

The trailer-type field sprayer (Figure 15) is the most common type used on farms in North America. It attaches easily to the tractor drawbar. The sprayer pump may be attached directly to the tractor PTO shaft. Some sprayers are equipped with a universal joint shaft that also permits operating the pump with the tractor PTO. The pump is then mounted on the sprayer hitch.

The trailer-type sprayer offers two primary advantages to an applicator spraying between other jobs: ease of attachment and greater volume. Also, the boom is farther away from the tractor, so there is less exposure of the operator to spray drift. However, much of the boom is usually out of the operator's

Figure 14 Sprayer with rear-mounted tank.

Figure 15 Trailer-type sprayer with tandem wheel.

view, and it is nearly impossible to see whether all nozzles are performing properly.

When barrels or small tanks are used, tank mounting is not a factor. Fiberglass or polyethylene tanks, however, must be properly mounted on a saddle that supports the tank over a large area. Without it, the weight of the liquid may break the tank as the sprayer bounces over obstructions or rough land.

Agitators

Agitators should be used to achieve uniform concentration of chemicals in the tank. The uniformity of active ingredients in a spray solution throughout the application period is highly influenced by how well the chemicals are mixed in the tank. An agitator initially mixes liquids, suspensions, and emulsions and keeps them mixed while spraying. The need for agitation and the type of agitation depend on the type of chemical applied. Most liquid concentrates, soluble powders, and emulsions require little agitation. Wettable powders and dry flowables, however, require sufficient agitation to prevent settling out. When using these formulations, the return flow coming from the bypass line of the pressure relief valve often does not provide sufficient agitation. Agitation can be accomplished with hydraulic jets or mechanical paddles. Figure 16 shows different types of agitators.

Mechanical agitators are propellers or paddles mounted on a shaft near the bottom of the tank. The shaft rotates at 100 to 200 rpm; excessive speeds cause foaming in some spray solutions. If foaming occurs, the agitator speed can be reduced by changing the drive ratio. Mechanical agitators require some maintenance, especially where shafts pass through tank walls.

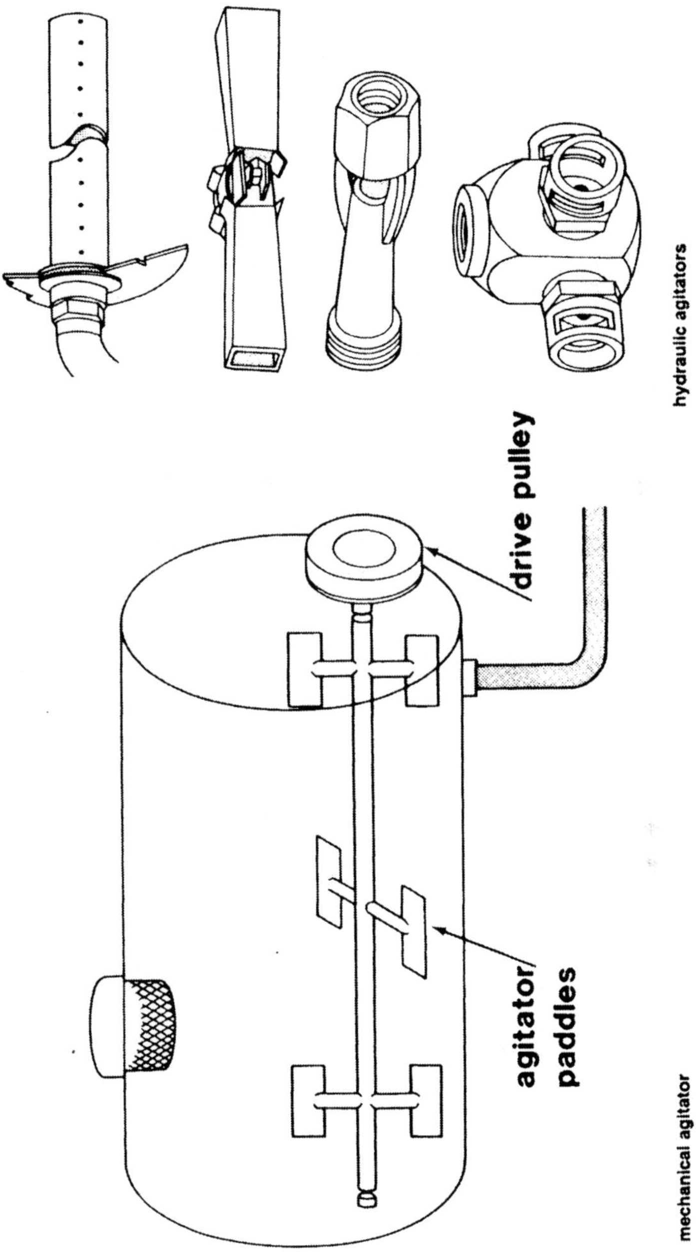

Figure 16 Types of agitators.

Hydraulic agitation is the most common means of agitation used on low-pressure boom sprayers. A small portion of the spray solution discharged from the pump is returned to the tank and discharged under pressure either through holes drilled in a pipe that runs the entire length of the tank or through agitator nozzles. Most hydraulic agitators use a special volume booster that increases the discharge from the nozzle two to three times. The amount of liquid returned to the liquid agitators from the pump should be about 2% to 3% of the tank volume for liquid chemicals and 4% to 5% for powders when no volume-booster nozzles are used.

Little research on hydraulic agitation systems has been directed toward recommendations on optimum equipment configurations. Researchers at the Prairie Agricultural Machinery Institute (PAMI) (1987) recommended jet agitation flow rates of 1.5 L/min per 100 L of tank capacity for emulsifiable concentrates. They also recommended jet flow rates of 3 L/min per 100 L of tank capacity for wettable powders. Folsom (1948) recommended jet diameters of one-seventeenth the distance the jet is required to flow. The required jet flow distance was reported as 75% of the tank diameter for vertical, cylindrical tanks. Similar results are not available for horizontal cylindrical tanks such as those used for most pesticide application equipment. Bode (1987) recommends a flow of 5 L/min for every 100 L of tank capacity for a simple jet agitation. If volume-booster nozzles are used, the agitation rate can be reduced to 2 to 3 L/min for every 100 L of tank capacity.

Hydraulic jet agitation is simple and effective as long as the device is installed correctly. Never connect the agitator to the bypass line of the pump and use a separate line from the pump output. As with mechanical agitators, foaming can occur if the liquid level in the tank is close to the tank bottom. To eliminate this, the liquid flow to the agitator should be reduced slowly as the tank empties.

Strainers and Screens

Strainers and screens remove undissolved clumps of pesticide formulation, sand, soil and other foreign material from the spraying system, protecting the pump and preventing nozzles from clogging. They are commonly used in four places on sprayers: (1) tank-filler screens, (2) suction strainers, (3) line strainers, and (4) nozzle strainers.

Nozzle strainers (filter screens) are placed just ahead of nozzle tips to screen the liquid and prevent plugging and excessive wear. Size the screens in the nozzle tips according to tip orifice, providing an aperture of the filter mesh screen smaller than the size of the nozzle orifice. Mesh refers to the number of openings per square inch of the screen material. Strainers with high mesh numbers have smaller openings than strainers with low mesh numbers. In other words, the larger the mesh number, the finer the screen. Most of the screens used in nozzles for crop spraying have 25- to 200-mesh openings. Smaller mesh screens are recommended when using nozzles with smaller orifices. In general, 100-mesh strainers are recommended for most nozzles with a flow rate below 0.8 L/min, and 50-mesh strainers for nozzles with a flow rate between 0.8 and 4 L/min. If a good line strainer is used, no nozzle strainer is needed for nozzles with flow rates above 4 L/min (Bode, 1987). For spraying abrasive materials, use a slotted strainer with openings equivalent to 16- to 50-mesh screens. Always check the manufacturer's recommendations for the specific nozzle.

Line strainers (Figure 17) prevent many of the small foreign materials in the spray liquid from entering and damaging the pump and/or nozzles. Location of the line strainers varies with the type of the pump. However, they are usually located between the pump and the nozzles. In a roller or diaphragm pump, as illustrated in Figure 10, a strainer in the intake line between the pump

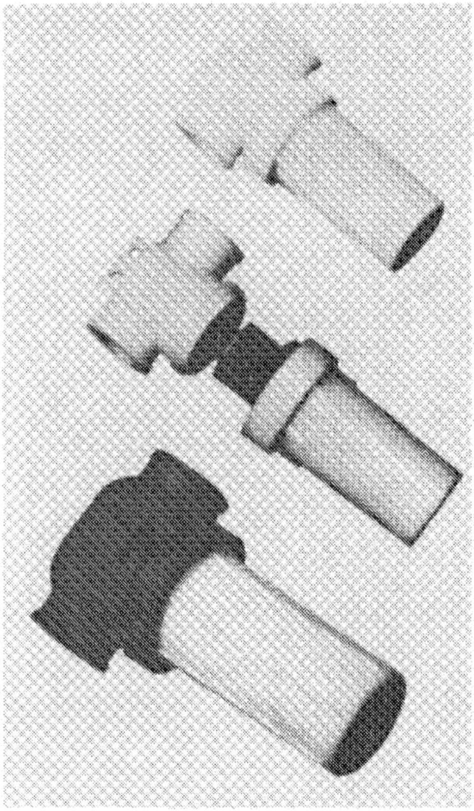

Figure 17 Line strainer.

and the tank is recommended. A 40- or 50-mesh strainer (usually 50) is adequate (Bode, 1987). Because of a centrifugal pump's large capacity, a strainer on the intake line is not required. As shown in Figure 12, a strainer in the spray line going from the pump to the boom is recommended to protect nozzles on the boom and agitation nozzles.

Position the line strainer to collect foreign materials at the bottom of the filter so blockage of liquid is unlikely even if some debris collects in the strainer.

Suction strainers are usually positioned near the bottom of the tank, connected to the end of the suction (intake) hose. They usually have aluminum bodies with stainless steel screens of 10- or 20-mesh. Suction strainers finer than 10 to 20 mesh may restrict liquid flow to the pump and plug easily, causing a pressure drop in the system and deterioration in performance of some pumps, especially centrifugal pumps. A suction strainer is not required for systems with a centrifugal pump unless water is pumped from a lake or a pond where the water contains large pieces of debris.

Tank-filler screens, which have relatively large openings, remove twigs, leaves, and other large foreign material when filling the tank. They also prevent large chunks of wettable powders in a solution from entering the tank.

Hoses, Couplings, and Fittings

Hoses, couplings, and fittings are necessary to convey the liquid through the sprayer. They should be strong and durable, able to withstand the peak pressure in the spraying system, and resistant to sunlight, oil, and corrosive chemicals.

Liquid pressure varies at different points on the sprayer. Select hoses strong enough to handle the chemicals at the required pressure. Also consider hose construction, composition, and size. Choose hoses rated for a working pressure higher than the expected operating pressure to provide a margin of safety and to prevent bursting from pressure surges. Peak pressures, much higher than average operating pressures, are common. These peak pressures usually occur when the spray boom is shut off.

It is extremely important that the hoses and fittings be in good condition. This will prevent a possible break and also protect the operator from being sprayed with chemicals. Because suction hoses are not pressurized, they do not burst. But they collapse if the inlet plugs. Plugging restricts the flow and "starves" a pump, causing decreased flow and damage to the pump or pump seals. Use special reinforced hose suction lines to prevent collapsing.

As liquid is forced through a hose, there will always be a certain amount of pressure drop due to friction between the liquid and the inside wall of the hose. The pressure drop is excessive when a large volume of liquid is forced through a small hose. Proper hose size varies with the size and application rate of the sprayer. Reducing the number of fittings, keeping the hoses as short and straight as possible, reduces the pressure drop in spray lines to a minimum. Suggested hose sizes for various pump flow rates are listed in Table 5.

Boom Stability

Boom stability is important for uniform spray application. Movements of the boom during field spraying cause uneven distribution of chemicals. This problem is accentuated as boom length increases. Ideally, the boom should be stable in all directions over its length. A boom must have adequate support. Support a boom longer than 15 m by wheels and tie rods or cables to keep it at a uniform height, particularly on rough or rolling land.

The most recent change in boom design is a special linkage of the boom to the spraying system that allows self-leveling of the boom when going over the rough ground. The self-leveling and stabilizing features of these new booms are entirely mechanical and do not involve hydraulic cylinders.

Flow Control Devices

The tank, agitator, pump, and nozzles are obviously important elements of a boom sprayer. However, without some flow control devices the system never operates properly.

Pressure regulators prevent possible pump damage by adjusting pressure within the operating limits of the pump. A *pressure relief valve* (Figure 18) usually has a spring-loaded ball, that opens when pressure exceeds a set level and allows excess fluid to return to the tank. The amount of tension on the ball is adjusted by turning a screw. Increasing the tension on the spring increases the liquid pressure going to the nozzles. The pressure relief valve must be large enough to allow the entire pump output to return to the tank when the boom is shut off. The pressure relief valve should be adjusted while the spray is discharged from the nozzles. When the flow to the nozzles is stopped, pres-

Table 5 Suggested Hose Sizes for Various Pump Flow Rates

Pump output (L/min)	Hose sizes (cm)	
	Suction	Pressure
0–3.9	1.25	.06
4–11.9	1.25	1.0
12–24.9	2.0	1.5
25–49.9	2.0	1.5
50–99.9	2.5	2.0
100–199.9	3.0	2.5
200–400	4.0	3.0

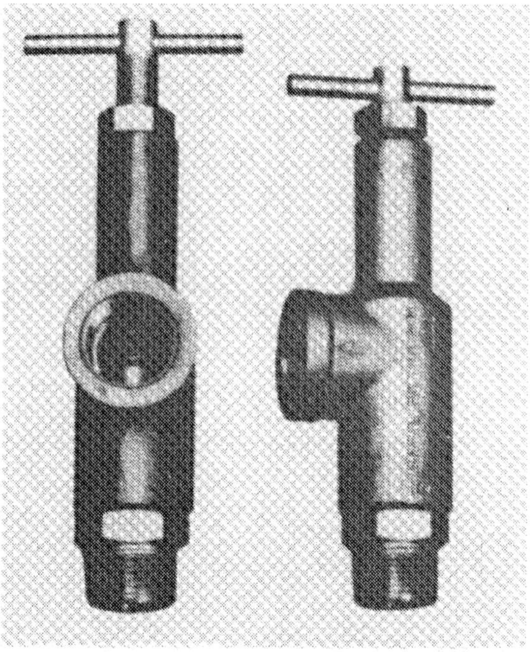

Figure 18 Pressure regulators.

sure in the system rises slightly and all the liquid is routed through the bypass. Because the output of a centrifugal pump can be shut off completely without pump damage, a pressure relief valve is not needed and a simple valve can be used to regulate spray pressure.

On sprayers that operate at pressures over 1400 kPa, use an *unloader valve* in place of a pressure relief valve to unload the pump when the boom is shut off. When using a hydraulic agitator, agitation flow rate may be insufficient when the valve is unloading.

Control valves are all valves other than the pressure regulator that regulate the flow of liquid. These control valves vary from a single valve that shuts off all the flow to multiple valves for individual outlet controls. Spray booms of low-pressure field crop sprayers often have two to three sections. Liquid flow in the boom can be directed to a portion of the boom, directed to the entire boom, or shut off completely by using a three-section boom control valve (Figure 19). With this valve it is possible to apply chemicals only from nozzles on the left, center, or right portion of the boom, or from any combination of these three sections. This is particularly useful when treating field edges when part of the boom is not required.

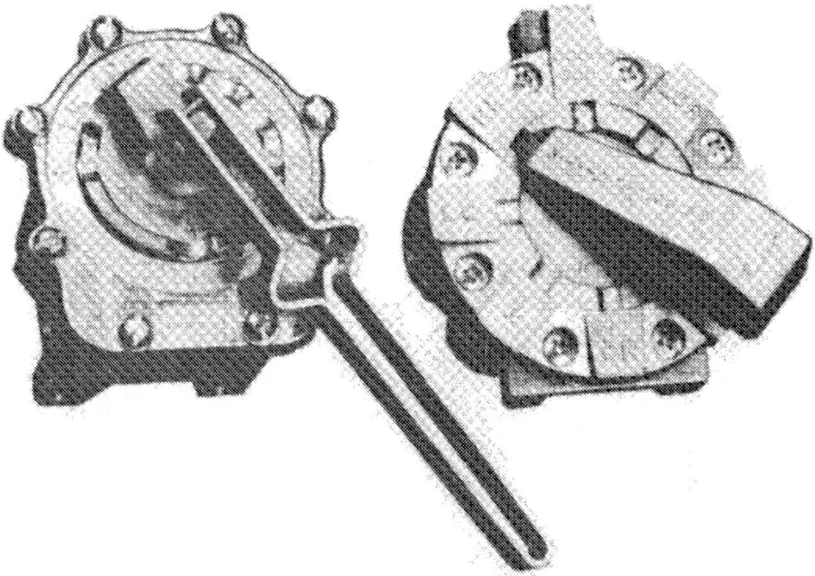

Figure 19 Boom flow control valves.

Flow control valves are usually of two types. Manually controlled valves are mounted in the hoses so that they are accessible by the tractor operator. Electronic control valves (Figure 20) may also have the control valve installed on the boom, but the switch box to activate these solenoid control valves is located within the operator's reach.

Pressure gauges (Figure 21) are important accessories of a spraying system because pressure affects the amount of liquid being sprayed and the size of droplets. To regulate spray pressure accurately, first, measure the spray. Use a pressure gauge with a range at least twice the expected operating pressure. A pulsation damper protects and increases the life of a pressure gauge and smooths pressure pulsations, thus making reading the gauge much easier.

There is usually a pressure drop in the spray line between the location of the pressure regulator valve and the nozzles. Therefore, pressure gauges should be located as close to the nozzles as possible, and the pressures at the nozzles are checked at least once to determine the magnitude of the pressure drop. An accurate pressure gauge is essential to proper calibration of sprayers. Therefore, compare the readings from your pressure gauge periodically with the readings form another calibrated gauge.

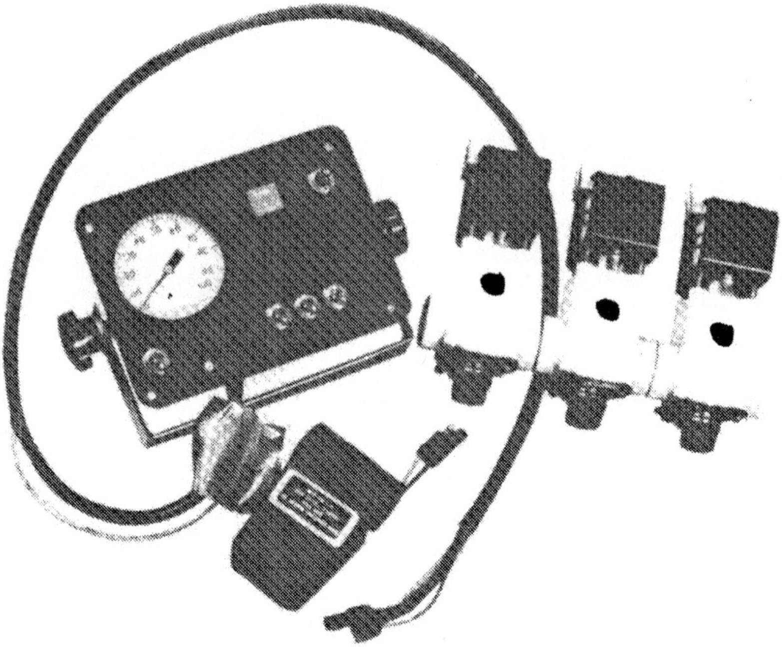

Figure 20 Electronic flow control valves.

Electronic monitors and flow rate control devices (Figure 22) are the latest additions to modern spraying systems. Some electronic monitors sense whether or not chemicals are applied by the nozzles. If flow through a nozzle stops for any reason, such as plugging or loss of pressure, a flashing light or buzzer alerts the operator. These monitors are especially useful for nozzles not visible to the operator or when spraying is combined with another operation, such as planting, that requires the operator's attention.

Other monitors can indicate various operating conditions of the sprayer. For example, a microprocessor in the monitor console can continuously calculate the application rate (liters per hectare). The console can display this information, along with other inputs such as travel speed, pressure, and flow rate. Some monitors supply the operator with information such as field capacity of the sprayer (hectares covered per hour), liters applied, liters remaining in the tank, total hectares covered, and total distance traveled. The sensors must be installed properly and the whole monitor unit calibrated before it will function.

Some monitors regulate application rate automatically. First, a desired application rate is entered in the monitor's memory. The controller receives the actual application rate from the monitor and continuously compares it with the

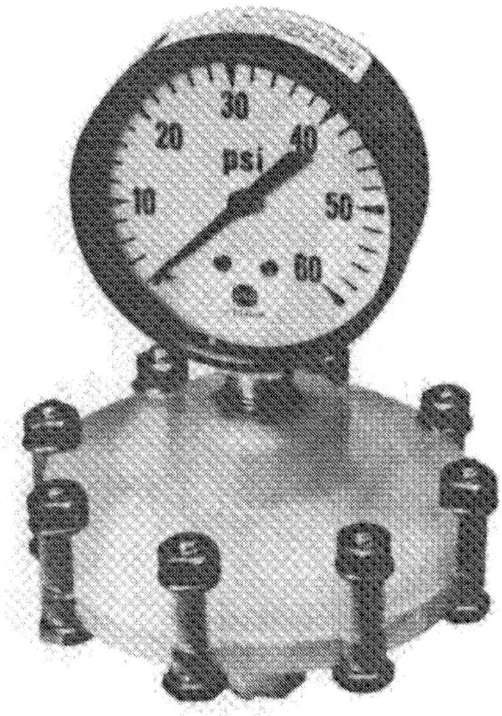

Figure 21 Pressure gauge with pulsation damper.

preset desired rate. If the difference (error) between the measured and the desired application rates is greater than a certain percentage (typically 1%) the controller adjusts application rate automatically (usually by adjusting pressure) to reduce the error. If the controller cannot reduce the error and if the error exceeds a certain amount (for example, 10%) the monitor signals the operator.

Marker systems

Marker systems are popular, especially among farmers doing custom spraying with large booms. Markers reduce skips and overlaps by indicating where to drive on each pass. Most marker systems use foam or dye. Markers may be continuous but are usually intermittent—about every 10 m. The foam or dye requires a separate tank and mix, a pump or compressor, a delivery to each end of the boom, and a control to select the desired boom end.

Maintenance

Maintenance is often performed only when parts fail and/or leaks occur. It is best, however, to follow a preventive maintenance program every year at the

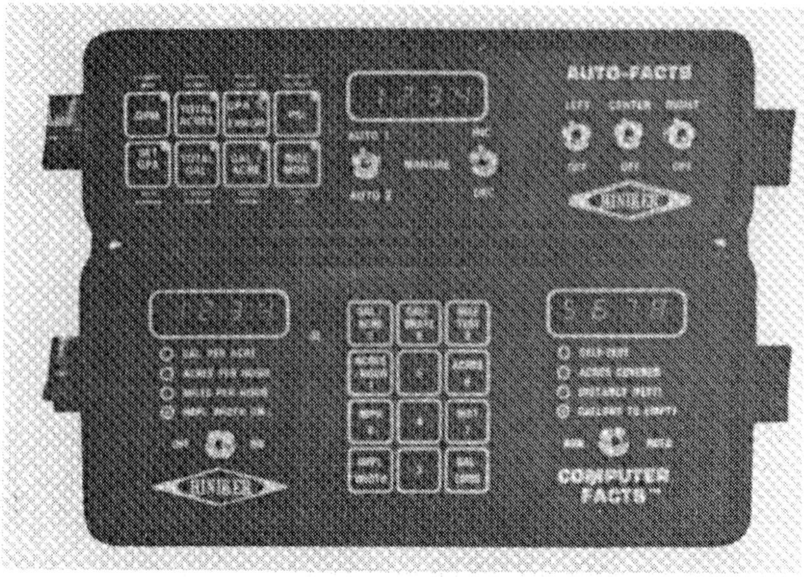

Figure 22 Electronic automatic control monitor.

end of the spraying season. Also, check the equipment carefully before spraying season begins. Periodically replace all components subject to wear. The following practices should reduce the failure of sprayer components and prolong useful life:

1. Flush a new sprayer before using it to remove metal chips and dirt resulting from manufacturing. Remove and clean all nozzles. Then flush the sprayer with clean water.
2. Use only those chemicals for which the sprayer is designed. For instance, liquid fertilizers corrode galvanized surfaces and parts made from copper, brass, bronze, and steel. Use sprayers made completely of stainless steel to apply fertilizers. Aluminum is satisfactory for some nitrogen fertilizers but not for mixed fertilizers.
3. Use clean water. Many troubles with sprayers can be traced to foreign materials in the water, which clog screens and wear out nozzles and pumps. If the water must be pumped from ditches, ponds, or lakes, filter as the tank is being filled.
4. Observe the output pattern of nozzles periodically. Streaks in the pattern indicate that the nozzles are partially plugged. Remove and clean nozzles with a soft brush for the tip and screen. Cleaning with a pin, a knife, or any other metallic object can completely change the spray pattern and capacity of the tip.

5. Inspect all strainers, screens, and nozzle tips after each day's spraying. Clean strainers daily. Partially clogged strainers create a pressure drop and reduce nozzle flow rate.
6. Thoroughly clean all the sprayer parts and drain all the liquids from the spray lines before taking the sprayer to an enclosed area for storage.

Operation Checklist

When spraying pesticides, the ultimate goal of an applicator should be to apply the right amount of chemical uniformly across the spray swath. Uneven spray coverage will result from use of misaligned or clogged nozzles, use of nozzles with different fan angles, or uneven height throughout the boom. These common problems result in streaks, untreated areas, or overapplication of chemicals as illustrated in Figure 23. The following list is a quick summary of guidelines to follow for an effective spraying job:

1. Select a recommended chemical for the specific spraying job.
2. Read and follow all directions on the pesticide label.
3. Mix the chemical in sufficient liquid carrier to get uniform recommended coverage over the sprayed area. Use clean water.
4. Consider the possibility of drift causing damage to plants, animals, or people adjacent to the sprayed area. Usually less drift occurs in early morning and late evening because there is less wind.

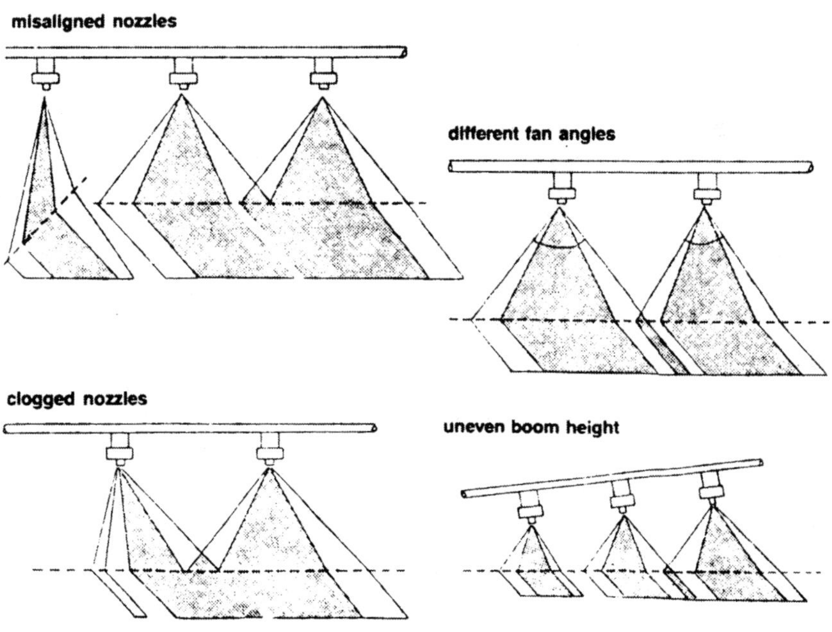

Figure 23 Common applicator errors.

5. Check whether the framework of the sprayer is secure and fastened to its carrier.
6. Make sure all vital parts of the sprayer are functioning properly. Check for leaks.
7. Select nozzles that provide the desired droplet size, flow rate, and pattern. Mount nozzles securely to desired positions.
8. Observe the nozzles periodically to detect clogging.
9. Operate the tractor at a uniform speed to ensure uniform application. Doubling speed halves the application rate.
10. Keep the proper height at all sections of the boom.
11. Set the pressure within the recommended range for precise application of the desired rate.
12. Do not run the pump without liquid in the tank.
13. Always calibrate the sprayer for application of the exact amount of chemical needed.
14. Maintain the system in peak condition by periodic inspections and repairs. Carry extra nozzles, washers, other spare parts, and tools for quick repairs in the field.
15. Clean the sprayer whenever pesticides are changed and at the end of the season.
16. Store the sprayer in a dry building.
17. Check all sprayer components carefully. Make note of any repairs that are needed. Order the required parts and make repairs before the spraying season starts.

Cleaning

Many pesticides cause rapid corrosion of metal parts in the spraying system. Pesticides should be washed from the whole system immediately after use. Cleaning a sprayer thoroughly not only increases its life but also reduces the chance of cross-contamination of chemicals and prevents crop injury.

Always try to end the day with an empty tank. If the same chemical is to be used the next day, flushing the sprayer with clean water is sufficient. If a different chemical will be used, then a more effective cleaning is recommended. Sometimes trace amounts of one pesticide can react with another, causing a loss in effectiveness of the second chemical and/or crop injury. Therefore, use a special cleaning solution whenever pesticides are changed or when the final cleaning at the end of the season is done.

The cleaning solution used depends on the pesticide used. Always check the pesticide label for specific cleaning instructions. A solution containing 1 kg of detergent for each 375 to 570 liters of water is sufficient for removing most pesticides. Flush the sprayer with clean water and add the cleaning solution to the tank. Agitate thoroughly and allow the water-detergent solution to circulate through the system for several minutes. Remove nozzles and flush the system twice with clean water. Clean nozzle tips and screens in a strong detergent solution or kerosene, using a soft brush such as an old toothbrush.

Some pesticide combinations (especially if oil is used) may produce a putty-type paste (buttering out) in the tank. Flushing out the residuals of such chemicals after each load may prevent accumulation. If water alone does not dissolve and remove the buildup, prepare a mixture of water and solvent such as kerosene, diesel fuel, or comparable light oil (1 liter solvent for 25 liters of water). Allow the paste to dissolve, then agitate, and flush. Next, flush with detergent and clean water as explained in the previous paragraph.

After rinsing the equipment with an oil-based solvent or a water-detergent solution, rinse with 1 liter of household ammonia or 1/4 liter of trisodium phosphate (TSP) in 100 liters of water. Circulate this solution and let a small amount flow through the nozzles. Keep the remainder of the solution in the system approximately 6 hours (or overnight), and then pump it out through the nozzle. Finally, remove the nozzles and strainers and flush the system twice with clean water.

Cleaning the outside of the sprayer components deserves equal attention. Remove compacted deposits with a bristle brush. Then flush the exterior parts of the equipment with water. A high-pressure washer can be used, if available. Wash the exterior of the equipment either in the field at an appropriate location or at a specially constructed concrete rinse pad.

Disposal of Rinsewater and Left-Over Tank Mixes

Disposal of rinsewater and left-over tank mixes requires much consideration. When one thinks of pesticidal wastes the obvious sources come to mind: left-over tank mixes and concentrates, empty containers, and chemicals that are unusable because they are obsolete, are unidentifiable, or have been frozen. While these wastes do present problems to users of chemicals, it is not the intent of this unit to address these disposal problems in detail.

In cleaning the spraying equipment to prolong its life, other less obvious wastes are generated. Consider these sources: pesticides left clinging to the exterior of spraying equipment; wash and rinse water used to clean spray or nurse tanks; spray remaining in hoses, booms, and tanks after the pump loses prime; and natural runoff from an area where spills occur as a result of mixing, where tank rinse water is dumped, or where vehicles are washed. Even though these sources represent fairly low levels of pesticides, repeated exposure at one location may contribute to groundwater contamination and runoff problems.

When cleaning a sprayer, select an area where it will not contaminate water supplies, streams, or crops, or other plants and where puddles will not be accessible to children, pets, livestock, or wildlife. Preferably, the area should have a wash rack or cement apron with a well-designed sump to catch contaminated wash water and pesticides. Spray the rinse water and the cleaning solution on the same field where the pesticide was applied in a manner consistent with the intended use of the chemical. Avoid discharging the whole cleaning solution in a small area.

The quickest and easiest way to rinse the spraying equipment and dispose of the waste safely is to carry a 200- to 400-L drum of fresh water with the spraying equipment. When spraying is finished, flush the system in the field. A concrete pad with a rinsate collection pit can be used to wash the interior and exterior of the sprayer without creating potential contamination of soil, groundwater, and surface water. Most importantly, the concrete rinse pad also helps contain pesticides that are spilled onto the sprayer or onto the ground during loading or mixing.

Sprayers are used only a few weeks in a year. When they are not in use, protect them against the harmful effects of snow, rain, sun, and strong winds. Moisture in the air, whether from snow, rain, or soil, rusts metal parts of unprotected equipment. The sun helps reduce moisture in the air, but it also causes damage. Ultraviolet light softens and weakens rubber materials such as hoses and tires and degrades some tank materials. The best protection from the environment is provided by storing sprayers in a dry building. Storing sprayers in a building allows for repairs at any time during the off-season regardless of weather. If storing in a building is not possible, provide some sort of cover. Remove the hoses, wipe them clean of oil, and store them inside a building. Hanging them over a nail or sharp object causes a permanent crease that reduces flow through the hose. Coil hoses around a basket or other large round object to prevent sharp bends. When storing trailer-type sprayers, put blocks under the frame or axle and reduce tire pressure during storage.

When it is time to store the sprayer, it is important to do the following:

1. Add 5 to 20 liters of lightweight oil such as diesel fuel, depending on the size of the tank, to the rinsing water before the final flushing. As water is pumped from the sprayer, the oil leaves a protective coating inside the tank, pump, hoses, and other parts.
2. To prevent corrosion, remove nozzle tips and strainers, dry them, and store them in a can of light oil such as diesel fuel or kerosene.
3. Drain all cleaning water from all parts to prevent freezing.
4. Pumps require special care. After draining the water, add a small amount of oil and rotate the pump four or five revolutions by hand to coat interior surfaces completely. Make sure that this oil is not going to damage rubber rollers in a roller pump or rubber parts in a diaphragm pump. Check the operator's manual. If oil is not recommended, pouring 1 tablespoon of radiator rust inhibitor in the inlet and outlet part of the pump

also prevents the pump from corroding. Another alternative is to use automotive antifreeze with rust inhibitor in the pump and other sprayer parts. This also protects against corrosion and prevents freezing in case all the water is not drained.

5. Cover openings so that insects, dirt, and other foreign material cannot get into the system.
6. Finally, check the sprayer for scratched spots. Touch up these areas with paint to eliminate corrosion.

B. Manual Hand Pump and Backpack Sprayers

Manual sprayers—hand pumps, hand cans, and back packs (Figure 24)—are designed for spot treatment and for spraying of other areas not suitable for power sprayers. Most of these sprayers do not have pressure gauges or pressure controls. The pressure drops continuously as you spray. Therefore, you need to repressurize the tank at frequent intervals. When spraying, either hold the nozzle steady at a constant height and walk back and forth, or swing the nozzle in a steady, sweeping, overlapping motion. Maintaining a uniform walk-

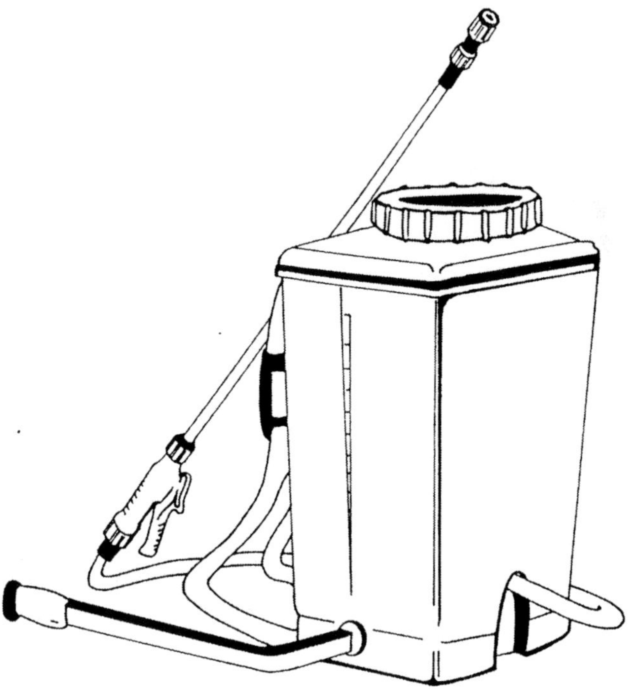

Figure 24 Manual backpack sprayer.

ing speed is essential to keeping the application rate relatively uniform throughout the operation.

Here are some hints for obtaining more uniform application when using manual sprayers:

1. Tie a weighted cord or chain to the wand near the nozzle. This can serve as a height gauge to maintain a set distance from the nozzle to the turf.
2. Walk a known space that should be covered by the sprayer. Practice until you can consistently spray this area with the correct amount of material. A ticking stopwatch or timer will improve pacing.
3. For hand pump sprayers, attach a pressure gauge to the spray wand and check the rate of pressure drop. Count the number of seconds needed for the pressure to drop 70 kPa. Count the number of pumps needed to return the pressure to the proper level.
4. Attach a gauge to the spray wand of backpack sprayers and determine how often you need to pump to keep the gauge pressure constant while spraying.

C. Hand Gun Nozzles

Hand gun nozzles (Figure 25) are used for applying herbicides and fertilizers on home lawns and small-size recreational turf grass areas. When using a hand gun nozzle, four factors are critical to delivering the correct rate uniformly over the application area: 1) constant pressure, 2) proper walking pace, 3) uniform hand/arm motion, and 4) relatively constant nozzle height and angle in reference to the ground.

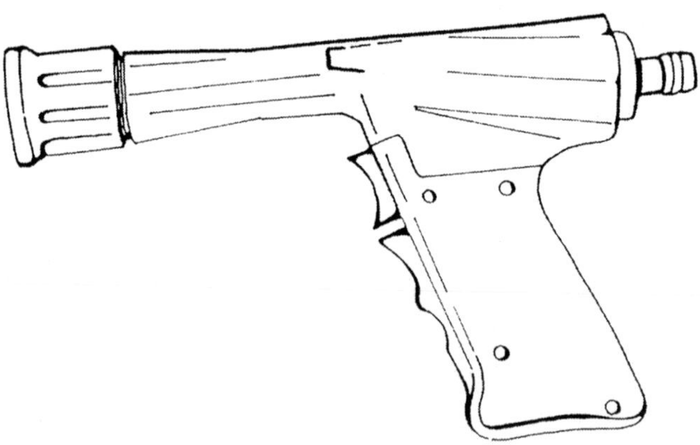

Figure 25 Hand gun nozzle.

D. Wiper Applicators

Although the idea of using wiper applicators to control weeds goes back to the early 1900s, their widespread use did not take place until the discovery and development of nonselective herbicides. Rather than using hydraulic nozzles to produce and transport small droplets to weed canopy, wiper applicators use the principle of actual physical contact with the weeds. These applicators have a limited use. They allow for the application of nonselective herbicides to tall weeds that cannot be controlled otherwise. They gained rapid acceptance in the Corn Belt states in the United States of America to control weeds in soybean fields when weeds were taller than soybeans.

Derting (1987) lists the advantages and disadvantages of wiper applicators as follows. The major advantages are: 1) selective application by mechanical device of nonselective herbicides, 2) practical control of species which were difficult to control with other methods, 3) efficient herbicide application by using only required quantities as opposed to broadcast spray, 4) elimination of drift, 5) reduced environmental hazard, 6) low cost of application equipment, 7) low maintenance requirements and cost, and 8) relative ease of use by unskilled personnel.

The three major disadvantages of wiper applicators are that 1) weed species that usually grow about the same height as field crops cannot be controlled, 2) application timing is restricted to postemergence use, and 3) formulations that are suitable to wiper applicators are restricted to water-soluble formulations. Wettable powders and flowable liquids cause flow blockage of wiper applicators.

Several types of wiper applicators are available commercially. Two of the most common types are the rope-wick and the roller.

Rope-wick applicators (Figure 26) consist of a long horizontal tube or pipe which is filled with a herbicide mixture, and a series of short braided ropes to wick herbicide solution from inside to outside the pipe where the herbicides can wipe onto tall weeds. Excessive restriction around the rope pieces due to small holes or a tight cap around the hose can substantially restrict solution flow through the ropes, resulting in poor weed control. Ideally, the rope should wick as fast as possible without dropping excessively. To obtain more rapid wicking, an air vent should be installed on the pipe. Pressurized rope wick applicators are also available to increase wicking efficiency.

Roller applicators work on the same basic principle as the rope-wick. Instead of using a series of short ropes, herbicide solution is trickled onto a carpet-covered roller that is rotated by a small motor as the applicator is driven across the field. The carpet on the early models of roller applicators was wetted manually by turning the liquid flow to the roller on and off to produce the desired wetness and to eliminate dripping. Some of the latest models are

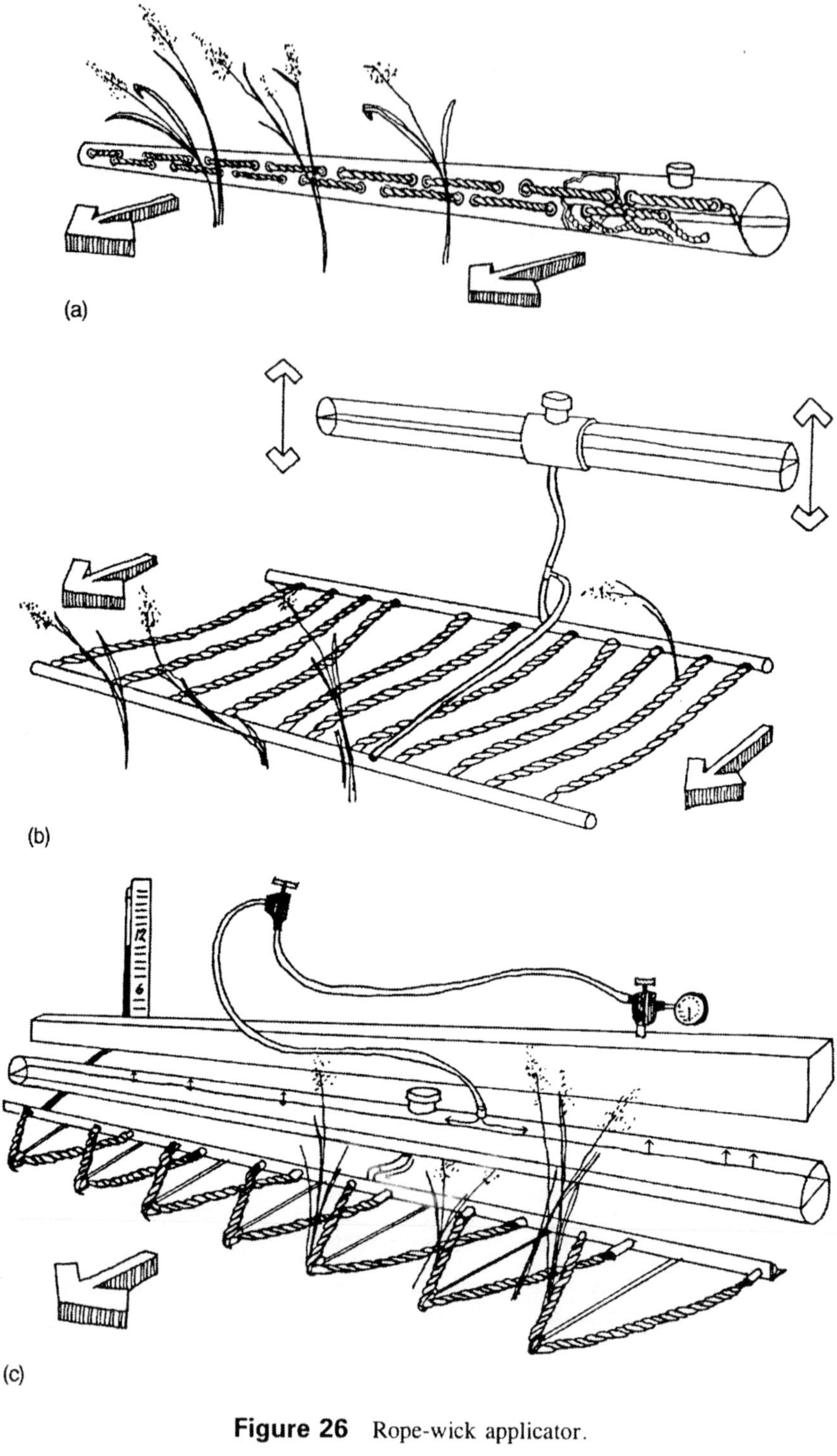

Figure 26 Rope-wick applicator.

equipped with electronic systems which sense the moisture level on the roller and automatically turn on the liquid flow to the roller.

E. Electrostatic Spraying

Transport of droplets to the spray target is greatly influenced by the size and velocity of droplets, dynamics of the spray vehicle, ambient winds, and spray targets. Small spray droplets which rely primarily on gravitational forces for transportation and deposition are highly susceptible to spray drift. Also, spray penetration into dense plant canopies is often inadequate. Electrostatic forces may become effective when droplets are smaller than 100 μm in diameter. Electrical charging of spray droplets has been investigated for use in pesticide application for several decades. Several researchers have investigated charged sprays for improving deposition efficiency on plants. Extensive reviews of research done on electrostatic atomizers are given by Hislop (1988) and by Gebhardt (1987). While many electrostatic sprayers have been tested extensively and provided favorable results in laboratory environments (Almekinders et al., 1992; Franz et al., 1987; Giles and Law, 1985; Law and Lane, 1981), field results have been inconsistent (Hislop, 1988). The deposition of charged droplets under field conditions varies significantly from laboratory applications. The general conclusion reached by many researchers is that electrostatic forces can be used to improve the deposit efficiency of small spray droplets. However, they also indicate that forces such as those exerted by air jets are needed in addition to those created by electrostatic fields to reduce drift losses and to increase penetration of charged droplets into plant canopies.

III. GRANULAR APPLICATION EQUIPMENT

Similar to the liquid application equipment, granular herbicide application equipment consists of a reservoir or hopper containing material to be applied, a mechanism to regulate flow rate, and a delivery or distribution system. Many applicators select granular products as part of their program or even as a complete program because of easy handling and low risk of environmental contamination. The popularity of dry granular herbicide formulations is expected to grow as a result of increased concerns associated with cleanup of accidentally spilled products and contamination of water sources with pesticides. Most granular spreaders are usually one of three types: drop (gravity), rotary (centrifugal), and pneumatic.

The typical *drop-type spreader* (Figure 27) has a full-length agitator mounted on the spreader axle. The agitator rotates over a series of openings in the bottom of the spreader hopper, and granules exit the hopper through these openings. The size of the openings can be adjusted to obtain different rates of application. Slight overlapping of the wheels with each pass is necessary to

Figure 27 Drop-type spreader.

ensure uniform coverage. Common problems associated with drop spreaders include skips, excessive overlap, nonuniform spread when turning corners, and clogging of openings in the bottom of the hopper when the ground is moist or when wet weeds wipe against the hopper opening during the application.

However, drop spreaders are generally more precise than rotary spreaders and deliver a better pattern. In addition, because the granules drop straight down a short distance, there is less chemical drift and pattern distortion. To prevent nonuniform spread when turning corners, always turn off the spreader while making turns. This means that you should treat the strips at both ends of the area first. This allows you to start in this strip and turn on the spreader as you cross the start line.

With *rotary spreaders* (Figure 28) granules drop out of one or more adjustable openings at the bottom of the hopper onto a rotating plate and are spread in a semicircular arc. Rotary spreaders cover a wider swath than drop spreaders. Therefore, they are preferred by the majority of commercial applicators. Common problems associated with rotary spreaders include the following:

1. Dry products with different granule sizes and weights do not spread uniformly.
2. Overlapping to obtain a uniform pattern varies with each product.

Figure 28 Rotary spreader.

3. Drop holes commonly clog.
4. Turning changes the rotating plate speed.

Although rotary spreaders are generally less precise than drop spreaders in terms of uniformity of distribution, with proper calibration and operation they provide satisfactory results. The problem with uniform distribution of granules can be corrected by proper overlapping of adjacent swaths. Always try to minimize drift of small particles when using rotary spreaders.

Small rotary spreaders used on lawn mowers throw material 1 to 1.5 m in front of the spreader. The operator is another 1 to 1.5 m behind the spreader hopper. This means that when the material finally reaches the ground, it is 2 to 3 m in front of the operator. It is essential that the operator leave a border along the edges of the area being treated so that there is enough room to maneuver the spreader for the next pass. This technique also eliminates the possibility of pesticides' reaching nontarget areas.

Much attention is currently being given to nontarget application of fertilizer and pesticides. Because rotary spreaders throw material 2 to 3 m wide, materials will likely be spread on sidewalks, driveways, and patios. To prevent such nontarget applications, operators must use extreme caution around nontarget areas. Some spreaders have shut-off valves to limit the spread of materials to the right of the operator. The use of this shut-off valve along with some commercially available reflectors that divert any other material will prevent nontarget applications to the right of the applicator.

Remember to open and close the spreader at the beginning and ending of each pass. Start walking before opening the lever and close it before stopping. Keeping the spreader open while turning around will cause misapplication because the impeller plate is driven by one wheel. Therefore, the impeller stops when turning in one direction but turns faster than normal in the other direction. Always hold the spreader handle at the proper height to keep the impeller level.

Do not operate spreaders traveling backward. The pattern with rotary spreaders will be skewed and the delivery rate of drop spreaders will be difficult.

Pneumatic applicators (Figure 29) use a combination of air, tubes, and deflectors to distribute the granules. Granular material is gravity fed and metered into a chamber and trapped in a high-speed air stream that takes the granules through delivery tubes to the discharge point. The deflector plates at the

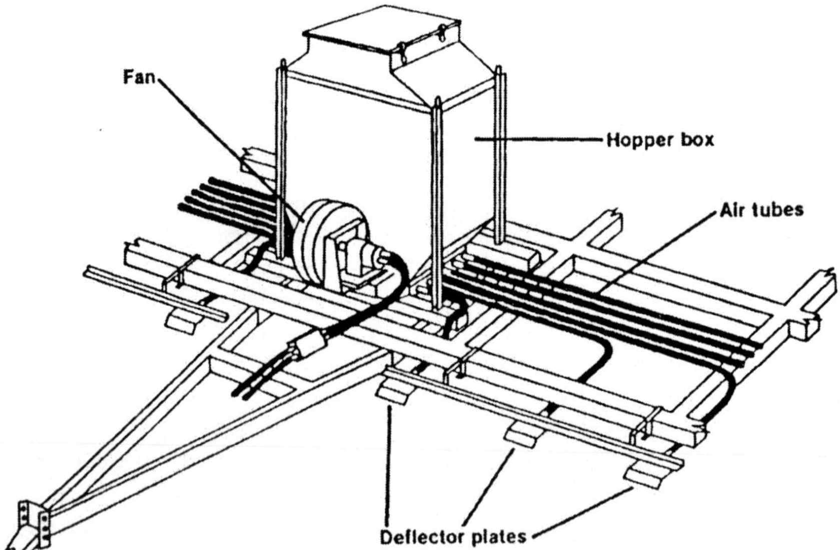

Figure 29 Pneumatic granular applicator.

discharge point help evenly distribute granules across the width of the boom or the tool to which the applicator is attached.

IV. INCORPORATION OF HERBICIDES INTO THE SOIL

Herbicides vary in their chemical and physical characteristics, such as vapor pressure. Some herbicides have a relatively high vapor pressure and volatilize rapidly from the soil surface. Incorporating them into the soil reduces surface loss through volatility.

Generally, weeds are most susceptible to herbicide activity when the weed seeds germinate and the seedlings are very young and "tender" (Knake, 1980). Weed seeds that are located near the soil surface are more likely to produce seedlings than those deeper in the soil. Therefore, the depth of placement and uniformity of herbicide concentration are important factors in achieving satisfactory results from herbicides incorporated into the soil. Generally the herbicides should be incorporated into the upper 6 cm.

How uniformly an incorporated herbicide is distributed in the soil is influenced by several factors, such as soil physical condition, soil moisture, type of equipment, and speed, direction, and depth of operation. Soil should be relatively dry and not too cloddy for proper mixing. Cloddy conditions prevent good horizontal distribution of herbicides. As the clods "melt down," weeds may grow from the untreated soil.

There are two common methods of herbicide incorporation: complete incorporation and shallow incorporation. The choice of incorporation type depends largely on the type of herbicide used and the type of weeds to be managed. If the weeds are annual grasses, which are most likely to germinate in the top 2 cm of the soil, shallow incorporation may do a better job. However, if the problem weeds germinate deep in the soil, a more complete, deeper incorporation will produce more effective weed control.

Almost all types of tillage implements except subsoilers and moldboard plows are used to incorporate herbicides. Usually a disk or a field cultivator is used for complete incorporation; shallow incorporation is accomplished with a combination disk-chisel plow type of equipment or "one-pass" incorporation equipment that has three or four types of tillage tools including a series of rolling baskets or spiked teeth.

A *tandem disk* with blade diameter less than 50 cm and blade spacing less than 20 cm is generally the preferred type of disk harrow for incorporation. It should be set to cut 10 to 15 cm deep and be operated 6 to 10 km/h.

Field cultivators with both the C-shank and Danish S-tine shanks can do an effective incorporation job. They should have at least three rows of sweeps spaced at intervals of 18 cm and staggered so no soil is left unturned. The sweeps should be at least 18 cm wide. Using chisels or points instead of sweeps is not recommended. The field cultivator should be operated level. If the rear rows of shanks operate deeper than the front rows, untreated soil may be brought to the surface, resulting in unsatisfactory weed control.

Whether incorporation is with a disk or field cultivator, pulling a leveling tool such as a tine drag, spike-tooth harrow at the rear will improve mixing action and will result in a better seed bed.

Combination tillage tools feature various sequences of disk gangs, C-tines, or Danish-tines with chisels or sweeps, ground-driven reels, spike-tooth harrows, and rolling baskets. These implements are usually referred to as one-pass incorporation equipment. They should be set to operate level and run at least 8 km/h. Incorporation depth usually doesn't exceed 8 cm.

Bode et al. (1979) and Thompson et al. (1981) used a fluorescent tracer material instead of granular herbicides to determine the patterns of incorporation with various tillage implements. The results of their study indicate that one pass with a disk or a field cultivator does not provide sufficient mixing of soil and herbicides. Two passes with these implements usually produce satisfactory distribution of herbicides. However, the risk of incorporating herbicides too deep is higher with a disk. A combination of disk and field cultivator does a better job of incorporation than two passes with a disk.

V. CALIBRATION OF EQUIPMENT

Calibration entails *all* the operations that ensure the correct amount of pesticide will be applied to the treated area. These operations include 1) calculating the amount of product, 2) mixing properly and safely, 3) adjusting equipment to deliver the desired rate uniformly across the spray swath, 4) determining effective swath/overlap, 5) checking accuracy during operation, and 6) detecting and correcting errors.

Application equipment must be calibrated periodically to achieve effective pest control and to reduce the potential for excessive pesticide residues remaining on sprayed surfaces. While applying too little pesticide may result in ineffective pest control, using of too much pesticide wastes money, may damage the crop, and increases the potential risk of contaminating groundwater and environment.

A survey of 53 private and municipal golf course pesticide applicators in Nebraska revealed that only one of six applicators applied pesticide carrier volumes within the acceptable error margin of $\pm 5\%$. Application errors ranged from 83% underapplication to 177% overapplication. Thirty-six percent of the applicators overapplied pesticides by more than 5% of their intended application rate. The applicators overapplied pesticides by an average of 19%. The cost of overapplication for groups using Daconil 2787 ranged from 12 cents per hectare to $16 per hectare. Faulty pressure gauges and improperly sized hoses and plumbing components are factors that contributed to inaccurate tank-mix application rates and nonuniform applications (Varner et al., 1990).

A survey of 152 private and commercial applicators in Nebraska revealed that only one of four farmers was applying pesticides within 5% of their intended application rate (Rider and Dickey, 1982). A follow-up field survey of 103 private pesticide applicators in Nebraska indicated that only 30% of the cooperators were applying herbicides within 5% of their intended application rate (Grisso et al., 1989). Twenty-six percent of the cooperators overapplied herbicides during a single application, with an average cost due to misapplication of $3.11/ha.

A survey conducted in South Carolina indicated that nearly 85% of applicators applied pesticides with errors greater than 10% when compared to intended rates (Wolak, 1989). Hofman and Hauck (1983) reported that only 11 of the 60 sprayers they calibrated in North Dakota were applying chemicals as the operator predicted. Calibration clinics conducted in Iowa and Ohio revealed that only 20% of sprayers calibrated were capable of applying a tank-mix within ±5% of the intended application rate (Ozkan, 1987). Similar findings were reported from calibration clinics conducted in Montana and Colorado. This limited data indicate that in general pesticide application accuracy continues to be a problem regardless of the geographic location. This problem, however, can be reduced to minimum with proper and frequent calibration of sprayers.

Sprayers should be calibrated several times a year. Changes in operating conditions and type of chemical used require a new calibration. Frequent calibration is even more important with liquid application because nozzles wear out with use, increasing the flow rate. For example, tests conducted by Reichard et al. (1991) indicated that the flow rate of 0.8 liter-per-minute capacity brass nozzle increased by 10% after only 3 hours of use with a wettable powder material.

The results of the survey conducted by Varner et al. (1990) indicated a positive correlation between application accuracy and frequency of calibration. Approximately 67% of the operators who calibrated before every spray operation had application errors below 5% while only 5% of those operators who calibrated their equipment less than once a year (once every 2, 3, 4 years) achieved the same degree of application accuracy.

A. Calibrating Liquid Application Equipment

Although there are many methods used to calibrate liquid application equipment, the methods described here are the most common. Choose a method that you find most comfortable.

For safety, use water instead of actual chemical mixtures when calibrating. However, some carriers, such as liquid fertilizers, are much denser than water and may cause the nozzle flow rate to vary from the rate obtained with water. In this case, determine the average nozzle output with water and use a conversion factor (usually listed in the nozzle manufacturer's catalogue).

To calibrate a sprayer, you need a measuring tape, a stopwatch or an ordinary watch with a second hand, and a measuring jar graduated in milliliters. A pocket calculator also will be handy. Regardless of the type of liquid application equipment you are calibrating, the calibration procedure requires checking the nozzle flow rate and the actual travel speed of the sprayer. Although the procedure is similar, steps involved in calibrating boom-type, manual backpack, and hand gun sprayers are presented separately.

Follow these steps when *calibrating low-pressure boom sprayers*:

Step 1. Fill the sprayer tank with water.

Step 2. Run the sprayer, inspect it for leaks, and make sure all vital parts function properly.

Step 3. Determine the spray width per nozzle (W) in centimeters. For broadcast application,

$$W = \text{nozzle spacing}$$

For band spraying

$$W = \text{bandwidth}$$

For directed spraying

$$W = \frac{\text{Row spacing (or bandwidth)}}{\text{Number of nozzles per row (or band)}}$$

Step 4. Determine travel speed (km/h). To determine the travel speed, measure a known distance. Use fence posts or flags to identify this distance. A distance over 50 m and a tank at least half full are recommended. Travel the distance determined at your normal spraying speed and record the lapsed time in seconds. Repeat this step and take the average of the two measurements. Use the following equation to determine the travel speed in km/h:

$$\text{Travel speed (km/h)} = \frac{\text{Distance (m)} \times 1000}{\text{Time (seconds)} \times 60}$$

Step 5. Determine nozzle flow rate (L/min). With the sprayer parked, run the sprayer at the same pressure level and catch the output from each nozzle in a measuring jar for 1 minute (or collect output for half a minute and then double the amount collected) to determine the nozzle flow rate in liters per minute. Check for clogs and clean the tip or replace any nozzle tips having flow rates 10% more or less than the average of the other nozzles checked and/or having obviously different fan angles or patterns. Repeat the process until the variation in discharge rate for all nozzles is within 10% of the average.

Step 6. Determine the actual application rate (L/ha) using the following equation:

$$L/ha = \frac{L/min \times 6000}{(km/h) \times W}$$

L/min: average nozzle flow rate in liters per minute
km/h: travel speed in kilometers per hour.
W: effective spray width per nozzle (calculated previously)
6000: constant to convert units to liters/hectare

Step 7. Compare the actual application rate with the recommended or intended rate. If the actual rate is more than 5% higher or lower than the recommended or intended rate, you must make adjustments in either the pressure or the travel speed or both. If these changes do not bring the application rate to the desired range, then you may have to select a new set of nozzles with smaller or larger openings. (Refer to Section VII, Useful Formulas, to determine the new travel speed and/or the pressure rate.)

Step 8. Recalibrate the sprayer. Repeat steps 5 through 7 until the application error is within 5% of the intended application rate.

The principles for *calibrating manual hand can and backpack sprayers* are essentially the same as those for boom sprayers outlined previously. Here are the steps involved in the calibration process:

Step 1. Measure and mark off an area equal to 100 square meters (such as 10 m × 10 m).

Step 2. Add a measured amount of water to the tank, spray the area, and then measure the amount of water remaining in the tank. The difference between the amount in the tank before and after spraying is the amount used per 100 square meters.

Step 3. Compare the measured rate with the intended or recommended rate. If the difference is greater than ±5% of the intended rate, make necessary adjustments and recalibrate the sprayer.

An alternative method is to record the time required to spray 100 square meters and later catch and measure the spray from the nozzle (or nozzles) used for the same period.

Calibrating a hand gun nozzle is similar to calibrating a backpack sprayer. Spray a known area and record the time required to cover this area. Then use a bucket to catch the spray from the hand gun for the period elapsed during spraying of the marked area. Determine the application rate by measuring the amount of liquid in the bucket.

B. Calibration of Granular Herbicide Application Equipment

Herbicide labels list the rate at which the product should be applied. Some labels also list the correct setting for some name-brand spreaders. However, these settings are just recommendations to provide a good reference point when calibrating spreaders. Seldom does a setting give the same application rate in all operating conditions and application sites. Variables such as particle size, particle density, particle shape, relative humidity, drop point of granules on the impeller plate, impeller speed, and ground speed all affect the application rate and the distribution pattern. Therefore, granular applicators should be calibrated for each type of granular product being applied and for change in operator, weather, and field conditions.

Before beginning to calibrate a granular applicator, be sure that it is clean and all parts are working properly. Always wear rubber gloves and other protective clothing to prevent contact with chemicals during cleaning and calibration of equipment and actual application of chemicals.

The primary purpose of calibration is to determine the actual delivery rate of a spreader regardless of its type. However, with rotary spreaders, it is also necessary to check and correct the distribution pattern.

Delivery rate refers to the average amount of product applied over a known area; it usually is expressed in kg/ha or kg/1000 ft^2. The easiest way for an operator to check the actual output rate is to collect and weigh the granules actually applied to a known area. You can do this by following one of the two methods explained next. Both methods are acceptable, but the first produces a more precise calibration.

Method 1

1. Spread out a plastic tarp of known size on the ground.
2. Operate the spreader at a known speed across the tarp.
3. Determine the area covered. If the swath width is wider than the tarp, the area covered is equal to the area of the tarp. If the swath is narrower than the tarp, multiply the swath width by the tarp length to determine the area covered. It is highly recommended you keep the tarp width wider than the swath width for more accurate measurement of the delivery.
4. Place all the granules on the tarp into a container and weigh them.

5. Divide the weight of granules collected by the area covered to determine the actual delivery rate in grams per square meter (g/m^2).
6. Compare the measured (actual) delivery rate with the rate recommended on the label. Make proper adjustments and recalibrate the spreader until you reach the desired accuracy (a variation within $\pm 5\%$ of the label recommendation is acceptable).

Method II

The second method is not as precise as the first one, but it is preferred by many operators because it does not require a tarp. However, the principle is the same. For calibrating, this area should be at least 50 m^2 for drop spreaders and 100 m^2 for rotary spreaders. Once you set up the known area that will be covered, follow the steps to complete calibration.

1. Fill the spreader to a known level marked on the inside of the hopper.
2. Operate the spreader over the known area already set up.
3. Weigh the amount of material needed to refill the spreader to the mark. Record the weight in grams.
4. Grams of product actually applied per square meter can be calculated as follows:

$$\frac{\text{Amount needed to refill (g)}}{\text{Area covered (m}^2\text{)}} = \text{Product applied (g/m}^2\text{)}$$

Drop-type spreaders provide fairly uniform *distribution of granules* across the swath width. The distribution is not affected much by product physical characteristics, weather conditions, etc. However, rotary spreaders are very sensitive to these variables, and patterns can be severely skewed if they are overlooked. Therefore, always check the distribution pattern of granules when using rotary spreaders.

The best method to check the distribution pattern is to lay out a row of shallow cardboard boxes on a line perpendicular to the direction of travel (Figure 30). Boxes with partitions will prevent chemical particles from rebounding. All boxes should have the same area and height (0.20 m^2 area, 3 to 5 cm tall). Space the boxes 30 cm apart, and make sure the row is 1.5 to 2 times the width of the anticipated effective swath.

To conduct the test, fill the spreader at least half full, set it at the label setting for rate and pattern, and make three passes in the *same* direction over the boxes. Be sure to turn on the spreader and reach proper travel speed before crossing the boxes. Put the material for each box into a test tube, vial, or small, narrow bottle. When the bottles are placed side by side in order, a plot of the pattern is visible. The effective swath width is twice the distance out from the center of the spreader to the point where the rate is one-half the average

Figure 30 Checking the distribution pattern of granules.

rate at the center of the pattern. For example, if the first three to four bottles for the center have material 4 cm deep and the bottles from the 3-meter positions (3 m left of the spreader centerline or 3 m right of the centerline) have material 2 cm deep, the effective swath width is 6 m. Therefore, you should be making passes 6 m apart to provide uniform coverage throughout the area covered.

In some cases, a rotary spreader's distribution pattern may not be uniform enough, even after all spreader adjustments have been made. A common recommendation in such cases has been to cut the rate of application in half to cover the field twice at right angles. However, research conducted by Parish (1986) revealed that cutting the rate in half and cutting the swath width in half while maintaining parallel swaths is more effective than the "right angle" method. The "half width" method is practical mainly for correcting severe pattern problems that remain after all mechanical adjustments to the spreader have been made.

There is always the question of which of the two modes of spreader operation—the continuous, back-and-forth mode or the circuitous, round-and-round mode—yields a more uniform spreader distribution. Parish (1988) analyzed the distribution pattern data from 90 spreader test runs involving four spreaders with several products and pattern settings. They found that when the continuous spreader pattern is generally acceptable, there is statistically no advantage to using the more cumbersome circular mode of operation. However, if the basic spreader pattern remains badly skewed after all the spreader adjustments are made, the circular mode may help improve the overlapping pattern. However, the half-rate, half-width method with the continuous mode is probably more effective in such cases. It is easier to use and requires little additional travel over the turf area.

IV. MIXING AND LOADING CHEMICALS

Mixing the pesticide thoroughly and carefully is an important step in good sprayer operation. Incomplete mixing results in varying application rates—too high at times, too low otherwise. Some chemicals, when mixed improperly with others, form a thick, mayonnaise-like mixture that will not spray properly and is difficult to clean out of the sprayer. Always read the label for proper mixing sequence, and test the compatibility of chemicals in a small container before mixing.

Generally, when preparing the spray mixture, fill the tank half full of water and pour in the correct amount of chemical while the pump is running. Then finish filling the tank. When using wettable powders, again fill the tank half full, make a slurry in a separate container or through an inductor system, and then add the slurry to the tank to ensure good mixing. Rinse the slurry container three times, pouring the rinsate into the tank. Finally, finish filling the tank with water.

You can spend hours in the field calibrating your sprayer to achieve the desired accuracy, but your efforts are wasted if you do not know how much chemical to put in the tank. In a survey conducted by Rider and Dickey (1982), 38% of the applicators failed to add the correct amount (within $\pm 5\%$ of the required amount) of chemical to the tank. In South Carolina, 76% of the applicators surveyed had tank-mix errors greater than 10% (Wolak, 1989).

The amount of chemical needed per tank depends on the recommended rate and the size of area you can treat per tank of spray. Calculations are the same whether you are using a manual backpack sprayer or a powered boom sprayer. Assuming that a boom sprayer is to be used, the following steps are involved in determining the right amount of pesticide to add to the tank: 1) Determine the area (hectares) to be treated per tank by dividing the tank capacity by the liter/hectare (L/ha) application rate; 2) find the amount of pesticide needed per tank by multiplying the number of hectares treated per tank with the amount of actual chemical to be applied per hectare.

Sometimes, chemical manufacturers give recommended rates in terms of active ingredients (AI) to be used per acre rather than the amount of total product per acre. In these cases, calculate the amount of material to be applied as shown in examples 2 and 3 for dry chemicals and liquid chemicals, respectively.

VII. USEFUL FORMULAS

1. To determine travel speed:

$$\text{Travel speed (km/h)} = \frac{\text{Distance (m)}}{\text{Time (sec)}} \times \frac{1 \text{ km}}{1000 \text{ m}} \times \frac{3600 \text{ sec}}{1 \text{ h}}$$

2. To calculate the resulting application rate (L/ha) for any change in travel speed (km/h):

$$(\text{L/ha})_2 = \frac{(\text{L/ha})_1 \times (\text{km/h})_1}{(\text{km/h})_2}$$

3. To determine the desired travel speed (km/h) when changing application rate (L/ha):

$$(\text{km/h})_2 = \frac{(\text{L/ha})_1 \times (\text{km/h})_1}{(\text{L/ha})_2}$$

4. To determine the resulting application rate (L/ha) when changing nozzle pressure:

$$(\text{L/ha})_2 = (\text{L/ha})_1 \times \sqrt{\frac{\text{kPa}_2}{\text{kPa}_1}}$$

5. To determine the desired pressure (kPa) that will produce a desired application rate (L/ha):

$$\text{kPa}_2 = \text{kPa}_1 \times \left(\frac{[\text{L/ha}]_2}{[\text{L/ha}]_1}\right)^2$$

6. To determine the resulting nozzle flow rate (L/min) when changing nozzle pressure:

$$(\text{L/min})_2 = (\text{L/min})_1 \times \sqrt{\frac{\text{kPa}_2}{\text{kPa}_1}}$$

$(\text{L/ha})_2$, $(\text{L/ha})_1$: desired and measured application rates, respectively
$(\text{km/h})_2$, $(\text{km/h})_1$: desired and measured travel speeds, respectively
kPa_2, kPa_1: desired and measured spraying pressures, respectively

ACKNOWLEDGMENT

The author thanks T. Camm for providing technical assistance and D. C. Yagich for typing the manuscript. The author also acknowledges the use of materials from Bode and Butler (1981), and Daum and Reed (1983).

REFERENCES

Almekinders, H., Ozkan, H. E., Reichard, D. L., Carpenter, T. G., and Brazee, R. D. (1992). Spray deposit patterns of an electrostatic atomizer, *Trans. ASAE, 35*: 1361-1367.

Akesson, N. B. and Yates, W. E. (1979). *Pesticide Application Equipment and Techniques,* FAO Agricultural Services Bulletin 38, Food and Agriculture Organization of the United Nations, Rome.

American Society of Agricultural Engineers. (1991). Procedure for measuring spray nozzle wear rate, in *ASAE Standard* S471, St. Joseph, Missouri, p. 262.

Bode, L. E. (1987). Spray application technology. *Methods of Applying Herbicides* (C. G. McWorter, and M. R. Gebhardt, eds.), WSSA Monograph 4, Weed Science Society of America, Champaign, Illinois, pp. 85–110.

Bode, L. E. and Butler, B. J. (1981). *Equipment and Calibration: Low-Pressure Sprayers*, Cooperative Extension Service Circular No. 1192, University of Illinois, Urbana-Champaign, Illinois.

Bode, L. E., Butler, B. J. and Wax. L. M. (1989). Herbicide incorporation: which works best? *Crops and Soils Magazine 31*: 17–20.

Daum, D. R. and T. F. Reed (1983). *Boom Sprayers*, The Northeast Regional Agricultural Engineering Service Publication 19, Cornell University, Ithaca, New York.

Derting, W. C. (1987). Wiper application, in *Methods of Applying Herbicides* (C. G. McWorter and M. R. Gebhardt, eds.), WSSA Monograph 4, Weed Science Society of America, Champaign, Illinois, pp. 207–229.

Folsom, R. G. and Ferguson, C. K. (1949). Jet mixing of two liquids, *Trans. Am. Soc. Mech. Engineers 71*: 73.

Franz, E., Reichard, D. L., Carpenter, T. G., and Brazee, R. D. (1987). Deposition and effectiveness of charged sprays for pest control, *Trans. ASAE 30*: 50–55.

Gebhardt, M. R. (1987). Electrostatic spraying, in *Methods of Applying Herbicides* (C. G. McWorter, and M. R. Gebhardt, eds.), WSSA Monograph 4, Weed Science Society of America, Champaign, Illinois, pp. 137–144

Giles, D. K. and Law, S. E. (1986). Space charge deposition of pesticide sprays onto cylindrical target arrays, *Trans. ASAE 28*: 658–664.

Grisso, R. D., Dickey, E. C., and Shulze, L. D. (1989). The cost of misapplication of herbicides, *Appl. Engineering Agric 5*: 344–347.

Hislop, E. C. (1988). Electrostatic ground-rig spraying: an overview, *Weed Technology* 2: 94–105.

Hofman V. and Hauck D. (1983). *Sprayer check in North Dakota*, ASAE Paper No. NCR-83-405, American Society of Agricultural Engineers, St. Joseph, Missouri.

Knake, E. L. (1987). Incorporating herbicides, in *Agronomy News: Crops and Soils*, No. 516, Cooperative Extension Service, University of Illinois, Urbana-Champaign.

Law, S. E. and Lane, M. D. (1981). Electrostatic deposition of pesticide spray onto foliar targets of varying morphology, *Trans ASAE 24*: 1441–1445, 1448.

Marer, P. J. (1988). *The Safe and Effective Use of Pesticides*, Pesticide Application Compendium 1, Division of Agriculture and Natural Resources Publication 3324, University of California, Oakland.

Mathews, G. A. (1982). *Pesticide Application Equipment*, Longman, Inc., New York.

McWorter, C. G. and Gebhardt, M. R. eds. (1987). *Methods of Applying Herbicides*, WSSA Monograph 4, Weed Science Society of America, Champaign, Illinois.

Ozkan, H. E. (1987). Sprayer performance evaluation with microcomputers, *Appl. Engineering Agric 3*: 36–41.

Ozkan, H. E., Reichard, D. L., and Ackerman K. D. (1992). Effect of orifice wear on spray patterns delivered by fan pattern nozzles, *Trans ASAE, 35*: 1091–1096.

Ozkan, H. E., Reichard, D. L., and Sweeney, J. S. (1992). Droplet size distributions across the fan-patterns of new and worn nozzles, *Trans ASAE, 35*: 1097–1102.

Parish, R. L. (1986). Evaluation of two methods of fertilizer spreader pattern correction, *Trans. ASAE 29*: 370–373.

Parish, R. L. (1988). An improved spreader pattern starts with the spreader, *ALA/ Maintenance* November, pp. 99–101.

Prairie Agricultural Machinery Institute. (1987). Evaluation Report 528. Vicon Model LS 2340T Field Sprayer.

Reed, T. (1984). *Wear Life of Agricultural Nozzles*, ASAE Paper No. AA84-001, American Society of Agricultural Engineers. St. Joseph, Missouri.

Reichard, D. L., Ozkan, H. E., and Fox, R. D. (1991). Nozzle wear rates and test procedures, *Trans. ASAE 34*: 2309–2316.

Rider, A. R. and Dickey, E. C. (1982). Field evaluation of calibration accuracy for pesticide application equipment, *Trans. ASAE, 25*: 258–260.

Wolak, F. J. (1989). Pesticide application accuracy survey in South Carolina. *Appl. Engineering Agric. 5*: 514–516.

7

Herbicide Formulations, Adjuvants, and Spray Drift Management

H. Erdal Ozkan
The Ohio State University, Columbus, Ohio

I. HERBICIDE FORMULATIONS

Only a small portion of chemicals in a herbicide product are responsible for controlling weeds. These active ingredients are usually diluted in inert (inactive) materials such as petroleum solvents, emulsifiers, powders, or granules. Inert materials change the characteristics of the active ingredients in many ways, for example, they may make them safer, easier to apply, and/or more convenient to handle. The mixture of active and inert ingredients is called a *herbicide formulation*. Inert materials may include chemicals such as wetting agents, spreaders, stickers, and extenders, all of which improve the effectiveness of an active ingredient.

The same active ingredient can be packaged and sold in several different kinds of formulations. The following factors should be considered when deciding which formulation to buy: 1) physical and biological characteristics of the target weed, 2) application equipment available and best suited for the job, 3) hazard of drift and runoff, 4) possible crop injury, 5) cost, 6) storage requirements, and 7) type of environment in which the application must be made (agricultural, aquatic, forest, urban, etc.).

While some formulations are mixed with water or oil for final application, some others such as baits, granules, and dusts can be applied without mixing with a liquid. Some products, especially those designed for the home owner, are sold in ready-to-use formulations.

The letters that follow, or are part of the brand name, are useful in identifying the formulation type. For example Dual 8E indicates that the herbicide is an emulsifiable concentrate (liquid), and each gallon of product contains 8 pounds of the active ingredient metolachlor. The amount of active ingredient in a dry formulation is usually expressed in terms of percentages. For example, Dual 25G indicates that the product is granular, and the active ingredient makes up 25% by weight of the formulated product. That is, 10 kg of Dual 25G contains 2.5 kg of the active ingredient metolachlor and 7.5 kg of inert materials.

In some cases, the letters following the brand name may indicate how the pesticide is intended to be used (for example, RP indicates range or pasture), what it is used for (for example, WK indicates use with weed killers), or special characteristics of the formulation (for example, LV indicates low volatility). A partial list of suffixes attached to formulation brand names and their descriptions is given in Table 1.

Although there are numerous types of pesticide formulations, as shown in Table 1, the majority of herbicides fall in one of the major formulation types described in the following sections. Detailed information on pesticide formulations can be obtained from a number of references (Van Valkenburg, 1973; McDonald, 1980; Marer, 1988; Ware, 1988).

A. Liquid Formulations

Emulsifiable Concentrates (EC or E): Some herbicides are applied through spraying in the form of water emulsions. An *emulsion* is one liquid dispersed throughout another liquid. An emulsifiable concentrate formulation usually contains one or several petroleum solvents and an emulsifier that allows the formulation to be mixed with water to form an opaque or milky emulsion. Advantages of emulsifiable concentrates include the following: 1) the price per pound of active ingredient is relatively low; 2) they are relatively easy to handle, transport, and store; 3) little agitation is required; 4) they are not abrasive and do not plug screens or nozzles. Some of the major disadvantages of emulsifiable concentrates are that 1) a high concentration makes it easy to overdose or underdose; 2) some crops are sensitive to solvents and additives used in ECs which may result in phytotoxicity; 3) they are readily absorbed through the skin of animals and humans; 4) solvents may cause rubber or plastic hoses, gaskets, and pump parts and surfaces to deteriorate; 5) they may cause pitting or discoloration of painted finishes; 6) they are flammable and corrosive.

Table 1 Suffixes of Chemical Brand Names

Suffix	Meaning
Describe the formulation:	
AF	Aqueous flowable
AS	Aqueous suspension
D	Dust
DF	Dry flowable
EC	Emulsifiable concentrate
Ec	Emulsifiable concentrate
ES	Emulsifiable solution
F	Flowable
FL	Flowable
G	Granular
OL	Oil-soluble liquid
P	Pelleted
PS	Pelleted
S	Soluble powder
SG	Sand granules
SL	Slurry
SP	Soluble powder
ULV	Ultra-low-volume concentrate
W	Wettable powder
WDG	Water-dispersible granules
WP	Wettable powder
Describe how a pesticide is used:	
GS	For treatment of grass seed
LSR	For leaf spot and rust
PM	For powdery mildew
RP	For range and pasture
RTU	Ready to use
SD	For use as a side dressing
TC	Termiticide concentrate
TGF	Turf grass fungicide
WK	To be used with weed killers
Describe characteristics of the formulation:	
BE	The butyl ester of 2,4-D
D	Ester of 2,4-D
K	Potassium salt of the active ingredient
LO	Low odor
LV	Low volatility
MF	Modified formulation
T	Triazole
2X	Double strength
Label for use in special locations:	
PNW	For use in the Pacific Northwest (i.e., Benlate PNW)
TVA	For use in the waterways of the Tennessee Valley Authority (i.e., Aqua-Kleen TVA)

Source: Marer (1988).

Flowables (F or L) contain active ingredients of insoluble solids, in which the finely ground active ingredients are mixed with a liquid, along with inert ingredients, to form a suspension. Flowables are mixed with water, along with emulsifiers, to form a concentrated emulsion. They are similar to EC or WP formulations in ease of handling and use. They require only moderate agitation and seldom clog spray nozzles if mixed thoroughly and agitated properly.

Solutions or water-soluble concentrates (S) contain active ingredients that dissolve readily and form a true solution when mixed with the appropriate carrier designated on the label. Once the solution is dissolved in the spray carrier, no further mixing or agitation is required. There are, however, a limited number of herbicides being formulated as solutions.

Ultra-low-volume concentrates (ULV) may approach 100% active ingredient. They are designed to be used as is or to be diluted with only small quantities of specified solvents. These special-purpose formulations must be applied with highly specialized equipment. The advantages and disadvantages of liquid concentrates are similar to those of emulsifiable concentrates.

Invert emulsions contain a water-soluble pesticide dispersed in an oil carrier. They contain small water droplets suspended in oil, as opposed to regular emulsions, in which water contains small droplets of oil. Invert emulsion concentrates have the consistency of mayonnaise and require a special kind of emulsifier that allows the pesticide to be mixed with a large volume of petroleum carrier, usually fuel oil. When applied, invert emulsions form large droplets that do not drift easily. Therefore, they are most commonly used in vegetation control along rights-of-way where drift to susceptible nontarget plants is a problem.

Microencapsulated formulations consist of dry and liquid pesticide particles enclosed in tiny plastic capsules which are mixed in water and sprayed in the same way other liquid formulations are (Figure 1). After spraying, the capsule slowly releases the pesticide. The encapsulation process can prolong the active life of the pesticide by providing timed release of the active ingredient. Since

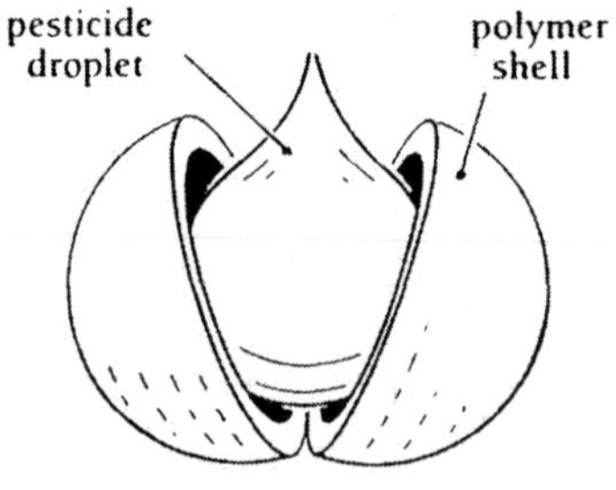

Figure 1 Microencapsulated formulations consist of pesticides enclosed in tiny plastic capsules. (From Marer, 1988).

they pose less dermal exposure hazard than other liquid formulations, microencapsulated formulations ensure that handling of highly toxic pesticides is safer for the applicator. They may, however, pose a significant hazard to bees. Microcapsules are about the same size as pollen grain and bees may mistakenly carry these capsules back to their hives.

B. Dry Formulations

Granular (G) formulations are produced by applying a liquid formulation of the active ingredient to particles of absorptive materials such as clay, corn cobs, or walnut shells. The active ingredient either coats the outside of the granules or is absorbed into them. The granule carrier is prepared in advance to a standard size and than the liquid formulation is added. The amount of active ingredient is relatively low, usually ranging from 1% to 15%.

Granular pesticides are most often used to apply chemicals to the soil to control weeds, nematodes, and insects living in the soil. Major advantages of granular formulations include the following: 1) they are ready to use, so no mixing is required; 2) the larger size and greater weight of granules eliminate drift; 3) hazards to the applicator and environment are minimized with granular formulations as a result of reduced dust and lack of spray mist; 4) they are more persistent in the environment than other formulations as a result of a slower release rate of the active ingredient.

Major disadvantages include the following: 1) they are more expensive than wettable powders or ECs relative to the active ingredient in the formulation; 2) incorporation into soil may be required; 3) moisture may be needed to activate pesticidal action; 4) they may be hazardous to nontarget species, especially waterfowl and other birds that mistakenly feed on the grainlike or seedlike granules.

Wettable powders (WP or W) are dry, finely ground formulations that resemble dust. Unlike dusts, however, they contain wetting and dispersing agents. Wettable powders usually must be mixed with water for application as a spray. A few products, however, may be applied either as a dust or as a wettable powder—the choice is left to the applicator.

Most wettable powders contain 15% to 80% active ingredient. Wettable powder particles do not dissolve in water. They settle out quickly unless constant agitation is used to keep them suspended. Wettable powders are among the most widely used formulations because they offer many advantages, such as 1) reasonable cost; 2) ease of storing, transporting, and handling; 3) lower phytotoxicity hazard than ECs and other liquid formulations; 4) ease of measurement and mixing; 5) less skin and eye absorption compared with ECs and other liquid formulations.

Major disadvantages of wettable powders include the following: 1) they pose an inhalation hazard to the applicator while pouring and mixing the concentrated powder; 2) they require adequate and constant agitation (usually mechanical) in the spray tank and quickly settle out if agitation is turned off; 3) they are abrasive to many pumps and nozzles, causing them to wear out quickly; 4) they are difficult to mix in very hard or very alkaline water; 5) they often clog nozzles and screens; and 6) their residues may be visible.

Dry flowables or water-dispersible granular (DF or WDG) formulations are like wettable powder formulations, except that the active ingredient is prepared as granule-sized particles. Water-dispersible granules must be mixed with water to be applied. Once in water, the granules break apart into fine powder. The formulation requires constant agitation to keep it suspended in water. Water-dispersible granules share the advantages and disadvantages of wettable powders, except that 1) they are more easily measured and mixed, 2) they do not present a dust problem and therefore pose less inhalation hazard to the applicator during pouring and mixing.

C. New Environmentally Safe Formulations

Water-soluble packets are used to reduce the mixing and handling hazards of herbicides. Preweighed amounts of wettable powder, soluble powder, or gel formulations are packaged in water-soluble plastic bags (Figure 2). When the bags are dropped into a filled spray tank, they dissolve and release their contents to mix with the water.

Effervescent tablets have been developed by some companies to eliminate the problem of container rinsing and to eliminate disposal problems altogether (Figure 3). The tablets can be broken into small parts or used whole, depending on the size of the field sprayed or the application rate recommended for different soil and crop conditions. Once the tablet is dropped into the sprayer tank, it dissolves in a short time. A moderate level of agitation is recommended to provide a uniform distribution of active ingredients released from the tablet throughout the spray tank.

II. ADJUVANTS

An *adjuvant* is a chemical added to a pesticide formulation or tank mix to improve mixing and application or enhance performance. With judicious use of adjuvants, the effectiveness of the pesticide compound may be increased as much as 5- or 10-fold (Van Valkenburg, 1982). Most pesticide formulations contain at least a small percentage of adjuvants. Wetting agents and spreaders are the adjuvants most frequently added by pesticide users. A small number of products, however, require avoiding adjuvants altogether. One should read the pesticide label before adding any surfactant to the spray mixture. Common adjuvants and their functions are as follows:

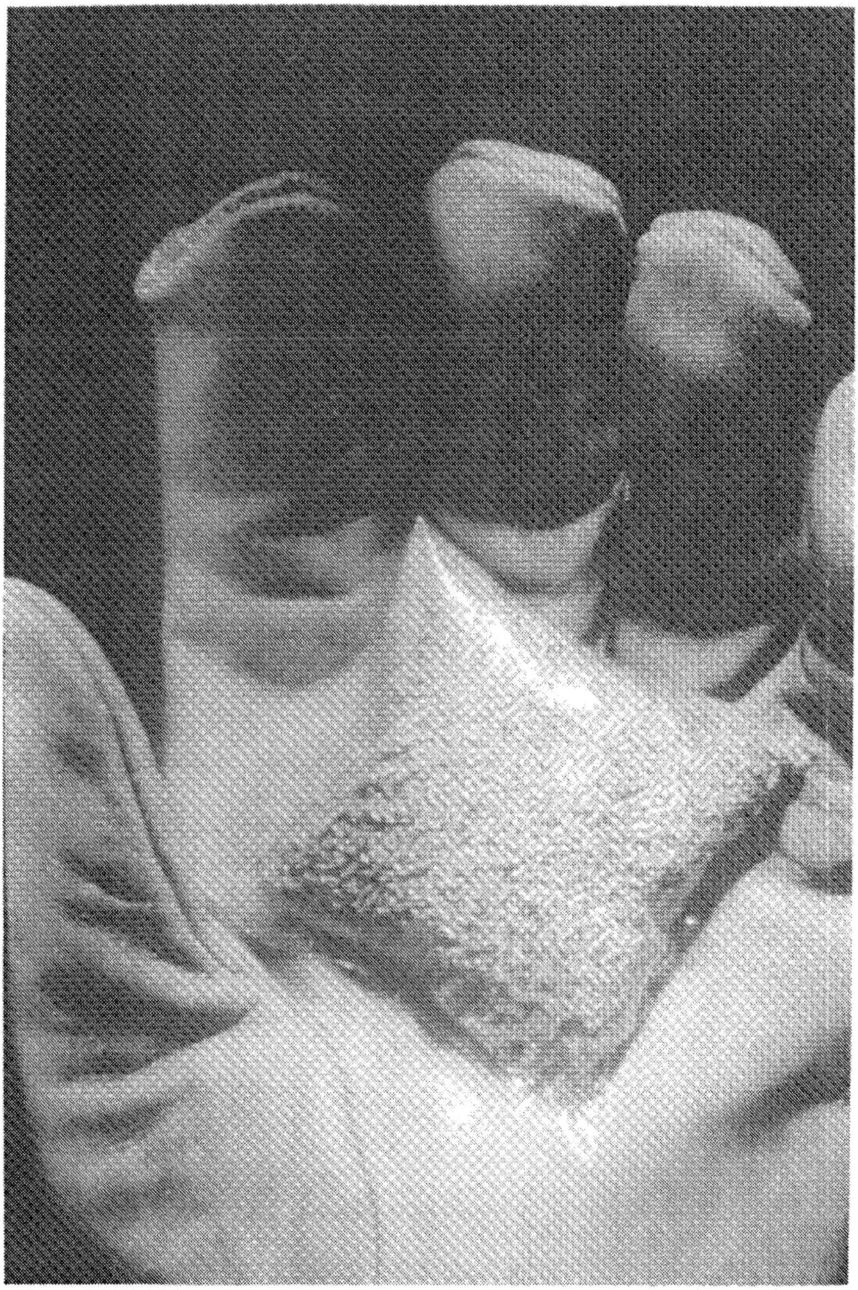

Figure 2 Water-soluble pouch. (Courtesy of Ciba Chemical Co.)

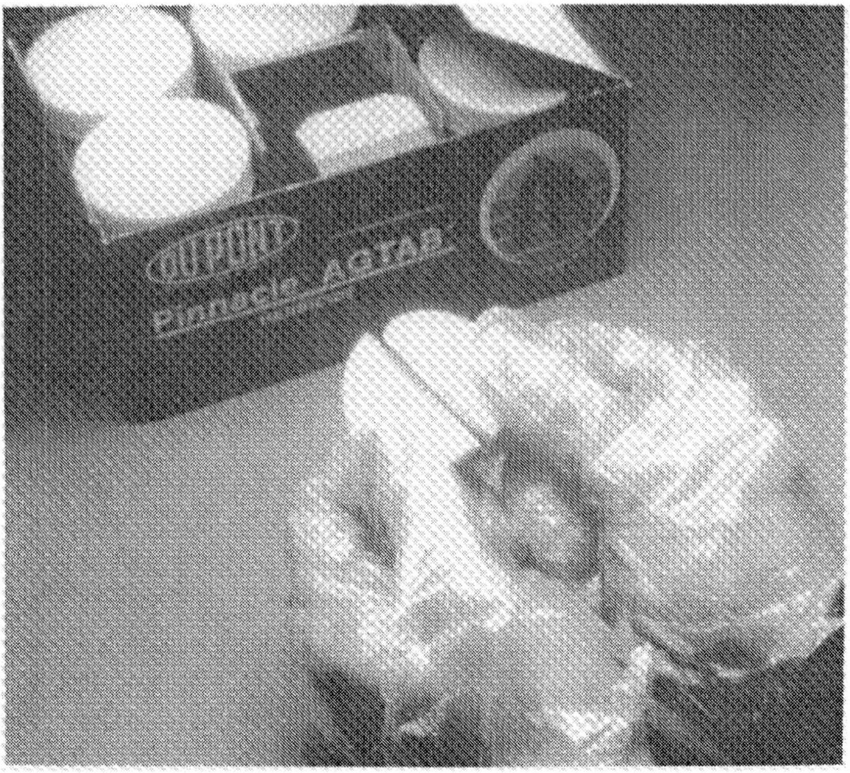

Figure 3 Effervescent tablet. (Courtesy of DuPont Agrichemical Products.)

Wettings agents allow wettable powders to mix with water.
Emulsifiers allow petroleum-based pesticides (ECs) to mix with water.
Invert emulsifiers allow water-based pesticides to mix with petroleum carrier.
Spreaders allow pesticide to form a uniform coating layer over the treated surface.
Stickers allow pesticide to stay on the treated surface.
Penetrants allow pesticide to get through the outer surface to the inside of the treated area.
Foaming agents reduce drift.
Thickeners reduce drift by increasing droplet size and reducing volume of spray contained in drift-prone droplets.
Safeners reduce the toxicity of a pesticide formulation to the pesticide handler or to the treated surface.
Compatibility agents aid in combining pesticides effectively.

Figure 3 Continued.

Buffers allow mixing of pesticides of different acidity or alkalinity.
Antifoaming agents reduce foaming of spray mixtures that require vigorous agitation.

Detailed information on adjuvants is given by Hodgson (1982) and Marer (1988).

III. SPRAY DRIFT MANAGEMENT

Spray drift occurs wherever liquid sprays are applied. Although complete elimination of spray drift is impossible, problems can be reduced to a minimum if chemicals are applied with the proper equipment under favorable weather conditions. Increased awareness of environmental quality and better understanding of the causes of spray drift can help operators make sounder judgments for safer, more efficient applications.

Drift has been defined as the movement of a pesticide through air, during or after application, to a site other than the intended site of application (Holst and Ellvanger, 1984). The affected area may be another crop, a garden, or any other vegetation within range of the pesticide movement. In most cases, this movement is limited to the edges of the field or fence row. However, under certain conditions, off-target movement may affect areas at greater distances from the field. Problems occur when this movement affects a sensitive crop or another person's property.

Drift is often associated with the physical movement of spray droplets away from the target site at the time of application. This type of drift, usually called *airborne drift*, results from factors associated with application methods and equipment. Airborne drift may be simply a spray swath displacement that results in targeted deposition of the spray droplet at a distance not more than 15 m from the point of spray release. Bode (1984) indicates that approximately one-half of the drift deposits occur in the first 8 m downwind and decrease very rapidly with distance. Small spray droplets, however, may travel several thousand meters before depositing into a nontarget area. Smaller airborne droplets may evaporate in the atmosphere and travel for greater distances. The resulting vaporized droplets usually do not create significant problems because they are dispersed over large areas. Airborne drift can be reduced to a minimum with good judgment and proper selection and operation of application equipment.

Drift sometimes occurs even days after the application is made. This type of drift, usually called *vapor drift*, is most commonly associated with volatilization (changing from a liquid to a gas) of pesticides, with subsequent movement from the target area. The vapor drift becomes a significant concern only if the pesticide applied is highly volatile and the atmospheric conditions become suitable for rapid vaporization of the pesticide.

When pesticide drift injury occurs on a sensitive crop, it is difficult to determine the relative contribution of airborne and vapor drift to overall off-target movement. Airborne drift is responsible for most drift damage, but, fortunately, it can be controlled.

Regardless of where it occurs and what causes it, drift is undesirable because

1. It results in inefficient use of application equipment and applicator time.
2. It may cause underapplication of chemicals and ineffective pest control, which lead to additional applications, reduced yield, and higher production costs.
3. It may result in overapplication if the applicator knowingly overapplies chemicals to compensate for drift losses and to ensure the desired level of control.
4. Losses and/or costly litigation may result if sensitive crops in adjacent fields are damaged.
5. Unintentional contamination of foodstuffs from unacceptable pesticide residues can result in mandatory destruction of the crop.
6. It may contribute to pollution of air and water resources.
7. It may affect the health and safety of susceptible human and livestock populations.

Drift is influenced by many factors that usually may be grouped into one of the following four categories: 1) spray characteristics, 2) equipment and application techniques used, 3) weather, and 4) operator care and skill.

A. Droplet Size

Spray droplet size is the most important factor affecting drift. Spray droplet diameters are measured in micrometers (μm). A micrometer is 1 millionth (10^{-6}) of a meter and is usually referred to as a *micron*. For reference, the thickness of a human hair or a sheet of paper is roughly 75 microns. The most widely used parameter of droplet size is *volume median diameter* (VMD), meaning one-half of the spray volume contains droplets smaller than the VMD, while the other half contains droplets larger than the VMD.

Small droplets can drift long distances because of their light weight. Research shows there is a rapid decrease in the drift potential of droplets greater than about 150 or 200 microns. Yates et al. (1985) measured drop size spectra of fan and cone nozzles in a wind tunnel. They concluded that, although drops as large as 400 microns may pose some drift hazard, drops smaller than 150 microns in diameter would generally pose the most serious drift hazard. Droplet size, where drift potential becomes insignificant, depends on wind speeds but lies in the range of 150 to 200 microns for wind speeds of 0.5 to 4 m/s (Bode, 1985).

Zhu et al. (1994) determined the drift distances of water droplets from field sprayers for several variables using a computational fluid dynamics computer program with a series of tests in a wind tunnel. The accuracy of this computer model was verified by Reichard et al. (1992). As shown in Figure 4, all water droplets 70 μm diameter or larger reached 0.5 m below point of discharge and deposited at 50% relative humidity and 20°C temperature. The mean drift distances of these droplets increased with increased wind velocity but decreased as initial droplet size increased. From the data on Figure 4 one can conclude that drift is far less likely to be a problem when spraying with droplets of 200 μm and larger.

Spray particles less then 50 μm in diameter remain suspended in the air indefinitely or until they evaporate. They should be avoided because there is no way to control deposition of these small droplets. Nozzles which produce a relatively high number of small droplets should not be used unless there are no other alternatives. For example, there is no need to use small droplets when applying soil-incorporated or systemic herbicides. However, small to medium size droplets are desirable when applying insecticides and fungicides because they provide better penetration into the canopy and better coverage. Good coverage is essential for insecticides and fungicides because of the small size of the target organism.

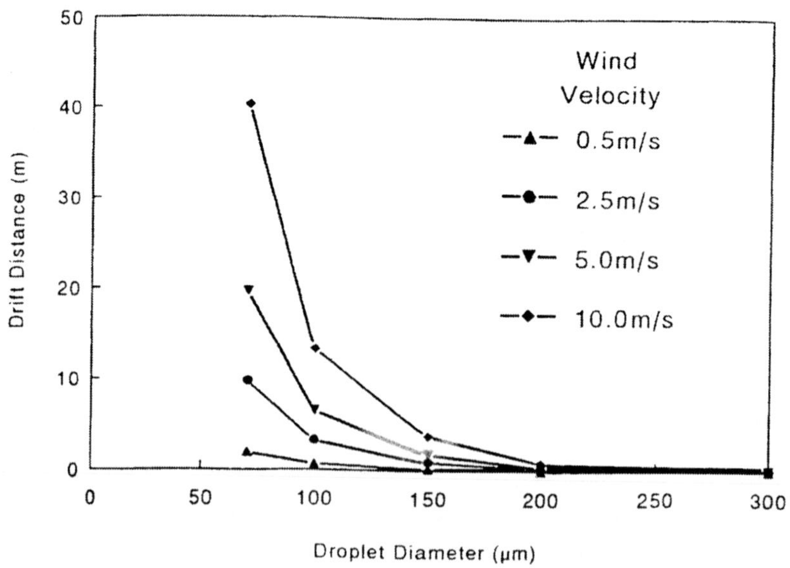

Figure 4 Effect of droplet diameter and wind velocity on mean drift distances of water droplets directed downward with 20 m/s velocity toward a target 0.5 m below point of discharge. Temperature, 20°C; relative humidity, 50%. (From Heping et al., 1994.)

Air-blast sprayers, which are commonly used to apply insecticides and fungicides in orchards, produce a relatively high number of small droplets. However, some of the droplets never reach their target because they are directed upward where there is no canopy. Therefore, spray drift is a more serious problem with air-blast sprayers. An option to decrease drift with these sprayers is to install a baffled extension manifold that directs the spray toward the tree canopy.

Most pesticide applications should attain a balance between drift reduction provided by large droplets and good coverage provided by small droplets. Generally, spray-droplet size should be no finer than necessary to do an effective job. Recommended droplet sizes for fungicides, insecticides, and herbicides are 150–200, 200–300, and 250–400 μm, respectively.

B. Nozzle Type and Size

Most current agricultural sprayers use hydraulic nozzles to meter and atomize the liquid into drops (Figure 5). Hydraulic nozzles produce a wide spectrum of droplet sizes ranging from less than 10 to more than 1000 μm.

Drift can be minimized with nozzles that produce relatively large droplets but still give sufficient penetration and coverage of the target. For example, flood-fan and wide-angle full-cone nozzles produce much fewer drift-prone

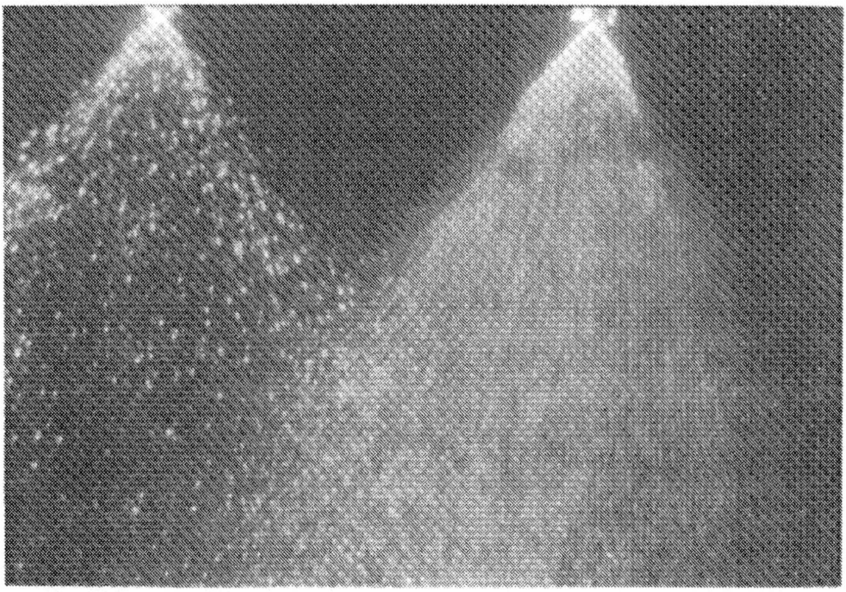

Figure 5 Nozzles with large orifices generally produce larger droplets.

droplets than hollow-cone nozzles and flat-fan nozzles. However, flood-fan and full-cone nozzles do not provide adequate control when applying contact herbicides, insecticides, and fungicides. We are likely to continue using other nozzles for application of these pesticides even though they increase the risk of drift. Fortunately, nozzle manufacturers have developed new "low-drift" fan-pattern and cone nozzles that are capable of producing relatively large droplets with adequate spray coverage. For example, drift experiments conducted by Ozkan et al. (1993) in a wind tunnel showed that, using the same spray mixture, drift deposits beyond 0.5 m were 34.6%, 15.3%, and 24.7% less from a nozzle which is designed to reduce drift (RF4) than those from a conventional fan-pattern nozzle (XR8004VH) at 2.0, 3.0, and 4.0 m/s wind velocities, respectively.

As the size of the nozzle opening increases, droplet size generally increases while the percentage of spray volume in small droplets decreases. As shown in Table 2, the percentage of spray volume contained in droplets smaller than 100 μm varied from 23.1% to 10.0% as the nominal flow rate of the nozzle increased from 0.8 to 3.0 L/min, respectively. Therefore, from the drift standpoint, it is safer to use nozzles with higher flow rates. However, smaller-capacity nozzles are often preferred because they require less volume of spray per hectare.

C. Nozzle Orientation

Orientation of nozzles is not critical for ground applications but plays an important role in reducing drift from aircraft applications. When a nozzle is pointed backward toward the tail of the aircraft, larger droplets are produced. The same nozzle produces medium droplets when pointed downward and smaller droplets when pointed into the airstream.

Table 2 Percentage of Spray Volume Contained in Droplets Smaller Than 100 μm from Hydraulic Fan-Pattern Nozzles

Nozzle ID[a] No.	Nozzle flow rate (L/min)	% Spray volume in droplets < 100 μm
XR8002VS	0.8	23.1
XR8004VS	1.5	15.6
XR8006VS	2.3	12.8
XR8008VS	3.0	10.0

[a]All nozzles have stainless steel tips and are manufactured by Spraying Systems Co.

D. Spray Height

Small spray droplets have low inertial energy, making them highly susceptible to drift. In addition, wind velocity is usually greater as height above the ground increases. Therefore, the closer the nozzle tips are to the ground, the smaller the likelihood of spray pesticide drift. However, lowering the spray boom without making adjustments in nozzle spacing may result in improper spray overlap, which will cause streaks of untreated areas, especially with flat-fan nozzles. This problem can be eliminated by using nozzles with a wider fan angle. However, nozzles with wider fan angles produce smaller drops than nozzles with narrow fan angles when operated at the same pressure and flow rate.

Computer simulations by Zhu et al. (1994) indicated that increased droplet discharge height resulted in increased drift distances. The mean drift distances of 50-μm-diameter and smaller water droplets were nearly constant with each wind velocity for the discharge height range of 0.25 to 3.0 m. This occurs because these droplets have short lifetimes and do not travel downward far enough to deposit before completely evaporating. However, as illustrated on Figure 6, increasing discharge height of 100 μm diameter droplets signifi-

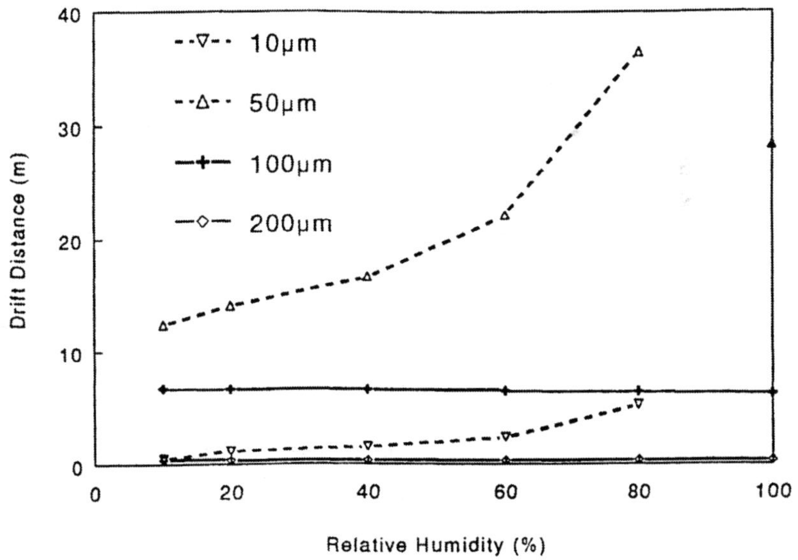

Figure 6 Effect of discharge height and droplet diameter on mean drift distances of water droplets directed downward with 20 m/s velocity through wind with velocity of 5.0 m/s at 20°C temperature and 50% relative humidity. *Dash line*, droplets completely evaporated before deposition. (From Heping et al., 1994.)

cantly affects their mean drift distances. Changes in discharge heights have less effect on mean drift distances as droplet size increases above 200 μm diameter.

E. Spray Pressure

Spray pressure exerted on hydraulic nozzles provides the energy that breaks a stream of water into droplets, and thus is a key factor in regulating spray-droplet size. Increased pressure generally causes a decrease in droplet size. One study of a flat-fan nozzle with 2.3-L/min nominal flow rate demonstrated that increasing the pressure from 1 kPa to 2 kPa increased the volume of droplets in the 100-μm and smaller range from 6% to 12%. Reducing pressure will reduce drift since larger droplets are formed, but operating nozzles below the recommended pressure will also reduce effective coverage and may result in poor distribution and improper spray overlap. The recommended pressure range for conventional flat-fan nozzles is about 1.4 to 2.4 kPa. However, nozzle manufacturers have introduced nozzles that can be used at pressures as low as 0.7 kPa with little reduction in coverage quality.

F. Chemical Formulation

The type of pesticide formulation is important in the reduction of drift. Some pesticide formulations are more volatile than others. For example, 2,4-D or mecoprop esters are susceptible to vapor drift, while 2,4-D or mecroprop amines are practically nonvolatile. Therefore, formulations with low volatility should be used to reduce vapor drift.

A courser spray can be achieved by increasing the viscosity of the spray mix. Increased viscosity results in an upward shift of the droplet spectrum so that there are fewer of the fine drops that are subject to drift. Many spray adjuvants can be added to a tank mix to increase its viscosity. In addition to drift control, most of these adjuvants provide spreading characteristics on the leaf surface and improve retention of pesticides on leaves. These products do not completely eliminate drift; rather they reduce the number of drift-prone droplets. To achieve satisfactory results from drift retardants, they should be mixed and applied according to label directions.

There are many commercially available drift control additives at varying prices. There have been some concerns about basic spray characteristics of nozzles. Ozkan et al. (1993) conducted a study to determine the effects of various drift retardant chemicals (Direct, Driftgard, Formula 358, Nalco-trol, and Target) on droplet size, spray pattern, and spray drift reduction. The results of this study indicated that, at recommended rates, Nalco-trol provided the largest increase in volume median diameters of droplets (63%) followed by Direct, Target, Driftgard, and Formula 358 with 59.8%, 38.9%, 28.6%, and 3.5%, respectively.

In general all drift retardants created patterns with greater volume toward the center than that of water only. This is likely to create uneven spray distribution across the spray pattern of a sprayer boom equipped with fan-pattern nozzles. Adjusting the boom height and/or nozzle spacing may eliminate this problem.

Tests conducted in a wind tunnel showed that cumulative deposits beyond 0.5 m downwind increased for all drift retardants when the wind velocity was increased from 2.0 to 4.0 m/s. The least amount of downwind deposit beyond 0.5 m at 4.0 m/s was with Nalco-trol, followed by Target, Direct, Driftgard, and Formula 358. In general, a comparison of the spray deposits from mixtures containing drift retardants on horizontal and vertical targets indicated that Nalco-trol and Target were the most effective in reducing spray drift, followed by Direct, Driftgard, and Formula 358. This order was similar to the results of tests conducted by Ozkan et al. (1992) to determine the reduction in spray volume in droplets less than 100, 150, and 200 μm for spray mixtures containing water with drift retardants. As indicated by the data presented in Table 3, in comparison to spraying water only, all drift retardants reduced portion of small droplets in the spray but at varying magnitudes. For instance, Nalco-trol had the highest reduction (67.8%) in the portion of spray volume in droplets less than 100 μm, followed by Target, Direct, Driftgard, and Formula 358 with reductions of 64.7%, 62.5%, 44.2%, and 30.4%, respectively.

G. Evaporation

With water carriers, spray droplets become smaller as they fall because of evaporation. Elliot and Wilson (1983) indicate that

> spray drops evaporate relatively slowly down to about 150 μm. As drops become smaller, the rapidly increasing ratio of surface area to volume results in an accelerating evaporation rate, so by the time a water drop has

Table 3 The Reduction in Spray Volume in Droplets Less Than 100, 150, and 200 mm for Spray Mixtures Containing Water with Drift Retardants Compared to Water Only

Liquid sprayed	% Reduction in spray volume contained in droplets		
	<100 μm	< 150 μm	<200 μm
Formula 358	30.4	28.3	23.1
Driftgard	44.2	40.9	34.5
Direct	62.5	63.2	55.8
Target	64.7	67.8	61.4
Nalco-trol	67.8	69.3	62.6

> reached 50 μm, it will evaporate to extinction before it has fallen a further 10–20 cm. For example, an initial 217 μm drop under the 59% relative humidity conditions will fall 23 m before evaporating to 111 μm, but needs only a further 3 m before becoming extinct.

For typical ground applications, droplets of 30 μm and less will completely evaporate to a residual core of pesticide before reaching the target. Droplets greater than 150 μm will have no significant reduction in size before deposition on the target. Evaporation of droplets between 30 and 150 μm is significantly affected by temperature, humidity, and other weather conditions (Bode, 1984).

H. Weather

Weather can critically influence the off-target movement of pesticides. Several factors associated with the microclimate at the application site can contribute to drift, including 1) wind velocity and direction, 2) relative humidity and temperature, and 3) atmospheric stability and inversions.

The impact that various weather factors have on drift can be related to the volume of droplets 150 μm or less in diameter. If these small droplets are eliminated, weather's effect on drift is reduced to a minimum.

We have very little control over the weather, but its effect on drift can be controlled. Factors such as droplet size and downward velocity, air turbulence, and spray boom height affect the distance a droplet travels before depositing on an object. However, *wind velocity* is usually the most critical factor of all meterological conditions affecting drift. The greater the wind speed, the farther off-target a droplet of a given size will be carried. The larger the droplet, the less it is affected by the wind and the faster it falls. High winds can cause even larger droplets to move off-target. Therefore, spraying operations should be stopped if wind speeds are excessively high.

Figure 7 illustrates the simulated effect of wind velocities up to 10 m/s on the mean drift distances for 70-, 100-, and 200-μ diameter water droplets at 50% relative humidity (Heping et al., 1994). Figure 7 shows that there is a nearly linear relationship between mean drift distances and wind velocity for each droplet size. The rate of change in drift distance with change in wind velocity is much greater with smaller than with larger diameter droplets.

Wind direction is as important as wind velocity in reducing the damage caused by drift. The presence of sensitive vegetation near the spray site, particularly downwind, is one of the first factors that should be evaluated but is often overlooked when beginning a spray application. Check in all directions from a spray site for sensitive vegetation, and be alert to changes in wind direction during application. If there are sensitive crops downwind, leave a buffer

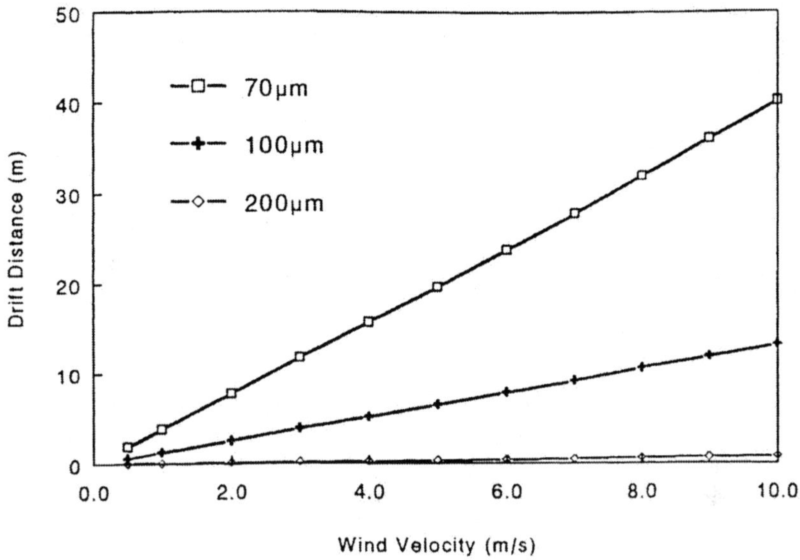

Figure 7 Effect of wind velocity and droplet diameter on mean drift distances of water droplets directed downward with 20 m/s velocity toward a target 0.5 m below point of discharge. Temperature, 20°C; relative humidity, 50%. (From Heping et al., 1994.)

strip of at least 40 meter or whatever the label recommends. Spray the buffer strip later, when the wind has shifted away from sensitive crops.

Relative humidity and temperature go hand-in-hand in affecting spray drift. While they generally are not as critical as wind velocity, they have a strong influence in some geographic regions or under certain meterological conditions. As a particle falls through the air, surface molecules of water evaporate into the atmosphere. This evaporation reduces the size and mass of the particle, enabling it to remain airborne longer and, under the right conditions, to drift farther from the application site. The rate at which water evaporates from the spray particles depends primarily on the ambient air temperature and relative humidity.

While evaporative loss of spray materials occurs under almost all atmospheric conditions, these losses are less pronounced under the environmental conditions that occur in the cooler parts of the day—early morning and late afternoon. The relative humidity is usually highest during these cool periods.

Temperature influences drift in other ways beyond its effect on evaporative loss of droplets. Pesticide volatility increases with increasing temperature. Temperature also influences atmospheric air turbulence, stability, and inversions—all of which, in turn, affect drift.

Figure 8 illustrates the simulated mean drift distances for 50-, 100-, and 200-μm-diameter water droplets with 5 m/s wind velocity, 50% relative humidity (RH), and ambient temperature of 10°, 20°, and 30°C (Heping et al., 1994). The curve for 50-μm droplets shows that the droplets completely evaporated before deposition, and the evaporation distance decreased as temperature increased. The curve for 100-μm-diameter droplets shows that drift distance before deposition increased with increased temperature. Temperature over the range of 10° to 30°C had little influence on drift distances of 200-μm diameter-droplets.

Pesticides are applied over a wide range of relative humidity (RH). Data from the computer simulation study conducted by Heping et al. (1994) indicate that all water droplets 50 μm diameter and smaller completely evaporated before they reached 0.5 m below the point of discharge with the ambient temperature and RH ranges of 20°–30°C and 20%–80%, respectively. With 5.0 m/s wind velocity and 20°C temperature, 50-μm-diameter water droplets drifted 15.0 and 36.5 m before completely evaporating when relative humidities were 20% and 80%, respectively. Figure 9 illustrates the effect of relative humidity on mean drift distances of various size water droplets directed downward with 20 m/s velocity toward a target 0.5 m below the discharge point for 5.0

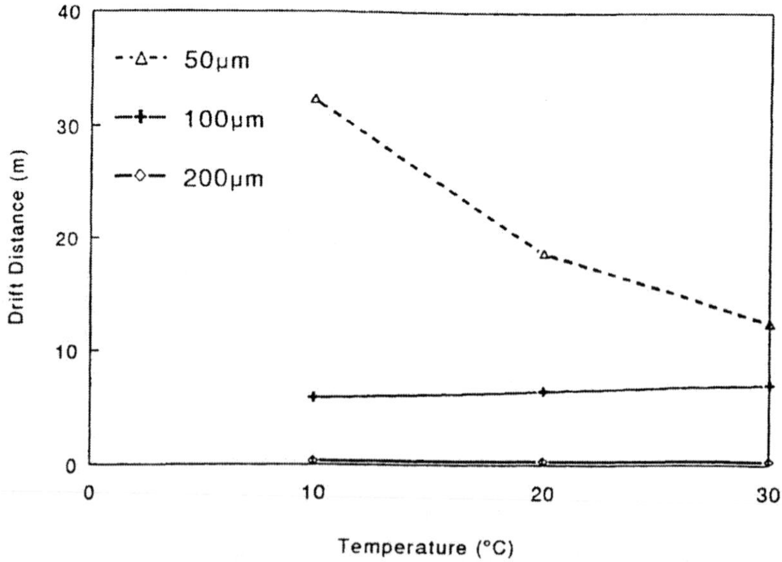

Figure 8 Effect of temperature and droplet diameter on mean drift distances of water droplets directed downward with 20 m/s velocity toward a target of 5.0 m below discharge point and through wind with velocity of 5.0 m/s at 50% relative humidity. Dash line, droplets completely evaporated before deposition. (From Heping et al., 1994).

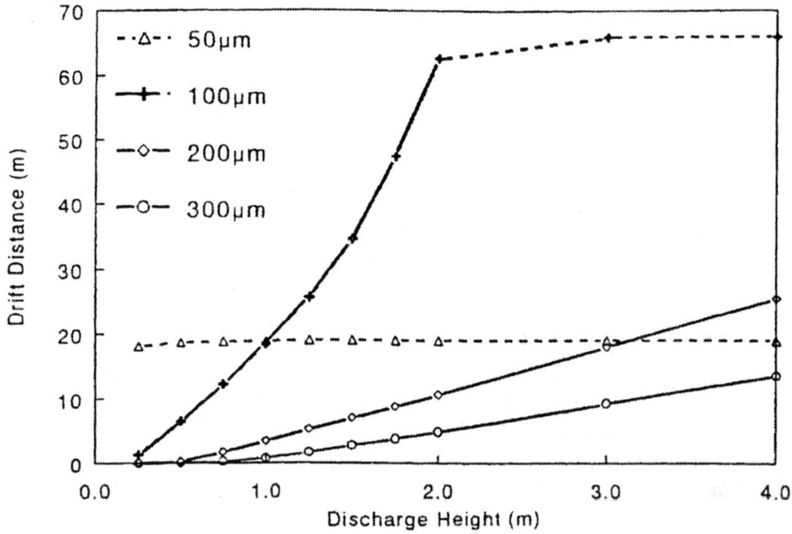

Figure 9 Effect of relative humidity and droplet diameter on mean drift distances of water droplets directed downward with 20 m/s velocity toward a target 0.5 m below point of discharge and through wind with velocity of 5.0 m/s at 20°C temperature. Dash line, droplets complete evaporated before deposition. (From Heping et al., 1994.)

m/s wind velocity and 20°C ambient temperature. As shown in Figure 9, while the RH affects drift distances of small droplets, there is no change in drift distance of 200-μm water droplets over the 10% to 80% range of RH.

Atmospheric stability is an important factor influencing drift. Under normal meterological conditions, the air temperature decreases by approximately 5°C per 300 m of height. Cool air tends to sink, displacing lower warm air and causing vertical mixing. As a warm air layer rises, suspended droplets rise with it and dissipate into the upper layers by normal air turbulence and vertical mixing. Under these conditions, the opportunity for crop injury at any specific off-target size is very small because the pesticide is dispersed and diluted into the atmosphere.

However, other problems may arise when the atmosphere is very stable. Under stable conditions, a warm air layer at some distance overhead may become a blanket, holding down cooler air underneath. This phenomenon is usually referred to as atmospheric inversion and can be recognized by observing a column of smoke (Figure 10). Particles suspended in the cool layer cannot move anywhere except laterally, possibly for several miles. Eventually, the suspended cloud may encounter a downdraft, forcing it back to earth and depositing it off-target, possibly over a sensitive crop.

Figure 10 Atmospheric inversion usually occurs early in the morning.

Inversions are part of a daily atmospheric cycle, occurring in the early-morning hours when the ground cools the air layer immediately above it. Inversions tend to dissipate during the middle of the day, when wind currents mix the air layers. Applicators concerned about these adverse spray conditions should wait until late afternoon or early evening to spray, when there is less chance of the atmosphere's being inverted and conditions are more favorable. Again, the best way to avoid drift associated with atmospheric inversions is to eliminate the formation of small particles (150 μm or smaller) from the spray effluent. If these particles are not in the spray emission, this weather-related phenomenon can be ignored.

I. Operator Skill and Care

Under a given spray situation, *any one* of the previously mentioned factors may be the most critical in reducing drift hazards. It is the applicator who determines this critical factor and takes precautions against it. By exercising good judgment regarding both equipment and weather factors relative to each application, applicators can minimize drift potential in nearly every case.

J. Equipment Modifications

Spray droplets from conventional sprayers deposit mainly on upper sides of horizontal surfaces because of gravity or on vertical surfaces by their velocity and movement in air currents around the target. Scientists and equipment companies have been exploring the possible use of other forces to increase application efficiency while reducing spray drift. Some new developments for increasing deposition efficiency of especially small droplets include partially or completely covered booms, air-assisted spraying, and electrostatic spraying.

The *shielded sprayer boom* has been shown to be helpful in reducing spray drift. Results from laboratory tests conducted at the University of Missouri (Smith et al., 1982) indicated that a mechanical shield could reduce spray drift deposit by up to 70%. Tests conducted under field conditions generally showed reduction in drift deposits of up to 65% with the shield. However, in 3 of the 11 tests, there was an increase in drift deposits with the shield. The researchers indicated that operating conditions, prevailing wind velocity, and size and configuration of the shield greatly influence the reduction of drift deposits. For example, a shield was most beneficial at the highest wind-speed/ground-speed combinations tested.

Shrouded (or hooded) booms have been shown to reduce drift. Some companies have developed simple shrouds (Figure 11), while others have improved the aerodynamics of airflow around the shield by placing an airfoil over the shield. Another advantage of a covered shield is that pesticides can be applied using small droplets that provide better coverage, and the volume of carrier needed can be reduced significantly. One drawback of a completely covered

Figure 11 Sprayers with shrouded booms significantly reduce spray drift. (Courtesy of Blumhardt Equipment Co.)

boom is that it restricts the operator's visual ability to monitor the nozzles for plugging. However, electronic monitors are being developed to detect changes in spray pattern. For instance, one company has developed a detector that continuously monitors the sound of the spray at each nozzle and warns the operator if nozzle sound levels vary from an ideal sound measurement taken earlier.

Air-assisted spraying has been shown to reduce drift deposits from boom sprayers (Almekinders et al., 1993; Quanquin, 1992; Walklate, 1991; Western and Hoslop, 1991; Young, 1991). The general concept with air-assisted spraying is to use air to replace part or all of the water as a pesticide carrier. Some air-assisted systems simply atomize the spray solution, while others use high-velocity airflow to transport the spray mixture to the target. This enables the applicator to use small droplets without increasing the risk of drift. Several companies manufacture and market sprayers that use a stream of air to direct the spray to the target by entraining droplets in a downward-moving air curtain. Figure 12 shows drift potential from a sprayer with and without the air-assist system. Air-assist spray systems generally improve penetration of spray material into the canopy of plant foliage and improve the coverage on the undersides of leaves.

Figure 12 Spray penetration efficiency from two identical sprayers: one utilizing the air-assist system (*right*) and the other one (*left*) with the air-assist system turned off. (Courtesy of Hardi, Inc.)

Electrostatic charging of small droplets has been considered as a possible way to increase the deposition of small droplets. An electrical field between the nozzle and the plant leaf is generated by charging droplets. The attraction between small, charged droplets and the plant canopy is expected to provide the additional force to move droplets toward the plants and resist drift. Although several studies with electrostatic spraying indicated reductions in drift deposits of up to 40%, electrostatic atomizers have not gained wide-scale acceptance mainly because of their high cost (Almekinders et al., 1992; Western and Hislop, 1991; Hislop, 1988; Franz et al., 1987; Sharp, 1984).

K. Management Strategies to Reduce Drift

Conscientious and experienced operators rarely get into serious trouble with drift damage because they understand drift and take steps to avoid it. Follow these management strategies to reduce drift:

1. Use nozzles that produce coarser droplets when applying pesticides on targets that do not require small, uniformly distributed droplets (e.g., soil-incorporated and systemic herbicides).
2. Keep the boom closer to the spray target.
3. Use nozzles with narrower spray-fan angles.
4. Keep spray volume up, and use nozzles with larger orifices.
5. Keep spray pressure down, and make sure pressure gauges are accurate.
6. Use spray additives and thickeners if necessary.
7. Follow label recommendations to prevent drift with highly volatile pesticides.
8. Avoid spraying on extremely hot and dry days, especially if sensitive vegetation is nearby.
9. Do not spray when conditions are favorable for an atmospheric inversion.
10. Do not spray when wind speeds are high enough to cause drift.
11. Avoid spraying near sensitive crops that are downwind. Leave a buffer strip of 50 to 100 ft, and spray the strip later when the wind shifts.
12. Be sure to keep good records and evaluate the results as efficient operators do for all applications.

Drift is undesirable for economic, environmental, and safety reasons. Efficient applicators do not spend money for pesticides to watch them drift away from their target fields. Today's chemicals are more potent and require more precise application. Unsatisfactory pest control can result if a significant portion of the chemical is lost in drift. This may require respraying the same field. You may even find yourself in court if spray drift damages sensitive crops in a neighbor's field.

The environmental effects of spray drift are equally costly and unacceptable. By reducing drift to a minimum, you can reduce the potential for pollution of streams, lakes, and other water supplies that could endanger fish and wildlife.

Regardless of how accurately an application is made, the possibility of drift is always present. You can minimize this possibility by selecting the right equipment and using sound judgment when applying pesticides. Your judgment can mean the difference between an efficient, economical application and one that results in drift, damaging nontarget crops and creating environmental pollution. Reducing spray drift not only improves application efficiency, but also reduces the risk of safety- and health-related problems caused by drift. Because it is impossible to eliminate drift altogether, always wear protective clothing when applying pesticides. A respirator is a must, especially if your tractor does not have a cab. If you have any doubts about a spraying job that might result in drift, wait until you no longer have that element of doubt. Your goal should be to eliminate off-target movement of pesticides, no matter how small it may be.

ACKNOWLEDGMENT

The author acknowledges that some of the information presented in this chapter related to herbicide formulations was adapted from *Applying Pesticides Correctly*, a study manual for private and commercial applicators, Ohio State University Extension Publication. The principal author is S. A. McDonald. The author thanks D. C. Yagich for typing the manuscript.

REFERENCES

Almekinders, H., Ozkan, H. E., Reichard, D. L., Carpenter, T. G., and Brazee, R. D. (1992). Spray deposit patterns of an electrostatic atomizer, *Trans. ASAE* 35(5): 1361–1367.

Almekinders, H., Ozkan, H. E., Reichard, D. L., Carpenter, T. G., and Brazee, R. D. (1993). Deposition efficiency of air-assisted charged sprays in a wind tunnel, *Trans. ASAE* 36(2): 321–325.

Bode, L. E. (1984). "Downwind drift deposits by ground applications. Proceedings of the Pesticide Drift Management Symposium," South Dakota State University, Brookings, p. 50.

Elliot, J. G. and Wilson, B. J., eds. (1983). *The Influence of Weather on the Efficiency and Safety of Pesticide Application*, BCPC Publications, Occasional Publication No. 3, Croydon, U.K., p. 32.

Franz, E., Reichard, D. L., Carpenter, T. G., and Brazee, R. D. (1987). Deposition and effectiveness of charged sprays for pest control, *Trans. ASAE* 30(1): 50–55.

Hodgson, R. H., ed. (1982). *Adjuvants for Herbicides*, Weed Science Society of America, Champaign, Illinois.

Holst, R. V. and Ellvanger, T. G. (1984). *Pesticide Assessment Guidelines*, Subdivision R, Pesticide Drift Evaluation. EPA Report No. EPA-540/9-84-002, NTIS Report No. PB84-189216, Washington, D.C.

Marer, P. J. (1988). *The Safe and Effective Use of Pesticides*, University of California, Division of Agriculture and Natural Resources Publication 3324, Oakland, pp. 96–107.

McDonald, S. A. (1980). *Applying Pesticides Correctly: A Guide for Private and Commercial Applicators*, North Carolina Extension Service Publication, North Carolina State University, Raleigh, pp. 24–28.

Ozkan, H. E., Reichard, D. L., Zhu, H., and Ackerman, K. D. (1992). *Effect of Drift Retardant Chemicals on Spray Drift, Droplet Size and Spray Pattern*, ASAE Paper No. 921613, American Society of Agricultural Engineers, St. Joseph, Missouri.

Ozkan, H. E., Reichard, D. L., Zhu, H, and Ackerman, K. D. (1993). Effect of drift retardant chemicals on spray drift, droplet size, and spray pattern, in *Pesticide Formulations and Application Systems, Vol. 13*, ASTM STP 1183 (P. D. Berger, B. N. Devisety, and F. R. Hall, eds.), American Society for Testing and Materials, Philadelphia, pp. 173–190.

Quanquin, B. J. (1992). *Less Drift, More Target with the Twin System*, ASAE Paper No. 921574, American Society of Agricultural Engineers, St. Joseph, Missouri.

Sharp, R. B. (1994). Comparison of drift from charged and uncharged hydraulic nozzles, *Proc. BCPC Pests and Diseases 3*: 1027–1031.

Smith, D. B., Harris, F. D., and Goering, C. E. (1982). Variables affecting drift from ground sprayers, *Trans. ASAE* 25(6): 1499–1503.

Van Valkenburg, J. W. ed. (1973). *Pesticide Formulations*, Marcel Dekker, Inc., New York.

Van Valkenburg, J. W. (1982). Terminology, classification and chemistry, in *Adjuvants for Herbicides* (R. H. Hodgson, eds.), Weed Science Society of America, Champaign, Illinois, p. 2.

Walklate, P. J. (1991). Pesticide drift from air-assisted orchard sprayers: a numerical simulation study, in *Air-Assisted Spraying in Crop Protection* (A. Lavers, P. Herrington, and E. S. E. Southcombe, eds.), BCPC Monograph No. 46, BCPC, Farnham, Surrey, U.K., pp. 61–68.

Ware, G. W. (1988). *Complete Guide to Pest Control*, Thompson Publications, Fresno, California, pp. 17–22.

Western, N. M. and Hislop E. C. (1991). Drift of charged and uncharged spray droplets from an experimental air-assisted sprayer, in *Air-Assisted Spraying in Crop Protection* (A. Lavers, P. Herrington and E. S. E. Southcombe, eds.), BCPC Monograph No. 46, BCPC, Farnham, Surrey, U.K. pp. 69–76.

Yates, W. E., Cowden, R. E., and Akesson, N. B. (1986). Drop size spectra from nozzles in high-speed airstream, *Trans. ASAE* 28(2): 405–410.

Young, B. W. (1991). A method for assessing the drift potential of hydraulic nozzle spray clouds, and the effect of air assistance, in *Air-Assisted Spraying in Crop*

Protection (A. Lavers, P. Herrington and E. S. E. Southcombe, eds.), BCPC Monograph No. 46, BCPC, Farnham, Surrey, U.K., pp. 77-86.

Zhu, H., Reichard, D. L., Fox, R. D., Brazee, R. D., and Ozkan, H. E. (1994). Simulation of drift of discrete sizes of water droplets from field sprayers, *Trans. ASAE* (in press).

8

Fate of Herbicides in the Environment

Patrick L. Havens
DowElanco, Indianapolis, Indiana

Gerald K. Sims
U.S. Department of Agriculture, Urbana, Illinois

Susan Erhardt-Zabik
DowElanco, Indianapolis, Indiana

I. INTRODUCTION

As soon as a herbicide is applied to its target, a number of processes immediately begin to remove the compound from the original site of application. This is the process of environmental fate. For the herbicide which is intercepted by plants, the chemical may be taken up by the plant itself (a process not addressed in this review), may be washed off by precipitation onto the soil, may undergo photodegradation on plant surfaces, or may volatilize back into the air (Figure 1). The herbicide which falls directly upon the soil or is washed onto the soil can undergo myriad processes which may be broken down into two main groupings: degradation and transport processes. Degradation processes include biological degradation by soil organisms and abiotic chemical and photochemical transformations. Transport of herbicides within the soil compartment can occur downward into the soil profile (leaching), across the soil surface (runoff), or into the air (volatilization). Each can be a combination of more fundamental processes including adsorption, convection, and diffusion.

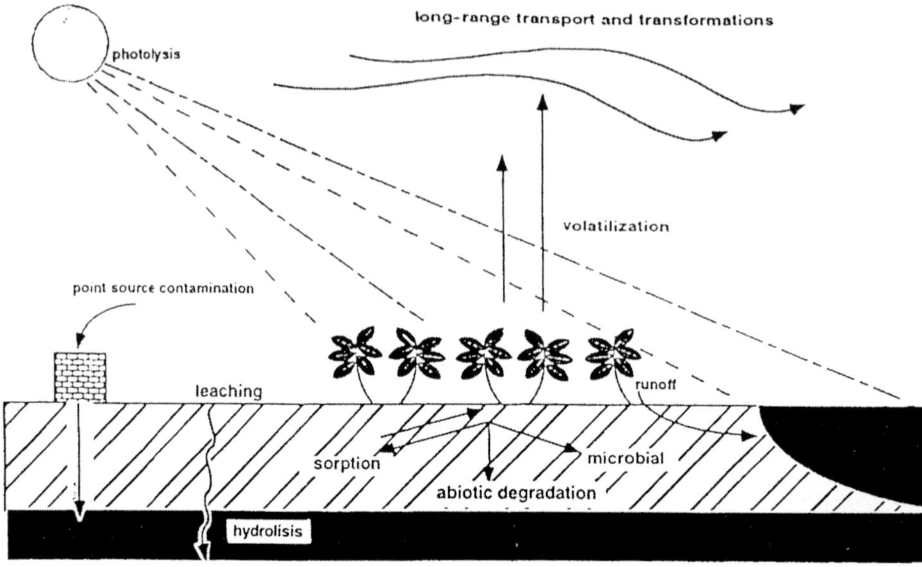

Figure 1 Potential routes of dissipation and degradation for herbicides. Note that each route is not independent of the others but forms an integrated whole.

Most herbicides are organic compounds and are therefore basically unstable in the environment. Inherent instability is essential to prevent these materials from accumulating in the environment as compounds are repeatedly used. Accumulation of pesticides not only poses environmental hazards, but prohibits rotation to sensitive crops. Unfortunately, not all herbicides exhibit optimal stability properties, and therefore numerous examples of insufficient control (too little stability) or carryover (too much stability) have been reported. Furthermore, persistence in the environment prolongs exposure of the materials to forces that can cause movement of the herbicide away from the application site. The extent of transport, with the exception of spray drift, is a function of both the stability and the physicochemical characteristics of the herbicide in the environment to which it is applied. The discussion presented in this chapter describes both transport and degradation processes and is meant as an introduction and not a comprehensive discussion for the reader. Suggested additional readings that provide in-depth discussions of these processes are included for the reader's information at the end of the chapter.

II. BIODEGRADATION

The environmental stability of a herbicide is strongly dependent on its tendency to decompose into an inactive or nontoxic form (or in some cases a more active or toxic form) as a result of chemical reactions that may involve living organisms. Biological reactions generally involve bacteria and/or fungi. Reactions not involving living organisms are abiotic and may involve light or catalytic surfaces such as soil clays. Some organisms (mostly fungi) are capable of producing catalytic proteins outside the cell (extracellular enzymes) that can partially degrade certain pesticides or incorporate them into large humuslike macromolecules. There is some argument as to whether such reactions should be considered biological or abiotic. Generally speaking, most abiotic reactions, with the exception of those induced by light in certain compounds, result in only partial modification of herbicides. Under natural conditions, complete degradation of pesticides to mineral constituents such as carbon dioxide, chloride, or phosphate nearly always requires biological degradation. However, inactivation or detoxification of a pesticide may require only a single reaction and often occurs via abiotic mechanisms. Whether inactivation or detoxification is adequate to remove risk associated with herbicides has been a subject of much debate; however, most agree that persistence of an inactive form of a herbicide poses less risk than persistence of the active form.

A. Nomenclature

Extensive and often confusing terminology has been developed to describe the environmental fate of organic compounds such as herbicides. Terminology becomes especially perplexing when biological transformations are being described. Some scientists have strong opinions about the precise usage of such terminology and others use many terms interchangeably, and thus the reader is often in the position of interpreting the intent of the author. For the benefit of the reader, we will discuss a few of the commonly used terms in the pesticide fate literature. *Degradation* generally refers to the process of chemical modification of pesticides by environmental forces, and sometimes man-made forces, whether those forces are biological or abiotic. The mechanism is often used as a descriptor (biological degradation, biodegradation, photochemical degradation, etc.). *Decomposition* is often used interchangeably with *degradation*, as is *dissipation*, although many use the latter term to denote the process of removal of the pesticide from the environment, without regard to whether the material is degraded in the process. Usually, the terms *degradation* and *decomposition* are used to imply modifications leading toward destabilization and ultimately mineralization (defined later) of a pesticide. (Occasionally, stabilizing reactions, such as incorporation into organic polymers, have been referred to as degradation.) *Inactivation* typically refers to the loss of pesticidal activity, through either structural modification of the pesticide or transfer to some unavailable compartment (such as irreversible sorption to soil colloids). *Detoxification* may refer to inactivation or more commonly indicate loss of ecological toxicity of a pesticide. Finally, *transformation* denotes a change in chemical structure of a pesticide (often used with a mechanistic modifier such as *biotransformation*), whereas *mineralization* refers to complete degradation to mineral constituents described previously. *Transformation* is often used in place of *degradation* or *decomposition* to describe minor modifications or reactions that do not lead to mineralization.

B. Role of Chemical Properties

Biological degradation of herbicides for the most part involves recognition of the herbicide by one or more enzymes. Complete mineralization of a herbicide to carbon dioxide may require the cooperation of more than one type of organism. It is important to remember that nearly all herbicides are *xenobiotic* substances: that is, they are foreign to biological systems. Since they are *anthropogenic* (man-made) and novel, they were not present during the evolution of present-day organisms. Recognition of a xenobiotic substance, such as a herbicide by an enzyme, requires either a certain sloppiness of enzymes (nonspecificity) or some adaptation of the enzymes due to mutation. It is clear that some enzymes, particularly those capable of hydroxylating aromatic rings, may recognize xenobiotic substances without prior exposure to the substance of the organisms that produces the enzymes. Peroxidases, phenol oxidases, and esterases from a variety of sources can transform certain herbicides, demonstrating that mutation is not necessary for recognition of xenobiotic substances by biological systems. In the environment, some substances are more rapidly degraded with each passing season, probably as a result of some type of microbial adaptation, suggesting that mutations important to pesticide degradation do in fact occur in nature. Whether or not organisms must change in order to degrade pesticides, it is unlikely that mutations resulting from natural selections will often result in rapid degradation of extremely resistant chemical structures. Unless they are particularly toxic to microorganisms, extremely resistant structures (which typically do not produce microbial growth) do not provide adequate selective pressure to give an advantage to mutants capable of degrading them. Thus mutants are not amplified by selective pressure and therefore may disappear without consequence.

The fact that huge differences in degradation rates exist among herbicides indicates either that some structures are easier for organisms to degrade or that the genetic information for degradation of some structures is not widely distributed among actively growing organisms. Some examples of persistent and biodegradable functional groups are shown in Table 1. More specifically, the addition of a chlorine to position 5 of the aromatic ring of 2,4-D, to produce 2,4,5-T results in much slower degradation of the phenoxy herbicide ring (Figure 2A). Similarly, methoxychlor, a relatively short-lived insecticide, differs from DDT, an extremely long-lived insecticide, only in the presence of two

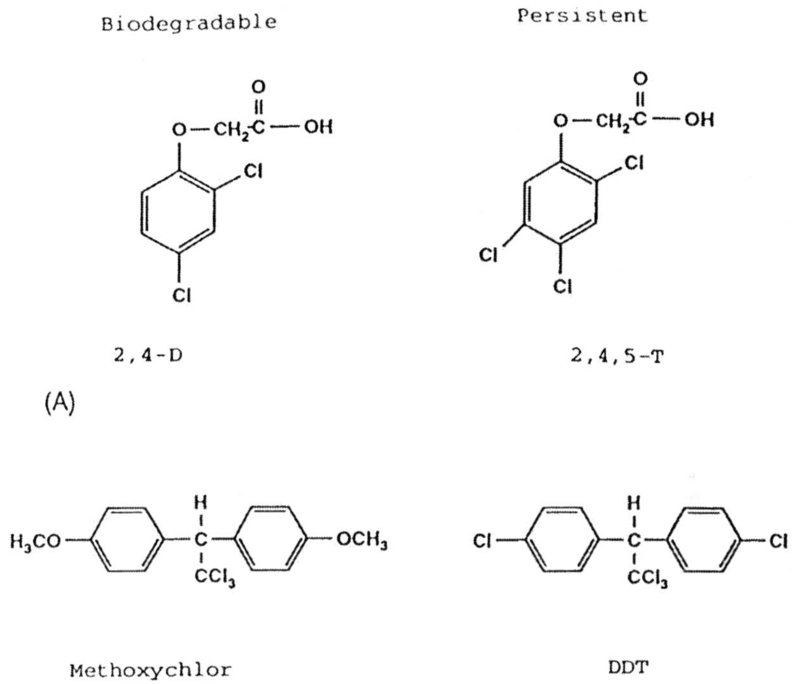

Figure 2 Effect of halogens on the biodegradability of pesticides. Addition of one chlorine (A) or replacement of methoxy groups with chlorines (B) significantly reduces biodegradability of two simple pesticides. (From Alexander, 1977; Atlas and Bartha, 1987.

methoxy (O-CH_3) groups in place of the chlorines present in DDT (Figure 2B). Numerous studies have been performed to examine the relationships between chemical structure and resistance to biological degradation. These have been summarized in Table 1. Generally, addition of functional groups that are difficult to remove by oxidative reactions usually increases the persistence of compounds in aerobic environments. Halogens such as chlorines and bromines fall into this category. Halogens withdraw electrons from adjacent carbons, rendering them less susceptible to electrophilic substitution and, in general, less susceptible to oxidation. As the number of halogen substituents increases, the relative persistence of a compound increases. It is worthwhile to note that the halogen-carbon bond is relatively uncommon in nature and may explain in part why removal of halogens from pesticides is often slow in the environment.

Table 1 Effect of Chemical Structure on Biodegradability of Aromatic and Aliphatic Compounds

Structure	Compound
Biodegradable	
Aromatic ring substituents	Carboxylic acids, phenolic hydroxyls, methoxyl and methyl groups
Aliphatic chain features	Straight unsubstituted saturated chains
Persistent	
Aromatic ring substituents	Nitro, amino, sulfonate, and halogen groups
Other chain features	Branched chains, halogenated chains

Removal of halogens from heavily halogenated compounds may actually be faster in anaerobic than aerobic environments. This may be attributed to activation toward reductive reactions imparted by numerous halogen substituents. Reductive dehalogenation appears to slow after a number of the halogens have been removed.

C. Aerobic Biodegradation Processes

Most microbial processes in the environment are strongly influenced by the availability of molecular oxygen. Many soil organisms are unable to function or compete under anaerobic conditions, and thus the influence of their presence diminishes when oxygen is withheld. In terrestrial environments, oxygen depletion usually results from limited oxygen supply (typically due to excess water) simultaneously with significant oxygen consumption (driven by carbon sources being degraded by microorganisms). Many reactions involved in pesticide degradation are carried out more rapidly by aerobic organisms, and some reactions actually require molecular oxygen as a substrate. Thus pesticide degradation, or at least certain steps in pesticide degradation, was once considered to be much slower in the absence of oxygen.

We will consider microbial degradation of a hypothetical herbicide to demonstrate some of the features inherent in aerobic biodegradation pathways. We will then examine the degradation of the herbicide 2,4-D in soil environments. The approach will be teliological (i.e., thinking from the organism's perspective) to promote understanding as to why reactions occur as they do.

Consider the hypothetical structure (Superkill) in Figure 3. The ester group (Figure 3a) on this compound is labile and thus should be easy to remove. It bears a resemblance to naturally occurring esters to which organisms are continuously exposed. Many organisms are capable of ester hydrolysis, so these

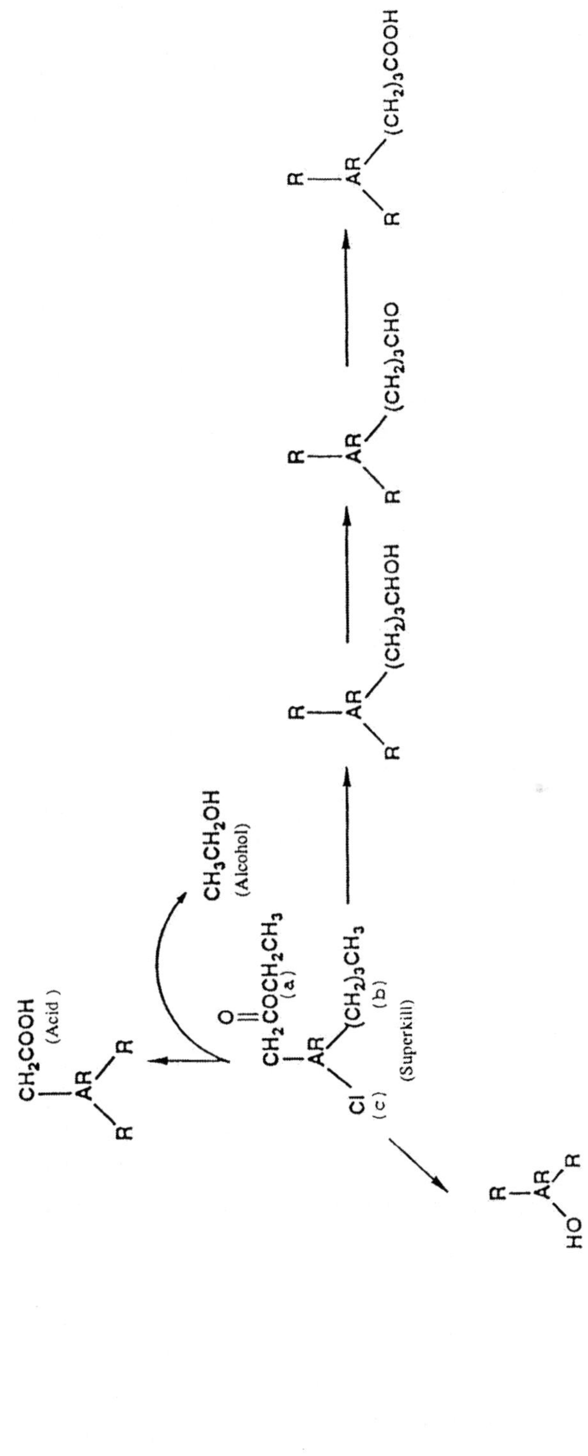

Figure 3 Possible aerobic transformations of a hypothetical pesticide structure: (a) the ester group; (b) the alkyl side chain; (c) chlorine. AR = aromatic ring.

reactions often occur without appreciable lag. Loss of such ester functions is thus often rapid in the environment. Hydrolytic cleavage of an ester bond releases a free acid and an alcohol as depicted in Figure 3.

Like the ester group, the alkyl side chain (Figure 3b) may be acted upon in a similar fashion. Typically, aliphatic hydrocarbons are converted to the corresponding acids via stepwise oxidation to alcohols and aldehydes. The resulting fatty acids are enzymatically "chopped up" to produce simple organic acids that can enter central metabolic pathways. Oxidation of the alkyl side chain of our hypothetical pesticide is shown in Figure 3. After stepwise oxidation, the side chain may be removed from the aromatic ring. Alternatively the side chains may persist until the aromatic ring is cleaved. Alkyl side chains are not necessarily toxic to microorganisms, and thus the consequences of removing them after ring fission are probably minor.

Chlorine (Figure 3c) is often removed from aromatic rings prior to ring fission although some organisms cleave chlorinated aromatic rings to release chlorinated aliphatic hydrocarbons. In some cases, release of chlorinated aliphatics is a kind of lethal synthesis leading to inhibition of central metabolic pathways by the chlorinated metabolites. Chlorinated forms of key metabolic intermediates sometimes bind irreversibly to important enzymes and block the host organism's central metabolism. When chlorine is removed from aromatic rings by aerobic organisms, it is typically replaced with a hydroxyl group, as shown in Figure 3, and as a result, loss of chlorine is sometimes a very slow process.

Perhaps the most ubiquitous theme in pesticide degradation by aerobic organisms is the manner in which aromatic rings are decomposed. Microbes tend to activate aromatic rings prior to ring cleavage by introducing hydroxyl groups. Hydroxyls are often incorporated by oxygenase enzymes which use molecular oxygen as a substrate. Such reactions are dependent upon molecular oxygen and may not proceed in an anaerobic environment, even if the organism is capable of growing anaerobically. Activation of aromatic rings by hydroxylation is a common theme in microbial metabolism of plant lignins and aromatic hydrocarbons. Enzymes responsible for hydroxylating pesticide ring systems are thus probably recruited from pathways for degradation of naturally occurring materials. An oxygenase may introduce a single hydroxyl (monooxygenase) or two hydroxyls (dioxygenase) into an aromatic ring. Oxygenase attack on a benzene ring is shown in Figure 4.

Aromatic rings may also be coupled rather than degraded in soil. Soil organisms, especially fungi, are thought to be involved in the formation of humic materials through oxidative coupling of simple aromatic compounds. In this process, aromatic structures are activated to form reactive intermediates which condense to form more stable oligomers and polymers. Some aromatic

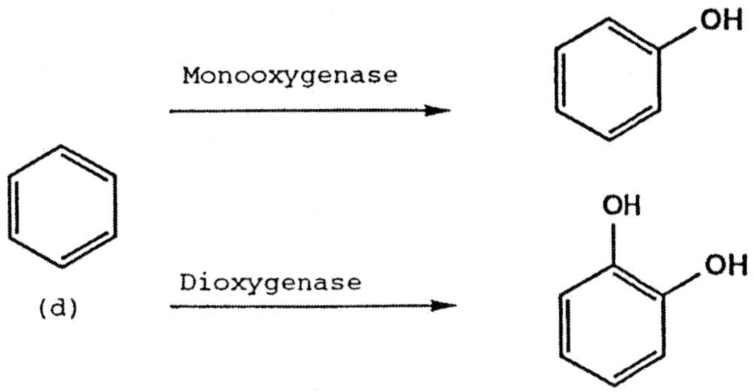

Figure 4 Two potential aerobic biodegradation pathways for aromatic ring systems. (d) benzene ring.

structures derived from pesticides, such as phenols and anilines, are particularly susceptible to oxidative coupling and often become incorporated into soil organic matter through such processes.

Mechanisms in degradation of the herbicide 2,4-D are remarkably well understood. Biochemical pathways have been worked out in numerous organisms, and the pattern of formation and decline of degradation products has been followed in numerous studies with soils. The compound 2,4-D is applied as either the free acid (active form) or an ester or salt (inactive) as depicted in Figure 5. Degradation of 2,4-D in soil is very similar to the degradation of the hypothetical herbicide discussed previously. The first reaction to take place is loss of the ester group (or salt). This step activates the herbicide. The herbicide is rapidly deactivated by loss of the ester group to form 2,4-dichlorophenol (DCP). DCP can undergo several fates in soil. The ring may be hydroxylated to activate it for ring fission, followed by loss of chlorine (Figure 5). The second chlorine is probably removed at a later point to yield a simple carboxylic acid capable of entering central metabolic pathways. Alternatively, the ring system may be oxidatively coupled to soil humic materials through the reactive phenolic hydroxyl group. Pesticide material incorporated into soil organic matter appears to be inactive at least for the short term. Long-term fate (especially possible release) of such bound residues is not well characterized for most pesticides.

D. Anaerobic Biodegradation Processes

Pesticide degradation does not stop in the absence of oxygen. Contrary to previous theories, anaerobic systems are often metabolically very active and uti-

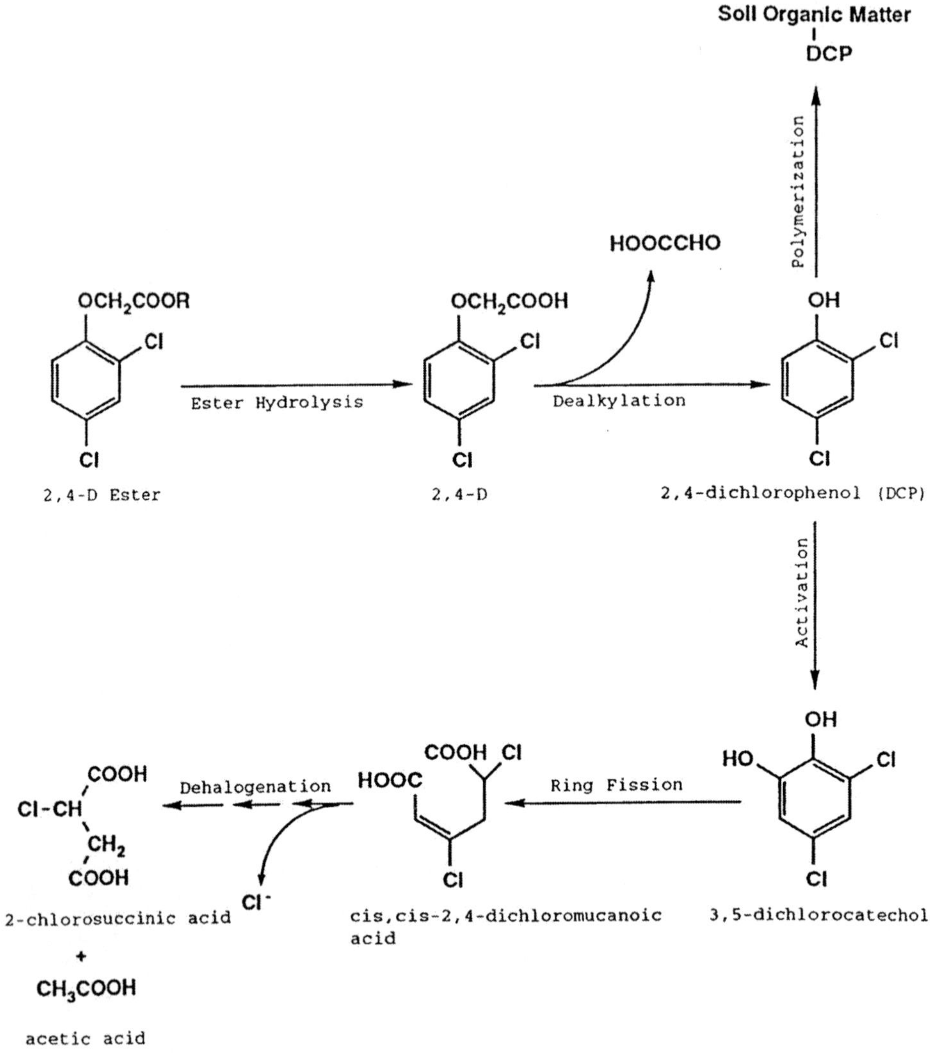

Figure 5 Biodegradation of a hypothetical ester of 2,4-D. Pathway as shown is abbreviated.

lize novel mechanisms to degrade pesticides. Anaerobic metabolism of pesticides is usually different from aerobic metabolism but is not necessarily either faster or slower. Among the most noteworthy anaerobic transformations of pesticides are reductive dechlorination and aromatic ring reduction. Reductive

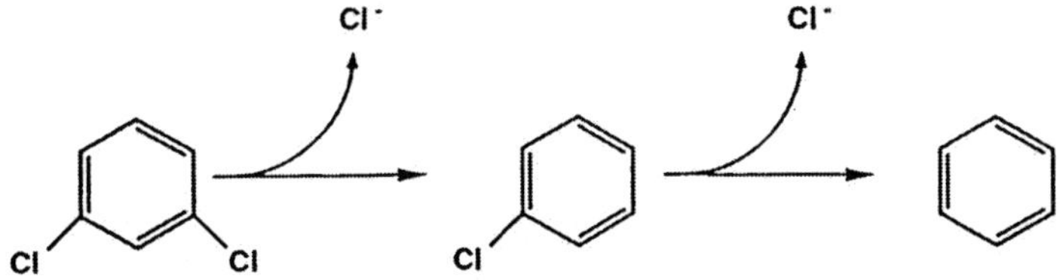

Figure 6 Reductive dechlorination of dichlorobenzene.

dechlorination results in removal of chlorine from the aromatic ring without the introduction of hydroxyls or removal of aromaticity (Figure 6). Reductive dechlorination appears to be fastest for compounds heavily substituted with chlorine and slows after several chlorines have been removed. It is a very important mechanism for decomposition of halogenated organic compounds in lake sediments. Aromatic ring reduction, in comparison, results in the loss of double bonds and thus dearomatization of the ring. The remaining double bonds are not involved in ring resonance and behave as isolated double bonds. Numerous natural compounds contain isolated double bonds, and organisms have evolved means to degrade them. The double bond can then be enzymatically hydrated to form hydroxyl groups (Figure 7). Subsequent oxidation introduces another double bond, which is then hydrated to open the ring structure.

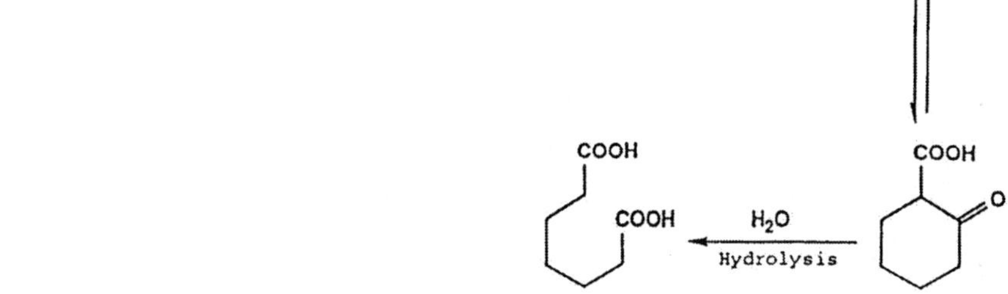

Figure 7 Anaerobic aromatic ring metabolism (reductive dearomatization).

III. ABIOTIC DEGRADATION

Although soil can exhibit intense biological activity, nonbiological, or *abiotic*, transformations may become dominant in some cases. Reactions can take place both in the homogeneous phases of soil, especially the water phase, and at the interfaces between phases. It must be stressed, however, that in most soil systems, biological and abiotic transformations occur simultaneously, and thus efforts to segregate individual processes can prove to be difficult.

Hydrolysis and photolysis are the two main classes of abiotic reactions. In addition, some other reactions can be important for some chemicals, including epoxidation, oxidation-reduction, and free radical reactions. These reactions for various classes of agrichemicals have been briefly reviewed by Armstrong and Konrad (1974).

A. Hydrolysis

Hydrolysis is defined as the chemical process of decomposition involving the splitting of a bond and the addition of the elements of water. Hydrolysis is the most prevalent reaction to occur in the liquid phase of soils, the soil water. The hydrolytic conversion of an ester to its acid is probably the most important hydrolysis reaction, although other functional groups can be transformed as well (Wolfe et al., 1990). Depending upon the molecule, the hydrolysis reactions can be acid-catalyzed, base-catalyzed, or neutral. The first two of this list are pH-dependent, while the third is pH-independent; thus the pH condition of the soil can be an important factor affecting the rate of degradation.

Hydrolysis reactions can also take place near the water-solid interface, where the surface properties of the solid phase often dominate. These reactions are catalyzed by components of the solid surface, including clays, organic matter, and minerals. As many surface reactions require previous sorption of the chemical before they occur (Burkhard and Guth, 1981), sorption may govern the kinetics of the reaction. Furthermore, the moisture content of the soil can drastically influence the rate of surface-catalyzed hydrolysis (Farmer and Aochi, 1987). The moisture content determines the degree of hydration of surface sites, which in turn affects the rate and mechanisms of the reactions which occur. Increased hydration can increase the efficiency of the reaction of cations and pesticides (Wolfe et al., 1990), while, through a different mechanism, hydrolysis rates may be enhanced at low water content (Saltzman et al., 1976; Getzin, 1981).

In natural water systems, hydrolysis is also a prevalent reaction and is greatly influenced by pH and temperature. The types of hydrolysis reactions are similar to those which can take place in soil, and as in soil, the particular chemical properties of the system and environmental conditions affect the contribution of hydrolysis to the degradation of the compound.

B. Photolysis

Photochemical degradation by sunlight is often the primary degradation route for herbicides present in exposed environmental compartments including soil and plant surfaces, water, and the atmosphere. All photolytic reactions begin with the absorption of light radiation by some chemical species which can then undergo a variety of chemical or physical processes. Photolysis is often viewed as a detoxifying mechanism, producing products of lower environmental significance than the parent molecule, although some notable exceptions have been observed (Miller and Hebert, 1987). Photolysis reactions can be grouped as direct or indirect.

In direct photolysis, the pesticide itself absorbs light energy and then reacts. The rate of photolytic reaction is directly correlated to the overlap of the absorption spectrum of the herbicide and the spectral distribution of sunlight. Many herbicides absorb light most readily in the near ultraviolet (UV) region; thus the intensity of natural sunlight (the irradiance) in this region will greatly influence photodegradation rates. Irradiance is dependent upon latitude, season, and time of day and can also be attenuated by ozone absorption, molecular scattering, and particulate diffusion (Wolfe et al., 1990). Direct photolysis reactions can be significant in any environmental compartment that receives light, depending upon the particular chemical and environmental conditions.

Indirect photolysis occurs when a chemical species other than the compound of interest absorbs light energy and initiates a series of reactions that eventually degrades the pesticide. These other chemical species are known as photosensitizers and can include naturally occurring organic and inorganic species, including humic materials, clay minerals, transition metals, ozone, and various free radicals produced by the interaction of sunlight with air and water.

In natural water systems, the dissolved humic materials present can serve as strong photosensitizers (Zepp et al., 1981; Burkhard and Guth, 1976). In addition, direct photodegradation may be important, depending upon the properties of the chemical (McCall and Gavit, 1986). On soil surfaces, there has been evidence of both direct and indirect photolysis of pesticides (e.g., Gohre and Miller, 1983; Choudhry and Webster, 1985; Burkhard and Guth, 1979; Klehr et al., 1983), although investigations have not been extensive. Since laboratory water systems are homogeneous and simple to characterize, photodegradation in water has been extensively studied in the laboratory (reviews include Choudhry and Webster, 1985, and Miller and Crosby, 1982). In the atmosphere, photodegradation is thought to be a major degradation pathway (Miller and Zepp, 1983), although very little is known about actual reaction rates and their significance in the dissipation of herbicides (Miller and Crosby, 1982). A further discussion of vapor phase transformation processes is presented later.

IV. TRANSPORT PROCESSES

Soil is a porous medium that consists of mineral and organic aggregates or peds, surrounded by pore space containing soil water and soil air. The peds themselves also contain pores, although they are much smaller than the interparticle pores. Soluble herbicides are primarily transported through the soil profile as solutes dissolved in the soil water while volatile compounds are carried in the soil air.

The description of transport can be broken down into a series of fundamental processes, all interacting closely and often simultaneously. Because of the complexity of the soil system, the interactions of the processes are not completely understood and, thus, a more simplified or empirical approach may be taken.

A. Sorption

Adsorption is defined as the accumulation of a chemical at either the soil-water or the air-water interface. The processes in effect may include true adsorption at the molecular scale onto a soil particle surface, precipitation from solution, or even covalent bonding to soil organic matter. It is often difficult to distinguish the separate processes, so they are often lumped into the more general heading of sorption (Koskinen and Harper, 1990). In general, all these processes contribute to the removal of the chemical from the soil water phase, rendering them less available for further transport, uptake by plants, or degradation.

The most commonly used measure of sorption in a soil system is the sorption constant, often termed K_d, which gives the ratio between the solid and liquid phases:

$$K_d = \frac{[C_s]}{[C_w]} \tag{1}$$

where C_s is the concentration in the solid phase (g/kg; soil particulate) and C_w is the concentration in the liquid phase (g/L; soil solution). This approach assumes that C_s depends linearly upon C_w, which for many herbicides may be assumed without significant error. To normalize the K_d values, they are often corrected for organic carbon content of the soil to obtain the quantity K_{oc}:

$$K_{oc} = K_d/f_{oc} \tag{2}$$

where f_{oc} is the fraction of organic carbon in the soil. This assumes that soil organic carbon has constant sorption capacity (Green and Karickoff, 1990).

Both of these approaches assume an instantaneous and complete equilibrium between soil and soil solution which in reality may never be attained. In the environment, sorption is a dynamic, highly complex process in which partitioning between soil solution and surfaces occurs continually. Chemicals can be associated with the solid phase by a wide range of physicochemical processes including van der Waal's forces, hydrogen bonding, dipole-dipole interactions, ion exchange, covalent bonding, protonation, ligand exchange, cation adsorption, and water bridging. These processes have been described in detail elsewhere (Koskinen and Harper, 1990). As herbicides exhibit a wide diversity of chemical properties, the relative importance of the various sorption processes will vary for each chemical. For example, for nonionic compounds, hydrophobic partitioning may be the dominant mechanism (Chiou et al., 1979), whereas for weakly acidic compounds the distribution of charged and neutral forms, and hence the local pH conditions, has been shown to be important (Fontaine et al., 1991).

B. Convection, Diffusion, and Dispersion

In convection, or mass transport, dissolved chemicals (solutes) move with the soil water through the soil profile. Especially in highly structured or fractured soils, mass flow can transport pesticides over large distances in relatively short time (the phenomenon of preferential flow). Steady-state flow can be described by Darcy's law (Enfield and Yates, 1990): the flux of water is proportional to the hydraulic conductivity of the soil and the hydraulic head. This approach assumes that all the soil pores are the same size—an obvious oversimplification. If this were true, the water, along with the dissolved chemical, would move down the soil profile as a sharp front, in "plug flow." This has been shown many times not to be the case (e.g., Letey and Farmer, 1974). The chemical front spreads out over the soil profile, a result of varying pore sizes and corresponding differences in local flow velocities. This occurrence can be conceptualized as an effect of mechanical dispersion and the convective flux, J_{CL}, that can be described (Hutson and Wagenet, 1989) as follows:

$$J_{CL} = -\theta D_M(q)\frac{dC_L}{dz} + qC_L \qquad [3]$$

where q is the volumetric water content, $D_M(q)$ is the dispersion coefficient, C_L is the concentration of solute in the liquid phase, and q is the macroscopic (Darcian) water flux.

Diffusion, on the other hand, is a slow process, moving solute molecules only short distances. Solute molecules move in solution driven by concentration gradients from areas of high to low concentration. Diffusion coefficients can be obtained in pure solutions in the laboratory, but in soil systems, the incomplete filling of pore spaces as well as the tortuous nature of the pores lead to lower "effective" diffusion coefficients. Diffusion rates are generally so much slower than convective rates that they can often be neglected in modeling solute flux in soils. In simplest terms, diffusion contributes further to the spreading of the solute front, albeit in a minor fashion in most cases (Green and Khan, 1987).

All of the effects discussed—convection/dispersion, sorption, and source/sink terms—can be combined into a single expression, termed the convection-dispersion equation (Wagenet and Hutson, 1986). Because it is a a partial differential equation, numerical techniques are usually required for its solution, as well as a large number of input parameters. Further refinements can be made to include such phenomena as nonequilibrium transport and degradation (Gamerdinger et al., 1990).

C. Runoff

As many herbicides are applied to, or wash onto, the surface of the soil, runoff can be a significant route of transport of herbicides from their initial site of application. Often, the ultimate sink for runoff is local surface waters such as ponds, lakes, and streams, where it is likely to be of practical as well as public concern (Merkle and Bovey, 1974). The Federal Water Pollution Control Act Amendments of 1972 (P.L. 92-500) require that states develop guidelines to help control water pollution from nonpoint sources, such as agricultural runoff.

Runoff includes not only the water leaving a field in surface drainage but also any dissolved or suspended material it may contain. Herbicides may be dissolved in the runoff water, suspended as particles (e.g., in dust or granular formulations), or sorbed to soil carried by runoff water. Although direct transport from the soil surface is the most important source of runoff, movement of water can expose deeper layers of soil, and if chemicals are present, they may be available for runoff. A number of factors affect the amount of chemical transported by runoff. These may be subdivided into climatic, soil, chemical, and management factors (Leonard, 1990); again, the site-specific combination of all these factors influences the amount of chemical lost by runoff.

Climatic Factors

The timing of runoff events governs the concentration of herbicide present in runoff water. Highest concentrations of active ingredients occur at the first significant runoff event after application. Concentrations decrease with time as the compound dissipates from soil and plant surfaces as a result of the environmental processes described in this chapter. Increased intensity of the rainfall event causes a greater volume of runoff water and can increase the energy for extraction of the compound from the soil matrix. On the other hand, an increase in intensity (before infiltration rate is surpassed) can leach the chemical into the soil profile, rendering it unavailable for runoff.

Soil Factors

Water tends to run off more quickly in finer-textured or compacted soils because of decreased infiltration rates. Topography is an important influence: increasing slope can increase the actual rate of runoff, increase the amount of entrained sediment, and increase the depth in the soil at which runoff has an effect. High organic matter content affects sorption of the chemical as well as infiltration rates, thus reducing runoff potential. In addition, water content is important, as runoff occurs when the surface soil becomes saturated.

Chemical Factors

The chemical properties of the applied herbicide and its formulation have a pronounced effect on the degree of loss through runoff. Wauchope (1978) attempted to correlate solubility with runoff potential and found that chemicals with water solubilities > 10 parts per million (ppm) were primarily transported as dissolved species in the water phase of runoff while more hydrophobic compounds were carried in entrained sediments. The same author found that wettable powder formulations are most prone to runoff (up to 5% of applied) because of easy washing off of the powder. In addition, more persistent compounds which remain near the surface after application, as well as compounds applied at high rates, are more prone to run off.

Management Factors

Erosion control practices on agricultural land can reduce runoff losses of compounds sorbed to suspended particles. However, these practices will have little effect upon the transport of soluble compounds unless the erosion control also reduces the volume of runoff water. Crop residues may also impede sediment transport, although increased loss of chemical may occur if the compound is washed directly into runoff water. In addition, the use of vegetative buffer strips around application areas can help to intercept entrained or dissolved chemicals through sediment deposition or plant sorption.

V. VOLATILIZATION OF HERBICIDES

A herbicide has the capacity to move into the atmosphere by two distinct processes. These are spray drift, which occurs during application of a herbicide formulation to a site, and volatilization, which usually occurs after the completion of application. Although spray drift can contribute significantly to the dissipation of a herbicide applied under nonideal conditions, this section focuses on the volatilization process, which can make a larger contribution to the dissipation of a herbicide. Volatilization is considered one of the primary pathways for herbicide dissipation from the site of a herbicide application. Under unfavorable conditions, losses that result from volatilization can reach 80% to 90%

within a few days although such rates are dependent upon climatic and microclimatic conditions (Taylor and Glotfelty, 1988).

Early work on volatilization was conducted on insecticides because of their environmental stability and concerns arising from bioconcentration in wildlife. However, the chemical and physical principles described later are identical for both insecticides and herbicides, thus eliminating the need for distinctions between the two. If appropriate herbicide examples are not available, insecticide examples from these early studies are used to develop a particular concept.

A. Basic Physical Process

The process of volatilization is controlled by two mechanisms: the evaporation of the active ingredient from the soil or plant surface to which it has been applied or has migrated and dispersion of the resulting vapor through the overlying atmosphere by diffusion and turbulent mixing. Though these two processes are different in character and are controlled by different chemical and environmental factors, they are not independent of one another and in general should be considered as an integrated whole.

The first process of evaporation involves a phase change from its initial state of solid or liquid to its corresponding vapor. Thus, one of the main factors controlling the rate of loss by volatilization is the rate at which surface residues evaporate. The rate of evaporation in turn is controlled by the vapor pressure of the exposed residues and is either the vapor pressure of the active ingredient or lower values to which it is reduced by interaction of the compound with surfaces. For residues no longer on the surface but drawn away to underlying layers by capillary action or diffusion after application, the rate of volatilization becomes dependent upon the rate of "back diffusion" of the material to the upper layers (Taylor and Glotfelty, 1988).

The second process of dispersion into the atmosphere acts through the turbulent flow of air over the evaporating surface. The flow of air continuously replaces the air around the evaporating surface and mixes and dilutes the herbicide vapors in the surrounding air. The movement and dispersion of vapors within the turbulent layer are considered rapid in relation to those at the evaporative surface.

The flow of air over plant and soil surfaces is generally turbulent except for a shallow stagnant layer of air directly above the evaporating surface where flow is regarded as laminar (Monteith, 1973). The actual rate of volatilization away from the evaporative surface through the stagnant layer is proportional to the molecular diffusion coefficient and the vapor pressure of the herbicide at the evaporating surface. The depth of the stagnant layer in turn is affected by both the surface geometry and the air flow of the system. Surface geometry or roughness alters the turbulent flow of air over the evaporative surface.

Thus an increase in either air flow or turbulence increases the rate of vaporization (Spencer et al., 1982) as vapors are removed from the evaporative surface.

Effect of Soil Organic Matter

Once a herbicide has been applied to a soil, the material either remains on the surface, diffuses into the soil by capillary action, or becomes sorbed to the soil matrix. The extent of sorption to the soil matrix is influenced by the carbon content of the soil, which in turn affects the rate of volatilization from the soil. Soil carbon content controls the concentration of pesticide present in the soil solution, and Spencer (1970) has demonstrated that the vapor density of a herbicide over soil is proportional to its concentration in the soil solution. In general, the higher the soil carbon content, the lower the concentration of herbicide in the soil solution. As a result of its reduced concentration in soil solution, the rate of volatilization for a herbicide from a soil with a high organic matter content is lower than that from one with a low organic matter content (Jury et al., 1980; Getzin, 1981; Glotfelty et al., 1984).

Empirically, the concentration of a herbicide in soil solution is equal to the ratio of its vapor pressure and water solubility. This concept has been further developed by Swann and coworkers (1982) as an empirical rate constant where the rate of volatilization of a herbicide (K_v) is related to its vapor pressure (p), water solubility (S), and soil adsorption coefficient (K_{oc}) by

$$K_v = Q(p/K_{oc}\ S) \qquad [4]$$

where Q is an empirically determined coefficient. Glotfelty and coworkers (1989) applied equation [4] to field generated data from the application of atrazine, simazine, alachlor and toxaphene to bare soil. The calculated rates of volatilization, with the exception of that of toxaphene, were within an order of magnitude of that observed in the field. It was suggested that toxaphene exhibited a higher rate of volatilization than calculated because of its nature as a complex mixture of many components of differing volatilities.

Effect of Soil Moisture

The moisture content of soil has been determined to be one of the most significant environmental parameters which influence the rate of herbicide volatilization (Nash, 1983). In general, herbicides volatilize more rapidly from moist than from dry soils. The reduced volatilization of a herbicide under dry conditions has been attributed to the exposure of additional adsorption sites on the soil by the evaporation of water from the soil surface (Spencer et al., 1982). The effect of low soil moisture content can be dramatic, effectively halting volatilization of a herbicide from a given soil. Increased binding of a herbicide to a soil has been estimated to occur once the soil moisture content decreases to one to three molecular layers of water on the surface of a soil particle. However, the point at which increased herbicide binding occurs varies and is also dependent upon soil composition, which, in turn, influences soil moisture capacity. The binding process is reversible, and with increased soil moisture content, such as after rainfall or dew fall, an increase in herbicide flux from the soil can be detected (White et al., 1977; Taylor et al., 1976; Harper et al., 1976; Grover et al., 1985).

This influence of soil moisture has been observed in both field and laboratory studies. Getzin (1981) found that chlorpyrifos-treated soils equilibrated at 91% relative humidity (RH) had a higher proportion of volatiles than the same soils equilibrated at 30% RH. A similar influence of soil moisture on the herbicide triallate has also been observed on sandy and silt loam soils (Jury et al., 1980). Once triallate had dissipated from the soil surface, the controlling mechanism for dispersion was convection or the upward movement of triallate from the bulk soil to the surface. The volatilization of triallate was influenced by the relative humidity over the sandy versus the silty loam soil. Observed differences between the two soils were attributed to the organic matter content of the soils and its influence on the concentration of triallate in the soil solution.

In the field, similar influences of soil moisture on herbicide flux have been observed. The flux of trifluralin from soil (Harper et al., 1976) decreased shortly after application when the surface moisture was low. After sundown, the surface water content of the soil increased as evaporative pressure decreased, resulting in a corresponding increase in trifluralin flux. An apparent inverted flux pattern from that anticipated, based on the energy input from sunlight, was observed, i.e., high flux at night and low flux during the day. This inverted pattern of volatilization has also been observed with alachlor and toxaphene (Glotfelty et al., 1989). However, when soil moisture is no longer limiting, volatilization increases proportionally with increased energy input in the form of sunlight (Glotfelty et al., 1989; Harper et al., 1976). In these instances, maximum flux of pesticide is detected at or around noon when energy input from sunlight and ambient temperature is at its maximum.

Effect of Temperature

The vapor pressure of a herbicide varies proportionally with temperature. Thus with an increase in ambient temperature, a corresponding increase in volatilization from soil or plant surfaces is anticipated. When sorption sites on the soil are saturated with herbicide, the vapor density or vapor pressure for that herbicide above the soil may be equivalent to that of the pure active compound (Spencer and Cliath, 1969) and varies proportionally with temperature. At lower concentrations, this temperature effect on the vapor pressure for a herbicide no longer occurs but is directly influenced by the energy of sorption to the soil matrix, becoming less well defined as a result (Spencer and Claith, 1970).

The preceding discussion only applies if soil moisture is not a limiting factor. As discussed previously, under low moisture conditions, increased temperature induces increased herbicide sorption to the soil matrix as additional sites are exposed by the evaporation of water (Spencer, 1970; Taylor and Glotfelty, 1988). As a result, herbicide volatilization may be reduced or halted until water displaces the sorbed herbicide from the soil particle.

Effect of Formulation

The manner in which a herbicide has been formulated affects its rate of volatilization from a site. Use of emulsifiable concentrates (ECs) results in direct application of the active ingredient to the soil, allowing for sorption to the soil matrix. In comparison, wettable powders (WPs), which carry the active ingredient on fine clay particulate, remain on the surface of the soil with limited sorption to the soil matrix. Glotfelty et al. (1989) reported that simazine and atrazine applied as WP formulations were displaced from the soil on dry days by the wind, as compared to toxaphene and alachlor applied as EC formulations. During the same sampling period, volatilization had dropped to a minimum at noon and increased again toward evening with increased moisture availability. Total losses of simazine and atrazine were significantly less than that observed for toxaphene and alachlor, though this may be attributed to their lower inherent vapor pressure. Dacthal, applied as a WP, exhibited similar behavior (Glotfelty et al., 1984), with 30% of the airborne material attributed to particulate displacement during windy periods.

A change in formulation can be utilized to reduce volatilization of a herbicide. This has been demonstrated by Turner et al. (1978) for chloropropham, which was applied to bare soil in two forms: EC and microencapsulated. The microencapsulated chloropropham exhibited much lower loss of material due to volatilization than the conventional EC. Additionally, the chloropropham microencapsulated formulation increased the efficacy of the herbicide by increasing its persistence.

B. Surface versus Incorporated Material

Evaporation of a surface-applied chemical is proportional to the gross area of the site. Under such conditions the herbicide is lost at a constant rate under constant conditions until significantly depleted. The rate of volatilization varies from compound to compound, depending upon the product of its molecular diffusion coefficient and saturation vapor density. Hartley and Graham-Bryce (1980) noted that the diffusion coefficient for gases varies inversely with the square root of the molecular weight. Volatilization rates have been shown to reflect this relationship under constant conditions.

For soil-incorporated herbicides, the rate of volatilization is controlled by their movement through the soil to the surface and ultimately the air. Soil-incorporated residues differ from surface residues in that the concentration in the soil can be up to an order of magnitude lower than that of those applied to the surface. The volatilization of incorporated residues is greatly restricted and is dependent upon the rate of their upward movement to the soil surface. The upward movement of herbicide residues is influenced by the mass flow of water (convection) and by diffusion. Diffusion, as previously discussed, is driven by concentration gradients and solute movement from areas of high to areas of low concentration. The concentration gradient in this case is inverted, with the active ingredient having a lower concentration on the surface of the soil than at the depth of incorporation. As a result, upward rather than downward movement is promoted.

C. Transformation and Loss Processes

Once a herbicide has evaporated from the soil surface, it may be acted upon by both chemical and physical processes in the atmosphere. The physical removal processes involve both wet and dry deposition of gas and particulate bound herbicide residues. The chemical processes, which can lead to inactivation of the herbicide, include photolytic and direct reactions with atmospheric components such as OH, HO_2, NO_3 radicals, O_3, N_2O_5, and gaseous acids.

The magnitude of deposition removal rates, whether wet or dry, is highly dependent upon the vapor pressure and aqueous solubilities of the chemical (Jury et al., 1987). The extent of dispersion also is dependent upon the particulate size of the material, whether the material is sorbed onto particulate or in its vapor state, and meteorological conditions. This subject has been extensively reviewed elsewhere (Glotfelty and Caro, 1975; Pitter and Baum, 1975; Junge, 1976).

As with dispersion mechanisms, the reactions that the herbicide undergoes in the atmosphere are also influenced by its physical state, i.e., vapor phase or sorbed. Often herbicides may not be able to absorb sunlight because of their chemical structure. In such instances, the chemical in the vapor phase is unlikely to react photolytically in sunlight. However, sorption of a chemical onto a particle may shift the compound's absorption spectrum such that it overlaps with the spectrum of sunlight, thus increasing the likelihood of photolytic decomposition.

A herbicide may react directly with atmospheric components such as O_3 and OH and NO_3 radicals. The atmospheric concentrations of these reactive species are known to vary with time of day, season, altitude, and degree of pollution in the atmosphere. The gas phase reactions of these species with various functional groups are shown in Table 2. These reactive species may also interact with sunlight to react with and alter a chemical through photosensiti-

Table 2 Major Atmospheric Reaction Pathways for Various Functional Groups Associated with Herbicides

Functional groups	Atmospheric reactive species
Amines	OH, HNO_3/HONO, NO_3
>S	OH, NO_3
-SH	OH, NO_3, O_3
Alkenes	OH, NO_3, O_3
$-CH_2$	OH, NO_3, O_3
$-CH_3$	OH
>CH	OH
Hetercyclic aromatics	OH, HNO_3/HONO

Source: Jury et al. (1987).

zation. In most instances, such interactions produce compounds of lesser toxicity than the parent, thus reducing their impact on the environment.

VI. SIMULATION OF ENVIRONMENTAL FATE PROCESSES

The preceding sections have described the soil, air, and aquatic environments, and, as can be readily seen, a complete understanding of the processes involved is far from the grasp of science. A useful approach to improving our understanding of the process occurring in nature is simulation modeling. In addition, increased public awareness of the impact of agriculture, warranted or not, has necessitated attempts to predict the fate of agrochemicals in the environment. Significant research on this multifaceted problem has taken place in recent years. Table 3 gives a short (and noncomprehensive) list of some of the models recently developed for predicting agrichemical fate.

Simulation models of environmental fate can be roughly divided into three groups: research models, management models, and screening models (Wagenet and Rao, 1990). The quantity and quality of input data, the degree of detail simulated, and the usefulness of results vary widely among model types. The user must be very clear about the intended application of the modeled results. The availability of input data remains a large problem for all modeling, although efforts have been put forth to gather the "best available" data on chemicals and soils (Wauchope et al., 1992; U.S. Department of Agriculture, 1991).

A. Research Models

Research models attempt to simulate fundamental transport and degradation processes in a quantitative fashion through application of basic concepts of

Table 3 A Short List of Agrochemical Fate Models

Model	Process simulated	Type
AF	Leaching	Screening
AGNPS	Runoff	Research
ANSWERS	Runoff	Management
ARM	Runoff	Management/screening
BAM	Leaching	Screening
CMLS	Leaching	Screening
CREAMS	Runoff/leaching	Management/research
EPIC	Runoff	Research
GLEAMS	Runoff/leaching	Management
GUS	Leaching	Screening
HILLS	Runoff	Research
LEACH	Leaching	Screening
LEACHM	Leaching	Research
MOUSE	Leaching	Screening
PESTAN	Leaching	Screening
PRM	Runoff	Screening
PRZM	Leaching/runoff	Management
SESOIL	Leaching	Management
SWERB	Runoff	Management
SWM	Runoff	Research
USLE	Runoff	Management
WEPP	Runoff	Research

Source: Ghadiri and Rose (1992), Thorlourn et al. (1992), Rao et al. (1985), and Wagenet and Rao (1990).

physics, chemistry, and biology. The number of inputs required is substantial and may include parameters that are difficult to measure. This input intensity often renders them cumbersome for routine management use. In addition, the numerical methods required for the solution of these models have, until recently, required the use of powerful mainframe computers. However, the recent advances in inexpensive computational power have made this less of a problem. Even so, the deterministic, site-specific nature of research models has made their use for management purposes a risky prospect at best.

In the soil environment, where a majority of this effort has been focused, the utilization of miscible displacement theory as described by the convection-dispersion equation (Section IVB) has been at the forefront of research models. For situations of constant water flow, analytical solutions can be obtained (e.g., van Genuchten and Wagenet, 1989) and have been validated through careful laboratory experiments (e.g., Gamerdinger et al., 1990). In simulation of actual field conditions, steady-state conditions rarely exist, and numerical solutions of transient water flow equations are required. In addition, most of these models are one-dimensional and cannot account for the spatial variability that can be significant within even a few meters.

Similarly, significant effort has been expended in the physical simulation of runoff. The appearance of runoff from a field is also a many-faceted process involving infiltration of water into the surface, the hydrography of overland flow, and the transport of sediments and dissolved species. Runoff models have recently been reviewed by Ghadiri and Rose (1992). Again, research-type models require large numbers of site-specific inputs, are computationally intensive, and cannot deal well with spatial variability.

It can be seen from this discussion that research models can be most successfully utilized to gain a better understanding of the physical processes in effect in the environment. It is dangerous to try to utilize these models for management purposes, a use for which they were not intended.

B. Management Models

Management models require fewer parameters, some of which can be estimated by "rules of thumb," and are usually less computationally intensive than research models. Management models are more empirically based than research models and generally predict only quantitative water and solute transport. However, they are deterministic, and a large number of iterations varying input parameters statistically may provide a good representation of field-scale transport of chemicals.

An application of this approach to regional assessments of leaching probability has been recently carried out (Tillotson and Fontaine, 1991; Laskowski et al., 1990). The authors used a Monte Carlo technique to sample actual distributions of input parameters from soil surveys for multiple executions of the PRZM model (Carsel et al., 1984). An output was a distribution showing the probability that a certain chemical would leach beyond a certain depth. This technique holds promise in the regulatory arena as well as in the identification of areas at highest risk for leaching.

Further expansion of stochastic analysis, where the soil-water system itself is not deterministically but statistically described, has been proposed by several workers (e.g., Bresler and Dagan, 1981). These researchers assume that all processes in the environment are too variable to be described physically and thus can only be described as probability distributions. Conceptually, the idea of modeling spatial variability using these techniques is attractive. However, validations and data bases describing variabilities in the environment are lacking (Wagenet and Rao, 1990).

C. Screening Models

Screening models are designed to describe purely relative behavior of chemicals in the environment (often termed a "benchmarking" approach). These models require relatively few input parameters and can usually be implemented easily, often as numerical indices. The user must realize that screening models should not be used to predict fate under field conditions, and that the models are only useful to group chemicals into classes that may behave similarly under identical environmental conditions. Barring these warnings, screening models can serve a useful purpose, especially in the regulatory process, to group new chemicals into general classes and select those chemicals which may be candidates for further monitoring or site-specific studies.

A number of screening models have been developed over the years to describe the relative risks that a chemical will reach groundwater. Some of these are based upon convective-dispersive transport concepts while others form indices from properties of the chemical and soil itself.

The benchmarking method of Jury et al. (1984) is based upon an analytical solution of the convection-dispersion equation and defines a travel time (T_c) for the solute to reach a certain depth in the soil profile without taking degradation into account.

In another approach, Laskowski et al. (1982) define an index LEACH, calculated as

$$\text{LEACH} = \frac{(S \times t_{1/2})}{(V_p \times K_{oc})} \qquad [5]$$

where S is the water solubility, $t_{1/2}$ is the half-life in days, V_p is the vapor pressure, and K_{oc} is the water-soil organic partition coefficient as described previously. The index has no real physical meaning, but it is intuitively consistent in terms of chemical properties. For example, water-soluble and persistent compounds would have a greater potential to leach to groundwater. Similarly, Gustafson (1989) proposed the groundwater ubiquity score (GUS), based only on K_{oc} and half-life. In a combination approach, Rao et al. (1985) proposed an attenuation factor (AF) that uses the transport approach of Jury and also takes degradation into account. Relative rankings produced by these four screening models appear in Table 4. The compounds are ranked from highest to lowest leaching potential: i.e., for the LEACH index, picloram has the greatest and trifluralin the lowest probability of leaching.

It can be seen from Table 4 that all models do not give equivalent results. For example, the first three indices indicate that 2,4-D has little leaching potential whereas the T_c index shows high leaching potential. This is an effect of 2,4-D's high water solubility (900 mg/L), used by all four models, but short

Table 4 Comparison of Some Common Screening Indices for Leaching Potential

	Ranking scheme			
Chemical	Leach[a]	AF[b]	GUS[c]	T_c[d]
Picloram	1	1	1	3
Diuron	8	2	7	11
Monuron	3	3	5	8
Terbacil	2	4	4	5
Cyanazine	11	5	6	6
Carbofuran	7	6	3	4
Aldicarb	5	7	2	1
Simazine	4	8	8	7
Atrazine	6	9	9	8
2,4-D	12	10	10	2
Linuron	9	11	11	12
Alachlor	10	12	12	10
Trifluralin	13	13	13	13

[a]Laskowski et al., 1982.
[b]Rao et al., 1985.
[c]Gustafson, 1989.
[d]Jury et al., 1984.

half-life (15 days), not accounted for in the calculation of T_c. However, the concept of a single half-life is overly simplistic in trying to compare the properties of different chemicals. Soil degradation rate is highly dependent upon soil and climatic conditions and, as such, cannot be normalized (like K_{oc} and K_d). Thus, using a reported half-life for environmental fate modeling is fraught with peril.

VII. CONCLUSIONS AND GENERAL CONSIDERATIONS

Pesticide users have come under public scrutiny recently because of concern over pesticide occurrence in natural waters. Furthermore, since pesticide users have greater exposure to pesticides than other groups, concern over health risks to pesticide users has also increased. Though the pesticide user may not need to know structure biodegradability relationships and complex photochemistry to use pesticides effectively, knowledge of why pesticides pose environmental risk can assist the applicator in making decisions that may determine the environmental impact of the use of pesticides.

Manufacturers choose new candidates for development on the basis of not only its efficacy but its environmental properties. Users should also include potential environmental impact in their choice of product lines and formulations. Once a product is chosen, the chemistry is fixed. How the product is used thus becomes the deciding factor in minimizing environmental impact. The user is limited in flexibility by approved use practices specified on the herbicide label. However, with some understanding of environmental properties of the herbicide to be applied, the user can have a significant effect on the environmental impact.

Some sites are more vulnerable than others, as discussed previously, because of the soil type and/or topography. Deep sands are more permeable than clay soils; thus groundwater is more vulnerable at such sites. Water wells are perhaps the most important features at any site, since contamination of groundwater through well-heads is instantaneous, typically severe, and, most likely, long-term and difficult to remedy. The user should determine the location of wells on the site and assess their condition. Spraying and handling should never be done near an uncapped wellhead. Proximity to surface water is an extremely important feature, since herbicides are commonly transported to streams in runoff water, in detached sediment, or as spray drift. Use of buffer zones (untreated areas) near surface water is an obvious choice. Soils that are plowed annually pose a greater risk for surface water contamination than those under long-term reduced tillage. There is some concern that reduced tillage may at least slightly increase the risk of groundwater contamination relative to annual plowing. This is because annual plowing breaks up natural macropores which facilitate rapid downward movement of water in undisturbed sites. Whether reduced tillage actually increases the risk of groundwater contamination by pesticides is not well understood, although decreased risk of surface water contamination as a result of reduced tillage has been clearly demonstrated.

Timing of application is probably the most important input the user has, and, unfortunately, is often not within the user's control. Most pesticide movement either to the surface or to the groundwater occurs in the first major storm event after application. Heavy losses are reported when application occurs immediately before a major storm. Light gentle rain lacking significant runoff may help move the compound into the soil micropores, thus reducing the effects of a subsequent heavy rainfall. This scenario holds not only for transport of chemicals in surface runoff, but also for movement of chemicals to groundwater via macropores. Applicators obviously cannot control the weather but can be aware of the relationship between timing of application and weather patterns.

Proper disposal of pesticides may be more important than proper use in many cases. Unused materials and tank mixes are concentrated forms of pesticides, which when applied to small areas or dumped on the soil surface may be much more toxic and likely to move than when disposed of properly. Disposal should follow recommended procedures. When choices exist, the user may wish to inquire which disposal technique causes the least impact. It should be noted that historical practices of rinsing tanks (especially disposing of rinsate on the soil surface near a well) have often been cited as contributing factors in groundwater contamination. These observations make it clear that management practices other than pesticide use may influence the environmental impact of pesticides.

It is evident from the discussion throughout this chapter that the environmental fate of herbicides is a complex and multidimensional combination of many interconnecting factors. Reseach to understand these factors on a fundamental basis continues at all levels of scientific inquiry. Through this understanding, we can more precisely gauge the true impact of agriculture on the environment around us and direct our efforts to minimizing potential adverse impacts.

APPENDIX

Pesticides Mentioned in Chapter

Common name	Chemical Name
2,4-D	(2,4-Dichlorophenoxy)acetic acid
Methoxychlor	2,2-bis(*p*-Methoxyphenyl)-1,1,1-trichloroethane
Atrazine	2-Chloro-*N*-ethyl-*N*′-(1-methylethyl)-1,3,5-triazine-2,4-diamine
Alachlor	2-Chloro-*N*-(2,6-diethylphenyl)-*N*-(methoxymethyl)acetamide
Chlorpyrifos	*O,O*-Diethyl-*O*-(3,5,6-trichloro-2-pyridiyl)phosphorothioate
Trifluralin	2,6-Dinitro-*N,N*-dipropyl-4-(trifluoromethyl)benzenamine
Chloropropham	1-Methylethyl-3-chlorophenylcarbamate
2,4,5-T	(2,4,5-Trichlorophenoxy)acetic acid
DDT	Dichlorodiphenyltrichloroethane
Simazine	6-Chloro-*N,N*′-diethyl-1,3,5-triazine-2,4-diamine
Toxaphene	Chlorinated camphene
Triallate	*S*-(2,3,3-Trichloro-2-propenyl)bis(1-methylethyl)carbamothioate
DCPA	Dimethyl 2,3,5,6-tetrachloro-1,4-benzenedicarboxylate
Picloram	4-Amino-3,5,6-trichloro-2-pyridinecarboxylic acid
Diuron	*N*′-(3,4-Dichlorophenyl)-*N,N*-dimethylurea
Monuron	*N*′-(3-Chlorophenyl)-*N,N*-dimethylurea
Terbacil	5-Chloro-3-(1,1-dimethylethyl)-6-methyl-2,4-(1*H*,3*H*)-pyrimidinedione
Cyanazine	2-[[4-Chloro-6-(ethylamino)-1,3,5-triazin-2-yl]amino]-2-methylpropionitrile
Aldicarb	2-Methyl-2(methylthio)propionaldehyde
Linuron	*N*′-(3, 4-Dichlorophenyl)-*N*-methyoxy-*N*-methylurea

REFERENCES

Armstrong, D. E. and Konrad, J. G. (1974). Nonbiological degradation of pesticides, in *Pesticides in Soil and Water*, (M. Guenzi, ed.) Soil Science Society of America, Madison, Wisconsin, p. 123.

Bresler, E. and Dagan, G. (1981). Convective and pore scale dispersive solute transport in unsaturated heterogeneous field, *Water Resour. Res. 17*: 1683.

Burkhard, N. and Guth, J. A. (1976). Photodegradation of atrazine, atraton and ametryne in aqueous solution with acetone as a photosensitizer, *Pestic. Sci. 7*: 65.

Burkhard, N. and Guth, J. A. (1979). Photolysis of organophosphorus insecticides on soil surfaces, *Pestic. Sci. 10*: 313.

Burkhard, N. and Guth, J. A. (1981). Chemical hydrolysis of 2-chloro-4,6-bis (alkylamino)-1,3,5-triazine herbicides and their breakdown in soil under the influence of adsorption, *Pestic. Sci. 12*: 45.

Carsel, R. F., Smith, C. N., Mulkey, L. A., Dean, D., and Jowis, P. (1984). *User's Manual for the Pesticide Root Zone Model (PRZM), Release I*, U.S. Environmental Protection Agency, EPA-600/3-84-109, Athens, Georgia.

Chiou, C. T., Peters, L. J., and Freed, V. H. (1979). A physical concept of soil-water equilibrium for nonionic organic compounds, *Science 206*: 831.

Choudhry, G. G. and Webster, G. R. B. (1985). Protocol guidelines for the investigations of photochemical fate of pesticides in water, air, and soils, *Res. Rev. 96*: 79.

Enfield, C. G. and Yates, S. R. (1990). Organic chemical transport to groundwater, in *Pesticides in the Soil Environment: Processes, Impacts, and Modeling* (H. H. Cheng, ed.), Soil Science Society of America, Madison, Wisconsin, p. 271.

Farmer, W. J. and Aochi, Y. (1987). Chemical conversion of pesticides in the soil-Water environment, in *Fate of Pesticides in the Environment* (J. W. Biggar and J. N. Sieber, eds.), Agricultural Experiment Station Pub. No. 3320, Division of Agriculture and Natural Resources, University of California, Oakland, California, p. 69.

Fontaine, D. D., Lehmann, R. G., and Miller, J. R. (1991). Soil adsorption of neutral and anionic forms of a sulfonamide herbicide, flumetsulam, *J. Environ. Qual. 20*: 759.

Gamerdinger, A. P., Wagenet, R. J., and van Genuchten, M. T. (1990). Application of two-site/two-region models for studying simultaneous nonequilibrium transport and degradation of pesticides, *Soil Sci. Soc. Am. J. 54*: 957.

Getzin, L. W. (1981). Dissipation of chlorpyrifos from dry soil surfaces, *J. Econ. Entomol. 74*: 707.

Ghadiri, H. and Rose, C. W. (1992). Sorbed chemical transport modeling, in *Modeling Chemical Transport in Soils* (H. Ghadiri and C. W. Rose, eds.), Lewis Publishers, Boca Raton, Florida, p. 15.

Glotfelty, D. E. (1981). *Atmospheric dispersion of pesticides from treated fields*, Ph.D. Dissertation, Univ. Maryland, College Park.

Glotfelty, D. E. and Caro, J. H. (1975). Introduction, transport and fate of persistent pesticides in the atmosphere, *Am. Chem. Soc. Symp. Ser. No. 17* (V. R. Deitz, ed.), American Chemical Society, Washington, D. C., p. 42.

Glotfelty, D. E., Leech, M. M., Jersey, J., and Taylor, A. W. (1989). Volatilization and wind erosion of soil surface applied atrazine, simazine, alachlor and toxaphene, *J. Agric. Food Chem. 37*: 546.

Glotfelty, D. E., Taylor, A. W., Turner, B. C., Zoller, W. H. (1984). Volatilization of surface applied pesticides from fallow soil, *J. Agric. Food Chem. 32*: 638.

Gohre, K. and Miller, G. C. (1983). Singlet oxygen generation on soil surfaces, *J. Agric. Food Chem. 31*: 1104.

Green, R. E. and Karickoff, S. W. (1990). Sorption estimates for modeling, in *Pesticides in the Soil Environment: Processes, Impacts, and Modeling*, (H. H. Cheng, ed.), Soil Science Society of America, Madison, Wisconsin, p. 79.

Green, R. E. and Khan, M. A. (1987). Pesticide movement in soil: mass flow and molecular diffusion, in *Fate of Pesticides in the Environment* (J. W. Biggar and J. N. Sieber, eds.), Agricultural Experiment Station Pub. No. 3320, Division of Agriculture and Natural Resources, University of California, Oakland, p. 87.

Grover, R., Shewchuk, S. R., Cessna, A. J., Smith, A. E., and Hunter, J. H. (1985). Fate of 2,4-D iso-octyl ester after application to a wheat field, *J. Environ. Qual. 14*: 203.

Gustafson, D. I. (1989). Groundwater ubiquity score: a simple method for assessing pesticide leachability, *Environ. Toxicol. Chem. 8*: 339.

Hamaker, J. W. (1975). The interpretation of soil leaching experiments, in *Environmental Dynamics of Pesticides* (R. Haque and V. H. Freed, eds.) Plenum Press, New York, p. 115.

Hance, R. J. (1967). The speed of attainment of sorption equilibria in some systems involving herbicides, *Weed Res. 7*: 29.

Harper, A, White, A. W. Jr., Bruce, R. R., Thomas, A. W. and Leonard, R. A.. (1976). Soil and microclimate effects on trifluralin volatilization, *J. Environ. Qual. 5*: 236.

Hartley, G. S. and Graham-Bryce, I. J. (1980). *Physical Principles of Pesticide Behavior*, Vol. I, Academic Press, New York.

Hutson, J. L. and Wagenet, R. J. (1989). Predicting the fate of herbicides in the soil environment, *Brighton Crop Protection Conf. 3*(9A-1): 1111.

Junge, C. E. (1975). Transport mechanisms for pesticides in the atmosphere, *Pure Appl. Chem. 42*: 95.

Jury, W. A., Farmer, W. J., and Spencer, W. F. (1984). Behavior assessment model for trace organics in soil. II. Chemical classification and parameter sensitivity, *J. Environ. Qual. 13*: 567.

Jury, W. A., Focht, D. D., and Farmer, W. J. (1987). Evaluation of pesticides groundwater pollution potential from standard indices of soil-chemical adsorption and biodegradation, *J. Environ. Qual. 16*: 422.

Jury, W. A., Grover, R., Specer, W. F., and Farmer, W. J. (1980). Modeling vapor losses of soil incorporated triallate, *J. Soil Sci. Soc. Am. 3*: 445.

Jury, W. A., Winer, A. M., Spencer, W. F., and Focht, D. D. (1987). Transport and transformation of organic chemicals in the soil-air-water ecosystem, *Rev. Environ. Contam. Toxicol. 99*: 120.

Klehr, M., Iwan, J., and Riemann, J. (1983). An experimental approach to the photolysis of pesticides adsorbed on soil: thidiazuron, *Pestic. Sci. 14*: 359.

Laskowski, D. A., Goring, C. A. I., McCall, P. J., and Swann, R. L. (1982). Terrestrial environment, in *Environmental Risk Analysis for Chemicals*, (R. A. Conway, ed.), Van Nostrand Reinhold, New York, p. 198.

Laskowski, D. A., Tillotson, P. M., Fontaine, D. D., and Martin, E. J. (1990). Probability modeling, *Phil. Trans. R. Soc. Lond. 329*: 383.

Leonard, R. A. (1990). Pesticide movement into surface waters, in *Pesticides in the Soil Environment: Processes, Impacts, and Modeling* (H. H. Cheng, ed.), Soil Science Society of America, Madison, Wisconsin, p. 303.

Letey, J. and Farmer, W. J. (1974). Movement of pesticides in soil, in *Pesticides in Soil and Water* (M. Guenzi, ed.), Soil Science Society of America, Madison, Wisconsin, p. 67.

McCall, P. J. and Gavit, P. D. (1986). Aqueous photolysis of triclopyr and its butoxyethyl ester and calculated environmental photodecomposition rates, *Environ. Toxicol. Chem. 5*: 879.

Merkel, M. G. and Bovey, R. W. (1974). Movement of pesticides in surface water, in *Pesticides in Soil and Water* (M. Guenzi, ed.), Soil Science Society of America, Madison, Wisconsin, p. 99.

Miller, G. C. and Crosby, D. G. (1982). Pesticides photoproducts: generation and significance, *J. Toxicol. Clin. Toxicol. 19*: 707.

Miller, G. C. and Hebert, V. R. (1987). Environmental photodecomposition of pesticides, in *Fate of Pesticides in the Environment* (J. W. Biggar and J. N. Sieber, eds.), Agricultural Experiment Station Pub. No. 3320, Division of Agriculture and Natural Resources, University of California, Oakland, p. 75.

Monteith, J. L. (1973). *Principles of Environmental Physics,* American Elsevier, New York, chapter 2.

Pitter, R. L. and Baum, E. J. (1975). Chemicals in the air: the atmospheric system and dispersal of chemicals, in *Environmental Dynamics of Pesticides* (R. Haque and V. H. Greed, eds.), Plenum Press, New York, p. 5.

Rao, P. S. C., Hornsby, A. G., and Jessup, R. E. (1985). Indices for ranking the potential for pesticide contamination of groundwater, *Soil Crop Sci. Soc. Florida Proc. 44*: 1.

Saltzman, S., Mingelgrim, U., and Yaron, B. (1976). Role of water in the hydrolysis of parathion and methylparathion on kaolinite, *J. Agric. Food Chem. 24*: 739.

Spencer, W. F. (1970). Distribution of pesticides between soil, water and air, in *Pesticides in the soil: Ecology, Degradation and Movement*, Michigan State University, East Lansing, p. 120.

Spencer, W. F., and Cliath, M. M. (1969). Vapor density of dieldrin, *Environ. Sci. Tech. 3*: 670.

Spencer, W. F. and Cliath, M. M.. (1970). Desorption of lindane from soil as related to vapor density, *Soil Sci. Soc. Am. Proc. 34*: 574.

Spencer, W. F., Farmer, W. J., and Jury, W. (1982). Review: behavior of organic chemicals at soil, air, water interfaces as related to predicting the transport and volatilization of organic pollutants, *Environ. Toxicol. Chem. 1*: 17.

Swann, R. L., McCall, P. J., and Unger, S. M. (1982). Volatility of pesticides From soil surfaces, in *Handbook of Chemical Property Estimation Methods* (W. J. Layman, W. F. Reehl, and D. H. Rosenblatt, eds.), CRC Press, Boca Raton, Florida, p. 16.

Taylor, A. W. and Glotfelty, D. E. (1988). Evaporation from soils and crops in *Environmental Chemistry of Herbicides,* Vol. I., (R. Grover, ed.), CRC press Boca Raton, Florida, p. 89.

Taylor, A. W., Glotfelty, D. E., Glass, B. L., Freeman, H. P., and Edwards, W. M. (1976). Volatilization of dieldrin and heptachlor from a maize field, *J. Agric. Food Chem. 24*: 625.

Thorburn, P., Shaw, R., and Gordon, I. (1992). Modeling salt transport in the landscape, in *Modeling Chemical Transport in Soils*, (H. Ghadiri and C. W. Rose, eds.), Lewis Publishers, Boca Raton, Florida, p. 145.

Tillotson, P. M. and Fontaine, D. D. (1991). *A User's Guide to GRASP*, Midland, Michigan, DowElanco, pp . 1–194.

Turner, B. C., Gloltfelty, D. E., Taylor, A. W., and Watson, D. R. (1978). Volatilization of microencapsulated and conventionally applied chloropham in the field, *J. Agron. 70*: 933.

United States Department of Agriculture, Soil Conservation Service. (1991). *State Soil Geographic Data Base (STATSGO) Data Users Guide*, USDA/SCS Miscellaneous Publication 1492, U.S. Government Printing Office, Washington, D.C.

van Genuchten, M. T. and Wagenet, R. J. (1989). Two-site/two-region models for pesticide transport and degradation: theoretical development and analytical solutions, *Soil Sci. Soc. Am. J. 53*: 1303.

Wagenet, R. J. and Hutson, J. L. (1986). Predicting the fate of nonvolatile pesticides in the unsaturated zone, *J. Environ. Qual. 15*: 315.

Wagenet, R. J. and Rao, P. S. C. (1990). Modeling pesticide fate in soils, in *Pesticides in the Soil Environment: Processes, Impacts, and Modeling*, (H. H. Cheng, ed.), Soil Science Society of America, Madison, Wisconsin, p. 351.

Wauchope, R. D. (1987). The pesticide content of surface water draining from agricultural fields: a review, *J. Environ. Qual. 7*: 459.

Wauchope, R. D., Buttler, T. M., Hornsby, A. G., Augustijn Beckers, P. W. M., and Burt, J. P. (1992). The SCS/ARS/CES pesticide properties database for environmental decision-making, *Rev. Environ. Contam. Toxicol. 123*: 1.

Wauchope, R. D. and Leonard, R. A. (1980). Maximum pesticides concentrations in agricultural runoff: a semiempirical prediction formula, *J. Environ. Qual. 9*: 665.

White, L. A. Jr., Harper, R. A., Turnball, L., and Turnbull, J. W. (1977). Trifluralin volatilization losses from a soybean field, *J. Environ. Qual. 6*: 105.

Wolfe, N. L., Mingelgrin, U., and Miller, G. C. (1990). Abiotic transformations in water, sediments, and soil, in *Pesticides in the Soil Environment: Processes, Impacts, and Modeling* (H. H. Cheng, ed.), Soil Science Society of America, Madison, Wisconsin, p. 103.

Zepp, R. G., Baughman, G. L., and Schlotzhauer, P. F. (1981). Comparison of photochemical behavior of various humic substances in water. I. Sunlight induced reactions of aquatic pollutants photosensitized by humic substances, *Chemosphere 10*: 109.

ADDITIONAL READING

Alexander, M. (1977). *Introduction to Soil Microbiology*, 2nd Edition, John Wiley and Sons, New York.

Atlas, R. M. and Bartha, R. (1987). *Microbial Ecology: Fundamentals and Applications*, 2nd Edition, Benjamin/Cummings, Menlo Park, CA.

Bollag, J.-M. and Liu, S.-Y. (1990). Biological transformation processes of pesticides, in *Pesticides in the Soil Environment: Processes, Impacts, and Modelling*, Soil Science Society of America, Madison, Wisconsin pp. 169–211.

Sims, G. K., Turco, R. F., and Buhler, D. D. (1992). Residue management impact on the environment, in *Managing Agricultural Residues* (P. Unger, ed.) CRC Press, Boca Raton, Florida.

Wolfe, N. L., Mingelgrin, U., and Miller, G. C. (1990). Abiotic transformations in water, sediments, and soil, in *Pesticides in the Soil Environment: Processes, Impacts, and Modelling*, Soil Science Society of America, Madison, Wisconsin, pp. 103–168.

9

Biological Weed Management

John Cardina
The Ohio State University, Wooster, Ohio

I. INTRODUCTION

This chapter focuses on the use of biological agents for the management of weeds. *Biological management* is a broad term for the exploitation of living organisms, or their products, to reduce or prevent the growth and reproduction of weeds. The organisms that have been used for biological weed management are diverse and include various types of animals as well as plants, fungi, and bacteria. Biological management seeks to manipulate one or more of these organisms, or agents, in ways that enhance processes that hinder the normal development of weeds at one or more phases of their life cycle.

A. The Context of Biological Management

Many reviews, books, and popular articles have been written about biological *control* of weeds with insects or plant pathogens. This is only one approach to biological *management*. Biological control is a type of biological weed management that uses organisms other than the associated crop to reduce the growth and adverse effects of weeds. Biological management differs from biological control in the nature of organisms used and in the objectives to be achieved by their use. Biological control involves the deliberate use of organisms such as insects or fungi to control weeds; biological management also involves the use of crop competition, allelopathy, resistant varieties, natural chemical agents, and other approaches. The difference in objectives is due to the different connotations of "control" and "management." The objective of biological control is often to obtain the same dramatic reduction in weed populations as may be obtained by chemical weed control. Biological weed management objectives are broader and include more subtle restrictions and constraints on weed populations resulting from the manipulation of organisms or the environment. Without minimizing the importance of biological control, it is important to recognize that biological management encompasses a diversity of agents and approaches that are equally important, though often underexploited.

History and familiarity with chemical weed control may lead practitioners to consider weed management only in the context of products applied to control weeds. Biological management tactics are not listed in the charts that show weed response to various commercial products that many growers use as a guide to weed control. Biological management does not rely primarily on commercial products that a grower can pick off a shelf and purchase. Whereas herbicides are labeled to indicate which weeds they will kill, few biological management tactics are guaranteed, by themselves, to kill a particular weed. Instead, biological approaches to weed management are part of an overall cropping system, consisting of many interacting components, designed to reduce the impact of weeds on potential crop productivity. The application of herbicides is rather simple, and their effects are often dramatically rapid. The implementation of biological management, however, is relatively complex and may require several years to reduce the impact of weeds on crops.

Biological management is more complex than chemical weed control because it requires 1) long-term planning, 2) multiple tactics, and/or 3) manipulation of cropping systems. It is also more complex than chemical control because it deals with variability in behavior and response of multiple organisms interacting with the environment. Moreover, biological management is oriented toward prevention of weed problems, as opposed to the therapeutic approach taken in chemical (and mycoherbicidal) weed management. Most weed control in modern agriculture may be considered "chemical intensive," and biological weed management may be considered both "biologically intensive" and "information intensive" in that it integrates knowledge of weed, crop, and soil ecology with biotechnology to arrive at intelligent approaches to reducing the impact of weeds on crops.

Biological management uses biological processes that are always at work and that can be increased and manipulated to the detriment of weeds. Biological stress is always happening at some level in weed populations that infest crops. Weeds are always constrained to some extent, by factors such as disease, arthropod feeding, predispersal and postdispersal seed predation by various animals, allelopathic effects, competition from neighboring plants, and other factors (Table 1). Biological management seeks to manipulate these factors to stress weeds throughout their life cycle and reduce their ability to survive, reproduce, and compete with crops.

B. Classifications Within Biological Management

Pesticides are commonly classified by properties such as chemical structure, mode of action, and selectivity. Attempting, by analogy, to classify biological management strategies in a similar fashion illustrates the complexity of biological approaches. Specific agents such as insects and pathogens can be classified taxonomically as well as by host range, type of damage caused (galls, necrosis, etc.), or part of the plant damaged. For many biological management tactics the mechanisms of weed suppression and modes of action are not known. The "agents" include plants, fungi, bacteria, insects, nematodes, fish, birds, and grazing animals that exert one or more physical, mechanical, or physiological stresses that are often poorly understood. Plants (the principal crop, cover crop, smother crop, rotation crop), nematodes, and some insects compete for limited resources. Large animals, leaf-feeding insects, and some fungi and bacteria are generally defoliators. Other mechanisms of weed suppression for biological agents include complex processes such as allelopathy, herbivory, parasitism, and predation. A single agent may affect weeds by more than one mechanism, such as a crop that competes for light and is allelopathic, or an insect that defoliates a weed and thereby reduces its competitiveness. Multiple biological agents with varying mechanisms of suppression are more effective than a single agent or a single mechanism of activity against weeds. Any attempt to reduce weed growth will influence the environment of all pests and natural enemies in the plant community, often resulting in dramatic shifts in populations.

There are many approaches, tactics, and agents that can be used in biological weed management, only a few of which are highlighted here. Appropriate biological management approaches vary according to specific grower situations, environmental conditions, and cultural practices. This chapter therefore examines general categories of biological management agents and examples of their adaptations. The emphasis is on the agents themselves and how they can be integrated into a cropping system. What follows is an overview of some of the approaches to biological weed management. The categories discussed below are somewhat artificial, and there may be considerable overlap among them, but with an understanding of the possibilities for biological management, growers can apply these concepts to their particular weed and crop management situation.

Table 1 Generalized Weed Life Cycle and Typical Biological Constraints Attributed to Disease, Arthropods, Plants, and Grazing Animals

	Potential biological constraints on weed populations			
Weed life cycle	Diseases	Arthropods	Plants	Grazers
Germinating seedling	Seed rot Dampling off Root rot	Many large and small consumers	Allelopathic and physical suppression	
Vegetative plant	Root rots and galls Stem rots and cankers Stem blights Vascular wilts Leaf spots, blights, and mosiacs Anthracnoses	Root and stem borers Phloem feeders Defoliators Leaf miners Disease transmitters	Allelopathic suppression Competition for light, water, nutrients	Defoliation
Reproductive plant (Annuals) (Perennials)	All the above plus: Flower infections Fruit rots Mildews Rusts Smuts	Root and stem borers Phloem feeders Defoliators Leaf miners Disease transmitters Bud and flower Feeders Predispersal seed predators	Competition for light, water, nutrients	Defoliation Consumption of flowers and Developing seeds
Vegetative Propagules	Rots and cankers Blights	Root and stem borers		Rooting of tubers
Seeds in soil	Seed rots Molds	Seed predators		

II. PLANTS

Plants provide a rich source of biological material that can be manipulated for biological weed management. Organisms have profound influences on one another when they exploit the same physical environment and compete for the same resources. Thus, it is logical that the manipulation of plants is of interest in the management of weeds. The management of harvestable crop plants is of central concern in their interaction with weeds, but auxiliary crops, such as cover crops, rotation crops, smother crops, and companion crops, are important resources that can be exploited in weed management.

The mechanisms by which these crops influence weed populations are not well understood but often are attributed to physical factors that are responsible for plant competition or to chemical factors that account for allelopathic effects. An important limitation of plants in weed management is general lack of selectivity. Most allelochemicals and cover crops exert their effects on crops as well as on weeds. Therefore, the main crop should be separated in time or space from an auxiliary crop used to suppress weeds. Some ways plants can be manipulated in weed management are discussed in the following sections.

A. Crop Competition

The first consideration in the defense of a crop against any pest is the crop itself. Although crops are rarely considered biological control agents, competition of crops with weeds is one of the principal ways that the growth of weeds is reduced in crop production systems. Proof of this is often observed where a planter malfunction leaves a section of row without the normal density of crop plants. Lacking competition from the principal crop, weed growth is abundant in these spots even though a herbicide or cultivation is applied uniformly to the entire field. The same response can be seen where an opening occurs in a pasture sward and lack of competition results in weed encroachment in spite of uniform grazing management. Crop competition may reduce weed biomass and seed production 80% or more (Minotti and Sweet, 1981). Because of its effect on weed seed production and return of seed to the seed bank, crop competition is important in the long-term management of weed populations.

Competitiveness of crops is determined by earliness of emergence, seedling disease resistance, relative growth rate, canopy architecture, root growth, production of allelochemicals, and tolerance of adverse environmental conditions. Relatively little attention has been given to development of crop varieties that tolerate weeds. One reason is that the "crop" is more than simply the plant itself; it also involves planting pattern, row spacing, fertility practices, and other cultural practices that influence competitiveness against weeds. These are discussed in more detail in Chapters 2 and 3. Herbicide resistance is one biological factor that makes some varieties more competitive where the appropriate herbicide is applied. The development of herbicide-resistant crops is, in a sense, a biological management tactic that offers the opportunity to substitute herbicides with less potential hazard for herbicides that pose greater potential danger. Since this approach relies on herbicide use, it is considered in more detail elsewhere (Chapter 5).

Principles of Crop Competition

Plants growing in close proximity may compete for one or more limiting resources required for growth, including light, soil water, and nutrients. Of secondary importance is competition for space and symbiotic organisms. Plants respond to the presence of neighbors by altering their growth form or growth rate in a way that increases their potential for resource capture. Allelopathy may also be a factor in crop competition with weeds, but well-documented examples are difficult to find. Thus, any crop characteristic or management practice that tilts the balance of competition toward the crop will reduce the proportion of resources available to weeds, thereby reducing their growth potential and impact on crop production.

Time is an important factor in crop competition and is described by two benchmarks. The first is the critical weed-free period, the length of time the crop must be free of competition to prevent yield loss. The second is the tolerable duration of competition, the length of time weeds can remain in a crop before they reduce crop yield. The interval between these two benchmarks is referred to as the critical period of weed control. Although these may vary with cultivar, environmental conditions, and cultural practices, they can be used as a guide to weed management. For example, the critical period of weed control began between the 6th and 9th leaf stages and ended about the 14th leaf stage in corn (Hall et al., 1992). In some cases, the difference of only a few days in emergence or canopy closure can determine whether weeds will significantly affect crop yields. Application of the critical weed-free period can be used in determining the duration of control required by a management tactic and in determining the time of seeding a companion crop so that competition will be minimized. The tolerable duration of competition can be used in timing the application of control measures. These critical periods vary over years and with growing conditions (Hall et al., 1992) and should be considered as general guidelines only.

Biological weed management requires the adoption of cultural practices that maximize crop competitiveness and yield potential. This is more important than trying to choose a crop or a cultivar on the basis of morphological characteristics or growth habit thought to be related to competitiveness. Crops are most competitive with weeds when cultural practices provide early emergence, rapid canopy closure, rapid root development, rapid growth rates, high harvest index, and freedom from insects and diseases. For example, fertility practices that enhance crop competitiveness may include a "starter" supply of P for crop establishment during cool periods and proper amounts of additional fertilizer applied in the crop root zone at a time it is most needed by the crop. Rapid, early crop emergence can be encouraged by proper planting time, planting depth, and starter fertilizer. Rapid canopy closure can be enhanced by choice of planting pattern in addition to factors that allow for early emergence. Crop root development can be increased by appropriate tillage, seed placement, fertilizer placement, and irrigation practices. Crop growth rate, harvest index, and freedom from pests are, of course, affected by many management practices. Competition between weeds and crops is complex, yet the basic concept in enhancing crop competitiveness is to maximize access to and capture of resources by the crop at the expense of the weeds. Most competition is for light, water, or nutrients, so any management practice that selectively promotes rapid crop canopy or root development will increase crop competitiveness with weeds. Although these practices may be considered cultural rather than strictly biological, their effect is to increase crop competition and it is this competition—not the cultural practice—that suppresses weed growth.

Choice of Crops

Many crops have competitive characteristics that can be exploited in weed management. Plant vigor, competitive ability, robustness, and tolerance of environmental extremes are among the traits that early agriculturalists selected for in developing crops. Modern breeding and variety development programs continue to favor these traits, while novel DNA transfer techniques have conferred traits such as disease and insect resistance that increase crop competitiveness indirectly. Many crops are less sensitive to shade than the weeds that commonly infest them (Shetty et al., 1982). For example, sorghum and cowpea are particularly effective competitors for light compared to yellow nutsedge (Keely and Thullen, 1978). Tall, rapidly growing crops like corn, sorghum, and sugarcane can compete effectively for light against weeds that are slow growing, late emerging, or shade sensitive. Crops such as soybean and peanut, that form a dense canopy when planted in appropriate row spacings, can be effective competitors with late emerging weeds. This is because light quality and quantity are affected as light passes through a dense crop canopy. As green light is reflected and red light is absorbed by the canopy, the light reaching the soil surface is enriched in the far-red, inhibiting the germination of many weeds. Weeds that germinate under the canopy receive little radiant energy for early growth and ideally are unable to penetrate the crop canopy to compete effectively.

Growers have adopted cultural practices that use crop competition to suppress weeds for crops where other weed control options are either limited or too expensive. Herbicide use in wheat, oats, and hay crops is minimal in many areas because these crops are strong competitors once they are established. However, it is impractical to rely on crop competition alone as an approach to weed management for any crop. Even in the examples just cited, some land preparation or preplant herbicide normally precedes planting to assure good crop emergence and to reduce weed competition during establishment. Potential yield losses will be economically intolerable if no effort is made to control weeds beyond preparing land and planting a crop.

Specialized cultural practices must be adopted to reduce weed effects on some crops, such as onions, garlic, carrots, and beets, that are relatively poor competitors. The potential exploitation of above- and below-ground resources is limited in these crops and in any other plant that is restricted in size, growth rate, or life span. For example, to prevent harvest losses, onions must be maintained free of weeds for at least 12 weeks, a longer period than for almost any other crop (Wicks et al., 1973).

Choice of Cultivars

Crop cultivars can vary considerably in their ability to compete with weeds. Many studies have been conducted to measure differences in competitiveness among crop cultivars. For example, 'Amsoy 71' soybeans were more competitive with both early and late emerging weeds than were 'Beeson' soybeans (Burnside, 1979). 'Bragg' and 'Semmes' soybeans, of intermediate maturity, were more competitive with johnsongrass and common cocklebur than were four other varieties differing in maturity (McWhorter and Hartwig, 1972). Additional examples of differences in weed competitiveness among cultivars are listed in Table 2.

There is little agreement regarding characteristics that contribute to differential competitiveness of crop cultivars. The competitiveness of eight soybean cultivars was associated with greater leaf area expansion and branch production during the first month of growth, but not with maturity group, height, or growth habit (Forcella, 1987). In wheat, more competitive cultivars produced greater biomass and grain yield but did not intercept more light and were not taller than less competitive cultivars (Ramsel and Wicks, 1988). However, other research has shown that competitiveness of wheat cultivars was associated with greater height (Appleby et al., 1976; Challaiah et al., 1986). Tall rice cultivars that produced greater dry matter and long leaves were more competitive than short, semidwarf, early maturing, and erect-leaf cultivars (Smith, 1988). The minimum height of upland rice cultivars necessary to suppress weeds adequately was estimated to be 1.0 to 1.15 m (Garrity et al., 1992).

It is difficult to generalize about competitiveness of cultivars because of the interaction between cultivars and cultural practices such as row spacing, fertility practices, or planting time. Different cultivars respond favorably, in competitiveness and productivity, to different management and environmental

Table 2 Examples of Crop Cultivars Showing Differences in Competitiveness Against Weeds

Crop	Weeds	Varietal ranking	Mechanism suggested	Reference
Alfalfa	Smooth bromegrass	ScMf3713 > Beaver = Rambler = Rangelander = Drylander = Peace = Anik	Greater fall dormancy and superior winter hardiness	Bittman et al., 1991
Barley	Wild oat	Fairfield > Betzes = Steptoe = Laurier = Bonanza (at high P) Betzes > Steptoe = Laurier = Bonanza (at low P)		Konesky et al., 1989
Bean, field	Hairy beggarticks, tropical spiderwort	Indeterminate bush or trailing > determinant bush	LAI, leaf size, growth rate	Wortmann, 1993
Cotton	Common cocklebur Tall morningglory Redroot pigweed Prostrate spurge	Normal leaf type, okra leaf type > super okra leaf type		Andres et al., 1974
	Spurred anoda	Deltapine > DES21326-04 = Stoneville 213		Chandler and Meredith, 1983
	Johnsongrass	1209-619-7 = TAMCOT SP37-H = Stoneville 213		Bridges and Chandler, 1988
Peanuts	Common cocklebur	F8143B > BL-8 > BL-10 > NC7	Longer duration of vegetative development, higher rate of partitioning of assimilates to pod development	Fiebig et al., 1991
Potato	Redroot pigweed,	Red Pontiac > Red Norland		Nelson and Giles, 1989
Rice	Barnyardgrass	Starbonnet > Nova 66 > Bluebelle	Time required for maturity	Smith, 1974
	Barnyardgrass	Newbonnet > Lemont	Height of rice	Stauber et al., 1991
	Goosegrass	B3622E-Tb-5-4-4 = Salumpikit = C894-21 = UPLRi7 > CR156-5021-20 = IR9782-111-2-1-2 = IR9101-124-11	Plant height and LAI	Garrity et al., 1992
Sorghum	Redroot pigweed Green foxtail Tall waterhemp Large crabgrass	RS671 = RS690 = AKS663 > AKS614 = NB505 = RS633	Rapidity of germination, emergence and early root and shoot growth	Burnside and Wicks, 1972

(continued)

Table 2 Continued

Crop	Weeds	Varietal ranking	Mechanism suggested	Reference
Soybean	Velvetleaf Tall waterhemp Large crabgrass	Amsoy 71 > Beeson	Shading	Burnside, 1979
	Tall waterhemp Green foxtail	Harosoy 63 = Amsoy = Corsoy > Hawkeye 63 = Shelby = Lindarin 63		Burnside, 1972
	Johnsongrass	Bragg > Davis = Semmes = Lee = Jackson = Hardee		McWhorter and Hartwig, 1972
	Common cocklebur	Semmes > Bragg = Davis = Lee = Jackson = Hardee		McWhorter and Hartwig, 1972
	Common cocklebur	Bragg > Lee = Hill = Semmes = Forrest		McWhorter and Barrentine, 1975
	Common cocklebur	Centennial > Forrest (for first 8 weeks after emergence) Forrest > Centennial (after 8 weeks from emergence)		Monks and Oliver, 1988
	Pitted morningglory	Govan > Bragg > Ransom	Earlier shade development	Murdock et al., 1986
Wheat, spring	Wild oat	Neepawa > HY355 > HY320		Kirkland and Hunter, 1991
Wheat, winter	Barnyardgrass	Centurk 78 > Bennett = Eagle	Not a result of shading; authors suggest allelopathy	Ramsel and Wicks, 1988
	Downy brome	Centura > Centurk 78 (at Lincoln) SD75284 > Centurk 78 (at North Platte)	Height of cultivar	Challaiah et al., 1986
	Wild oat	WH147 = HD2285 > HD2009 = WH291 = S308		Balyan et al., 1991
	Annual grasses	Lancota > Homestead	Not height, light interception	Wicks et al., 1986

conditions. Therefore, cultivars must be compared where conditions enable them to exhibit their competitive potential. Researchers have studied cultivars varying in plant height, leaf shape, canopy architecture, root morphology, or other characteristics that might confer differences in competitiveness against weeds. Tall, large leaved, wide canopied cultivars are not necessarily more competitive, and more competitive cultivars are not necessarily higher yielding (Weaver et al., 1978; Bridges and Chandler, 1988; Challaiah et al., 1986). For example, indeterminant soybean cultivars are more bushy and are about 50% taller than semidwarf cultivars. These characteristics make the indeterminants more competitive when both cultivars are planted in wide row spacings. In narrow row spacings and high densities, the yield of indeterminant cultivars is reduced by lodging, yet semidwarf cultivars stand and yield well. Cooper (1977) demonstrated that indeterminant cultivars were most productive (570 kg/ha) when planted in 76-cm row spacings using 555,800 seeds/ha, whereas the yield of semidwarf soybean cultivars was highest (870 kg/ha) when planted in 17-cm row spacings using 741,000 seeds/ha.

Multiple Cropping

The practice of growing two or more crops on the same field in a year, or multiple cropping, is a centuries-old method of maximizing the productivity of land. One advantage of multiple cropping is that by keeping a vegetative cover on the soil most of the year, solar radiation is intercepted by crops at the expense of weeds. One or more crops in the system may also produce allelochemicals that may contribute to weed control. Because the time land lies idle is minimized, there is little opportunity for establishment and unhindered growth of weeds.

Sequential cropping and relay intercropping are multiple cropping methods adaptable to mechanized agriculture that are common in grain production areas of the United States. Cropping systems where principal summer crops are planted into winter annual cover crops or living mulch crops, as discussed later, are examples of sequential cropping. In relay intercropping systems, a second crop is planted before harvesting the first crop. An example is direct-seeding soybeans into standing wheat. In systems developed in Ohio, winter wheat is planted in a skip-row pattern to facilitate interplanting of soybeans (Jeffers, 1987). The wheat is harvested at 22% to 25% moisture to provide early "release" of soybeans and to prevent interference of the soybeans with wheat harvest. Full-season soybean varieties are interplanted in early May to minimize damage to wheat. A vigorous stand of wheat is needed to provide adequate shade to compete with weeds and to suppress early growth of soybeans. Since the wheat canopy is already well formed during the peak period of annual weed emergence in May and June, light attenuation by the crop canopy helps reduce weed emergence and early growth. After wheat harvest, the soybean canopy fills in rapidly and provides season-long competition. The principal weed problems in this system are perennial broadleaf weeds that become established in wheat and may attain considerable growth before soybeans develop a complete canopy.

Few studies have been conducted to examine weed populations in relay cropping systems. In a 2-year study at Rodale Research Center weed biomass was reduced in intercropped soybeans compared to monoculture soybeans (Peters, 1992). An intercropped barley-soybean system without herbicides provided 70% to 93% control of fall panicum, redroot pigweed, and giant ragweed (Prostko and Ilnicki, 1988).

Intercropping (two or more crops grown on the same field at the same time) is often practiced in highly advanced cropping systems in areas of the world where temperature and moisture do not limit plant growth most of the year. The implications of weed infestations and interference in intercropping systems are particularly important in tropical and subtropical countries, where hand weeding is the most common method of weed control for small-scale farmers. The success of intercropping depends on minimizing competition among crop species for available light, water, and nutrients by the choice of crops, spatial patterns, and planting schedules. Plants with dissimilar rooting patterns or canopy architecture exploit resources from different niches and therefore compete less with each other than do plants with very similar morphologies and growth habits. Crops and planting patterns are chosen in intercropping systems to exploit the soil and aerial environment for crop production and to leave little opportunity for weed establishment. Thus, crop species diversity permits more thorough spatial and temporal exploitation of resources, which contributes to more effective resource capture by crops at the expense of weeds. The extended cropping period in many tropical and subtropical environments may allow indigenous biological control agents to maintain higher population levels (Litsinger and Moody, 1976). Ideally, crops with different life cycles can be included in a system so that weed populations are interrupted at different times during a rotation. In addition, rotations are designed so that no group of weeds develops without disturbance. The crop sequence must be chosen to allow for rapid development of a persistent canopy and for vigorous crop growth to limit light penetration to the soil surface and to maintain competition against weeds. The duration of the growing season and of crop rotation is long compared with that of annual crop production in temperate climates. Therefore, limiting weed seed production is particularly important to reduce weed buildup and to prevent interference from multiple generations of weeds within a rotation.

The diversity of soils, crops, and cropping systems on which intercropping is practiced in traditional agricultural systems makes generalization about biological weed management practices difficult. For example, the Medje of Zaire grow about 80 varieties of 30 food crop species (Miracle, 1967), and in

Abakaliki, Nigeria, more than 50 crop species are grown in intercropped mixtures within a field (Okigbo and Greenland, 1977). Complex intercropping systems may include crops for a variety of uses, including staple foods, fruits, vegetables, condiment plants, ornamental plants, and gourds as well as hedges, trees, and fences to support climbing beans or yams. Plants are often grown on mounds with planting times and spatial arrangements that ensure adequate crop cover to compete with weeds.

The effect of multiple cropping systems on weeds has generally been attributed to the efficiency with which the crops intercept light and thereby reduce the energy available for weed growth. Intercropping can be most effective in weed suppression if crops are chosen for their efficiency in shading weeds. Multiple crop canopies can be manipulated to create desired shade levels that suppress grass and broadleaf weeds (Shetty et al., 1982). Cowpea and mung bean develop ground cover rapidly and can suppress weed growth, decrease weeding costs, and increase hand-weeding efficiency in intercrops with sorghum (Shetty and Krantz, 1980).

B. Allelopathy

Allelopathy is an adverse effect on plants caused by biochemicals that are produced by a living plant or by the microbial degradation of a plant. Allelopathy has been implicated as one means by which weeds interfere with crops. There is also evidence that allelopathic crops might be able to reduce weed emergence and growth. Selection of allelopathic crop germplasm, use of allelopathic rotational or companion crops, and production of natural product herbicides are among the ways that allelopathy may be exploited for weed management.

The use of allelopathic cover crops, rotation crops, or companion crops is an appealing biological weed management tactic, especially in reduced tillage and no-tillage farming systems. Crops like rye have been planted to compete with weeds, cover the soil in the winter, and improve soil tilth when they are plowed or disked in the spring. Part of the weed suppression may be due to physical impedance of seedling growth or to shading. These effects are difficult to separate from chemical inhibition due to the action of allelochemicals produced by the plant or during degradation of the plant residue. Toxins that inhibit weeds are contained in many plant residues or are produced by microbes in the degradation of crop residues that remain in the field. Table 3 lists some examples of mulch, cover, and rotation crops as well as principal crops where weed suppression by allelopathy has been demonstrated or strongly implicated. The discussion that follows deals with cases where the primary means of weed suppression is attributed to allelopathy, even though other factors may also be involved.

Table 3 Examples of Mulch, Cover, or Rotation Crops Where Weed Suppression Has Been Attributed to Allelopathy

Allelopathic crop	Management	Weeds suppressed	Associated crop	Reference
Barley	Cover	Chickweed		Overland, 1966
		Shepherd's-purse		
Beet	Rotation	Corn cockle		Funke, 1941
Cassava	Principal crop	Pigweed spp.		Altieri and Doll, 1978
		Ivyleaf red morningglory		
		Large crabgrass		
Cucumber	N/A	Wild-proso millet		Lockerman and Putmnam, 1978
		Barnyardgrass		
		Redroot pigweed		
		White mustard		
Hairy vetch	Cover	Goosegrass, stinkgrass	Maize	Teasdale et al., 1991
Horsebean	Companion	Not specified	Maize	Putnam and Duke, 1978
Lupine	Rotation	Common lambsquarters		Rice, 1983
		Redroot pigweed		
Maize	Rotation	Common lambsquarters		Rice, 1983
		Redroot pigweed		
Oat	Rotation	Wild mustard		Fay and Duke, 1977
		Redroot pigweed		

Oat	Rotation	Corn poppy		Martin and Rademacher, 1960
Oat, Rye	Rotation	Wild mustard		Martin and Rademacher, 1960
Oat, Rye	Cover	Broadleaf weeds	Tobacco	Shilling et al., 1986
Rye	Rotation	False chamomile		Martin and Rademacher, 1960
Rye	Cover	Barnyardgrass Redroot pigweed		Barnes and Putnam, 1983
Rye, wheat, sorghum barley, sudangrass	Cover	Various weeds	Large seeded legumes, perennial tree crops,	Putnam, 1988
Sorghum		Common purslane Smooth crabgrass		Putnam and DeFrank, 1983
Sorghum	Rotation	Various weeds	Snap beans	
Sunflower	Rotation	Horseweed Hairy coneflower Japanese brome		Wilson and Rice, 1968
Sunflower	Principal crop	Broadleaf weeds		Leather, 1983
Sweet potato	Principal crop	Yellow nutsedge		Harrison and Peterson, 1986
Wheat	Rotation	Ivyleaf morningglory Velvetleaf		Steinsiek et al., 1982

Allelopathic Crops

A direct way to use allelopathy is to plant crops that produce allelopathic substances that suppress the germination or development of neighboring weeds. Putnam and Duke (1974) screened 526 accessions of cucumber in sand culture and identified superior weed suppressing types whose leachates were more toxic to indicator species. Subsequent research indicated that selected cucumber accessions did not inhibit weeds consistently in the field (Lockerman and Putnam, 1979). In a study of 3000 oat accessions, researchers identified one accession, PI266281, that inhibited wild mustard growth and produced up to three times as much of the allelochemical scopoletin as the standard oat cultivar 'Gary' (Fay and Duke, 1977). Allelopathic potential in rice germplasm was studied by Lin and coworkers (1992). They identified five accessions that reduced weight of aquatic weeds 80% to 91% compared to commercial cultivars and suppressed weeds in a band up to 48 cm wide. Superior weed suppression has been attributed to allelopathy for cultivars or accessions of sunflower (Leather, 1983), sweet potato (Harrison and Peterson, 1986), soybeans (James et al., 1988; Rose et al., 1984), and sorghum (Alsaadawi et al., 1986). There are no published references on attempts to incorporate allelopathy into commercial cultivars.

The choice of rotation crop can greatly affect weed populations in subsequent seasons, a phenomenon that has been attributed to allelopathy. For example, yellow foxtail populations were lower in flax preceded by flax than in flax preceded by oats, wheat, corn, or soybeans (Kommendahl and Linck, 1957). The germination of common chickweed and shepherd's-purse was inhibited by previous crops of barley (Overland, 1966). Rye is a crop that has often been associated with suppression of growth of succeeding crops and weeds (Rice, 1983). Including rye in a rotation may help control annual broadleaf weeds and seedlings of perennials such as dandelion (Phillips and Young, 1973). Sorghum crops have also been reported to reduce seedling growth of broadleaf weeds in subsequent crops (Einhellig and Leather, 1988). Prior cropping with a sorghum-sudangrass hybrid or with foxtail millet has been shown to suppress broadleaf weeds and enhance alfalfa establishment (Forney et al., 1985). Sunflower-oat rotations have resulted in decreased grass and broadleaf weed infestations (Leather, 1983).

The allelopathic potential of decaying plant residues has received considerable attention in weed management. Putnam et al. (1983) planted sorghums, wheat, rye, oats, and barley; killed them when they reached a height of 40 to 50 cm; and left their residues on the soil surface. Up to 95% control of certain weeds was obtained for 30 to 60 days. Populations of common purslane and smooth crabgrass were reduced 70% and 98%, respectively, by sorghum residues without injury to larger seeded vegetables (Putnam and DeFrank, 1983). Sunflower planted in early spring and incorporated into the soil after 45 days of growth reduced weed density 55% in soybeans and 81% in cotton, an effect attributed to allelopathy (Semidey et al., 1992). Hairy vetch and crimson clover debris have allelopathic potential against weeds such as pitted morningglory, with little impact on corn or cotton (White et al., 1989).

Management Considerations

An important aspect of the use of allelopathy in weed management is the potential for crop injury, especially for small seeded vegetable crops. For example, lettuce is very susceptible to residues of barley, oats, rye, sorghum, sudangrass, and wheat (Putnam and DeFrank, 1983). Stands of radish were reduced by all these crop residues except those of rye, whereas tomato stands were reduced by residues of barley, sorghum, sorghum × sudangrass, and wheat. Larger seeded vegetables, such as beans, cucumbers, and peas, were not adversely affected by these crop residues. Rye residues have been shown to reduce the growth of many crops including corn, soybean, tomato, lettuce, and tobacco (Putnam, 1988). In 4 years of studies at two locations, rye residue reduced stands and yields of no-till corn and soybeans, especially where previous crop residues were present (Eckert, 1988). Residues of corn, soybean, and grass hay reduced germination and early growth of oat, soybean, and corn (Martin et al., 1990).

Management can be directed toward increasing the production and timing the release of allelochemical(s) to obtain maximum weed suppression. Many factors influence the quantity of secondary metabolites, including allelochemicals produced by plants. In general, any factor that induces stress will increase secondary metabolite production. The most commonly discussed stress factors include light (quality, intensity, duration), mineral deficiencies, drought, temperature extremes, and plant growth regulators (Aldrich, 1984; Rice, 1983). Some plants produce toxic defensive chemicals in response to defoliation by herbivores (Rhoades, 1985). For example, alkaloid levels in lupine species increased by a factor of 4 within hours after partial defoliation (Wink, 1983).

Stress factors that may be most easily manipulated in weed management are nitrogen deficiency, water stress, and defoliation. Rice (1983) published data showing that the combined stress of low nitrogen and water increased isochlorogenic acid concentration in sunflower by a factor of 16. The manipulation of stress factors for increased allelochemical production has received virtually no attention from weed scientists; thus one can only speculate on the importance of seeding rates, species mixtures, and defoliation management in allelopathy.

Allelochemicals enter the environment of affected weeds by volatilization, leaching, exudation, and decomposition (Aldrich, 1984; Rice, 1983). Volatile inhibitors may be most important in dry environments (Rice, 1983). Volatile emissions from fresh residues of the winter legume crops berseem clover, hairy vetch, and crimson clover greatly inhibited seed germination of small seeded vegetables (Bradow and Connick, 1990). Water-soluble toxins may leach from above- or below-ground parts of living and dead plants. Whether toxins leach from plants or are actively exuded is a matter of dispute. It is difficult to determine whether toxic substances associated with the decomposition of plant residues are plant derived or are by-products of decomposition. Nevertheless, in dealing with allelochemicals some rudimentary concepts regarding toxic compounds in the environment can be borrowed from familiar herbicide management principles. To inhibit seedling emergence, water soluble toxins in plant residues must be moved into the weed seed germination zone by rain or mechanical incorporation. Volatile inhibitors may be lost to the atmosphere if not incorporated and may be lost more readily from moist soil than from dry soil. The fate of allelochemicals in the soil environment is a function of the chemical and physical properties of the compound and the amount released into the environment over time.

The potential of allelopathy in weed management may be underexploited. More effort to integrate allelopathic crops, cover crops, or companion crops into comprehensive weed management strategies is needed. The effectiveness of allelopathy for weed management will undoubtedly vary with soil type and environmental conditions, particularly as they affect microbial breakdown of plant residues and fate of allelochemicals.

C. Cover Crops

Cover crops are used in weed management because of their physical (competition) and chemical (allelopathy) effects on weeds. Cover crops are most competitive when they form a dense plant population that initiates growth early, produces rapid ground cover, and maintains a high leaf area for a sufficient duration to exploit physical resources at the expense of weeds. Cover crops, or their residue, may contain allelochemicals that suppress weed germination, emergence, and early growth. They may alter the quantity and quality of light reaching weed seeds and seedlings. In addition, cover crops may contribute to weed management by physically impeding weed emergence and by providing a habitat for beneficial insects and other organisms. Often it is not known what mechanism is most important, and in many cases cover crop suppression of weeds is probably due to a combination of these factors.

The quantity and quality of light reaching the soil surface after passing through a cover crop or its residue can influence weed germination and seedling growth. Fall planted cover crops exhibit rapid growth in the spring during the period of germination of many summer annual weeds, some of which require light for germination (Wesson and Wareing, 1967). A dense cover crop canopy can greatly reduce the quantity of light received by weed seeds on or just below the soil surface, thereby inhibiting germination of those species with a light requirement. Since a canopy of leaves "filters out" red and blue light and transmits far-red light, the spectral composition of light passing through the cover crop canopy tends to be enriched in far-red light (Teasdale and Daughtry, 1993). As indicated previously, far-red light inhibits germination in phytochrome controlled seeds (Taylorson and Borthwick, 1969). Cover crop inhibition of seed germination may therefore be due to a change in light quality as well as to a reduction in the amount of light passing through the canopy. This may be most important for small weed seeds, which are more likely to have a light requirement for germination than are large seeds. The amount of light that passes through killed cover crop residue is generally greater than the amount that passes through a living cover crop and the red/far-red ratio is more similar to that of unfiltered light (Teasdale and Daughtry, 1993). The result may be the promotion of germination of previously inhibited weed seeds after the cover crop is killed. There is some evidence that for some species, seeds whose germination is inhibited by canopy-filtered light may acquire dormancy that inhibits germination after canopy removal (Fenner, 1980).

Cover crops and their residues can alter temperature and moisture conditions of the seed microsite, reducing or delaying emergence. Hairy vetch and rye cover crop residues reduced temperature and temperature fluctuation sufficiently to delay weed seed germination in studies by Teasdale and Mohler (1993). They reported higher soil moisture under the cover crop residue that could interact with cooler temperatures to delay germination early in the season or maintain soil moisture at more favorable levels during dry periods.

Cover crops have been used for many years in cropping systems to provide soil cover during periods when the principal annual crops do not occupy the land. The potential benefits of cover crops, aside from weed suppression, include protecting soil against wind and water erosion, adding organic matter, improving soil structure, and providing forage to livestock (Lal et al., 1991). Some nitrogen fixed by legume cover crops will be available to the crop that follows.

Establishment

Establishment of cover crops is an important management constraint, especially in continuous annual cropping systems where it is not practical or economical to interrupt crop production to provide time for cover crop establishment. Knowledge of critical weed-free periods can be used as a guide to appropriate timing of cover crop establishment. A cover crop can be sown into an existing crop early in the season as long as the cover crop emerges after the critical period of weed interference. Rye, ryegrass, wheat, and oat can be established by broadcasting seed into with hairy vetch, bigflower vetch, and crimson clover in regions where these species are adapted (Hargrove, 1991). Overseeding subterranean clover into standing rice has been a successful method of cover crop establishment in certain areas of Uruguay. Seeding after crop harvest, either no-till or into prepared ground, is successful in temperate climates only if there is adequate time for seedling establishment and development of soil cover before growth is halted by cold weather (Triplett, 1986). Hairy vetch can be planted as late as late October in the southeastern United States (Owsley and Hargrove, 1991), late September in the mid-South (Duck and Tyler, 1991), mid-August in Ohio (Cardina and Gallagher, unpublished data), and mid-July in Wisconsin (Stute and Posner, 1993). In cool temperate regions, companion seeding with spring crops and seeding after harvest of small grains or short season crops such as canning pea are ways to integrate this cover crop into a rotation.

Where herbicides are used in a system with cover crops, the effect of residual herbicide activity on cover crop establishment must be considered. Harmful residues may interfere with cover crop establishment after the application of triazine herbicides in corn, sulfonylurea herbicides in wheat or soybeans, and imidazolinone herbicides in soybeans. Integrated systems can be designed wherein postemergence herbicides with low soil activity are used in the principal crop, leaving minimum residues that restrict the use of cover crops.

Management for Weed Suppression

Any management technique that favors growth of the cover crop over that of weeds will aid in weed suppression. A fall-planted cover crop such as hairy vetch or rye is capable of rapid dry matter accumulation in spring and early summer. A dense cover crop can suppress winter annuals or seedling perennials and biennials by competition for light, water, and nutrients and may occupy space that emerging seedlings need for resource capture. However, few plant competition studies have been conducted to demonstrate the competitiveness of cover crops against weeds in field crops.

The greater the cover crop residue the greater the potential for physical obstruction of seedling emergence. Teasdale et al. (1991) showed a linear relationship between cover crop residue and weed suppression: 300 g/m^2 of hairy vetch or rye residue, covering over 90% of the soil, was needed to reduce weed density 78%. Physical obstruction is commonly seen in fields that have been cropped for several years with high residue crops such as corn and grain sorghum. Lifting up undecayed stalks in the spring often reveals etiolated seedlings that germinated beneath—but were unable to penetrate—the residue. Obstruction of weed emergence also occurs where thick stands of cover crops or their residues are present during periods of weed emergence. The most susceptible weeds are those with small seeds and thus small energy reserves that might be expended before emergence through crop residue. In no-tillage cropping systems where residues accumulate on the soil surface, some weed seeds falling on thick residue are not able to make contact with the soil. These seeds are more vulnerable to predation or exposure to temperature extremes than are seeds that reach the relatively protected environment of the soil seed bank (Chepil, 1946). This may be more important for large seeds that are more likely to be trapped on the surface residue than for small seeds that move more easily through crevices to the soil surface. These physical effects of cover crops can be viewed as modifications of potential "safe sites" (Harper, 1977) for seed survival, germination, and emergence.

Managing cover crops can be difficult, and growers should recognize the potential drawbacks of including cover crops in a rotation. In spite of many benefits, the use of cover crops in weed management remains a compromise involving the potential for delayed planting, delayed crop emergence, allelopathic effects on the principal crop, and increased pest pressure (including insects and slugs). Properly managed cover crops can effectively suppress emergence of seedling weeds but are less effective in suppressing regrowth of perennials. There are even some indications that the far-red light-enriched environment under a cover crop canopy may favor regrowth of some perennial weeds (Aldrich, 1984).

Limiting Interference

Most cover crops must be killed or suppressed to minimize interference with establishment or growth of the principal annual crop that may be seeded directly into the cover crop or its residue. Winter and spring oats overseeded in the fall at northern latitudes greater than about 38° N can be expected to winterkill, thereby eliminating the need for herbicidal control (Putnam et al., 1983). Mustard has been used as a fast growing cover crop to smother weeds in fall planted small grains, with the expectation that severe frosts will kill the mustard and prevent it from competing with the crop (Parish, 1990). Otherwise, glyphosate, glufosinate, or other broad spectrum herbicides can be used to kill most plants used as cover crops. Subterranean clover, hairy vetch, and crimson clover were controlled more effectively with glufosinate or paraquat than with glyphosate in studies by Griffin and Dabney (1990). Before flowering, hairy vetch is difficult to control with glyphosate or mowing, but after flowering it is easily killed (Hoffman et al., 1993). Hairy vetch in early bloom has been suppressed by planting no-tillage corn followed by an early postemergence application of 2,4-D in a 30-cm band over the corn row (Hoffman et al., 1993). Other approaches to cover crop suppression are possible. Hairy vetch, for example, is easily suppressed by physical damage from tractor tires, planting equipment, or heavy turf rollers. Research is being conducted to develop a device that will "undercut," or sever the roots of a cover crop just below the soil surface, leaving the relatively undisturbed top growth to die (Wood, 1992).

Living Mulch, Companion Crops, and Smother Plants

Several attempts have been made to intercrop a low-growing cover crop, or "living mulch," to provide weed control in summer annual crops such as corn or soybeans (Elkins et al., 1983; Enache and Ilnicki, 1990; Hartwig, 1983). Early attempts were not successful because of excessive competition with principal crops and lack of herbicides to suppress the living mulch and control weeds (Kurtz et al., 1952; Robinson and Dunham, 1954). In more recent attempts, with a variety of grass and legume species as living mulches, competition with the principal crop remained a problem to the extent that the living mulch system often required as much or more herbicide than would be used in a standard management system without a cover crop or living mulch crop (Berkowitz et al., 1986; Hartwig, 1983; Regnier and Stoller, 1981).

One exception is a system of subterranean clover as a living mulch in field corn that has been developed in New Jersey (Enache and Ilnicki, 1990). Growth of corn and growth of subterranean clover appear to be well synchronized at this location: the subterranean clover reseeds itself, thereby eliminating the need for yearly reestablishment, and senesces early in the spring, thereby allowing for corn establishment without competition from the living mulch. Weed biomass in a field infested with ivyleaf morningglory and fall panicum was reduced by the living mulch without the application of herbicides (Enache and Ilnicki, 1990). Another successful example is the production of reduced tillage corn in a living mulch of stylo in tropical West Africa (Lal, 1975; Wilson et al., 1982).

The use of companion small grain crops seeded with forage legume crops was previously very common, and one of the principal benefits of this practice was control of weeds by the companion crop (Schmid and Behrens, 1972). Companion crops are used almost exclusively with spring seeded forages and in areas of adequate rainfall. The most common small grain for simultaneous seeding is oats. Red clover can be sown during early spring into fall sown winter wheat or barley, but rye is too competitive with the clover.

In temperate climates cover crops are generally planted in the fall to provide soil cover during winter and early spring. An alternative approach has been to plant a cover crop in early spring, prior to planting the principal summer crop. Researchers in Minnesota have seeded a dwarf *Brassica* species at rates of 530 and 2120 seeds/m^2 in early spring as a "smother plant" prior to corn planting (DeHaan et al., 1991). The concept is that the rapid cycling *Brassica* germinates and grows rapidly in cool spring conditions, and senesces before the critical period of competition that corn tolerates. Reductions in weed biomass of 41% and 65% have been recorded, while corn yield reductions have ranged from 0% to 91% (DeHaan et al., 1991). The mechanisms of weed reduction may be competition for light and physical impedance of weed emergence. A similar approach using sunflower is discussed in the section on allelopathy.

III. GRAZING ANIMALS

The use of grazing animals to manage weeds is often overlooked, as livestock are not generally considered to be biological control agents. However, weed populations can be managed by using many different grazing animals, including cattle, sheep, and goats, as well as geese and rabbits (Combellack, 1989; Hope-Simpson, 1940; Scifres, 1981).

In agronomic crops the potential uses of grazing in weed management are few and not well documented. Geese have reportedly been used successfully for the selective removal of seedling grasses in cotton and strawberry (Scifres, 1981; National Academy of Sciences, 1968). Sheep have been used experimentally to control weeds in seedling alfalfa (Bell and Guerrero, 1992) and to reduce seed production by fall panicum in silage corn late in the season (McClure, 1991). Selectivity is the limiting factor in this approach, as the grazing animal must have a strong preference for the target weeds to minimize damage to the crop.

Selective grazing is more applicable to weed management in pastures and rangeland, where selectivity is not so critical and where secondary effects of grazing, such as trampling and fouling, can be tolerated. Selectivity and dietary requirements vary among animal types and among individual animals, depending on sex and age. Food preferences change with quantity, quality, and botanical composition of available forage over a season and in response to grazing. Different animals prefer different plant types, different parts of the same plant, or different growth stages of the same plant. In general, cattle prefer grasses to forbs and shrubs, making them the least suitable for weed management in grass dominated pastures and rangeland. Sheep prefer forbs, including some weeds and some desirable pasture plants. Goats are the most selective and versatile animal for brush control in grazing lands. There is overlap among animals in plant species preference that is dictated by seasonal variation in forage availability and quality.

Weeds consumed by grazing animals have been listed by Vallentine (1971) and Dellow et al. (1987). In a review of effects of grazing on weed suppression, Scifres (1981) provides a list of 38 species suppressed by selective grazing. Table 4 contains a partial list of a broad variety of weeds suppressed by grazing.

A. Effects of Grazing

Different types of animals cause different types of damage to plants. Specific plant parts may be removed or entire plants may be uprooted. Plants may be killed if grazed when they are seedlings. Species with epigeal germination are more susceptible to damage by defoliation and trampling than are species with hypogeal germination. The effect of grazing may be to delay flowering and seed

Table 4 Examples of Weeds That Have Been Managed with the Aid of Grazing Animals

Target weed	Crop and situation	Animal	Reference
Blackberry	Pasture	Goats	Dellow et al., 1987
Briar, sweet	Pasture	Goats	Dellow et al., 1987
Cheat	Wheat for pasture and grain	Cattle	Koscelny and Pepper, 1990
Dandelion	Rangeland	Sheep	Vallentine, 1971
Dyer's woad	Rangeland	Sheep	West and Farah, 1989
Garlic, wild	Pasture	Sheep	Runnels & Schaffner, 1931
Gorse	Pasture	Goats	Radcliffe, 1990
Plantain	Pasture	Sheep	Scifres, 1981
Ragwort	Pasture	Sheep	Parsons, 1973
Ryegrass	Wheat after pasture	Sheep	Reeves & Smith, 1975
Spurge, leafy	Pasture	Lambs	Walker et al., 1992
St. Johnswort	Rangeland	Sheep and cattle	Vallentine, 1971 Campbell & Dellow, 1984
Sunflower	Pasture	Sheep and cattle	Scifres, 1981
Thistle, Illyrian	Pasture	Goats	Campbell & Holst, 1987
thistle, saffron	Pasture	Goats	Pierce, 1987
tussock, poa	Rangeland	Goats	Campbell et al., 1984
various	Rangeland	Sheep	Walker et al., 1992
	Alfalfa pasture	Sheep	Bell and Guerrero, 1992

set or to reduce the number of flowers and seed formed. Grazing may affect competition among plants in a pasture by changing the proportion of light captured or the number of growing points in competing plants. Regrowth from defoliated plants may be reduced if emerging buds are removed repeatedly by grazing.

B. Management Considerations

One aim of grazing management is to balance the nutritional needs of the animals with the quality of available forage. Nutritional needs are very high for calves, lambs, and high-producing milk cows. Ewes and beef cows have medium nutritional needs, whereas needs of goats are comparatively low. Animals with the greatest nutritional needs should be allowed to graze forage of the highest quality, leaving those with lower needs to consume remaining forage of lesser quality, including weeds (Blaser, 1986). This can be done in rotational grazing systems where first grazers select leafy forage high in protein, energy, and digestibility and low in fiber. As the pasture is consumed the proportion of stemmy, less palatable plant material increases. Allowing "second grazers" with lower nutritional needs to graze this material, even forcing them to consume relatively unpalatable forage, is a way of suppressing growth of undesirable weeds, brush, and woody plants while furnishing the nutrition of grazing animals. Heavy stocking rates for brief periods may be effective in suppressing some species, particularly annual weeds; lighter stocking rates and continuous grazing may be more effective against other species.

A flexible approach may be mixed grazing of cattle with sheep and/or goats, the proportion being adjusted to match changing botanical composition over the season. For example, about 60% sheep graze with cattle on common broomweed-infested areas in northwest Texas (Scifres, 1981). The sheep selectively graze on broomweed seedlings and young plants during the period of early vegetative growth from mid-March to mid-May. Intensive grazing by sheep early in this period may be more effective than moderate grazing pressure over the entire 2 months. In contrast, sheep grazed leafy spurge for about 4 years before this persistent perennial was effectively suppressed (Johnston and Peake, 1960).

The much maligned goat reigns as the most versatile and effective biological brush (undesirable shrub or small tree) management agent among all domestic animals because of its preference for woody plants (Scifres, 1981). "Goating" may be more widely applicable than might be suggested by current practices. Not only will goats consume foliage, young twigs, and bark of many woody plants, they are particularly well adapted to rough terrain and are more adaptable in their dietary habits than other livestock. Goats are especially effective in controlling sprout regrowth after removal of mature brush. Spanish goats consume more browse than do Angora goats, but both are used for brush control. Stocking rate and duration of grazing must be managed according to the amount and botanical composition of the brush. Continuous grazing of rangeland by 0.83 to 1.25 goats/ha resulted in an 85% reduction in cover of mixed brush species after 5 years (Merill and Taylor, 1976). Twelve to 20 goats per hectare may in some cases achieve desired brush control within a month (Scifres, 1981).

Another way to use grazing animals for weed management is in establishment of desirable range or pasture species. Goats were used to control woody species after removal of undesirable oak species in the improvement of grazing lands in Texas (Darrow and McCully, 1959). Intensive grazing by sheep suppressed native grassland (*Sporobuols* spp. and *Cynodon plectostachyum*) during no-tillage establishment of stylo in experiments in Ghana (Cardina, 1977). Sheep grazed pastures twice: once before seeding and once to suppress grass regrowth after seeding until legume emergence, to permit legume establishment and improve pasture quality.

IV. INSECTS

Insects are important regulators of all plant populations, including weeds. The deliberate exploitation of insects to reduce weed populations dates back to 1795, when the cochineal insect *Dactylopius ceylonicus* was imported to India from Brazil to control the pricklypear (Goeden, 1988). Since that time there have been many examples of successful management of weed populations using one or more insects, and researchers continue to discover and develop additional insects as biological control agents. The utility of insects is due to their great variety, host specialization, mobility, and ability to establish stable populations in dynamic ecosystems.

Three ways insects can be used in weed management, in order of disturbance of the agroecosystem, are 1) enhancement, 2) augmentation, and 3) introduction. Enhancement involves increasing the amount of damage caused by existing insect populations, with little change in insect numbers. Augmentation means increasing the size of existing insect populations as a means of increasing feeding damage. Introduction, sometimes referred to as classical biological control, usually involves obtaining a natural enemy from the native range of a weed and introducing that insect enemy into areas where the alien weed is abundant because of lack of natural enemies. The applicability of this method is not restricted to alien weeds or to alien insects, and successful use of insects in weed management may involve a combination of enhancement, augmentation, and introduction in conjunction with other methods of stressing target weeds (Harris, 1980).

A. Enhancement

Insect damage to weeds can be increased by enhancing the acceptability of weeds to insects through integration of cultural practices. Addition of fertilizer to plants can change their acceptability to insects and can affect insects feeding on plants (Rodriguez, 1960). Leaves are often more succulent after applications of nitrate, which makes them more attractive and a better food source for insects. For example, the insect *Cactoblastis cactorum* (Berg.) more readily attacked and destroyed pricklypear plants after N-fertilizer was applied (Wilson, 1960). Addition of P and Ca to mats of alligatorweed that escaped initial control by the flea beetle *Agasicles hydrophila* resulted in increased insect feeding and fecundity (Maddox and Rhyne, 1975). The population of many insects, especially those with high reproductive ability, is limited by the nutritional value of the host plant, so correcting mineral deficiencies may improve the effectiveness of some biological control agents.

Herbicides, even at sublethal doses, can alter plant susceptibility to insects. This is sometimes observed in crops, particularly after hormone-type herbicides are applied and insect feeding increases (Oka and Pimentel, 1976). This effect has been used to increase damage to alligatorweed by the beetle *Agasicles hygrophila* (Durden et al., 1975).

A novel enhancement approach is increasing insect damage by coating larvae with a herbicide so that insects visiting a weed and dispensing small amounts of the herbicide will compound feeding damage with herbicide injury. Such an approach was taken by Quimby and Frick (1985), using third-instar larvae of the nutsedge moth (*Bactra verutana* Zeller) coated with the herbicide glyphosate or bentazon. The herbicide coated larvae reduced purple nutsedge and yellow nutsedge dry weights about 80% and 50%, respectively, compared with 53% and 24% for uncoated larvae. Using a similar approach, larvae of the American waterhyacinth borer moth (*Arzama densa* Walker) were coated with 2,4-D amine and gibberellic acid, nearly doubling the damage caused to water hyacinth (O'Leary, 1983).

Another enhancement method is the use of microorganisms to increase the damage caused by insects. The scentless plant bug (*Niesthrea louisianica* Sailer) is a native insect that reduces velvetleaf seed numbers and viability by penetrating the wall of immature seed capsules and consuming the contents of developing seeds (Kremer and Spencer, 1989). Seed pathogenic fungi applied to immature velvetleaf seed capsules were introduced into developing seeds by the probing and penetrating activity of the scentless plant bug. The combination of insect feeding and microbial infection reduced velvetleaf seed viability 98%. A similar approach is being studied to enhance the activity of root-boring larvae attacking leafy spurge.

Success with enhancement is limited by the difficulty in achieving selective suppression of weeds with techniques that increase the acceptability of plants to insects. The use of herbicide coated larvae helps to achieve selectivity because the host range of these insects is limited and little damage is caused by infrequent visitation on nonhosts. The combination of insects with pathogenic fungi or bacteria is a novel and effective way of achieving selectivity that merits further study. Merely manipulating growing conditions to increase susceptibility of weeds to insects is rarely effective and requires intimate understanding of insect behavior and weed host biology. Very little research has been conducted to support the practical application of this approach.

B. Augmentation

Augmentation is used where insect damage to weeds is limited because the insect population is too small. There are several ways to increase numbers of endemic insects that damage weeds. They are based on knowledge of life cycles of weed feeding insects and on efforts to remove constraints that reduce the amount of damage they inflict on weed populations.

Management of insecticides is an important consideration for augmenting weed feeding insect populations. Routine application of insecticides, as in-row or broadcast applications of systemic or contact materials, may protect weeds from feeding damage by endemic insects. This was observed when withholding insecticides allowed populations of the sawfly (*Schizorcerella pilicornis* Holmgren) to increase to levels sufficient to destroy common purslane (Norris and Ayres, 1981). The biomass of ducksalad (*Heteranthera limosa*), an annual broadleaf weed in rice, was reduced 58% to 87% by natural populations of the waterlily aphid (*Rhopalosiphum nymphaeae*) when insecticides were withheld (Oraze and Grigarick, 1992).

The population of weed feeding insects is often low because of a lack of overwintering sites that are not disturbed by cultural practices such as tillage, harvest operations, or herbicide applications. One approach to insect augmentation is to provide alternative overwintering sites for some insects, particularly where fields are tilled in the fall, by allowing weeds or other materials on which insects survive to persist in hedgerows or adjacent fields.

Another approach is to reduce the activity of parasites and predators responsible for mortality of these insects. Activity of the scale insect *Dactylopius opuntiae* on pricklypear increased when low rates of DDT were applied to kill a scale predator (Annecke et al., 1969). Although populations of insect pests of crops have gone "out of control" when their predators were killed by insecticides, a similar effect on pests of weeds has rarely been observed.

The manipulation of alternate hosts as a means of increasing insect numbers has been studied by scientists in Indiana and Ohio (Meyers et al., 1991). One objective of their research is to increase the population of velvetleaf seed beetle by providing alternate hosts during the early season when velvetleaf is not yet in bloom. The alternate host might be planted in field borders or in strips through fields and would senesce, or be mowed, when velvetleaf flowers. Conceptually, the enlarged beetle population would move to velvetleaf, its preferred host. This strategy is still experimental, but it is one that could be duplicated for other insects that attack weeds at any stage of their life cycle. Reducing the production of viable weed seed is one of many strategies that can be integrated into a system in an attempt to achieve long-term reductions in weed populations.

Insect numbers can be augmented by direct field releases of endemic insects that have been collected, conserved, or reared. The rationale is to increase the insect population in areas where the weed host is a problem and where the current insect population is small and ineffective. Depending on the biology of the insect and its host, releases can be made in large numbers to achieve rapid, inundative destruction of the weed population; otherwise, releases can involve relatively few insects that essentially "inoculate" a field with the intention of establishing a population level that will maintain itself with little additional manipulation. An example is the release of laboratory-reared nutsedge moth larvae and adults to suppress purple nutsedge in cotton (Frick and Chandler, 1978). Natural populations of the nutsedge moth remain at a low level until late in the season and are therefore ineffective in suppressing purple nutsedge. To augment the population earlier in the season, insects were reared (Frick et al., 1983) and released into purple nutsedge-infested fields at a rate of five to six larvae per plant. Periodic releases of larvae, when properly timed, suppressed purple nutsedge so that it was no longer competitive with cotton.

Examples of inoculative augmentation include the release of the pigweed fleabeetle [*Disonycha glabrata* (Fab.)] for control of pigweeds (Balsbaugh et al., 1981) and release of the scentless plant bug to destroy velvetleaf seeds (Kremer and Spencer, 1989). The pigweed fleabeetle is a native insect that was collected from a fallow field in the southern United States (Mississippi) and moved to a northern state (North Dakota) in early June. At this time pigweed is still a seedling in the North, and normal migration of the fleabeetle has not begun. The fleabeetle adults were released by hand into fields infested with pigweeds, but the insect failed to overwinter in the cooler northern climate. To augment scentless plant bug populations, methods were developed for culturing the insect (Jones et al., 1985; Spencer, 1986), which was released at sites in five states from Iowa to New York, where seed germination was reduced from 95% to 55%. The destruction of velvetleaf seeds by this insect was increased substantially when they were released in combination with microorganisms, as described previously.

It is common to find insects that are considered crop pests feeding on weeds, especially where the weed and crop are closely related. However, where a crop is not a host of the insect, and no crop host is likely to be grown nearby, it may be possible to use an insect against weed hosts. For example, the tobacco budworm (*Heliothis virescens* [F]) is a tobacco pest that will also feed on several weeds, including morningglories. Quimby and coworkers (1982) sprayed tobacco budworm eggs on flowering morningglories growing on levees in a field of rice, a crop that is not a tobacco budworm host. The larvae fed on morningglory fruits and reduced seed production, demonstrating that a carefully managed crop pest could safely and effectively suppress weeds in a nonhost crop.

An advantage of augmentation is that weeds are controlled only where the insect is introduced and damage is unlikely in areas where the weed is beneficial. Another advantage is that control can be relatively rapid and the technique can be used with annual as well as biennial and perennial weeds. However, the cost of rearing and releasing organisms is high and depends on the area where control is needed and the number of releases required per season.

C. Introduction of Self-Sustaining Insects

Insects that attack a weed in its native habitat have been introduced into areas where the weed has become troublesome because natural enemies are absent. Until recently this approach was virtually synonymous with "biological weed control." This "classical" approach to biological control involves searching for natural enemies of the weed in the area of its origin; importation, quarantine, and rearing of the insect; extensive tests to ensure the safety of useful plants; and introduction of the insect where the weed is troublesome. Over 500 releases of exotic invertebrates, most of them insects, have been reported worldwide for biological control of weeds since the late nineteenth century (Julien, 1992). The development of classical biological weed control programs using insects has been described extensively (Harris, 1971; Andres et al., 1976; Andres, 1981; Harris, 1993). The importation of exotic insects is regulated by various government agencies, such as the Animal and Plant Health Inspection Service (APHIS) in the United States, and insect introductions have generally been under the auspices of government or university researchers. Therefore, this approach to biological weed management is not one that growers can easily employ at their discretion.

The long history of efforts to introduce exotic insects points to several factors that influence the effectiveness of this approach. Success has usually been achieved with weeds that belong to plant families that are of relatively little economic importance. Most have been naturalized perennial weeds that dominate areas of low disturbance and compete with desirable plants whose growth is favored by the stress of the weed. Most successful cases have involved highly host-specific insects and often multiple insects or insects acting in concert with plant pathogens. The introduced insect should be able to disperse and locate the hosts and prevent reproduction of the weeds, if not kill a large percentage of them outright. The insect should, of course, be adapted to the weed and to the environment it inhabits. The life cycle of the insect must be synchronized with that of the weed so that the insect is able to survive periods when the plant is dormant and to reinfest the weed when conditions are favorable for plant growth. In cases of successful control, no further manipulation has been necessary after insects were introduced into new areas.

The result of classical biological control efforts has been to reduce the density of the weed host but not to eradicate it. A "balance" is achieved between the weed and the insect predator such that the activity of both appears low. The ability of injured perennial weeds to regrow helps to secure the survival of the predator. Because this approach relies on introduction of quarantined insects from relatively remote geographical (usually transoceanic) locations, the activity of the insect is not hindered by its native parasites and predators. However, insects colonizing weeds have eventually been affected by locally adapted parasites and predators (Harper, 1977). Introduction of insects that feed on plants related to the target weed has been proposed as a way to increase the effectiveness of biological control and to use biological control technology against native weeds (Pimentel, 1991).

The mechanism of action of weed-feeding insects may include destruction of leaves, flowers, stems, or seeds; tunneling and feeding in leaves, stems, or roots; or formation of galls. In many cases plant metabolism is disrupted by the destruction of the weed's photosynthetic capability, energy (carbohydrate) storage, or ability to transfer metabolites to growing points. However, the reduction in weed growth may not be attributable directly to the activity of the insect. Neighboring plants provide competitive stress and can fill in gaps where the weed is destroyed. Insects often transmit pathogenic agents that can move through the plant vascular tissue to cause further destruction. For example, *Cactoblastis cactorum* acts on pricklypears primarily by feeding damage, but also by transmittal of a bacterial rot.

The principal advantages of classical biological control are that the insects reproduce and disperse on their own and that the cost of control, after the initial release, is relatively low. Since this approach depends on self-dispersing and self-sustaining agents, the cost of control is independent of the area over which control is desired. The principal disadvantage is that the organism may spread to areas where the weed is considered beneficial, leading to conflicts of interest.

Successful insect introductions for biological control of weeds have been reviewed previously (National Academy of Sciences, 1968; Harper, 1977; Andres, 1981; Huffaker, 1971). Julien et al. (1984) surveyed attempts at biological control of weeds prior to 1980 and listed 171 species of insects, 71% of which became established. A fairly comprehensive list of insect releases for biological weed control is given by Julien (1992). Examples of attempts at biological control of weeds with insects are listed in Table 5. Details of two recent attempts at classical biological control involve tansy ragwort and leafy spurge and are described in the following sections.

Tansy Ragwort

Tansy ragwort has been the object of biological control efforts since 1936 when insects were introduced from England into New Zealand, where they became established and caused minor reductions in seed head production (Julien, 1992). The weed is a poisonous composite of European origin, and it infests rangeland and pastures in coastal areas from California to Canada. The cinnabar moth (*Tyria jacobaea*), imported from France into California in the 1960s, strips the weed of leaves and feeds on flower heads, severely limiting seed production. The ragwort fleabeetle (*Longitarsus jacobaeae*), imported later from Italy, attacks the crown, stem, and roots of tansy ragwort and remains active later in

Table 5 Examples of Weeds for Which Insect Biological Control Agents Have Been Developed

Target weed	Insect	Area of importation	Comments	Reference
Alligatorweed	*Agasicles hygrophila*	USA, Thailand, New Zealand	Fleabeetle feeds on stem and leaves	Buckingham et al., 1983
	Amynothrips andersoni	USA	Requires repeated release	Coulson, 1977
	Vogtia malloi	USA, New Zealand	Stem-boring moth, effective	Coulson, 1977
Hydrilla	*Bagous affinis* *Hydrellia balciunasi* *H. pakistanae*	Florida	Weevil infests tuber; leaf-mining flies; release over 3+ months	Buckingham, 1988
Knapweeds	*Sphenoptera jugoslavica*	USA, Canada	Buprestid beetle mine taproot	Story, 1985
	Urophora affinis *U. gadrifasciata*	USA, Canada	Seed head gall fly larvae feed on flower	Nechols et al., 1992
Puncturevine	*Microlarinus lareynii* *M. lypriformis*	Western U.S., Papua New Guinea, Canada	Seed and stem feeding weevils	Huffaker et al., 1983
Starthistle	*Chaetorellia australis* *Bangasternus orientalis*	Western U.S.	Destroys developing seed	Nechols et al., 1992
			50+% Seed reduction	Nechols et al., 1992
St. Johnswort	*Chrysolina guadrigemina*	Western U.S.	Feeds on foliage; >99% control	National Academy of Sciences, 1968
Thistle, Canada	*Urophora cardui*	USA, Canada, New Zealand	Gall fly, uncertain effectiveness	Peschken, 1979
Thistle, Russian	*Coleophora klimeschiella* *C. parthenica*	Hawaii, Western U.S., Canada	Larvae hollow out leaves	Goeden et al., 1987
Waterhyacinth	*Sameodes albiguttalis*	Australia; USA	Larvae feed in young leaf petioles	Wapshere et al., 1989

the season than the cinnabar moth larvae. In parts of California the two insects have controlled over 95% of tansy ragwort (Hawkes and Johnson, 1978). Similar results were obtained after introduction of the two insects in pastures in Oregon.

Insects alone are not completely responsible for the control of tansy ragwort. On well-drained sites in the northern range of the weed the short growing season and slow regrowth delay regeneration and subject the plant to winterkill. Thus the interaction of insects and environmental stress determines the success of biological control of this weed (Harris, 1980).

Leafy Spurge

About a dozen insects have been introduced into the United States and Canada to control leafy spurge, an aggressive, poisonous perennial that has infested over a million hectares of rangeland in North America. A native of Europe and Asia, it was introduced in the 1820s into the United States, where no natural enemies were present to keep the population in check. The focus of current biological control efforts is on five insects, each of which attacks and weakens the weed in a different way (Department of Agriculture, 1989). Three species of root mining fleabeetles (*Aphthona flava, A. cyparissiae* and *A. czwalinae*) lay eggs on leafy spurge stems. The larvae feed on primary and secondary roots, and the adults feed on leaves. The cerambyciid long-horned beetle (*Oberea erythrocephala*) adult feeds on the stem, girdling it, and the larvae bore into the stem, crown, and large roots. Larvae of a cecidomyiid gall midge (*Bayeria* sp.) feed on shoots and induce gall formation in shoot tips and can produce up to five generations in a growing season. These insects are being reared and released by state and federal cooperators, and though dramatic results have not been observed, effects may not be evident for several years. Efforts have been made to limit the use of insecticides that might kill the introduced agents. In addition, grazing land must be managed so that competition from desirable species can add stress to insect-damaged leafy spurge and can replace the weed as it deteriorates.

V. PLANT PATHOGENS

Weed populations are often constrained by diseases caused by fungi, bacteria, viruses, and viruslike agents. Among the classes of plant pathogens, fungi have received the greatest attention in efforts to control weeds. The activity of these organisms can be enhanced to reduce the potential growth of weed populations and to reduce the impact of weeds. In some cases it has been possible to isolate, culture, formulate, and disseminate fungal propagules as mycoherbicides. Many other pathogens attacking weeds have not been successfully developed into biological control agents or have not been studied from a biological control perspective. Several reviews and texts detail the history, development, prospects, and technical aspects of the use of plant pathogens for biological control of weeds (Charudattan, 1990; Charudattan and Walker, 1982; Quimby, 1986; TeBeese, 1991; TeBeest and Templeton, 1985; Templeton et al., 1986). Charudattan (1991), Watson (1991), and Julien (1992) provide comprehensive lists of pathogens that have been used or evaluated for biological control of weeds. The list of weeds and pathogens in Table 6 shows the many types of weeds that have been studied.

Plant pathogens have been used in weed management for 1) enhancement of existing disease levels, 2) augmentation of pathogen populations to induce an epiphytotic, and 3) one-time introduction of exotic agents in a classical biological control approach, with the intent that the organism will be self-perpetuating without further manipulation.

Regardless of the approach, plant disease requires a susceptible host, a virulent pathogen, and a favorable environment. Enhancement involves manipulation of host susceptibility or environmental conditions to promote disease but may be limited by lack of a virulent pathogen. Likewise the success of augmentation and introduction of exotic agents is determined by host susceptibility and environmental conditions. All three factors must be considered, and possibly manipulated in concert, to promote disease. For example, release of high-quality, infective propagules can be timed to coincide with host susceptibility and favorable environmental conditions to provide the essential factors for disease development. Approaches that manipulate the amount of inoculum are based on experience with diseases of crop plants that shows that avoidance or removal of primary inoculum by sanitation, rotation, or exclusion can constrain the development of disease. To create a disease epidemic in susceptible weed populations the introduction or augmentation of inoculum overcomes this crucial constraint and, under appropriate environmental conditions, can lead to suppression of weed populations below economic levels.

One of the advantages of plant pathogens over other biological agents is specificity. Many organisms that incite disease have a very narrow host range, some restricted to a single weed species. In some cases this specificity may be a hindrance to commercial development because of industry reluctance to develop products that control only one or two weeds.

Effective pathogens are those that reproduce at the expense of nutrients that would otherwise be used by the weed for growth, maintenance, and reproduction. The process of parasitism for a typical fungal pathogen begins with germination of spores or growth of vegetative structures. Upon contact with the plant surface the organism forms an appressorium, from which an infection peg grows, piercing the cuticle and cell wall. In some cases, the fungus grows through intercellular space before entering a cell. Most fungi penetrate the host directly, through intact plant surfaces; most fungi, and all bacteria, can also

Table 6 Examples of Weeds for Which Potentially Useful Pathogens Have Been Studied

Target weed	Pathogen	Crop and situation	Approach[a]	Comments	Status[b]	Reference
Anoda, spurred	*Alternaria macrospora*	Row crops	a	Limited by dew period	E, L, p	Walker and Sciumbato, 1979
Bindweed, field	*Phomopsis convolvulus*	Row crops	a	Limited by weed growth	E. L, p	Ormeno-nunez, et al., 1988
Blackberry	*Phragmidium violaceum*	Untilled fields	i	Introduced from Europe	A	Oehrens, 1977
Bluegrass, annual	*Xanthomonas campestris*	Lawns, turf	a	Bacteria applied prior to wounding	E, F, g	Baragona and White, 1992
Cocklebur, spiny	*Colletotrichum orbiculare*	Row crops and pastures in Australia	a a	Needs dew within 4 hr of inoculation	E, F, g	Auld et al., 1988
Goosegrass	*Pyricularia grisea*	Row crops	a	Limited by dew period	E,p	Figliola et al., 1988
Gourd, Texas	*Fusarim solani*	Annual field crops	a	Soil-borne fungus	E, F, g	Boyette et al., 1984
Jimsonweed	*Alternaria crassa*	Row crops	a	Limited by dew period	E, F, p	Boyette et al., 1991
Koster's curse	*Colletotrichum gloeosporioides*	Forests in Hawaii	i	Introduced from Panama	E, F, g	Trujillo et al., 1986
Pamakani, Hamakua	*Entyloma ageratinae*	Pastures in Hawaii	i	Introduced from Jamaica	A	Trujillo, 1985
Sesbania, hemp	*Colletotrichum truncatum*	Annual field crops	a	In invert emulsion	F,g	Boyette et al., *1993*
Sicklepod	*Alternaria cassiae*	Row crops in South	a	Inconsistent results	E, F, p	Walker and Boyette, 1985
Thistle, Canada	*Pseudomonas syringae*	Field crops, pasture	a	Wide host range	F, g	Johnson and Wyse, 1992

[a]a = augmentation of native pathogen population, usually as a mycoherbicide; i = introduction of a self-sustaining foreign pathogen in the sense of "classical" biological control.
[b]A = available for practical use; E = experimental; F = field trials have been conducted; L = laboratory studies only have been conducted; g = good degree of success; p = poor prospects of eventual development.

penetrate the host through wounds; and some fungi and bacteria enter the host through natural openings such as stomata or lenticels. The steps from germination to penetration require free water or at least high relative humidity to prevent desiccation. There are two common modes of infection by which pathogens contact and obtain nutrients from host cells: 1) the microbe produces a toxin that kills host tissue in advance of colonization; or 2) the microbe colonizes host tissue and drains the host of energy or unique metabolites, resulting in cell death. The many phytotoxic fungal or bacterial compounds that have been characterized show a high level of host selectivity, though few toxins from weed pathogens have been studied. Once the host has been infected, many fungi spread to all plant tissues, growing through or between cells.

The nature of the damage that pathogens cause varies with the type of organism in question, and this may determine which management approach is most appropriate. Specialized pathogens may exploit specific plant niches and are characterized as parasites of roots, crowns, stems, old leaves, leaf mesophyll, young leaves, ovaries, seeds, and other plant parts. The types of disease are also diverse and include rots, galls, rusts, necroses, smuts, and wilts.

A. Enhancement

Few studies have been conducted to investigate methods of enhancing the effects of endemic diseases of weeds. The reason is not a shortage of fungal or bacterial pathogens or of infected weeds. One reason may be that cultural or environmental manipulations that enhance weed diseases will likely increase diseases of crops. Cultural practices that may influence weed disease pressure include the type, amount, and timing of fertilizer application; the amount and timing of irrigation; and the application of low rates of certain herbicides or growth regulators. Fertilizers, especially nitrate, can stimulate growth of succulent leaves that may be more susceptible to infection than leaves of unfertilized plants. Irrigation can contribute to disease because many pathogens are dispersed in water and because periods of free moisture are usually required for pathogen germination and infection. It may be possible to plant alternate hosts as a means of increasing the spread and survival of inoculum. For example, Peeper (1984) suggested that planting alternate hosts in the buttercup family (*Ranunculaceae*) could be used to increase stress on *Bromus* spp. from leaf rust (*Puccinia rubigo-vera agropyri* [Erikss.] Arth.). Low rates of herbicides or growth regulators may enhance disease development by lowering host resistance to infection (Altman and Campbell, 1977).

B. Augmentation

The augmentation approach can be used when disease is limited by lack of effective inoculum at a time when the weed is susceptible and the environment is conducive to disease development. Commonly, a massive, or inundative, dose of inoculum is applied to incite a level of disease that may virtually destroy the weed host in a relatively short time. Inoculative augmentation is also appropriate where disease development is limited by a lag period for inoculum buildup and distribution. In this case, it is necessary to apply only a small amount of inoculum to initiate a destructive level of disease earlier in the growing season than would normally occur. The augmented pathogen may be an endemic organism that is not normally at a level that incites adequate damage to the weed.

Formulations of fungi applied as inundative inoculum in a manner similar to that of chemical herbicides have been termed "mycoherbicides." Fungi from over 100 taxonomic groups have been investigated as agents for inundative or inoculative release to control weeds. Three have been developed and registered as mycoherbicides for commercial use in North America (discussed later). These are commonly pathogens that coevolved with their weed hosts, resulting in a stable, balanced relationship. But they do not normally produce self-sustaining epidemics of sufficient severity to cause major damage to weeds. These pathogens are generally soil-borne or mucoid-spored fungi that are poorly disseminated. This is why inundative doses of inoculum are often needed to incite disease at a level that can be highly destructive. Pathogens with wet (sticky) spores and small spores are better able to adhere to and hydrate on plant surfaces and thus are less limited by environmental conditions than are pathogens with large, dry spores. Much research emphasis has been put on facultative parasites or facultative saprophytes, although obligate parasites have been used successfully.

Suitability of pathogens for development as mycoherbicides depends on criteria similar to those used in herbicide development. Of primary concern is effectiveness in controlling an economically important weed. This is determined by how readily disease is incited and how rapidly an epidemic is developed. Virulence has been optimized by selecting for the most active, adapted strains of the pathogen. Effectiveness of the pathogen at endemic levels does not necessarily indicate the type of epidemic that might be incited when inoculum is applied at the appropriate time and dose. Speed of infection and disease development is critical because environmental conditions suitable for these processes may be brief and because the host weed may be susceptible to the pathogen only at restricted growth stages. A pathogen is not successful if infected weeds are able to "grow out of" effects of a disease.

Host specificity is an important consideration. Weed pathogens have been applied to potential hosts under optimum conditions for infection and disease development to determine host specificity. This is done to assure the safety of crops and other desirable plants as well as to determine the susceptibility of plants taxonomically related to the weed host. Experience with the biological control of northern jointvetch has demonstrated the importance of testing a pathogen on several varieties within a crop species, as these may vary considerably in susceptibility (TeBeest, 1988).

Most pathogens under development as mycoherbicides have host ranges limited to one, or at most a few, economically important weeds. Many mycoherbicide pathogens are strains or pathotypes of a single species that infect multiple weed hosts. A specific pathotype of *Phytophthora palmivora*, for example, has been developed for control of stranglervine. There are several subspecies of the fungal pathogen *Colletotrichum gloeosporiodes* that are fairly specific to various weed hosts. Attempts have been made either to enhance the efficacy of host-specific pathogens so as to increase the number of weeds affected, or to restrict the host range of broad-host pathogens while maintaining their virulence against economically important weeds (Sands et al., 1990).

Other attributes of pathogens suitable for development as mycoherbicides include the ability to grow and produce viable inoculum in culture and the ability of this inoculum to survive in storage and infect plants after application. The organism must be stable in culture, without loss or virulence or change of host specificity. Mycoherbicides must be safe to humans and livestock, and because this is a concern with any pesticide, the sale of mycoherbicides is regulated by government agencies such as the U.S. Environmental Protection Agency.

Susceptibility of weeds to a given dose of inoculum varies with weed growth stage. Susceptible growth stages may correspond to periods of low energy status, such as newly autotrophic seedlings, or to stages of growth when cuticles are not well developed or when concentrations of certain metabolites are limited. For example, *Alternaria cassiae* can infect and kill young sicklepod seedlings, but equivalent spore densities applied to older plants will result in a few infections but not plant death or significant reduction in growth. Waterhyacinth is highly susceptible to the leaf spot fungus *Cercospora rodmanii* but is able to tolerate loss of diseased tissue due to its clonal growth habit.

One advantage of pathogen augmentation is the ability to time the release of inoculum when environmental conditions are optimum for disease development. However, the effectiveness of weed control is often limited by epidemiological factors. The processes of pathogen germination and infection, in particular, often require specific environmental conditions. Dry fungal spores or mycelia applied to a host weed must be hydrated by dew or other forms of free water so that germination and host infection can proceed. In addition, growth of some fungi is inhibited by exposure to some wavelengths of solar radiation. Minimum periods of free water at a narrow range of temperature are often difficult to achieve under field conditions. The free water requirement can be overcome somewhat by formulation components that prevent drying and encourage moisture retention, or by application of inoculum during periods of high humidity, at night, or in sprinkler irrigation water. Substances that reduce adverse effects of solar radiation can be added to formulations.

Mycoherbicides are generally applied in the same manner as herbicides. Because the "active ingredient" is a living organism, some extra care is required in shipping, storage, and handling. Sprayers used for mycoherbicide application must be cleaned out carefully, because residues of active or inert ingredients of conventional pesticides may be toxic to the fungi. This is especially true for a sprayer that is also used for fungicide application. To attain uniform spread of inoculum, surfactants have been added to some mycoherbicide formulations. Since some surfactants are toxic to microorganisms, only the specifically recommended surfactant should be used. Researchers have added various substances to liquid formulations in efforts to improve mycoherbicide efficacy, including sugars, gums, gels, oils, and waxes. Sugars are used in an effort to enhance spore viability or modify spore germination, whereas most other additives have been used to overcome limitations of moisture requirements. For example, sorbitol added to spray mixes increased spore survival and efficacy of *Colletotrichum coccodes* on velvetleaf (Wymore and Watson, 1986). Quimby and coworkers (1988) developed an invert emulsion containing wax and paraffin oil to retard evaporation, mixed with lecithin as an emulsifier. This formulation has been used with several fungal pathogens to reduce or eliminate the lengthy period of free water normally required for spore germination and penetration.

Chemical herbicides have been used in conjunction with mycoherbicides to increase the number of weeds controlled or to increase the severity of disease. The herbicide acifluorfen can be added to mixtures of the fungal pathogen of northern jointvetch to control hemp sesbania at the same time (Templeton et al., 1984). A sequential application of 2,4-D improved the activity of *Fusarium lateritium* against velvetleaf, but mixing the herbicide with the fungus reduced spore survival and level of disease (Boyette and Walker, 1985). The cotton defoliant thidiazuron has been added to mixtures of *Colletotrichum coccodes* to improve control of velvetleaf (Hodgson et al., 1988). Several researchers have successfully combined fungal pathogens as a means of increasing the number of weeds killed in a single mycoherbicide application. The facultative parasitic fungus *Colletotrichum orbiculare* infected Noogoorabur via nonlethal leaf and stem lesions caused by the rust fungus *Puccinia xanthii*, causing severe symptoms and plant death (Morin et al., 1993).

The mycoherbicide approach has several advantages over other uses of pathogens. One is that it uses native organisms so the need for overseas searches and quarantine is eliminated. Another is that it can be used against annuals in repeated applications. The effects are relatively rapid, so there is no lag time for inoculum buildup, and after application the pathogen population is expected to return to its low, endemic level.

However, the mycoherbicide approach is essentially a technological rather than an ecological one and is not without drawbacks. Practical constraints, which are common to other pathogen approaches, include lack of available pathogens that are safe enough to use, difficulty in overcoming environmental limitations, specificity of organisms to only one or two weeds, complications where fungicides are used, and possible conflicts of interest with desirable plants. Biological challenges include the potential for resistance development in weed populations, possible altered host range of mass reared and genetically altered pathogens, and inability to stop an infection in desirable plants if it should occur. The similarity between the use of mycoherbicides and use of conventional herbicides with which growers have experience can be seen as an advantage. Both tactics focus on control rather than long-term management of weeds; they are short-term solutions to long-term weed problems that can be managed by manipulation of cultural practices and other biological components of the cropping system.

In contrast to other biological management approaches, mycoherbicides have been considered sufficiently appealing as an alternative and supplement to chemical weed control to attract industry to develop and register some as commercial products. Others, though less attractive commercially, exhibit biological potential and could be developed with public support or by small entrepreneurs. Following are examples of three commercially registered mycoherbicides and two organisms that may be available by other means.

Colletotrichum gloeosporoides f.sp. aeschynomene

Spores of this anthracnose fungus have been sold under the trade name Collego* for control of northern jointvetch in rice and soybeans. This product is a wettable powder containing dried spores produced by liquid fermentation and has been labeled for application at a rate of 3.1×10^{10} viable spores/ha. The product can be applied to northern jointvetch that is at least 30 cm tall in soybeans that have begun to flower or in rice at any stage of growth. To provide a high-moisture environment, soybean fields should be irrigated and rice fields flooded just before application. Late afternoon applications are most effective. To prevent seed production, applications to northern jointvetch should be made before bloom, which corresponds to the time pods are forming on lower nodes of soybeans or rice heads are emerging from the boot. This mycoherbicide can be mixed with the herbicide acifluorfen, but not with fungicides, insecticides, or other herbicides. The fungus was given the *forma specialis* designation *aeschynomene* because it was originally thought to have a host range restricted to seven species of the *Aeschynomene* genus. Subsequent studies showed that

*Ecogen Inc., Langhorne, PA 19047.

21 additional species in eight other genera were susceptible, including several cultivars of peas. However, susceptibility differed greatly and only northern jointvetch was killed by the fungus (TeBeest, 1988). The history, development, and registration of this product have been reviewed extensively (Smith, 1986; TeBeest, 1991; TeBeest and Templeton, 1985; Templeton et al., 1986; Templeton et al., 1984; Templeton and Trujillo, 1981).

Phytophthora palmivora

A pathotype of *Phytophthora palmivora* has been formulated as a liquid suspension of chlamydospores for control of stranglervine in Florida citrus under the trade name DeVine*. This organism was developed by Ridings and coworkers and was the first mycoherbicide to be registered (Ridings, 1986; Ridings et al., 1976). The fungus, a native of Florida, kills seedling and adult stranglervine, a native of South America, at application rates of 1.2 L/ha. The fungus can be applied from May through September after the weed has germinated or is actively growing. A root infection is incited, and although control may appear to spread slowly the first year, control of 90% or more can be obtained for 2 years or more after application. In spite of initial concerns over susceptibility to nontarget crop plants and potential for development of new strains of *Phytophthora*, the use of this organism is considered safe in selective, site-specific applications in the citrus growing areas of Florida.

Colletotrichum gloeosporiodes f.sp. malvae

BioMal† is the trade name given to a formulation containing the fungal pathogen *C. gloeosporioides* f.sp. *malvae* for control of round-leaved mallow in lentil and flax crops in Canada. Both weed and pathogen are endemic to Canada. The fungus incites stem lesions on eight other *Malva* species, with low infection rates on velvetleaf, venice mallow, and safflower (Mortensen, 1988). Environmental conditions necessary for infection include 16 to 20 hr of free water at 20°C to 25°C, so applications must be made in conjunction with irrigation or be carefully timed to coincide with periods of high relative humidity.

Puccinia canaliculata

The native rust fungus *Puccinia canaliculata* is a macrocyclic, heteroecious organism that normally attacks yellow nutsedge late in the season, after the weed has flowered and produced many new tubers. When uredospores of this fungus are released in early spring, high levels of disease inhibit flowering and tuber formation and can result in dehydration of existing tubers and plants

*Abbott Laboratories, North Chicago, IL 60064.

†Philom Biol Inc., Saskatoon, Saskatchewan.

(Phatak et al., 1983). Common cocklebur and sunflower are alternate hosts supporting the relatively undestructive aecial stage of the fungus, but no other hosts are known. The fungus is an obligate parasite that cannot be cultured on artificial media, but uredospores can be collected from infected plants and stored for later application in water with appropriate surfactants. Yellow nutsedge stands were reduced 54% and tuber production 75% when the pathogen was released in Georgia (Phatak et al., 1983). One of the challenges facing biological control of perennial weeds like yellow nutsedge is variable susceptibility of populations from different geographical locations. Considerable variation in phenological and morphological characteristics, herbicide sensitivity, and isozyme pattern has been found among yellow nutsedge populations from diverse locations. Not only do yellow nutsedge populations vary in their susceptibility to *P. canaliculata*, but strains of the pathogen also vary in their ability to incite disease on weed populations from different areas. For example, the Georgia isolate of *P. canaliculata* does not attack yellow nutsedge from Maryland, and Maryland isolates are ineffective in Georgia (Bruckart et al., 1988).

Cercospora rodmanii

One of several pathogens attacking waterhyacinth, *C. rodmanii* is a highly specific fungus native to Florida. The pathogen incites leaf spots, leaf necrosis, and secondary root rot, resulting in reduced plant height, density, and biomass (Freeman and Charudattan, 1984). Infections can be established by applying 2.2 mg of mycelium/m^2 to waterhyacinth mats in May or early June when daytime temperatures range from about 18°C to 30°C. Leaf spots are evident 16 days after application and in 3 to 7 months plants appear pale, weakened, and brown-spotted. Multiple applications of the fungus should be used when conditions support slow growth of waterhyacinth, such as in the early phase of seasonal growth. Application in combination with other agents that retard plant growth increases effectiveness. For example, two weevils (*Neochetina bruchi* and *N. eichhorniae*) and a moth (*Sameodes albiguttalis*) are among more than 70 species of arthropods that feed on waterhyacinth. Combinations of arthropods with *C. rodmanii* have controlled 99% of waterhyacinth within 7 months (Charudattan, 1986). Reduced rates of 2,4-D, diquat, glyphosate, or bensulfuron have also been used to manage waterhyacinth suppressed by *C. rodmanii.*

C. Introduction of Self-Sustaining Pathogens

In the use of self-sustaining pathogens, a pathogen is introduced into susceptible weed populations where the organism is allowed to spread without further manipulation. The aim is that the organism will become established and thrive in its new environment. The activity of the pathogen is intended to provide sustained regulation of the weed at low population levels but is not intended to lead to weed eradication. The unintentional association of bacteria (*Erwinia* spp.) and fungi (*Colletotrichum gloeosporioides* and *Fusarium oxysporum*) with the moth *Cactoblastis cactorum*, that was introduced into Australia in 1925 for control of pricklypear, was probably the first introduction of plant pathogens in biological control of weeds (Goeden, 1988). Holes bored by *Cactoblastis* larvae provided means of entry by the fungi and within 10 years 90% of the original pricklypear population was destroyed. Modern quarantine regulations have eliminated unintentional importation of pathogens with introduced insects, but since 1970 more than 30 pathogens have been introduced intentionally to various geographical areas as classical biological weed control agents (Julien, 1992).

Pathogens selected for introduction must be effective in controlling the weed, be safe for desirable plants, and be able to survive and spread with minimal manipulation. Effectiveness is determined by the nature of the organism itself, including virulence of the particular strain of a pathogen that is selected, as well as the suitability of other factors necessary for disease development. Safety is determined by host specificity requirements for the area where a pathogen is introduced. Extensive and rigorous evaluations in quarantine have been conducted to determine the safety of pathogens before they are introduced for control of weeds. Host specificity is based on the specialization of a pathogen and its coevolved host. Often, fairly specific chemical signals or nutritional needs must be met for a pathogen to recognize its host, to produce structures capable of infecting host tissue, to obtain nutrients from the host, and to develop propagules by which it is spread throughout the host population. Strains of a pathogen may vary considerably in their ability to incite epidemics. Disease buildup may require considerable time before significant effects on the weed are observed and the effectiveness of the inoculative agent is evident. It may be very difficult to halt the spread of an agent after it is released, so the safety of organisms must be assured.

Most self-sustaining pathogens are rusts, smuts, or dry-spored foliar fungi. These organisms are capable of rapid production of wind-disseminated spores which incite disease quickly and result in rapid disease spread. In addition, propagules of these organisms may survive periods of environmental stress, maintain their virulence, and continue to incite disease when suitable conditions prevail. As a result of stringent host specificity requirements, many successful candidate biological control pathogens have been limited to obligate parasites. Although these are generally considered to cause relatively little damage to their weed hosts, successful biological control of rush skeletonweed and others demonstrates the potential of obligate parasites for effective and long-term weed management.

Although few bacteria have been investigated for biological control of weeds, recent research has identified rhizobacteria that infect and interfere with the growth of several annual broadleaf weeds (Kremer et al., 1990) and of downy brome (Gealy et al., 1992). The bacteria invade and multiply in the intercellular space in roots, where they may disrupt cell walls, produce toxins, or interfere with plant hormones. In contrast to foliar pathogens, indigenous rhizobacteria do not require specific environmental conditions for infection and disease development. The bacteria colonize the microenvironment surrounding roots and possibly seeds and may be ideal self-perpetuating biological control agents. They do not provide high levels of control by themselves, but when used in combination with other stress factors they may be an important component of an integrated biological management system (Kremer and Schulte, 1989).

Conventional wisdom implies that the ideal weed for biological control, with the classical approach, is an introduced species that has spread on land or waterways of little economic value to result in extensive, dense populations, unrestricted by natural enemies. However, the classical approach has also shown potential against some weeds of cultivated fields, including rush skeletonweed, common cocklebur, common ragweed, and common groundsel, in areas of intensive cropping (Watson, 1991). Research projects on this approach to biological control have emphasized weeds for which alternative control measures are lacking or are very expensive, giving sufficient economic incentive to justify the research effort. Research has also focused on weeds that have no value as sources of nectar, food for wildlife, or other benefits that may create conflicts of interest. Introduced weeds are separated from pathogens of their native habitat and may lose some disease resistance over time. For this reason weeds may be particularly susceptible to a pathogen introduced into its new environment. Therefore, success with biological control is favored by low genetic variability in weed populations, as the rate and extent of disease epidemics are determined by the degree of resistance in the weed population.

Many plant pathogens have rather specific environmental requirements for growth, infection, propagule production, and survival. The success of self-sustaining organisms is determined by their ability to adapt to the new environment to which they are introduced. It is unlikely that the environment where the organism is introduced will be identical to that of the native habitat. Of particular concern are periods of drought, cultivation, grazing, or competition in the new environment that may disrupt the life cycle of the organism at a time when it is especially susceptible to stress.

The success of pathogens introduced for weed control has been due to increased stress from competition by associated plants or from adverse environmental conditions. For example, common groundsel was less competitive with lettuce when the weed was infected by a rust fungus. The disease did not result in weed mortality but allowed the crop to be more competitive, especially at high densities (Paul and Ayres, 1987). Examples of two weeds that have been targets of classical biological control with plant pathogens follow; additional examples are listed in Table 6.

Rush Skeletonweed

A relatively innocuous weed in the Mediterranean region to which it is native, introduced rush skeletonweed stands have spread to millions of hectares of cereal crops in Australia and rangeland in the western United States. Several organisms have been introduced to control this weed, including the rust fungus *Puccinia chondrillina*. This pathogen destroys rush skeletonweed seedlings and rosettes, damages flowering shoots, and reduces seed production. The pathogen is a highly specialized organism that infects and has all stages of its life cycle on a narrow-leaf biotype of the weed. Strains of *P. chondrillina* specific to other biotypes are under investigation (Hasan, 1981). Optimum conditions for infection are dark periods with at least 6 hr of free water and temperatures of 8°C (Blanchette and Lee, 1981). Effective control is limited by spore dispersal because the rush skeletonweed is adapted to semiarid environments where conditions for disease development occur only briefly. In Idaho, spore inoculum has been disseminated by aerial application of 2 g of uredospores/ha (Lee, 1986). Establishment and effectiveness of the rust may require several years, with significant damage appearing on rosettes infested in fall or spring up to 2 years after spore release. The development of *P. chondrillina* is inhibited by soil disturbance and crop cultivation, but it is self-sustaining in undisturbed conditions. Effectiveness of control is increased where rush skeletonweed is in competition with other plant species such as subterranean clover (Lee, 1986).

Musk Thistle

Musk thistle and related *Carduus* species introduced into North America have become aggressive weeds in grazing lands. A 1974 survey showed that about 10% of U.S. counties had economically important infestations of musk thistle (Dunn, 1976). Among the organisms studied for biological control of musk thistle are strains of the rust fungus *Puccinia carduorum*, imported from Eurasia. Optimum conditions for infection are temperatures of 17°C to 24°C and a minimum of 4 hr of free water (Politis and Bruckart, 1986). Young plants are particularly susceptible to the rust; the main effect is a decrease in plant growth. The fungus was released in southwest Virginia in 1987 and has become established in the surrounding area (Baudoin, 1992). Host specificity studies indicated that limited infections also occurred on artichoke and some *Circium* species, but these species are not considered to be at risk where *P. carduorum* is introduced for musk thistle control.

VI. INTEGRATING BIOLOGICAL AND OTHER APPROACHES TO WEED MANAGEMENT

Natural populations of plants are usually regulated by interactions among several biotic and abiotic agents (Harper, 1977). Even when a single natural enemy is a key factor contributing to the control of a pest population, a combination of interrelated factors is usually involved in constraining most natural populations (Pimentel, 1991). Single-tactic approaches often have not been effective in long-term weed management, and reliance on a single tactic has sometimes resulted in weed shifts, replacement with more competitive species, and development of resistant populations. This is a problem not only for chemical control, but also for reliance on a single organism for biological control. Therefore, the integration of multiple tactics, including a diversity of biological agents and approaches, favors the effectiveness and stability needed for long-term management of weeds.

In nearly all cases of successful classical biological control of weeds, more than one organism interacts with other stresses to suppress the weed population. There is no reason why competition from desirable plants, allelopathy, and grazing animals could not be integrated with insects or pathogens, as well as with compatible cultural, mechanical, and chemical techniques, to suppress weeds in various cropping systems. Integrated management can make use of techniques that by themselves are considered ineffective but contribute to long-term weed suppression when used together with other methods. For this reason, techniques or biological agents should not be rejected as unworthy of study simply because they do not provide weed control comparable to that achieved with herbicides. For example, water hyacinth is not successfully managed by a single insect or pathogen, but the interaction of a pathogen with arthropods or low doses of herbicides provides an effective management approach. It is possible that the effectiveness of other biological management agents could be enhanced if they were used in combination with other agents or compatible weed control techniques.

The aim of the integrated approach is to manage the weed population at an acceptable level by reducing 1) the number of propagules, 2) the number of emerging seedlings, and 3) the survival and competitiveness of emerged weeds. This involves both preventive methods and therapeutic methods.

Prevention makes use of all appropriate and compatible biological and cultural techniques to achieve an acceptable level of weed suppression. Biological agents such as insect predators or fungal pathogens can be used to reduce the production of weed seeds or vegetative propagules. By themselves these agents may not have a large impact on the weed population, but they contribute to weed management by attacking a crucial phase of the life cycle of most weeds. Insects and pathogens can also be used to decrease the survival of seeds and vegetative parts. The integration of the scentless plant bug and pathogenic fungi to attack velvetleaf seed (Kremer and Spencer, 1989) and the use of grazing animals to consume weeds before corn harvest (McClure, unpublished observation) are examples of novel and successful uses of this approach. The number of emerging seedlings can be reduced by several types of biological agents, including principal or auxiliary crop plants (through allelopathy or physical effects), foliar or soil-borne seedling pathogens, or insects. The survival and competitiveness of emerged seedlings can be reduced by methods that increase crop competitiveness as well as by all manner of insect and microbial agents. Therapeutic methods are appropriately employed when preventive methods have failed to reduce the weed population to an acceptable level. Biological therapeutic methods include inundative releases of insects or pathogens, as in the use of mycoherbicides.

Designing an integrated system can often be complicated by conflicting goals. For example, the need to reduce the weed population to an economically acceptable level may result in a subthreshold population that produces seed or vegetative propagules that will be a problem in crops the following season. This goes counter to the goal of reducing production of propagules as a means of long-term management. An important conflict for classical biological control involves balancing the needs of the biological control agent with the desire to destroy the weed on which the biological control agent depends for sustenance. An example is *Rhinocyllus conicus* Froel., a weevil whose larvae feed in the receptacle of thistles in the genus *Carduus*, thereby reducing seed formation. Conventional management of thistles (mowing) disrupts emergence of *R. conicus* unless mowing is delayed until senescence (Tipping, 1991). Obvious conflicts for the use of weed pathogens arise where fungicides are employed to reduce the impact of disease on crops or where herbicides stunt or burn crops and thereby decrease crop competitiveness.

Long-term planning is essential in developing integrated management systems which emphasize biological methods. Planning can help to clarify management objectives and can alleviate some conflicts among goals. Knowledge of the biology of the agroecosystem must serve as the basis for planning the system. Computer models can be used to simulate the influence of long-term management strategies on weed population dynamics. Projections of 5 years or more are helpful to predict how weeds will respond to management efforts and to plan possible responses to the dynamics of the system. For example, the history and biology of herbicide use have demonstrated that continued use of a single control tactic favors resistance development in certain types of weed populations and cropping systems. A similar response should be anticipated for any control tactic, especially if it is highly effective. The shift to broadleaf biotypes of rush skeletonweed where the rust fungus *Puccinia chondrillina* has been released demonstrates the vulnerability of host-specific biological control agents. A knowledge of ecological succession can be used to predict the response of a system to changes in tillage practices and their impact on weed populations. For example, pioneer annuals would be expected to increase where tillage is increased, whereas biennial and perennial weeds would be expected to become more troublesome as tillage is reduced. By developing a plan to anticipate and prevent future weed problems over an entire farm unit, a comprehensive weed management system can be developed that is based on a true integration of approaches and that emphasizes ecologically based biological weed management.

APPENDIX A

List of Common and Scientific Names of Weeds

Common name	Scientific name	BAYER code[a]
Alligatorweed	*Alternanthera philoxeroides* (Mart.) Griseb.	ALRPH
Anoda, spurred	*Anoda cristata* (L.) Schlecht.	ANVCR
Barnyardgrass	*Echinochloa crus-galli* (L.) Beauv.	ECHCG
Beggarticks, hairy	*Bidens pilosa*	BIDPI
Bindweed, field	*Colvolvulus arvensis* L.	CONAR
Blackberry	*Rubus* spp.	
Bluegrass, annual	*Poa annua* L.	POAAN
Brome, downy	*Bromus tectorum* L.	BROTE
Brome, Japanese	*Bromus japonicus* Thunb. ex Murr.	BROJA
Broomweed, common	*Gutierrezia dracunculoides* (DC.) Blake	GUEDR
Bur, Noogoora	*Xanthium occidentale* Bertol.	N/A
Chamomile, false	*Matricaria maritima* L.	MATMA
Chickweed, common	*Stellaria media* (L.) Vill.	STEME
Cockle, corn	*Agrostemma githago* L.	AGOGI
Cocklebur, common	*Xanthium strumarium* L.	XANST
Cocklebur, spiny	*Xanthium spinosum* L.	XANSP
Coneflower, hairy	*Rudbeckia laciniata* L.	RUDLA
Crabgrass, large	*Digitaria sanguinalis* (L.) Scop.	DIGSA
Crabgrass, smooth	*Digitaria ischaemum* (Schreb. ex Schweig.) Schreb. ex Muhl.	DIGIS
Dandelion	*Taraxacum officinale* Weber in Wiggers	TAROF
Ducksalad	*Heteranthera limosa* (Sw.) Willd.	HETLI
Foxtail, green	*Setaria viridis* (L.) Beauv.	SETVI
Foxtail, yellow	*Setaria glauca* (L.) Beauv.	SETLU
Garlic, wild	*Allium vineale* L.	ALLVI
Goosegrass	*Eleusine indica* (L.) Gaertn.	ELEIN

Appendix A Continued

Common name	Scientific name	BAYER code[a]
Gorse	*Ulex europaeus* L.	ULEEU
Gourd, Texas	*Cucurbita texana* (Scheele) Gray	CUUTX
Groundsel, common	*Senecio vulgaris* L.	SENVU
Horseweed	*Conyza canadensis* (L.) Conq.	ERICA
Hydrilla	*Hydrilla verticillata* (L.f.) Royle	HYLLI
Jimsonweed	*Datura stramonium* L.	DATST
Johnsongrass	*Sorghum halpense* (L.) Pers.	SORHA
Jointvetch, northern	*Aeschynomene virginica* (L.) B.S.P.	AESVI
Knapweeds	*Centaurea* spp.	
Koster's curse	*Clidemia hirta* (L.) D. Don	N/A
Lambsquarters, common	*Chenopodium album* (L.)	CHEAL
Mallow, round-leaved	*Mava pusilla* Sm.	MALPU
Mallow, Venice	*Hibiscus trionum* L.	HIBTR
Millet, wild-proso	*Panicum miliaceum* L.	PANMI
Morningglory, ivyleaf	*Ipomoea hederacea* (L.) Jacq.	IPOHE
Morningglory, ivyleaf red	*Ipomoea hederifolia* L.	IPOHF
Morningglory, pitted	*Ipomoea lacunosa* L.	IPOLA
Morningglory, tall	*Ipomoea purpurea* (L.) Roth	PHBPU
Mustard, white	*Brassica hirta* Moench	SINAL
Mustard, wild	*Brassica kaber* (DC.) L.C.Wheeler	SINAR
Nutsedge, purple	*Cyperus rotundus* L.	CYPRO
Nutsedge, yellow	*Cyperus esculentus* L.	CYPES
Oat, wild	*Avena fatua* L.	AVEFA
Pamakani, hamakua	*Ageratina riparia* (Regel) K. & R.	N/A
Panicum, fall	*Panicum dichotomiflorum* Michx.	PANDI
Pigweed	*Amaranthus* spp.	
Pigweed, redroot	*Amaranthus retroflexus* L.	AMARE
Plantain	*Plantago* spp.	
Poa tussock	*Poa labillardieri* L.	N/A
Poppy, corn	*Papaver rhoeas* L.	PAPRH
Pricklypear	*Opuntia* spp.	
Purslane, common	*Portulaca oleracea* L.	POROL
Ragweed, common	*Ambrosia artemisifolia* L.	AMBEL
Ragweed, giant	*Ambrosia trifida*	AMBTR
Ragwort, tansy	*Senecio jacobaea* L.	SENJA
Sesbania, hemp	*Sesbania exaltata* (Raf.) Rydb. ex A.W.Hill	SEBEX
Shepherd's-purse	*Capsella bursa-pastoris* (L.) Medicus	CAPBP
Sicklepod	*Senna obtusifolia* L.	CASOB
Skeletonweed, rush	*Chondrilla juncea* L.	CHOJU
Spiderwort, tropical	*Commelina benghalensis* L.	COMBE

(*continued*)

Appendix A Continued

Common name	Scientific name	BAYER code[a]
Spurge, leafy	*Euphorbia esula* L.	EPHES
Spurge, prostrate	*Euphorbia humistrata* Engelm. ex Gray	EPHHT
Starthistle, yellow	*Centaurea solstitialis* L.	CENSO
Stranglervine	*Morrenia odorata* (H. & A.) Lindl.	MONOD
Sweet briar	*Rosa rubiginosa* L.	ROSRB
Thistle, Canada	*Cirsium arvense* (L.) Scop.	CIRAR
Thistle, Illyrian	*Onorpordium illyricum* N/A	
Thistle, musk	*Carduus nutans* L.	CRUNU
Thistle, Russian	*Salsola iberica* Sennen & Pau	SASKR
Thistle, saffron	*Carthamus lanatus*	N/A
Velvetleaf	*Abutilon theophrasti* Medicus	ABUTH
Waterhemp, tall	*Amaranthus tuberculatus* (Moq.) J.D.Sauer	AMATU
Waterhyacinth	*Eichhornia crassipes* (Mart.) Solms	EICCR

[a]N/A indicates that the Bayer code is not listed in *Composite List of Weeds*.

APPENDIX B

List of Common and Scientific Names of Crops

Common name	Scientific name
Alfalfa	*Medicago sativa* L.
Artichoke	*Cyanara scolumus* L.
Barley	*Hordeum vulgare* L.
Beans	*Phaseolus* spp.
Beet	*Beta vulgaris* L.
Carrot	*Dacus carota* L.
Cassava	*Manihot esculenta* Crantz
Clover, berseem	*Trifolium alexandrinum* L.
Clover, crimson	*Trifolium incarnatum* L.
Clover, red	*Trifolium pratense* L.
Clover, subterranean	*Trifolium subterraneum* L.
Corn	*Zea mays* L.
Cotton	*Gossypium hirsutum* L.
Cowpea	*Vigna unguiculata* (L.) Walp.
Cucumber	*Cucumis sativus* L.
Flax	*Linum usitatissimum* L.
Garlic	*Allium sativum* L.
Horsebean	*Vicia faba* L. var. *major* L.

Appendix B Continued

Common name	Scientific name
Lettuce	*Lactuca sativa* L.
Lentil	*Lens culinaris* Medic.
Lupine	*Lupinus* spp.
Mung bean	*Phaseolus aureus* Roxb.
Mustard	*Brassica* spp.
Oat	*Avena sativa* L.
Onion	*Allium cepa* L.
Pea	*Pisum sativum* L.
Peanut	*Arachis hypogaea* L.
Radish	*Raphanus sativus* L.
Rice	*Oryza sativa* L.
Rye	*Secale cereale* L.
Ryegrass	*Lolium multiflorum* Lam.
Safflower	*Carthamus tinctorius* L.
Sorghum	*Sorghum bicolor* (L.) Moench
Soybean	*Glycine max* (L.) Merr.
Stylo	*Stylosanthes gyianensis* (Aubl.)
Sudangrass	*Sorghum bicolor* (L.) Moench
Sugarcane	*Saccharum officinarum* L.
Sunflower	*Helianthus annuus* L.
Sweet potato	*Ipomoea batatas* Lam.
Tomato	*Lycopersicon esculentum* Mill.
Tobacco	*Nicotiana tabacum* L.
Vetch, bigflower	*Vicia grandiflora* Scop.
Vetch, hairy	*Vicia villosa* Roth.
Wheat	*Triticum aestivum* L.

APPENDIX C

List of Common and Chemical Names of Herbicides

Common name	Chemical name
Acifluorfen	5-[2-Chloro-4-(trifluoromethyl)phenoxy]-2-nitrobenzoic acid
Bensulfuron	2-[[[[[(4,6-Dimethoxy-2-pyrimidinyl)amino]carbonyl] amino]sulfonyl]methyl]benzoic acid
Bentazon	3-(1-Methylethyl)-(*H*)-2,1,3-benzythiadiazin-4(3*H*)-One 2,2-dioxide
Diquat	6,7-Dihydrodipyrido[1,2-α:2′1′-*c*]pyrazinediium ion
Glufosinate	2-Amino-4-(hydroxymethylphosphinyl) butanoic acid

(*continued*)

Appendix C Continued

Common name	Chemical name
Glyphosate	*N*-(Phosphonomethyl) glycine
Paraquat	1,1'-Dimethyl-4,4'-bipyridinium ion
Sulfosate (SC-0224)	Trimethylsulfonium carboxymethylaminomethyl-phosphonate
2,4-D	(2,4-Dichlorophenoxy)acetic acid
Growth regulator	
Thidiazuron	*N*-Phenyl-*N'*-1,2,3-thiadiazol-5-yl urea
Insecticide	
DDT	Dichloro diphenyl trichloroethane

REFERENCES

Aldrich, R. J. (1984). *Weed-Crop Ecology: Principles in Weed Management*, Brenton Publishers, North Scituate, Massachusetts.

Alsaadawi, I. S., Al-uqaili, J. K., Alrubeaa, A. J., and Al-Hadithy, S. M. (1986). Allelopathic suppression of weeds and nitrification by selected cultivars of *Sorghum bicolor* (L.) *Moench*, *J. Chem. Ecol. 12*: 200–204.

Altieri, M. A. and Doll, J. D. (1978). The potential of allelopathy as a tool for weed management in crop fields. *PANS 24*: 495–502.

Altman, J. and Campbell, L. C. (1977). Effect of herbicides on plant diseases. *Annu. Rev. Phytopohthol. 15*: 361–385.

Andres, L. A. 1981. Insects in the biological control of weeds, in *CRC Handbook of pest management in agriculture*, Vol. II, (Pimentel, D., ed.), CRC Press, Boca Raton, Florida, pp. 337–344.

Andres, L. A., Davis, C. J., Harris, P., and Wapshere, A. J. (1976). Biological control of weeds, in *Theory and practice of biological control.* (Huffaker, C. B. and Messenger, P. S., eds.), Academic Press, New York.

Andries, J. A, Douglas, A. G., and Cole, A. W. (1974). Herbicide, leaf type, and row spacing response in cotton, *Weed Sci. 22*: 496–499.

Annecke, D. P., Karney, M., and Burger, W. A. (1969). Improved biological control of the prickly-pear, *Opunita megacantha* Salm Dyck, in South Africa through use of an insecticide, *Phytophylactia 1*: 9–13.

Appleby, A. P., Olson, P. D., and Colert, D. R. (1976). Winter wheat yield reduction from interference by Italian ryegrass, *Agron. J. 68*: 463–466.

Auld, B. A., McRae, C. F., and Say, M. M. (1988). Possible control of *Xanthium spinosum* by a fungus, *Agric. Ecosyst. Environ. 21*: 219–223.

Balsbaugh, E. U., Jr., Frye, R. D., Scholl, C. G., and Anderson, A. W. (1981). Insects for weed control: status in North Dakota, *N. Dakota Farm Res. 39*(3): 3–7.

Balyan, R. S., Malik, R. K., Panwar, R. S., and Singh, S. (1991). Competitive ability of winter wheat cultivars with wild oat (*Avena ludoviciana*), *Weed Sci. 39*: 154–158.

Baragona, J. S. and White, J. C. (1992). Colonization of the bacterial biocontrol agent *Xanthomonas campestris* in annual bluegrass (*Poa annua* L.), *Abstr. Weed Sci. Soc. Am.* 23: 50.

Barnes, J. P. and Putnam, A. R. (1983). Rye residues contribute weed suppression in no-tillage cropping systems, *J. Chem. Ecol. 9*: 1045–1057.

Baudoin, A. (1992). Personal communication. Department of Plant Pathology, Physiology and Weed Science, Blacksburg, Virginia.

Bell, C. E. and J. N. Guerrero. (1992). Sheep grazing for weed control in seedling alfalfa, *Res. Prog. Rep. West. Soc. Weed. Sci.* VII/5-VII-6.

Berkowitz, A. R., Wien, H. C., and Chabot, B. F. (1986). Competition for water between beans and interplanted perennial grass cover crops, *Agron. Abstr. 86*: 107.

Bittman, S., Waddington, J., and McCartney, D. H. (1991). Performance of alfalfa strains grown in mixture with smooth bromegrass as affected by management, *Can. J. Plant Sci. 71*: 1029–1037.

Blanchette, B. L. and Lee, G. A. (1981). The influence of environmental factors in infestation of rush skeletonweed *Chondrilla juncea* by *Puccinia chondrillina*, *Weed Sci. 29*: 364–367.

Blaser, R. E. (1986). Forage-animal management systems, *Virginia Agric. Exp. Sta. Bull.* 86-87.

Boyette, C. D., Quimby, P. C., Bayson, C. T., Egley, G. H., and Fulgham, F. E. (1993). Biological control of hemp sesbania (*Sesbania exaltata*) under field conditions with *Colletotrichum truncatum* formulated in an invert emulsion, *Weed Sci. 41*: 497-500.

Boyette, C. D. and Walker, H. L. (1985). Evaluation of *Fusarium lateritium* as a biological herbicide for controlling velvetleaf (*Abutilon theophrasti*) and prickly sida (*Sida spinosa*), *Weed Sci. 34*: 106–109.

Boyette, C. D., Templeton, G. E., and Oliver, L. R. (1984). Texas gourd (*Cucurbita texana*) control with *Fusarium solani* f.sp. *cucurbitae*, *Weed Sci. 32*: 649–653.

Boyette, C. D., Weidemann, G. J., TeBeest, D. O., and Quimby, P. C. (1991). Biological control of jimsonweed (*Datura stramonium*) with *Alternaria crassa, Weed Sci. 39*: 678-681.

Bradow, J. M. and Connick, W. J. (1990). Volatile seed germination inhibitors from plant residues, *J. Chem. Ecol. 16*: 645–666.

Bridges, D. C. and Chandler, J. M. (1988). Influence of cultivar height on competitiveness of cotton (*Gossypium hirsutum*) with johnsongrass (*Sorghum halpense*), *Weed Sci. 36*: 616–620.

Bruckart, W. L., Johnson, D. R., and Frank, J. R. (1988). Bentazon reduces rust-induced disease in yellow nutsedge, *Cyperus esculentus, Weed Tech.* 2: 299–303.

Buckingham, G. R. (1988). Reunion in Florida: hydrilla, a weevil, and a fly, *Aquatics 10*: 19–25.

Buckingham, G. R., Boucias, D., and Theriot, R. F. (1983). Reintroduction of the alligatorweed fleabeetle (*Agasicles hygrophila* Selman and Vogt) into the United States from Argentina, *J. Aquatic Plant Manage. 21*: 101–102.

Burnside, O. C. (1972). Tolerance of soybean cultivars to weed competition and herbicides, *Weed Sci. 20*: 294–297.

Burnside, O. C. (1979). Soybean (*Glycine max*) growth as affected by weed removal, cultivar, and row spacing, *Weed Sci. 27*: 562–565.

Burnside, O. C. and Wicks, G. A. (1972). Competitiveness and herbicide tolerance of sorghum hybrids, *Weed Sci. 20*: 314–316.

Campbell, M. H. and Dellow, J. J. (1984). Control of St. John's wort (*Hypericum perforatum*) by insects and/or grazing animals, *Proc. Seventh Aust. Weeds Conf. 1*: 109–112.

Campbell, M. H. and Holst, P. J. (1987). Control of Illyrian thistle, *Onoprdum illyricum*, by goats, *Proc. Eighth Aust. Weeds Conf. 1*: 24–26.

Campbell, M. H., Holst, P. J., Keys, M. J., Dellow, J. J., and Duncan, M. R. (1984). The effect of goats in controlling *Poa labillardieri* and trees, *Proc. Seventh Aust. Weeds Conf. 1*: 118–120.

Cardina, J. (1977). Establishment of improved legumes into native grassland on the Accra Plains, Special Pub. Animal Res. Inst., Accra, Ghana.

Challaiah, Burnside, O. C., Wicks, G. A., and Johnson, V. A. (1986). Competition between winter wheat (*Tricicum aestivum*) cultivars and downy brome (*Bromus tectorum*), *Weed Sci. 34*: 689–693.

Chandler, J. M. and Meredith, W. R. (1983). Yields of three cotton (*Gossypium hirsutum*) cultivars as influenced by spurred anoda (*Anoda cristata*) competition, *Weed Sci. 31*: 303–307.

Charudattan, R. (1986). Integrated control of waterhyacinth (*Eichhornia crassipes*) with a pathogen, insects, and herbicides, *Weed Sci. 34*(Suppl. 1): 26–30.

Charudattan, R. (1990). Prospects for biological control of weeds by plant pathogens, *Fitopatol. Bras. 15*: 13–19.

Charudattan, R. (1991). The mycoherbicide approach with plant pathogens, in *Microbial Control of Weeds*, (TeBeest, D. O., ed.), Chapman and Hall, New York, pp. 24–57.

Charudattan, R. and Walker, H. L. (1982). *Biological Control of Weeds with Plant Pathogens*, John Wiley & Sons, New York.

Chepil, W. S. (1946). Germination of weed seeds. I. The influence of tillage treatments on germination, *Sci. Agric. 26*: 347–357.

Combellack, J. H. (1989). The importance of weeds and the advantages and disadvantages of herbicide use, *Plant Prot. Quarterly 4*: 14–37.

Cooper, R. L. (1977). Response of soybean cultivars to narrow rows and planting rates under weed-free conditions, *Agron. J. 69*: 89–92.

Coulson, J. R. (1977). *Biological Control of Aligatorweed, 1959-1972: A Review and Evaluation*, USDA Tech. Bull. 1547, Washington, D.C.

Darrow, R. A. and McCully, W. G. (1959). Brush control and range improvement in the post oak-blackjack oak area of Texas, *Tex. Agric. Exp. Sta. Bull. 942*.

DeHaan, R. L., Wyse, D. L, Ehlke, N. J., Maxwell, B. D., and Putnam, D. H. (1991). Weed control in corn with a *Brassica spp.* smother plant, *Proc. N. Centr. Weed. Sci. Soc. 46*: 8.

Dellow, J. J., Mitchell, T., Johnston, W., Hennessey, G., and Gray, P. (1987). Large area blackberry, *Rubus fruticosus* agg., control using grazing goats, *Proc. Eighth Austr. Weeds Conf. 1*: 70.

Department of Agriculture. (1989). Biological control of leafy spurge, Program Aid No. 1435. Washington, D.C.

Duck, B. N. and Tyler, D. D. (1991). Adaptation of legume species as cover crops in no-till systems, in *Cover Crops for Clean Water*, (Hargrove, W. L. ed.), Soil and Water Conservation Society Ankeny, Iowa, pp. 130–131.

Dunn, P. H. (1976). Distribution of *Cauduus nutans, C. acanthoides, C. pycnocephalus,* and *C. crispus* in the United States, *Weed Sci. 24*: 21–23.

Durden, W. C., Blackburn, R. D., and Gangstad, E. O. (1975). Control program for alligatorweed in southeastern states, *Hyacinth Control 13*: 27–30.

Eckert, D. J. (1988). Rye cover crops for no-tillage corn and soybean production, *J. Prod. Agric. 1*: 207–210.

Einhellig, F. A. and Leather, G. A. (1988). Potential for exploiting allelopathy to enhance crop production, *J. Chem. Ecol. 14*: 1829–1844.

Elkins, D. I., Frederking, D., Marashi, R., and McVay, B. (1983). Living mulch for no-till corn and soybeans, *J. Soil Water Cons. 26*: 431–433.

Enache, A. J. and Ilnicki, R. D. (1990). Weed control by subterranean clover (*Trifolium subterraneum*) used as a living mulch, *Weed Tech. 4*: 534–538.

Fay, P. K. and Duke, W. B. (1977). An assessment of allelopathic potential in *Avena* germplasm, *Weed Sci. 25*: 224–228.

Fenner, M. (1980). The inhibition of germination of *Bidens pilosa* seeds by leaf canopy shade in some natural vegetation types. *New Phytol. 84*: 95–101.

Fiebig, W. W., Shilling, D. G., and Knauft, D. A. (1991). Peanut genotype response to interference from common cocklebur, *Crop Sci. 31*: 1289–1292.

Figliola, S. E., Camper, N. D., and Ridings, W. H. (1988). Potential biological control agent for goosegrass (*Elucine indica*), *Weed Sci. 36*: 830–835.

Forcella, F. (1987). Characteristics associated with highly competitive soybeans, *Agron. Abstr. 87*: 111.

Forney, D. R., Foy, C. L., and Wolf, D. D. (1985). Weed suppression in no-till alfalfa (*Medicago sativa*) by prior cropping of summer-annual forage grasses, *Weed Sci. 33*: 490–497.

Freeman, T. E. and Charudattan, R. (1984). *Cercospora rodmanii Conway, a Biocontrol Agent for Waterhyacinth*, Technical Bull. 842, Florida Agricultural Experiment Station, University of Florida, Gainesville.

Frick, K. E. and Chandler, J. M. (1978). Augmenting the moth *Bactra verutana* in field plots for early-season suppression of purple nutsedge (*Cyperus rotundus*), *Weed Sci. 26*: 703–710.

Frick, K. E., Hartley, G. G., and King, E. G. (1983). Large-scale production of *Bactra verutana* [Lep:Tortricidae] for biological control of nutsedge, *Entomophaga 28*: 107–116.

Funke, G. L. (1941). Essai de phytosociologie experimentale, *Bull. Soc. Hist. Nat. Toulouse 76*: 19–21.

Garrity, D. P., Movillon, M., and Moody, K. (1992). Differential weed suppression ability in upland rice cultivars, *Agron. J. 84*: 586–591.

Gealy, D. R., Gurusiddaiah, S., Kennedy, A. C., and Ogg Jr., A. G. (1992). Effects of phytotoxins from *Pseudomonas fluorescens* strain D7 on seed germination and seedling growth of downy brome (*Bromus tectorum* L.), *Abstr. Weed Sci. Soc. Am. 32*: 51.

Goeden, R. D. (1988). A capsule history of biological control of weeds, *Biocontrol News and Information 9*: 55–61.

Goeden, R. D., Ricker, D. W., and Muller, H. (1987). Introduction, recovery and limited establishment of *Coloephora klimeschiella* (Lepidoptera: Coleophoridae) on Russian thistles, *Salsola australis*, in southern California, *Environ. Entomol. 16*: 1027–1029.

Griffin, J. L. and Dabney, S. M. (1990). Preplant-postemergence herbicides for legume cover-crop control in minimum tillage systems, *Weed Technol. 4*: 332–336.

Hall, M. R., Swanton, C. J., and Anderson, G. W. (1992). The critical period of weed control in grain corn (*Zea mays*), *Weed Sci. 40*: 441–447.

Hargrove, W. L. (1991). *Cover Crops for Clean Water*, Soil and Water Conservation Society, Ankeny, Iowa.

Harper, J. (1977). *Population Biology of Plants*, Academic Press, New York.

Harris, P. (1971). Current approaches to biological control of weeds, *Commonwealth Inst. Biol. Control Tech. Commun. 4*: 67–76.

Harris, P. (1980). Stress as a strategy in the biological control of weeds, in *Biological control in crop protection*. (Papavizas, G. C., ed.), BARC Symp. No. 5. London: Granada.

Harris, P. (1993). Effects, constraints, and the future of weed biocontrol, *Agric. Ecosyst. Environ. 46*: 289-303.

Harrison, H. F., Jr. and Peterson, J. K. (1986). Allelopathic effects of sweet potatoes (*Ipomoea batatas*) on yellow nutsedge (*Cyperus esculentus*) and alfalfa (*Medicago sativa*), *Weed Sci. 34*: 623-627.

Hartwig, N. L. (1983). Crownvetch: a perennial legume "living mulch" for no-tillage crop production, *Proc. Northeast Weed Sci. Soc.* Suppl. 37: 28–38.

Hasan, S. (1981). A new strain of the rust fungus *Puccinia chondrillina*, *Ann. Appl. Biol. 72*: 257–263.

Hawkes, R. B. and Johnson, G. R. (1978). *Longitarsus jacobaeae* aids moth in the biological control of tansy ragwort, *Proc. IV Int. Symp. Biol. Control Weeds 4*: 193.

Hodgson, R. H., Wymore, L. A., Watson, A. K., Snyder, R. H., and Collette, A. (1988). Efficacy of *Colletotrichum coccodes* and thidiazuron for velvetleaf (*Abutilon theophrasti*) control in soybean, *Weed Tech. 2*: 473–480.

Hoffman, M. L., Regnier, E. E., and Cardina, J. (1993). Weed and corn responses to a hairy vetch (*vicia villosa*) cover crop, *Weed Tech. 7*: 594–599.

Hope-Simpson, J. (1940). Studies of the vegetation of the English chalk. IV. Late stages in succession leading to chalk grassland, *J. Ecol. 28*: 386–402.

Huffaker, C. B. (1971). *Biological Control*, Plenum Press, New York.

Huffaker, C. B., Hamai, J., and Nowierski, R. M. (1983). Biological control of puncturevine, *Tribulus terrestris* in California after twenty years of activity of introduced weevils, *Entomophaga 28*: 387–400.

James, K. L., Banks, P. A., and Karnok, K. J. (1988). Interference of soybean, *Glycine max*, cultivars with sicklepod, *Cassia obtusifolia, Weed Technol. 2*: 404–409.

Jeffers, D. L. (1987). Multiple cropping, in *The soybean in Ohio*, (J. Beuerlein, ed.). Ohio Coop. Ext. Serv. Bull. *741*: 90–95.

Johnson, D. R. and Wyse, D. L. (1992). Biological control of weeds with *Pseudomonas syringae* pv. *Tagetis*, *Proc. N. Central Weed Sci. Soc. 47*: 16.

Johnston, A. and Peake, R. W. (1960). Effect of selective grazing by sheep on the control of leafy spurge (*Euphorbia esula* L.), *J. Range Manage. 13*: 192.

Jones, W. A., Jr., Walker, H. E., Quimby, P. C., and Ouzts, J. D. (1985). Biology of *Niesthrea louisianica* (Hemiptera:Rhopalidae) on selected plants, and its potential for biocontrol of velvetleaf, *Abutilon theophrasti* (Malvaceae), *Ann. Entomol. Soc. Am. 78*: 326–330.

Julien, M. H. (1992). *Biological Control of Weeds: A World Catalogue of Agents and Their Target Weeds*, 3rd Edition, C.A.B. International, Wallingford, U.K.

Julien, M. H., Kerr, J. D., and Chan, R. R. (1984). Biological control of weeds: an evaluation, *Protection Ecol.* 7: 3–25.

Keely, P. E. and Thullen, R. J. (1978). Light requirement of yellow nutsedge (*Cyperus esculentus*) and light interception by crops, *Weed Sci. 26*: 10–16.

Kirkland, K. J. and Hunter, J. H. (1991). Competitiveness of Canada prairie spring wheats with wild oats (*Avena fatua* L.), *Can. J. Plant Sci. 71*: 1089–1092.

Kommendahl, T. and Linck, A. J. (1957). The ecological effects of different preceding crop plants on *Setaria glauca* in flax, *Proc. Minn. Acad. Sci. 25*: 91–94.

Konesky, D. W., Siddiqi, M. Y., and Glass, A. D. M. (1989). Wild oat and barley interactions: varietal differences in competitiveness in relation to phosphorus supply, *Can. J. Bot. 67*: 3366–3371.

Koscelny, J. A. and Pepper, T. F. (1990). Effect of grazing cheat (*Bromus secalinus*)-infested wheat (*Triticum aestivum*) on both grasses, *Weed Tech. 4*: 565–568.

Kremer, R. J., Begonia, M. F. T., Stanley, L., and Lanham, E. T. (1990). Characterization of rhizobacteria associated with weed seedlings, *Appl. Environ. Microbiol. 56*: 1649–1655.

Kremer, R. J. and Schulte, L. K. (1989). Influence of chemical treatment and *Fusarium oxysporum* on velvetleaf (*Abutilon theophrasti*), *Weed Tech. 3*: 369–374.

Kremer, R. J. and Spencer, N. R. (1989). Impact of a seed-feeding insect and microorganisms on velvetleaf (*Abutilon theophrasti*) seed viability, *Weed Sci. 37*: 211–216.

Kurtz, T., Melsted, S. W., Bray, R. H., and Breland, H. L. (1952). Further trials with intercropping of corn in established sods, *Soil. Sci. Soc. Am. Proc. 16*: 282–285.

Lal, R. (1975). Role of mulching techniques in tropical soil and water management, *IITA Tech. Bull. 1.*

Lal, R., Regnier, E., Eckert, D. J., Edwards, W. M., and Hammond, R. (1991). Expectations of cover crops for sustainable agriculture, in *Cover Crops for Clean Water*, (Hargrove, W. L. ed.), Soil and Water Conservation Society, Ankeny, Iowa, pp. 1–11.

Leather, G. R. (1983). Sunflowers (*Helianthus annus*) are allelopathic to weeds, *Weed Sci. 31*: 37–42.

Lee, G. A. (1986). Integrated control of rush skeletonweed (*Chondrilla juncea*) in the Western U.S., *Weed Sci. 34*(Suppl. 1): 2–6.

Lin, J., Smith, R. J., Jr., and Dilday, R. H. (1992). Allelopathic activity of rice germplasm on ducksalad and redstem, *Abstr. Weed Sci. Soc. Am. 32*: 54.

Litsinger, J. A. and Moody, K. (1976). Integrated pest management in multiple cropping systems, in Multiple cropping (Papendick, R. I., Sanchez, P. A., and Triplett, G. B., eds.), *Am. Soc. Agron. Special Pub. 27.*

Lockerman, R. H. and Putnam, A. R. (1979). Evaluation of allelopathic cucumbers (*Cucumis sativus*) as an aid to weed control, *Weed Sci. 27*: 54–57.

Maddox, D. M. and Rhyne, M. (1975). Effects of induced host-plant mineral deficiencies on attraction, feeding and fecundity of the alligatorweed flea beetle, *Environ. Entomol. 4*: 682–686.

Martin, V. L., McCoy, E. L., and Dick, W. A. (1990). Allelopathy of crop residues influences corn seed germination and early growth, *Agron. J. 82*: 555–560.

Martin, P. and Rademacher, B. (1960). Studies on the mutual influences of weeds and crops, *Symp. Br. Ecol. Soc. 1*: 143–152.

McWhorter, C. G. and Barrentine, W. L. (1975). Cocklebur control in soybeans as affected by cultivars, seeding rates, and methods of weed control, *Weed Sci. 23*: 386–390.

McWhorter, C. G. and Hartwig, E. E. (1972). Competition of johnsongrass and cocklebur with six soybean varieties, *Weed Sci. 20*: 56–59.

Merrill, L. B. and Taylor, C. A. (1976). Take note of the versatile goat, *Rangeman's J. 3*: 74–76.

Meyers, H. B., Gibb, T. J., Edwards, C. R. and Cardina, J. (1991). In search of alternative host plants of *Althaeus folkertsi* (velvetleaf seed beetle), *Proc. Indiana Acad. Sci. 107*: 46.

Minotti, P. L. and Sweet, R. D. (1981). Role of crop competition in limiting losses from weeds, in *CRC Handbook of pest management in agriculture,* Vol. II, (Pimentel, D., ed.), CRC Press, Boca Raton, Florida, pp. 351–367.

Miracle, M. P. (1967). *Agriculture of the Congo Basin*, University of Wisconsin Press, Madison.

Monks, D. W. and Oliver, L. R. (1988). Interaction between soybean (*Glycine max*) cultivars and selected weeds, *Weed Sci. 36*: 770–774.

Morin, L., Auld, B. A., and Brown, J. F. (1993). Synergy between *Puccinia xanthii* and *Colletotrichum orbiculare* on *Xanthium occidentale, Biol. Control 3*: 296–310.

Mortensen, K. (1988). The potential of an endemic fungus, *Colletotrichum gloeosporiodes*, for biological control of round-leaved mallow (*Malva pusilla*) and velvetleaf (*Abutilon theophrasti*), *Weed Sci. 36*: 473–478.

Murdock, E. C., Banks, P. A., and Toler, J. E. (1986). Shade development effects on pitted morningglory (*Ipomoea lacunosa*) interference with soybeans (*Glycine max*), *Weed Sci. 34*: 711–717.

National Academy of Sciences. (1968). *Principles of Plant and Animal Pest Control*, Vol. 2. Pub. 1597, Washington, D.C.

Nechols, J. R., Andres, L. A., Beardsley, J. W., Goeden, R. D., and Jackson, C. G. (1992). *Biological Control in the U.S. Western Region: Accomplishments and Benefits of Regional Research Project W-84 (1964–1989)*. University of California Division of Agriculture and Natural Resources, Berkeley.

Nelson, D. C. and Giles, J. F. (1989). Weed management in two potato (*Solanum tuberosum*) cultivars using tillage and pendimethalin, *Weed Sci. 37*: 228–232.

Norris, R. F. and Ayres, D. R. (1981). Reduction of biological control of common purslane (*Portulaca oleracea* L.) by the purslane sawfly (*Schizorcerella pilicornia* Holmgren) through use of insecticides, *Abstr. Weed Sci. Soc. Am. 21*: 6.

Oehrens, E. (1977). Biological control of the blackberry through the introduction of rust, *Phragmidium violaceum* in Chile, *FAO Plant Prot. Bull. 25*: 26–28.

Oka, I. N. and Pimentel, D. (1976). Herbicide (2,4-D) increases insect and pathogen pests on corn, *Science 193*: 239–240.

Okigbo, B. N. and Greenland, D. J. (1977). Intercropping systems in tropical Africa, in Multiple cropping (Papendick, R. I., Sanchez, P. A., and Triplett, G. B., eds.), *Am. Soc. Agron. Special Pub. 27*: 63–101.

O'Leary, P. F. (1983). The biological control of water hyacinth, *Eichornia crassipes* (Mart.) Solms, with *Arzama densa* Walker, M. S. Thesis, Delta State Univ., Cleveland.

Oraze, M. J. and Grigarick, A. A. (1992). Biological control of ducksalad (*Heteranthera limosa*) by the waterlily aphid (*Rhopalosiphum nymphaeae*) in rice (*Oryza sativa*), *Weed Sci. 40*: 333–336.

Ormeno-nunez, J., Reeleder, R. D., and Watson, A. K. (1988). A foliar disease of field bindweed (*Convolvulus arvensis*) caused by *Phomopsis convolvulus*, *Plant Dis. 72*: 338–342.

Overland, L. (1966). The role of allelopathic substances in the "smother crop" barley, *Am. J. Bot. 53*: 423–432.

Owsley, C. M. and Hargrove, W. L. (1991). New cool-season legumes for use in conservation tillage, in *Cover Crops for Clean Water* (Hargrove, W. L. ed.), Soil and Water Conservation Society, Ankeny, Iowa.

Parish, S. (1990). A review of non-chemical weed control techniques, *Biol. Agric. Hort. 7*: 117–137.

Parsons, W. T. (1973). *Noxious Weeds of Victoria*, Inkata Press, Melbourne, Australia.

Paul, N. D. and Ayres, P. G. (1987). Effects of rust infection of *Senecio vulgaris* on competition with lettuce, *Weed Res. 37*: 431–441.

Peeper, T. F. (1984). Chemical and biological control of downy brome (*Bromus tectorum*) in Wheat and alfalfa in North America, *Weed Sci. 32* (Suppl. 1): 18–25.

Peschken, D. P. (1979). Biological control of weeds in Canada with the aid of insects and nematodes, *Zeitschrift fur Angewandte Entomol. 88*: 1–16.

Peters, S. E. (1992). Personal communication, Rodale Research Center, Kutztown, Pennsylvania.

Phatak, S. C., Sumner, D. R., Wells, H. D., Bell, D. K., and Glaze, N. C. (1983). Biological control of yellow nutsedge with the indigenous rust fungus *Puccinia canaliculata, Science 219*: 1446–1447.

Phillips, S. H. and Young, H. M., Jr. (1973). *No-tillage farming*, Reiman, Milwaukee.

Pierce, J. R. (1987). The use of goats for control of saffron thistle, *Carthamus lanatus* L., *Proc. Eighth Aust. Weed Conf.* 189–197.

Pimentel, D. (1991). Diversification of biological control strategies in agriculture, *Crop Prot. 10*: 243–253.

Politis, D. J. and Bruckart, W. L. (1986). Infection of musk thistle by *Puccinia carduorum* as influenced by conditions of dew and plant age, *Plant Dis. 70*: 288–290.

Prostko, E. and Ilnicki, R. D. (1988). Residual weed control in intercropped soybeans, *Proc. Northeast Weed Sci. Soc. 42*: 22–23.

Putnam, A. R. (1988). Allelolpathy: problems and opportunities in weed management, in *Weed Management in Agroecosystems: Ecological Approaches* (Altieri, M. A. and Liebman, M. eds.), CRC Press, Boca Raton, Florida, pp. 77–88.

Putnam, A. R. and DeFrank, J. (1983). Use of phytotoxic plant residues for selective weed control, *Crop Prot. 2*: 173–181.

Putnam, A. R., DeFrank, J., and Barnes, J. P. (1983). Exploitation of allelopathy for weed control in annual and perennial cropping systems, *J. Chem. Ecol. 9*: 1001–1009.

Putnam, A. R. and Duke, W. B. (1974). Biological suppression of weeds: evidence for allelopathy in accessions of cucumber, *Science 185*: 370.

Putnam, A. R. and Duke, W. B. (1978). Allelopathy in agroecosystems. *Annu. Rev. Phytopathol. 16*: 431–451.

Quimby, P. C., Jr. (1986). Microbial control of weeds symposium, *Weed Sci. 34*(Suppl. 1): 1–53.

Quimby, P. C., Jr. and Frick, K. E. (1985). Evaluation of herbicide-coated larvae of *Bactra verutana* (Lep:Tortricidae) to control nutsedges (*Cyperus rotundus* L. and *C. esculentus* L.), *Entomophaga 30*: 287–292.

Quimby, P. C., Jr., Fulgham, F. E., Boyette, C. D., and Connick, W. J., Jr. (1988). An invert emulsion replaces dew in biocontrol of sicklepod—a preliminary study, in Hovde, D. A. and Beestman, G. B., eds. Pesticide formulations and application systems, Vol. 8. pp. 264–270. American Society of Testing and Materials, Philadelphia.

Quimby, P. C., Jr., Miller, T. C., and McWilliams, J. M. (1982). Male-sterile *Heliothis* suppresses *Ipomoea* in rice, *Proc. S. Weed Sci. Soc. 35*: 297.

Radcliffe, J. E. (1990). Gorse control by goats: effective strategies in Canterbury, *For. Res. Inst. Bull. 155*: 144–149.

Ramsel, R. E. and Wicks, G. A. (1988). Use of winter wheat (*Triticum aestivum*) cultivars and herbicides in aiding weed control in an ecofallow corn (*Zea mays*) rotation, *Weed Sci. 36:* 394–398.

Reeves, T. G. and Smith, I. S. (1975). Pasture and cultural methods for the control of annual ryegrass (*Lolium rigidum*), *Aust. J. Exp. Agric. An. Husb. 15*: 527–530.

Regnier, E. E. and Stoller, E. W. (1981). A living mulch for weed control in soybeans, *Proc. N. Cent. Weed Control Conf. 36*: 24–25.

Rhoades, D. F. (1985). Offensive-defensive interactions between herbivores and plants: their relevance in herbivore population dynamics and ecological theory, *Am. Nat. 125*: 205–238.

Rice, E. L. (1983). *Allelopathy*, 2nd Edition, Academic Press, New York.

Ridings, W. H. (1986). Biological control of stranglervine in citrus: a researcher's view, *Weed Sci. 34*(Suppl. 1): 31–32.

Ridings, W. H., Burnett, H. C., Schoulties, C. L., and El-Gholl, N. E. (1976). Biological control of milkweed vine in Florida citrus groves with a pathotype of *Phytophthora citrophthora* (T. E. Freeman, ed.), *Proc. IV Int. Symp. Biol. Control Weeds* 224–240.

Robinson, R. G. and Dunham, R. S. (1954). Companion crops for weed control in soybeans, *Agron. J. 46*: 278–281.

Rodriguez, J. G. 1960. Nutrition of the host and reaction to pests, in Biological and chemical control of plants and animal pests, *Am. Assoc. Advan. Sci. 61*: 149–167.

Rose, S. J., Burnside, O. C., Specht, J. E., and Swisher, B. A. (1984). Competition and allelopathy between soybeans and weeds, *Agron. J. 76*: 523–528.

Runnels, H. A. and Schaffner, J. H. (1931). Manual of Ohio weeds, *Ohio Agric. Exp. Sta. Bull.* 475.

Sands, D. C., Ford, E. J., and Miller, R. V. (1990). Genetic manipulation of broad host-range fungi for biological control of weeds, *Weed Tech. 4*: 471–474.

Schmid, A. R. and Behrens, R. (1972). Herbicides vs. oat companion crops for alfalfa establishment, *Agron. J. 64*: 157–159.

Scifres, C. J. (1981). Selective grazing as a weed control method, in *CRC Handbook of pest management in agriculture*, Vol. II. (Pimentel, D., ed.), CRC Press, Boca Raton, Florida, pp. 369–381.

Semidey, M. Frans, R. E., and Oliver, L. R. (1992). Weed management in cotton and soybean using allelopathic sunflower, *Abstr. Weed Sci. Soc. Am. 32*: 54.

Shetty, S. V. R. and Krantz, B. A. (1980). Weed research at ICRISAT (International Crops Research Institute for the Semi-Arid Tropics), *Weed Sci. 28*: 451–454.

Shetty, S. V. R., Sivakumar, M. V. K., and Ram, S. A. (1982). Effect of shading on the growth of some common weeds of the semi-arid tropics, *Agron. J. 74*: 1023–1029.

Shilling, D. G., Worsham, A. D., and Danehower, D. A. (1986). Influence of mulch, tillage and diphenamid on weed control, yield and quality in no-till flue-cured tobacco (*Nicotania tabacum*), *Weed Sci. 34*: 738–744.

Smith, R. J., Jr. (1974). Competition of barnyardgrass with rice cultivars, *Weed Sci. 22*: 423–426.

Smith, R. J., Jr. (1986). Biological control of northern jointvetch *Aeschynomene virginica* in rice (*Oryza sativa*) and soybeans (*Glycine max*): a researcher's view, *Weed Sci. 34*(Suppl. 1): 17–23.

Smith, R. J. (1988). Weed thresholds in southern U.S. Rice, *Oryza sativa, Weed Tech. 2*: 232–241.

Spencer, N. R. (1986). Velvetleaf (*Abutilon theohrasti* Medik.) seed production reduced by an insect, *Abstr. Weed Sci. Soc. Am. 26*: 56.

Stauber, L. G., Smith, R. J., Jr., and Talbert, R. E. (1991). Density and spatial interference of barnyardgrass (*Echinochloa crus-galli*) with rice (*Oryza sativa*), *Weed Sci. 39*: 163–168.

Steinsiek, J. W., Oliver, L. R., and Collins, F. C. (1982). Allelopathic potential of wheat (*Triticum aestivum*) straw on selected weed species, *Weed Sci. 30*: 495–497.

Story, J. M. (1985). Status of biological weed control in Montana, *Proc. 6th Int Symp. Biol. Control Weeds* 837–842.

Stute, J. K. and Posner, J. L. (1993. Legume cover crop options for grain rotations in Wisconsin, *Agron. J. 85*: 1128–1132.

Taylorson, R. B. and Borthwick, H. A. (1969). Light filtration by foliar canopies: significance for light-controlled weed seed germination, *Weed Sci. 17*: 48–51.

Teasdale, J. R., Beste, C. E., and Potts, W. E. (1991). Response of weeds to tillage and cover crop residues, *Weed Sci. 39*: 195–199.

Teasdale, J. R. and Daughtry, C. S. T. (1993). Weed suppression by live and desiccated hairy vetch (*Vicia villosa*), *Weed Sci. 41*: 207–212.

Teasdale, J. R. and Mohler, C. L. (1993). Light transmittance, soil temperature, and soil moisture under residue of hairy vetch and rye, *Agron. J. 85*: 673–680.

TeBeest, D. O. (1988). Additions to the host range of *Colletotrichum gloeosporiodes* f.sp. *aeschynomene, Plant Dis. 72*: 16–18.

TeBeest, D. O. (1991). *Microbial Control of Weeds*, Chapman and Hall, New York.

TeBeest, D. O. and Templeton, G. E. (1985). Mycoherbicides: progress in the biological control of weeds, *Plant Dis. 69*: 6–10.

Templeton, G. E., Smith, R. J., Jr., and TeBeest, D. O. (1986). Progress and potential of weed control with mycoherbicides, *Rev. Weed Sci. 2*: 1–14.

Templeton, G. E., TeBeest, D. O., and Smith, R. J., Jr. (1984). Biological weed control in rice with a strain of *Colletotrichum gloeosporioides* (Penz.) Sacc. used as a mycoherbicide, *Crop Prot. 4*: 409–422.

Templeton, G. E. and Trujillo, E. E. (1981). The use of plant pathogens in the biological control of weeds, in *CRC Handbook of Pest Management in Agriculture*, Vol. II (Pimentel, D., ed.), CRC Press, Boca Raton, Florida, pp. 345–350.

Tipping, P. W. (1991). Effects of mowing or spraying *Carduus thoermeri* on *Rhinocyllus conicus, Weed Tech. 5*: 628-631.

Triplett, G. B. (1986). Crop management practices for surface-tillage systems, in *No-Tillage and Surface-Tillage Agriculture* (Sprague, M. A. and Triplett, G. B. eds.), John Wiley & Sons, New York, pp. 149–182.

Trujillo, E. E. (1985). Biological control of hamakua pamakani with *Cercosporella* sp. in Hawaii, (E. S. Delfosse, ed.), *Proc. VI Int. Symp. Biol. Control Weeds*, 661–671.

Trujillo, E. E., Latterell, F. M., and Rossi, A. E. (1986). *Colletotrichum gloeosporioides*, a possible biological control agent for *Clidemia hirta* in Hawaiian forests, *Plant Dis. 70*: 974–976.

Vallentine, J. F. (1971). *Range Development and Improvements*, Brigham Young University Press, Provo, Utah.

Walker, H. L. and Boyette, C. D. (1985). Biocontrol of sicklepod (*Cassia obtusifolia*) in soybeans (*Glycine max*) with *Alternaria cassiae, Weed Sci. 33*: 212–215.

Walker, J. W., Hemenway, K. G., Hatfield, P. G., and Glimp, H. A. (1992). Training lambs to be weed eaters: studies with leafy spurge, *J. Range Manage. 45*: 245–249.

Walker, H. L. and Sciumbato, G. L. (1979). Evaluation of *Alternaria macrospora* as a potential biocontrol agent for spurred anoda (*Anoda cristata*): host range studies, *Weed Sci. 27*: 612–614.

Wapshere, A. J., Delfosse, E. S., and Cullen, J. M. (1989). Recent developments in biological control of weeds, *Crop Prot. 8*: 227–250.

Watson, A. K. (1991). The classical approach with plant pathogens, in *Microbial Control of Weeds* (TeBeest, D. O., ed.) Chapman and Hall, New York, pp. 3–23.

Weaver, J. B., Miller, J., and Weaver, D. (1978). Tolerance and weed control response of three cotton genotypes to three levels of two dinitroaniline herbicides, *Univ. Georgia Agric. Exp. Sta. Res. Bull.* 213.

Wesson, G. and Wareing, P. F. (1967). Light requirements of buried seed, *Nature 213*: 600–601.

West, N. E. and Farah, K. O. (1989). Effects of clipping and sheep grazing on dyers woad, *J. Range Manage. 42*: 5–10.

White, R. H., Worsham, D., and Blum, U. (1989). Allelopathic potential of legume debris and aqueous extracts, *Weed Sci. 37*: 674–679.

Wicks, G. A., Johnston, D. N., Nuland, D. S., and Kinbacher, E. J. (1973). Competition between annual weeds and sweet Spanish onions, *Weed Sci. 21*: 436–439.

Wicks, G. A., Ramsel, R. E., Nordquist, P. T., Schmidt, J. W., and Challaiah. (1986). Impact of wheat cultivars on establishment and suppression of summer annual weeds, *Agron. J. 78*: 59–62.

Wilson, F. (1960). A review of the biological control of insects and weeds in Australia and Australian New Guinea, *Commonwealth Inst. Biol. Control Tech. Commun.* 1.

Wilson, G. F., Lal, R., and Okigbo, B. N. (1982). Effects of cover crops on soil structure and on yield of subsequent arable crops grown under strip tillage on an eroded Alfisol, *Soil Tillage Res. 2*: 233–250.

Wilson, R. E. and Rice, E. L. (1968). Allelopathy as expressed by *Helianthus annuus* and its role in old-field succession, *Bull. Torrey Bot. Club 95*: 432–448.

Wink, M. (1983). Wounding-induced increase of quinolizidine alkaloid accumulation in lupin leaves, *Z. Naturforsch 38*: 905–909.

Wood, R. W. (1992). Personal communication, Department of Agricultural Engineering Ohio State University, Columbus, Ohio.

Wortmann, C. S. (1993). Contribution of bean morphological characteristics to weed suppression, *Agron. J. 85*: 840–843.

Wymore, L. A. and Watson, A. K. (1986). An adjuvant increases survival and efficacy of *Colletotrichum coccodes*, a mycoherbicide for velvetleaf (*Abutilon theophrasti*). *Phytopathol. 76*: 1115–1116.

10

Weed Management Systems for Oil Seed Crops

John W. Wilcut and Alan C. York
North Carolina State University, Raleigh, North Carolina

David L. Jordan
Louisiana State University Agricultural Center, St. Joseph, Louisiana

I. INTRODUCTION

Cotton, peanut, and soybean are three of the most widely planted agronomic broadleaf crops in the United States. Cotton is grown primarily in the southern region of the country from Virginia and Tennessee in the north, south to Florida, and west to Arizona and California. Cotton is grown on approximately 5.7 million hectares annually in the United States (Supka et al., 1992). The major cotton production areas are in the Southeast, the Midsouth, the Southwest, and Arizona and California. Upland cotton is grown in the southeastern and Midsouth regions, whereas Pima cotton and upland cotton are grown in the Far West.

Peanut is grown on approximately 810,000 hectares in the United States with a farm value exceeding $1 billion annually. Seven states (Georgia, Alabama, Texas, North Carolina, Virginia, Oklahoma, and Florida) account for 98% of the production. Peanut also is grown in New Mexico, South Carolina, and Mississippi. The Virginia–North Carolina area grows the large-seeded Virginia-type peanut; the southeastern area (Alabama, Florida, Georgia, Mississippi, and South Carolina) grows Virginia-, Spanish-, and runner-type peanut; and the southwestern area (Texas, Oklahoma, and New Mexico) grows Spanish- and runner-type, with New Mexico growing mostly Valencia-type peanut (Wilcut et al., 1994a).

Soybean is grown on approximately 25 million hectares in the United States and is used in numerous products for both animal and human consumption. Soybean is grown in numerous states in the United States with the major hectarage in the midwestern, midsouthern, and southeastern regions.

Pest management is a key to management strategy for profitable cotton, peanut, and soybean production. Adequate control of weeds is important in obtaining optimum yield and quality of these crops.

II. LOSSES AND COSTS DUE TO WEEDS

Weed management systems in cotton consist primarily of herbicides and cultivation. Cost of weed management systems currently employed in cotton can vary from $55 to $165/hectare for a total cost of approximately $410 million annually in the United States (Chandler and Cooke, 1992). Estimated loss from the 10 most common weeds in U.S. cotton was approximately $190 million (Chandler and Cooke, 1992). Weeds and other foreign matter can impair grade (Perkins et al., 1984). From 1961 to 1986, approximately $9.4 million was lost annually as a result of classification of cotton bales as "grassy" (Chandler and Cooke, 1992). Because of the limited competitiveness of cotton and the necessity for clean, uncontaminated fiber, weed management systems which maintain essentially a weed-free crop are necessary.

Herbicide costs in peanut range from $37 to $124/ha with a net cost to U.S. producers in excess of $70 million annually (Bridges, 1992; Bryson, 1989). Weeds also increase the need for additional primary and secondary tillage, with a net loss to producers of $7 to $20/ha due to yield reductions and $7 to $62/ha due to quality reductions. Finally, harvesting efficiency was reduced and yield losses increased in harvesting weedy fields as a result of physical obstruction by the weed biomass (Wilcut et al., 1994a).

Yield losses incurred by weeds in soybean can be substantial. Crop loss in the United States using best management practices was estimated as high as $800 million (Bridges, 1992).

III. WEED INTERFERENCE AND COMPETITION

Crop competition is an effective but often underused weed management option. The basic strategy is to follow practices that result in rapid crop growth and canopy closure so as to provide maximum shading of weeds. Any practice that enhances the chances of the crop's dominating the site is beneficial in weed management (Walker, 1982). Manipulating row width may allow the crop to become dominant.

Historically, row width was set as the minimum distance needed for the passage of draft animals. These distances were frequently carried over with minimal modification into mechanized production practices. It is obvious that reducing row spacing would reduce the amount of time required for the crop to "canopy over" the row middles. Planting in narrow rows is the most effective means of achieving rapid canopy closure, particularly in soybean, when the crop can be drilled into the seedbed. Early canopy closure allows full utilization of sunlight by the crop and reduces late-season weed emergence and subsequent competition and interference with harvest (Yelverton and Coble, 1991).

Herbicides are used primarily for early-season weed management in cotton, peanut, and soybean. Later in the season, as herbicides dissipate to noneffective levels, the crop itself may keep weeds in check by successfully competing for light. Use of good cultural practices, including seedling disease control and early-season insect control, will help to establish a more vigorous crop that will compete more successfully with the weeds. In cotton, this vigorous early-season crop growth is necessary to obtain a height differential between cotton plants and weeds. This height differential is essential for postemergence-directed (POST-directed) herbicide applications and also enhances cotton's ability to compete with later-emerging weeds. Cultural practices that aid in weed management include proper fertilization and liming, good seedbed preparation, use of high-quality seed, precise control of seeding depth, and planting into warm soil. The use of in-furrow insecticides such as aldicarb which also suppress nematodes is recommended in cotton and peanut.

Decreased row spacings in cotton may reduce weed interference by allowing more rapid canopy closure (Bryson, 1990; Coble and Byrd, 1992; Rogers et al., 1976). Because of the limited competitiveness of cotton, fields must be maintained free of weed interference for 8 to 10 weeks after emergence to prevent yield reduction (Byrd and Coble, 1991; Coble and Byrd, 1992). Narrow rows are primarily beneficial for late-season weed control. One of the major limitations preventing producers from shifting to narrow rows has been the lack of appropriate harvesting equipment. Until recently, spindle pickers were set on wide-row spacings (90 to 110 cm) which limited feasibility of narrow-row cotton production.

Inconsistent results have been seen in peanut with reduced row spacings (Cardina et al., 1987; Colvin et al., 1985b; Wehtje et al., 1984). Most row spacing studies in peanut, with or without weed competition, have shown either no differences in peanut yield or only minor increases with reduced row spacings. It has been suggested that narrow row patterns may serve to increase the peanut's ability to compete with weeds such that herbicide inputs could be reduced relative to those required in traditionally spaced rows. Although twin-row patterns with reduced row spacings influenced weed growth and peanut yield in a positive manner, it was not consistent and herbicide inputs could not be reduced. In Georgia and Alabama, herbicide inputs could be reduced when weed populations were low (Cardina et al., 1987). If weed populations were high, increased herbicide inputs were required to maintain profitability. Planting in narrow rows reduced the potential for cultivation as an early-season control option and increased reliance on herbicides.

Numerous studies in soybean document the advantages of narrow rows in decreasing the deleterious effects of weeds on soybean (Stoller et al., 1987). Soybean canopy closure decreased weed resurgence after POST herbicide application (Yelverton and Coble, 1991).

The degree to which weeds cause problems is related to the competitiveness of the weed with the crop, the quality and purity required of the processed crop in relation to the ability of the weed to impair quality, the propagation of the weed, and the ability to control the weed economically.

Weeds that infest and cause yield reductions requiring control practices vary across the United States. The most troublesome weeds in cotton include various *Ipomoea* morningglory species and perennial nutsedge species (Table 1). The most common weed species in cotton are crabgrass species and pigweed species. The most common weeds in peanut are crabgrass species and Florida beggarweed (Table 2). Florida beggarweed and yellow nutsedge were the most troublesome species in a majority of the peanut-producing states. In soybean, sicklepod was listed as the most troublesome weed in 7 of the 28 states, and these states were found predominantly in the southern United States (Table 3). Velvetleaf and foxtail species were listed as most common or most troublesome 6 and 10 times, respectively. These two weeds were found predominantly in the midwestern states. Morningglory was listed as the most common weed in four states and the most troublesome in five states (data not shown). Although not listed as most troublesome, pigweed species and Palmer amaranth were listed as the most common weeds in six states.

IV. MANAGEMENT STRATEGIES

A. Cultivation

Cultivation is of primary benefit in controlling annual weed species. Vegetative reproductive organs of perennial weeds are frequently spread by cultivation. Perennials are not always uprooted by cultivation, and weeds may resprout from the vegetative reproductive organs.

Use and effectiveness of cultivation and other tillage practices are dependent upon the topography of land, location of weeds in relation to the crop row,

Table 1 Most Common and Troublesome Weeds in Cotton Listed by State

State	Common	Troublesome
Alabama	Pitted morningglory	Velvetleaf
Arkansas	Annual grasses	*Ipomoea* morningglories
Arizona	Palmer amaranth	Purple nutsedge
California	Not listed	Not listed
Florida	Crabgrass species	Common cocklebur
Georgia	Sicklepod	Yellow nutsedge
Kansas	Pigweed species	Not listed
Louisiana	Johnsongrass	Johnsongrass
Missouri	Large crabgrass	Pitted morningglory
Mississippi	Southern crabgrass	Bermudagrass
North Carolina	Large crabgrass	*Ipomoea* morningglories
New Mexico	Pigweed species	Purple nutsedge
Oklahoma	Pigweed species	*Ipomoea* morningglories
South Carolina	Crabgrass species	*Ipomoea* morningglories
Tennessee	Large crabgrass	*Ipomoea* morningglories
Texas	Palmer amaranth	Purple nutsedge
Virginia	Large crabgrass	Nutsedge species

and size of crop and weeds. On highly erodable soils, tillage may not be an option. Additionally, if weeds are present within the crop row, cultivation is of little benefit. In situations where the crop is extremely small or large, cultivation may be of limited benefit.

Cultivation is considered a necessary component of a good weed management program in cotton (Snipes et al., 1984; Snipes and Mueller, 1992b; Snipes

Table 2 Most Common and Troublesome Weeds in Peanut Listed by State

State	Common	Troublesome
Alabama	Florida beggarweed	Florida beggarweed
Florida	Crabgrass species	Florida beggarweed
Georgia	Florida beggarweed	Yellow nutsedge
North Carolina	Large crabgrass	Yellow nutsedge
New Mexico	Pigweed species	Spurred anoda
Oklahoma	Crabgrass species	Carolina horsenettle
South Carolina	Crabgrass species	Florida beggarweed
Texas	Palmer amaranth	Yellow nutsedge
Virginia	Large crabgrass	Bermudagrass

Table 3 Most Common and Troublesome Weeds of Soybean Listed by State

State	Common	Troublesome
Alabama	Sicklepod	Sicklepod
Arkansas	Annual grass	Sicklepod
Delaware	Canada thistle	Morningglory
Florida	Crabgrass species	Sicklepod
Georgia	Common cocklebur	Sicklepod
Iowa	Giant foxtail	Giant foxtail
Illinois	Velvetleaf	Giant ragweed
Indiana	Giant foxtail	Velvetleaf
Kansas	Pigweed species	Common cocklebur
Kentucky	Johnsongrass	Johnsongrass
Louisiana	Common cocklebur	Morningglory
Maryland	Giant foxtail	Giant ragweed
Michigan	Pigweed species	Common lambsquarters
Minnesota	Foxtail species	Woolly cupgrass
Missouri	Giant foxtail	Velvetleaf
Mississippi	Sicklepod	Johnsongrass
North Carolina	Large crabgrass	Sicklepod
North Dakota	Foxtail species	Kochia
Nebraska	Pigweed species	Velvetleaf
New Jersey	Velvetleaf	Morningglory
Ohio	Giant foxtail	Canada thistle
Oklahoma	Pigweed species	Morningglory
Pennsylvania	Common cocklebur	Horsenettle
South Carolina	Palmer amaranth	Sicklepod
South Dakota	Green foxtail	Common milkweed
Tennessee	Common cocklebur	Sicklepod
Texas	Palmer amaranth	Morningglory
Wisconsin	Giant foxtail	Velvetleaf

Source: Bridges (1992).

et al., 1992). Depth should be shallow (2 to 5 cm) to prevent damaging the cotton roots and bringing up untreated soil. Cultivation is most effective when weeds are small. It also improves cotton growth in tight or crusted soils. Cultivation alone will not provide acceptable weed control under most conditions (Snipes and Mueller, 1992b). At least one cultivation usually is needed to maximize cotton yield. Although POST-directed herbicide applications sometimes do not increase yields, they are needed to reduce weed biomass, especially if preemergence (PRE) herbicides are banded. Reductions in weed biomass improve crop quality and harvestability.

Peanut generally is cultivated after land plaster and/or granular insecticide applications are made at pegging (start of fruit set underground). Cultivation in peanut is considered to be a fairly exacting process. Cultivation should be done early (before mid-June) or not at all to prevent injury to peanut vines since vine injury can increase the incidence of soil-borne diseases (Porter et al., 1982). Even an early season cultivation can be damaging if not done properly. Many peanut pods are formed on the cotyledonary lateral branches. Movement of soil onto these branches and around the base of the plant causes physical damage and enhances development of stem and pod diseases (Porter et al., 1982; Wilcut et al., 1994a). For example, cultivation that is not precise and "nondirting" (cultivation where soil is not moved onto the peanut stems or foliage) may increase the occurrence of *Sclerotium rolfsii* Sacc (various common names include stem rot, white mold, and southern blight), sclerotinia blight caused by *Sclerotinia minor* Jagger, and sometimes *Sclerotinia sclortiorum* (Lib.) de Bary. These diseases can cause catastrophic yield reductions (Porter et al., 1982).

Research has been conducted to determine weed management, peanut yield, and economic returns of weed management systems that included cultivation only, herbicides only, and a combination of herbicides with one or two timely cultivations (Bridges et al., 1984; Wilcut et al., 1987a; Wilcut et al., 1987b). Generally cultivation or herbicides alone failed to provide adequate weed control or high peanut yield. Systems that used herbicides with cultivation provided the greatest levels of weed control, peanut yield, and net returns (Bridges et al., 1984; Wilcut et al., 1987a; Wilcut et al., 1987b). Intensive cultivation, however, also can result in significant peanut yield reductions (Hauser et al., 1974).

Cultivation is an effective and economical method of weed control in soybean (Miller and Griffin, 1994; Poston et al., 1992; Shaw et al., 1991; Steckel et al., 1990). Cultivation alone seldom provides sufficient weed control when moderate to heavy populations of weeds exist. Utilizing cultivation enables herbicides to be applied on a band (Poston et al., 1992). Precision cultivation coupled with reduced-rate technology has allowed producers in Arkansas to save approximately $17 per hectare (Baldwin and Hall, 1993).

Cultivation of cotton, peanut, and soybean, when feasible, should be shallow (2 to 5 cm) to prevent damaging roots and breaking through any residual herbicide layer and bringing up untreated soil and weed seed.

Tillage during the fall and winter is an effective method of reducing populations of perennial weeds such as rhizomatous johnsongrass, bermudagrass, and quackgrass. Tillage exposes rhizomes and stolons to cold temperatures which reduce viability and help reduce populations. Regardless of the crop, these practices can be implemented during the fall or winter, if the soil is not frozen, to reduce populations.

B. Crop and Herbicide Rotation

Crop rotation should be an integral component of a weed management program. It allows use of herbicides from different chemical families and with different modes of action. Crop rotation is not only agronomically advantageous, it can prevent the buildup of problem weeds and diseases, avoid the development of weed resistance to herbicides, and help to keep the overall weed population low (Wilcut et al., 1994a).

A major benefit of rotating soybean with grain sorghum or corn in the southern states is the relative ease of controlling sicklepod with atrazine, 2,4-D, and dicamba. Reducing the sicklepod population in grass crops is beneficial to management in cotton, peanut, and soybean.

Common cocklebur resistance to imazaquin after repeated use in soybean has been reported.* A plausible way to reduce or delay resistance of common cocklebur to imazaquin and possibly other inhibitors of acetolactate synthase (enzyme that is essential for synthesis of the branch chain amino acids isoleucine, leucine, and valine) would be use of bentazon (a photosynthetic inhibitor) or crop rotation; use of atrazine, 2,4-D, or dicamba in corn; or fluometuron and MSMA in cotton. The basic concept of chemical rotation, which is often facilitated by crop rotation, is critical for delaying or preventing development of herbicide resistance in weeds (Gressel and Segel, 1990).

Crop rotation also results in healthier soybean that compete better with weeds. Monoculture peanut and soybean production often results in buildup of diseases and nematodes; crop rotation is effective in reducing incidence of disease (Porter et al., 1982).

Some herbicides used in soybean may persist and damage rotational crops. Imazaquin injured cotton and corn planted the year following application to soybean (Barnes et al., 1989; Monks and Banks, 1991; Renner et al., 1988). However, rice was not injured by clomazone, imazaquin, imazethapyr, or chlorimuron applied the previous year to soybean (Johnson et al., 1992). Wheat and clover crops rotated with soybean may be injured when clomazone or norflurazon is applied to soybean (Chemical Pharmaceutical, 1994). In other studies, chlorimuron, clomazone, imazaquin, and imazethapyr applied to soybean did not injure corn or wheat planted after soybean (Krausz et al., 1992). In a cotton and peanut rotation, fluometuron did not injure peanut (York et al., 1994). Carryover to rotational crops is dependent upon the herbicide, the sensitivity of the rotational crop, the environmental and edaphic conditions of the growing seasons, as well as the rate and method of herbicide application (Monks and Banks, 1991; Renner et al., 1988). All of these factors should be considered during herbicide selection for any crop.

*Byrd, J. D., Jr, 1993, Mississippi State Univ., personal communication.

C. Biological Control

Biological weed control generally employs insects, plant pathogens, or nematodes. Biological weed control has had only minor commercial success in cotton, peanut, and soybean because most biological control agents are generally effective on only one weed species. Most agronomic crops are infested with several weed species at densities exceeding economic thresholds.

Although *Alternaria cassia* has been developed for sicklepod control (Walker and Riley, 1982), it has not been produced commercially. Until biological control agents that provide relatively rapid and consistent control of a multiple weed complex are developed, the likelihood of widespread use in agronomic crops is small. However, if a biological agent were developed that provided acceptable and consistent control of troublesome weeds (i.e., sicklepod, yellow and purple nutsedge), it would likely be readily accepted.

D. Herbicide Selection

Selecting a herbicide program, especially in soybean, has become a confusing exercise because of the number of herbicides, prepackaged mixtures, and tank mixes available. In determining the most appropriate and cost-effective herbicide program to use, one must consider the spectrum of weeds present, tillage system, method of application, crop tolerance relative to edaphic conditions, possible rotation restrictions, and cost of the herbicide. Selecting a herbicide program is not a simple matter. However, following five logical steps may eliminate confusion in this process.

Step 1

To plan a weed management program wisely one must accurately identify and/or predict what weeds will be present in a given field. Herein lies the value of scouting in-season and in the fall and weed mapping. Weeds present in the fall have set seed and most likely will be the predominant problems the following season. Knowledge of the weeds present allows producers to choose from preplant-incorporated (PPI), PRE, POST, and POST-directed herbicide options. Depending upon the weed complex, herbicide options may be limited to one method of application. Timeliness of application and management practices are extremely important for POST management.

Most fields, regardless of the crop, contain a complex of annual broadleaf and grass weeds. In peanut and soybean, numerous herbicides are available for POST control of most species. In cotton, however, POST applications of currently available herbicides for control of broadleaf weeds are not recommended because they are too injurious when applied POST to cotton (Byrd and York, 1987a; Guthrie and York, 1989; Snipes and Byrd, 1994). Therefore, growers must rely on soil-applied herbicides for early-season weed control.

Although numerous POST herbicides are available in peanut and soybean, certain weeds are controlled better with soil-applied herbicides or a combination of soil- and POST-applications. For example, POST herbicides are generally inadequate for sicklepod control in soybean. A program consisting of metribuzin applied PRE coupled with chlorimuron applied early POST in soybean is effective. Without knowledge of sicklepod as a problem, producers may not apply metribuzin; consequently adequate weed control may be more difficult to obtain. Similarly, Florida beggarweed is difficult to control in peanut, and a complete herbicide program employing PPI, PRE, and POST herbicides often is necessary for season-long control (Wilcut et al., 1994a).

Some weeds can be effectively controlled with POST herbicides, and if a given field contains patches of weeds (Wiles et al., 1992b), producers may forgo soil application of herbicide to the complete field in favor of treating only those areas infested. Control of annual grasses POST is an alternative to soil-applied herbicide programs in cotton, soybean, and peanut (Byrd and York, 1987b; Grichar and Boswell, 1986; Nastasi et al., 1986; Wilcut et al., 1990b). Total POST programs also are preferred in high organic soils where soil-applied herbicides are either not effective or cost prohibitive because of the rate required.

After determining which weeds will most likely cause problems the following season, herbicide options can be narrowed significantly. Consultants, extension personnel, and dealers usually have weed response charts available with recommendations of herbicides that are most effective for a specific weed problem. Compile a list of herbicides and tank mixes that will control the various weeds anticipated. Ideally, this should be accomplished on a field by field basis. The ultimate goal is to determine the best herbicide program for each individual field. If use of specific herbicide programs for individual fields is not practical, at least group fields with similar weed problems. Herbicides are too expensive to try to come up with a single blanket treatment to handle all the different weeds likely to be found in different fields on large farms.

In the list of possible herbicide options, initially include all registered products and tank mixtures for PPI, PRE, POST, and POST-directed application regardless of cost, carryover potential, or soil type limitations. The presence of certain weeds may limit herbicide options to only one application method or only one or two products or tank mixtures. For most situations, however, several herbicide options are available.

Step 2

Reduce the list of herbicide options by considering the rotational restrictions on certain products. Eliminate those herbicides that have the potential to carry-over and damage the crop to be planted the following season. This information can be found on the herbicide labels and/or in extension publications.

Step 3

Consider your soil texture and organic matter content. Use of certain soil-applied herbicides can be risky on coarse-textured or low organic matter soils, or fields with variable soils. For example, linuron can cause severe injury on coarse-textured soils. In contrast, trifluralin and vernolate are not cost-effective on soils with high organic matter (>5%). Eliminate those choices that don't fit your soil type. On these soils, graminicides (POST herbicides that control annual and perennial grasses) and POST herbicides for broadleaf weeds such as bentazon, chlorimuron, or acifluorfen may be more effective and economical. This information can be obtained from herbicide labels or from extension publications.

Step 4

The method of herbicide application is important as herbicides can be applied PPI, PRE, POST, or POST-directed. Some herbicides have more than one application method and the weeds may be susceptible to more than one method of application. If the list of herbicide options includes more than one application method, decide which method is best for a particular situation. The method of application choice will vary with the producer, depending on equipment and management input limitations. With several broad-spectrum soil-applied broadleaf herbicides and prepackaged mixtures now available for use in peanut and soybean, application of herbicides for grass and broadleaf weed control at planting may be most appropriate. By correctly matching herbicides to specific weed complexes, the need for POST or POST-directed herbicides may be eliminated in some cases. In contrast, soil-applied herbicides are always necessary in cotton. Depending upon the weed spectrum, herbicides applied to the soil at planting as well as POST-directed herbicides may be necessary. Additionally, environmental conditions affecting herbicide activation may limit the effectiveness of soil-applied herbicides. In cases where soil-applied herbicides are not activated by rainfall or irrigation, POST herbicides may be necessary.

Consider benefits and limitations of PPI versus PRE application. PRE herbicides require rainfall for activation. PPI application provides insurance against a performance failure due to lack of rainfall or irrigation. On the other hand, incorporating herbicides may require more time. Also, proper incorporation is critical. Poor incorporation may result in poor weed control and/or crop injury. Some PPI herbicides cost less than PRE herbicides; however, the extra expense of incorporation should be considered. Also, consider the savings possible with banded PRE herbicides used in conjunction with cultivation (Poston et al., 1992).

For some soils and some weeds, POST herbicides may be more economical than soil-applied herbicides. This is especially true for the good manager who can get acceptable control with reduced application rates (Baldwin and Hall, 1993). However, unless weeds are small, weather conditions are ideal, and application variables such as pressure, volume, nozzle selection, and adjuvant usage are optimum, reduced rates may lead to disappointing results.

When applying POST herbicides, one should consult the label for adjuvant selection. Most POST and POST-directed herbicide labels recommend the use of a spray adjuvant. Spray adjuvants are any nonpesticidal materials added to a herbicide to improve its effectiveness. The most commonly used types of adjuvants are nonionic surfactants and crop oil concentrates.

A major advantage of using POST herbicides is the ability to wait and see what weed problems are present before determining whether the herbicide treatment is justified. Proper timing of application is one of the keys to a successful POST herbicide program. Too many farmers allow their weeds to get too big before they spray. About 5 to 7 days after planting, start checking the fields and continue on a 3- to 5-day interval. If ensuring timely scouting and application will be difficult, inclusion of a soil-applied herbicide program at planting may be the best option.

Annual weeds can be controlled with a total POST program. If the herbicides are properly applied (at the correct rates and times), POST management can be successful. The major disadvantage of a total POST program is the need for a higher level of management ability. Identification of seedling weeds so that the proper herbicide or tank mix can be selected is critical. Timing of application is also critical, and adverse weather may delay application beyond the proper time.

In many cases, the optimum times to apply both the graminicide and broadleaf herbicides will coincide. When this occurs, a tank-mix application can save a trip across the field. With most tank mixes of graminicides and broadleaf herbicides, reduced grass control (antagonism) is possible (Grichar et al., 1991; Holshouser and Coble, 1990; Jordan and York, 1989; Jordan et al., 1989; Vidrine, 1989). To overcome this problem, labels for some POST graminicides suggest increased rates, and this practice can reduce antagonism. The increased rate may not be necessary under ideal conditions (small grass that is actively growing with good soil moisture), but it is suggested for consistent control. Sequential applications can eliminate or reduce antagonism. If sequential applications are made, the graminicide rate is not increased.

If sequential applications are made, the recommended waiting interval between applications of the graminicide and broadleaf herbicides varies, depending upon which herbicides are used and which herbicide is applied first (see the herbicide labels for details). However, the following are general guidelines for peanut and soybean: 1) If the POST graminicide is applied first, the broadleaf herbicide can be applied 24 hours later; 2) if bentazon is applied first, the graminicide can be applied 24 hours later; 3) if chlorimuron is applied first, delay graminicide application for at least 3 days; 4) if imazaquin or imazaquin-containing premixes or imazethapyr is applied first, delay graminicide application for at least 5 days; 5) if acifluorfen, acifluorfen plus bentazon, fomesafen, or lactofen is applied first, delay grass herbicide application 7 days or until the grass resumes active growth with development of new leaves. In cotton, apply graminicides at least 24 hours before fluometuron or MSMA or wait at least 5 days if applying after MSMA or fluometuron. Always refer to the respective herbicide labels for specific guidelines.

Use of spray additives can reduce antagonism. Addition of ammonium sulfate and/or replacement of crop oil concentrate with BCH 815085 (Dash) reduced or eliminated the antagonism observed between sethoxydim and bentazon (Jordan and York, 1989). However, there are few other herbicide combinations in which adjuvants overcome antagonism.

Step 5

The final consideration is price of a particular herbicide or mixture of herbicides that effectively controls specific weed complexes. Choosing the most cost-effective option will provide the greatest profit.

V. WEED MANAGEMENT IN COTTON

A. Introduction

Effective weed management is one of the many necessary components in a successful cotton production program. Good weed management is more difficult to achieve in cotton than in corn, peanut, or soybean because cotton grows more slowly and because there are fewer herbicide options for cotton. Unfortunately, cotton is one of the few agronomic crops with no suitable POST herbicide for annual broadleaf weed control (Byrd and York, 1987a; Guthrie and York, 1989; Snipes and Byrd, 1994).

Better weed management is needed in cotton than in corn, peanut, or soybean because cotton is not a strong early season competitor with weeds. Lower weed densities will reduce cotton yield more than corn or soybean yield (Byrd and Coble, 1991; Coble and Byrd, 1992; Zimdahl, 1980). Weeds also interfere more with harvesting operations in cotton, and weeds may reduce lint grades because of the presence of trash or green stain. Additional ginning operations to remove foreign material from cotton also reduce fiber quality. Weed problems are often greater in the southern United States, partly because of the warm, humid growing conditions, which accelerate herbicide breakdown and stimulate rapid weed germination and growth.

Weed management in cotton is most effective when an integrated approach to weed management is used. A weed management program for each field is developed by using an integrated approach. All available weed management methods and crop rotations are combined in a coordinated strategy.

B. Soil-Applied Herbicides

Except for fields known to have very low infestations of annual grasses, application of a soil-applied grass control herbicide is recommended. The dinitroaniline herbicides (pendimethalin and trifluralin), clomazone, and norflurazon are the products currently available for PPI application.

Pendimethalin and trifluralin are very similar products and basically control the same weeds with very minor differences. Incorporated applications of pendimethalin or trifluralin provide good control of all common annual grasses including Texas panicum. These two herbicides also control certain small-seeded broadleaf weeds such as Florida pusley, pigweed species, and common lambsquarters.

Cotton normally has good tolerance of pendimethalin and trifluralin. However, an excessive application rate or deep incorporation may delay early season growth and possibly reduce yield or delay maturity. This crop damage is more likely to occur on very sandy soils and under cool, wet conditions.

For best results, pendimethalin or trifluralin should be incorporated into the top 5 to 8 cm of the final seedbed. Where cotton is to be planted flat, one pass with a power-driven tiller or two passes with either a finishing disk or a field cultivator result in good incorporation. If cotton is to be planted on beds, incorporation on the bed with a power-driven tiller or a properly adjusted rolling cultivator may be preferred because of the shallow incorporation pattern. Except for extremely sandy soils, incorporation with a disk before bedding is adequate. On sand soils, the deep herbicide distribution from incorporation with a disk before bedding has been shown to reduce cotton yield (York and Tucker, 1985).

According to herbicide labels, incorporation of trifluralin and pendimethalin can be delayed for 24 hr and 7 days, respectively, after application (Chemical Pharmaceutical, 1994). However, incorporation as soon as possible after application is suggested for best results. Consult the respective labels for specific instructions on application rates for various soils and methods of incorporation. Pendimethalin also is labeled for PRE application as a tank mix with fluometuron. Although pendimethalin provides good annual grass control when properly incorporated, research in North Carolina and Georgia sometimes has shown poor control with PRE application of pendimethalin on fields with moderate to heavy grass pressure (Wilcut and York, unpublished data). Pendimethalin applied PRE can provide good weed control, but rainfall soon after application and before weed germination and emergence is essential.

Norflurazon controls all common annual grasses except Texas panicum. It also controls or suppresses pigweed species, common lambsquarters, prickly sida, common ragweed, spurred anoda, spotted spurge, tropic croton, velvetleaf, and nutsedge. Norflurazon can be applied PPI or PRE, or it can be split into two applications, with half of the recommended rate applied PPI and the other half applied PRE (Chemical Pharmaceutical, 1994). A PPI or split application generally is preferred, especially for annual broadleaf and nutsedge control (McLean et al., 1993; Snipes, 1991; York, 1994). Norflurazon as a split application will improve control of large-seeded broadleaf weeds (especially annual morningglory species and sicklepod) when used in conjunction with fluometuron (McLean et al., 1993). Control frequently is 10% to 20% higher with this combination than with fluometuron applied PRE alone.

Damage from norflurazon carryover to rotational crops planted after cotton can be significant (Kendig and Talbert, 1991). Therefore, one should consider crop rotation plans carefully before using norflurazon. Only cotton, peanut, or soybean can be planted the year after a norflurazon application (Chemical Pharmaceutical, 1994). Corn, grain sorghum, small grains, and tobacco planted after cotton treated with norflurazon may be damaged.

Because of rotational limitations and because it is more expensive than pendimethalin or trifluralin, use of norflurazon is suggested for specific situations. In particular, norflurazon application is suggested where nutsedge is expected to be a problem, prickly sida, spotted spurge, spurred anoda, or velvetleaf is expected (Snipes, 1991; Wilcut et al., 1993; York, 1994). Norflurazon is often applied in tank mixture with trifluralin or pendimethalin for improved annual grass control.

Clomazone received registration for PPI and PRE application to cotton in early 1993. Clomazone controls numerous annual grasses and broadleaf weeds. It does not provide adequate control of Texas panicum or pigweed species (Wilcut and Richburg, 1994b). Most producers apply clomazone with pendimethalin or trifluralin for better annual grass control. Pitted morningglory, prickly sida, spurred anoda, tropic croton, and velvetleaf are controlled by clomazone (Harrison and Hays, 1992; Jordan et al., 1993e; Smith et al., 1994). Clomazone is more effective applied PRE than PPI and takes very little rainfall for activation. However, off-site movement from volatility is worse with PRE application. Restrictions for PRE application are significant in regard to buffer areas around sensitive vegetation (Chemical Pharmaceutical, 1994). Clomazone can be applied PRE in reduced tillage systems (Harrison and Hayes, 1992). Greater pitted morningglory control with clomazone plus fluometuron applied PRE than with standard herbicide programs including dinitroaniline herbicides and fluometuron has been reported (Jordan et al., 1993e). However, improvement in morningglory control has not been consistent (Wilcut and Richburg, 1994b). To be used in cotton, clomazone must be applied with either disulfoton or phorate, and these insecticides must be in direct contact (in-furrow) with the cotton seed (Smith et al., 1994; York et al., 1991a). When it is applied without an organophosphate insecticide, excessive seedling injury may result.

Metolachlor is registered for PRE application in cotton in some states. Metolachlor does not control Texas panicum (Wilcut et al., 1994a). However, it normally controls other annual grasses and pigweed species. Additionally, yellow nutsedge may be suppressed. Metolachlor should not be used on coarse-textured, low organic matter soils. Use of metolachlor on these soils may lead to severe cotton injury if heavy rainfall is received shortly after planting (Brown et al., 1993).

Use of a PRE herbicide to control broadleaf weeds is always advisable. In the absence of a PRE broadleaf herbicide, broadleaf weeds often will be too large to control by the time the cotton is large enough to apply POST-directed herbicides. As previously mentioned, no POST herbicide is currently available for broadleaf weed control that does not delay cotton maturity and/or reduce yields.

PRE herbicides can be banded over the row or broadcast. Because the row middles normally will be cultivated, banding of PRE herbicides reduces cost. The PRE herbicide should be applied as soon as possible after planting and always before cotton emergence. Application of PRE herbicides to cotton that is beginning to emerge can result in severe crop injury or death. Ideally, planting and spraying are done in the same operation.

PRE herbicide choices for broadleaf weed control include diuron, fluometuron, and prometryn. Growers should read herbicide labels carefully for specific directions on application rates, soil type limitations, application methods, and rotational restrictions (Chemical Pharmaceutical, 1994).

Fluometuron is the herbicide of choice for southeastern and Midsouth cotton production. Both diuron and fluometuron have good activity on small-seeded broadleaf weeds, but fluometuron is more effective on large-seeded broadleaf weeds such as common cocklebur, morningglory species, sicklepod, and smartweed (*Polygonum spp.*) (McLean et al., 1993; Wilcut et al., 1993). Diuron also is less active than fluometuron under low rainfall conditions (Wilcut, unpublished data).

Cyanazine is labeled for PRE application to cotton in some areas of the Cotton Belt. However, label restrictions essentially eliminate its PRE use in a majority of the Coastal Plain soils of the Southeast. Cotton has very limited tolerance of PRE applications of cyanazine, and there are very few cotton fields in the southeastern region where cyanazine can safely be used as a PRE application. Cyanazine and norflurazon can be applied in some states for spurge control, although cotton injury may occur.

If replanting is necessary, it is best to run the planter back in the original drill without soil preparation if soil conditions permit. In this case, do not apply any additional residual herbicides. If emerged weeds are present, paraquat can be applied for burndown. If it is necessary to rework the seedbed, use shallow tillage such as light disking. Do not rebed land without first disking old beds down or unacceptable injury may occur (York, personal observations). Do not apply additional PPI herbicides. If the original PRE herbicide was broadcast, do not apply more. If the original PRE herbicide was banded, a second PRE banded application at the minimum rate for the soil type is advised.

For good results with a PRE application of fluometuron, 2.0 cm or more of rainfall is required within 7 to 10 days after application. If the amount of rainfall needed to activate the fluometuron is not received, the field can be shallowly cultivated to remove emerged weeds and to mix the fluometuron into the soil. Unfortunately, it is impossible to cultivate in the row where incorporation is needed most.

Fluometuron can be tank mixed with trifluralin and incorporated before planting. Incorporation will eliminate the need for rainfall to activate the herbicide. If sufficient rainfall for herbicide activation is not received during the first 2 weeks after planting, an incorporated application of fluometuron will perform much better than a PRE application. If rainfall is sufficient for activation, a PRE application of fluometuron will always be the superior treatment. Incorporating fluometuron is insurance against performance failure in case rainfall is inadequate. In return, there will be somewhat less weed control if fluometuron is incorporated deep with a disk. A shallow incorporation (less than 3 cm) will provide the most consistent control.

C. Postemergence Herbicides.

Postemergence Over-the-Top Herbicides

It is recommended that all POST broadleaf herbicides be applied as POST-directed sprays. On some occasions, however, an over-the-top application is inevitable. Because weeds may grow faster than cotton, the height differential necessary for a directed application may never be achieved.

DSMA and MSMA may be applied POST (over-the-top) to cotton that is 8 to 15 cm tall and before the first square stage. Only one over-the-top application should be made. Rates for POST application are approximately one-half the rates for POST-directed application. Over-the-top application of DSMA or MSMA is suggested only for control of cocklebur or nutsedge, and then only if POST-directed application is not possible. Crop maturity and yields may be reduced by POST applications of DSMA and MSMA (Guthrie and York, 1989; Snipes and Byrd, 1994).

In Alabama and Missouri, 0.56 kg/ha of fluometuron applied over-the-top of cotyledon cotton is recommended, mainly for morningglory control (Snipes and Byrd, 1994). Fluometuron also may be applied over-the-top of cotton that is 8 to 15 cm tall. Fluometuron will control a broader spectrum of weeds than DSMA or MSMA, but it is more damaging to cotton. Over-the-top application of fluometuron may reduce cotton yield 20% or more and often delays maturity 1 week or more (Guthrie and York, 1989; Snipes and Byrd, 1994). Over-the-top application of fluometuron is suggested only for salvage situations. Crop injury and potential for yield reduction increase as the application rate increases. Rates greater than 1.1 kg ai/ha are not advised.

Clethodim, fenoxaprop-P, fluazifop-P, quizalofop-P, and sethoxydim are POST herbicides (graminicides) registered for grass control in cotton. These graminicides, if properly used, provide excellent control of annual and perennial grasses but are ineffective on nutsedge or broadleaf weeds. Cotton at all growth stages is tolerant of these herbicides.

Graminicides may be used to remove annual grasses that have escaped control with soil-applied herbicides or may be used as the primary grass control treatment. Research has shown that management programs using these herbicides in the absence of soil-applied grass control herbicides can result in lint yields and grades, and net profits equal to those from management programs using PPI herbicides (Byrd and York, 1987b; Nastasi et al., 1986). Because POST graminicides provide no residual control and because cotton is slow to form a canopy, two applications often are needed in moderately and heavily infested fields for acceptable season-long grass control (Byrd and York, 1987b; Nastasi et al., 1986).

Except for fields known to have a very low infestation of annual grasses, eliminating soil-applied grass control herbicides and relying entirely upon POST herbicides for annual grass control generally is not recommended. An annual grass management program that relies totally on POST herbicides requires carefully timed applications and a higher level of management skill (Buhler and Burnside, 1984). In addition, small-seeded broadleaf weeds such as Florida pusley, pigweed species, and common lambsquarters may be more troublesome than when a soil-applied grass control herbicide is used (Wilcut et al., 1993).

When using the POST graminicides, follow label directions for application rates, application methods, use of adjuvants, and optimum grass size for treatment. Labels for the POST graminicides do not allow tank mixing with broadleaf herbicides in cotton as antagonism may occur (Byrd and York, 1987a).

Bermudagrass and johnsongrass can be controlled in cotton with clethodim, fluazifop-P, quizalofop-P, and sethoxydim. Fenoxaprop-P controls johnsongrass but not bermudagrass (Wilcut et al., 1993). Consult product labels for application rates, application methods, optimum grass size for treatment, and adjuvant usage. A soil-applied grass control herbicide, as previously discussed, should also be applied in most situations.

Sometimes two applications are needed for johnsongrass control but the need for two is much more likely for bermudagrass control. Two herbicide applications plus timely cultivations usually are required for good control of common bermudagrass. Proper timing of application is important. Cultivation will enhance control. To prevent putting the bermudagrass under stress, do not cultivate for 7 days before or after herbicide application. Cultivation within 7 days of application may reduce control. Good spray coverage is necessary. For bermudagrass, better spray coverage with the second application may be achieved if the spray solution is semidirected under the cotton foliage. See the section on perennial grass control in soybean for fall management systems for johnsongrass and common bermudagrass.

Postemergence-Directed Herbicides

As discussed, a PRE application of broadleaf herbicide (fluometuron, diuron, norflurazon, prometryn) is always recommended. In addition, fields with a moderate to heavy infestation of broadleaf weeds normally require cultivation and one or more POST-directed herbicide applications (Snipes et al., 1984; Snipes and Mueller, 1992; Snipes et al., 1992). Directed application of all POST broadleaf herbicides is strongly recommended. As discussed earlier, POST over-the-top treatments may reduce yield and delay maturity.

If a POST-directed broadleaf herbicide is needed, it often will be needed when the cotton is in the 8- to 15-cm size range. Most broadleaf weeds that escape soil-applied herbicides will emerge with or shortly after the cotton. Directing herbicides to cotton that is 8 to 15 cm tall requires precision equipment, careful equipment adjustment, and patient operators. However, application to small weeds is the key to success with POST-directed herbicides, and delaying application until the cotton is larger usually results in poor control.

Recommended herbicides for directed application to 8- to 15-cm-tall cotton include fluometuron, DSMA, MSMA, and fluometuron plus DSMA or MSMA. If cocklebur is the only broadleaf weed present, DSMA or MSMA alone is sufficient unless arsenical-resistant cocklebur populations are present. Otherwise, fluometuron mixed with either DSMA or MSMA will provide more effective control of a broader spectrum of broadleaf weeds (Jordan et al., 1992; Jordan et al., 1993c). Always add a nonionic surfactant when using fluometuron POST.

In most fields, a PRE application of fluometuron, an early POST-directed application of fluometuron tank mixed with DSMA or MSMA, and cultivation will provide sufficient broadleaf weed control. Fields with a heavy infestation of broadleaf weeds, especially morningglory or sicklepod, may need a late POST-directed herbicide application to control those weeds emerging in the row after the early POST-directed treatment (York and Wilcut, unpublished data).

After cotton is at least 15 to 20 cm tall, one can choose from several herbicides or tank mixes. Options include DSMA, MSMA, fluometuron plus DSMA or MSMA, lactofen, lactofen plus DSMA or MSMA, diuron plus DSMA, cyanazine plus MSMA, oxyfluorfen, oxyfluorfen plus MSMA, oxyfluorfen plus cyanazine, and prometryn plus MSMA or DSMA. DSMA or MSMA alone is sufficient for cocklebur. Mixtures of MSMA with the other herbicides often will improve control and broaden the spectrum of control (Jordan et al., 1992; Jordan et al., 1993c). All of these treatments must be directed, never applied POST over-the-top of cotton. Also, do not apply treatments containing DSMA or MSMA after first bloom.

For effective POST-directed application of herbicides to small cotton, precision equipment is needed plus a flat surface across the drill area must be created at planting and preserved until at least the early POST-directed application has been made.

Various types of equipment are available for applying POST-directed herbicides. Fender-type applicators are very effective for making precision-directed application to small cotton. Nozzles are mounted on the fender frame and adjusted to cover a 30- to 40-cm band under the cotton. The combined output from all nozzles should be the equivalent of 6900 to 9400 ml per hectare per 2.5 cm of band width. Adjust the nozzles to give good weed coverage but minimum contact with the cotton leaves.

With early POST-directed applications to 8- to 15-cm cotton, it is impossible to cover the weeds without allowing the herbicide to contact the cotton. The important thing is to confine the herbicide to the lower half or less of the cotton seedlings. Any injury from DSMA, fluometuron, or MSMA applied in this manner will be temporary. For late POST-directed applications to larger cotton, such as cyanazine plus MSMA, contact no more than the lower 5 to 10 cm of the cotton plant (refer to chapter for calibration techniques). Do not allow the herbicide spray to contact leaves on the sympodia (fruiting branches).

D. Nutsedge Control

In fields where yellow and/or purple nutsedge is expected to be a problem, norflurazon should be incorporated before planting or used in a split application (half of the rate applied PPI, the other half applied PRE) (York, 1994). Research in Georgia and North Carolina (Wilcut and York, unpublished data) has shown norflurazon applied PPI to be most effective on nutsedge, whereas norflurazon applied PRE is not effective for nutsedge management. Norflurazon normally provides only 50% to 70% control of nutsedge, but it is the only soil-applied cotton herbicide that affects nutsedge. Follow the norflurazon treatment with cultivation and a POST application of DSMA or MSMA. Two applications of MSMA or DSMA normally are required for acceptable nutsedge control (York, 1994). These treatments are more effective if preceded by a norflurazon treatment. POST-direct the MSMA or DSMA if possible. If at least half of the nutsedge plant can be contacted by the spray, POST-directed application of MSMA or DSMA will be more effective than POST over-the-top application because higher rates can be directed (York, unpublished data).

Nutsedge does not tolerate shade. If nutsedge is suppressed until the cotton gets above it, further treatment normally is not necessary. Nutsedge management starts in rotational crops that have more herbicide options.

E. Reduced Tillage

Cotton traditionally has been produced under conventional-tillage methods. However, compliance with government programs will necessitate that cotton grown on erodable soils be produced under a reduced-tillage program. Research indicates that reduced-tillage and even no-tillage systems can be as productive as conventional-tillage systems (Brown and Whitwell, 1985; Bryson and Keeley, 1992; Elmore and Williford, 1993; Keeley et al., 1992; Keeling et al., 1991; Keeling et al., 1992).

Reduced-tillage cotton production requires careful management of weeds because cultivation may not be a management option. No-till cotton may be planted into the residue of a previous crop or into a cover crop. Although a legume cover crop will provide part or all of the cotton's nitrogen requirements, small grain cover crops generally are preferred (White and Worsham, 1990; York and Worsham, 1992). Small grains are easier and more economical to establish, they can be planted later in the fall, they are easier to kill than legumes, and they produce more residue than legumes (Hurst, 1992; York and Worsham, 1992). Additionally, it is easier to manage the nitrogen fertilization of cotton planted in small grain residues than in legume residues.

It is essential that the cover crop and any winter annual weeds or newly emerged summer annual weeds be killed prior to planting cotton. Either glyphosate or paraquat can be used for this purpose. Suggested application rates vary by weed or cover crop species and size. Small grains, especially wheat, are more susceptible if paraquat is applied before tilling or after the boot stage. High rates of glyphosate or two applications of paraquat are necessary for kill of legume cover crops.

The cover crop normally should be desiccated at least 2 weeks ahead of planting. Controlling the cover crop at this time will prevent soil moisture depletion by the cover crop and may make planting easier. Additionally, killing the cover crop ahead of planting allows additional burndown herbicide to be applied at planting, if needed, to kill any recently emerged weeds or any streaks of the cover crop that may have been missed during the first burndown herbicide application.

Interest in stale seedbed production has increased in the Midsouth. Dinitroaniline herbicides are applied in the fall or early winter and the land is bedded. Early preplant herbicide applications are made in February or March to kill winter annual weeds. Research has shown promising results with oxyfluorfen applied in combination with paraquat or glyphosate (Crawford and Leake, 1991; McClelland et al., 1992; McGee et al., 1992). Oxyfluorfen provides residual control of some winter annual weeds and enhances burndown of some species.

A major drawback of reduced-tillage cotton is the inability to incorporate dinitroaniline herbicides (pendimethalin or trifluralin) for annual grass and small-seeded broadleaf weed control, and norflurazon for nutsedge control (Keeling et al., 1991; Keeling et al., 1992; Wilcut et al., 1993; York, 1994). The result is increased reliance on PRE herbicides which require sufficient rainfall or irrigation for activation. Rainfall within the first week or 10 days after planting usually is necessary for activation. Incorporated herbicides are somewhat effective even without activating rainfall. With the exception of trifluralin, herbicides registered for use in conventionally-planted cotton also are available for reduced-tillage cotton. Lack of cultivation in no-till cotton also places greater reliance on chemical weed control. Recently, no-till or high-residue cultivators have been developed for cultivating no-till cotton. Additionally, hooded sprayers have been developed primarily to allow application of glyphosate to control weeds in the row middles. This sprayer essentially allows producers to cultivate the row middles chemically.

F. Summary and Future Strategies

Probably the most promising weed management tool since the registration of the POST graminicides is the development of transgenic cottons resistant to bromoxynil or glyphosate applied POST (Jordan et al., 1993d; Wilcut et al., 1992; Wilcut et al., 1995). Equally promising is the development of pyrithiobac sodium, which also can be applied POST over-the-top of cotton with little or no crop injury (Jordan et al., 1993a; Jordan et al., 1993b). Both pyrithiobac sodium and bromoxynil control common cocklebur, morningglories, and pigweeds when applied POST, and these weeds are among the most troublesome broadleaf weeds in cotton (Byrd and Coble, 1991; Chandler and Cooke, 1992; Coble and Byrd, 1992). Bromoxynil is also effective for bristly starbur, coffee senna, common lambsquarters, spurred anoda, tropic croton, and velvetleaf. Early application to small pigweed species and prickly sida is critical for success with bromoxynil. Pyrithiobac sodium will have restrictions for certain rotational crops (Jordan et al., 1993f; Eastin et al., 1995). Glyphosate will control most grasses and broadleaf weeds in glyphosate-resistant cotton plus suppress nutsedge species.

These herbicides will provide producers with greater flexibility in their weed management programs and reduce reliance on soil-applied herbicides. However, they are not yet registered for cotton, and in the case of bromoxynil-

and glyphosate-resistant cottons, the gene conferring resistance must be incorporated into agronomically desirable cultivars.

The increased interest in reduced-tillage cotton has brought about the need to evaluate the herbicide performance under differing tillage and cover crop regimens, the benefits of cover crops in terms of allelopathic effects, and the interrelationship of weed management and other pest and agronomic disciplines (Bryson, 1993; Crawford and Leake, 1992; Derkens et al., 1991; Elmore and Williford, 1993; Hurst, 1992; Keeley et al., 1992; Keeling et al., 1992; Liebl and Worsham, 1983; Liebl et al., 1992; Shilling et al., 1985).

I. WEED MANAGEMENT IN PEANUT

A. Introduction

Compared to many other crops, peanut presents several unique features that must be taken into consideration in developing weed management systems. Peanut requires a fairly long growing season; approximately 140 to 160 days depending on cultivar and geographical region. This long growing season allows more time for weed problems to develop. Peanut has a prostrate growth habit and a relatively shallow canopy that is slow to shade the row middles, which allow weeds to be more competitive (Wilcut et al., 1994a). Additionally, the fruit develops underground on pegs (gynophores) which originate from the stems that grow parallel to the soil surface. Thus herbicides cannot be applied to the base of the plant (POST-directed), in contrast to more upright crops such as cotton, corn, and soybean (Brecke and Colvin, 1991; Wilcut et al., 1994). Also, the prostrate growth habit and pattern of fruit development restrict cultivation to an early season management option only (Brecke and Colvin, 1991; Wilcut et al., 1994a).

B. Soil-Applied Herbicides

Herbicides can be applied to the soil either before or immediately after peanut planting. The intent is to prevent weeds from becoming established. Herbicides from four classes are used in this way: 1) the dinitroaniline herbicides (benefin, ethalfluralin, pendimethalin, and trifluralin); 2) the chloroacetamide herbicides (alachlor and metolachlor); 3) a thiocarbamate herbicide (vernolate); 4) an imidazolinone herbicide (imazethapyr); and 5) norflurazon.

In the mid- to late-1970s, research results (Buchanan et al., 1978) indicated that the dinitroaniline herbicides provided similar levels of weed control and crop tolerance. However, trifluralin (the most widely used dinitroaniline herbicide in cotton and soybean) can injure the pegging and fruiting process of runner-type peanut in the southeastern region. Trifluralin greatly increased the number of pegs, not only around the base of the peanut plant, but also on the lateral branches of a Spanish cultivar. Most of these pegs either were nonfruiting or terminated in small, misshapen, and immature embryos (Buchanan et al., 1978). There was a high percentage of severely malformed pods and nuts at harvest. These trends have been confirmed by other researchers. As a result, trifluralin is used only in Texas and Oklahoma (Chemical Pharmaceutical, 1994; Wilcut et al., 1994a).

Dinitroaniline herbicides control annual grasses, including Texas panicum, and small-seeded broadleaf weeds such as Florida pusley, common lambsquarters, and pigweed species. The dinitroaniline herbicides are the only soil-applied herbicides registered for use in peanut that will provide full-season control of Texas panicum (Wilcut et al., 1994a). Dinitroaniline herbicides do not adequately control large-seeded broadleaf weeds (Buchanan et al., 1982; Wilcut et al., 1994a).

Alachlor and metolachlor control annual grasses, except Texas panicum and shattercane, and control small-seeded broadleaf weeds such as Florida pusley and pigweed species, and yellow nutsedge (Buchanan et al., 1982; Wilcut et al., 1989; Wilcut et al., 1994a). They also suppress large-seeded broadleaf species including bristly starbur, common ragweed, Florida beggarweed, and sicklepod (Wilcut et al., 1994a). Alachlor suppresses broadleaf weeds better than metolachlor, while metolachlor provides better yellow nutsedge control. Both of these herbicides can be applied either PPI, PRE, or at ground cracking (GC). Additionally, metolachlor can be applied at layby.

The chloracetamide herbicides are commonly used for yellow nutsedge control and suppression of large-seeded broadleaf weeds. Suppression of large-seeded broadleaf weeds was perceived as critical for successful weed management, since no herbicide prior to registration of imazethapyr in 1990 provided residual control of these weeds. All POST herbicides available for use prior to imazethapyr registration provided no appreciable soil activity. Chlorimuron has soil residual activity, but when applied 60 days after crop emergence (see later section on chlorimuron) it is primarily a POST herbicide (Wilcut et al., 1994a).

The use of both alachlor and metolachlor during the past decade has not been without controversy. Sporadic peanut injury attributed to metolachlor has been observed by growers and extension personnel in the southeastern region (Cardina and Swann, 1988). Peanut emergence and seedling growth were delayed. Severely injured peanut may fail to emerge from the soil. Cardina and Swann (1988) examined growth and yield of peanut as influenced by PPI applications of metolachlor at 2.2 to 6.7 kg/ha at two Georgia locations on Tifton loamy sands (fine-loamy, siliceous, thermic Plinthic Paleudults) containing less than 1% organic matter and a pH of 6.0 or 6.2. Metolachlor delayed peanut emergence and reduced peanut growth when irrigation followed planting. In 1 year of the study at one location, irrigated peanut yielded less than nonirrigated peanut only when metolachlor was applied at 6.7 kg/ha, a rate well above the registered use rate.

A field study conducted in Alabama investigated both alachlor and metolachlor applied at registered use rates as well as excessive rates (Wehtje et al., 1988). Across the 3-year study, at least a twofold safety factor existed between the maximum registered rate and the rate necessary for peanut injury. During 1 year of the study, metolachlor at 9.0 kg/ha and alachlor at 13.4 kg/ha reduced yields. Metolachlor may have a slightly greater propensity to damage peanut under conditions of extensive leaching and/or slow peanut emergence. Additionally, peanut is more tolerant of alachlor than metolachlor. However, in 12 years of field trials in North Carolina (York, unpublished data) and 7 yr of field trials in Virginia and Georgia (Wilcut, unpublished data), peanut has never been injured from alachlor or metolachlor, including metolachlor applied 2.2 kg/ha PPI followed by 2.2 kg/ha GC.

Vernolate is the only thiocarbamate herbicide registered for use in peanut. Vernolate is extremely volatile and should be immediately incorporated into the soil. It frequently is mixed with a dinitroaniline herbicide (Wilcut et al., 1994a). Vernolate controls annual grasses, except Texas panicum, as well as small-seeded broadleaf weeds such as common lambsquarters, Florida pusley, and pigweed species. Vernolate provides only about 4 to 5 weeks of control. It is primarily used for yellow and purple nutsedge control. Until imazethapyr was registered in 1990, vernolate was the only herbicide registered for purple nutsedge control in peanut.

Thiocarbamate herbicides such as vernolate can reduce leaf cuticle formation. Sicklepod cuticle thickness was decreased 35% on the petiole and 15% on the upper leaf surface by EPTC (Wilkinson and Hardcastle, 1969). As a result, POST herbicides may work more effectively on some weeds. For example, burgherkin control with alachlor plus naptalam plus dinoseb applied at GC was improved when vernolate was applied PPI (Buchanan et al., 1982). In Virginia, sequential applications of paraquat provided better common ragweed control with vernolate applied PPI than without vernolate PPI (Wilcut and Swann, 1990). However, peanut yield also was reduced when vernolate was applied PPI and followed with sequential applications of paraquat.

Imazethapyr, an imidazolinone herbicide, may be applied PPI, PRE, GC, or POST (Chemical Pharmaceutical, 1994). Imazethapyr also can be used as split applications of the aforementioned application methods. Imazethapyr applied PPI or PRE controls many troublesome weeds such as coffee senna, common lambsquarters, *Ipomoea* and smallflower morningglory species, pigweed species including Palmer amaranth, prickly sida, purple and yellow nutsedge, spurred anoda, and wild poinsettia (Richburg et al., 1993; Wilcut et al., 1991a; Wilcut et al., 1991b; Wilcut et al., 1994b; Wilcut et al., 1994c). Common lambsquarters control is better and more consistent with PPI or PRE applications; POST applications usually provide poor control. Split applications (PPI or PRE applications followed by a GC or early POST application) also are very effective for the aforementioned weed species. Control of most species is better and more consistent with soil applications or a GC application on very small weeds (generally less than 3 cm tall). Common ragweed, sicklepod, tropic croton, and Florida beggarweed were not controlled with imazethapyr (Wilcut et al., 1994a).

Norflurazon was registered for PRE application in peanut in 1993. The major fit for norflurazon in peanut is for Florida beggarweed control in the Southeast. Norflurazon is the only registered soil applied herbicide in peanut that provides adequate control of this troublesome weed (McLean et al., 1994). Readers should refer to the cotton section for other weeds that norflurazon controls.

C. Postemergence Herbicides

Paraquat, paraquat plus 2,4-DB, paraquat plus bentazon, and paraquat plus bentazon plus 2,4-DB mixtures are currently the POST standards for broadleaf weed control in southeastern peanut production (Wilcut et al., 1994a). Paraquat provides good to excellent control of Florida beggarweed and sicklepod (Wilcut et al., 1989; Wilcut et al. 1990b).

Although paraquat applied POST to peanut injures the foliage, peanut rapidly recovers under good growing conditions and yield is unaffected. Peanut tolerance of paraquat does not appear to be cultivar dependent (Wehtje et al., 1991). However, making two applications of paraquat has occasionally reduced yields (Knauft et al., 1990; Wilcut and Swann, 1990) Paraquat can be applied from crop emergence until 28 days after emergence (Chemical Pharmaceutical, 1994). Paraquat applied after this 28-day period increased the chance of significant yield reductions (Wehtje et al., 1986). It is important to manage thrips early in the season. Yield reductions and maturity delays are sometimes seen where injurious POST herbicides like paraquat or acifluorfen are applied and then additional stress from thrips feeding occurs (Herbert et al., 1991).

Bentazon plus paraquat mixtures control more broadleaf weed species than paraquat alone. Bentazon plus paraquat mixtures control bristly starbur, coffee senna, prickly sida, and smallflower morningglory (Wilcut et al., 1994a). Paraquat alone does not control these weed species. Bentazon also reduces paraquat-induced foliar injury to peanut by reducing paraquat absorption into peanut foliage (Wehtje et al., 1992). Although paraquat absorption also was reduced in several weed species, including Florida beggarweed, sicklepod, and Texas panicum, control of these species was not reduced unless the application was made to weeds larger than the size recommended for application (generally 5 cm or taller) (Wilcut et al., 1994a). Paraquat reduces bentazon absorption into several weed species because of rapid destruction of cell membranes by paraquat. Apparently very little absorption of either herbicide is needed for good control since the weeds controlled by either herbicide are extremely sensitive to the respective herbicide. Control of species not highly sensitive to either herbicide, most notably sicklepod, will be reduced if applications are not made to small weeds.

Paraquat and paraquat plus bentazon and/or 2,4-DB mixtures also are used by some producers in the Virginia–North Carolina region but not to the same extent as in the rest of the southeastern United States. Paraquat does not adequately control common lambsquarters, common ragweed, prickly sida, spurred anoda, and tropic croton (Wilcut and Swann, 1990; Wilcut, 1991b). These weeds are commonly found in Virginia and North Carolina and to a lesser extent in the rest of the southeastern United States (Bryson, 1989; Wilcut et al., 1994a). The best application timing for paraquat or paraquat mixtures in the southeastern region is about 2 to 3 wk after GC (Wilcut et al., 1989; Wilcut et al., 1990b).

Paraquat plus imazethapyr as a GC application provided excellent control of broadleaf and nutsedge species in Georgia (Wilcut et al., 1994b). There were no significant interactions for a paraquat plus imazethapyr mixture in regard to peanut injury. Paraquat plus imazethapyr provided better full-season control of many weed species than paraquat plus bentazon because of the residual control from imazethapyr. Bentazon and imazethapyr frequently control the same species, including bristly starbur, coffee senna, prickly sida, smallflower morningglory, spurred anoda, and yellow nutsedge. Neither bentazon nor imazethapyr provides acceptable control of Florida beggarweed or sicklepod (Wilcut et al., 1994a). Paraquat controls both species.

Peanut is tolerant of 2,4-DB applied POST for broadleaf weed control (Buchanan et al., 1982). A single application of 2,4-DB at 0.9 ae kg/ha to peanut during the reproductive stage (the most herbicide-sensitive period) reduced both yield and market grade (Ketchersid et al., 1978). However, repeated applications of 0.45 kg/ha had no effect. The normal use rate is 0.28 kg/ha. This herbicide is used for controlling broadleaf weeds including common cocklebur, *Ipomoea* morningglory species, sicklepod, and smallflower morningglory (Buchanan et al., 1982). Smallflower morningglory is more tolerant of 2,4-DB than *Ipomoea* morningglory species (Wilcut et al., 1994a). Pitted morningglory is the most 2,4-DB-tolerant *Ipomoea* morningglory species (Barker et al., 1984). 2,4-DB applied POST following alachlor applied PRE provides near complete bristly starbur control (Wilcut et al., 1994a).

2,4-DB commonly is mixed with other POST broadleaf herbicides. It increases control of many broadleaf species, particularly if the broadleaf weeds are larger than recommended size for treatment with herbicides such as acifluorfen and bentazon (Buchanan et al., 1982; Wilcut et al., 1994a). 2,4-DB also is mixed with foliar fungicides to reduce the expense of making two separate applications.

Acifluorfen applied POST is widely used in the Virginia–North Carolina and southwestern peanut regions of the United States. Acifluorfen controls common lambsquarters, common ragweed, eclipta, jimsonweed, pigweed species, Pennsylvania smartweed, and tropic croton (Buchanan et al., 1982; King and Oliver, 1992; Wilcut and Swann, 1990; Wilcut, 1991b; Wilcut et al., 1994a; York et al., 1994). Only small cocklebur are controlled with acifluorfen. Acifluorfen does not adequately control bristly starbur, coffee senna, Florida beggarweed, prickly sida, sicklepod, or spurred anoda (Buchanan et al., 1982; Wilcut et al., 1994a). Acifluorfen at rates used POST in peanut does not provide residual control.

Bentazon applied POST controls bristly starbur, coffee senna, common cocklebur, common lambsquarters, prickly sida, smallflower morningglory, spurred anoda, velvetleaf, and yellow nutsedge (Buchanan et al., 1982; Wilcut et al., 1994a). Common ragweed can be controlled with bentazon and crop oil. Bentazon does not control Florida beggarweed, *Ipomoea* morningglory species, pigweed species, or sicklepod (Buchanan et al., 1982; Wilcut et al., 1994a). Bentazon does not provide residual control.

Acifluorfen plus bentazon is a standard mixture in the Virginia–North Carolina and southwestern regions. Acifluorfen plus bentazon mixtures control most broadleaf weed species, except Florida beggarweed and sicklepod (Wilcut et al., 1994a). Acifluorfen plus bentazon often is applied in a commercially available prepackaged 2:1 ratio of bentazon to acifluorfen (Chemical Pharmaceutical, 1994). Rates of acifluorfen and bentazon in this mixture may need to be adjusted for certain troublesome weed species such as common lambsquarters, common ragweed, prickly sida, spurred anoda, velvetleaf, and yellow nutsedge. The bentazon rate needs to be increased for consistent control of prickly sida, spurred anoda, velvetleaf, and yellow nutsedge, while the acifluorfen rate should be increased for control of large (>6 cm) common ragweed and common lambsquarters.

Imazethapyr applied POST provides the broadest spectrum and most consistent control when applied within 10 days of weed emergence (Wilcut et al., 1994a). Imazethapyr is the only POST herbicide to control both yellow and purple nutsedge (Richburg et al., 1993). Control is most effective when imazethapyr is applied to the soil or to yellow nutsedge that is no more than 5 to 10 cm tall (Wilcut et al., 1994a). Other research has reported good nutsedge control with imazethapyr applied PPI plus EPOST.

Imazethapyr provides excellent full-season control (greater than 90%) of bristly starbur, coffee senna, *Ipomoea* morningglory species, prickly sida, smallflower morningglory, and spurred anoda (Wilcut et al., 1994a; Wilcut et al., 1994b). Coffee senna, prickly sida, and spurred anoda can be controlled if imazethapyr is applied when the weeds are in the two-leaf growth stage or smaller (Wilcut et al., 1994a). Larger weeds are not controlled by this treatment. Bristly starbur less than 4 cm tall is extremely sensitive to imazethapyr (Wilcut et al., 1994a). However, bristly starbur taller than 4 cm is not controlled by imazethapyr. Imazethapyr will kill only the terminal growing point of taller bristly starbur.

Chlorimuron applied POST provides effective late-season Florida beggarweed control (Wehtje et al., 1993b; Wilcut et al., 1994a). Chlorimuron can be applied from 60 days after crop emergence until 45 days prior to harvest (Chemical Pharmaceutical, 1994). Applying chlorimuron prior to 60 days after crop emergence increases the potential for significant yield reductions. If weeds are actively growing with good soil moisture, chlorimuron also may control bristly starbur, morningglory species including smallflower morningglory, sicklepod, and yellow and purple nutsedge. However, late POST-applied chlorimuron generally controls only Florida beggarweed (Wehtje et al., 1993). Sicklepod can be controlled with chlorimuron but greatest control results from chlorimuron applications at the two- to three-leaf stage of development.

Decreased alachlor use without other herbicides for early season suppression of Florida beggarweed may result in reduced chlorimuron activity on Florida beggarweed (Wilcut et al., 1994a). Chlorimuron consistently controls Florida beggarweed less than 25 cm tall. Without some early season suppression, Florida beggarweed often is larger than 25 cm when chlorimuron is applied, resulting in inconsistent control. Irrigated peanut should be watered 4 or 5 days before chlorimuron application to improve efficacy if the weeds are moisture stressed. Additionally, the potential for yield reduction from chlorimuron is increased if peanut is moisture stressed when chlorimuron is applied (Wilcut, unpublished data).

Pyridate was registered for use in peanut in late 1992. Peanut has excellent tolerance of pyridate with no phytotoxicity symptoms evident from POST application (Hicks et al., 1990). Pyridate is a POST herbicide with no soil residual activity. It should be applied with 2,4-DB for the control of a broader spectrum of weeds (Hicks et al., 1990; Wilcut et al., 1994a). Pyridate plus 2,4-DB controls cocklebur, common lambsquarters, pigweed species, and morningglory species. Small prickly sida, sicklepod, and spurred anoda (less than 3 cm tall) also are controlled by pyridate. Pyridate will control 15-cm tall common lambsquarters under good growing conditions. Pyridate alone is not effective on sicklepod, does not control common ragweed, and will only suppress yellow nutsedge growth. Cotton and soybean are extremely sensitive to pyridate.

Fenoxaprop-P and sethoxydim applied POST control annual and perennial grasses. Sethoxydim controls crabgrass species and bermudagrass better than fenoxaprop-P (Grichar and Boswell, 1986; Grichar and Boswell, 1989; Wilcut, 1991a). Two applications of sethoxydim are often required for control of the perennial grasses johnsongrass and bermudagrass (Chemical Pharmaceutical, 1994; Wilcut, 1991a). Both fenoxaprop-P and sethoxydim are most active if the grass weeds are not moisture-stressed when treated. The sethoxydim registration label recommends the use of a crop oil concentrate for improved grass control compared to control from sethoxydim and a nonionic surfactant (Chemical Pharmaceutical, 1994). BCH 81508 S, $(NH_4)_2SO_4$, and Urea ammonium nitrate can be added with crop oil for more consistent control (Jordan and York, 1989; York et al., 1990).

D. Reduced Tillage

Peanut traditionally has been grown in a well-prepared seedbed. Relatively little research has been conducted in peanut compared to other agronomic crops using minimum-tillage production practices (Worsham, 1985). The lack of interest was due to a perceived need to bury crop residues to reduce disease problems (Colvin et al., 1985a). Many producers and researchers also have considered well-prepared seedbeds as essential for high peanut yields.

Weed management practices for minimum-tillage peanut were studied in the early 1980s. Effective weed control and peanut yields equivalent to those in hand-weeded control plots were obtained in minimum-tillage systems with several herbicide programs in Alabama (Colvin et al., 1985a; Wilcut et al., 1987c).

Most studies concerning minimum-tillage peanut production were conducted within the last decade. Most researchers believe that the likelihood of reduced and inconsistent yield with minimum-tillage production does not result from poor weed control or increased disease incidence (Colvin et al., 1985a; Grichar and Boswell, 1986; Grichar and Boswell, 1987; Wilcut et al., 1987c; Wilcut et al., 1990a; Wilcut et al., 1994a). Rather, it was the subtle additional benefits of tillage, i.e., improved seedling performance and ease of pegging and harvesting operations that provided the consistently better yields in conventional-tillage systems.

E. Summary and Future Strategies

AC 263,222 is an imidazolinone herbicide very similar to imazethapyr. AC 263,222 controls every weed in peanut controlled by imazethapyr. Additionally, AC 263,222 controls Florida beggarweed and sicklepod (Wilcut et al., 1994a; Wilcut et al., 1994c). These two broadleaf weeds are the most troublesome weeds in the southeastern production area, which generates approximately 60% of the U.S. peanut crop. AC 263,222 controls purple and yellow nutsedge better and more consistently than imazethapyr (Richburg et al., 1993; Richburg et al., 1994). The registration of this herbicide will provide peanut producers with a management tool for some of the most troublesome weeds in peanut. However, carryover to cotton and certain vegetables may be a problem (York and Wilcut, 1995).

VII. WEED MANAGEMENT IN SOYBEAN

A. Introduction

The judicious and safe application of herbicides has become the most effective and widely used method for controlling weeds in soybean. Numerous herbicides are available, and combinations of these herbicides can control the majority of weeds in soybean. However, shrinking profit margins limit the number of herbicide applications a grower can economically justify. Matching herbicides with specific weed complexes is important.

B. Soil-Applied Herbicides

Soil-applied broadleaf herbicides such as flumetsulam, imazaquin, imazethapyr, linuron, metribuzin, or prepackaged mixtures of these herbicides suppress some annual grasses. With light to moderate infestations of susceptible species, one of these broadleaf herbicides plus cultivation for annual grass control may be adequate. Where moderate to heavy grass infestations or nonsusceptible species are expected, or where cultivation is not an option, a grass herbicide application is recommended.

PPI herbicide options for annual grasses include ethalfluralin, pendimethalin, trifluralin, and vernolate. Alachlor, dimethenamid, and metolachlor may also be incorporated, but they are generally applied PRE.

Ethalfluralin, pendimethalin, and trifluralin are very similar and can basically be used interchangeably. They provide good to excellent control of all the common annual grasses including Texas panicum. They also provide good control of certain small-seeded broadleaf weeds such as carpetweed, common lambsquarters, common purslane, Florida pusley, and pigweed species. Generally, control is season-long and carryover is not likely to be a problem at normal use rates.

The labels for ethalfluralin, pendimethalin, and trifluralin specify the maximum waiting period between application and incorporation (24 hours for trifluralin, 2 days for ethalfluralin, and 7 days for pendimethalin) (Chemical Pharmaceutical, 1994). It is best to incorporate these herbicides as soon as possible after application. A delay in incorporation can result in a significant herbicide loss.

Vernolate controls most annual grasses except broadleaf signalgrass and Texas panicum. Small-seeded broadleaf weeds such as carpetweed, common lambsquarters, common purslane, Florida pusley, and pigweed species also are controlled (Wilcut et al., 1994a). Vernolate normally provides control for about 4 to 5 weeks. Combined with a competitive crop canopy and cultivation if needed, this length of control should be sufficient in soybean. Vernolate should be incorporated immediately after application. A short delay in incorporation (15 minutes or longer) can result in significant volatilization losses and reduced weed control.

Clomazone provides good to excellent season-long control of all common annual grasses except shattercane and Texas panicum, and it controls several troublesome broadleaf weeds including prickly sida, spurred anoda, tropic croton, velvetleaf, and wild poinsettia (Griffin, 1986; Jordan et al., 1993e; Smith et al., 1994; Westberg et al., 1989). Clomazone does not control Florida pusley, carpetweed, or pigweed species. Off-site movement due to vapor drift has been observed with clomazone, particularly with PRE application. Incorporation greatly reduces the risk of vapor drift.

PRE herbicides available for annual grass control include alachlor, clomazone, dimethenamid, metolachlor, and pendimethalin. Alachlor, dimethenamid, and metolachlor control annual grasses except Texas panicum, seedling johnsongrass, and shattercane (Wilcut et al., 1989; Wilcut et al., 1994a). Alachlor, dimethenamid, and metolachlor also control carpetweed, Florida pusley, and pigweed species but are usually inadequate for common lambsquarters. In conventional-tillage systems, control with alachlor and the microencapsulated formulation of alachlor is similar. However, the microencapsulated formulation is superior for reduced-tillage production systems (Wilson et al., 1988).

Clomazone applied PRE will control the common annual grasses except shattercane and Texas panicum. Clomazone also controls several problem broadleaf weeds. Control is generally superior with PRE compared to PPI application. In spite of the excellent weed control clomazone provides, its use as a PRE application is subject to important limitations because of the potential for drift and volatilization (Chemical Pharmaceutical, 1994).

Pendimethalin is registered for PRE application alone or as a tank mixture with certain broadleaf herbicides. Pendimethalin normally controls weeds better when applied PPI than when applied PRE. In contrast to PPI applications, pendimethalin applied PRE often will not provide good or consistent control of moderate to heavy annual grass infestations.

As with annual grasses, annual broadleaf weeds can be controlled with numerous soil-applied and POST herbicides. The best application method depends upon the weed species present and compatibility with the total production system.

Soil-applied grass control herbicides also control certain broadleaf weeds. Ethalfluralin, pendimethalin, trifluralin, and vernolate control annual grasses plus certain small-seeded broadleaf weeds such as carpetweed, common lambsquarters, common purslane, Florida pusley, and pigweed species. If the predominant weed problem is annual grasses plus pigweed species, common lambsquarters, or both, one of these PPI herbicides plus cultivation will be adequate.

Alachlor, dimethenamid, and metolachlor control Florida pusley and suppress pigweed species. Clomazone controls common lambsquarters, common ragweed, jimsonweed, prickly sida, spurred anoda, tropic croton, velvetleaf, and wild poinsettia (Griffin, 1986; Jordan et al., 1993e; Smith et al., 1994; Westberg et al., 1989). Clomazone is also effective on Pennsylvania smartweed but does not control ladysthumb smartweed (York, unpublished data). Clomazone does not control carpetweed and pigweed species.

Metribuzin normally provides adequate control of many annual broadleaf weeds. Cocklebur and morningglories are not controlled adequately. There are several prepackaged mixtures of metribuzin with herbicides for grass control (i.e., trifluralin and metolachlor). Several prepackaged mixtures of metribuzin plus chlorimuron are available. These mixtures provide better control of sicklepod and morningglories than metribuzin alone (Bruff and Shaw, 1992; Green et al., 1988; Lanie et al., 1994a).

Soybean have a narrow margin of tolerance for metribuzin. One should use extreme care when selecting the application rate of a product that contains metribuzin. This is especially important in fields with varying soil types. There are several restrictions on the use of metribuzin. Specific restrictions vary, depending upon application methods and tank mixtures. In general, this herbicide should not be used on sands or loamy sands with less than 1% organic matter or soil with less than 0.5% organic matter (Chemical Pharmaceutical, 1994). Certain cultivars of soybean are very susceptible to metribuzin injury. Before applying a product containing metribuzin to a new cultivar, one should check with a dealer or extension agent to determine potential sensitivity to metribuzin. Increased soybean injury also will be observed when metribuzin is used in conjunction with soil-applied organophosphate insecticide/nematicides; see the product labels for details.

Linuron controls common ragweed, common lambsquarters, and pigweed species but is inadequate for control of most other broadleaf weeds (Moseley and Hagood, 1990). Linuron will not adequately control cocklebur and morningglories. Linuron should not be used on sand, loamy sand, or gravelly soils or on any soil with less than 0.5% organic matter. Linuron is safer than metribuzin on light-textured soils. However, injury may occur if excessive rates are applied or if heavy rainfall is received shortly after planting. Do not use on soils with more than 3% organic matter.

Imazaquin can be applied alone or in various PPI or PRE tank mixtures. The most consistent and broadest spectrum control is with imazaquin tank mixed with ethalfluralin, pendimethalin, or trifluralin and incorporated. Morningglory control is improved 10% to 15% with preplant-incorporated application of imazaquin with the aforementioned herbicides (Wilcut and York, unpublished data).

More rainfall is needed to activate a PRE application of imazaquin than is required for most PRE herbicides. For good control, 2 to 3 cm of rainfall is needed within 7 to 10 days after application. Generally imazaquin gives best weed control in years with ample rainfall. For this reason, incorporated applications perform more consistently.

Soil-applied imazaquin controls most broadleaf weeds (Mills and Witt, 1989; Poston et al., 1992; Oliver et al., 1993). Major exceptions are hemp sesbania, sicklepod, spurred anoda, tropic croton, and volunteer cowpea (Wesley et al., 1989a; Wesley et al., 1989b). Morningglory control is variable but usually adequate, especially when followed by cultivation.

Soybean tolerance of imazaquin is normally good. However, slight to severe crop stunting has occasionally been noted, especially with very wet (water logged) conditions shortly after planting. To reduce the potential for crop stunting, avoid deep incorporation and shallow planting, especially on sandy soils.

Soil-applied imazethapyr provides good to excellent control of many broadleaf weeds, including jimsonweed, *Ipomoea* morningglories, coffee senna, pigweed species, prickly sida, smallflower morningglory, spurred anoda, velvetleaf, and wild poinsettia (Cantwell et al., 1989; Wilcut et al., 1994a). Imazethapyr does not control common ragweed, Florida beggarweed, sicklepod, and tropic croton. Imazethapyr also suppresses or controls purple and yellow nutsedge as a soil application (Richburg et al., 1993). Imazethapyr controls some annual grasses; however, an application of a dinitroaniline or chloroacetamide herbicide provides more consistent and better control (Cantwell et al., 1989; Wilcut et al., 1991b; Wilcut et al., 1994a).

C. Postemergence Herbicides

Herbicide options for POST control of annual grasses include clethodim, fenoxaprop-P, fluazifop-P, quizalofop-P, and sethoxydim. With few exceptions, these POST graminicides control all common annual grasses. Sethoxydim does not control crowfoot grass. Although fenoxaprop-P controls most annual grass species, generally it is the least effective of the POST graminicides.

Two applications of all POST graminicides may be made; however, one properly timed application in combination with a narrow row spacing (<50 cm) or cultivation should be sufficient for annual grasses. The optimum grass size for application depends upon the product being used and the species of grass. Application to larger grasses, especially if under drought stress, may result in insufficient control. Cultivation will enhance control, but to prevent stressing the grass, do not cultivate for 7 days before or after application.

All common annual broadleaf weeds can be controlled with POST herbicides. However, weed species vary in their response to different herbicides, and proper weed identification and herbicide selection are essential for satisfactory results.

In addition to proper herbicide selection, timeliness of application and use of proper application methods are essential. The most common cause of poor results with POST herbicides is application when the weeds have grown beyond the optimum size for treatment (DeFelice et al., 1989; King and Oliver, 1992). As a general rule, POST broadleaf herbicides should be applied when weeds are in the two- to four-leaf stage or are 5 to 8 cm tall. This application normally occurs within 2 weeks after planting.

Weather conditions also are important to success with POST herbicides. For best results, weeds should be actively growing and not under moisture stress. When they are under moisture stress, they have thicker cuticles and less herbicide is absorbed into the plant.

Acifluorfen controls many annual broadleaf weeds, with the exception of Florida beggarweed, prickly sida, sicklepod, spurred anoda, and velvetleaf (King and Oliver, 1992; Monks et al., 1993; Wilcut et al., 1994a). Acifluorfen normally causes some soybean leaf crinkling and leaf bronzing or leaf burn. Addition of crop oil concentrate or higher rates of surfactant increases the amount of leaf burn. This injury is temporary and the soybean recover quickly and grow normally.

Bentazon gives excellent control of cocklebur, jimsonweed, and smartweed and good control of prickly sida, spurred anoda, velvetleaf, and giant ragweed (King and Oliver, 1992; Wilcut et al., 1994a). Although control may be inconsistent, bentazon plus crop oil concentrate usually controls small common ragweed and common lambsquarters and crop tolerance is excellent. A tank mixture of bentazon plus acifluorfen is a very effective and broad-spectrum POST treatment (Wilcut et al., 1994a). The best rate of each herbicide to use in the tank mixture depends upon the weed species and weed size; see the product labels for details. Several commercial prepackaged mixtures are available.

Chlorimuron applied POST will control most broadleaf weeds (King and Oliver, 1992). Major exceptions are coffee senna, common lambsquarters, prickly sida, and tropic croton (Monks et al., 1993; Wilcut et al., 1994a). Control of spurred anoda and velvetleaf also may not be adequate. Chlorimuron is the POST herbicide of choice for sicklepod control. Sicklepod control is best when flumetsulam, imazaquin, metribuzin, or metribuzin plus chlorimuron mixtures are soil-applied and followed with chlorimuron applied POST and cultivation as needed (Shaw et al., 1991; York, unpublished data). Soybean tolerance of chlorimuron normally is good, but some leaf distortion and yellowing and minor crop stunting may be noted, especially when chlorimuron is applied to soybean under stress conditions. The soybean will recover and grow normally.

Fomesafen controls many common broadleaf weeds; major exceptions are common lambsquarters, Florida beggarweed, prickly sida, sicklepod, spurred anoda, and velvetleaf. Unless treated when smaller than 2 cm (two-leaf stage or earlier), the aforementioned weeds will not be controlled. Soybean tolerance of fomesafen is normally good. Although fomesafen has some crop rotational restrictions, carryover problems to agronomic crops have rarely been observed.

Imazaquin applied POST controls cocklebur and pigweed species (King and Oliver, 1992; Risley and Oliver, 1991). Control of other broadleaf weeds will be inadequate (see the section on sicklepod control). Soybean tolerance of imazaquin applied POST is normally excellent. Imazaquin can be applied in a POST mixture with acifluorfen. This mixture controls cocklebur and pigweed plus other weeds normally controlled by acifluorfen, including common lambsquarters, common ragweed, jimsonweed, *Ipomoea* morningglories, and tropic croton. However, this mixture should not be used for sicklepod control as acifluorfen may reduce imazaquin's control of sicklepod (Wesley and Shaw, 1992; Westberg and Coble, 1992).

Imazethapyr applied POST controls bristly starbur, cocklebur, pigweed species, jimsonweed, smartweed, and velvetleaf (Klingaman et al., 1992; Wilcut et al., 1994a). Morningglory control is usually acceptable but can be inconsistent. This herbicide has become the standard in the Midwest for velvetleaf control (Cantwell et al., 1989; Lueschen and Hoverstad, 1991). Small coffee senna, prickly sida, and spurred anoda (fewer than two true leaves), and bristly starbur and wild poinsettia (less than 3 cm tall) can be controlled with imazethapyr applied POST (Wilcut et al., 1994a).

In addition to controlling certain broadleaf weeds, imazethapyr applied POST usually gives adequate control of broadleaf signalgrass, foxtail species, seedling johnsongrass, and shattercane and sometimes gives adequate control of rhizome johnsongrass. Imazethapyr applied POST also controls small yellow nutsedge (less than 5 cm tall) and purple nutsedge (Richburg et al., 1993). Imazethapyr applied POST does not control common lambsquarters, common ragweed, Florida pusley, hemp sesbania, sicklepod, and tropic croton (Wilcut et al., 1994a).

Lactofen controls many annual broadleaf weeds. Major exceptions are common lambsquarters, sicklepod, and spurred anoda (Wilcut et al., 1994a). Morningglory control may be inconsistent (Higgins et al., 1988). Lactofen is a very good herbicide for POST control of common ragweed. Lactofen normally causes more leaf distortion and leaf burn than acifluorfen or fomesafen; crop injury is sometimes severe. However, soybean usually will recover and grow normally.

Thifensulfuron applied POST controls common lambsquarters and pigweed species (Monks et al., 1993). Thifensulfuron will not provide adequate control of cocklebur, common ragweed, *Ipomoea* morningglories, and sicklepod. Injury to soybean can be significant although soybean usually recovers. A mixture of chlorimuron plus thifensulfuron provides broader-spectrum control than either herbicide alone.

A common practice in many regions is to use 2,4-DB at 0.035 kg/ha with other POST broadleaf herbicides to improve *Ipomoea* morningglory species control (Barker et al., 1984; Kapusta et al., 1986; York et al., 1991b). Soybean is tolerant and yield reductions were not observed in full-season or double-crop soybean with tank mixtures of acifluorfen, chlorimuron, imazaquin, or imazethapyr.

Although not widely used, POST-directed herbicide applications are an effective and economical tool to control late-emerging weeds or weeds that have escaped at-planting or early season POST herbicides. Directed applications are also very effective as a component of a sicklepod management program; see the section on sicklepod control.

Herbicides and tank mixes labeled for POST-directed application to soybean include linuron, metribuzin, paraquat, 2,4-DB, and certain mixtures of these herbicides. A surfactant should be added according to label directions to all these herbicides and tank mixes except 2,4-DB applied alone.

For POST-directed treatments to be effective, the weeds must be shorter than the crop. The soybean should be at least 20 cm tall (30 cm for metribuzin) and the weeds should be 8 to 10 cm tall or less, depending upon the herbicide and application rate selected. Precision application is necessary to prevent unacceptable crop injury. The spray solution should contact no more than the lower one-fourth to one-third (or about 8 cm) of the soybean plant. Nozzles should be mounted in a manner that allows good control of the spraying height. Never attempt to use booms with drop nozzles. Mount the nozzles on a cultivator frame or use a directed-spray rig. Shielded sprayers are particularly effective. Use regular flat fan or off-center flat fan nozzles.

Depending upon the herbicide chosen, directed sprays will control a broad spectrum of annual grasses and broadleaf weeds. 2,4-DB is effective on cocklebur and morningglory; treatment is suggested when weeds are 8 cm tall or less, although much larger weeds can be killed if spray coverage is good on the lower portion of the weeds. Paraquat controls pigweed species and annual grasses up to 8 cm. Metribuzin controls most species of broadleaf weeds up to 8 cm tall and annual grasses up to 2 cm tall. Linuron is effective on most annual broadleaf weeds and grasses up to 10 cm tall. A directed application of linuron or metribuzin plus 2,4-DB is very effective on a wide range of weed species.

D. Sicklepod Management

Sicklepod is a very serious soybean weed and appears to be spreading throughout the southeastern and midsouthern sections of the United States. Although sicklepod is not very competitive on a per-plant basis, heavy infestations can reduce soybean yields significantly. In addition, a price dockage may occur because of foreign matter (sicklepod seed) in the harvested product.

Several herbicides effectively control sicklepod. A single herbicide application plus cultivation often gives adequate control in light to moderately infested fields. Adequate control of heavy infestations normally requires two herbicide applications. Herbicide programs that have given acceptable control of moderate to heavy sicklepod infestations through midseason include flumetsulam tank mixed with trifluralin applied PPI and followed with chlorimuron applied POST if needed; metribuzin alone or in tank mixtures with trifluralin, metolachlor, or chlorimuron applied at planting followed by cultivation and a POST application of chlorimuron or imazaquin; or imazaquin applied at planting followed by cultivation and a POST application of imazaquin or chlorimuron (Poston et al., 1992; Shaw et al., 1991).

Any of the programs described will give adequate control for 6 to 8 weeks. After that time, sicklepod can reinfest the crop unless the rows are lapped. The late-emerging sicklepod has little effect on soybean yield but increases the amount of foreign matter in the harvested seed, produces seed to infest the field further, and may delay harvest of early-maturing soybean varieties.

The use of a POST-directed application of linuron plus 2,4-DB or of metribuzin plus 2,4-DB is highly beneficial in a sicklepod management program. Good to excellent season-long sicklepod control has been obtained with the previously described soil-applied herbicides at planting followed by cultivation and a POST-directed application of linuron or metribuzin plus 2,4-DB. POST application of chlorimuron or imazaquin followed by cultivation and a POST-directed application of linuron or metribuzin plus 2,4-DB has also given excellent season-long management.

Planting soybean in narrow rows (narrow enough to prohibit cultivation) in fields heavily infested with sicklepod can be risky. While adequate control may be obtained with metribuzin or metribuzin tank mixtures or commercial prepackages at planting followed by a POST application of chlorimuron or with imazaquin at planting followed by a POST application of imazaquin, the lack of the option to cultivate makes this program risky. Research has shown cultivation to be the most cost-effective component of a sicklepod management program (Miller and Griffin, 1994; Poston et al., 1992; Shaw et al., 1991).

Strive for good sicklepod control in rotational crops. For example, in corn, apply atrazine at planting. Follow with cultivation and a layby application of 2,4-D amine plus a surfactant, dicamba, or ametryn plus a surfactant. After corn harvest, disk the field to reduce seed production by weeds that have escaped treatment.

E. Perennial Weed Management

Grasses

The major perennial grasses infesting soybean are bermudagrass, johnsongrass, and quackgrass. Although graminicides are available for their control, several cultural control options are beneficial in decreasing populations. Implementing such options may become more important, especially with the development of graminicide-resistant johnsongrass biotypes (Barrentine et al., 1992).

Except in severely infested fields, bermudagrass does not cause major yield reductions. However, heavy infestations of johnsongrass and quackgrass can reduce yields significantly. Regardless of the program chosen to control perennial weeds, it is important to maintain control in rotational crops and to control the weeds in adjoining areas to prevent reinfestation of treated fields. Herbicide options for johnsongrass in corn include nicosulfuron and primisulfuron. Additionally, perennial grasses can be controlled along field borders and fence rows and around stump piles and other obstructions with POST applications of glyphosate or graminicides.

Although not absolutely necessary, a tillage program before planting soybean will aid in perennial grass control. Using a moldboard plow or chisel plow followed by several diskings spaced at 4- to 6-week intervals during the fall and winter is most effective. This tillage cuts the rhizomes and stolons into small pieces, thus stimulating sprouting the following spring, and also brings some of the rhizomes to the soil surface, where winter weather may kill them. If winter tillage is not possible, tillage just before planting is still of some benefit.

Although applications of POST graminicides are effective on johnsongrass, recent reports indicate that repeated use of POST graminicides has resulted in johnsongrass resistance to sethoxydim, quizalofop-P, and fluazifop-P (Barrentine et al., 1992). Controlling johnsongrass with rope wick or recirculating applicators and glyphosate is a viable option for control. A major disadvantage to this method of application is that some degree of competition already has occurred prior to this treatment.

Light infestations of johnsongrass can be controlled in soybean with glyphosate applied with a wick bar or other type of wiper applicator. For heavy infestations, one of the control programs outlined later is preferred. If johnsongrass populations resistant to the POST graminicides are present, then glyphosate applied with a wick bar, etc., may be the only POST option. If resistant johnsongrass populations are nearby, glyphosate applied with a wick or recirculating sprayer will reduce potential for resistant populations to become established. In addition, high rates of trifluralin are beneficial for management.

The johnsongrass should be a minimum of 15 cm above the soybean canopy and the wiper applicator should be carried at least 6 cm above the top of the soybean canopy. Better results will be obtained when more of the weed is exposed to the herbicide solution.

Johnsongrass can be controlled in soybeans with POST applications of clethodim, fenoxaprop-P, fluazifop-P, quizalofop-P, or sethoxydim (Vidrine, 1989). There are no soil-type limitations nor significant rotational restrictions with use of the graminicides, and they do not injure soybean. If good control is obtained in soybean, the rhizome johnsongrass population in the following year should be reduced 85% or more. However, a control program for seedling johnsongrass will be necessary in the following crop.

It is best to begin the herbicide program in soybean with an incorporated application of ethalfluralin, pendimethalin, trifluralin, or mixtures thereof. These herbicides will control seedling johnsongrass and give some suppression of rhizome johnsongrass.

Apply the recommended rate of one of the POST graminicides when the johnsongrass is actively growing and at the proper size for treatment. Application of herbicides to johnsongrass under drought stress may result in inadequate control. In most cases, one application will be sufficient. However, a second application may be made if necessary to control regrowth of newly emerged plants. Do not mix any POST broadleaf herbicide with graminicides as antagonism will likely occur.

Alternatively, a POST application of imazethapyr can be considered when both johnsongrass and broadleaf weeds controlled by imazethapyr are present. Imazethapyr will give good initial control of johnsongrass, in some cases, it alone will provide adequate control. In others, enough regrowth will occur to justify a later application of a POST graminicide. Imazethapyr should be applied when the broadleaf weeds are within the recommended growth stages and before the johnsongrass is more than 20 to 25 cm tall. Generally the POST graminicides are more effective for rhizome johnsongrass control than imazethapyr. Imazethapyr is more effective on seedling johnsongrass than on rhizome johnsongrass.

If properly applied, a fall application of glyphosate will effectively control rhizome johnsongrass. Glyphosate normally is applied after corn harvest but may also be applied to set-aside land. Corn should be harvested 6 weeks before the first killing frost. After corn harvest, rotary mow the field to remove old weed foliage and excess trash and to stimulate regrowth of the johnsongrass foliage. Do not till. Allow the johnsongrass to regrow until it has reached the boot to head stage. This will normally take about 3 to 5 weeks.

For best results, the johnsongrass must be actively growing and not under drought stress. There should be at least 7 days between application and the first killing frost. Delay tillage for 10 days after treatment. After the 10-day waiting period, the field can be tilled and small grain or a cover crop planted. If seedling johnsongrass emerges, tillage is suggested before the seedlings exceed 2 weeks of age to prevent new rhizome production.

An integral component of the fall glyphosate program is control of seedling johnsongrass the following season. There should be no need to repeat the glyphosate treatment for several years if seedling johnsongrass is controlled in all crops in the rotation. To control seedling johnsongrass in soybean, use a PPI application of ethalfluralin, pendimethalin, or trifluralin and cultivate as needed. In corn, butylate or EPTC applied PPI or nicosulfuron or primisulfuron applied POST will provide control.

Although fall glyphosate application is preferred, spring application is also an option. With spring application, delay land preparation and allow the johnsongrass to grow undisturbed until the boot stage is reached. Apply glyphosate, wait 7 to 10 days, and then incorporate ethalfluralin, pendimethalin, or trifluralin. Spring glyphosate application delays soybean planting until early to mid-June, which reduces yield potential.

Bermudagrass can be controlled with POST applications of clethodim, fluazifop-P, quizalofop-P, and sethoxydim (Grichar and Boswell, 1989; Wilcut, 1991a). Fenoxaprop-P does not control bermudagrass. One application of any of the four graminicides in combination with cultivation and/or a row spacing that results in early canopy closure will usually be sufficient to prevent soybean yield reductions. However, two applications can be made if necessary. Fluazifop-P, quizalofop-P, and clethodim provide similar levels of control with adequate but lesser control from sethoxydim.

Apply one of the POST graminicides when the bermudagrass is actively growing and at the proper size for treatment. See product labels for the most appropriate growth stage (Chemical Pharmaceutical, 1994). Do not tank mix the POST graminicides with a POST broadleaf herbicide. If a second application is needed, better spray coverage will be obtained if drop nozzles are used to semidirect the herbicide under the soybean foliage.

Cultivation will enhance control. However, do not cultivate for 7 days before or after herbicide application. Additionally, avoid moving bermudagrass from ends of fields with cultivators. Continue bermudagrass suppression in rotational crops.

Bermudagrass may also be controlled with fall applications of glyphosate. The program for bermudagrass is similar to that previously described for johnsongrass. Corn should be harvested at least 3 weeks before the first killing frost. Do not till. When the bermudagrass is 20 to 25 cm tall or when some seedheads can be seen, apply glyphosate. Glyphosate must be applied at least 14 days before the first killing frost. Delay tillage for 14 days after application. Small grains or a cover crop can be sown if desired. The following spring, plant soybean in row spacings of less than 50 cm. Spot treatment with any of the POST graminicides except fenoxaprop-P is also an option for bermudagrass escapes. Suppression must be continued in the following crop.

Broadcast applications of glyphosate are expensive. To reduce costs, spot spray glyphosate only on infested areas. A 2% solution can be used in hand sprayers for treatment of small spots.

Broadleaf Weeds and Sedges

Perennial broadleaf weeds such as bigroot morningglory, common milkweed, curly dock, hemp dogbane, Carolina and robust horsenettle, maypop passionflower, and trumpetcreeper are normally more prevalent in reduced-tillage and no-tillage systems than in conventional-tillage systems. Perennial broadleaf weeds cannot be controlled in soybean. The only way to control these weeds is to plant corn in rotation with soybean and follow an effective management program in corn.

To control these weeds in corn, apply atrazine at planting. When the corn is 30 to 50 cm tall, post-direct 2,4-D amine with a surfactant. After corn harvest and before frost, retreat infested areas with a tank mix of 2,4-D amine plus dicamba with a surfactant. Alternatively, apply glyphosate at this stage of growth. Fall tillage also is helpful in reducing the buildup or spread of these perennial weeds.

Nutsedge is not very competitive with soybean. Cultivation alone will usually reduce competition enough to prevent yield reduction. Depending upon other crops in the rotation, however, you may want to follow a good nutsedge control program in soybean to reduce problems with this weed in the rotational crops.

If nutsedge is detected during fall scouting, it is important to determine whether it is yellow or purple nutsedge since management programs for the two species may differ. Vernolate and imazethapyr are the only soil-applied herbicides that control purple nutsedge. Imazethapyr can also be applied PRE or POST for purple nutsedge control (Richburg et al., 1993). Preplant incorporation of metolachlor or vernolate is effective on yellow nutsedge. Adequate control of light infestations of yellow nutsedge may be obtained with a PRE application of metolachlor; however, control is more consistent when the metolachlor is incorporated. Alachlor suppresses yellow nutsedge to some extent, but control is usually less than with metolachlor.

Yellow nutsedge can be controlled with POST applications of bentazon, chlorimuron, or imazethapyr. Bentazon is applied with a crop oil concentrate when yellow nutsedge is 15 to 20 cm tall. If needed, a second application can be made at the same rate 7 to 10 days later. Alternatively, apply chlorimuron with a surfactant or imazethapyr with a crop oil concentrate when nutsedge is 5 to 10 cm tall.

F. Salvage Treatments and Harvest Aid Herbicide Options

For optimum yields, weeds must be controlled for the first 4 to 5 weeks after soybean planting. Weeds that emerge later have little or no direct effect on yield but may interfere with harvesting and add foreign matter to the harvested product.

Although the manager should strive for good early season weed control, there are situations in which a salvage treatment is necessary. Remember that salvage treatments are not a substitute for good early season weed control.

2,4-DB can be applied POST to soybean from 7 to 10 days before bloom until midbloom. Application before or after this period may result in severe crop injury. Application to soybeans under drought stress also will increase injury. This is a risky treatment and should be considered only as a salvage operation to remove cocklebur and morningglory when harvesting would otherwise be impossible.

Paraquat can be applied for weed desiccation to facilitate harvest. This treatment may be of value for early maturing soybeans that are ready for harvest before frost kills weeds. Observe all precautions on the label as applications made too early can be injurious (Chemical Pharmaceutical, 1994).

G. Reduced Tillage

Reduced-tillage systems are effective and economical for soybean production, and adequate weed control can usually be obtained (Kapusta and Krausz, 1993; Lewis, 1985; Sims and Guethle, 1992; Witt, 1991; Wilson and Worsham, 1988). Advantages of reduced tillage include reduced erosion on sloping land, lower energy consumption due to fewer trips across fields, maintenance and increase of soil organic matter and improvement in soil structure and tilth, greater moisture conservation, and reduced weed infestations (Witt, 1991). Potential disadvantages include initial investment in appropriate planting equipment, inability to use tillage or cultivation as a weed control tool, possible increases in herbicide inputs to offset reductions in tillage, increased reliance on PRE herbicides which require rainfall for activation, and changes in pest populations including weeds, insects, and disease (Witt, 1991). Even with these potential problems, technology is available for successful reduced-tillage soybean production. Although the term *conservation tillage* has numerous meanings, a basic premise is the elimination of tillage. However, changes in tillage often mandate changes in weed management strategies.

Three basic reduced-tillage systems used for soybean production in the United States are double cropping wheat and soybean, full-season soybean planted into residues of corn or some other rotational crop, and stale-seedbed production, in which soybean is planted into a previously established seedbed with little or no tillage at the time of planting. These systems have similarities and differences in terms of weed management. All require use of nonselective herbicide(s) to control emerged vegetation at planting. Glyphosate and paraquat are used extensively for this purpose. Selection of these herbicides is dependent upon the target species and size. Paraquat is less effective in controlling Pennsylvania smartweed and horseweed than glyphosate (Crawford et al., 1992; McClelland et al., 1992). In most reduced tillage situations, either glyphosate or paraquat is used to ensure sufficient burndown of existing vegetation. Glyphosate is better on larger annual weed species and on species like cutleaf eveningprimrose, horseweed, and, of course, perennial weed species.

Glyphosate and paraquat do not provide residual control. Research has shown that these burndown herbicides can be applied with residual-type herbicide(s), including chlorimuron, imazaquin, imazethapyr, and metribuzin (Lanie et al., 1994b; Moseley and Hagood, 1990; Moseley and Hagood, 1991).

Double-cropped soybean into wheat stubble and full-season soybean planted into crop mulch or cover crop are similar in weed management strategies. Both systems may require use of nonselective herbicides and both may benefit from allelopathic effects of residues.

Small grain and other cover crop residues provide weed control with allelochemicals (Barnes and Putnam, 1983; Liebl and Worsham, 1983; Shilling et al., 1985; White and Worsham, 1990). Rye suppressed growth of redroot pigweed, common lambsquarters, and common ragweed (Shilling et al., 1985). Wheat, barley, oats, grain sorghum, and Sudangrass also have been used to control or suppress weeds (Worsham, 1991).

Timely desiccation of the cover crop is necessary for establishment of soybean. Destruction of cover crops too late resulted in excessive residues, lower than optimal stands, and subsequently lower soybean yield (Liebl et al., 1992). Additionally, herbicides in no-tillage systems may be less effective than in conventional tillage if soil contact is necessary for activity (Buhler and Daniel, 1988; Triplett and Lythe, 1972). Vegetation or excessive residues may intercept herbicides and prevent them from reaching the soil. However, residues do not always reduce efficacy of residual herbicides (Witt, 1991).

The benefits of cover crops in soybean weed management have not been fully researched. The extent to which plant residues can consistently replace PRE herbicides is still uncertain. Rye mulch has provided sufficient weed control to eliminate the necessity of using PRE or POST herbicides in soybean (Worsham, 1991). Additionally, Sudangrass residues suppressed several broadleaf weeds in soybean sufficiently that herbicides were not necessary (Smeda et al., 1993).

In the Mississippi delta region of the United States, especially on alleuvial clay soils, seedbeds are prepared in the fall or early spring as weather permits, and soybeans are planted into these beds with minimum disturbance of the soil. Stale seedbed systems have been as successful as conventional tillage systems (Bruff and Shaw, 1992; Lanie et al., 1994a; Lanie et al., 1994b; Oliver et al., 1993). Tillage of these clay soils at planting is undesirable because difficulties in establishing soybean occur if primary tillage is performed at planting. Vegetation in the form of cool-season weeds or emerging warm-season weeds must be controlled at or prior to planting with glyphosate or paraquat. Additionally, residual herbicides can be applied to control specific weeds prior to planting, thereby reducing excessive vegetation at planting. Mixtures of glyphosate or paraquat with imazaquin, chlorimuron, and/or metribuzin provide effective burndown and residual weed control (Moseley and Hagood, 1990; Moseley and Hagood, 1991; Bruff and Shaw, 1992).

Fall and early preplant applications of cyanazine prevented horseweed, Pennsylvania smartweed, field pennycress, giant foxtail, and fall panicum from becoming established at planting. Although early preplant herbicides provided control of vegetation at planting, season-long control did not always occur and supplemental herbicide applications were necessary to maintain control.

Herbicide selection in stale-seedbed soybean and conservation-tillage soybean with or without use of cover crops will be dependent upon the weed complex prior to planting, and the weed spectrum anticipated after planting. Numerous herbicide options are available for selecting an appropriate weed management system in soybean.

Changes in tillage practices generally result in weed shifts. Annual grasses are more prevalent in no-till production than in conventional, while populations of velvetleaf (Buhler and Daniel, 1988), pitted morningglory, tall morningglory, common ragweed, and common lambsquarters (Shilling et al., 1985) are less prevalent. In the southeastern region, Florida pusley is a major problem in reduced-tillage fields where dinitroaniline or chloroacetamide herbicides are not used (Wilcut, unpublished data). Green foxtail densities were greater in chisel plow and no tillage corn than in conventional or ridge-till corn (Buhler, 1992). In the same study horseweed were present only in no-till and ridge-till systems.

Additionally, reduced tillage is thought to increase populations of perennial weeds. Anticipating weed shifts and developing management systems to control new weeds will be necessary as tillage practices are altered (Staniforth and Wiese, 1985).

H. Summary and Future Strategies

Because of the large soybean hectarage throughout the world, developing herbicides for soybean continues to be profitable for industry, and new and more effective herbicides will be available. Recently, a gene expressing resistance to glyphosate has been incorporated into soybean. Glyphosate may provide total vegetation control, including control of most grass and broadleaf weeds, in soybean. Glyphosate generally provides adequate control of *Ipomoea* morningglory species and sicklepod, especially if the weeds are not under moisture stress at the time of application (Doss and York, 1995). However, with a genetically tolerant soybean cultivar, many weeds can be controlled with increased rates of glyphosate.

One major advantage of glyphosate is in herbicide resistance management. Glyphosate offers a herbicide option with a mechanism of action distinctly different from that of other herbicides used in soybean. Because glyphosate controls both grass and broadleaf weeds, use in herbicide rotation with graminicides and broadleaf herbicides may suppress and delay weed resistance.

In addition to glyphosate-tolerant soybean, a soybean cultivar tolerant of the sulfonylurea herbicides has been developed (Forney et al., 1989). Although this cultivar is not genetically resistant, a greater margin of tolerance of sulfonylurea herbicides exists. The major advantage of this tolerance is the ability to use thifensulfuron and chlorimuron in tank mixtures at higher rates for more acceptable control of common lambsquarters, sicklepod, and velvetleaf, without soybean injury. Although control of specific weeds can be enhanced by increasing the herbicide rate, there is debate among weed scientists as to the benefits of such a weed management tool in light of the recent increase in incidence of weed resistance to ALS (acetolactate synthese) inhibiting herbicides.

A significant data base for the competitiveness of individual weeds, their economic impact, and predictive models for economic threshold implementation has been developed (Coble and Mortenson, 1992; Keisling et al., 1984). A major goal of weed scientists is to provide growers with information on whether or not herbicide applications are economically justified. Many fields contain populations of weeds that economically justify herbicide use. However, light to moderate weed infestations frequently occur, and herbicide applications may not be needed to maximize profit. Additionally, increased concern for herbicide impact on the environment has generated interest in management tools that reduce herbicide use. Computer models have been developed in North Carolina (Wilkerson et al., 1991) and Arkansas (Baldwin and Hall, 1993). These decision aid or expert systems allow producers to determine whether the weed population in a given field is sufficient to warrant herbicide application to prevent economic loss.

The expert system developed in North Carolina (HERB) ranks the competitive ability of individual weeds relative to that of common cocklebur, which is considered the most competitive weed with soybean in North Carolina (Wilkerson et al., 1991). A competitive index, based on a particular weed's competitive ability with soybean relative to common cocklebur, is assigned to each weed. The index indicates the deleterious effect of the weed on yield if allowed to compete throughout the season. A total competitive load for a given field is then calculated. From the total competitive load, yield loss if weeds are not controlled can be determined. On the basis of the predicted yield loss from the weed complex, cost of implementing a control practice, and projected yield (weed-free) and price, an economic decision can be made.

The Arkansas program (SWC) also estimates yield loss as a result of weed competition. This decision aide predicts yield loss from a complex of weeds by subtracting the percentage yield loss from the most competitive weed and then repeating the same procedure for the remaining weeds in the complex (Keisling et al., 1984). As with HERB, projected yield, soybean price, and cost of control are necessary in making an economically based decision. HERB provides, in terms of the data provided, the most economical option available. SWC allows growers to select from a list of economical herbicide options.

The popularity of both systems has increased. However, weeds are generally patchy in their distribution (Wiles et al., 1992b), and thus the decision models do not always recommend herbicide use accurately (Wiles et al., 1992a). Additionally, these models were developed on the basis of competition data within the respective states. Development of threshold data in other states and modifications of these systems may increase accuracy. Currently these models do not consider variety, row spacing, planting date, weather conditions, and other factors that influence the magnitude of yield reductions.

These models do not consider problems caused by weeds left in the field when populations are below the economic injury level, which later mature and set seed (Forcella et al., 1993). However, future models will most likely address this aspect. Additionally, harvesting efficiency is not considered in these models.

Many states are evaluating these programs and modifying inputs and weed species for that particular region. As these decision aids increase in accuracy and acceptance, a more integrated approach to weed management which would increase soybean profitability may be possible.

In addition to the system developed for soybean, similar systems are being developed for cotton and peanut (Barbour and Bridges, 1992). Development of these systems may lead to a more integrated approach to weed management in these crops. A major limitation to such a program in cotton is the lack of safe and effective POST control options for broadleaf weeds. Producers must rely on soil-applied and POST-directed herbicides to control broadleaf weeds;

because these treatments often are preventative, they do not allow use of thresholds to predict the need for herbicide applications.

IX. CONCLUSIONS

Weed management in these and other crops has progressed from hand weeding to mechanical control and more recently to heavy reliance on chemical weed control. The concept of integrated weed management has increased in popularity over the last few years. Although the concept has been utilized by weed scientists for many years, weed management has generally been equated to chemical control. It is becoming more and more evident that to be profitable and "sustainable," all control options need to be integrated for successful weed management. Along with the judicious use of herbicides, cultivation, crop and herbicide rotation, and establishment of a competitive crop should be employed. Specific production systems may limit the use of some control options such as cultivation in a no-till system. However, developing best management practices for an overall farm system will be the most profitable and sustainable approach.

APPENDIX

List of Certain Weeds Mentioned in Chapter

Common name	Scientific name	BAYER code
Amaranth, Palmer	*Amaranthus palmeri* S. Wats.	AMAPA
Anoda, spurred	*Anoda cristata* (L.) Schlecht.	ANVCR
Beggarweed, Florida	*Desmodium tortuosum* (Sw.) DC.	DEDTO
Crabgrass species	*Digitaria* spp.	DIGZZ
Croton, tropic	*Croton glandulosus* var. septentrionalis Muell.-Arg.	CVNGS
Jimsonweed	*Datura stramonium* L.	DATST
Johnsongrass	*Sorghum halepense* (L.) Pers.	SORHA
Lambsquarters, common	*Chenopodium album* L.	CHEAL
Morningglories, *Ipomoea*	*Ipomoea* spp.	IPOZZ
Nutsedge, purple	*Cyperus rotundus* L.	CYPRO
Nutsedge, yellow	*Cyperus esculentus* L.	CYPES
Panicum, Texas	*Panicum texanum* Buckl.	PANTE
Poinsettia, wild	*Euphorbia heterophylla* L.	EPHHL
Pusley, Florida	*Richardia scabra* L.	RCHSC
Ragweed, common	*Ambrosia artemisiifolia* L.	AMBEL
Senna, coffee	*Cassia occidentalis* L.	CASOC
Sesbania, hemp	*Sesbania exaltata* (Raf.) Rydb. ex A.W. Hill	SEBEX
Sicklepod	*Cassia obtusifolia* (L.) Irwin and Barneby	CASOB
Sida, prickly	*Sida spinosa* L.	SIDSP

REFERENCES

Baldwin, F. L. and Hall, S. (1993). University of Arkansas soybean weed control program SWC user's guide. Cooperative Extension Service, Little Rock, Arkansas.

Barbour, J. C. and Bridges, D. C. (1992). Modeling peanut yield losses due to weeds, *Proc. S. Weed Sci. Soc. 45*: 298.

Barker, M. A., Thompson, L., Jr., and Godley, F. M. (1984). Control of annual morningglories (*Ipomoea* spp.) in soybeans (*Glycine max*), *Weed Sci. 32*: 813–818.

Barnes, C. J., Goetz, A. J., and Lavy, T. L. (1989). Effects of imazaquin residues on cotton (*Gossypium hirsutum*), *Weed Sci. 37*: 820–824.

Barnes, J. P. and Putnam, A. R. (1983). Rye residues contribute weed control suppression in no-tillage cropping systems, *J. Chem. Ecol. 9*: 1045–1057.

Barrentine, W. L., Snipes, C. E., and Smeda, R. J. (1992). Herbicide resistance confirmed in johnsongrass biotypes, *Mississippi Agr. Forest. Exp. Sta. Res. Rep. 17*(5): 1–5.

Brecke, B. J. and Colvin, D. L. (1991). Weed management in peanuts, in *CRC Handbook of Pest Management in Agriculture*, Vol. 3, 2nd edition (D. Pimentel, ed.), CRC Press, Boca Raton, Florida, pp. 239–259.

Bridges, D. C. (1992). *Crop Losses Due to Weeds in Canada and the United States—1992,* Weed Science Society of America, Champaign, Illinois.

Bridges, D. C., Walker, R. H., McGuire, J. A., and Martin, N. R. (1984). Efficiency of chemical and mechanical methods for controlling weeds in peanuts (*Arachis hypogaea*), *Weed Sci. 32*: 584–591.

Brown, S. M., Dowler, C. C., Vencill, W. K., and Wilcut, J. W. (1993). Cotton and nutsedge response to Dual (metolachlor) herbicide in Georgia, *Proc. Beltwide Cotton Conf. 3*: 1520.

Brown, S. M. and Whitwell, T. (1985). Weed control programs for minimum-tillage cotton, *Weed Sci. 33*: 843–847.

Bruff, S. A. and Shaw, D. R. (1992). Tank-mix combinations for weed control in stale seedbed soybean (*Glycine max*), *Weed Tech. 6*: 45–51.

Bryson, C. T. (1989). Economic losses due to weeds in the southern states, *Proc. S. Weed Sci. Soc. 42*: 385–392.

Bryson, C. T. (1990). Differential growth of bermudagrass (*Cynodon dactylon*) biotypes in cotton (*Gossypium hirsutum*), *Weed Tech. 4*: 612–614.

Bryson, C. T. (1993). Population dynamics and seedbank of weeds in various cotton tillage production systems, *Proc. Beltwide Cotton Conf. 3*: 1521.

Bryson, C. T. and Keeley, P. E. (1992). Reduced-tillage systems, in *Weeds of Cotton: Characterization and Control* (C. G. McWhorter and J. R. Abernathy, eds.), The Cotton Foundation, Publisher, Memphis, Tennessee, pp. 323–363.

Buchanan, G. A., Hauser, E. W., and Street, J. (1978). Response of peanuts to dinitroaniline herbicides, *Proc. S. Weed Sci. Soc. 31*: 105–114.

Buchanan, G. A., Murray, D. S., and Hauser, E. W. (1982). Weeds and their control in peanuts, in *Peanut Science and Technology* (H. E. Pattee and C. T. Young, eds.), *American Peanut Research and Education Society,* Yoakum, Texas, pp. 209–249.

Buhler, D. D. (1992). Population dynamics and control of annual weeds in corn (*Zea mays*) as influenced by tillage systems, *Weed Sci. 40*: 241–248.

Buhler, D. D. and Burnside, O. C. (1984). Effect of application factors on postemergence phytotoxicity of fluazifop-butyl, haloxyfop methyl, and sethoxydim, *Weed Sci. 32*: 573–583.

Buhler, D. D. and Daniel, T. C. (1988). Influence of tillage systems on giant foxtail and velvetleaf density and control in corn, *Weed Sci. 36*: 642–647.

Byrd, J. D., Jr. (1991). Report of the 1990 cotton weed loss committee, *Proc. Beltwide Cotton Conf. 3*: 949.

Byrd, J. D., Jr. and Coble, H. D. (1991). Interference of selected weeds in cotton (*Gossypium hirsutum*), *Weed Tech. 5*: 263–269.

Byrd, J. D., Jr. and York, A. C. (1987a). Interactions of fluometuron and MSMA with fluazifop and sethoxydim, *Weed Sci. 35*: 270–276.

Byrd, J. D., Jr. and York, A. C. (1987b). Annual grass control in cotton (*Gossypium hirsutum*) with fluazifop, sethoxydim, and selected dinitroaniline herbicides, *Weed Sci. 35*: 388–394.

Cantwell, J. R., Liebl, R. A., and Slife, F. W. (1989). Imazethapyr for weed control in soybeans (*Glycine max*), *Weed Tech. 3*: 596–601.

Cardina, J., Mixon, A. C., and Wehtje, G. R. (1987). Low-cost weed control systems for close-row peanuts (*Arachis hypogaea*), *Weed Sci. 35*: 700–703.

Cardina, J. and Swann, C. W. (1988). Metolachlor effects on peanut growth and development, *Peanut Sci. 15*: 57–60.

Chandler, J. M. and Cooke, F. T. Jr. (1992). Economics of cotton losses caused by weeds. in *Weeds of Cotton: Characterization and Control* (C. G. McWhorter and J. R. Abernathy, eds.), The Cotton Foundation, Reference Book Series Number 2, Memphis, Tennessee.

Coble, H. D. and Byrd, J. D. Jr. (1992). Interference of weeds with cotton, in *Weeds of Cotton: Characterization and Control* (C. G. McWhorter and J. R. Abernathy, eds.), The Cotton Foundation, Reference Book Series Number 2, Memphis, Tennessee, pp. 73–82.

Coble, H. D. and Mortensen, D. (1992). The threshold concept and its application to weed science, *Weed Tech. 6*: 191–195.

Colvin, D. L., Wehtje, G. R., Patterson, M., and Walker, R. H. (1985a). Weed management in minimum-tillage peanuts (*Ararchis hypogaea*) as influenced by cultivar, row spacing, and herbicides, *Weed Sci. 33*: 233–237.

Colvin, D. L., Walker, R. H., Patterson, M. G., Wehtje, G., and McGuire, J. A. (1985b). Row pattern and weed management effects on peanut production, *Peanut Sci. 12*: 22–27.

Crawford, S. H. and Leake, K. D. (1992). Preplant weed control in conservation tillage cotton: approaches and options, *Proc. Beltwide Cotton Conf. 3*: 1319.

DeFelice, M. S., Brown, W. B., Aldrich, R. J., Sims, B. D., Judy, D. T., and Guethle, D. R. (1989). Weed control in soybeans (*Glycine max*) with reduced rates of postemergence herbicides, *Weed Sci. 37*: 365–374.

Derkens, D. A., Swanton, C. J., and Thomas, A. G. (1991). Weed community changes over time in reduced tillage systems, *Abstr. Weed Sci. Soc. Am. 31*: 120.

Doss, L. G. and York, A. C. (1995). Roundup ready soybean response to glyphosate and acetolactate synthase inhibitor combinations. *Proc. S. Weed Sci. Soc. 48*: (in press).

Eastin, E. F., Wilcut, J. W., and Richburg, J. S., III. (1995). Vegetable rotational response to Staple carryover from cotton, *Proc. Beltwide Cotton Conf.* (in press).

Elmore, C. D. and Williford, J. R. (1993). Tillage and cover crop effects on cotton yield and weed control, *Proc. Beltwide Cotton Conf. 3*: 1523.

Forney, D. R., Sebastian, S., Fader, G. M., and Ulrich, J. (1989). Development of suflonylurea-resistant soybeans and assoicated weed control options, *Weed Sci. Soc. Am. Abstr. 29*: 176.

Forcella, F. K., Oskoui, E., and Wagner, S. W. (1993). Application of weed seedbank ecology to low-input crop management, *Ecol. Appl. 3*: 74–83.

Gossett, B. J., Murdock, E. C., and Toler, J. E. (1992). Resistance of Palmer amaranth (*Amaranthus palmeri*) to the dinitroaniline herbicides, *Weed Tech. 6*: 587–591.

Green, J. M., Obrigawitch, T. T., Long, J. D., and Hutchinson, J. M. (1988). Metribuzin and chlorimuron mixtures for preemergence broadleaf weed control in soybeans (*Glycine max*), *Weed Tech. 2*: 355–363.

Gressel, J. and Segel, L. A. (1990). Modelling the effectiveness of herbicide rotations and mixtures as strategies to delay or preclude resistance, *Weed Tech. 4*: 186–198.

Grichar, W. J. (1991). Sethoxydim and broadleaf herbicide interaction effects on annual grass control in peanuts (*Arachis hypogaea*), *Weed Tech. 5*: 321–324.

Grichar, W. J. and Boswell, T. E. (1986). Postemergence grass control in peanut (*Arachis hypogaea*), *Weed Sci. 34*: 587–590.

Grichar, W. J. and Boswell, T. E. (1987). Comparison of no-tillage, minimum, and full-tillage cultural practices on peanuts, *Peanut Sci. 14*: 101–103.

Grichar, W. J. and Boswell, T. E. (1989). Bermudagrass (*Cynodon dactylon*) control with postemergence herbicides in peanut (*Arachis hypogaea*), *Weed Tech. 3*: 267–271.

Grichar, W. J., Colburn, A. E., and Bauman, P. A. (1991). Control of yellow nutsedge (*Cyperus esculentus*) with postemergence metolachlor applications, *Am. Peanut Res. Educ. Soc. 23*: 55.

Griffin, J. L. (1986). Wild poinsettia control in soybeans, *Proc. S. Weed Sci. Soc. 39*: 475.

Guthrie, D. S. and York, A. C. (1989). Cotton (*Gossypium hirsutum*) development and yield following fluometuron postemergence applied, *Weed Tech. 3*: 501–504.

Harrison, M. A. and Hayes, R. M. (1992). Utilization of Command with selected in-furrow insecticides in no-till and conventional cotton, *Proc. S. Weed Sci. Soc. 45*: 44.

Hauser, E. W., Dowler, C. C., Jellum, M. D., and Cecil, S. R. (1974). Effects of herbicide-crop rotation on nutsedge, annual weeds, and crops, *Weed Sci. 22*: 172–176.

Herbert, D. A., Jr., Wilcut, J. W., and Swann, C. W. (1991). Effects of various postemergence herbicide treatments and tobacco thrips (*Frankliniella fusca*) injury on peanut yields in Virginia, *Peanut Sci. 19*: 91–94.

Hicks, T. V., Wehtje, G. R., and Wilcut, J. W. (1990). Weed control in peanuts (*Arachis hypogaea*) with pyridate, *Weed Tech. 4*: 493–495.

Higgins, J. M., Whitwell, T., Murdock, E. C., and Toler, J. E. (1988). Recovery of pitted morningglory (*Ipomoea lacunosa*) and ivyleaf morningglory (*Ipomoea heder-*

aceae) following applications of acifluorfen, fomesafen, and lactofen, *Weed Sci. 36*: 345-353.

Holhouser, D. L. and Coble, H. D. (1990). Compatibility of sethoxydim with five postemergence broadleaf herbicides, *Weed Tech. 4*: 128–133.

Hurst, H. R. (1992). Cotton lay-by herbicides as wheat, vetch, and winter weeds as cover crops, *Proc. Beltwide Cotton Conf. 3*: 1308–1312.

Johnson, D. H., Talbert, R. E., Beaty, J. D., Grey, C. B., and Smith, R. J. (1992). Rice response following imazaquin, imazethapyr, chlorimuron, and chlomazone use, *Proc. S. Weed Sci. Soc. 45*: 371.

Jordan, D. L. and York, A. C. (1989). Effects of ammonium fertilizers and BCH 81508 S on antagonism with sethoxydim and bentazon mixtures, *Weed Tech. 3*: 450–454.

Jordan, D. L., McClelland, M. R., and Frans, R. E. (1992). MSMA effect on postemergence directed cotton herbicides, *Arkansas Farm Res. 41*(6): 6–7.

Jordan, D. L., Frans, R. E., and McClelland, M. R. (1993a). Cotton (*Gossypium hirsutum*) response to DPX-PE350 applied postemergence, *Weed Tech. 7*: 159–162.

Jordan, D. L., Frans, R. E., and McClelland, M. R. (1993b). Influence of application rate and timing on efficacy of DPX-PE350 applied postemergence, *Weed Tech. 7*: 216–219.

Jordan, D. L., McClelland, M. R., Frans, R. E., and Kendig, J. A. (1993c). Effect of MSMA on entireleaf morningglory (*Ipomoea hederacea* var. *integriuscula*) control with postemergence herbicides, *Weed Tech. 7*: 36–41.

Jordan, D. L., Smith, M. C., McClelland, M. R., and Frans, R. E. (1993d). Weed control with bromoxynil applied alone and with graminicides, *Weed Tech. 7*: 835–839.

Jordan, D. L., York, A. C., McClelland, M. R., and Frans, R. E. (1993e). Clomazone as a component in cotton (*Gossypium hirsutum*) herbicide programs, *Weed Tech. 7*: 202–211.

Jordan, D. L., Johnson, D. H., Johnson, W. G., Kendig, J. A., Frans, R. E., and Talbert, R. E. (1993f). Carryover of DPX-PE350 to grain sorghum (*Sorghum bicolor*) and soybean (*Glycine max*) on two Arkansas soils, *Weed Tech. 7*: 645–649.

Kapusta, G., Jackson, L. A., and Mason, D. S. (1986). Yield response of weed-free soybeans (*Glycine max*) to injury from postemergence broadleaf herbicides, *Weed Sci. 34*: 304–307.

Kapusta, G. and Krausz, R. F. (1993). Weed control and yield are equal in conventional, reduced-, and no-tillage soybean (*Glycine max*) after 11 years, *Weed Tech. 7*: 443–451.

Keeley, P., Thullen, R., Carter, L., and Chesson, J. (1992). Control of weeds in cotton with winter cover crops, *Proc. Beltwide Cotton Conf. 3*: 1304–1307.

Keeling, J. W., Henniger, C. G., and Abernathy, J. R. (1992). Integrated weed management systems for conservation tillage cotton, *Proc. Beltwide Cotton Conf. 3*: 1319.

Keeling, J. W., Siders, K. T., and Abernathy, J. R. (1991). Palmer amaranth (*Amaranthus palmeri*) control in a conservation tillage system for cotton (*Gossypium hirsutum*), *Weed Tech. 5*: 137–141.

Keisling, T. C., Oliver, L. R., Crowley, R. H., and Baldwin, F. L. (1984). Potential use of response surface analysis for weed management in soybean (*Glycine max*), *Weed Sci. 32*: 552–557.

Kells, J. J. and Wanamarta, G. (1987). Effect of adjuvant and spray volume on quackgrass (*Agropyron repens*) control with selective postemergence herbicides, *Weed Tech. 1*: 129–132.

Kendig, J. A. and Talbert, R. E. (1991). Wheat recovery from clomazone and norflurazon injury, *Proc. S. Weed Sci. Soc. 44*: 352.

Ketchersid, M. L., Boswell, T. E., and Merkle, M. G. (1978). Effects of 2,4-DB on yield and pod development in peanuts, *Peanut Sci. 5*: 35–39.

King, C. A. and Oliver, L. R. (1992). Application rate and timing of acifluorfen, bentazon, chlorimuron, and imazaquin, *Weed Tech. 6*: 526–534.

Klingaman, T. E., King, C. A., and Oliver, L. R. (1992). Effect of application rate, weed species, and weed stage of growth on imazethapyr activity, *Weed Sci. 40*: 227–232.

Knauft, D. A., Colvin, D. L., and Gorbet, D. W. (1990). Effect of paraquat on yield and market grade of peanut (*Arachis hypogaea*) genotypes, *Weed Tech. 4*: 866–870.

Krausz, R. F., Kapusta, G., and Krake, E. L. (1992). Soybean (*Glycine max*) and rotational crop tolerance to chlorimuron, clomazone, imazaquin, and imazethapyr, *Weed Tech. 6*: 77–80.

Lanie, A. J., Griffin, J. L., Vidrine, P. R., and Reynolds, D. B. (1994a). Herbicide combinations for soybean (*Glycine max*) planted in stale seedbed, *Weed Tech. 8*: 17–22.

Lanie, A. J., Griffin, J. L., Vidrine, P. R., and Reynolds, D. B. (1994b). Weed control with non-selective herbicides in soybean (*Glycine max*) stale seedbed culture, *Weed Tech. 8*: 159–164.

Lewis, W. M. (1985). Weed control in reduced-tillage soybean production. In *Weed Control in Limited-Tillage Systems* (A. F. Wiese, ed.), Weed Science Society of America, Monograph 2, Champaign, Illinois, pp. 41–50.

Liebl, R. A., Simmons, F. W., Wax, L. M., and Stoller, E. W. (1992). Effect of rye (*Secale cereale*) mulch on weed control and soil moisture in soybean (*Glycine max*), *Weed Tech. 6*: 838–846.

Liebl, R. A. and Worsham, A. D. (1983). Inhibition of pitted morningglory and certain other weed species by phytotoxic components of wheat straw, *J. Chem. Weed Ecol. 9*: 1027–1043.

Lueschen, W. E. and Hoverstad, T. R. (1991). Imazethapyr for weed control in no-till soybean (*Glycine max*), *Weed Tech. 5*: 845–851.

McGee, J. W., Buchert, K. P., and Walton, L. C. (1992). Results from fallowbed applications of Goal 1.6E herbicide in mid-south cotton, *Proc. Beltwide Cotton Conf. 3*: 1303.

McClelland, M. R., Jordan, D. L., Frans, R. E., and Keisling, T. C. (1992). Weed management in minimum tillage cotton, *Proc. S. Weed Sci. Soc. 45*: 31.

McLean, H. S., Richburg, J. S., III, and Wilcut, J. W. (1993). Zorial interactions with Cotoran or Karmex for broadleaf weed control in Georgia cotton, *Proc. Beltwide Cotton Conf. 3*: 1528.

McLean, H. S., Wilcut, J. W., and Richburg, J. S. III. (1994). Norflurazon and metolachlor systems for weed management in Southeastern peanut, *Proc. S. Weed Sci. Soc. 47*: 228–229.

Miller, D. K. and Griffin, J. L. (1994). Comparison of herbicide programs and cultivation for sicklepod (*Cassia obtusifolia*) control in soybean (*Glycine max*), *Weed Tech. 8*: 77–82.

Mills, J. A. and Witt, W. W. (1989). Efficacy, phytotoxicity, and persistence of imazaquin, imazethapyr, and clomazone in no-till double-crop soybean (*Glycine max*), *Weed Sci. 37*: 353–359.

Monks, C. D. and Banks, P. A. (1991). Rotational crop response to chlorimuron, clomazone, and imazaquin applied the previous year, *Weed Sci. 39*: 629–633.

Monks, C. D., Wilcut, J. W., and Richburg, J. S., III. (1993). Broadleaf weed control in soybean (*Glycine max*) with chlorimuron plus acifluorfen or thifensulfuron mixtures, *Weed Tech. 7*: 317–321.

Moseley, C. M. and Hagood, E. S., Jr. (1990). Reducing herbicide inputs when establishing no-till soybeans (*Glycine max*), *Weed Tech. 4*: 14–19.

Moseley, C. M. and Hagood, E. S., Jr. (1991). Decreasing rates of nonselective herbicides in double-crop no-till soybeans (*Glycine max*), *Weed Tech. 5*: 198–201.

Nastasi, P., Frans, R. E., and McClelland, M. R. (1986). Economics and new alternatives in cotton (*Gossypium hirsutum*) weed management programs, *Weed Sci. 34*: 634–638.

Oliver, L. R., Klingaman, T. E., McClelland, M., and Bozsa, R. C. (1993). Herbicide systems in stale seedbed soybean (*Glycine max*) production, *Weed Tech. 7*: 816–823.

Perkins, H. H., Jr., Ethridge, D. E., and Bragg, C. K. (1984). Fiber, in *Cotton* (R. J. Kohel and C. F. Lewis, eds.), Agron. Monograph 24. Amer. Soc. Agron. Madison, Wisconsin, 437–509.

Porter, D. M., Smith, D. H., and Rodriquez-Kabana, R. (1982). Peanut plant diseases, in *Peanut Science and Technology* (H. E. Pattee and C. T. Young, eds.), American Peanut Research and Education Society, Yoakum, Texas, 326–410.

Poston, D. H., Murdock, E. C., and Toler, J. E. (1992). Cost-efficient weed control in soybean (*Glycine max*) with cultivation and banded herbicide applications, *Weed Tech. 6*: 990–995.

Renner, K. A., Meggitt, W. F., and Penner, D. (1988). Effect of soil pH on imazaquin and imazethapyr adsorption to soil and phytotoxicity to corn (*Zea mays*), *Weed Sci. 36*: 78–83.

Richburg, J. S., III, Wilcut, J. W., and Wehtje, G. R. (1993). Toxicity of foliar and/or soil applied imazethapyr to purple and yellow nutsedge (*Cyperus rotundus* and *C. esculentus*), *Weed Tech. 7*: 900–905.

Richburg, J. S., III, Wilcut, J. W., and Wehtje, G. R. (1994). Toxicity of AC 263,222 to purple (*Cyperus rotundus*) and yellow nutsedge (*C. esculentus*), *Weed Sci. 42*: 398–402.

Risley, M. A. and Oliver, L. R. (1991). Efficacy of imazaquin on various weed species, *Weed Sci. 39*: 243–250.

Rogers, N. K., Buchanan, G. A., and Johnson, W. C. (1976). Influence of row spacing on weed competition with cotton, *Weed Sci. 24*: 410–413.

Shaw, D. R., Newsom, L. J., and Smith, C. A. (1991). Influence of cultivation timing on chemical control of sicklepod (*Cassia obtusifolia*) in soybean (*Glycine max*), *Weed Sci. 39*: 67–72.

Sherman, M. E., Thompson, L., Jr., and Wilkinson, R. E. (1983). Sicklepod (*Cassia obtusifolia*) management in soybeans (*Glycine max*), *Weed Sci. 31*: 622–627.

Shilling, D. G., Liebl, R. A., and Worsham, A. D. (1985). Rye and wheat mulch: the suppression of certain broadleaf weeds and the isolation and identification of phytotoxins, in *The Chemistry of Allelopathy* (A. C. Thompson, ed.), ACS Symp. Ser. No. 268. Amer. Chem. Soc. Washington, D.C.

Sims, B. D. and Guethle, D. R. (1992). Herbicide programs in no-tillage and conventional-tillage soybeans (*Glycine max*) double cropped after wheat (*Triticum aestivum*), *Weed Sci. 40*: 255–263.

Smeda, R. J. (1993). Potential of spring-seeded sorghum-Sudangrass for weed management in soybean, *Proc. S. Weed Sci. Soc. 46*: 57.

Smith, M. C., Bagwell, R. D., Tugwell, N. P., and Frans, R. E. (1994). Efficacy of clomazone with disulfoton and aldicarb, *Proc. Beltwide Cotton Conf. 3*: 1685–1688.

Snipes, C. E. (1991). Preemergence velvetleaf (*Abutilon theophrasti*) control in cotton (*Gossypium hirsutum*), *Weed Tech. 5*: 142–146.

Snipes, C. E., Colvin, D. L., Patterson, M. G., and Crawford, S. H. (1992). Cotton (*Gossypium hirsutum*) yield response to cultivation timing and frequency, *Weed Tech. 6*: 31–35.

Snipes, C. E. and Byrd, J. D., Jr. (1994). Influence of fluometuron and MSMA on cotton yield and fruiting characteristics, *Weed Sci. 42*: 210–215.

Snipes, C. E. and Mueller, T. C. (1992). Cotton (*Gossypium hirsutum*) yield response to mechanical and chemical weed control systems, *Weed Sci. 40*: 249–254.

Snipes, C. E., Walker, R. H., Whitwell, T., Buchanan, G. A., McGuire, J. A., and Martin, N. R. (1984). Efficacy and economics of weed control methods in cotton (*Gossypium hirsutum*), *Weed Sci. 32*: 95–100.

Stainforth, D. W. and Wiese, A. F. (1985). Weed biology and its relationship to weed control in limited-tillage systems, in *Weed Control in Limited-Tillage Systems* (F. Wiese, ed.), Weed Science Society of America, Monograph Number 2, Champaign, Illinois, pp. 15–25.

Steckel, L. E., DeFelice, M. S., and Sims, B. D. (1990). Integrating reduced rates of postemergence herbicides and cultivation for broadleaf weed control in soybeans (*Glycine max*), *Weed Sci. 38*: 541–545.

Stoller, E. W., Harrison, S. K., Wax, L. M., Regnier, E. R., and Nafziger, E. D. (1987). Weed interference in soybeans (*Glycine max*), *Rev. Weed Sci. 3*: 155–181.

Supka, J. R., Anderson, C. G., and Mayfield, W. D. (1992). Trends in cotton production: history, culture, mechanization, and economics. in *Weeds of Cotton: Characterization and Control* (C. G. McWhorter and J. R. Abernathy, eds.), The Cotton Foundation Reference Book, Series Number 2, Memphis, Tennessee, pp. 9–41.

Triplett, G. B., Jr. and Lyth, G. D. (1972). Control and ecology of weeds in continuous grown corn without tillage, *Weed Sci. 20*: 453–457.

Vidrine, P. R. (1989). Johnsongrass (*Sorghum halepense*) control in soybeans (*Glycine max*) with postemergence herbicides, *Weed Tech. 3*: 455–458.

Walker, R. H. (1982). Crop manipulation in integrated weed management systems, *Weed Sci. 30*(Suppl. 1): 17–24.

Walker, H. L. and Riley, J. A. (1982). Evaluation of *Alternaria cassiae* for biocontrol of sicklepod (*Cassia obtusifolia*), *Weed Sci. 30*: 651–654.

Wehtje, G. R., McGuire, J. A., Walker, R. H., and Patterson, M. G. (1986). Texas panicum (*Panicum texanum*) control in peanuts (*Arachis hypogaea*) with paraquat, *Weed Sci. 34*: 308–311.

Wehtje, G., Walker, R. H., Patterson, M. G., and McGuire, J. A. (1984). Influence of twin rows on yield and weed control in peanuts, *Peanut Sci. 11*: 88–91.

Wehtje, G. R., Wilcut, J., and Hicks, V. (1988). Relative tolerance of peanuts to alachlor and metolachlor, *Peanut Sci. 15*: 53–56.

Wehtje, G. R., Wilcut, J. W., McGuire, J. A., and Hicks, T. V. (1991). Foliar penetration and phytotoxicity of paraquat as influenced by peanut cultivar, *Peanut Sci. 18*: 67–71.

Wehtje, G. R., Wilcut, J. W., and McGuire, J. A. (1992). Influence of bentazon on the phytotoxicity of paraquat to peanuts (*Arachis hypogaea*) and associated weeds, *Weed Sci. 40*: 90–95.

Wehtje, G. R., Wilcut, J. W., and McGuire, J. A. (1993). Absorption, translocation, and phytotoxicity of chlorimuron and 2,4-DB mixtures in peanut (*Arachis hypogaea*) and selected weed species, *Weed Sci. 41*: 347–352.

Wesley, R. A., Jr., Shaw, D. R., and Barrentine, W. L. (1989a). Application timing of metribuzin, chlorimuron, and imazaquin for common cocklebur (*Xanthium strumarium*) control, *Weed Tech. 3*: 364–368.

Wesley, R. A., Jr., Shaw, D. R., and Barrentine, W. L. (1989b). Incorporation depths of imazaquin, metribuzin, and chlorimuron for common cocklebur (*Xanthium strumarium*) control in soybeans (*Glycine max*), *Weed Sci. 37*: 596–599.

Wesley, M. T. and Shaw, D. R. (1992). Interactions of diphenylether herbicides with chlorimuron and imazaquin, *Weed Tech. 6*: 345–351.

Westberg, D. E. and Coble, H. D. (1992). Effect of acifluorfen on the absorption, translocation, and metabolism of chlorimuron in certain weeds, *Weed Tech. 6*: 4–12.

Westberg, D. E., Oliver, L. R., and Frans, R. E. (1989). Weed control with clomazone alone and with other herbicides, *Weed Tech. 3*: 678–685.

White, R. H. and Worsham, A. D. (1990). Control of legume cover crops in no-till corn (*Zea mays*) and cotton (*Gossypium hirsutum*), *Weed Tech. 4*: 57–62.

Wilcut, J. W., Wehtje, G. R., and Patterson, M. G. (1987a). Economic assessment of weed control systems for peanuts (*Arachis hypogaea*), *Weed Sci. 35*: 433–437.

Wilcut, J. W., Wehtje, G. R., and Walker, R. H. (1987b). Economics of weed control in peanuts (*Arachis hypogaea*) with herbicides and cultivations, *Weed Sci. 35*: 711–715.

Wilcut, J. W., Wehtje, G. R., Colvin, D. L., and Patterson, M. G. (1987c). Economic assessment of herbicide systems for minimum-tillage peanuts, *Peanut Sci. 14*: 83–86.

Wilcut, J. W., Wehtje, G. R., Cole, T. A., Hicks, T. V., and McGuire, J. A. (1989). Postemergence weed control systems without dinoseb for peanuts (*Arachis hypogaea*), *Weed Sci. 37*: 385–391.

Wilcut, J. W. and Swann, C. W. (1990). Timing of paraquat applications for weed control in Virginia-type peanuts (*Arachis hypogaea*), *Weed Sci. 38*: 558–562.

Wilcut, J. W., Wehtje, G. R., and Cole, T. A. (1990a). Evaluation of herbicide systems in minimum- and conventional-tillage peanuts, *Weed Sci. 38*: 243–248.

Wilcut, J. W., Wehtje, G. R., Hicks, T. V., and Cole, T. A. (1990b). Postemergence weed management systems for peanuts, *Weed Tech. 4*: 76–80.

Wilcut, J. W. (1991a). Efficacy and economics of common bermudagrass (*Cynodon dactylon*) control in peanut (*Arachis hypogaea*), *Peanut Sci. 18*: 106–109.

Wilcut, J. W. (1991b). Tropic croton (*Croton glandulosus*) control in peanut (*Arachis hypogaea*), *Weed Tech. 5*: 795–798.

Wilcut, J. W., Walls, F. R., Jr., and Horton, D. N. (1991a). Imazethapyr for broadleaf weed control in peanuts (*Arachis hypogaea*), *Peanut Sci. 18*: 26–30.

Wilcut, J. W., Walls, F. R., Jr., and Horton, D. N. (1991b). Weed control, yield, and net returns using imazethapyr in peanuts (*Arachis hypogaea*), *Weed Sci. 39*: 238–242.

Wilcut, J. W., Coble, H. D., Murdock, E. C., and Snipes, C. E. (1992). Weed response to bromoxynil in Georgia, North and South Carolina, and Mississippi cotton, *Proc. Beltwide Cotton Conf. 3*: 1317.

Wilcut, J. W., York, A. C., and Jordan, D. L. (1993). Weed management in reduced-tillage southeastern cotton, in *Conservation Tillage Systems for Cotton 1993: A Review of Research and Demonstration Results from Across the Cotton Belt* (M. McClelland, T. Valco, and R. E. Frans eds.), Arkansas Agricultural Experiment Station Special Rep. 160 and National Cotton Council, Memphis, Tennessee, pp. 29–35.

Wilcut, J. W. and Richburg, J. S., III. (1994a). Reduced tillage weed management in Georgia cotton, *Proc. Beltwide Cotton Conf. 3*: 1699.

Wilcut, J. W. and Richburg, J. S., III. (1994b). An evaluation of clomazone systems in Georgia cotton, *Proc. Beltwide Cotton Conf. 3*: 1702.

Wilcut, J. W., York, A. C., and Wehtje, G. R. (1994a). Interaction and control of weeds in peanut (*Arachis hypogaea*), *Rev. Weed Sci. 6*: 177-205.

Wilcut, J. W., Richburg, J. S., III, Eastin, E. F., Wiley, G. R., Walls, F. R. Jr., and Newell, S. (1994b). Imazethapyr and paraquat systems for weed management in peanut (*Arachis hypogaea*), *Weed Sci. 42*: 601–607.

Wilcut, J. W., Richburg, J. S., III, Wiley, G., Walls, F. R., Jr., Jones, S. R., and Iverson, M. J. (1994c). Imidazolinone herbicide systems for peanut (*Arachis hypogaea* L.), *Peanut Sci. 21*: 23–28.

Wilcut, J. W., Coble, H. D., York, A. C., and Monks, D. W. (1995). The niche for herbicide resistant crops in U.S. agriculture, in *Herbicide Resistant Crops: Agricultural, Environmental, Economic, Regulatory, and Technical Aspects* (S. O. Duke, ed.), Marcel Dekker, Inc., New York, (in press).

Wiles, L. J., Wilkerson, G. G., Gold, H. J., and Coble, H. D. (1992a). Modeling weed distribution for improved postemergence control systems, *Weed Sci.* 546–553.

Wiles, L. J., Oliver, G. W., York, A. C., Gold, H. J., and Wilkerson, G. G. (1992b). Spatial distribution of broadleaf weeds in North Carolina soybean (*Glycine max*) fields, *Weed Sci. 40*: 554–557.

Wilkinson, R. E. (1988). Carbamothiates, in *Herbicides: Chemistry, Degradation, and Mode of Action*. Vol. 3 (P. C. Kearney and D. D. Kaufman, eds.), Marcel Dekker, Inc., New York, pp. 245–300.

Wilkerson, G. G., Modena, S. A., and Coble, H. D. (1991). HERB: Decision model for postemergence weed control in soybeans, *Agron. J. 83*: 413–417.

Wilkinson, R. E. and Hardcastle, W. S. (1969). EPTC effects on sicklepod petiolar fatty acids, *Weed Sci. 17*: 335–338.

Willard, T. S. and Griffin, J. L. (1993). Soybean (*glycine max*) yield and quality responses associated with wild poinsettia (*Euphorbia heterophylla*) control programs, *Weed Tech. 7*: 118–122.

Wilson, H. P., Hines, T. E., Hatzios, K. K., and Doub, J. P. (1988). Efficacy comparisons of alachlor and metolachlor formulations in the field, *Weed Tech. 2*: 24–27.

Wilson, J. S. and Worsham, A. D. (1988). Combinations of nonselective herbicides for difficult to control weeds in no-till corn, *Zea mays*, and soybeans, *Glycine max*, *Weed Sci. 36*: 648–652.

Witt, W. W. (1991). Impact of continuous no-tillage on weed management, *Proc. S. Weed Sci. Soc. 44*: 70–72.

Worsham, A. D. (1985). No-till tobacco (*Nicotiana tabacum*) and peanuts (*Arachis hypogaea*), in *Weed Control in Limited-Tillage Systems* (A. F. Wiese, ed.), Weed Science Society of America, Champaign, Illinois, pp. 101–126.

Worsham, A. D. (1991). Allelopathic cover crops to reduce herbicide input, *Proc. S. Weed Sci. Soc. 44*: 58–69.

Wrubel, R. P. and Gressel, J. (1994). Are herbicide mixtures useful for delaying the rapid evolution of resistance? A case study, *Weed Tech. 8*: 635–648.

Yelverton, F. L. and Coble, H. D. (1991). Narrow row spacing and canopy formation reduces weed resurgence in soybean (*Glycine max*), *Weed Tech. 5*: 169–174.

York, A. C. and Tucker, M. R. (1985). Nitrification inhibitor evaluation in cotton. II. Inhibitor incorporation depth and N placement. *Agron. J. 77*: 407–411.

York, A. C., Jordan, D. L., and Wilcut, J. W. (1990). Effects of $(NH_4)_2SO_4$ and BCH 81508 S on efficacy of sethoxydim, *Weed Tech. 4*: 76–80.

York, A. C., Jordan, D. L., and Frans, R. E. (1991a). Insecticides modify cotton (*Gossypium hirsutum*) response to clomazone, *Weed Tech. 6*: 796–800.

York, A. C., Wilcut, J. W., Keene, M. M., and Walls, F. R., Jr. (1991b). Soybean (*Glycine max*) response to postemergence herbicide mixtures containing 2,4-DB, *Weed Tech. 5*: 43–47.

York, A. C. and Worsham, A. D. (1992). Weed management systems in no-till vs conventional cotton, *Proc. S. Weed Sci. Soc. 45*: 33–34.

York, A. C. (1994). Nutsedge management in cotton, *Proc. Beltwide Cotton Conf. 3*: 1698.

York, A. C., Wilcut, J. W., Swann, C. W., Jordan, D. L., and Walls, F. R., Jr. (1995). Efficacy of imazethapyr in peanut (*Arachis hypogaea*) as affected by time of application, *Weed Sci. 43*: (in press).

York, A. C. and Wilcut, J. W. (1995). Potential for cadre and pursuit applied to peanuts to carry over to cotton in North Carolina and Georgia, *Proc. Beltwide Cotton Conf.* (in press).

Zimdahl, R. L. (1980). *Weed Crop Competition: A Review*, International Plant Protection Center, Oregon State Univ., Corvallis.

Zimdahl, R. L. (1991). *Weed Science: A Plea for Thought*, United States Department of Agriculture-Cooperative State Research Service, Fort Collins, Colorado.

11

Weed Management Systems for Grain Crops

William W. Donald
U.S. Department of Agriculture, Columbia, Missouri

E. Ford Eastin
University of Georgia, Tifton, Georgia

I. CEREAL PRODUCTION PRACTICES

Weed management systems in cereals are designed to create conditions which allow cereals to reach their potential yield within limits imposed by cereal genotype and environment. Managing weeds minimizes yield losses due to weeds rather than increasing the theoretical yield potential. While entomologists and plant pathologists have developed pest management strategies for individual pests, farmers and weed scientists have been concerned with developing weed management for several grass and broadleaf weed species at a time. Current weed control measures were developed under the assumption that controlling weeds maximizes yield. However, maximizing yields may not always maximize profits because profitability of cereal production depends on fluctuating crop price, input costs, and weed density and distribution, among other factors. Currently used weed control methods may sometimes conflict with other goals, such as maximizing farmer profit, minimizing environmental degradation, or managing other crop pests. Nevertheless, some farming systems which were developed to minimize soil erosion and improve water quality, such as no-tillage, would not be feasible without herbicides for weed management.

Weed control methods used in cereals are restricted by cereal production practices. Dryland cereal production practices have been summarized (Cook and Veseth, 1991; Donald, 1990b; Dregne and Willis, 1983) and must be understood to implement weed control measures fully. Winter and spring wheat production practices, weeds, and weed control measures differ in several respects and have been reviewed for various regions in North America (Nalewaja and Arnold, 1970; Donald, 1990b). Weed management in conservation (no-till and reduced till) tillage systems also was reviewed for winter wheat (Wicks, 1985b) and spring wheat (Donald and Nalewaja, 1990).

The relative proportion of different cereals produced in the United States is presented in Figure 1. Cereals are grown in regions which are unsuitable for growing other, more profitable field crops, such as corn or soybean. Most spring-sown barley, durum wheat, and oat are produced like spring wheat in the same geographic regions (Donald, 1990b; Donald and Nalewaja, 1990) and have many common weed problems (Fay, 1990), although fewer herbicides* are registered for these crops than for spring wheat (Table 1). Production practices were reviewed for wheat (Donald, 1990b; Donald and Nalewaja, 1990), barley (Baldridge et al., 1985; Briggs, 1978), and oat (Helm and McMullen, 1989). Spring barley is grown in most of the United States, but some spring barley is sown in fall or winter in the Southwest and harvested in spring (Baldridge et al., 1976). More winter barley is grown in the South and Southeast than elsewhere.

II. WEED MANAGEMENT IN WINTER WHEAT

A. Weeds of Winter Wheat

Different weeds infest the diverse regions where winter wheat is grown, according to expert opinion (Appleby and Morrow, 1990; Banks, 1990; Mitich and Kyser, 1990a and b; Peeper and Wiese, 1990; Peters, 1990; Wicks and Smika, 1990). Weed problems differ between various cereal producing regions of North America (Fay, 1990), and many winter annual weeds infesting winter wheat are not troublesome where spring wheat, barley, and oat are grown. In the United States estimated dollar losses due to weeds in winter wheat were summarized by region on the basis of expert opinion (Bridges, 1992). Weeds of wheat that were intentionally or unintentionally introduced into North America and major routes of introduction have been summarized (Donald, 1990c). Many of the

*Since herbicide registrations and labels are constantly changing, current labels and information available through state agricultural extension service and state agricultural experiment stations should be consulted for the most recent information and recommendations.

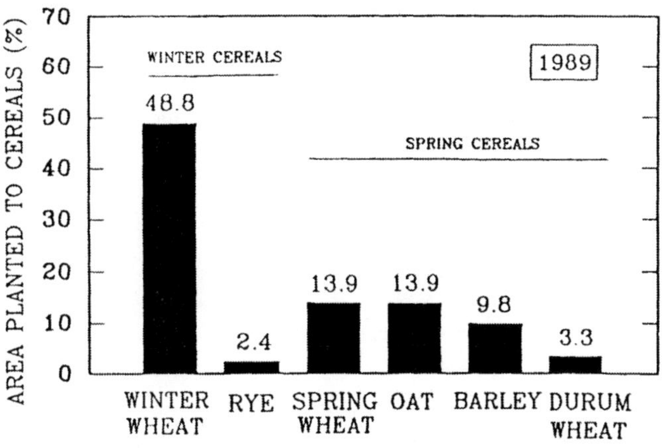

Figure 1 The relative proportion of acreage planted to small grains in the United States in 1989. (From Agricultural Statistics, 1990.)

most troublesome weeds infesting cereals in North America were intentionally introduced as herbs, as ornamentals, or for forage by settlers, whereas others were accidentally introduced in imported seed, hay, livestock, and ship ballast. The distribution and spread of downy brome (Douglas et al., 1990; Mack, 1981; Morrow and Stahlman, 1984) and jointed goatgrass (Donald and Ogg, 1991) have been reconstructed from herbaria and expert opinion. But information on the rate of spread and avenues of spread of weeds in North America is limited.

Winter annual grass weeds, such as cheat, downy brome, Italian ryegrass, jointed goatgrass, and volunteer rye, are more difficult to manage in winter wheat than are winter annual broadleaf weeds (Wiese, 1983). Fewer effective, selective herbicides are registered for grass weed management in winter wheat than for broadleaf weed management (Table 1). Most summer annual broadleaf weeds, such as kochia or Russian thistle, which emerge in spring do not concern winter wheat producers if cereal canopy closure is complete early in spring. Wild buckwheat is an exception because it emerges before winter wheat canopy closure in some parts of the Great Plains.* When disease, insects, winter kill, drought, or, less commonly, carryover herbicide residues thin winter wheat growth or stands, weeds can grow and fill in gaps. Late weed growth in gaps may not reduce yield directly but can interfere with harvesting, reduce grain quality, and result in dockage.

Summer annual weeds are best adapted to the spring wheat life cycle (Cook and Veseth, 1991; Donald and Nalewaja, 1990). Summer annual grass weeds,

*Dr. T. Peeper, 1993. Oklahoma State Univ., personal communication.

Table 1 Herbicides and Prepackaged Herbicide Mixtures Registered for Use in Small Grains by the United States Environmental Protection Agency in 1993[a]

Application time	Herbicide	Winter wheat	Winter barley	Spring wheat	Durum wheat	Spring barley	Oat	Rye	Sorghum
Early preplant	Alachlor + atrazine								X
	Atrazine + metolachlor								X
	Cyanazine + (2,4-D or paraquat; atrazine; metolachlor; alachlor)	X							X
	Glyphosate	X	X	X		X	X		X
	Glyphosate + dicamba	X	X	X		X	X		X
	Glyphosate + 2,4-D	X	X	X	X	X	X	X	X
	Metribuzin	X							
	Picloram	X	X	X	X	X	X		
Early preplant incorporated	Triallate (fall)			X	X	X			
Preplant	Alachlor + atrazine								X
	Atrazine								X
	Atrazine + cyanazine								X
	Atrazine + metolachlor								X
	Cyanazine								X
	Cyanazine + (alachlor, atrazine, metolachlor, propachlor, or propazine)								X
	Dicamba	X	X	X		X	X		X
	Glyphosate	X	X	X		X	X		X
	Metolachlor								X
	Paraquat	X	X	X		X			X

Preplant incorporated	Alachlor								X
	Alachlor + atrazine								X
	Atrazine								X
	Atrazine + metolachlor								X
	Cyanazine + (alachlor, atrazine, metolachlor, propachlor, or propazine)								X
	Diclofop	X							
	Metolachlor								X
	Triallate	X	X	X	X	X			
	Trifluralin	X				X			
	Triallate + trifluralin	X	X	X	X	X			
Postplant incorporated	Triallate	X	X	X	X	X			
	Trifluralin	X	X	X	X	X			
	Triallate + trifluralin			X	X	X			
Preemergence	Alachlor								X
	Alachlor + atrazine								X
	Alachlor + glyphosate								X
	Atrazine								X
	Atrazine + metolachlor								X
	Atrazine + propachlor								X
	Chlorsulfuron	X						X	
	Chlorsulfuron + metsulfuron	X							
	Cyanazine + (alachlor, atrazine, metolachlor, propachlor, or propazine)								X
	Dicamba	X	X	X		X	X		
	Diclofop	X							
	Diuron	X	X				X		X

(*continued*)

Table 1 Herbicides and Prepackaged Herbicide Mixtures Registered for Use in Small Grains by the United States Environmental Protection Agency in 1993[a]

Application time	Herbicide	Winter wheat	Winter barley	Spring wheat	Durum wheat	Spring barley	Oat	Rye	Sorghum
	Glyphosate	X	X	X		X	X		X
	Linuron								X
	Metolachlor								X
	Paraquat	X	X	X		X	X		X
	Propachlor								X
	Triallate	X	X	X		X			
Postemergence	Alachlor + atrazine (surface applied)								X
	Atrazine								X
	Atrazine + bentazon								X
	Atrazine + bromoxynil								X
	Atrazine + dicamba								X
	Atrazine + propachlor								X
	Bentazon								X
	Bromoxynil	X	X	X		X	X	X	X
	Bromoxynil + MCPA	X	X	X		X	X	X	
	Chlorsulfuron	X	X	X	X	X	X		
	Chlorsulfuron + metsulfuron	X	X			X	X		
	Clopyralid	X	X	X		X	X		
	Clopyralid + MCPA	X	X	X		X	X		
	Dicamba	X	X	X		X	X		X
	Diclofop	X	X	X	X	X			

	Difenzoquat	X	X	X	X	X			
	Diuron	X	X				X		X
	Fenoxaprop + MCPA		X						
	Fenoxaprop + 2,4-D + MCPA			X					
	Fenoxaprop + MCPA + Thifensulfuron + tribenuron	X		X					
	Imazamethabenz	X	X	X	X	X			
	Linuron (directed)								X
	MCPA	X	X	X	X	X	X	X	
	Metribuzin	X	X			X			
	Metsulfuron	X	X	X	X	X			
	Picloram	X	X	X		X	X		
	Propanil			X	X	X	X		
	2,4-D	X	X	X		X	X	X	X
	Triasulfuron	X	X	X	X	X			
	Tribenuron	X	X	X	X	X			
	Thifensulfuron + tribenuron	X	X	X	X	X			
Postemergence incorporated (layby treatment)	Pendimethalin								X
	Trifluralin								X
Preharvest	Diquat (dessicant)								X
	2,4-D	X	X	X		X	X	X	

[a]Other combinations of these individual herbicides or prepackaged herbicide mixtures are registered uses; consult the registration label. Registered uses may be restricted to particular states or parts of states. Consult the current registration label before use.

such as wild oat and foxtail species, concern spring wheat producers (Donald and Nalewaja, 1990; Fay, 1990; Hunter et al., 1990) more than summer annual broadleaf weeds do because there are several effective, registered postemergence herbicides for broadleaf weed management in spring wheat. Winter annual weeds are generally not a problem in spring wheat because mechanical seedbed preparation in spring usually controls small emerged winter annuals. Perennial weeds, such as Canada thistle or field bindweed, can reduce wheat yields in both spring- and fall-sown cereals but are usually localized problems (see later discussion).

B. Weed Competition, Yield Losses, and Reduced Grain Quality

Competition of weeds with winter wheat has been reviewed (Zimdahl, 1980; Zimdahl, 1990) (Table 2). In general, those broadleaf and grass weeds which have a winter annual or perennial growth habit compete more effectively with winter wheat than summer annual weeds (Peeper and Wiese, 1990; Wicks and Smika, 1990; Zimdahl, 1990). However, the extent to which weeds reduce crop yield is not well enough understood to predict crop yield loss for any small grain or sorghum accurately. Information on the extent to which weeds limit crop yield and profitability also is not "packaged" in a form that cereal farmers can use to make economic choices between alternative weed management strategies and, consequently, has had little impact on year-to-year weed management decisions of farmers.

Weeds limit cereal yield potential in arid regions partially because they increase evapotranspiration (Greb, 1983) and compete with winter wheat for limited soil moisture reserves stored in the soil profile (Cook and Veseth, 1991; Dregne and Willis, 1983). In regions where the winter wheat–fallow rotation is practiced for soil moisture conservation (Donald and Nalewaja, 1990; Dregne and Willis, 1983; Peeper and Wiese, 1990; Wicks and Smika, 1990), weeds growing in fallow can deplete soil moisture reserves before winter wheat planting and limit the potential yield of the fall-planted winter wheat crop, even if weeds are well controlled in winter wheat itself. Consequently, weeds must be controlled during the entire winter wheat–fallow rotation to minimize depletion of limited soil moisture reserves and subsequent crop yield potential. Weeds also reduce soil moisture reserves before planting dryland sorghum in spring in 2-year sorghum-fallow or 3-year winter wheat–fallow–sorghum rotations.

While weed scientists have been most concerned with relating weed density to reduced crop yield, weed dockage (i.e., chiefly green residue and weed seeds) may also limit cereal grain quality as reflected in grain grading standards. For example, U.S. Department of Agriculture grain grading standards stipu-

Table 2 Selected Studies Documenting the Impact of Density of Individual Weeds on the Yield of Small Grains, Rice, and Sorghum

Crop	Weed	Location	Reference
Winter wheat	Blue mustard	Texas	Cited by Zimdahl, 1980
	Downy brome		Cited by Zimdahl, 1980
	Fiddleneck	Oregon	Cited by Wiese, 1983
	Rigid and Italian ryegrass	Pacific Northwest	Cited by Zimdahl, 1980
	Tansymustard	Texas	Cited by Wiese, 1983
Spring wheat	Canada thistle	North Dakota	Donald and Khan, 1992
	Green foxtail		Cited by Zimdahl, 1980
	Quackgrass		Cited by Zimdahl, 1980
	Wild buckwheat	Canada	Holm and Kirkland, 1986; cited by Zimdahl, 1980
	Wild mustard	Canada	Holm and Kirkland,1986; cited by Zimdahl, 1980; Wiese, 1983
		North Dakota	Cited by Wiese, 1983
	Wild oat	North Dakota	Bell and Nalewaja, 1968 Cited by Wiese, 1983
	Yellow foxtail		Cited by Zimdahl, 1980
Spring barley	Canada thistle	Canada	O'Sullivan et al., 1982
	Common chickweed	England	Scragg and McKelvie, 1976
	Hempnettle	England	Scragg and McKelvie, 1976
	Jerusalem artichoke	Canada	Wall et al., 1986 Wall and Friesen, 1989
	Quackgrass	England	Scragg and McKelvie, 1976
	Tartary buckwheat	Canada	Remy et al., 1985
	Wild oat	North Dakota	Bell and Nalewaja, 1968
		England	Cussans and Wilson, 1975
		Canada	Dew, 1972
		England	Wilson and Peters, 1982
		Idaho	Morishita and Thill,1988a,b
		Quebec	Rioux, 1982
Grain sorghum	Common milkweed	Nebraska	Cramer and Burnside, 1982
	Smooth pigweed Palmer amaranth	Texas	Graham et al., 1988

(continued)

Table 2 Continued

Crop	Weed	Location	Reference
	Tall waterhemp	Kansas	Feltner et al., 1969b
	Volunteer sorghum	Kansas	Vesecky et al., 1973
	Yellow foxtail	Kansas	Feltner et al., 1969a
Rice	Barnyardgrass	Arkansas	Smith, 1974
		Arkansas	Smith, 1988
		Arkansas	Stauber et al., 1991
		Japan	Noda, 1973
		Japan	Noda et al., 1968
	Bearded sprangletop	Arkansas	Smith, 1983a
		Arkansas	Smith, 1988
	Ducksalad	Arkansas	Smith, 1968
		Arkansas	Smith, 1988
	Eclipta	Arkansas	Smith, 1988
	Hemp sesbania	Arkansas	Smith, 1968
		Arkansas	Smith, 1988
	Junglerice	Philippines	Mercado and Talatala,1977
	Northern jointvetch	Arkansas	Smith, 1968
		Arkansas	Smith, 1988
	Red rice	Arkansas	Kwon et al., 1991a,b
		Louisiana	Pantone and Baker, 1991
		Arkansas	Smith, 1988
	Spreading dayflower	Arkansas	Smith, 1984
		Arkansas	Smith, 1988

late that winter wheat should be graded as "garlicky" if it contains two bulblets of wild garlic/kg grain (Gast et al., 1990). Winter wheat is severely discounted if it is garlicky and some elevators may reject contaminated winter wheat. Wild garlic is primarily a major problem for winter wheat producers in the Corn Belt (Peters, 1990) and some southern states. Crop management practices also can influence how much weeds contribute to dockage in winter wheat. For example, dockage in winter wheat from cheat depended on crop seeding date and herbicide effectiveness in Oklahoma (Ferreira et al., 1990b). If winter wheat was seeded early in September for winter forage, dockage from cheat in winter wheat grain was greater than if winter wheat was planted later. Grain standards also limit the amount of jointed goatgrass seed permitted to contaminate winter wheat grain (Donald and Ogg, 1991).

C. Winter Wheat Grown for Both Forage and Grain

Cattle are often grazed on winter wheat fields in the Southern and Central Great Plains during winter months (Koscelny and Peeper, 1990a). Consequently, the same winter wheat crop can provide income from both forage and grain for mixed cereal-livestock farmers. However, optimal winter wheat planting dates for maximum forage production are earlier than for maximum grain production. Improperly timing winter wheat grazing can inadvertently encourage weeds, such as cheat, to build up at the expense of winter wheat grain yield and quality. For example, fall grazing in Oklahoma reduced winter wheat foliage more than cheat (52% versus 28%), leading to increased dockage in winter wheat due to cheat (9%) and reduced yield. Grazing did not influence cheat control or winter wheat yields when superimposed on metribuzin treatments (Koscelny and Peeper, 1990b). When metribuzin did not damage winter wheat itself, it reduced cheat biomass enough that winter wheat replaced cheat.

Registration labels for certain herbicides used on cereals, including winter wheat, either forbid herbicide use where grazing is planned or impose restrictions on the interval between herbicide application and later grazing (Table 3). Such restrictions are designed to minimize the chance that herbicide residues will be present in animal products. A portion of the hectares planted to winter wheat and spring-sown cereals, such as oat, is undersown with small-seeded legumes as a method of forage legume establishment (as "companion crops"). In this way, farmers can get cash income from harvested grain and also establish a perennial forage crop at the same time. Herbicide registration labels indicate which herbicides can be safely applied to such small-seeded legume and cereal interseedings. For example, difenzoquat, imazamethabenz-methyl, MCPA, and tribenuron are registered for use on established small-seeded legumes in a winter wheat cover crop. Other herbicides, such as diuron, metribuzin, and trifluralin, can be applied to winter wheat undersown with legumes only as dormant-season treatments.

D. Nonchemical Weed Management Methods

Whether winter wheat is used for grain, silage, hay, grazing, green crop, or cover determines the profitability of the weed control measures farmers use. Management which enhances vigorous winter wheat growth suppresses weed growth both in the growing crop and after harvest (Wicks et al., 1989), minimizing the need for herbicides. Planting competitive varieties with high wheat shoot density, planting at optimum dates, fall fertilization, fallowing, and rotation with spring-sown crops, such as corn, reduce winter annual weed seed production, soil seed banks, and later infestations and enhance winter wheat

Table 3 Restrictions on Use of Herbicide-Treated Winter Wheat, Spring Wheat, or Sorghum for Livestock Following In-Crop Treatment for Weed Control According to the 1993 U.S. Environmental Protection Agency registration labels[a]

Herbicide	Restrictions on use for livestock
Alachlor	Do not graze or harvest forage for 70 days after treatment.
Alachlor + atrazine	Do not graze or harvest forage for 70 days after treatment.
Alachlor + glyphosate	Do not graze or harvest forage for 70 days after treatment.
Atrazine	Do not graze or feed forage within 21 days after treatment.
Atrazine + bentazon	Do not graze or feed treated forage for 21 days after treatment.
Atrazine + bromoxynil	Do not graze or cut for feed for 30 days after treatment.
Atrazine + dicamba	Do not graze or harvest for feed until the grain sorghum is mature.
Atrazine + metolachlor	Do not graze or feed forage for 30 days after treatment.
Atrazine + propachlor	Do not graze for 21 days after treatment. Do not graze or feed sorghum forage, silage or fodder (stubble) to dairy animals.
Bentazon	Do not graze for 12 days after treatment.
Bromoxynil	Do not graze treated small grains for 30 days after treatment. Do not cut treated sorghum for feed, fodder, or graze within 30 days after treatment.
Bromoxynil + MCPA	Do not graze or feed treated grass to livestock.
Chlorsulfuron	No registration restrictions on grazing.
Chlorsulfuron + metsulfuron	No registration restrictions on grazing.
Clopyralid	Do not graze dairy cattle for 14 days after treatment. Remove meat animals from treated areas 7 days before slaughter unless 2 or more weeks have passed since application. Do not cut treated grass for hay within 30 days after application.
Clopyralid + MCPA	Do not graze dairy animals or meat animals on forage or graze treated areas within 1 week before slaughter. Do not harvest hay from treated areas.

Table 3 Continued

Herbicide	Restrictions on use for livestock
Cyanazine	No restrictions on the registration label for grazing or feeding grain sorghum or winter wheat.
Cyanazine + metolachlor	There are no label restrictions regarding livestock use, although it is not registered for use on forage sorghum.
Dicamba	Do not graze or harvest for livestock feed before small grains or sorghum are mature. If small grains or sorghum are used for pasture or hay, remove meat animals 30 days before slaughter. There are waiting periods for lactating animals, but not for nonlactating animals. Waiting periods between treatment and grazing vary from 7 to 60 days depending upon dicamba rate. Waiting periods between treatment and hay harvest vary from 37 to 90 days depending upon dicamba rate.
Diclofop	Do not graze or harvest forage, hay or straw before grain harvest.
Difenzoquat	Do not graze or feed cut forage, hay, or silage before harvest. Wheat and barley grain and straw can be fed to livestock after harvest.
Diquat	No restrictions on the registration label regarding grazing or feeding treated crops to livestock.
Diuron	No restrictions on the registration label regarding grazing or feeding treated crops to livestock.
Fenoxyprop + MCPA	No restrictions on the registration label regarding grazing or feeding treated crops to livestock.
Fenoxyprop + 2,4-D + MCPA	No restrictions on the registration label regarding grazing or feeding treated crops to livestock.
Fenoxyprop + MCPA + thifensulfuron + tribenuron	Do not graze or feed forage or hay to livestock (harvested straw may be used as bedding).
Glyphosate	Wait 14 days before grazing or harvesting for livestock.
Glyphosate + dicamba	Do not feed or use for forage for 8 weeks after application.

(continued)

Table 3 Continued

Herbicide	Restrictions on use for livestock
Glyphosate + 2,4-D	Do not feed or use for forage for 8 weeks after application.
Imazamethabenz	Do not graze or cut treated forage for silage or hay (harvest wheat and barley straw may be used as bedding).
Linuron	Do not graze or feed 3 months after postemergence application to grain sorghum.
MCPA	Do not graze or forage on treated areas within 7 days of slaughter.
Metolachlor	No restrictions on the registration label regarding grazing or feeding treated crops to livestock.
Metribuzin (Lexone)	No restrictions on the registration label regarding grazing or feeding treated crops to livestock.
Metribuzin (Sencor)	Wait 14 days before grazing wheat. Wait until crop maturity to graze or harvest barley for livestock feed.
Metsulfuron	No restrictions on the registration label regarding grazing or feeding treated crops to livestock.
Paraquat	Do not graze, cut or harvest for forage or hay.
Pendimethalin	No registration label restrictions regarding grazing or feeding treated crops to livestock.
Picloram	Do not cut for feed 2 weeks after treatment. Do not graze lactating animals for 2 weeks after treatment. Meat animals should be removed from treated areas 3 days before slaughter.
Propachlor	Do not graze or feed to dairy animals.
Propachlor + atrazine	Do not graze or feed to livestock for 21 days after treatment. Do not graze or feed sorghum forage, silage or fodder (stubble) to dairy animals.
Propanil	Do not graze or cut for green chop feed.
2,4-D	Do not graze dairy cattle or meat animals being finished for slaughter for 7 to 14 days after treatment, depending upon formulation (see label). Remove meat animals from treated areas 3 to 7 days before slaughter (see label) unless 2 weeks has elapsed after treatment. Do not cut treated grass for hay within 30 days after treatment.

Table 3 Continued

Herbicide	Restrictions on use for livestock
Triallate	Do not graze livestock after treatment.
Triallate + trifluralin	No restrictions on the registration label regarding grazing or feeding treated forage to livestock.
Triasulfuron	No restrictions on the registration label regarding grazing or feeding treated forage to livestock.
Tribenuron	Do not graze or feed forage or hay to livestock. Harvested straw may be used for bedding or feed.
Thifensulfuron + tribenuron	Do not graze or feed forage or hay to livestock. Harvested straw may be used for bedding or feed.
Trifluralin	Do not graze or feed forage or hay to livestock.

Note: Herbicides and rates are based on current labels or expected labels. Before using a herbicide, refer to a current label for appropriate rates.
[a]When "grain sorghum (milo)" is specified, forage sorghum should not be treated. These restrictions minimize the chance that illegal herbicide residues will be present in meat or dairy products.

competitiveness with weeds (Wicks et al., 1989). Winter wheat is also more competitive with weeds than spring wheat, presumably because of early canopy development and closure in spring.

Some winter wheat cultivars are more competitive with weeds than others. For example, cheat competing with 'Pioneer 2157' reduced yields less than with 14 other cultivars in Oklahoma (Koscelny et al., 1990). In Nebraska, some winter wheat varieties suppressed barnyardgrass more than others both in winter wheat (Ramsel and Wicks, 1988) and after harvest (Ramsel and Wicks, 1988; Wicks et al., 1989).

Unfortunately, the basis for varietal differences in winter wheat competitiveness with weeds is too poorly understood to be routinely used as a screening criterion in winter wheat breeding programs or to be recommended by extension personnel. In one study, varietal differences in winter wheat competitiveness were unrelated to differences in interrow shading (Ramsel and Wicks, 1988). Differences in varietal competitiveness with weeds has not been extensively studied for other cereals or sorghum.

As a rule of thumb, weeds which emerge before or with the crop tend to reduce yields more than those weed cohorts which emerge progressively later after crop emergence (Zimdahl, 1980, 1990). Optimum winter wheat planting date can enhance crop competitiveness with weeds because it allows the crop to emerge and grow more quickly and uniformly before most weeds emerge. For example, in Nebraska, late fall-planted winter wheat failed to tiller sufficiently to compete with weeds as well as earlier planted wheat (Wicks et al., 1989). Both the timing of seeding and forage removal influenced wheat grain yield, cheat biomass, and herbicide efficacy for managing cheat in winter wheat in Oklahoma (Ferreira et al., 1990b). Winter wheat planted at normal seeding dates (October) yielded more than when seeded early (September) for forage production or late (November) for cheat management. Delaying seeding until November greatly reduced cheat populations. Seeding date and herbicides also influenced dockage in winter wheat due to cheat.

In Oklahoma, decreasing winter wheat row width from 23 to 8 cm increased weed-free wheat yield in two of three locations but did not increase the yield of cheat-infested winter wheat in 6 of 10 experiments (Koscleny et al., 1990). Doubling winter wheat seeding rates from 265 to 530 seed/m increased winter wheat yield in competition with cheat. However, increasing seeding rate only partially controlled cheat, although it reduced the amount of cheat seed harvested. In other research, increasing winter wheat density above the recommended density for the locale improved wild oat control, but control was still incomplete and unsatisfactory (Martin et al., 1987).

Diseases, insects, winter-kill, drought, or, occasionally, herbicide damage (whether from misapplication, drift, or carryover) may thin winter wheat stands (Cook and Verseth, 1991). When winter wheat stands are sparse or winter wheat foliage grows slowly in spring, summer annual weeds can emerge and develop more vigorously in spring than when the winter wheat canopy closes quickly and completely. When canopies are thin or not completely closed, spring herbicide treatment may be required to control weeds.

E. Herbicides for Weed Control

Herbicides registered for use on winter wheat, many of which can be combined, are summarized in Table 1. More herbicides are registered for use on winter or spring wheat than for other cereals (Table 1). Most cereal herbicides are applied postemergence and there are more herbicides for broadleaf weed management than for grass weed management. Herbicides would probably be used more if effective, selective herbicides were marketed for controlling the grass weeds cheat, downy brome, Italian ryegrass, jointed goatgrass, and volunteer rye in winter wheat.*

The winter wheat treated with herbicides and other pesticides is summarized for 1991 in Table 4. Even though there were several registered

*Dr. T. Peeper, 1993. Oklahoma State Univ., personal communication.

Table 4 Frequency and Extent of Pesticide Use in Winter Wheat in the Major Production States in 1990

Pesticide	Area treated, %	Number of applications	Rate, kg/ha	Total applied, × 1000 kg
Herbicides				
2,4-D	15.6	1.02	0.41	1090
Bromoxynil	1.6	1.00	0.21	55
Chlorsulfuron	12.6	1.00	0.01	33
Clopyralid	<1.0			
Dicamba	3.7	1.05	0.12	80
Diclofop-methyl	<1.0			
Diuron	<1.0			
Glyphosate	1.3	1.13	0.50	121
Imazamethabenz	<1.0			
MCPA	2.5	1.05	0.43	181
Metribuzin	<1.0			
Metsulfuron	<1.0			
Paraquat	<1.0			
Picloram	<1.0			
Terbutryn	<1.0			
Thifensulfuron	3.1	1.02	0.01	7
Triallate	<1.0			
Tribenuron	3.1	1.02	0.01	6
Trifluralin	<1.0			
Insecticides and Fungicides				
Benomyl	<1.0			
Chlorpyrifos	1.0	1.00	0.48	78
Dimethoate	1.2	1.08	0.28	58
Disulfoton	1.0	1.00	0.75	122
Esfenvalerate	<1.0			
Ethyl parathion	<1.0			
Malathion	<1.0			
Mancozeb	<1.0			
Methomyl	<1.0			
Methyl parathion	<1.0			
Phorate	<1.0			
Propiconazole	1.9	1.00	0.11	35
Thiophanate-methyl	<1.0			
Triadimefon	<1.0			

Source: NASS, SDA, Agricultural Chemical Usage, March, 1991

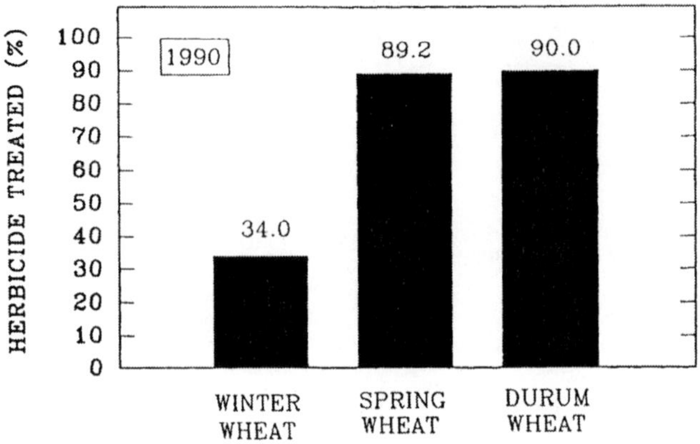

Figure 2 The relative proportion of acreage planted to winter, spring, and durum wheat that was treated with herbicide in 1990. (From Agricultural Chemical Usage, 1991.)

winter wheat herbicides, only a third of the planted winter wheat was routinely treated, compared to greater than 95% for spring wheat (USDA, 1990, 1991; Delvo, 1989) (Figures 2 and 3). In 1990, 16.2 million hectares of winter wheat were harvested, but only 34%, 4% and 3% of that was treated with herbicides, insecticides, and fungicides, respectively (USDA, 1991). Nevertheless, the total herbicide-treated winter wheat roughly equaled that of spring wheat in the United States (Figure 3), largely because so many more hectares of winter wheat than spring wheat were planted (Figure 1). The percentage of the win-

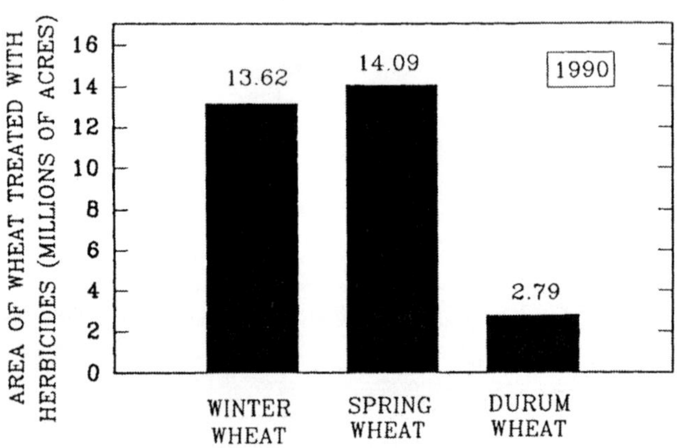

Figure 3 The acreage of winter, spring, and durum wheat treated with herbicide in 1990 (From Agricultural Chemical Usage, 1991.)

ter wheat treated ranged from 10% to 85%, depending upon the state. Consequently, some states treated no more winter wheat with herbicide than they did in the early 1950s (Osteen and Szemdra, 1989). Herbicides were used less in winter wheat than in spring-sown cereals, and nonchemical weed management measures were used more. Tables listing which weeds of winter wheat are controlled by individual herbicides have been compiled for each major wheat production region (Donald, 1990b), and current tables are available from state extension personnel.

F. Combinations of Nonchemical Management and Herbicides for Weed Control

Sometimes it is advantageous to manage weeds in winter wheat or other cereals by combining nonchemical cultural practices with application of herbicides, as has been suggested for downy brome in winter wheat (Wicks, 1984). Crop rotation for weed management often involves herbicide rotation, by default. Rotation of crops with different life cycles (fall-sown and spring-sown crops) is advantageous for preventing the buildup of certain weeds, but there are few examples supporting this assertion for crops alone without herbicide treatment (Black and Siddoway, 1976a,b). Rotation of broadleaf crops with grass or cereal crops also may be advantageous. Finally, rotation of crops with different times of planting can be used to manage weeds. Crop rotation is preferable to continuous cropping, because it limits weed shifts and reduces the likelihood that diseases and insects will increase to levels which will reduce yield.

Winter annual grasses that are especially difficult to manage in winter wheat include cheat, downy brome, Italian ryegrass, jointed goatgrass, volunteer rye, wild garlic, and wild oat (Donald, 1990b). The biology and control of certain weeds in winter wheat have been reviewed: Canada thistle (Donald, 1990a), downy brome (Morrow and Stahlman, 1984; Peeper, 1984), jointed goatgrass (Donald and Ogg, 1991), and wild oat (Jones, 1976; Smith, 1984; Smith and Hsiao, 1984). Management strategies for winter wheat weeds are discussed at greater length elsewhere (Donald, 1990b).

Winter·annual bromegrasses, such as cheat and downy brome, are major problems of winter wheat in the Central and Southern Great Plains and Pacific Northwest (Wicks and Smika, 1990; Peeper and Wiese, 1990). Reduced tillage for crop residue management to reduce wind and water erosion has created a favorable environment for grass weed growth in winter wheat. In Oklahoma, cheat and downy brome densities were greater in continuous winter wheat with a stubble-mulched surface than in a nonmulched, no-till soil surface (Dao, 1987). Removing residue without tillage reduced cheat and downy brome infestations (Dao, 1987).

Massee and Higgins (1977) and Wicks (1984) suggested using a combination of cultural practices for managing downy brome. Tillage is recommended to maintain soil moisture in fallow for the later-sown winter wheat crop, but mechanical fallowing often destroys surface crop residue used for wind erosion control. Rod weeding just before planting winter wheat after mid-September was recommended to allow some downy brome to emerge before being killed by tillage. Other methods were faulted: deep moldboard plowing killed downy brome and buried seed so deeply that fewer seedlings emerged, but control was never complete; the temperatures from burning wheat residue in fall did not kill surface-lying downy brome seed but exposed the soil to erosion. Harrowing or stubble mulch tillage (i.e., undercutting with sweeps) did not completely control downy brome because its fibrous roots rerooted and plants regrew after rains. Herbicidal control of downy brome was reviewed (Peeper, 1984; Wicks, 1984), but herbicides have not been widely adopted for downy brome control because of their potential for damaging winter wheat (e.g., metribuzin) and their inconsistent weed control (Peeper and Morrow, 1990). Metribuzin is recommended for use on winter wheat fields with a history of heavy cheat infestations. According to the registration label, metribuzin should be applied to only certain winter wheat varieties and only when they have at least three to four tillers and four secondary roots greater than 5 cm long. Because herbicides for downy brome control require soil moisture for activation, they have performed erratically in arid regions with sporadic rainfall.

Wild garlic is a problem in winter wheat in the Corn Belt and some southern states because it reduces grain quality and, to a lesser extent, yield (Peters, 1990). Traditionally, postemergence 2,4-D was used to manage wild garlic (Rhodes and Shelby, 1991). Although 2,4-D did not prevent yield loss from wild garlic, it caused wild garlic scapes to grow downward so that aerial bulblets were less likely to be harvested during combining and did not contribute to dockage in winter wheat (Rhodes and Shelby, 1991). Recently, postemergence sulfonylurea herbicides, such as thifensulfuron plus tribenuron, were shown to control wild garlic effectively, increase yields, and decrease dockage in winter wheat (Gast et al., 1990). These sulfonylurea herbicides controlled offset bulbs better when applied in late March to early April than in late April (Gast et al., 1990). According to the registration label, thifensulfuron plus tribenuron should be applied when wild garlic is 2.5 to 5 cm tall but less than 30 cm tall (Rhodes and Shelby, 1991).

Phenoxy herbicides, chiefly 2,4-D and MCPA, are widely used for winter annual and summer annual broadleaf weed control in winter wheat (Table 4). Other postemergence herbicides have been marketed either as alternatives to phenoxy herbicides or as tank mixes with phenoxy herbicides to broaden the spectrum of weed control. For example, wild buckwheat is tolerant of phenoxy herbicides but can be controlled with picloram or picloram plus 2,4-D (Heering and Peeper, 1991), among other registered herbicides. Likewise, common chickweed is well controlled by mixtures of 2,4-D plus thifensulfuron, tribenuron, or thifensulfuron + tribenuron but not by 2,4-D alone (Harrison and Beuerlein, 1989). Synergistic interactions between 2,4-D and sulfonylurea herbicides were observed on common chickweed and may be exploited on other weeds for reducing application rates without sacrificing weed control. Reducing application rates of many sulfonylurea herbicides would be advantageous to reduce the chance of carryover damage to later-sown rotational crops.

Sometimes fall-applied herbicides can be used to control winter annual broadleaf weeds better and increase yields more than spring-applied herbicides in winter wheat. In Canada, fall-applied herbicides controlled field pennycress and flixweed better than spring-applied herbicides (Blackshaw, 1990). The residual herbicides chlorsulfuron, dicamba plus nonresidual 2,4-D, metribuzin, metsulfuron, and picloram applied in fall increased winter wheat yields most because they were persistent and controlled both fall- and spring-germinating flushes of field pennycress and flixweed. Bromoxynil, bromoxynil plus MCPA, MCPA, thifensulfuron, and 2,4-D applied in fall did not persist long in soil and failed to control spring flushes of either weed. Some sulfonylurea herbicides can carry over to damage rotational crops (Table 5) and registration labels caution against the increase of sulfonylurea-resistant weed populations after repeated use of sulfonylurea herbicides over several years.

More soybeans in Illinois, Indiana, Kentucky, and Ohio are being double cropped after harvesting winter wheat using no-till planting methods (Lewis, 1976). Selecting cereal and soybean varieties with short enough life cycles to match the local length of the growing season is critical for successful double cropping. Winter barley, an alternative to winter wheat, matures early, making it a good candidate double crop with soybean in some regions, such as the Carolinas (Lewis, 1976), but soybeans are most commonly double cropped into winter wheat stubble. The success of double cropping depends on available soil moisture after winter wheat harvest. Small etiolated weeds, such as clover, common ragweed, foxtails, large crabgrass, and Pennsylvania smartweed, which have not competed well with winter wheat may survive and then grow after winter wheat harvest (Triplett, 1978). It is critical to plant double cropped soybean soon after winter wheat harvest to suppress later regrowth of these previously suppressed weeds. If much time elapses between winter wheat harvest and soybean planting, these weeds grow quickly unless they are controlled with postemergence herbicides. Glyphosate or paraquat can be used to manage weeds growing after winter wheat harvest but before soybean emergence (Triplett, 1978). Winter wheat herbicides must be carefully selected to prevent herbicide carryover damage to double cropped soybean. Short residual herbicides, such as 2,4-D, dicamba, 2,4-D plus dicamba, MCPA, and bromoxynil, can be

Table 5 Restrictions on Rotational Crops That Can Be Planted in the Growing Season After Treatment or the Recropping Intervals After Treating Cereals In-Crop with Various Herbicides, According to 1993 U.S. Environmental Protection Agency Registration Labels[a]

Herbicide	Restrictions on recropping interval in the growing season after treatment
Alachlor	Corn, peanuts, peas, sorghum (milo), soybeans, or sunflower may be planted 1 year after application.
Alachlor + atrazine	Corn, peanuts, sorghum (milo), or soybean may be planted 1 year after application
Alachlor + glyphosate	Corn, cotton, dry beans, peanuts, sorghum (milo), soybeans or sunflower may be planted 1 year after application.
Atrazine	Corn or sorghum may be planted within 1 year after application. Do not plant any other crops for 1 year after application.
Atrazine on furrow irrigated bedded production	Do not replant sorghum for 8 months after application
Atrazine + bentazon	Do not plant sugar beet or sunflower in the following growing season or oat in calcareous soils
Atrazine + bromoxynil	Residual atrazine may injury crops rotated within the last year. Contact local extension agents or perform a field bioassay before planting susceptible rotational crops.
Atrazine + dicamba	If applied after June 10th, do not rotate to crops other than corn or sorghum 1 year after treatment.
	Plant only corn, peanuts, sorghum, or soybeans 1 year after treatment.
	Small grains can be planted 10 months after treatment, otherwise wait for 18 months before replanting to other crops.
Atrazine + metolachlor	Corn, cotton, peanuts, sorghum, and soybeans may be planted in the spring after application.
	If applied after June 10, only corn or sorghum may be planted in the following year.
	Do not plant spring-wheat, small grains, small seed legumes, sugar beets, tobacoo, or vegetables (including dry beans) in the year following application.
Atrazine + propachlor	Corn, sorghum, and soybeans may be grown 1 year after treatment.
Bentazon	No restrictions on the registration label.
Bromoxynil	All rotational crops can be planted one year after application.

Table 5 Continued

Herbicide	Restrictions on recropping interval in the growing season after treatment
Bromoxynil + MCPA	No restrictions on the registration label.
Chlorsulfuron	Recommended for use only where wheat or barley are grown because residues can persist for 2 to 3 years. Recropping intervals depend on crop, application rate, soil pH, cumulative precipitation, and site (see label).
Chlorsulfuron + metsulfuron	Recommended for use only where wheat or barley are grown because residues can persist for 2 to 3 years. Recropping intervals depend on crop, application rate, soil pH, cumulative precipitation, and site (see label).
Clopyralid	Barley, corn, oat, grasses, sugar beets, and wheat may be planted any time after treatment. Do no plant alfalfa, asparagus, canola, cole crops, grain sorghum, onions, popcorn, safflower, sweet corn, and strawberries for 12 months after application. Do not plant dry beans, soybeans, and sunflower for 12 to 18 months after application depending upon soil organic matter and precipitation. Other crops, including peas, lentils, potatoes, and broadleaf crops for seed, should not be planted for 18 months after treatment to avoid phytotoxicity.
Clopyralid + MCPA	Do not plant wheat, barley, oat, grasses or sugar beets for 12 months after treatment. Do not plant alfalfa, asparagus, canola, cole crops, grain sorghum, onions, popcorn, safflower, sweet corn, and strawberries for 12 months after application. Do not plant dry beans, soybeans, and sunflower for 12 to 18 months after application depending upon soil organic matter and precipitation. Peas, lentils, potatoes, and broadleaf crops for seed should not be planted for 18 months after treatment to avoid phytotoxicity.
Cyanazine	Only corn, sorghum or soybeans may be planted 1 year after treatment. If cyanazine or combinations with cyanazine are applied after June 1, only corn and sorghum can be planted 1 year after treatment. Small grains may be planted 15 months after treatment and all other crops can be planted 18 months after application.
Cyanazine + metolachlor	If the treated crop is lost, corn (field, silage, and sweet) and Concept-treated grain sorghum may be replanted immediately. Wait 4.5 months before planting barley,

(*continued*)

Table 5 Continued

Herbicide	Restrictions on recropping interval in the growing season after treatment
	oat, rye or wheat. Barley, buckwheat, cotton, oat, peanuts, pod crops, potatoes, rice, root crops, rye, safflowers, sorghum, and soybeans may be planted 9 months after treatment. Clover may be planted 18 months after treatment.
Dicamba	Corn, sorghum, soybean, and wheat may be planted in the spring following treatment with 2 lb ai/A dicamba. Check the registration label for restrictions at higher application rates. Following normal harvest of these crops or one year after treatment with dicamba at registered rates, any rotational crop may be grown.
Diclofop	There are no registration restrictions on recropping.
Difenzoquat	There are no registration restrictions on recropping.
Diquat	There are no registration restrictions on recropping.
Diuron	Wait 4 months for band application to 6 months for broadcast application to replant following grain sorghum treatment. Cotton or corn can be replanted at any time after treatment.
Fenoxyprop + MCPA	No rotational crop restrictions on registration label.
Fenoxyprop + 2,4-D + MCPA	No rotational crop restrictions on registration label.
Fenoxyprop + MCPA + thifensulfuron + tribenuron	Do not plant any crop other than wheat or barley 60 days after treatment.
Glyphosate	Do not feed or forage before 8 weeks after treatment.
Glyphosate + dicamba	Apply 15 days before planting wheat, barley, oat or sorghum. Do not plant any crop other than barley, corn, oat, sorghum, or wheat for 3 months after application or until glyphosate + dicamba disappear from the soil.
Glyphosate + 2,4-D	Do not plant crops, other than barley, corn, oat, rye, sorghum, or wheat, for 3 months after treatment.
Imazamethabenz	Barley, corn, edible bean, safflower, soybean, sunflower, and wheat may be grown in the growing season after treatment. Other crops can be planted 15 months after treatment, except sugar beets, which can be planted 20 months after treatment.
Linuron	Any crop can be planted 4 months after treatment, except for cereals. Only barley, oat, rye, and wheat can be planted 4 months after treatment.

Table 5 Continued

Herbicide	Restrictions on recropping interval in the growing season after treatment
MCPA	No rotational crop restrictions on the registration label.
Metolachlor	Barley, oat, rye or wheat may be planted within 4.5 months of application. Any crop registered for metolachlor may be planted in the following spring After 8 months any crop can be planted.
Metribuzin (Lexone)	Crops registered for the rate used may be replanted in case of crop failure, but may not be retreated within the same growing season. In the Pacific Northwest, wait 4 months after applying 0.5 lb ai/A metribuzin to plant to barley or wheat. If 2/3 lb ai/A has been applied, wait 8 months to plant barley or wheat. In areas treated with up to 1 1/3 lb ai./A, wait 10 months to plant barley or wheat.
Metribuzin (Sencor)	Wait 4 months to recrop to alfalfa, asparagus, barley, corn, forage grasses, potatoes, sainfoin, soybeans, sugarcane, tomatoes, or wheat. Wait 8 months to recrop to barley, cotton, lentils, peas, and rice. Wait 10 months to recrop to other crops except root crops. Wait 18 months to recrop to onions, sugar beets, and other root crops.
Metsulfuron	Although metsulfuron is recommended on land dedicated to wheat and barley production, other rotational crops can be grown following metsulfuron application. Detailed instructions for recropping to alfalfa, barley, Conservation Reserve Program grasses, corn, cotton, dry beans, durum wheat, flax, grain sorghum, proso millet, safflower, soybean, spring wheat, sunflower, winter wheat and other crops can be found on the registration label and depend on geographic location, soil pH, and moisture.
Paraquat	There are no registration label restrictions on recropping interval.
Pendimethalin	Winter wheat or winter barley can be planted 90 days after treatment in irrigated sorghum or 120 days after treating dryland sorghum. Any crop can be planted after 1 year.
Picloram	May be used on land to be planted to barley, grass, oat, or wheat or which will be fallowed in the year after

(continued)

Table 5 Continued

Herbicide	Restrictions on recropping interval in the growing season after treatment
	treatment. Do not rotate to nonregistered crops for 36 months or until a field bioassay shows that damaging residues no longer persist.
Propachlor	Corn, sorghum, and soybeans may be grown 1 year after treatment.
Propanil	No restrictions on rotational crops on the registration label.
2,4-D	Wait 3 months or until the herbicide has dissipated before planting any crop.
Triallate	Do not plant oat in the growing season after treatment.
Triallate + trifluralin	Sugar beets, red beets or spinach should not be planted for 14 months after treatment. If less than 20 inches of water are used to produce these crops, do not plant them for 20 months after treatment. Sorghum, proso millet, corn or oat should not be planted for 16 months after treatment.
Triasulfuron	Although triasulfuron is recommended on land dedicated to wheat and barley production, other rotational crops can be grown following triasulfuron application subject to a field bioassay. Detailed instructions for recropping to grain sorghum, soybean, spring wheat, winter wheat and other crops without a field bioassay can be found on the registration label and depend on geographic location, soil pH, and moisture.
Tribenuron	Wait 60 days before recropping to crops other than barley or wheat.
Thifensulfuron + tribenuron	Wait 60 days before recropping to crops other than barley or wheat.
Trifluralin	See registration label for detailed instructions on rotational intervals. 14 months before planting sorghum (milo) proso millet, corn and oat after spring application or 16 months after a fall application.

Note: Herbicides and rates are based on 1993 labels or expected labels. Before using a herbicide refer to a current label.

[a]Restrictions help avoid the chance of crop damage from phytotoxic residues and minimize the change of illegal herbicide residues in harvested nonregistered crops. Label restrictions are closely paraphrased.

applied in winter wheat, but the interval between treatment and soybean planting is critical to prevent soybean damage (Lewis, 1976).

G. Postharvest and Fallow Weed Management

Postharvest or fallow weed management is critical for soil moisture storage and successful crop production in arid regions (Wicks, 1986). Herbicides such as cyanazine, 2,4-D, glyphosate, and paraquat have been used to control weeds in fallow and reduce the number of tillage operations needed for fallow weed management (Wiese et al., 1986). When volunteer wheat or weeds were allowed to grow postharvest, they reduced soil moisture storage (Hoefer et al., 1981) and yields of later planted crops (Wicks, 1986). In the Southern Great Plains, weeds should not be allowed to grow for more than 17 days to prevent reducing yields of later-planted crops (Unger, 1983).

Tillage (mechanical fallow) or herbicides (chemical fallow) can be used to limit soil moisture depletion by postharvest volunteer cereal (Table 6) and weed growth. Fallowing by undercutting weeds with sweep-type implements limits evapotranspiration yet keeps plant residues on the soil surface, limiting wind and water erosion (Willis et al., 1983). Chemical fallow with herbicides limits erosion by keeping the soil surface covered with residue. The timing of herbicide application after harvest is critical for successful chemical fallow. In Nebraska, atrazine applied postemergence to winter wheat stubble for weed management after mid-July was more heavily infested with weeds 9 months later than residue sprayed earlier with atrazine (Wicks et al., 1989).

Weeds in fallow also can be suppressed by postharvest winter wheat residue itself if the residue is dense and cut tall (Wicks et al., 1989). When wheat was harvested in Nebraska in mid-July, weeds were controlled 56% by wheat residue 7 weeks after harvest (Hoefer et al., 1981). In Minnesota, winter wheat and winter rye residue provided residual weed control of common lambsquarters and redroot pigweed and fair control of green and yellow foxtail in later-sown soybean (Robinson and Dunham, 1954). Winter rye was more effective than winter wheat for weed management in soybeans sown into the residue. Soybean yields were not reduced by the cereals, and rye and wheat died in August. This interseeding may be less successful in drier regions because early spring regrowth of winter cereals limits soil moisture reserves and, consequently,reduces soybean germination. Winter wheat has advantages as a soil cover for no-till crop production (Moschler et al., 1967). However, the order of preference was rye > wheat = oat > barley because of winter hardiness and susceptibility to being killed by atrazine and paraquat (Moschler et al., 1967).

Table 6 Registered Herbicides for Control of Volunteer Wheat in Field Crops and Fallow According to the 1993 U.S. Environmental Protection Agency Labels

Herbicide	Corn	Cotton	Chemical fallow	No-till before corn, grain sorghum, soybean emergence	Grain sorghum	Soybean
Atrazine	X		X		X	
Clethodim		X				X
Clomazone			X			X
Cyanazine	X	X	X		X	
EPTC	X					
Fluazifop		X				X
Fluazifop + fenoxyaprop						X
Fluazifop + fomesafen						X
Glyphosate			X	X		
Glyphosate + 2,4-D			X			
Metribuzin						X
Paraquat			X	X		
Pronamide			X			
Quizalofop						X
Sethoxydim		X				X

III. WEED MANAGEMENT IN SPRING WHEAT

A. Competition and Yield Loss Assessment

The surveyed distribution of weeds of spring wheat has been summarized for the Northern Great Plains (Dexter et al., 1981; Fay, 1990) and the neighboring Prairie Provinces of Canada (Hunter et al., 1990) (Table 7). Competition and yield loss assessment for weeds of spring wheat are summarized in Table 2 (Zimdahl, 1990). Estimated dollar losses in the United States due to weeds in spring wheat were summarized by region, based on expert opinion (Bridges, 1992). Secondary effects of weeds on spring wheat, such as decreased grain quality, have not been extensively studied. Wild oat, a major weed of spring wheat in the Northern Great Plains, reduced yield, but not percentage grain protein or seed size, important grain quality factors (Bell and Nalewaja, 1968). In fact, spring wheat percentage protein increased as yield was reduced by increasing wild oat densities, probably a stress effect. Tartary buckwheat reportedly darkened spring wheat flour and reduced its value (Remy et al., 1985).

B. Nonchemical Weed Management Methods

Nonchemical methods for weed management in spring wheat were recently reviewed (Donald and Nalewaja, 1990), including control of wild oat (Hunter, 1983) and Canada thistle (Donald, 1990a). Measures to prevent weed seed from being spread between fields include planting weed-free crop seed and cleaning grain drills, harvesting equipment, and combines between fields (Donald and Nalewaja, 1990).

Weed control is a major justification for crop rotation (Kirby, 1980). More research is needed to characterize better the relative competitiveness of spring wheat with weeds versus other spring-sown cereals or row crops. For example, spring-sown barley was more competitive with wild oat than spring wheat in North Dakota, as determined by smaller yield reductions at comparable wild oat densities (Bell and Nalewaja, 1968). Forages, barley, and fall rye were more competitive with weeds than spring wheat (Holm and Kirkland, 1986).

Spring wheat competitiveness with weeds can be enhanced by changing planting practices (i.e., early spring planting, decreasing row spacing, or increasing plant populations). For example, increasing wheat population increased wheat yields in competition with wild mustard (Holm and Kirkland, 1986). Some extension agents suggest that farmers grow competitive spring wheat cultivars as a weed management measure; however, they would be hard-pressed to name particular varieties or growth characteristics that make some varieties of spring wheat more competitive than others. More research is needed on this possibility for managing weeds in all cereals.

The timing of primary tillage can be used to manage weeds in spring wheat. Fall tillage or summer fallow controls winter annual weeds (Holm and

Table 7 Comparison of the Frequency of the Most Common Weeds of Spring Wheat[a]

Weed	Spring wheat 1978	Spring wheat 1979	Barley 1978	Barley 1979	Oat 1978	Oat 1979
Green foxtail	94	95	98	99	89	95
Wild oat	67	67	63	68	73	45
Wild buckwheat	54	66	53	86	65	77
Redroot pigweed	52	67	33	56	34	61
Yellow foxtail	17	27	-	25	15	31
Kochia	26	27	28	30	16	42
Russian thistle	22	33	16	-	25	57
Common lambsquarters	20	40	37	35	36	73
Canada thistle	11	17	15	32	7	15
Field bindweed	9	19	9	10	12	24
Wild mustard	12	39	13	32	19	54
Night-flowering catchfly	8	8	8	20	-	-
Volunteer sunflower	10	15	16	11	13	-
Perennial sowthistle	8	-	21	9	14	14
Prostrate pigweed	6	15	4	-	-	19
Russian thistle	-	-	16	16	26	54
Quackgrass	-	-	4	8	-	15
Wild rose	3	11	-	-	5	-

Source: Dexter et al. (1981).

[a](N = 859 in 1978 and N = 89 in 1979), barley (N = 303 in 1978 and N = 100 in 1979), and oat (N = 176 in 1978 and N = 74 in 1979) in North Dakota.

Kirkland, 1986), as does preplant primary tillage (i.e., chisel plowing) and secondary tillage for seedbed preparation (i.e., field cultivation-harrowing) in spring (Cook and Veseth, 1991). Although cultivation is not usually suggested for solid-seeded spring cereals or spring wheat, postseeding harrowing until spring wheat is 12.5 to 15 cm tall has successfully controlled summer annual broadleaf weeds, such as mustards, stinkweed, and Russian thistle, in Canada (Holm and Kirkland, 1986).

C. Herbicides for Weed Control

The major herbicides registered for use in spring wheat are summarized in Table 1; many of them can be applied as mixtures. Of the 6.4 million hectares planted to spring wheat in 1991, 89% and 7% were treated with herbicides and insecticides, respectively (USDA, 1991). The percentages of planted hectares treated with particular herbicides in spring wheat and durum wheat are summarized in Tables 8 and 9. In 1989, herbicide-treated spring wheat varied from a low of 71% in Montana to a high of 97% in Minnesota (Delvo, 1989). Ninety percent of the durum wheat was treated with herbicides (125,550 hectares, chiefly in North Dakota). Except for soil-applied triallate and trifluralin for grass control, most herbicides are applied postemergence to spring wheat and other spring-sown cereals (Table 1). Consequently, they are well adapted to no-till production practices. A history of the early introduction of phenoxy herbicides (Kirby, 1980) and wild oat herbicides (Banting, 1984) and a summary of herbicide use for wild oat control in Canada (Smith and Hsiao, 1984) have been published. While spring wheat is not generally grazed or harvested for forage in the Northern Great Plains (Donald and Nalewaja, 1990), farmers who graze spring cereals should be aware of restrictions on the interval between herbicide application and grazing or cutting (Table 3). Tables that indicate which weeds of spring wheat are controlled by individual herbicides have been constructed for each major wheat production region (Donald, 1990b), and current summaries are available from state extension agents.

Cereal yields are reduced less if summer annual weeds, such as wild oat, emerge progressively later after spring wheat or barley emergence (Donald and Nalewaja, 1990). Consequently, timely planting according to regional seeding recommendations minimizes spring cereal yield losses due to weeds. Early, timely application of postemergence herbicides is critical to kill weeds when they are young and prevent yield losses in spring wheat. "Time-of-removal" experiments in spring wheat show that weeds must be removed within 2 to 4 weeks of crop emergence to prevent initial yield loss (Zimdahl, 1980). Unlike hand hoeing or cultivation of row crops, there is a lag between herbicide application and later weed death during which weeds can continue to compete with cereals (Gillespie and Nalewaja, 1988).

Table 8 Frequency and Extent of Pesticide Use in Spring Wheat in the Major Production States in 1990

Pecticide	Area treated, %	Number of applications	Rate, kg/ha	Total applied, × 1000 kg
Herbicides				
2,4-D	53.4	1.03	0.38	1359
Bromoxynil	8.6	1.03	0.31	181
Chlorsulfuron	3.3	1.14	0.02	3
Clopyralid	2.0	1.00	0.11	15
Dicamba	21.3	1.00	0.09	125
Diclofop	3.6	1.00	0.92	212
Difenzoquat	2.9	1.00	0.68	127
Fenoxyprop	3.3	1.00	0.09	18
Glyphosate	<1.0			
Imazamethabenz	2.4	1.00	0.39	61
MCPA	32.4	1.06	0.34	755
Metsulfuron	<1.0			
Picloram	<1.0			
Thifensulfuron	5.3	1.00	0.01	4
Triallate	4.3	1.00	1.13	312
Tribenuron	10.2	1.00	0.01	7
Trifluralin	5.4	1.00	0.56	193
Insecticides and Fungicides				
Carbaryl	1.6	1.15	0.92	108
Carbofuran	1.0	1.00	0.26	16
Ethyl parathion	<1.0			
Malathion	<1.0			
Mancozeb	<1.0			
Methyl parathion	3.1	1.00	0.44	86
Propiconazole	1.7	1.00	0.15	16

Source: NASS, SDA, Agricultural Chemical Usage, March 1991.

Properly timed postemergence herbicide application influenced the profitability of herbicide use in spring wheat in North Dakota (Gillespie and Nalewaja, 1988). Control of wild oat and wild mustard and profitability were greatest when diclofop plus bromoxynil was applied at the two-leaf stage compared with the three- to five-leaf stage or boot stage of spring wheat. Delaying herbicide treatment reduced economic returns, and controlling wild oat without controlling wild mustard was less profitable than controlling both weeds.

Table 9 Frequency and Extent of Pesticide Use in Durum Wheat in the Major Production States in 1990

Pesticide	Area treated, %	Number of applications	Rate, kg/ha	Total applied, × 1000 kg
Herbicides				
2,4-D	50.0	1.00	0.46	289
Bromoxynil	< 1.0			
Chlorsulfuron	< 1.0			
Dicamba	24.6	1.00	0.09	28
Diclofop	5.2	1.00	0.81	53
Difenzoquat	< 1.0			
Glyphosate	< 1.0			
Imazamethabenz	< 1.0			
MCPA	34.3	1.04	0.40	177
Methsulfuron	< 1.0			
Thifensulfuron	< 1.0			
Triallate	13.4	1.00	1.03	174
Tribenuron	10.4	1.07	0.01	1
Trifluralin	30.6	1.00	0.45	175
Insecticides and Fungicides				
Carbaryl	< 1.0			
Carbofuran	< 1.0			
Malathion	< 1.0			
Methyl parathion	< 1.0			

Note: Herbicides and rates are based on current labels or expected labels. Before using an herbicide refer to a current label for appropriate rates.
Source: NASS, SDA, Agricultural Chemical Usage, March 1991.

Combining crop management practices which suppress weeds with use of herbicides helped improve weed management in spring wheat. For example, N-fertilizer placement, spring wheat row spacing, and wild oat herbicides in no-till spring wheat all influenced wild oat control (Reinertsen et al., 1984). Fewer wild oat plants emerged when ammonium nitrate was band applied below the spring wheat seed than when it was surface-applied. Narrow row spacing (20 cm) increased wheat yields and shoot dry weight more than wider spacings (40 cm) in competition with wild oat. Wild oat densities increased from 68 to 95 plants/m as crop spacing was increased from 20 to 40 cm. Triallate reduced wild oat populations more than difenzoquat or no herbicide but did not reduce wild oat shoot dry weight any more than difenzoquat.

Most herbicides used for grass or broadleaf control in spring wheat are applied postemergence and, consequently, are well suited for use in reduced- or no-tillage farming systems (Table 1). Herbicide formulation may influence the efficacy of soil-applied herbicides, such as triallate, in reduced or no-till crop production (Carlson and Morrow, 1986). Granular triallate before or at planting without incorporation adequately controlled wild oat in no-till spring wheat in Washington. The emulsifiable formulation of triallate was not as effective as the granular formulation. Preplant incorporation of both triallate formulations controlled wild oat better than surface-applied treatments, but surface treatment with both formulations still controlled wild oat well.

Weed management strategies for specific weeds of spring wheat are discussed at greater length elsewhere (Donald and Nalewaja, 1990). Even though most spring wheat herbicides are applied postemergence, some postemergence herbicides have residual soil activity. For example, chlorsulfuron, metsulfuron, and triasulfuron plus surfactant applied at planting controlled annual broadleaf weeds (kochia and wild mustard) for the entire growing season in North Dakota in place of glyphosate, if applied when spring wheat was no-till planted (Donald and Prato, 1991). Residual suppression of kochia and wild mustard was observed in the following growing season. However, repeated use of residual sulfonylurea herbicides may lead to development of herbicide-resistant weeds and persistent phytotoxic residues may limit rotational crop options.

IV. WEED MANAGEMENT IN BARLEY

A. Weed Competition

Barley is grown for different purposes in different regions in North America. It is marketed as a feed grain in Canada (Baldridge et al., 1985), but it is sold chiefly for malting in the Northern Great Plains. Barley is also used for forage or grazing, but farmers must observe grazing or feeding restrictions on herbicide registration labels, as noted for winter and spring wheat (Table 3). How barley will be used and grain grading standards influence the profitability of weed management strategies for barley.

The major weeds of barley in the Northern Great Plains have been surveyed (Table 7), but weed competition and the impact of weeds on barley yield and grain quality have not been studied as much as for wheat (Zimdahl, 1980, 1990) (Table 2). Estimated dollar losses in the United States due to weeds in barley, based on expert opinion, were summarized by region (Bridges, 1992). Little has been published on the impact of weeds on barley grain quality although weeds can reduce quality. For example, Tartary buckwheat reportedly decreased barley malting quality (Remy et al., 1985).

B. Nonchemical Weed Management Methods

Crop rotations are not as rigid for spring-sown barley (Baldridge et al., 1985) as they are for winter wheat. Typical rotations include alternate barley/fallow, row crop/barley, and continuous barley. Barley can be rotated with winter- and spring-sown crops to prevent the buildup of weeds and other pests. In the Northwestern Great Plains, a 3-year winter wheat or spring wheat–barley–fallow rotation is often used (Baldridge et al., 1985). Other typical rotations include 4-year winter wheat–oilseed (generally sunflower or safflower)–barley–fallow, safflower–barley–winter wheat rotation, and winter wheat–barley (winter wheat planted no-till into barley residue). Herbicide treatment in one crop may provide residual control in later crops. For example, trifluralin applied in sunflower in the Northern Great Plains controlled foxtails in spring barley and spring wheat in the following growing season (Black and Siddoway, 1976a,b). The mechanism of residual control with trifluralin was not determined, although it probably included both crop life cycle rotation and herbicide rotation.

Selection of competitive crops is often suggested as a weed management strategy. However, there are few reports comparing barley competitiveness with weeds to competitiveness of other crops in the absence of herbicides. In Canada, barley was more competitive with green foxtail than flax (Rahman and Ashford, 1972) and more competitive with wild oat than either canola or spring wheat (O'Donovan, 1988). Barley is reportedly more competitive than spring wheat with the following weeds: quackgrass (Williams, 1977), Tartary buckwheat (Remy et al., 1985), wild buckwheat (Pavlychenko, 1940), wild mustard (Pavlychenko, 1940), and wild oat (Bell and Nalewaja, 1968; Pavlychenko, 1940). Continuous monoculture is discouraged for producing all field crops because of the potential for buildup of weed, insect, and disease problems. An irrigated barley weed management system including herbicides allowed fewer barnyardgrass and redroot pigweed plants to survive than in corn, pinto bean, or sugar beet in Colorado (Schweizer et al., 1988). The reasons barley was more competitive with particular weeds than other cereals are not known and deserve further research.

Crop management practices, such as planting date, can drastically influence barley yields in competition with weeds, as observed for spring wheat. The interval between barley and weed emergence can impact the magnitude of barley yield losses due to weeds. In fall-tilled and no-tilled barley in Quebec, delaying spring seeding past the local optimum dates (May 15 to June 22) decreased yields, in part, because annual weeds increased (Barnett et al., 1984). As planting was delayed on a fall-prepared seedbed, subsequent barley yield decreased unless the seedbed was harrowed to control weeds immediately before late planting.

The effectiveness of planting date as a weed management strategy depended on weed and crop emergence relative to one another. In Alberta, when barley emerged 1 week before wild oat, barley yields were greater than when wild oat emerged 1 week before barley (McBeath et al., 1970). The later wild oat emerged after spring wheat or barley emergence, the less yield was reduced in Canada. In England, quackgrass and hempnettle did not reduce the yield of early-planted barley, but hempnettle reduced yield if barley planting was delayed (Scragg and McKelvie, 1976).

The presence of weeds, their density, and their emergence behavior must be known when considering planting date as a weed management strategy for any cereal. For example, delayed seeding for wild oat control may favor the buildup of later emerging species, such as green and yellow foxtails, in the Northern Great Plains. Also, weed-free barley yields will be reduced if planting is delayed past the optimum local planting date.

Past research on increased barley seeding rate to suppress weeds in cereals was briefly reviewed (Zimdahl, 1980) and has been known since the 1930s (Pavlychenko, 1940). In England, seeding spring barley at 180 kg/ha reduced wild oat and quackgrass competition more than seeding at 90 kg/ha at two row widths (10 and 20 cm) (Cussans and Wilson, 1975). Barley seeding rate decreased wild oat growth more than row width.

The influence of fertilization on weed competition with barley and spring wheat was reviewed (Zimdahl, 1980). Early research showed that fertilizer improved yields only when weeds (e.g., wild oat) were controlled. Spring barley or wheat yields were not increased by broadcast fertilization with N-P in competition with wild oat in North Dakota. Wild oat took advantage of surface-applied fertilizer better than cereal crops.

C. Herbicides for Weed Control

Currently registered barley herbicides and herbicide combinations are summarized in Table 1. There are few soil-applied herbicides for use in barley, as was noted for spring wheat. Herbicide efficacy for weed management can be improved by barley competitiveness. For example, barley competition enhanced trifluralin control of green foxtail compared with control by trifluralin alone in Canada (Rahman and Ashford, 1972). Weed management with herbicides is similar in barley, durum wheat, oat, and spring wheat.

V. WEED MANAGEMENT IN OAT

The major weeds of oat in the Northern Great Plains have been surveyed (Table 7). Estimated dollar losses in the United States due to weeds in oat were summarized by region, based on expert opinion (Bridges, 1992). Because there were fewer effective registered herbicides and herbicide combinations for oat than for other cereals (Table 1), oat should not be planted on weedy fields. Fields with wild oat should be avoided since wild oat cannot be selectively controlled in oat with herbicides (Helm and McMullen, 1989) and nonchemical weed management is ineffective. The best management strategy for wild oat control is to rotate to crops on which registered wild oat-control herbicides can be safely used, rather than to plant oat.

Good weed management in preceding crops is important for subsequent oat production. Fallowing in the growing season before planting oat can be used for reducing weed infestations in oat in the Northern Great Plains and elsewhere (Mitich and George, 1981). In nonirrigated oat, fallow tillage should be delayed until fall rains stimulate weed germination, before controlling weeds with tillage (Mitich and George, 1981). In irrigated oat, irrigation can be used to stimulate weed germination before starting tillage (Mitich and George, 1981). The "antidote" or "safener" strategy has been tried on oat for wild oat control with wild oat herbicides that would normally damage oat but has not been successful enough for commercial use (Chang et al., 1974).

VI. WEED MANAGEMENT IN SORGHUM

Sorghum production practices have been summarized (Bennett et al., 1990; McKinney, 1992) and weed management practices were briefly reviewed for conventional tillage (Phillips, 1970; Wiese, 1983) and conservation tillage (no-till and reduced-till) production systems (Wiese et al., 1985). Weed problems in sorghum differ between production regions. Estimated dollar losses in the United States due to weeds in sorghum were summarized by region on the basis of expert opinion (Bridges, 1992).

A. Weed Competition

Most research on sorghum competition with weeds was conducted in the Central Great Plains with irrigated and nonirrigated sorghum populations of 250,000 plants/ha and 75,000 to 100,000 plants/ha, respectively (Table 2). Damage thresholds from weeds for sorghum were briefly reviewed (Swann, 1980), but this competition information is not used by farmers in making weed control decisions, as noted for other cereals. There is essentially no published information on how weeds impact sorghum grain quality. In the only published study on the impact of weeds on sorghum harvesting efficiency, mixed weed populations which were killed and desiccated by freezing did not influence harvesting efficiency (Burnside et al., 1969). However, combining had to be delayed to allow weeds to dry.

"Time of removal" experiments for different weeds were used to estimate the interval from sorghum emergence to control of weeds before yields were reduced. These studies demonstrated that weeds emerging within the first few weeks after crop emergence reduced yields more than later emerging weeds (Burnside and Wicks, 1967), as observed for small grains. When sorghum was grown in 40-cm rows and kept weed-free for only 1 to 3 weeks after planting, it yielded as much as sorghum kept weed-free for the entire growing season in Nebraska. If sorghum infested with smooth pigweed, foxtail, and crabgrass was kept free of weeds for 4 weeks after planting, later emerging weeds did not reduce yields. Sorghum yields were progressively reduced as the period between planting and weeding increased from 4 to 8 weeks. In later research in Nebraska, sorghum yields were reduced 4%, 12%, and 18% when mixed populations of weeds were removed 3, 4, and 5 weeks after sorghum emergence (Burnside and Wicks. 1969). If weeds were allowed to grow for 2 weeks before being controlled, yields were not reduced. This information is valuable in planning weed management measures with either cultivation or herbicides.

B. Nonchemical Weed Management Methods

The profitability of the best weed management methods depends on yield potential, the weeds that are present, and weed density (Greer, 1990). Herbicides are best used to supplement traditional nonchemical weed management measures, not to substitute for them. Nonchemical methods, such as cultivation, may be sufficient for controlling light weed infestations (Bauman and Weaver, 1991).

Crop rotation has been suggested as a way to manage weeds in sorghum, but the efficacy of particular rotations on individual weeds has not been well documented (Bauman and Weaver, 1991). Good weed management in the previous rotational crop has been demonstrated to improve weed control in sorghum in Georgia (Diawara and Banks, 1990). Combining crop rotation with herbicide rotation may control weeds in late-planted sorghum better than crop rotation alone.

Weed shifts were observed in a cultivated soybean-sorghum rotation in Nebraska (Burnside, 1978). Green foxtail, large crabgrass, and tall waterhemp decreased as problems in atrazine-treated sorghum over time, but velvetleaf and Pennsylvania smartweed increased. Rotation to winter wheat or soybean treated with trifluralin was used to manage shattercane in sorghum planted in the subsequent growing season (Wiese et al., 1977). Common milkweed has been managed by rotating to alfalfa for several years before planting sorghum (Wiese et al., 1977). Both alfalfa's perennial life cycle and repeated mowing management probably reduced common milkweed populations in later-planted sorghum. Field sandbur has been managed in sorghum by rotation to winter wheat (Wiese et al., 1977).

Row spacing influenced grain sorghum competitiveness with weeds in Oklahoma (Smith et al., 1990) and Nebraska (Burnside and Wicks, 1969). Sorghum planted in narrow rows (61 cm) was more competitive with barnyardgrass, large crabgrass, and Texas panicum than sorghum in wide rows (91 cm) (Smith et al., 1990). In Nebraska, the yield of narrow row (50 cm) sorghum competing with mixed weed populations also was greater than that of wide row (100 cm) sorghum (Burnside and Wicks, 1969), but row spacing did not influence yield when kept weed-free by hand weeding. Narrow row spacing in nonirrigated, nontilled sorghum yielded well in competition with weeds (green foxtail, tall waterhemp, and large crabgrass) (Burnside, 1977). Postemergence atrazine reduced mixed weeds more than increasing plant population did, but sorghum population was correlated with observed weed control.

C. Weed Management in Various Sorghum Cropping Systems

Herbicides registered for use on sorghum are summarized in Table 1, although many more mixtures of these herbicides are registered. Young, volunteer sorghum can be controlled with postemergence paraquat or glyphosate before emergence of rotational crops (Wiese and Chenault, 1987).

Weed Management in Conventional Sorghum

Combinations of management practices, such as crop rotation, tillage, and herbicide application, offer the most potential for weed management in conventional sorghum. Wiese et al. (1977) suggested that total reliance on herbicides in sorghum has caused weed shifts and encouraged the buildup of certain weed species on farms. In Kansas and Nebraska, pigweeds present in the 1950s were controlled by 2,4-D and propazine, leading to increases of annual grass weeds, such as fall panicum, witchgrass, field sandbur, and green foxtail (Houston and Kimbrough, 1991; Swann, 1980; Wiese et al., 1977), which were not controlled by these broadleaf herbicides. In continuous sorghum, low rates of atrazine alone encouraged the buildup of severe infestations of stinkgrass (Chamberlain et al., 1970). Many, but not all, grasses can be suppressed in sorghum with atrazine, cyanazine, propachlor, and linuron (Wiese et al., 1977), although these are generally considered to be broadleaf herbicides. Control of broadleaf weeds probably allowed field sandbur to increase in Kansas sorghum fields (Wiese, 1983), and Texas panicum has increased elsewhere, in part because it is resistant to atrazine and alachlor (Swann, 1980).

Grass Weed Management

Annual grass weeds can be controlled before sorghum planting by mechanical seedbed preparation. If preemergence herbicides, such as alachlor or metolachlor, are not applied at planting for controlling grasses, grass weeds emerging after sorghum can be controlled by cultivation, or by pendimethalin or trifluralin applied as a postemergence-incorporated lay-by treatment (Table 1). These dinitroaniline herbicides have been used successfully on weeds, such as Texas panicum, but should be applied only after sorghum brace roots are well developed (Swann, 1980). Linuron plus surfactant as a postdirected application is used for controlling both grasses and broadleaf weeds.

Weedy sorghum and other grasses are major problems for sorghum producers, especially if crop rotation is not practiced. There are five types of weedy sorghums: tall mutants, off-types, silage types, rhizomatous grassy types, and nonrhizomatous grassy types. Tall mutants are 30 to 60 cm taller than grain sorghum hybrids but appear normal otherwise. The presence of numerous tall mutants in sorghum fields suggests that foundation seed was impure. Off-type sorghum or off-color heads do not look like the planted hybrid and off-type sorghum suggests that pollen from some other sorghum interfered with normal pollination. Silage-type sorghum is tall and thick stemmed. Because silage-type sorghum matures after grain sorghum, yield losses are likely in infested fields. Rhizomatous grassy-type sorghum is tall, tillers excessively, and produces sterile seedheads that look like johnsongrass. Rhizomes of this sorghum type are not as large or well established as johnsongrass and tend to winter-kill. Nonrhizomatous grassy types (shattercane) are tall and also tiller excessively. Their loose open heads produce viable seed which shatter and are very persistent in the soil seed bank (Burnside et al., 1977). Many of these weedy sorghums were introduced as contaminants in sorghum seed.

Hand roguing or spot herbicide treatment (e.g., glyphosate or paraquat) of small infestations of both rhizomatous and nonrhizomatous grassy sorghum is recommended to limit the buildup of weed populations. Because shattercane is difficult to control in sorghum, it is preferable to manage it by crop rotation and use of effective herbicides in rotational crops (Wiese et al., 1977; Wiese, 1983), in addition to cultivation. In dryland areas, shattercane can be managed by rotating to winter wheat or alfalfa for 3 years or more (Wiese, 1983). Shattercane also can be controlled in soybean with dinitroaniline herbicides, such as trifluralin. Rotation to corn allows the use of herbicides on shattercane that would normally damage sorghum. Farm machinery should be cleaned before moving between fields to prevent infesting of new fields (transferring seed from field to field). Limiting grazing on infested fields will also minimize the spread of weed-type sorghums from one field to another in manure.

Johnsongrass, a perennial rhizomatous sorghum species, is difficult to control selectively in sorghum (Greer, 1990). Fall plowing is recommended to bring rhizomes up to the soil surface, where they will dry and freeze over winter. Multiple cultivations during winter expose rhizomes to killing low temperatures and desiccation, but cultivation alone is not completely effective on johnsongrass (Obrigawitch et al., 1990). Cultivation after johnsongrass emerges in spring can stimulate uniform emergence under favorable environmental conditions in the Southern Great Plains (Greer, 1990). Short residual, nonselective, postemergence herbicides applied in fall also controlled johnsongrass if there was sufficient fall foliage growth to treat. Glyphosate is recommended in both fall and spring before planting sorghum, usually when johnsongrass is at least 30 cm tall (Greer, 1990). Glyphosate applied in mid-June, late July, or mid-September 1 year before planting sorghum totally controlled weeds in no-till sorghum in Texas (Brown et al., 1988). Glyphosate applied in mid-June controlled weeds better than at other times. In other research, fall-applied glyphosate controlled johnsongrass better than preemergence metolachlor or pendimethalin applied in sorghum. Fall plowing or spring disking more than 7 days after glyphosate treatment provided enough time for the herbicide to translocate to roots and also to prepare a seedbed. Preemergence metolachlor or alachlor controlled seedling johnsongrass in spring in sorghum (Greer, 1990), whereas atrazine was ineffective (Houston and Kimbrough, 1991).

Perennial Broadleaf Weed Management

Postemergence herbicides for annual broadleaf control in sorghum (Table 1) failed to control perennial weeds adequately. For example, 2,4-D did not adequately control common milkweed in sorghum (Wiese et al., 1977). If one does not wish to rotate to a perennial crop, such as intensively managed alfalfa, for controlling common milkweed (Wiese et al., 1977), fall-applied glyphosate at ≥ 1.1 kg/ha can be used before harvest to control common milkweed in the following year (Gigax and Burnside, 1976). However, this may require preharvest spot treatment of common milkweed patches growing in late-harvested row crops.

Nonselective translocated herbicides have been studied as fall-applied treatments for controlling honeyvine milkweed in sorghum planted 1 year after treatment (Claassen and Moshier, 1989). The following herbicide treatments suppressed this perennial for 9 months and increased the yield of sorghum planted in the following growing season: glyphosate at 3.4 kg/ha, glyphosate plus dicamba at 1.7 plus 0.6 kg/ha, glyphosate plus 2,4-D at 1.7 plus 1.1 kg/ha, picloram at 0.28 kg/ha, and all combinations of picloram at 0.14 and 0.28 kg/ha plus 2,4-D at 0.06 to 1.1 kg/ha. However, honeyvine milkweed had regrown enough that these fall-applied herbicides did not prevent yield losses of winter wheat planted 13 months after treatment. Honeyvine milkweed had regrown enough after 24 months to require retreatment.

Field bindweed is a continuing problem in sorghum and cereals in the Great Plains. Dicamba at ≥ 4.5 kg/ha applied 1 year before planting controlled field bindweed in sorghum in Colorado (Schweizer and Swink, 1971). However, dicamba residues were too persistent and phytotoxic to most other irrigated crops to allow their recommendation. Unfortunately, dicamba at lower rates and mixes of 2,4-D and dicamba were less effective for managing field bindweed than dicamba at higher rates. Combinations of herbicide, cultivation, and crop rotation were much more effective than reliance on dicamba or 2,4-D alone for field bindweed control. Glyphosate applied after the previous crop harvest or before planting can also be used to manage field bindweed.

Weed Management in No-Till Sorghum

Data from early research indicated that no-till sorghum in the Great Plains had several limitations: inconsistent weed control (Phillips, 1969), herbicide carry-over damage to succeeding crops (Phillips, 1969), and lack of profitability. However, recent research on no-till sorghum weed control has been more successful. In a winter wheat–fallow–sorghum rotation in Nebraska, atrazine applied alone after winter wheat harvest did not adequately control volunteer wheat, downy brome, or barnyardgrass, which depleted soil moisture reserves (Wicks et al., 1988). However, atrazine plus paraquat controlled weeds better and reduced weed biomass while maintaining greater crop residues for erosion control. This combination treatment increased soil moisture conservation and subsequent no-till sorghum yields compared with conventional seedbed preparation.

Preplant or planting-time "burndown" herbicides are often mixed with soil-residual herbicides for season-long weed control in no-till sorghum (Lewis et al., 1988). Either paraquat or glyphosate can be combined with residual herbicides, such as atrazine or a mixture of atrazine plus metolachlor, for longer broadleaf weed control or broadleaf plus grass control, respectively. As expected, paraquat alone did not control weeds for the entire growing season because it has only postemergence contact activity (Wicks and Grabouski, 1986). Disking for seedbed preparation and 2,4-D application before cultivation controlled broadleaf weeds, but not grasses, although Russian thistle escaped, presumably because it is a summer annual that emerges over an extended period later in the growing season.

Early preplant herbicides play an important role in no-till sorghum production by maintaining good weed control, reducing the chance of herbicide damage to sorghum, and increasing sorghum yields. Fall-applied herbicides, such as atrazine or cyanazine, controlled emerged winter annual weeds, such as henbit in Texas (Baumann and Weaver, 1991). Cyanazine may also be applied in late fall and up to 30 days before sorghum planting. In Nebraska, triazine herbicides were applied 41 and 25 days before no-till sorghum planting in standing wheat stubble (Wicks, 1985a). Cyanazine applied 41 days before planting did not damage sorghum but caused 18% injury when applied at planting (Wicks, 1985a). Metolachlor plus 2,4-D applied before planting was superior to alachlor plus 2,4-D for grass control. In Nebraska, no-till sorghum treated with paraquat, atrazine, and terbutryn yield 33% more than sorghum produced on a tilled seedbed (Wicks and Grabouski, 1986). Herbicides applied 14 days before planting gave more consistent control than when applied at planting.

Preemergence herbicides applied after sorghum planting are restricted for use to certain states (Table 1). Sorghum seed must be treated with safeners if either alachlor and metolachlor is used (Greer, 1990). Although these herbicides controlled small-seeded annual grasses well, Texas panicum and seedling Johnsongrass escaped control (Greer, 1990). Propachlor was not as effective for grass control as alachlor or metolachlor in Oklahoma (Greer, 1990). If these preemergence herbicides are not activated by rainfall within several days of planting, light cultivation may be needed to control emerging weeds (Greer, 1990).

Some selective postemergence herbicides can be applied until sorghum is 10 to 15 cm tall. Postemergence herbicides are useful for weed management if preemergence herbicides fail or are not applied (Greer, 1990). Postemergence atrazine must be applied to grass seedlings when they are small, 1 cm tall or less, and control is inferior to that achieved on broadleaf weeds. Addition of appropriate surfactants, crop oil concentrates, and crop oils made grass control more consistent with atrazine, but only for small grasses less than 5 cm tall (Swann, 1980). Atrazine is safest on sorghum when it is treated after the three-leaf stage (Swann, 1980).

If only broadleaf weeds are present, postemergence 2,4-D or dicamba can be applied. Postemergence 2,4-D can be applied later than some other herbicides (Wiese, 1983). Weeds in sorghum can be treated only after sorghum is 10 cm tall (measured at the point of emergence of leaves from the whorl) but before it is 38 cm tall (seedhead formation). High clearance drop nozzles should be used to apply 2,4-D after sorghum is 25 cm or more tall.

Weed Management in Double Cropping

The short growing season in the northern United States limits double cropping for grain, but double cropped forage sorghum can be grown after cereals (Okoli et al., 1984). In Wisconsin, conventionally planted spring oat can be followed by no-till corn, soybean, and grain sorghum for forage. However, corn had several advantages over sorghum in this climate. Canning peas are harvested early and can also be followed by minimum-till sorghum (Ndon et al., 1982). However, phytotoxic carryover residues of trifluralin applied to peas reduced sorghum stands 44% and yield 17% (Ndon et al., 1982). Herbicides in the preceding double crop must be chosen to minimize the chance of carryover damage to sorghum. Either atrazine plus paraquat or paraquat plus 2,4-D in reduced-till sorghum controlled green foxtail and common lambsquarters very well.

Rescue Herbicide Treatments and Sorghum Desiccation

"Salvage" or "rescue" herbicide treatments can be applied late in sorghum development to control large weed escapes that might interfere with harvest. Such treatments aid in harvest, but generally do not improve sorghum yields. Nonselective herbicides, such as paraquat and glyphosate, have been applied to tall-growing weeds in sorghum with a recirculating sprayer (Carlson and Burnside, 1981) or rope-wick applicator (Rhodes and Shelby, 1991). Spray drift and splash caused unacceptable grain sorghum injury with the recirculating sprayer (Carlson and Burnside, 1981).

Interest in preharvest desiccation of sorghum dates back to the 1950s (Bovey, 1969). Desiccants are applied at physiological maturity (i.e., black layer formation). At this time, grain moisture content is 30% to 40%, whereas sorghum can be safely stored only below 14%. Properly timed preharvest desiccation is advantageous because it aids in harvesting by drying green sorghum leaves and weeds without hastening sorghum maturity. Paraquat and diquat were shown to be effective desiccants in the late 1960s (Bovey, 1969; Gigax and Burnside, 1976). Sodium chlorate also can be used at moisture contents below 25% (Houston and Kimbrough, 1991). Preharvest glyphosate hastened seed drying without affecting sorghum seed germination or sorghum lodging (Gigax and Burnside, 1976). Preharvest glyphosate was superior to paraquat or sodium chlorate because these latter two treatments did not adequately kill the lower sorghum leaves (Bovey et al., 1975a). There were added weed control benefits of preharvest desiccation with glyphosate, especially for perennial weeds. Common milkweed regrowth was reduced in the growing season after applying fall-applied glyphosate to sorghum (Gigax and Burnside, 1976). Emerged johnsongrass, cogongrass, and browntop panicum present at harvest were killed by glyphosate (Bovey et al., 1975a).

Weed Management in Ratoon Cropping

Ratoon cropping of sorghum is not widely practiced in the United States because the growing season is usually too short. Ratoon sorghum can be grown when the growing season exceeds 225 days, but grain sorghum varieties with good regrowth after first harvest must be selected (Banks and Duncan, 1983). Successful weed control in first crop sorghum improved weed control in ratoon (second crop) in Georgia. Soil disturbance by hoeing after the first crop was harvested stimulated volunteer sorghum and annual grass weed growth. The method of stalk removal also influenced sorghum seed shattering and the incidence of volunteer sorghum in the ratoon crop. After the first harvest, nonselective herbicides controlled weeds for the subsequent ratoon sorghum crop. However, glyphosate applied to the stalks of the first crop caused some chlorosis and stunting of the subsequent ratoon crop. Metolachlor plus propazine controlled volunteer sorghum and grass weeds in ratoon crop sorghum when applied preemergence at planting and followed at harvest by metolachlor, cyanazine, or pendimethalin.

VII. WEED MANAGEMENT IN RICE

Rice is a major food crop in the world, grown under a wide range of cultural and environmental conditions; however, all of the rice grown in the United States is flooded, with many variations on when and how flooding is applied. Weed management in rice has been extensively reviewed (Chisaka, 1977; Eastin, 1983; Eastin, 1991; Moody, 1991; Smith et al., 1977). Weed management in rice consists of a combination of chemical, cultural, and preventative measures designed to reduce their occurrence and/or competitiveness. In the United States herbicides play a major role in weed management. Without herbicides, direct and indirect effect of weeds would account for a 70% reduction in rice yield (Abernathy, 1981). Even with present weed management, including use of herbicides, weeds account for an average 17% yield reduction in United States rice (Chandler, 1981). Estimated dollar losses in the United States due to weeds in rice has been summarized on the basis of expert opinion (Bridges, 1992).

A. Weed Competition

Weed competition with rice has been documented to reduce yields not only in the United States (Smith, 1968; Smith, 1970; Smith 1983b; Smith, 1988; Smith et al., 1977; Zimdahl, 1980) but elsewhere in the world (Zimdahl, 1980; Chisaka, 1977) (Table 1). Weeds typically compete with rice for light, nutrients, space, and water. However, competition for water is not as critical as with most nonirrigated crops since all United States rice is flooded for at least part of the growing season.

In addition to reducing yield, weeds can increase rice lodging, which interferes with harvesting operations and further reduces grain quality and yield. Seed from weeds such as dayflower, hemp sesbania, morningglory, and red rice are also very difficult to separate from the rice grain, thereby decreasing quality. Many weeds harbor disease organisms and insects which can attack rice (Eastin, 1983).

Integrated weed management systems depend on the weeds present or expected to be present. Weed histories of individual fields are very helpful in developing weed management systems to address particular problems. An integrated weed management system depends on the interaction of several types of weed control measures and requires a higher level of weed management than the use of herbicides alone. Various components of a weed management system are discussed in the sections that follow.

B. Nonchemical Weed Management Methods

Crop Rotation

Continuous rice culture tends to increase infestations of difficult-to-control weed species and soil-borne disease problems which may be present. In much of the southern United States, a rotation involving 1 year in rice and 2 years in other crops is normal. This rotation can be rice/soybean/soybean, rice/sorghum/soybean, rice/cotton/soybean, or a number of other crop options. There have been cases in Texas of effectively managing johnsongrass in sorghum by growing rice in the preceding 2 years, offsetting levees the second year in order to control johnsongrass, which is killed by flooding. Weed management techniques are available in several other crops which can be rotated with rice, such as cotton, sorghum, and soybeans, to control many problem weeds in rice, thereby helping to deplete the weed seed reserve. Many weeds which are difficult to control in rice, such as alligatorweed, red rice, and sprangletop, can be effectively controlled in soybeans or other broadleaf crops. Aquatic alligatorweed is partially controlled by lack of adequate moisture in rotational crops, while many grasses and broadleaf weeds are controlled by cultivation and/or herbicides used in the rotational crop.

The potential carryover of herbicides from one crop to others, such as atrazine from sorghum to rice, should be considered in planning crop rotations. Very few, if any, rice herbicides pose any significant carryover hazards to rotational crops. Many weeds can be controlled with crop rotations combined with other good weed management practices that otherwise would not be controlled in a continuous monoculture.

Prevention

If a weed species is not present in a field, the best management technique is to prevent its introduction. This requires good sanitation, particularly in using clean equipment. Equipment coming from a weedy field must be cleaned before entering a clean field; otherwise mud and debris on equipment can carry weed seed into clean fields.

The use of weed-free seed for planting cannot be overemphasized. In most states, even certified rice seed can contain some red rice seed; therefore, not only should the seed be certified but its source should be known in order to determine whether the seed producer had any red rice. Other weed seed can also be present with rice seed. The main consideration in preventative weed management is to do the necessary things to prevent introducing new weed seed. For a detailed discussion of preventative weed management refer to Chapter 3.

Cultural Practices

Many cultural practices used in rice production influence weeds and their management. Most cultural practices have been developed to optimize rice production; however, some have evolved specificially as weed management practices.

Precision grading of rice fields allows better water management by allowing shallower flooding and use of smaller levees, allowing a larger percentage of field area to be used to grow rice. In level fields with many high and low spots water is too deep for rice in low spots and too shallow for weed control in high spots. Precision grading improves weed management through better control of water depth. It also allows water to be applied quickly and more uniformly.

The method of seedbed preparation is determined by the anticipated seeding method. Rice can be 1) water seeded, using presprouted rice; 2) water seeded, using dry rice seed; 3) dry seeded, broadcast on the soil surface; or 4) drilled with a grain drill. Seeding method often determines the type of weeds present in rice. If rice is seeded on dry ground followed by flushing for emergence and moisture until flooding, terrestrial weeds predominate early in the growing season. Aquatic weeds predominate when rice is water seeded, particularly when the water is never completely drained. Each of these situations requires a different weed management strategy.

A dense rice stand can outcompete some weed species. Seeding rates resulting in rice stands of 160 to 200 healthy plants per square meter (15–20 per square foot) are optimum for both yield and competition with weeds. Current, widely grown semidwarf rice cultivars are less competitive with weeds than traditional cultivars; therefore, early season weed management is more critical for semidwarf cultivars than for conventional cultivars.

Planting date influences the type of weeds and difficulty of control. If rice is planted too early, it does not compete well with weeds because of slow crop growth in cool weather. Early planting also favors weeds such as barnyardgrass, while late planting encourages dayflower more than early planting. Planting date must be chosen for its effect on not only weeds and their management, but also yield potential. Planting too late or too early decreased yield even if rice was weed-free.

Fertilizer can significantly affect weeds and their control. Very early phosphorus application generally stimulates weed seed germination and growth while phosphorus applied into flooded fields stimulates growth of algae and other aquatic weeds. To reduce this weed stimulation, phosphorus is usually applied to rotational crops 1 year before planting rice. If this was not done, only enough phosphorus is used for the desired yield goal, since excess phosphorus not only stimulates weeds but reduces rice yields. Nitrogen timing can influence weed growth. Nitrogen is generally split between two or three applications; this practice not only is better for rice but eliminates excess nitrogen available to stimulate early weed growth. By timing fertilizer applications to time of flooding and herbicide applications, fertilizer and herbicide use is more efficient and effective.

Water management, a major cultural practice for rice, can also be used for weed management. Not only the timing of flooding is important; the depth of flooding is also critical in weed control. However, in practice there must be a compromise between the use of shallow water for rice growth and deep water for weed control. While shallow flooding is better for rice growth and yield, deep flooding (10–20 centimeters) controls many terrestrial weeds, such as young barnyardgrass. However, if applied early in the growing season, flooding may cause rice internodes to elongate, resulting in a weakened plant with a tendency to lodge before harvest. Early flooding also stimulates growth of aquatic weeds such as algae, alligatorweed, and ducksalad, and waterprimrose, which may necessitate draining fields and allowing the soil to dry. This stresses rice and allows terrestrial weeds, such as barnyardgrass, to germinate. While water management cannot be used as the only early season weed management practice, understanding its effect allows it to be used in a total weed management program. Because older rice tolerates deeper flooding than many weeds, deeper flooding can be used to reduce or prevent terrestrial weed germination.

Mechanical cultivation is not feasible in United States rice production. Hand weeding, or roguing, of specific weeds, such as red rice, in seed rice fields is practiced on limited hectares; however, this is not economically feasible in production fields.

Biological Weed Management

A biological control for northern jointvetch using the fungus *Colletotrichum gloeosporioides* species Asaeschynomene has been developed. This fungus is specific for northern jointvetch. Moreover, there are regions where United States rice is grown where it is feasible to use this biological control late in the growing season even after the time when most herbicides normally would be applied.

C. Herbicides for Weed Management

Even with available nonchemical weed management practices, herbicides are considered the backbone of any rice weed management program in the United States (Table 10). Rice herbicides, like all pesticides, must be properly used to prevent lack of weed control, crop injury, and/or damage to the environment. Since herbicide registrations and labels are constantly changing, current labels and information available through state agricultural extension services and state agricultural experiment stations should be consulted for current information and recommendations.

Essentially all United States rice is treated with at least one herbicide during the growing season. Barnyardgrass, junglerice, dayflower, sprangletop, red rice, and aquatics are the major weeds of United States rice. Aquatics include alligatorweed, waterprimroses, algae, ducksalad, and several other common weeds.

Table 10 Herbicides Registered for Use on Rice in 1992 in the United States

Time of application	Herbicide
Preplant	Molinate
	Thiobencarb
Preemergence	Quinclorac
	Thiobencarb
Early postemergence	Acifluorfen + surfactant
	Bentazon
	Propanil
	Propanil + molinate
	Propanil + pendimethalin
	Propanil + thiobencarb
	Quinclorac
Midpostemergence	Bromoxynil
	Fenoxaprop
Postflooding	Acifluorfen + surfactant
	Bensulfuron methyl
	Bentazon
	Molinate
Late postemergence	MCPA
	Triclopyr
	2,4-D

Note: Herbicides and rates are based on current labels or expected labels. Before using an herbicide refer to a current label for appropriate rate and species controlled.

Propanil has been the standard herbicide in rice for many years. It is a postemergence herbicide that, while used primarily for barnyardgrass and junglerice control, is very effective on many other common weeds in rice. Since propanil has no soil residual, weeds can reinfest rapidly after application, however. When using propanil without a residual herbicide, a second application is generally needed to control weeds until rice is flooded. While propanil controls barnyardgrass, addition of other herbicides expands the spectrum of weed control and/or provides residual activity. Addition of pendimethalin or thiobencarb to propanil not only provides residual control of barnyardgrass but increases sprangletop control. Addition of molinate to propanil, while not providing residual control, helps control emerged sprangletop and dayflower. Propanil interacts with carbamate and organophosphate insecticides and can injure rice. Therefore, these insecticides should not be applied with or within a few days of application of propanil.

Preemergence thiobencarb controls barnyardgrass well when applied to dry seeded rice; however, it does not control broadleaf signalgrass and several problem broadleaf weeds which propanil will control as a postemergence herbicide. Preemergence quinclorac controls barnyardgrass and many other grasses and broadleaf weeds well in rice, but it is ineffective on sprangletop.

Fenoxyprop-ethyl applied to four- to five-leaf rice before flooding will control many grass weeds. This herbicide controls larger barnyardgrass than propanil or quinclorac. It also controls sprangletop but does not completely control broadleaf weeds. Also, if rice is too young when treated, fenoxyprop-ethyl can injure it.

Acifluorfen or bentazon can be used postemergence to control several broadleaf weeds. Bentazon is particularly effective on dayflower and yellow nutsedge while acifluorfen controls hemp sesbania well. Just before flooding, bromoxynil can be applied to control many broadleaf weeds. Fenoxaprop can also be used at this time.

After rice is flooded, molinate granules can be used for control of barnyardgrass, dayflower, and sprangletop while bensulfuron-methyl can be used to control many aquatic broadleaf weeds. Bentazon or acifluorfen can be used either preflooding or postflooding.

After rice has started tillering but before panicle initiation, 2,4-D, MCPA, or triclopyr can be used for control of broadleaf weeds that have escaped previous herbicide applications.

Red rice is particularly difficult to manage in rice since it is the same species as the rice crop. At present no herbicides completely control red rice in rice; however, there are some management techniques that can be used to suppress it. Red rice is best controlled by rotating to a crop, such as cotton, sorghum, or soybeans, in which herbicides for red rice control are available. High rates of alachlor, metolachlor, or trifluralin in soybeans or cotton, or propazine or atrazine in sorghum, or atrazine in corn control red rice. Usually 2 years of this program will reduce red rice to a level at which a rice crop can be produced.

Several weed management practices can be combined to provide fair to good red rice suppression in rice, if red rice infestations are sparse. Cultural practices primarily involve use of continuous or pinpoint flooding combined with use of preplant-incorporated molinate or preplant, surface-applied thiobencarb before flooding. Presprouted rice is then broadcast by air into flooded fields, and flooded fields are either drained for a short time or not drained to prevent red rice from germinating. Continuous flooding inhibits germination of red rice. Table 10 lists the herbicides registered for use on rice in 1992 in the United States.

VIII. CONTROL OF VOLUNTEER CEREALS AND SORGHUM

Volunteer cereals are occasionally weeds in some rotational crops or fallow. Volunteer cereals may have to be controlled before planting no-till row crops, or if thin spring cereal stands warrant recropping to later-sown crops (Chow, 1983; Donald, 1990b; Wicks et al., 1988).

Although tillage for mechanical fallow or seedbed preparation kills seedling winter wheat and other cereals before planting other crops, large established plants are controlled better with herbicides. In no-tillage systems, herbicides are substituted for tillage for controlling volunteer cereals. In no-tillage systems, nonselective herbicides, such as glyphosate and paraquat, serve this role but must be applied before crop emergence. In fact, no-tillage farming systems would probably not be possible for erosion control without nonselective postemergence herbicides for contact weed control. Volunteer cereals can be selectively controlled in the major field crops with several postemergence grass herbicides (Table 6) (Beardmore and Linscott, 1989; Chow, 1983; Friesen and Wall, 1990; Weston, 1990).

The optimum time and rate of glyphosate or paraquat to use to control volunteer cereals have been evaluated so that only the lowest rates and least costly treatments can be used before crop emergence in no-till farming systems. Postemergence glyphosate or paraquat at 0.25 kg/ha adequately controlled winter wheat (Wiese and Chenault, 1987). Paraquat at 0.3 to 0.6 kg/ha applied postemergence reduced winter wheat biomass 84% when applied at the one- to three-leaf stage but was less effective on larger plants (Anderson and Nielsen, 1991). Glyphosate is generally more effective than paraquat for controlling large, established winter wheat and other cereals.

Postemergence atrazine plus either glyphosate or paraquat adequately controlled both established volunteer winter wheat and downy brome before planting of no-till sorghum (Wicks et al., 1988). In contrast, preemergence atrazine alone did not control these species well (Wicks et al., 1988). Combinations of atrazine with either glyphosate or paraquat can also be used for weed control in no-till corn.

Various surfactants have been added to glyphosate in an attempt to improve volunteer cereal control with lower application rates (O'Sullivan et al., 1981). Many surfactants and other additives, such as ammonium sulfate, increase glyphosate phytotoxicity to weeds and volunteer cereals (McWhorter and Derting, 1985; O'Keeffe and Makepeace, 1985; Turner, 1985). Directions for using spray tank additives with glyphosate are included in the registration label. Glyphosate applied in low spray volumes was more phytotoxic to cereals than when applied in higher volumes, allowing lower application rates to be used for controlling volunteer cereals (O'Sullivan et al., 1981).

Residues of rye, barley, and wheat killed with either paraquat or glyphosate suppressed weed growth for at least part of the growing season in no-till farming systems. For example, glyphosate-killed cereal residue prevented weed growth for 45 days after treatment in spring, but weeds escaped by 60 days after treatment in Kentucky (Weston, 1990). Herbicide-killed rye and wheat residues suppressed weeds longer than barley residues (Weston, 1990). Corn can be no-till planted into herbicide-killed barley, rye, oat, and spring wheat (Moschler et al., 1967). Paraquat plus atrazine successfully killed weeds and volunteer barley growing in barley stubble before no-till planting. Volunteer rye, oat, and spring wheat were easier to kill than barley with either atrazine or paraquat alone. Herbicide-killed cereal residues have potential for managing weeds in no-tillage farming systems in some locations, but more research is required to ensure that such systems control weeds as predictably as herbicides do now.

IX. CEREAL CROP DAMAGE FROM HERBICIDES

Yield-reducing crop damage from herbicides can occasionally limit the potential benefit of herbicides for reducing yield loss due to weed competition. Registered herbicides (Table 1) can sometimes damage cereals on particular soils, under abnormal or extreme environmental conditions, if they are improperly applied, or if the application is improperly timed relative to cereal growth stage. Common application errors include incorrect sprayer calibration and overlapping spray swaths, especially if sprayers are not turned off when turning near field borders.

A. Damage to Grain Crops from Registered Herbicides

Winter and Spring Wheat Damage

Damage to winter and spring wheat from registered herbicides has been reviewed for the following herbicides: diclofop (O'Sullivan, 1990), difenzoquat (Donald, 1990e), propanil (Eberlein, 1990), picolinic acid herbicides (Gillespie, 1990; Heering and Peeper, 1991; Martin et al., 1989; Ogg and Young, 1991), substituted ureas, triazines, and triazinones (Baker and Peeper, 1990; Ratliff et al., 1991), and sulfonylurea herbicides (Donald, 1990d; Ferreira et al., 1990a).

The damage that phenoxy herbicides can cause wheat has been studied most and early studies on these herbicides have shaped subsequent research. Several factors influenced the phytotoxicity of postemergence phenoxy herbicides to spring wheat: herbicide formulation, application rate, growth stage of cereals when treated, cereal variety, and seasonal environment (Shaw et al., 1955). Injury symptoms from 2,4-D and other phenoxy herbicides were unique and diagnostic (Olson et al., 1951; Shaw et al., 1955). Symptoms varied with the rate and date of 2,4-D application; they included leaf rolling, an erect growth habit, reduced tillering, failure of heads to emerge from the boot, emergence of heads laterally through rolled leaves, stunting, variable numbers of seed per seedhead, and delayed maturity (Shaw et al., 1955). Both phenoxy and benzoic acid herbicides injured plant parts undergoing meristematic activity at the time of treatment more than nongrowing tissue (Friesen et al., 1964; Lemerle et al., 1986). Consequently, treatment timing determined which of the injury symptoms named appeared. Yields are not always reduced even if shoots become injured by phenoxy herbicides. Yield losses were greatest if 2,4-D was applied soon after emergence (Olson et al., 1951). Yield also was reduced if phenoxy herbicides were applied between spring wheat emergence and tillering. The safest period to apply phenoxy herbicides was from the three-leaf stage until just before the boot stage (Olson et al., 1951; Shaw et al., 1955). Yield losses resulted from phenoxy injury during seedhead formation (Olson et al., 1951) or flowering (Shaw et al., 1955). Such late treatments usually are unwarranted for preventing yield loss due to weeds but may expedite harvesting.

Properly timed herbicide application to match tolerant growth stages of spring and winter wheat is critical to prevent damage or yield loss from phenoxy herbicides, such as 2,4-D and MCPA (Ivany et al., 1990; Martin et al., 1989; Schroeder and Banks, 1989); benzoic acid herbicides, such as dicamba (Martin et al., 1989); picolinic acid herbicides, such as clopyralid and picloram (Gillespie, 1990; Heering and Peeper, 1991; Martin et al., 1989; Ogg and Young, 1991); and triazinones, such as metribuzin (Blackshaw, 1990).

Commercially available cultivars of winter and spring wheat differ in their susceptibility to certain registered herbicides. Some normally tolerant varieties may be damaged if herbicide rates above the registered use rate are applied. Varietal tolerance is particularly critical for metribuzin on winter wheat (Baker and Peeper, 1990; Ratliff et al., 1991) and difenzoquat on spring wheat (Donald, 1990e). However, visible damage from herbicides may not always decrease spring wheat yield (Ivany and Nass, 1984). Current herbicide registration labels should be consulted because herbicide use may be limited to particular varieties.

Barley Damage

Both barley yields and malting quality can be reduced if phenoxy or benzoic acid herbicides are improperly timed in relation to barley growth stage or if high rates are applied (Lemerle et al., 1986; Martin et al., 1989; Olson et al., 1951; Shaw et al., 1955), as observed for spring wheat. Phenoxy and benzoic acid herbicide (Friesen et al., 1964; Lemerle et al., 1986; Briggs, 1978; Shaw et al., 1955) and picloram (Chang and Foy, 1971) injury symptoms were also similar to those in spring wheat.

Treatment timing with phenoxy and benzoic acid herbicides influenced crop yield, crop quality, and injury symptoms observed. When 2,4-D was applied to winter barley at various growth stages (Olson et al., 1951; Shaw et al., 1955), barley tolerated 2,4-D best between tillering and the two-joint stage, compared to the five-leaf stage, booting, or full bloom. As plants aged, fewer symptoms were observed on the main stem, but tillers exhibited increased injury (Friesen et al., 1964). Dicamba, applied later, injured tillers and had a greater impact on yield because seed production from tillers was more important to total yield than seed production from the main stem. Differences were observed in the response of barley varieties to herbicides, as noted for wheat (Clay et al., 1988; Hageman and Behrens, 1981; Lemerle et al., 1986; Miller et al., 1978).

Sorghum Damage

Atrazine can occasionally decrease sorghum yield by reducing stand and stunting surviving plants when applied preemergence or postemergence and burning leaves when applied postemergence (Swann, 1980). Atrazine damage to sorghum was rate dependent (Burnside and Moomaw, 1975). The registered rate range for preemergence- or postemergence-applied atrazine varies with soil texture, pH, organic matter, and location in order to minimize the risk of sorghum damage (Houston and Kimbrough, 1991). Irrigation after treatment caused more damage from atrazine, especially on sandy soils, because it leached into the root zone and sorghum roots absorbed it (Chamberlain et al., 1970). Fall plowing reduced the likelihood of atrazine injury more than spring plowing at one of two locations in Nebraska (Burnside and Moomaw, 1975), presumably by reducing root uptake of atrazine when sorghum was small.

Lower atrazine rates were registered for postemergence-applied than for preemergence-applied atrazine, decreasing the likelihood of crop damage (Houston and Kimbrough, 1991). In fact, in certain regions, atrazine can only be applied postemergence because of the potential for sorghum damage from preemergence treatment. Postemergence atrazine was less likely to damage sorghum if roots were well established when plants were treated. As sorghum developed, it became more tolerant to postemergence atrazine (Chamberlain et al., 1970; Swann, 1980).

Metolachlor or alachlor can occasionally damage sorghum; injury symptoms include reduced stand (Roeth et al., 1983; Simkins et al., 1980), stunting (Simkins et al., 1980), delayed maturity (Simkins et al., 1980), and decreased yield (Simkins et al., 1980). Like atrazine injury, metolachlor or alachlor injury depended on application rate and soil characteristics (texture, clay, organic matter, and pH) (Houston and Kimbrough, 1991), even if sorghum seed was treated with a safener to prevent damage. For example, alachlor damaged unsafened sorghum grown on coarse textured soils but did not damage sorghum when soil organic matter was more than 3% (Houston and Kimbrough, 1991).

Soil moisture also modified sorghum tolerance to alachlor and metolachlor (Ketchersid et al., 1981). Sorghum was not damaged by alachlor or metolachlor if the soil surface remained dry until sorghum emerged (Ketchersid et al., 1981), presumably reducing herbicide uptake by the emerging sorghum shoot. Alachlor or metolachlor was more likely to damage sorghum when shoots emerged under moist conditions in the field, conditions enhancing herbicide uptake. Herbicide incorporation modified sorghum tolerance. Metolachlor was less damaging applied preemergence than preplant-incorporated (Roeth et al., 1983).

Since the 1960s, seed-applied safeners have been tested on sorghum to protect it from thiocarbamate or chloroacetamide herbicides without success in reducing crop damage (Eastin, 1972). More recently introduced seed safeners make use of metolachlor and alachlor on grain sorghum commercially possible (Devlin et al., 1983; Foy and Witt, 1990; Wiese et al., 1986). Seed-applied flurazole (Screen) is used for alachlor, and cyometrinil (Concept II) is used for metolachlor (Roeth et al., 1983). Cyometrinil-treated grain and forage sorghum seed are commercially available for use with metolachlor and atrazine plus metolachlor. These safeners act by enhancing alachlor or metolachlor degradation in sorghum (Fuerst and Gronwald, 1986; Zama and Hatzios, 1986).

Sometimes sorghum has been damaged by postemergence 2,4-D (Ross, 1970), especially if sorghum was small when treated (Wiese, 1983). Injury symptoms included inhibited root development followed by later lodging, stalk brittleness, stunting, leaf curling, head sterility, and reduced yield. Registered broadcast application of 2,4-D is limited to sorghum that is less than 20 cm tall; directed or drop nozzle application is suggested thereafter so that the leaf whorl of sorghum is not treated (Houston and Kimbrough, 1991). If weeds emerge with sorghum, farmers sometimes treat sorghum when it is too young, risking damage. If sorghum emerges before weeds, herbicide application at the wrong time is less likely (Wiese, 1983). Treatment of sorghum at the boot, flowering, and early dough stages can reduce yields.

Trifluralin and pendimethalin are registered for controlling grass and small-seeded broadleaf weeds only as postemergence lay-by treatments in sorghum. Sorghum generally was not damaged by postemergence- (lay-by)-applied trifluralin incorporated with a rolling cultivator or sweep after sorghum roots were well developed (Banks et al., 1978). This herbicide-incorporating implement threw treated soil into the rows to control weeds within the row (Banks et al., 1978). The selectivity of registered lay-by treatment and unregistered preplant-incorporated trifluralin to sorghum was positional (Barrentine and Warren, 1971). Consequently, sorghum roots of very young seedlings growing into trifluralin-treated soil can be stunted, limiting later sorghum shoot growth. In fact, sorghum has been used as a bioassay specie to test for the persistence of the dinitroaniline herbicides isopropalin, nitralin, and trifluralin (Brewer et al., 1982; Hamilton, 1979; Nelson et al., 1983; Romanowski and Libik, 1978).

B. Damage to Grain Crops from Herbicide Drift and Carryover Applied to Other Crops

Winter Wheat

On occasion, winter wheat can be damaged by persistent residues of herbicides applied to previously grown rotational crops. Triazine herbicides, most commonly atrazine applied to corn or sorghum, have damaged later sown winter wheat and other cereals, especially after drought (Wiese et al., 1986). Drought also increased the chance of damage to winter wheat from atrazine applied for weed management in "chemical fallow" (14 month fallow, pH 7.4 and 1.3% organic matter) (Anderson, 1984). Atrazine used for chemical fallow may be more damaging to winter wheat planted on eroded hilltops because soil organic matter is often lower there than elsewhere and hilltops may be more arid, leading to greater atrazine persistence.

Herbicides other than atrazine reportedly can carry over to damage rotational winter wheat. Norflurazon applied to cotton damaged winter wheat planted 14 months after application (Keeling and Abernathy, 1989). Imazapyr applied in fallow for field bindweed management (not now registered for use) damaged winter wheat planted 122 days later in Texas (Schoenhals et al., 1990).

The potential for herbicide carryover damage should be considered when planning crop rotations including winter wheat and other cereals. Herbicide registration labels should be consulted for appropriate regional waiting periods for planting winter wheat after herbicide application, to prevent damaging rotational winter wheat. For example, short residual herbicides, such as cyanazine, alachlor, and metolachlor, should be substituted for atrazine in spring when planting corn or sorghum before fall-sown winter wheat (Wiese et al., 1986). Triazine herbicides persist longer when environmental conditions, such as limited soil moisture, reduce microbial degradation of triazines or alkaline soil pH limits chemical hydrolysis of triazines to nonphytotoxic degradation products.

Barley Damage

Atrazine residues have damaged barley planted a year after atrazine-treated corn (Baldridge et al., 1985; Brinkman et al., 1980a). Spring barley or spring wheat tolerated atrazine carryover better than oat in Wisconsin (Brinkman et al., 1980a). In other research, picloram carryover did not reduce barley germination, but it inhibited subsequent barley growth (Chang and Foy, 1971). Also, ethofumesate applied to sugar beet at commercial rates persisted and damaged later sown winter wheat and barley in Colorado (Schweizer, 1975). Residues of diuron, prometryn, and trifluralin applied to safflower did not persist to damage barley in Arizona (Hamilton, 1979). Barley tolerance to trifluralin is not surprising because it is registered for use on barley.

Oat Damage

Oat has been damaged by atrazine applied 1 year earlier to corn (Baldridge et al., 1985; Brinkman et al., 1980a,b). Oat grain quality (100-seed weight and groat protein percentage) were reduced when atrazine damage was severe (Brinkman et al., 1980b). Tillage system modified oat damage from persistent residues of atrazine in Nebraska (Burnside and Wicks, 1980). Oat grown in reduced tillage were less damaged than conventionally tilled oat 1 year after atrazine application. Different oat varieties were damaged to different extents by atrazine in Wisconsin (Brinkman et al., 1980b; Smith and Buchholtz, 1964). However, differential varietal tolerance was not great enough to warrant breeding for atrazine resistance (Brinkman et al., 1980b). In northern states such as Wisconsin, it is advisable to replant something other than oat if atrazine is applied 1 year earlier (Smith and Bucholtz, 1964). Granular formulations of atrazine persisted longer than other formulations (Bucholtz, 1965), and environmental conditions reducing atrazine breakdown enhanced the likelihood of residual damage to oat 1 year after atrazine application to corn (Bucholtz, 1965; Brinkman et al., 1980a,b). Consequently, atrazine carryover damage is more likely in cold or arid environments. In western Nebraska, EPTC and trifluralin also carried over to damage oat when incorporated with a rotary-hoe-like implement or disk (Robison and Fenster, 1968).

Sorghum Damage

Sorghum can be damaged by drift, especially of sulfonylurea herbicides, such as nicosulfuron or primisulfuron, applied to corn (Rhodes, 1990). Herbicides can also persist to damage rotational sorghum. In Wisconsin, the stand and yield of sorghum planted after harvesting trifluralin-treated canning peas were reduced 44% and 17%, respectively (Ndon et al., 1982). In Nebraska, sorghum planted after soybean treated with trifluralin was damaged by phytotoxic trifluralin residues (Burnside, 1978). Damage from trifluralin was less after moldboard plowing than on no-till plots. Trifluralin applied to winter-planted safflower in Arizona persisted longer in some years to damage bioassay sorghum planted in soil collected in the following spring (Hamilton, 1979).

Persistent residues of chlorsulfuron and some other sulfonylurea herbicides applied to cereals can damage sorghum (Peterson and Arnold, 1986). Typical sulfonylurea herbicide injury symptoms on sorghum include stunting, purple leaf coloration, and reduced yield. Herbicide-delayed maturity could reduce yield during years of early frost. Sorghum yield was reduced up to 60% when planted 15 months after chlorsulfuron in Colorado and Kansas; no reductions were observed after 27 months (Sutherland and Long, 1987). Similar chlorsulfuron persistence and degrees of crop damage have been observed elsewhere (Peterson and Arnold, 1986; Wiese et al., 1988). Sorghum yields were not decreased when planted 12 to 26 months after picloram application in Texas, depending upon application rate (Bovey et al., 1975b).

Sorghum is damaged by thiocarbamate herbicides, such as EPTC and butylate, which are registered for weed control in corn (Oliver et al., 1968). While residues of these herbicides usually do not persist from one year to the next at damaging concentrations, sorghum may be injured if it is planted into EPTC-treated soil soon after herbicide application. This situation might arise during replanting sorghum soon after failure of a corn crop to emerge.

Herbicides applied to cotton have carried over to damage sorghum under certain circumstances. In Tennessee, fluometuron damaged sorghum planted 3 to 9 weeks after treatment (Jackson et al., 1978). This situation might arise if sorghum were replanted after poor cotton emergence. Sorghum was successfully planted 3 weeks after band or broadcast application of fluometuron, but a 9 week delay was required after a broadcast application (Jackson et al., 1978). In Texas, preplant-incorporated and preemergence-applied norflurazon applied to cotton damaged sorghum planted 14 months later (Keeling et al., 1989). Preplant-incorporated norflurazon was more persistent than preemergence norflurazon, as measured by sorghum damage.

Sorghum has been injured by imidazolinone herbicides. In Texas, imazapyr in fallow controlled field bindweed (89%) for 1 year (Schoenhals et al., 1990), but sorghum planted the spring after application was injured (15%). In Virginia, imazapyr on pea and snap bean injured sorghum more than 50% when planted in the following growing season (Vencill et al., 1990). Consequently, this use is currently not registered. Imazethapyr residues can persist to damage sorghum the year after application to soybean.

C. Crop Damage to Other Crops from Herbicides Applied to Cereals

Spring Wheat Herbicides

Herbicides applied to winter or spring wheat can damage other crops, either as drift or because phytotoxic residues persist to damage later-sown rotational crops (Table 6). For example, drift from volatile phenoxy or benzoic acid herbicides (2,4-D, MCPA, dicamba, and picloram) has damaged sunflower, soybeans, and sugar beets, as well as shelterbelt trees, in the Northern Great Plains (Donald and Nalewaja, 1990). Carryover of the imidazolinone herbicide imazamethabenz injured sugar beet, lentil, and rape nearly 14 months after application in Montana (Fellows et al., 1990). Imazamethabenz persistence was greater in soils with lower pH. The sulfonylurea herbicide chlorsulfuron limited which rotational crops could be successfully grown for up to 7 years after application at two locations in Alberta (Moyer et al., 1990). Moyer et al. (1990) suggested that the need for long recropping intervals was due to Alberta's low temperature and high soil pH, which lengthened chlorsulfuron persistence. Because of the potential for carryover damage from chlorsulfuron and the development of sulfonylurea herbicide-resistant weeds, use of chlorsulfuron has been severely restricted compared to its initial registration label. Other shorter-residual sulfonylurea herbicides have replaced chlorsulfuron for use on cereals in many regions.

Sorghum Herbicides

Drift of 2,4-D applied to sorghum damaged cotton more than soybean (Houston and Kimbrough, 1991). In contrast, dicamba drift from applications to sorghum damaged soybean more than cotton. Bentazon is a useful substitute for 2,4-D or dicamba on sorghum where the chance of drift to neighboring crops poses problems of crop damage (Houston and Kimbrough, 1991).

Residual atrazine from applications to sorghum has damaged subsequently planted crops, as noted earlier for cereals (Table 6). In Nebraska, atrazine damaged rotational soybean (Burnside, 1978) and winter wheat (Burnside and Schultz, 1978). In a 6-year study, atrazine carryover was not a major problem in no-till but was a problem in conventional tillage in Nebraska, using oat as a bioassay plant in the field (Burnside and Wicks, 1980). In the Texas panhandle, alachlor and metolachlor could be used on sorghum without damaging later-sown winter wheat, but atrazine damaged rotational wheat (Wiese et al., 1986).

APPENDIX A

List of Crops and Weeds in Chapter

Common name	Scientific name	BAYER code
Crops		
Alfalfa	(*Medicago sativa* L.)	MEDSA
Barley	(*Hordeum vulgare* L.)	HORVS
Canola	(*Brassica campestris* L.)	BRSRA
Clover	(*Trifolium* sp.)	TRFXX
Corn	(*Zea mays* L.)	ZEAMX
Cotton	(*Gossypium hirsutum* L.)	GOSHI
Dry beans	(*Phaseolus vulgaris* L.)	PHSVX
Durum wheat	(*Triticum durum* Desf.)	TRZDU
Flax	(*Linum usitatissimum* L.)	LIUUT
Lentil	(*Lens culinaris* Medik.)	LENCU
Oat	(*Avena sativa* L.)	AVESA

(continued)

Appendix A Continued

Common name	Scientific name	BAYER code
Pea	(*Pisum sativum* L.)	PIBST
Pinto bean	(*Phaseolus vulgaris* L.)	PHSVX
Potato	(*Solanum tuberosum* L.)	SOLTU
Rape	(*Brassica napus* L.)	BRSNW
Rice	(*Oryza sativa* L.)	ORYSA
Rye	(*Secale cereale* L.)	SECCE
Safflower	(*Carthamus tinctorius* L.)	CDUCA
Sorghum	(*Sorghum bicolor* (L.) Moench.)	SORVU
Soybean	(*Glycine max* L.)	GLXMA
Sugar beet	(*Beta vulgaris* L.)	BEAVA
Sunflower	(*Helianthus annuus* L.)	HELAN
Spring wheat	(*Triticum aestivum* L.)	TRZVX
Winter wheat	(*Triticum aestivum* L.)	TRZVX
Weeds		
Alligatorweed	(*Alternanthera philoxeroides* (Mart.) Griseb.)	ALRPH
Barnyardgrass	(*Echinochloa crusgalli* (L.) Beauv.)	ECHCG
Bearded sprangletop	(*Leptochloa fascicularis* (Lam.) Gray)	LEFFA
Blue mustard	(*Chorispora tenella* (Pallas) DC.)	COBTE
Browntop panicum	(*Panicum fasciculatum* Sw.)	PANFA
Canada thistle	(*Cirsium arvense* (L.) Scop.)	CIRAR
Cheat	(*Bromus secalinus* (L.))	BROSE
Cogongrass	(*Imperata cylindrica* (L.) Beauv.)	IMPCY
Common chickweed	(*Stellaria media* (L.) Vill.)	STEME
Common hempnettle	(*Galeopsis tetrahit* L.)	GAETE
Common lambsquarters	(*Chenopodium album* L.)	CHEAL
Common milkweed	(*Asclepias syriaca* L.)	ASCSY
Common ragweed	(*Ambrosia artemisiifolia* L.)	AMBEL
Dayflower	(*Commelina* sp.)	COMXX
Downy brome	(*Bromus tectorum* L.)	BROTE
Ducksalad	(*Heteranthera limosa* (Sw.) Willd.)	HETLI
Eclipta	(*Eclipta prostrata* L.)	ECEPA
Fall panicum	(*Panicum dichotomiflorum* Michx.)	PANDI
Fiddleneck	(*Amsinckia* sp.)	AMSXX
Field bindweed	(*Convolvulus arvensis* L.)	CONAR
Field pennycress	(*Thlaspi arvensis* L.)	THLAR
Field sandbur	(*Cenchrus incertus* M.A. Curtis)	CCHIN
Flixweed	(*Descurainia sophia* (L.) Webb. ex Prantl)	DESSO
Green foxtail	(*Setaria viridis* (L.) Beauv.)	SETVI
Hemp sesbania	(*Sesbania exaltata* (Raf.) Rydb. ex A. W. Hill)	SEBEX

Appendix A Continued

Common name	Scientific name	BAYER code
Henbit	(*Lamium amplexicaule* L.)	LAMAM
Honeyvine milkweed	(*Ampelamus albidus* (Nutt.) Britt.)	AMPAL
Italian ryegrass	(*Lolium multiflorum* Lam.)	LOLMU
Jerusalem artichoke	(*Helianthus tuberosus* L.)	HELTU
Johnsongrass	(*Solanum halepense* (L.) Pers.)	SORHA
Jointed goatgrass	(*Aegilops cylindrica* Host)	AEGCY
Junglerice	(*Echinochloa colona* (L.) Link)	ECHCO
Kochia	(*Kochia scoparia* (L.) Scrad.)	ECHSC
Large crabgrass	(*Digitaria sanguinalis* (L.) Scop.)	DIGSA
Morningglory	(*Ipomoea* sp.)	IPOXX
Nightflowering catchfly	(*Silene noctiflora* L.)	MELNO
Northern jointvetch	(*Aeschynomene virginica* (L.) B.S.P.)	AESVI
Palmer amaranth	(*Amaranthus palmeri* S. Wats.)	AMAPA
Pennsylvania smartweed	(*Polygonum pensylvanicum* L.)	POLPY
Perennial sowthistle	(*Sonchus arvensis* L.)	SONAR
Quackgrass	(*Elytrigia repens* (L.) Nevski)	AGRRE
Red rice	(*Oryza sativa* L.)	ORYSA
Redroot pigweed	(*Amaranthus retroflexus* L.)	AMARE
Rigid ryegrass	(*Lolium rigidum* Gaudin)	LOLRI
Russian thistle	(*Salsola iberica* Sennen & Pau)	SASKR
Shattercane	(*Sorghum bicolor* (L.) Moench)	SORVU
Smooth pigweed	(*Amaranthus hybridus* L.)	AMACH
Sprangletop	(*Leptochloa* sp.)	LEFXX
Spreading dayflower	(*Commelina diffusa* Burm. f.)	COMDI
Stinkgrass	(*Eragrostis cilianensis* (All.) E. Mosher)	ERACN
Stinkweed	(*Pluchea camphorata* (L.) DC.)	PLUCA
Tall waterhemp	(*Amaranthus tuberculatus* (Moq.) J.D. Sauer)	AMATU
Tansymustard	(*Descurainia* sp.)	DESXX
Tartary buckwheat	(*Polygonum tartaricum* L.)	POLTA
Texas panicum	(*Panicum texanum* Buckl.)	PANTE
Velvetleaf	(*Abutilon theophrasti* Medik)	ABUTH
Waterprimrose	(*Ludwigia* sp.)	LUDXX
Wild buckwheat	(*Polygonum convolvulus* L.)	POLCO
Wild garlic	(*Allium vineale* L.)	ALLVI
Wild mustard	(*Brassica kaber* (DC.) L.C. Wheeler)	SINAR
Wild oat	(*Avena fatua* L.)	AVEFA
Wild onion	(*Allium canadense* L.)	ALLCA
Wild rose	(*Rosa* sp.)	ROSXX
Witchgrass	(*Panicum capillare* L.)	PANCA
Yellow foxtail	(*Setaria glauca* (L.) Beauv.)	SETLU
Yellow nutsedge	(*Cyperus esculentus* L.)	CYPES

APPENDIX B

List of Pesticides in Chapter

Common name	Chemical name
Herbicides	
Acetochlor	2-chloro-*N*-(ethoxymethyl)-*N*-(2-ethyl-6-methylphenyl)acetamide
Acifluorfen	5-[2-chloro-4-(trifluoromethyl)phenoxyl]-2-nitrobenzoic acid
Alachlor	2-chloro-*N*-(2,6-diethylphenyl)-*N*-(methoxymethyl)acetamide
Atrazine	6-chloro-*N*-ethyl-*N'*-(1-methylethyl)-1,3,5-triazine-2,4-diamine
Barban	4-chloro-2-butynyl-*m*-chlorocarbanilate
Bensulfuron	methyl 2-[[[[[(4,6-dimethoxy-2-pyrimidinyl)amino]carbonyl]-amino]sulfonyl]methyl]benzoate
Bentazon	3-(1-methylethyl)-(1*H*)-2,1,3-benzothiadiazin-3(3*H*)-one 2,2,-dioxide
Bromoxynil	3,5-dibromo-4-hydroxybenzonitrile
Butylate	*S*-ethyl bis(2-methylpropyl)carbamothioate
CGA 131036	2-(2-chloroethoxy)-*N*-[[4-methoxy-6-methyl-1,3,5-triazin-2-yl)-amino]carbonyl]benzenesulfonamide
CGA 92194	α-[(1,3-dioxolan-2-ylmethoxy)imino]benzeneacetonitrile
Chlorsulfuron	2-chloro-*N*-[[(4-methoxy-6-methyl-1,3,5-triazin-2-yl)amino]-carbonyl]benzenesulfonamide
Clopyralid	3,6-dichloro-2-pyridinecarboxylic acid
Cyanazine	2-[[4-chloro-6-(ethylamino)-1,3,5-triazin-2-yl]amino]-2-methyl-propanenitrile
Cyometrinil	(*Z*)-α[(cyanomethoxy)imino]benzeneacetonitrile
Diallate	*S*-(2,3-dichloro-2-propenyl)bis(1-methlethyl)carbamothioate
Dicamba	3,6-dichloro-2-methoxybenzoic acid
Diclofop	(±)-2-[4-(2,4-dichlorophenoxy)phenoxy]propanoic acid
Difenzoquat	1,2-dimethyl-3-5-diphenyl-1*H*-pyrazolium
Dipropetryn	2-ethylthio-4,6-bis(isopropylamino)-*s*-triazine
Diquat	6,7-dihydrodipyrido[1,2-α:2',2',1'-*c*]pyrazinediium ion
Diuron	*N'*-(3,4-dichlorophenyl)-*N,N*-dimethylurea
DPX-R9674	Chlorsulfuron + metsulfuron
EPTC	*S*-ethyl dipropyl carbamothioate
Ethofumesate	(±)-2-ethoxy-2,3-dihydro-3,3-dimethyl-5-benzofuranyl methanesulfonate
Fenoxyprop	(±)-2[4-[(6-chloro-2-benzoxazolyl)oxy]phenoxy]propanoic acid
Flamprop-methyl	methyl *N*-benzoyl-*N*-(3-chloro-4-fluorophenyl)-*D*-alaninate
Fluazifop	(*R*)-2-[4-[[5-(trifluoromethyl)-2-pyridinyl]oxy]phenoxy]prop-anoate acid
Fluometuron	*N,N*-dimethyl-*N'*-[3-(trifluoromethyl)phenyl]urea
Flurazole	phenylmethyl 2-chloro-4-(trifluoromethyl)-5-thiazolecarboxy-late
Glyphosate	*N*-(phosphonomethyl)glycine

Appendix B Continued

Common name	Chemical name
Haloxyfop	(±)-2-[4-[[3-chloro-5-(trifluoromethyl)-2-pyridinyl]oxy]phenoxy]propanoic acid
Imazamethabenz	(±)-2-[4,5-dihydro-4-methyl-4-(1-methylethyl)-5-oxo-1*H*-imidazol-2-yl]-4(and 5)-methylbenzoic acid
Imazapyr	(±)-2-[4,5-dihydro-4-methyl-4-(1-methylethyl)-5-oxo-1*H*-imidazol-2-yl]-3-pyridinecarboxylic acid
Imazethyapyr	2-[4,5-dihydro-4-methyl-4-(1-methylethyl)-5-oxo-1*H*-imidazol-2-yl]-5-ethyl-3-pyridinecarboxylic acid
Isopropalin	4-(1-methylethyl)-2,6-dinitro-*N*,*N*-dipropylbenzenamine
Linuron	*N'*-(3,4-dichlorophenyl)-*N*-methoxy-*N*-methylurea
MCPA	(4-chloro-2-methylphenoxy)acetic acid
Methazole	2-(3,4-dichlorophenyl)-4-methyl-1,2,4-oxadiazolidine-3,5-dione
Metolachlor	2-chloro-*N*-(2-ethyl-6-methylphenyl)-*N*-(2-methoxy-1-methylethyl)acetamide
Metribuzin	4-amino-6-(1,1-dimethylethyl)-3-(methylthio)-1,2,4-triazin-5(4*H*)-one
Metsulfuron	methyl 2-[[[[(4-methoxy-6-methyl-1,3,5-triazin-2-yl)amino]carbonyl]amino]sulfonyl]benzoate
Molinate	*S*-ethyl hexahydro-1*H*-azepine-1-carbothioate
MON-4606 (flurazole)	phenylmethyl 2-chloro-4-(trifluoromethyl)-5-thiazolecarboxylate
Naphthalic anhydride	napththalene-1,8-dicarboxylic acid anhydride
Nicosulfuron	2-[[[[(4,6-dimethoxy-2-pyrimidinyl)amino]carbonyl]amino]sulfonyl]-*N*,*N*-dimethyl-3-pyridinecarboxamide
Nitralin	4-methylsulphonyl-2,6-dinitro-*N*,*N*-dipropylaniline
Norflurazon	4-chloro-5-(methylamino)-2-(3-(trifluoromethyl)phenyl)-3-(2*H*)-pyridazinone
Oryzalin	4-(dipropylamino)-3,5-dinitrobenzenesulfonamide
Paraquat	1,1'-dimethyl-4,4'-bipyridinium ion
Pendimethalin	*N*-(1-ethylpropyl)-3,4-dimethyl-2,6-dinitrobenzenamine
Picloram	4-amino-3,5,6-trichloro-2-pyridinecarboxylic acid
Primisulfuron	methyl 2-[[[[[4,6-bis(difluoromethoxy)-2-pyrimidinyl]amino]carbonyl]amino]sulfonyl]benzoate
Prometryn	*N*,*N'*-bis(1-methylethyl)-6-(methylthio)-1,3,5-triazine-2,4-diamine
Propachlor	2-chloro-*N*-(1-methylethyl)-*N*-phenylacetamide
Propanil	*N*-(3,4-dichlorophenyl)propanamide
Propazine	2-chloro-4,6-bis(isopropylamino)-*s*-triazine
Quinchlorac	3,7-dichloro-8-quinolinecarboxylic acid

(*continued*)

Appendix B Continued

Common name	Chemical name
R-25788	2,2-dichloro-*N*,*N*-di-2-propenylacetamide
Sethoxydim	2-[1-(ethoxyimino)butyl]-5-[2-(ethylthio)propyl]-3-hydroxy-2-cyclohexen-1-one
Sodium chlorate	sodium chlorate
2,4-D	(2,4-dichlorophenoxy)acetic acid
Terbutryn	2-(tert-butylamino)-4-(ethylamino)-6-(methylthio)-*s*-triazine
Thifensulfuron (Thiameturon)	methyl 3-[[[[(4-methoxy-6-methyl-1,3,5-triazin-2-yl)amino] carbonyl]amino]sulfonyl]-2-thiophenecarboxylate
Thiobencarb	*S*-[(4-chlorophenyl)methyl]diethylcarbamothioate
Triallate	*S*-(2,3,3-trichloro-2-propenyl)bis(1-methylethyl)carbamothioate
Triasulfuron	2-(2-chloroethoxy)-*N*-[[(4-methoxy-6-methyl-1,3,5-triazin-2-yl) amino]carbonyl]benzenesulfonamide
Tribenuron	methyl 2-[[[[(4-methoxy-6-methyl-1,3,5-triazin-2-yl)methyl-amino]carbonyl]amino]sulfonyl]benzoate
Triclopyr	[(3,5,6-trichloro-2-pyridinyl)oxy]acetic acid
Trifluralin	2,6-dinitro-*N*,*N*-dipropyl-4-(trifluoromethyl)benzenamine
Insecticides	
Carbaryl	1-naphthyl methylcarbamate
Carbofuran	2,3-dihydro-2,2-dimethyl-7-benzofuranyl methylcarbamate
Chlorpyrifos	*O*,*O*-diethyl *O*-(3,5,6-trichloro-2-pyridinyl)phosphorothioate
Dimethoate	*O*,*O*-dimethyl *S*-methylcarbamoylmethyl phosphorodithioate
Disulfoton	*O*,*O*-diethyl *S*-[2-(ethylthio)ethyl]phosphorodithioate
Esfenvalerate	(*S*)-α-cyano-3-phenoxybenzyl (*S*)-2-(4-chlorophenyl)-3-methyl-butyrate
Ethyl parathion	*O*,*O*-diethyl *O*-(4-nitrophenyl)phosphorothioate, ethyl ester
Malathion	*O*,*O*-dimethyl phosphorodithioate of diethyl mercaptosuccinate
Methomyl	*S*-methyl *N*-[(methylcarbamoyl)oxy]thioacetimidate
Methyl parathion	*O*,*O*-diethyl *O*-(4-nitrophenyl)phosphorothioate, methyl ester
Phorate	*O*,*O*-diethyl *S*-[(ethylthio)methyl]phosphorodithioate
Fungicides	
Benomyl	methyl 1-(butylcarbamoyl)benzimidazol-2-ylcarbamate
Mancozeb	Zn, Mn ethylene bisdithiocarbamate
Propiconazole	1-[[2-(2,4-dichlorophenyl)-4-propyl-1,3-dioxolan-2-yl]methyl]-1*H*-1,2,4-triazole
Thiophanate-methyl	dimethyl [(1,2-phenylene)bis-(iminocarbonothioyl)]bis [carbamate]
Triadimefon	1-(4-chlorophenoxy)-3,3-dimethyl-1-(1*H*-1,2,4-trizol-1-yl)-2-butanone

REFERENCES

Abernathy, J. R. (1981). Estimated crop losses by weeds with nonchemical management, in *Handbook of Pest Management in Agriculture* (D. Pimentel, ed.), Vol. 1, CRC Press, Boca Raton, Florida, pp. 159–167.

Anderson, R. L. (1984). Chemical fallow in the Central Great Plains, *Colorado State Univ. Exp. Stn. Bull. 588S.*

Anderson, R. L. and Nielsen, D. C. (1991). Winter wheat (*Triticum aestivum*) growth stage effect on paraquat bioactivity, *Weed Technol. 5*: 439–441.

Appleby, A. P. and Morrow, L. A. (1990). The Pacific Northwest, in *Systems of Weed Control in Wheat in North America* (W. W. Donald, ed.), Weed Science Society of America, Champaign, Illinois, pp. 200–232.

Baker, T. K. and Peeper, T. F. (1990). Differential tolerance of winter wheat (*Triticum aestivum*) to cyanazine and triazinone herbicides, *Weed Technol. 4*: 569–575.

Baldridge, D. E., Brann, D. E., Ferguson, A. H., Henry, J. L., and Thompson, R. K. (1985). Cultural practices, in *Barley* (D. C. Rasmusson, ed.), American Society of Agronomy, Madison, Wisconsin, pp. 457–482.

Banks, J. C., Addison, D. A., Hicks, R. D., McNeill, K. E., Warner, L. C., and Webster, H. L. (1978). Trifluralin for weed control in grain sorghum and corn, *Proc. S. Weed Sci. Soc. 31*: 102–103.

Banks, P. A. and Duncan, R. R. (1983). Weed-control evaluations in ratoon-cropped grain sorghum (*Sorghum bicolor*), *Weed Sci. 31*: 254–258.

Banks, P. A. (1990). The Southeast, in *Systems of Weed Control in Wheat in North America* (W. W. Donald, ed.), Weed Science Society of America, Champaign, Illinois, pp. 182–191.

Banting, J. D. (1984). History of the development of wild oat herbicides, in *Wild Oat Symposium Proceedings*. Vol. 2. *Use and Mode of Action of Wild Oat Herbicides* (A. E. Smith and A. I. Hsiao, eds.), Canadian Plains Proc. 12. Agriculture Canada, Regina Res. Stn., Saskatchewan, pp. 11–16.

Barnett, G. M., Rioux, R., and Comeau, J. E. 1984. Soil preparation, herbicides, and seeding dates for barley production, *Can. J. Plant Sci. 64*: 1–8.

Barrentine, W. L. and Warren, G. F. (1971). Differential phytotoxicity of trifluralin and nitralin, *Weed Sci. 19*: 31–37.

Baumann, P. A. and Weaver, D. N. (1991). *Suggestions for Weed Control in Sorghum*, Tex. Agric. Ext. Serv.

Beardmore, R. A. and Linscott, D. L. (1989). Relative susceptibility of oats (*Avena sativa*) and wheat (*Triticum aestivum*) to fluazifop, haloxyfop, and sethoxydim, *Weed Technol. 3*: 76–81.

Bell, A. R. and Nalewaja, J. D. (1968). Competition of wild oat in wheat and barley, *Weed Sci. 16*: 505–508.

Bennett, W. F., Tucker, B. B., and Maunder, A. B. (1990). *Modern Grain Sorghum Production*, Iowa State University, Ames, pp. 63–81.

Black, A. L. and Siddoway, F. H. (1976a). Dryland cropping sequences within a tall wheatgrass barrier system, *J. Soil Water Cons. 31*: 101–105.

Black, A. L. and Siddoway, F. H. (1976b). Dryland cropping sequences within a tall wheatgrass barrier system, *Proc. Regional Saline Seep Control Symp., Montana State Univ. Coop. Ext. Serv. Bull. 1132*: 154-163.

Blackshaw, R. E. (1990). Control of stinkweed (*Thlaspi arvense*) and flixweed (*Descurainia sophia*) in winter wheat (*Triticum aestivum*), *Can. J. Plant Sci. 70*: 817-824.

Bovey, R. W. (1969). Effects of foliarly applied desiccants on selected species under tropical environment, *Weed Sci. 17*: 79-83.

Bovey, R. W., Miller, F. R., and Bauer, J. R. (1975a). Preharvest desiccation of grain sorghum with glyphosate, *Agron. J. 67*: 618-621.

Bovey, R. W., Miller, F. R., Bauer, J. R., and Meyer, R. E. (1975b). Growth of sorghum and soybeans in picloram-treated soil, *Agron. J. 67*: 433-436.

Brewer, F., Lavy, T. L., and Talbert, R. E. (1982). Effect of three dinitroaniline herbicides on rice (*Oryza sativa*) growth, *Weed Sci. 30*: 153-158.

Bridges, D. C., ed. (1992). *Crop Losses Due to Weeds in the United States - 1992*, Weed Science Society of America, Champaign, Illinois.

Briggs, D. E. (1978). *Barley*, Chapman and Hall, New York, pp. 292-369.

Brinkman, M. A., Langer, D. K., and Harvey, R. G. (1980a). Response of barley, spring wheat, and oats to atrazine, *Crop Sci. 20*: 319-322.

Brinkman, M. A., Langer, D. K., Harvey, R. G., and Hardie, A. R. (1980b). Response of oats to atrazine, *Crop Sci. 20*: 185-189.

Brown, S. M., Chandler, J. M., and Morrison, J. E., Jr. (1988). Glyphosate for Johnsongrass (*Sorghum halepense*) control in no-till sorghum (*Sorghum bicolor*), *Weed Sci. 36*: 510-513.

Buchholtz, K. P. (1965). Factors influencing oat injury from triazine residues in soil, *Weeds 13*: 362-367.

Burnside, O. C. (1977). Control of weeds in non-cultivated, narrow-row sorghum, *Agron. J. 69*: 851-854.

Burnside, O. C. (1978). Mechanical, cultural, and chemical control of weeds in a sorghum-soybean (*Sorghum bicolor*) - (*Glycine max*) rotation, *Weed Sci. 26*: 362-369.

Burnside, O. C. and Moomaw, R. S. (1975). Sorghum growth as affected by annual applications of atrazine, *Weed Sci. 23*: 494-498.

Burnside, O. C. and Schultz, M. E. (1978). Soil persistence of herbicides for corn, sorghum, and soybeans during the year of application, *Weed Sci. 26*: 108-115.

Burnside, O. C. and Wicks, G. A. (1967). The effect of weed removal treatments on sorghum growth, *Weed Sci. 15*: 204-207.

Burnside, O. C. and Wicks, G. A. (1969). Influence of weed competition on sorghum growth, *Weed Sci. 17*: 332-334.

Burnside, O. C. and Wicks, G. A. (1980). Atrazine carryover in soil in a reduced tillage crop production system, *Weed Sci. 28*: 661-666.

Burnside, O. C., Wicks, G. A., and Fenster, C. R. (1977). Longevity of shattercane seed in soil across Nebraska, *Weed Res. 17*: 139-143.

Burnside, O. C., Wicks, G. A., Warnes, D. D., Somerhalder, B. R., and Weeks, S. A. (1969). Effect of weeds on harvesting efficiency in corn, sorghum, and soybeans, *Weed Sci. 17*: 438-441.

Carlson, D. R. and Burnside, O. C. (1981). Use of the recirculating sprayer to control tall weed escapes in crops, *Weed Sci. 29*: 174-179.

Carlson, S. J. and Morrow, L. A. (1986). Control of wild oat with triallate in a spring wheat conservation tillage system, *Can. J. Plant Sci. 66*: 181-184.

Chamberlain, E. W., Becton, A. J., and LeBaron, H. M. (1970). Tolerance of sorghum to postemergence applications of atrazine, *Weed Sci. 18*: 410-412.

Chandler, J. M. (1981). Estimated losses of crops to weeds, in *Handbook of Pest Management in Agriculture,* Vol. 1 (D. Pimentel, ed.), CRC Press, Boca Raton, Florida, pp. 95-109.

Chang, I. and Foy, C. L. (1971). Effect of picloram on germination and seedling development of four species, *Weed Sci. 19*: 58-64.

Chisaka, H. (1977). Weed damage to crops: yield loss due to weed competition, in *Integrated Control of Weeds* (J. D. Fryer and S. Matsunaka, eds.), University of Tokyo Press, Tokyo. p. 1-16.

Chow, P. N. P. (1983). Herbicide mixtures containing BAS 9052 for weed control in flax (*Linum usitatissimum*), *Weed Sci. 31*: 20-22.

Claassen, M. M. and Moshier, L. J. (1989). Honeyvine milkweed (*Ampelamus albidus*) control in a winter wheat (*Triticum aestivum*): grain sorghum (*Sorghum bicolor*) rotation, *Weed Technol. 3*: 345-348.

Clay, S. A., Thill, D. C., and Cochran, V. L. (1988). Response of spring barley (*Hordeum vulgare*) to herbicides, *Weed Technol. 2*: 68-71.

Cook, R. J. and Veseth, R. J. (1991). *Wheat Health Management,* American Phytopathological Society, St. Paul, Minnesota.

Cramer, G. L. and Burnside, O. C. (1982). Distribution and interference of common milkweed (*Asclepias syriaca*) in Nebraska, *Weed Sci. 30*: 385-388.

Cussans, G. W. and Wilson, B. J. (1975). "Some effects of crop row width and seed rate on competition between spring barley and wild oat *Avena fatua* L. or common couch *Agropyron repens* (L.) Beauv," Proceedings European Weed Research Society Symposium. Status and Control of Grassweeds in Europe. p. 77-86.

Dao, T. H. (1987). Crop residues and management of annual grass weeds in continuous no-till wheat (*Triticum aestivum*), *Weed Sci. 35*: 395-400.

Delvo, H. W. (1989). *Agricultural Resources: Situation and Outlook* Report, USDA Econ. Res. Serv. AR-13, U.S. Government Printing Office, Washington, D.C.

Devlin, D. L., Moshier, L. J., Russ, O. G., and Stahlman, P. W. (1983). Antidotes reduce injury to grain sorghum (*Sorghum bicolor*) from acetanilide herbicides, *Weed Sci. 31*: 790-795.

Dew, D. A. (1972). An index of competition for estimating crop loss due to weeds, *Can. J. Plant Sci. 52*: 921-927.

Dexter, A., Nalewaja, J. D., Rasmusson, D. D., and Buchli, J. (1981). Survey of wild oats and other weeds in North Dakota, 1978 and 1979, *N. Dakota Res. Rep. 79.*

Diawara, M. M. and Banks, P. A. (1990). Weed control in barley (*Hordeum vulgare*) - no-till grain sorghum (*Sorghum bicolor*) production, *Weed Sci. 38*: 50-53.

Donald, W. W. (1990a). Management and control of Canada thistle (*Cirsium arvense*), *Rev. Weed Sci. 5*: 193-250.

Donald, W. W., ed.(1990b). *Systems of Weed Control in Wheat in North America,* Weed Science Society of America, Champaign, Illinois.

Donald, W. W. (1990c). Introduction, in *Systems of Weed Control in Wheat in North*

America (W. W. Donald, ed.), Weed Science Society of America, Champaign, Illinois, pp. 1–10.

Donald, W. W. (1990d). Difenzoquat, in *Systems of Weed Control in Wheat in North America* (W. W. Donald, ed.), Weed Science Society of America, Champaign, Illinois, pp. 298–321.

Donald, W. W. (1990e). Sulfonylurea herbicides, in *Systems of Weed Control in Wheat in North America* (W. W. Donald, ed.), Weed Science Society of America, Champaign, Illinois, pp. 423–477.

Donald, W. W. and Khan, M. (1992). Yield loss assessment for spring wheat (*Triticum aestivum*) infested with Canada thistle (*Cirsium arvense*), *Weed Sci. 40*: 590–598.

Donald, W. W. and Nalewaja, J. D. (1990). Northern Great Plains, in *Systems of Weed Control in Wheat in North America* (W. W. Donald, ed.), Weed Science Society of America, Champaign, Illinois, pp. 90–127.

Donald, W. W. and Ogg, A. G., Jr. (1991). Biology and control of jointed goatgrass (*Aegilops cylindrica*), a review, *Weed Technol. 5*: 3–17.

Donald, W. W. and Prato, T. (1991). Profitable, effective herbicides for planting-time weed control in no-till spring wheat (*Triticum aestivum*), *Weed Sci. 39*: 83–90.

Douglas, B. J., Thomas, A. G., and Derksen, D. A. (1990). Downy brome (*Bromus tectorum*) invasion into southwestern Saskatchewan, *Can. J. Plant Sci. 70*: 1143–1151.

Dregne, H. E. and Willis, W. O., eds. (1983). *Dryland Agriculture*, American Society of Agronomy, Madison, Wisconsin.

Eastin, E. F. (1972). Evaluation of a sorghum seed treatment to prevent injury from acetanilide herbicides, *Agron. J. 64*: 556–557.

Eastin, E. F. (1983). Weed, disease, insect interactions in rice, in *Weed Control in Rice*, International Rice Research Institute, Los Banos, Laguna, Philippines, pp. 213–218.

Eastin, E. F. (1991). Weed management in southern U.S. rice, in *CRC Handbook of Pest Management in Agriculture*, Vol. III (C. Pimentel, ed.), CRC Press, Inc., Boca Raton, Florida, pp. 329–338.

Eberlein, C. V. (1990). Propanil, in *Systems of Weed Control in Wheat in North America* (W. W. Donald, ed.), Weed Science Society of America, Champaign, Illinois, pp. 374–391.

Fay, P. K. (1990). A brief overview of the biology and distribution of weeds of wheat, in *Systems of Weed Control in Wheat in North America* (W. W. Donald, ed.), Weed Science Society of America, Champaign, Illinois, pp. 33–50.

Fellows, G. M., Fay, P. K., Carlson, G. R., and Stewart, V. R. (1990). Effect of AC 222,293 soil residues on rotational crops, *Weed Technol. 4*: 48–51.

Feltner, K. C., Hurst, H. R., and Anderson, L. E. (1969a). Yellow foxtail competition in grain sorghum, *Weed Sci. 17*: 211–213.

Feltner, K. C., Hurst, H. R., and Anderson, L. E. (1969b). Tall waterhemp competition in grain sorghum, *Weed Sci. 17*: 214–216.

Ferreira, K. L., Baker, T. K., and Peeper, T. F. (1990a). Factors influencing winter wheat (*Triticum aestivum*) injury from sulfonylurea herbicides, *Weed Technol. 4*: 724–730.

Ferreira, K. L., Peeper, T. F., and Epplin, F. M. (1990b). Economic returns from cheat

(*Bromus secalinus*) control in winter wheat (*Triticum aestivum*), *Weed Technol. 4*: 306–313.

Fuerst, E. P. and Gronwald, J. W. (1986). Induction of rapid metabolism of metolachlor in sorghum (*Sorghum bicolor*) shoots by CGA-92194 and other antidotes, *Weed Sci. 34*: 354–361.

Foy, C. L. and Witt, H. L. (1990). Seed protectants safen sorghum (*Sorghum bicolor*) against chloroacetamide herbicide injury, *Weed Technol. 4*: 886–891.

Friesen, G. H. and Wall, D. A. (1990). Improving the efficiency of sethoxydim in flax, *Weed Tech. 4*: 749–753.

Friesen, H. A., Baenziger, H., and Keys, C. H. (1964). Morphological and cytological effects of dicamba on wheat and barley, *Can. J. Plant Sci. 44*: 288–294.

Gast, R. E., Liebl, R. A., and Slife, F. W. (1990). Wild garlic (*Allium vineale*) control with thifensulfuron and DPX-L5300, *Weed Technol. 4*: 592–597.

Gigax, D. R. and Burnside, O. C. (1976). Chemical desiccation of grain sorghum, *Agron. J. 68*: 645–649.

Gillespie, G. R. (1990). Picloram and clopyralid, in *Systems of Weed Control in Wheat in North America* (W. W. Donald, ed.), Weed Science Society of America, Champaign, Illinois, pp. 477–488.

Gillespie, G. R. and Nalewaja, J. D. (1988). Economic control of weeds in wheat, *Triticum aestivum, Weed Technol. 2*: 257–261.

Graham, P. L., Steiner, J. L., and Wiese, A. F. (1988). Light absorption and competition in mixed sorghum-pigweed communities, *Agron. J. 80*: 415–418.

Greb, B. W. (1983). Water conservation: Central Great Plains, in *Dryland Agriculture* (H. E. Dregne and W. O. Willis, ed.), Am. Soc. Agron., Madison, Wisconsin, pp. 57–72.

Greer, H. A. L. (1990). Weed control in grain sorghum, *Oklahoma State Univ. Ext. Facts No. 2763.*

Hageman, L. H. and Behrens, R. (1981). Response of small-grain cultivars to chlorsulfuron, *Weed Sci. 29*: 414–420.

Hamilton, K. C. (1979). Persistence of annual applications of herbicides in winter-planted safflower (*Carthamus tinctorius*), *Weed Sci. 27*: 655–657.

Harrison, S. K. and Beuerlein, J. E. (1989). Effect of herbicide mixtures and seeding rate on soft red winter wheat (*Triticum aestivum*) yield, *Weed Technol. 3*: 505–508.

Heering, D. C. and Peeper, T. F. (1991). Winter wheat (*Triticum aestivum*) response to picloram and 2,4-D, *Weed Technol. 5*: 317–320.

Helm, J. and McMullen, M. S. (1989). Oat production in North Dakota, *N. Dakota State Univ. Ext. Serv.* 14 AGRI-1-5.

Hoefer, R. H., Wicks, G. A., and Burnside, O. C. (1981). Grain yields, soil water storage, and weed growth in a winter wheat-corn-fallow rotation, *Agron. J. 73*: 1066–1071.

Holm, F. A. and Kirkland, K. J. (1986). Annual broadleaf weed control in wheat in western Canada, in *Wheat Production in Canada: a Review* (A. E. Slinkard and D. B. Fowler, eds.), Proceedings Canadian Wheat Production Symposium, Univ. Saskatchewan, Saskatoon, pp. 375–390.

Houston, W. and Kimbrough, E. L. (1991). Grain and forage sorghum weed control, *Mississippi State Univ. Coop. Ext. Serv. Info. Sheet 803.*

Hunter, J. H. (1983). Cultural control of wild oats, in *Wild Oat Symposium Proceedings*, Vol. 1. (A. E. Smith, ed.), Canadian Plains Proceedings 12, Agriculture Canada Regina Res. Stn., Saskatchewan, pp. 43–52.

Hunter, J. H., Morrison, I. N., and Rourke, D. R. S. (1990). The Canadian Prairie Provinces, in *Systems of Weed Control in Wheat in North America* (W. W. Donald, ed.), Weed Science Society of America, Champaign, Illinois, pp. 51–90.

Ivany, J. A. and Nass, H. G. (1984). Effect of herbicides on seedling growth, head deformation, and grain yield of spring wheat cultivars, *Can. J. Plant Sci. 64*: 25–30.

Ivany, J. A., Nass, H. G., and Sanderson, J. B. (1990). Effect of time of application of herbicides on yield of three winter wheat cultivars, *Can. J. Plant Sci. 70*: 605–609.

Jackson, A. W., Jeffery, L. S., and McCutchen, T. C. (1978). Tolerance of soybeans (*Glycine max*) and grain sorghum (*Sorghum bicolor*) to fluometuron residue, *Weed Sci. 26*: 454–458.

Jones, D. P. (1976). *Wild Oats in World Agriculture*, Agricultural Review Council, London.

Keeling, J. W. and Abernathy, J. R. (1989). Preemergence weed control in a conservation tillage cotton (*Gossypium hirsutum*) cropping system on sandy soils, *Weed Technol. 3*: 182–185.

Keeling, J. W., Lloyd, R. W., and Abernathy, J. R. (1989). Rotational crop response to repeated applications of norflurazon, *Weed Technol. 3*: 122–125.

Ketchersid, M. L., Norton, K., and Merkle, M. G. (1981). Influence of soil moisture on the safening effect of CGA-43089 in grain sorghum (*Sorghum bicolor*), *Weed Sci. 29*: 281–287.

Kirby, C. (1980). *The Hormone Weedkillers*, BCPC Publications, Croydon, England.

Koscelny, J. A. and Peeper, T. F. (1990a). Effect of grazing cheat (*Bromus secalinus*)-infested wheat (*Triticum aestivum*) on both species. *Weed Technol. 4*: 565–568.

Koscelny, J. A. and Peeper, T. F. (1990b). Herbicide-grazing interactions in cheat (*Bromus secalinus*) infested-winter wheat (*Triticum aestivum*), *Weed Sci. 38*: 532–535.

Koscelny, J. A., Peeper, T. F., Solie, J. B., and Solomon, S. G., Jr. (1990). Effect of wheat (*Triticum aestivum*) row spacing, seeding rate, and cultivar on yield loss from cheat (*Bromus secalinus*), *Weed Technol. 4*: 487–492.

Kwon, S. L., Smith, R. J., Jr., and Talbert, R. E. (1991a). Interference of red rice (*Oryza sativa*) densities in rice (*O. sativa*), *Weed Sci. 39*: 169–174.

Kwon, S. L., Smith, R. J., Jr., and Talbert, R. E. (1991b). Interference durations of red rice (*Oryza sativa*) in rice (*O. sativa*), *Weed Sci. 39*: 363–368.

Lemerle, D., Kidd, C. R., and Read, B. J. (1986). Tolerances of barley cultivars to post-emergence herbicides, *Aust. J. Exp. Agric. 26*: 383–392.

Lewis, W. M. (1976). Double cropping soybeans after winter small grains, in *World Soybean Research: Proceedings of the World Soybean Research Conference* (L. D. Hill, ed.), Interstate Printers and Publishers, Inc., Danville, Illinois, pp. 44–52.

Lewis, W. M., Worsham, A. D., and York, A. C. (1988). Weed management, *in Conservation Tillage for Crop Production in North Carolina* (M. G. Cook and W. M. Lewis, eds.), N. C. Agric. Exp. Serv. 1-89-5M-TWK-190067, pp. 28–39.

Mack, R. N. (1981). Invasion of *Bromus tectorum* L. into western North America: an ecological chronicle, *Agro-ecosystems 7*: 145–165.

Martin, D. A., Miller, S. D., and Alley, H. P. (1989). Winter wheat (*Triticum aestivum*) response to herbicides applied at three growth stages, *Weed Technol. 3*: 90–94.

Martin, R. J., Cullis, B. R., and McNamara, D. W. (1987). Prediction of wheat yield loss due to competition by wild oats (*Avena* spp.), *Aust. J. Agric. Res. 38*: 487–499.

Massee, T. W. and Higgins, R. E. (1977). Downy brome (cheatgrass) control in a dryland winter wheat-fallow rotation, *Univ. Idaho Coop. Ext. Serv. Current Info. Ser. 371.*

McBeath, D. K., Dew, D. A., and Friesen, H. A. (1970). Competition between barley and wild oats as affected by nitrogen, barban, and time of seeding, *Can. J. Plant Sci. 50*: 541–550.

McKinney, N. V. (1992). Arkansas grain sorghum production handbook, *Univ. Arkansas Coop. Ext. Serv. Handbook AGR 10.*

McWhorter, C. G. and Derting, C. W (1985). Methods of application of glyphosate, in *The Herbicide Glyphosate* (E. Grossbard and D. Atkinson, eds.), Butterworths, Boston, pp. 241–260.

Mercado, B. L. and Talatala, R. L. (1977). Competitive ability of *Echinochloa colonum* L. against direct-seeded lowland rice, *Proc. Asian-Pacific Weed Sci. Soc. 6*: 161–165.

Miller, S. D., Nalewaja, J. D., Pudelko, J., and Adamczewski, K. A. (1978). Difenzoquat for wild oat (*Avena fatua*) control, *Weed Sci. 26*: 571–576.

Mitich, L. and George, M. (1981). Weed control in oats, *Univ. California Leaflet 21254.*

Mitich, L. W. and Kyser, G. B. (1990a). Wheat production practices in relation to weed control in California, in *Systems of Weed Control in Wheat in North America* (W. W. Donald, ed.), Weed Science Society of America, Champaign, Illinois, pp. 233–257.

Mitich, L. W. and Kyser, G. B. (1990b). Wheat production and weed control in Arizona, in *Systems of Weed Control in Wheat in North America* (W. W. Donald, ed.), Weed Science Society of America, Champaign, Illinois, pp. 257–266.

Moody, K. (1991). Weed management in rice, in *CRC Handbook of Pest Management in Agriculture*, Vol. III. (C. Pimentel, ed.), CRC Press Inc., Boca Raton, Florida, pp. 301–328.

Morishita, D. W. and Thill, D. C. (1988a). Factors of wild oat (*Avena fatua*) interference on spring barley (*Hordeum vulgare*) growth and yield, *Weed Sci. 36*: 37–42.

Morishita, D. W. and Thill, D. C. (1988b). Wild oat (*Avena fatua*) and spring barley (*Hordeum vulgare*) growth and development in monoculture and mixed culture, *Weed Sci. 36*: 43–48.

Morrow, L. A. and Stahlman, P. W. (1984). The history and distribution of downy brome (*Bromus tectorum*) in North America, *Weed Sci. 32*(Suppl.): 2–7.

Moschler, W. W., Shear, G. M., Hallock, D. L., Sears, R. D., and Jones, G. D. (1967). Winter cover crops for sod-planted corn: their selection and management, *Agron. J. 59*: 547–551.

Moyer, J. R., Esau, R., and Kozub, G. C. (1990). Chlorsulfuron persistence and response of nine rotational crops in alkaline soils of southern Alberta, *Weed Technol. 4*: 543–548.

Nalewaja, J. D. and Arnold, W. E. (1970). *Weed Control Methods, Losses and Costs Due to Weeds, and Benefits of Weed Control in Wheat and Other Small Grains*, FAO Int. Conf. Weed Control., Weed Sci. Soc. Am., Champaign, Illinois, pp. 48–64.

Ndon, B. A., Harvey, R. G., and Scholl, J. M. (1982). Weed control in double cropped corn, grain sorghum, or soybeans minimum-till planted following canning peas, *Agron. J. 74*: 266–269.

Nelson, J. E., Meggitt, W. F., Penner, D., and Ladlie, J. S. (1983). The influence of environmental factors on oryzalin activity, *Weed Sci. 31*: 752–758.

Noda, K. (1973). Competitive effects of barnyardgrass (*Echinochloa crus-galli*) on rice, *Proc. Asian-Pacific Weed Sci. Soc. 4*: 145–150.

Noda, K., Ozawa, K., and Ibaraki, K. (1968). Studies on the damage to rice plants due to weed competition, *Kyushu Agric. Exp. Stn. Bull 13*: 345–376.

Obrigawitch, T. T., Kenyon, W. H., and Kuratle, H. (1990). Effect of application timing on rhizome Johnsongrass (*Sorghum halepense*) control with DPX-V9360, *Weed Sci. 38*: 45–49.

Ogg, A. G., Jr., and Young, F. L. (1991). Effects of preplant treatment interval and tillages on herbicide toxicity to winter wheat (*Triticum aestivum*), *Weed Technol. 5*: 291–296.

O'Keefe, M. G. and Makepeace, R. J. (1985). Efficacy of glyphosate in arable situations, in *The Herbicide Glyphosate* (E. Grossbard and D. Atkinson, eds.), Butterworths, Boston, pp. 418–434.

Okoli, P. S. O., Drolsom, P. N., and Scholl, J. M. (1984). Forage production and weed control in a double-cropping program, *Agron. J. 76*: 363–366.

Oliver, L. R., Prendeville, G. N., and Schreiber, M. M. (1968). Species differences in site of root uptake and tolerance to EPTC, *Weed Sci. 16*: 534–537.

Olson, P. J., Zalik, S., Breakey, W. J., and Brown, D. A. (1951). Sensitivity of wheat and barley at different stages of growth to treatment with 2,4-D, *Agron. J. 43*: 77–83.

Osteen, C.D. and Szmedra, P. I. (1989). Agricultural pesticide use trends and policy issues, *USDA Econ. Res. Serv. Agric. Econ. Rep. 622.*

O'Sullivan, P. A. (1990). Diclofop, in *Systems of Weed Control in Wheat in North America* (W. W. Donald, ed.), Weed Science Society of America, Champaign, Illinois, pp.321–345.

O'Sullivan, P. A., O'Donovan, J. T., and Hamman, W. M. (1981). Influence of non-ionic surfactants, ammonium sulphate, water quality and spray volume on the phytotoxicity of glyphosate, *Can. J. Plant Sci. 61*: 391–400.

O'Sullivan, P. A., Kossatz, V. C., Weiss, G. M., and Dew, D. A. (1982). An approach to estimating yield loss of barley due to Canada thistle, *Can. J. Plant Sci. 62*: 725–731.

Pantone, D. J. and Baker, J. B. (1991). Reciprocal yield analysis of red rice (*Oryza sativa*) competition in cultivated rice, *Weed Sci. 39*: 42–47.

Pavlychenko, T. K. (1940). *Investigations Relating to Weed Control in Western Canada*, Imperial Bureau Pastures and Forages Crops, Aberystwyth, Wales, Great Britain, Herbage Pub. Serv. Bull. 27.

Peeper, T. F. (1984). Chemical and biological control of downy brome (*Bromus tectorum*) in wheat and alfalfa in North America, *Weed Sci. 32*(Suppl.): 18-26.

Peeper, T. F. and Morrow, L. A. (1990). Substituted ureas, triazines, and triazinones, in *Systems of Weed Control in Wheat in North America* (W. W. Donald, ed.), Weed Science Society of America, Champaign, Illinois, pp. 404-422.

Peeper, T. F. and Wiese, A. F. (1990). Southern Great Plains, in *Systems of Weed Control in Wheat in North America* (W. W. Donald, ed.), Weed Science Society of America, Champaign, Illinois, pp. 158-181.

Peters, E. J. (1990). The Midwest and Northeast, in *Systems of Weed Control in Wheat in North America* (W. W. Donald, ed.), Weed Science Society of America, Champaign, Illinois, pp. 191-199.

Peterson, M. A. and Arnold, W. E. (1986). Response of rotational crops to soil residues of chlorsulfuron, *Weed Sci. 34*: 131-136.

Phillips, W. M. (1969). Dryland sorghum production and weed control with minimum tillage, *Weed Sci. 17*: 451-455.

Phillips, W. M. (1970). *Weed Control Methods, Losses and Costs Due to Weeds, and Benefits of Weed Control in Grain Sorghum*, FAO Int. Conf. Weed Control, Weed Sci. Soc. Am., Champaign, Illinois, pp. 101-108.

Rahman, A. and R. Ashford 1972. Trifluralin for green foxtail control in barley and flax, *Weed Sci. 20*: 160-163.

Ramsel, R. E. and Wicks, G. A. (1988). Use of winter wheat (*Triticum aestivum*) cultivars and herbicides in aiding weed control in an ecofallow corn (*Zea mays*) rotation, *Weed Sci. 36*: 394-398.

Ratliff, R. L., Carver, B. F., and Peeper, T. F. (1991). Expression of metribuzin sensitivity in winter wheat (*Triticum aestivum*) populations, *Weed Sci. 39*: 130-133.

Reinertsen, M. R., Cochran, V. L., and Morrow, L. A. (1984). Response of spring wheat to N fertilizer placement, row spacing, and wild oat herbicides in a no-till system, *Agron. J. 76*: 753-756.

Remy, E. A. de. St., O'Donovan, J. T., Tong, A. K. W., O'Sullivan, P. A., Sharma, M. P., and Dew, D. A. (1985). Influence of tartary buckwheat (*Fagopyrum tataricum*) density on yield loss of barley (*Hordeum vulgare*) and wheat (*Triticum aestivum*), *Weed Sci. 33*: 521-523.

Rhodes, G. N., Jr. (1990). Weed control in grain sorghum. Univ. Tenn. Agric. Ext. Serv. PB 1190. 4 pp.

Rhodes, G. N., Jr., and Shelby, P. P. (1991). Weed control in wheat. Univ. Tenn. Agric. Ext. Serv. PB 1229. 4 pp.

Rioux, R. (1982). La mesure de l'interference du chiendent dans l'orge, *Can. J. Plant Sci. 62*: 183-188.

Robinson, R. G. and Dunham, R. S. (1954). Companion crops for weed control in soybeans, *Agron. J. 46*: 278-281.

Robison, L. R. and Fenster, C. R. (1968). Residual effects of EPTC and trifluralin incorporated with different implements, *Weed Sci. 16*: 415-417.

Roeth, F. W., Burnside, O. C., and Wicks, G. A. (1983). Protection of grain sorghum (*Sorghum bicolor*) from chloroacetanilide herbicide injury, *Weed Sci. 31*: 373-379.

Romanowski, R. R. and Libik, A. W. (1978). Soil persistence of isopropalin, nitralin, and trifluralin, *Weed Sci. 26*: 258-261.

Ross, W. M. (1970). Culture and use of grain sorghum, *USDA Agric. Res. Serv. Agric. Handbook 385.*

Schoenhals, M. G., Wiese, A. F., and Wood, M. L. (1990). Field bindweed (*Convolvulus arvensis*) control with imazapyr, *Weed Technol. 4*: 771–775.

Schroeder, J. and Banks, P. A. (1989). Soft red winter wheat (*Triticum aestivum*) response to dicamba and dicamba plus 2,4-D, *Weed Technol. 3*: 67–71.

Schweizer, E. E. (1975). Crop response to soil applications of ethofumesate, *Weed Sci. 23*: 409–413.

Schweizer, E. E. and Swink, J. F. (1971). Field bindweed control with dicamba and 2,4-D, and crop response to chemical residues, *Weed Sci. 19*: 717–721.

Schweizer, E. E., Lybecker, D. W., and Zimdahl, R. L. (1988). Systems approach to weed management in irrigated crops, *Weed Sci. 36*: 840–845.

Scragg, E. B. and McKelvie, A. D. (1976). The effect of certain species of weeds on the grain yield of spring barley, *Ann. Appl. Biol. 83*: 335–338.

Shaw, W. C., Willard, C. J., and Bernard, R. L. (1955). The effect of 2,4-dichlorophenoxyacetic acid (2,4-D) on wheat, oats, barley and the legumes underseeded in these crops, *Ohio Agric. Exp. Stn. Res. Bull. 761.*

Simkins, G. S., Moshier, L. J., and Russ, O. G. (1980). Influence of acetamide herbicide applications on efficacy of the protectant CGA-43089 in grain sorghum (*Sorghum bicolor*), *Weed Sci. 28*: 646–649.

Smith, A. E., ed. (1984). *Wild Oat Symposium Proceedings*, Vol. 1, Can. Plains Proc. 12, Agriculture Canada Regina Res. Stn., Saskatchewan.

Smith, A. E. and Hsiao, A. I., eds. (1984). *Wild Oat Symposium Proceedings. Use and Mode of Action of Wild Oat Herbicides*, Vol. 2, Can. Plains Proc. 12, Agriculture Canada Regina Res. Stn. Saskatchewan.

Smith, B. S., Murray, D. S., Green, J. D., Wanyahaya, W. M., and Weeks, D. L. (1990). Interference of three annual grasses with grain sorghum (*Sorghum bicolor*), *Weed Technol. 4*: 245–249.

Smith, D. and Buchholtz, K. P. (1964). Oat variety responses to triazine herbicides, *Crop Sci. 4*: 223–225.

Smith, R. J., Jr. (1968). Weed competition in rice, *Weed Sci. 16*: 252–254.

Smith, R. J., Jr. (1970). Weed control methods, losses and cost due to weeds and benefits of weed control in rice, in *FAO International Conference on Weed Control*, Weed Science Society of America, Champaign, Illinois, pp. 24–37.

Smith, R. J., Jr. (1974). Competition of barnyardgrass with rice cultivars, *Weed Sci. 22*: 423–426.

Smith, R. J., Jr. (1983a). Competition of bearded sprangletop (*Leptochloa fascicularis*) with rice (*Oryza sativa*), *Weed Sci. 31*: 120–123.

Smith, R. J., Jr. (1983b). Weeds of major economic importance in rice and yield losses due to weed competition, in *Weed Control in Rice*, International Rice Research Institute, Los Banos, Laguna, Philippines, pp. 19–36.

Smith, R. J., Jr. (1984). Competition of spreading dayflower (*Commelina diffusa*) with rice (*Oryza sativa*), *Weed Sci. 32*: 116–119.

Smith, R. J., Jr. (1988). Weed thresholds in southern U.S. rice, *Oryza sativa, Weed Technol. 2*: 232–241.

Smith, R. J., Jr., Flinchum, W. T., and Seaman, D. E. (1977). *Weed Control in U.S. Rice Production*, Handbook No. 497, USDA, Washington, D.C.

Stauber, L. G., Smith, R. J., Jr., and Talbert, R. E. (1991). Density and spatial interference of barnyardgrass (*Echinochloa crus-galli*) with rice (*Oryza sativa*), *Weed Sci. 39*: 163–168.

Sutherland, P. L. and Long, J. H., Jr. (1987). Comparative tolerance of no-till sorghum and proso millet to soil-applied chlorsulfuron as related to cropping interval, *Appl. Agric. Res. 2*: 330–336.

Swann, C. W. (1980). Principles and practices of weed control in corn and grain sorghum, *Univ. Georgia Coop. Ext. Serv. Bull. 834.*

Todd, B. G. and Derksen, D. A. (1986). Perennial weed control in wheat in western Canada, in *Wheat Production in Canada: A Review* (A. E. Slinkard and D. B. Fowler, eds.), Proc. Can. Wheat Production Symp. Univ. Saskatchewan, Saskatoon. p. 391–404.

Triplett, G. B, Jr. (1978). Weed control for doublecrop soybeans planted with the no-tillage method following small grain harvest, *Agron. J. 70*: 577–581.

Turner, D. J. (1985). Effects on glyphosate performance of formulation, additives and mixing with other herbicides, in *The Herbicide Glyphosate* (E. Grossbard and D. Atkinson, eds.), Butterworths, Boston, pp. 221–240.

Unger, P. W. (1983). Water conservation: Southern Great Plains, in *Dryland Agriculture* (Dregne, H. E. and W. O. Willis, eds.), Am. Soc. Agron., Madison, Wisconsin, pp. 35–55.

USDA. (1990). *Agricultural Statistics 1990.* U.S. Government Printing Office, Washington, D.C.

USDA. (1991). *Agricultural Chemical Usage 1990 Field Crops Summary.* U.S. Dep. Agric. Nat. Agric. Stat. Serv. Econ. Res. Serv. Ag., U.S. Government Printing Office, Washington, D.C.

Vencill, W. K., Wilson, H. P., Hines, T. E., and Hatzios, K. K. (1990). Common lambsquarters (*Chenopodium album*) and rotational crop response to imazethapyr in pea (*Pisum sativum*) and snap bean (*Phaseolus vulgaris*), *Weed Technol. 4*: 39–43.

Vesecky, J. F., Feltner, K. C., and Vanderlip, R. L. (1973). Wild cane and forage sorghum competition in grain sorghum, *Weed Sci. 21*: 28–32.

Wall, D. A. and Friesen, G. H. (1989). Volunteer Jerusalem artichoke (*Helianthus tuberosus*) interference and control in barley (*Hordeum vulgare*), *Weed Technol. 3*: 170–172.

Wall, D. A., Kiehn, F. A., and Friesen, G. H. (1986). Control of Jerusalem artichoke (*Helianthus tuberosus*) in barley (*Hordeum vulgare*), *Weed Sci. 34*: 761–764.

Weston, L. A. (1990). Cover crop and herbicide influence on row crop seedling establishment in no-tillage culture, *Weed Sci. 38*: 166–171.

Wicks, G. A. (1984). Integrated systems for control and management of downy brome (*Bromus tectorum*) in cropland, *Weed Sci.* 34(Suppl. 1): 26–31.

Wicks, G. A. (1985a). Early application of herbicides for no-till sorghum (*Sorghum bicolor*) in wheat (*Triticum aestivum*) stubble, *Weed Sci. 33*: 713–716.

Wicks, G. A. (1985b). Weed control in conservation tillage systems: small grains, in

Weed Control in Limited-Tillage Systems (A. F. Wiese, ed.), Weed Science Society of America, Champaign, Illinois, pp. 72-92.

Wicks, G. A. (1986). Herbicide applications on wheat and stubble for no-tillage corn, *Agron. J. 78*: 843–848.

Wicks, G. A. and Grabouski, P. H. (1986). Weed control in no-till sorghum (*Sorghum bicolor*), *Weed Sci. 34*: 577–581.

Wicks, G. A., Popken, D. H., and Lowry, S. R .(1989). Survey of winter wheat (*Triticum aestivum*) stubble fields sprayed with herbicides after harvest in 1986, *Weed Technol. 3*: 244–254.

Wicks, G. A. and Smika, D. E. (1990). Central Great Plains, in *Systems of Weed Control in Wheat in North America* (W. W. Donald, ed.), Weed Science Society of America, Champaign, Illinois, pp. 127–157.

Wicks, G. A., Smika, D. E., and Hergert, G. W. (1988). Long-term effects of no-tillage in a winter wheat (*Triticum aestivum*)-sorghum (*Sorghum bicolor*)-fallow rotation, *Weed Sci. 36*: 384-393.

Wiese, A. F. (1983). Weed control, in *Dryland Agriculture* (E. H. Dregne and W. O. Willis, eds.), Am. Soc. Agron., Madison, Wisconsin, pp. 463–488.

Wiese, A. F., Burnside, O. C., Phillips, W. M., and Eastin, E. F. (1977). Controlling tough weeds in sorghum, *Weeds Today 8*(2): 26–28.

Wiese, A. F. and Chenault, E. W. (1987). Controlling weeds and volunteer crops during fallow periods, *Texas Agric. Exp. Stn. B-1568.*

Wiese, A. F., Unger, P. W., and Allen, R. R. (1985). Reduced tillage in sorghum, in *Weed Control in Limited-Tillage Systems* (A. F Wiese, ed.), Weed Science Society of America, Champaign, Illinois, pp. 51-61.

Wiese, A. F., Unger, P. W., Allen, R. R., Jones, O. R., and Harman, W. L. (1986). Minimum tillage systems for the Texas panhandle, *Texas Agric. Exp. Stn. MP-1597.*

Wiese, A. F., Wood, M. L., and Chenault, E. W. (1988). Persistence of sulfonylureas in Pullman clay loam, *Weed Technol. 2*: 251–256.

Williams, E. D. (1977). Growth of seedlings of *Agropyron repens* L. Beauv. and *Agrostis gigantea* Roth. in wheat and barley: effect of time of emergence, nitrogen supply and cereal seed rate, *Weed Res. 17*: 69–76.

Willis, W. O., Bauer, A., and Black, A. L. (1983). Water conservation: Northern Great Plains, in *Dryland Agriculture* (H. E. Dregne and W. O. Willis, eds.), Am. Soc. Agron., Madison, Wisconsin, pp. 73–88.

Wilson, B. J. and Peters, N. C. B. (1982). Some studies of competition between *Avena fatua* L. and spring barley. I. The influence of *A. fatua* on yield of barley, *Weed Res. 22*: 143–148.

Zama, P. and Hatzios, K. K. (1986). Effects of CGA-92194 on the chemical reactivity of metolachlor with glutathione and metabolism of metolachlor in grain sorghum (*Sorghum bicolor*), *Weed Sci. 34*: 834–841.

Zimdahl, R. L. (1980). *Weed-Crop Competition: A Review*, Int. Plant Protection Ctr., Oregon State Univ., Corvallis, Oregon, pp. 31-39, 49–52.

Zimdahl, R. L. (1990). The effect of weeds on wheat, in *Systems of Weed Control in Wheat in North America* (W. W. Donald, ed.), Weed Science Society of America, Champaign, Illinois, pp. 11–33.

12

Weed Management Systems for Pastures and Hay Crops

Albert E. Smith
University of Georgia, Griffin, Georgia

Lionel D. Martin
Curtin University of Technology, Northam, Western Australia, Australia

I. INTRODUCTION

The authors of this chapter present the principles basic to weed management in pasture and hay cropping systems. Weed management strategies included in this chapter are common to the United States and Australia. However, this broad spectrum of information on a multitude of weeds found in numerous crops should be helpful to scientists and forage-crop managers throughout the world.

A weed has been defined as a plant growing where it is not desired, or a plant out of place. This definition holds true for weeds in row crops, horticulture, forestry, and other situations. However, in pastures, a "weed" is not always so obvious, as most of the weeds, at some growth stage, have forage value and contribute to the productivity of the pasture. If it is assumed that the greatest net returns are obtained from a pasture that comprises only sown species, then any invading plant regardless of its niche in the ecosystem that ultimately reduces net return from the pasture is an undesirable species or a weed (Wells, 1974).

Certain plants may be described as weeds or undesirable species at one time or in some situations, but as useful plants at another time or in other situations. For example, in southern Australia hare barley, though of poor nutritive value and troublesome to sheep in the spring, provides useful feed over vast areas in the autumn and winter. Similarly, in the southern United States little barley furnishes excellent forage in dormant bermudagrass stands during the early spring. However, as the plants mature and develop seedheads, the resulting awns cause irritation to the grazing animals' eyes. Annual *Medicago* species are important pasture legumes in southern Australia, but the burrs of many species are difficult to remove from wool (Michael, 1970).

Weeds are undesirable in pastures and hay crops because they interfere with desirable pasture plants, mainly by competing with them for soil moisture, nutrients, light, and space, thus reducing their yield (Harper, 1977; Klingman, et al., 1982; Ross and Lembi, 1985).

Some weeds interfere with desirable pasture plants through potential allelopathic effects (Smith, 1991a; Stachon and Zimdahl, 1980). Allelopathy, the direct and indirect effect of one plant on another through the production of chemical compounds that escape into the environment (Rice, 1984), is gaining importance as a process through which weeds in pastures and hay crops interfere with the desired pasture plants (Smith, 1990; Putnam and Weston, 1986; Rice, 1984). This is exemplified by the long list of pasture weed species with alleged allelopathic potential (Appendix C).

Many pasture weeds are unpalatable and are avoided by the grazing animals. Examples of weeds in this category are Californian burr, curly dock, horehound, horseweed, johnsongrass, Noogoora burr, quackgrass, spiny cocklebur, various species of thistles, wild artichoke, and wild onion (Australian Wool Corporation, 1989; Peters and Linscott, 1988).

Some weeds lower the quality of animal products. For example, Californian burr, common St. Johnswort, distaff thistle, dock, filaree, hare barley, Noogoora burr, red sorrel, spiny cocklebur, and three-cornered jack lower the quality of wool (Australian Wool Corporation, 1989; Lamp and Collet, 1989). Bitter sneezeweed, capeweed, dogfennel, fennel, heal-all, hedge mustard, Indian mustard, Indian sweet clover, mayweed chamomile, pennyroyal, peppercress, sand rocket, shepherd's purse, swinecress, tansy ragwort, wall rocket, and wild garlic taint milk, lowering the quality (Australian Wool Corporation, 1989; Smith, 1977, Lamp and Collet, 1989).

Other weeds cause physical injury to grazing animals. Examples are horsenettle, little barley, stinging nettle (Matthews, 1977), filaree, a range of thistle species, three-cornered jack, wild artichoke (Australian Wool Corporation, 1989), and bromegrasses (Peters and Linscott, 1988).

Still other weeds are poisonous and are capable of causing heavy livestock losses. Examples are buttercups, buttercup oxalis, blessed milkthistle, brackenfern, cape tulips, capeweed, common S. Johnswort, European helioptrope, filaree, goatsrue, largeleaf lantana, Patterson's curse, puncture vine, red sorrel, and tansy ragwort (Australian Wool Corporation, 1989; Lamp and Collet, 1989).

II. METHODS OF WEED MANAGEMENT

It has been estimated that every kilogram of herbaceous weeds grown in a pasture reduces the production of the desirable forage by an equivalent amount (Klingman, 1956; Klingman and McCarthy, 1958; Peters and Lowance, 1969). Therefore, proper management of weeds in pastures can dramatically improve pasture production, forage utilization, and animal production and ultimately has direct economic consequences.

The three fundamental objectives of weed management in pastures and hay crops, as in other ecosystems, are prevention, eradication, and management (Anderson, 1983; Ross and Lembi, 1985).

The main objective of prevention is the exclusion of weeds from areas not yet infested, or stopping their spread from infested areas to clean pastures. Prevention is practiced mainly through laws and their enforcement, such as federal and state seed laws and those established by local or regional weed districts. However, management is necessary to prevent the introduction of weeds into the pasture and hay fields as well as to isolate the infested areas. Tropical soda apple was introduced into Florida and is slowly infesting the pastures, primarily through seed distribution in animal dung (Mullahey et al., 1993). It is important for managers of noninfested farms and surrounding states to be aware of the potential infestation from the introduction of animals that carry the seed inside. A simple cleaning out period for the animal will prohibit the movement of soda apple seed and will prevent infestation of additional hectarage. Smutgrass can be introduced into the pasture ecosystem through hay infested with smutgrass seed (Smith, 1982).

The purpose of eradication is the elimination of a weed species after it has become established in an area. Because of the difficulty and high cost, eradication is usually attempted only in the case of new, small-scale infestations. Weed management is suppressing the weed populations to or below a predetermined level. This is a containment approach which implies that the target weed may still have a predetermined effect on pasture production or stock health, but the benefit to cost ratio is acceptable. Thus the concept of an economic or action threshold for weeds is central to weed management in the pasture ecosystem (Radosevich and Holt, 1987). Though economic threshold models have been used widely to make decisions on weed control in row crops, less work has been done to extend threshold models to pasture weed management decisions (Field and Daly, 1990).

The various methods of weed management in pastures and hay crops can be broadly classified as 1) preventive and cultural, 2) mechanical, 3) chemical, and 4) biological. The ideal weed management program is one which integrates these methods of weed management.

A. Mechanical Weed Management Methods

Mechanical methods are the oldest and still often the main ways in which weeds in pastures and hay crops are managed. They range from hand hoeing and pulling to mowing and tillage operations.

Hand Hoeing or Pulling

Hand hoeing and pulling methods are the most primitive types of weed management. They are still the major methods for managing weeds in pastures in developing countries, where labor is readily available and cheap and farm sizes are small. Hand hoeing or pulling is useful on scattered weeds that have appeared for the first time in pastures and that are particularly noxious. This method is highly effective in controlling annual and biennial weeds but is only effective on perennial weeds if the perennial is never allowed to grow undisturbed for more than about 2 weeks. The objective is to force constant regrowth without sufficient time to replenish the underground food reserves (Lorenzi and Jeffery, 1987; Ross and Lembi, 1985).

Tillage

Tillage as a method of weed management in pastures and hay crops, unlike in row crops, is limited to operations during seedbed preparation. Tillage prior to the establishment of pastures is very effective in controlling most of the perennial and annual weed seedlings that have emerged early in the season. Soil disturbance by tillage induces weed seed germination, which can be controlled by further tillage or by chemical means. Grain fields seeded with grass or clover frequently develop a crop of annual or biennial weeds after the grain is harvested. One or two harrowings of the stubble field will destroy many of the weed seedlings and give the pasture plants an initial advantage in competition with weeds.

Mowing

Mowing controls weeds by removing tops prior to seed formation, by depleting underground food reserves, and by favoring the growth habits of competitive pasture species. Mowing is relatively ineffective on weeds that can grow and produce seed below the normal cutting height. For mowing to prevent seed formation, the weed species must have a relatively tall growth habit, and mowing must be done prior to pollination and fertilization. Mowed plants generally survive and regenerate new tops. However, repeated mowing is required to prevent seed production. Mowing generally kills tall annual weeds with one or two passes, provided the buds on the stem are above the cutting height. When repeated frequently, mowing results in the depletion of food reserves and the gradual elimination of the perennial weeds. In most cases, the first mowing should be done before middle-to-late bud stage. Additional mowing probably will be required at 2- to 3-week intervals. Another advantage of mowing is that it stimulates the germination of underground buds of creeping perennials, thus depleting their numbers. Close mowing will induce the underground system to develop closer to the soil surface than it normally would, potentially resulting in better control when tillage is subsequently used or if the underground systems are sensitive to freezing temperatures. When mowing is coupled with other management practices such as grazing and fertilization, it results in a competitive advantage for the pasture species, which aids in smothering weeds (Ross and Lembi, 1985). Mowing can also be used to remove unpalatable or inedible weeds left after stock have selectively grazed a pasture, to prevent them from gaining a competitive advantage over the more desirable species in the pasture (Intaka Press, 1987).

B. Cultural Weed Management Methods

Cultural methods of weed management utilize practices common to good farm management. Practices such as use of weed-free seed and clean equipment fall under the category of preventive weed management. Other cultural methods such as crop rotation, burning, pasture plant competition, and pasture management can be incorporated into farm management programs to reduce many pasture weed problems.

Crop Rotation

Though crop rotation is not available as an option for weed management in permanent pastures, it is a very useful method in the case of temporary pastures, hay fields, and alfalfa fields, which can be rotated with other crops. Rotations can be used effectively to control perennial-broadleaf weeds, in alfalfa fields, during the cereal phase of the rotation.

Burning

In pastures, fire is mainly used for removing litter, controlling insects, and controlling some woody species. Noxious weeds which have been allowed to mature seed are sometimes cut and burned on the spot to prevent the spreading of their seeds. Patches of dodder growing in clover and alfalfa fields can be prevented from spreading to other parts of a farm by burning in place. Weedy pastures and meadows are sometimes burned with the hope of destroying weeds. Burning is more effective in destroying the underground parts of perennial weeds than any other method of removing their tops. However, frequent use of burning pastures leads to loss of many of the more desirable pasture species, development of fire resistant weed flora, and loss of surface organic matter, therefore increasing soil erosion and loss of soil fertility (Intaka Press, 1987).

Pasture Plant Competition

A vigorous pasture competes more effectively with weeds than a weak pasture. The competitive ability of a crop species depends on its density; time of emergence relative to the weeds; rate of shoot and root growth; canopy height and disposition of leaves; size, distribution, and activity of the root system; and speed of recovery from mowing and/or grazing. Any plant attribute or management practice that favors the pasture species enhances its competitive ability. The competitive ability of the pasture vegetation can be improved by use of well-adapted species and cultivars, soil preparation, optimum sowing date and rate, fertilizer application, liming (adjustment of pH), irrigation, drainage, effective insect and disease control, and optimum grazing management. All of these will not only increase the pasture yield, but also reduce the incidence and competitive effects of weeds (Intaka Press, 1987).

The other methods of weed management such as the use of herbicides, mowing, and grazing management are of limited value if a highly competitive pasture cover is not maintained. In these situations weeds emerge and reestablish in the gaps created by the weeds that were controlled. Pasture cover plays a major role in preventing germination of weed seed (Gorski et al., 1978; Silvertown, 1980) and establishment of weed seedlings (Hardine, 1985). For example, germination of blessed milkthistle, bull thistle, Canada thistle, mayweed chamomile, musk thistle, and tansy ragwort was markedly reduced under a canopy (Phung and Popay, 1981).

Liming and Fertilization

Liming and fertilization are particularly effective in reducing the weed populations in pastures over time. For example, in Connecticut applications of nitrogen, phosphorus, and potassium on a sparse stand of colonial bentgrass resulted in a shift to denser sod and a consequent reduction in weeds. Some weed species are favored by fertilization and improved grazing management. For example, common chickweed and curly dock are severe problems only on fertile soils (National Academy of Sciences, 1968). Broomsedge can be removed from a pasture by following liming with intense grazing pressure.

Drainage

Pastures are frequently poorly drained, and native weeds, which thrive in wet soils, are very common. Various species of buttercups, hardshack, horsetails, rushes, and many others are characteristic weeds of poorly drained, undergrazed pasture land. By improving the drainage, conditions are made less favorable for these weeds and more favorable for the grasses and legumes to compete with, and finally exclude, many weeds (National Academy of Science, 1968; Matthews, 1982).

Pasture Management

Weeds, in pastures, are usually a reflection of the climate, soil, and previous management of the area. The annual weeds germinate when the soil is at least partially bared by seasonal conditions or after overgrazing, mowing, or burning. Even slight changes in grazing pressure or the time at which livestock are turned into or removed from a pasture can have far reaching cumulative effects on the species composition over a few years. Trends toward weed dominance should be recognized as early as possible and steps taken to reverse the buildup by effectively grazing them when young and palatable, mowing after grazing, using chemicals, or reducing the amount of bare soil in which they can germinate.

Grazing Management

Grazing involves not only a reduction in leaf area of plants with a consequent reduction in growth rates, but also a reduction in the population of primary growing points (tillers, stolons, rhizomes, and roots). In addition, grazing modifies the microenvironment as a result of the presence of dung, urine, and trampling and the dispersal of propagules (seeds, fruits). Further, the effects of grazing on grassland communities are not simple summations of their effects on individual species. Species in pastures differ in germination, developmental patterns, and nutritive value such that their interrelationships may be profoundly altered by grazing. These very complex interactions often determine whether or not weeds emerge and become a problem in the pasture ecosystem (Medd et al., 1987).

The continued removal of the tops of weeds by grazing animals and close mowing prevents seed formation and also gradually weakens the underground parts. Sheep, goats, hogs, and cattle are effective in controlling many weed species. Grazing animals should not be turned into pastures too early in the growing season before the forage has made sufficient growth. Grazing, as a method of weed management, is generally more effective under a system of rotational rather than continuous grazing or set stocking. In the latter, animals are allowed to roam over a large area and choose the more desirable forage and leave the weeds to become rank and unpalatable.

Goats are used to control weeds in pastures in Australia (Holst, 1980; Holst and Campbell, 1987), New Zealand (Lambert, et al., 1981; Holgate, 1983), and several developing countries (Knight, 1965; Wilson, 1957). Goats control weeds by preventing flowering and subsequent seed dispersal, by preferentially grazing and thus placing the weeds at a competitive disadvantage to desirable species, by causing mechanical damage through ringbarking or destroying branches, and by a combination of these (Holst, 1980). Goats are browsers and opportunistic feeders that graze many plants that are either completely or partly avoided by sheep and cattle. Weeds that are most palatable to goats are generally the most easily controlled. In Australia, marked preference shown by goats for Eastern baccharis, European barberry, European blackberry, gorse, kudzu, and sweetbriar rose can result in complete control of these species. Curly dock, poa tussock, rushes, and white horehound are moderately attractive to goats. Thistles such as blessed milkthistle, bull thistle, Canada thistle, marsh thistle, and Italian thistle are eaten by goats during flowering and seeding. Control of these weeds is responsive to goat grazing pressure and seasonal effects (Holst and Campbell, 1987). Cogongrass, serrated tussock, and St. Johnswort are only controlled if the weed constitutes a small portion of the total pasture, for example, in a degenerating pasture or where traditional methods have reduced the density of weeds. Control is effected by the pasture species' competing with partially defoliated weeds (Campbell et al., 1979). Of the common pasture weeds, spiny cocklebur and brackenfern are rarely eaten by goats (Holst and Campbell, 1987).

Pasture Manipulation

Manipulation of the composition of the pasture by using low rates of herbicides alone and in combination with grazing, in order to improve the dominance of the desirable species by suppressing the undesirable weed species, is a recommended pasture management practice in Australia (Dickinson and Stephenson, 1990; Dellow and McDonald, 1991). Manipulation of the composition of the pasture is particularly important as some species which are desirable during their early stages of growth early in the season become undesirable as they mature. For example, annual grasses, particularly hare barley (in Australia) and little barley (in the United States), provide good-quality herbage early in the season when the forage component of the pasture is dormant but become undesirable later in the season as they compete with the desirable pasture species. Finally, at maturity the awned seeds cause injury to grazing animals' eyes and lower the quality of wool. Similarly, several broadleaf species contribute substantially to the herbage production early in the season when they are young and palatable, but later in the season become less palatable and also suppress the growth of the forage species.

C. Biological Weed Management

Biological weed management is the use of natural enemies to reduce weed populations to economically acceptable levels. The natural enemies used as biocontrol agents are mainly insects, plant pathogens, and nematodes. These biocontrol agents are very selective in their food preferences; therefore, biological methods cannot provide the broad-spectrum control that can be achieved with other methods such as tillage and herbicides. Their greatest value has been in situations in which a single aggressive weed is the major problem. Biological control is more suitable for perennial weeds growing in relatively undisturbed habitats like pastures and rangeland, as the biocontrol agent has a chance to build up its population over an extended period. It is more economical and feasible on these large areas than mechanical or chemical methods (Ross and Lembi, 1985).

The outstanding example of biological weed control is the control of prickly pear cactus, which infested over 30 million hectares of rangeland in Australia by the cactus moth, *Cactoblastis cactorum*, in the 1920s–1930s, and the control of St. Johnswort, a poisonous range weed which infested over 1 million hectares in California and Oregon, by the leaf eating beetle *Chrysolina quadrigemina* in 1944. More recent examples of successes of biological control of pasture weeds include rush skeleton weed in Australia using the rust *Puccinia Chondrillina*, musk thistle in Virgina and Georgia using the weevils *Rhinocyllus conicus* and *Ceuthorhynchidius horridus*, and tansy ragwort in California using the halticid flea beetle, *Longitarsus jacobaeae*, and the arctiid moth, *Tyria jacobaeae* (Schroeder, 1983; Wapshere 1982).

Although there have been some outstanding successes in biological control of pasture weeds and there are vigorous research programs being conducted in several countries (primarily in the United States, Canada, Australia, and New Zealand) (Schroeder, 1983), it will be some time before biocontrol agents are used for weed management in pastures with the frequency of either mechanical or chemical methods. However, biological control will have a place in the future as we move from controlling weeds to managing weeds in pastures, especially in situations where other methods cannot be applied, are not economical, or are environmentally undesirable.

D. Chemical Weed Management

Application of herbicides to manage vegetation in pastures and hayfields is now commonplace, and managers who contemplate the use of herbicides must be well versed in the properties and proper uses of these sophisticated chemicals. Therefore, it is important that the information in Chapter 5 be understood prior to developing the weed management system involving the application of herbicides. Herbicides are chemicals that kill plants or retard the rate and extent of their normal growth. Solutions of copper nitrate, iron sulfate, potassium salts, sulfuric acid, and other common chemicals were discovered to be plant control agents in the late 1800s. However, it wasn't until after World War II that chemicals which are selective in weed control became available. The growth regulating properties of the phenoxyalkanoic acid chemicals (2,4-D and

2,4,5-T) were reported in the early 1940s (Hamner and Tukey, 1944). These discoveries introduced an era of rapid growth in the agricultural chemical industry and gave rise to the discipline of weed science.

The new growth-regulating chemicals have three notable properties:

1. *They are highly selective.* Many herbaceous broadleaf and woody plants could be killed without damage to desirable grasses.
2. *They are systemic in action.* They are absorbed by plants and move internally to a site of action and disrupt vital physiological processes.
3. *Small amounts are required for weed control.* They are plant growth regulators.

Discovery of the herbicidal properties of the phenoxyacetic acids marked the beginning of the herbicide phase of the chemical era of agriculture (Hamner and Tukey, 1944). Since that time literally thousands of chemicals have been studied as potential herbicides and hundreds have been registered by the United States Environmental Protection Agency for use in the management of weeds in agricultural production systems. Because of the unique properties of the phenoxyalkanoic herbicides, many are presently utilized for managing the weeds in pastures and hay crops.

The registration of herbicides for managing weeds in the pasture ecosystem has not progressed as rapidly as in other cropping systems. This is primarily due to the limited use of herbicides in pastures compared to other crops, so that the monetary return on investment required for registration has been considered marginal by manufacturers of the herbicides. However, there are chemicals registered for use in most pasture and hay crops and the use of these herbicides will be discussed under the specific management systems. Table 1 lists the herbicides registered for use in pasture and hay crops in the United States. It is realized that some practices discussed in this chapter may not be registered for the reader's intended use; therefore, it is important to ascertain the registration of each herbicide before use.

III. WEED MANAGEMENT SYSTEMS FOR COOL SEASON PERENNIAL FORAGE CROPS

Cool season forage crops are predominant in the northern United States (Sheaffer and Marten, 1986), British Isles (Haggar et al., 1985), Europe (Dietl, 1982), temperate parts of Australia (Carpenter et al., 1990), New Zealand (Daly, 1990), Japan (Nemoto, 1982), and the temperate regions of South Africa (Wells and Stirton, 1982) and South America. Cool season forage crops produce the bulk of their feed during the winter-spring period and have the ability to tolerate frost and cold. They may be either pure grass pastures, as is most common in the northern United States (Sheaffer and Marten, 1986), or mixtures of grass and legumes as in New Zealand (Harris, 1990) and Australia (Carpenter et al., 1990). The principal cool season grasses are perennials such as bluegrasses, canary grasses, orchard grass, perennial ryegrass, smooth bromegrasses, tall fescue, and timothy. The cool season perennial legume species include red clover, strawberry clover, and white clover (Carpenter et al., 1990; Johnson, 1990; Langer, 1990).

Weeds are a continuing threat to newly seeded pastures. The reserves of weed seeds in the soil vastly outnumber the pasture seeds sown; therefore, as a pasture plant weakens or dies, it is likely to be replaced by a weed. Additionally, the vegetative reproductive organs of certain perennial weeds reside in the soil and must be managed prior to planting the crop. Hence, measures should be taken to minimize weed invasion during crop establishment. It is important for the landowner to consider the long-term effects of the inadequate management of weeds when deciding to convert land use to forage crops. Quite often, a simple management scheme or an application of a particular herbicide before planting the crop will save dollars in later years. All perennial weeds should be managed before seedbed preparation. Failure to manage such weeds as common bermudagrass, horsenettle, wild garlic, or wild onion before planting the crop will result in problems without a solution during the crop's establishment. Ecological principles of natural vegetation establishment indicate that Mother Nature strives to establish natural vegetation in all ecosystems. Such famous ideas as "nature abhors a vacuum" and "vegetation trends proceed toward climax" allow us to relate to the battle against Mother Nature in her thrust to establish natural species. The prepared seedbed opens the vacuum for the weeds and the abhorrence of that vacuum by a perennial weed results in complicated management decisions which generally are costly to the landowner. In certain instances the only decision may be to repeat the seedbed preparation and planting practices.

A. Seedbed Preparation

Chemical methods for managing the annual and perennial broadleaf weeds in the pasture are effective and economical. However, concern about the management of broadleaf weeds prior to seedbed preparation is as important as concern about the management of perennial grasses prior to seedbed preparation. Similarly, there are chemicals that are effective in managing grass weeds in legume crops; however, management of the perennial grass and broadleaf weeds prior to the preparation of the legume seedbed is very important.

Several herbicides can be used to manage common bermudagrass and johnsongrass prior to preparing the seedbed. Glyphosate is a relatively nonselective, broad-spectrum herbicide. It is applied as a postemergence foliar treatment and is readily translocated throughout the aerial and underground plant

Table 1 List of Herbicides Registered for Use, in the United States on Pasture and Hay Crops. The Herbicide Labels Should Be Read and Understood Prior to Using the Herbicides

Crop	Herbicide Name		Application type	Toxicity category
	Common	Trade		
Alfalfa	Benefin	Balan	Preplant incorporated	IV
(establishment)	EPTC	Eptam	Preplant incorporated	III
(first year)	2,4-DB	several	Postemergence	III
	Bromoxynil	Buctril	Postemergence	II
	Pronamide	Kerb	Postemergence	III
	Sethoxydim	Poast	Postemergence	II
(Established)	Sethoxydim	Poast	Postemergence	II
	Paraquat	Gramoxone Extra	Postemergence, after cut	II
(Established dormant)	Metribuzin	several	Post- and premergence	III
	Hexazinone	Velpar	Post- and premergence	III
	Pronamide	Kerb	Postemergence	III
	Paraquat	Gramoxone Extra	Postemergence	II
	2,4-DB	several	Postemergence	III

Clovers	EPTC	Eptam	Preplant incorporated	III
	Benefin	Balan	Preplant incorporated	IV
	Pronamide	Kerb	Postemergence	III
	2,4-DB	several	Postemergence	III
(+ Grass)	2,4-D	several	Postemergence	II, III
Perennial	Diuron	Karmex	Premergence to sprigged	III
grass	2,4-D	several	Postemergence	II, III
	Dicamba	Banvel	Postevergence	III
	2,4-D+picloram	Grazon P+D	Postemergence	IV
	2,4-D+dicamba	Weedmaster	Postemergence	III
	Triclopyr	Remedy	Postemergence	I, III
	2,4-D+triclopyr	Crossbow	Postemergence	I, III
	Hexazinone	Velpar	Post- and premergence	III
	Glyphosate	Roundup	Postemergence	II
	Tebuthiuron	Spike 20P	Post- and premergence	III
	Metsulfuron	Ally	Postemergence	III

parts (Anderson, 1983). It is phytotoxic to most annual, biennial, and perennial herbaceous plant species and certain woody species. It is strongly adsorbed by soil colloids, and it has little or no phytotoxicity following soil application. It will take several applications of glyphosate during the growing season prior to seeding the forage crop to control common bermudagrass and johnsongrass adequately. Often it is beneficial to till the area subsequent to the desiccation of the vegetation in response to the first application of glyphosate and prior to the second herbicide application.

Management of perennial broadleaf weeds prior to preparation of the planting bed can be accomplished by a single or several postemergence applications of dicamba, 2,4-D, glyphosate, picloram, and triclopyr. These herbicides are most effective when applied postemergence to the foliage of the weeds, and they are readily transported throughout the plant. These herbicides are organic acids and are formulated as esters or salts. The majority of the salt formulations are various substituted amine salts, whereas the ester forms are derived from several high molecular weight organic alcohols. When the salt formulation is used, it is suggested to include a surfactant in the aqueous mixture to increase wetting of and absorption through the leaf surface. The ester formulations are, generally, more effective than the salt formulations. However, ester formulations are more volatile than salt formulations and can result in damage to other nontarget species such as crops on adjoining land. Additionally, all of these herbicides except glyphosate have residual activity in the soil, and the length of that activity varies depending on the climate. Therefore, these herbicides must be applied to the area early enough to allow for adequate dissipation of the residue prior to seeding the forage crop.

All of these herbicides, except glyphosate, are selective in the management of broadleaf weeds. Dicamba, picloram, and triclopyr are more effective in the management of perennial broadleaf weeds than glyphosate and 2,4-D. Some formulations of triclopyr and dicamba include small quantities of 2,4-D since it has been reported to improve the transport of other herbicides through the plant. These herbicides are most effective when applied during the early growth stages of the weed and when the plant is not stressed. Upon entry into the plant, these herbicides are translocated in the symplast along with the photosynthate. If the weeds are under drought stress, translocation is reduced and the resulting effects are less than desirable.

Sethoxydim is a selective herbicide for grasses (graminicide) that must also be applied postemergence to the foliage of the grass. It will control both annual and perennial grasses. However, some grasses are more complicated to manage than others. Johnsongrass is much easier to manage with sethoxydim than is common bermudagrass. It may take several applications to manage common bermudagrass adequately before preparing the seedbed. It is necessary to include a nonphytotoxic oil in the aqueous mixture to obtain maximum weed management.

B. Weed Management During the Establishment of Cool-Season Forage Grasses

It is important to give the forage-crop seedlings every advantage possible in the competition with weeds that will germinate with the forage species. Several cultural practices are important to weed management at the time of seeding the forage crop. This includes use of the recommended forage species and cultivars, optimum seeding rate that results in uniform stand, sowing time that avoids particular weed problems, crop rotations that reduce the prevalence of weed seed in the soil, thorough seedbed preparation, fertilization, liming, and drainage. Early grazing or mowing can be used to manage tall-growing weeds, such as common lambsquarters. However, low-growing weeds such as common chickweed may benefit from mowing. Early grazing by sheep is very effective in reducing the competition from volunteer grasses and broadleaf species in legume based Mediterranean pastures in South and Western Australia. However, the success of these methods depends on the extent to which the sown species are subsequently able to dominate the pastures.

There are no herbicides labeled for use as preemergence treatments to seedbeds seeded to cool season forage grasses. Although some herbicides are reported to be effective in the management of some weeds and result in only limited damage to the forage grass seedlings, this use has not been registered by the EPA (Peters et al., 1989; Linscott and Vaughan, 1990; Malik, 1990). The importance of weed management during preparation of the seedbed is well known. Low rates of 2,4-D can be utilized for the management of small broadleaf weeds following grass seedling emergence and development beyond the two-leaf stage of growth. In most areas of pasture production this management is adequate for good establishment. The major grass weeds that may cause some problems are Italian ryegrass and little barley, since these annuals germinate about the same time as the seeded cool-season perennial forage grasses and respond well to high fertility. Early controlled grazing of the areas may help the perennial grasses overcome the competition from the annual weeds.

C. Weed Management in Established Stands of Cool-Season Forage Grasses

The most important management strategy for weeds in a perennial grass system is a combination of practices that maintain a good stand of grass. Allowing no voids to develop in the ground cover of the pasture forage is the best way to restrict invasion of the forage system by weed species. This requires a year-long management program and is most important in cool-season forage systems during the warm months of summer. Mismanagement during periods of heat and drought stress will result in stand reduction of the cool-season species. The resulting voids in ground cover are soon filled with warm-season annuals such as bitter sneezeweed, crabgrass, goosegrass, and pigweed. Crabgrass is especially competitive with the cool-season forage grass species and will respond to summer rainfall with a flush of growth. The forage grass is soon lost from the pasture.

Presently there are no herbicides labeled for use on cool-season forage grasses that effectively manage grass weeds. There are several strategies that are effective in relieving the pressure of competition with the grass weeds. If the weed is taller than the forage grass, wiper applications of glyphosate will suppress and sometimes kill the weed. It is important that the wiper applicator not come in contact with the forage plants. Glyphosate is nonselective and will damage or kill the plants that it contacts. If applied during the seedling stage of the grass weed, sometimes 2,4-D will suppress the weed. However, seldom are the effects adequate to justify the application solely for the management of the grass weeds. Metsulfuron can be applied to established fescue for the management of bahiagrass in the pasture.

Broadleaf weeds can be effectively managed by timely applications of 2,4-D, dicamba, triclopyr, and picloram. Also, glyphosate can be applied through a wiper applicator if there is adequate difference in height of the weed over the forage species. It is important to apply these herbicides during the early growth stages of the annual weeds for maximum uptake of the herbicide. When managing perennial broadleaf weeds in the pasture ecosystem, it is important to apply these herbicides during the early flowering stage of growth of the weed since during this growth stage translocation of the photosynthate is toward the storage organs, which also serve as the perennating organs. It is important to achieve maximum absorption and translocation of these compounds in perennial weeds for satisfactory management.

Dicamba, triclopyr, picloram, and metsulfuron can be safely applied to manage certain brush species that occur in pastures. It is important to read the label for the herbicides to determine the brush species on which each herbicide is effective, the requirement for the use of a surfactant, the timing of herbicide application, and the conditions for the herbicide use.

D. Renovation of Endophyte-Infected Fescue Pastures

The importance of tall fescue as a pasture grass is diminished when it is infected with the fescue endophyte. Tall fescue is the most important and widely grown pasture, turf, and conservation grass in the temperate regions of the world as well as in the southeastern United States. Much of the fescue is infected with an endophyte that causes fescue toxicosis symptoms in grazing animals (Bacon et al., 1977; Sampson, 1933; Stuedmann and Hoveland, 1988). The endophyte is seed transmitted and will die if the seed is stored 1 or 2 yrs (Bacon et al., 1977). Cultivars have been developed that are endophyte free. During a 4-yr grazing trial in Alabama, steer average daily gain on tall fescue with a 5% level of endophyte infection was 0.83 kg but only 0.45 kg on grass that was 94% infected (Hoveland et al., 1980).

Although much research has been conducted on the endophyte infection in fescue and its influence on animals, little is known about the level of control necessary to renovate the pasture effectively to "fungus-free" fescue. Results of studies indicate a linear relationship between endophyte level of infection and reduction of steer gains (Pederson et al., 1986; and Stuedmann et al., 1986). Stuedmann et al. (1986) observed a significant linear relationship between average daily gain and average percentage of endophyte among tall fescue treatments, indicating that for each 10% increase in endophyte frequency, average daily gain decreases 55 g. There are several strategies for managing the endophyte infected fescue pasture for improved animal performance. A second forage species can be interseeded into the fescue stand, allowing for a dilution of the infected foliage available to the herbivore. Also, the infected pasture can be renovated and seeded with fungus-free fescue or another forage species. Whether the endophyte is transferred from infected plants to noninfected plants through trampling and animal saliva is unknown. Until there is definitive information on the level of infection that can be tolerated, the assumed control level should be at least 90% for good to excellent control.

Prior to initiating the renovation, it is important to manage the seed head formation during the year of renovation. Reinfestation of the area from seed dropped during the renovation year is possible, and steps must be taken to manage the seed drop. This can be accomplished by mowing or spray topping. The infected fescue can be killed with one to several applications of paraquat or glyphosate. Research has indicated that paraquat applied in the early fall a month prior to seeding the fungus-free fescue will allow adequate time to determine the need for a second application to produce adequate kill of the infected fescue. Sequential applications of paraquat in the fall are generally required to kill more than 90% of the tall fescue stand consistently. Glyphosate applications were not as effective in managing the infected fescue as paraquat applications (Smith 1989). However, sequential applications of glyphosate may be recommended for use in pasture renovation when the pasture includes other perennial weeds that cannot be killed with paraquat.

E. Weed Management in Cool-Season Forage Legumes

There are fewer herbicides labeled for use in the cool-season forage legumes than in other types of forage crops. Therefore, weed management prior to preparing the seedbed is especially important. The use of a companion crop for managing weeds is of greater benefit to the cool-season forage legumes than to the warm-season legumes. Companion crops such as the cereal grains will help fill the void while the small-seeded legume species are germinating and becoming established.

There are several herbicides registered for use on small-seeded forage legume species at seeding. Benefin and EPTC are registered in the United States for application followed by light mechanical incorporation prior to planting the legume seed. It is important to incorporate these herbicides for maximum effectiveness. Both herbicides will effectively control grass weeds, and EPTC is more effective in the control of certain broadleaf weeds than benefin. These herbicides cannot be used when including a grass companion crop for the legume. Broadleaf weeds that escape these treatments can be managed by treatments with 2,4-DB.

In the subtropical regions of the United States the cool-season legume species are interseeded into established grass pastures to improve the forage quality and lengthen the season of production. The success of interseeding the small-seeded legume species into an established stand of grass has been variable. The chances of successful interseeding are dependent on managing the competition from the established forage grass and diseases. In fact the forage grass becomes the undesirable competitor with the interseeded species. Mechanical destruction of a bermudagrass sod prior to seeding the legume species has been successful. We have found that disking the grass pastures several times is better than single disking. A common rule of thumb is to disk for at least 50% destruction of the sod prior to seeding the area with the legume. Mechanical destruction of the cool-season grasses, such as fescue sod, must be carefully controlled since fescue does not vegetatively propagate and the stand will not fill in.

Two herbicides can be used to strip kill the sod over the planted row. Paraquat will effectively manage the cool-season grasses, such as fescue and bluegrass, and can be applied in a 10-cm swath over the row at seeding. Glyphosate will similarly kill the sod of warm-season perennial grasses such as bermudagrass in a strip over the seeded row at planting. Following germination of the legume seed, foliage of the grasses can be managed by mowing to reduce shading by the established forage grass.

F. Weed Management in Grass-Legume Mixtures

Cool-season grass species are often interseeded with small-seeded legume species for several reasons. A primary purpose of including the grass species is to serve as a companion crop. To reduce weed germination, a companion crop of cereal grains, peas, or other crops is seeded with the small-seeded legume.

Generally, the companion crop will germinate and develop faster than the small-seeded legume and will furnish an early ground cover that will interfere with weed seed germination. Research in many northern and northeastern states has shown that companion crops such as barley and oats may injure slow-starting forage species as much as weeds. Nevertheless, many farmers consider it advantageous to grow a crop of grain, silage, or pasturage while the legume is becoming established. If a companion crop is used while establishing the legume, it is important to remove the forage as early in the spring as possible and treat the regrowth with sethoxydim to reduce further competition during spring growth.

Additionally, certain perennial-grass species are planted with legumes for permanent grazing. The mixture has advantages over a monospecies forage crop of improved 1) forage quality, 2) distribution of forage quality and production over the growing season, 3) efficient utilization of nitrogen produced by the legume, and 4) weed management. Generally, there is no program of herbicide use that will assist in the management of weeds during the establishment of the mixed species grazing system. The herbicides that can be used in legume seedbeds will destroy the grass, and the herbicides that can be applied to young grass stands will harm or destroy the legume component. However, the major challenge is not in weed management but in the management of the composition of both species, simultaneously, in the grazing system. There will be a tendency to utilize management strategies that are beneficial to one component in the mixed pasture, and as a result of that action the favored component will become the dominant species, potentially leading to a loss of the second component from the grazing system. For example, the fertility regimen necessary for maintaining the grass component in the mix is not beneficial to the legume. The management program necessary to maintain both species in the mix will differ from region to region, and in the best of conditions it is a complicated program.

As a result of competition between the two forage species, it is necessary to adjust the foliage accumulation of both species in order to limit the degree of competition. Grazing intensity can be managed by season to manage the composition of the forage. Certain herbicides can be used for spray topping. Applications of low rates of graminicides and certain growth-regulating chemicals will restrict the formation of seed heads on most cool-season grass species. It is during the culm elongation and seed head formation stages of grass development that shading of the legume species will result. Generally, the best method for increasing the composition of the grass component is to increase the quantity of fertilizer since most cool-season grass species respond rapidly to this management practice. There is a potential for using low rates of some of the broadleaf weed herbicides (i.e., 2,4-D, dicamba, etc.) for suppression of the legume component. However, this has been found to be very risky and a slight mistake may result in complete loss of the legume species from the forage mix.

G. Weed Management in Annual Legume Based Pastures

Another type of cool-season pasture is the annual legume based pasture of the Mediterranean regions. These pastures are predominant in the wheat-sheep farming systems common in South and Western Australia and the southeastern United States (Madin and Bowran, 1989). In these pastures, the only sown species is the legume component, and all other volunteer grasses and broadleaf species are considered the undesirable vegetation. The principal legume species in annual legume based pastures are barrel medic, burr medic, murex medic, serradella, strand medic and subterranean clover (Gillespie and Nicholas, 1991) in Australia and arrowleaf clover and subterranean clover in the United States.

During establishment of annual legume based pastures, weeds can be managed by cultivating the land prior to establishment. Mowing the pasture at the appropriate stage of development will reduce seed set of many grasses and some broadleaf species. However, for these mechanical methods to be effective they must be carried out in conjunction with other methods of weed management.

Benefin and EPTC can be used as preplant incorporated herbicides for the control of seedling grasses during the establishment of most annual legumes. Pronamide and 2,4-DB are used for the control of grass and broadleaf weeds, respectively, after seedling establishment. Generally, the high cost of pronamide prevents its use except in high value seed crops. A disadvantage to the use of pronamide is the high potential for residual carryover into the following crop. Selective graminicides such as sethoxydim are being used in annual clovers especially during the establishment year.

Pasture topping or spray topping, the use of sublethal doses of knockdown herbicides, such as glyphosate and paraquat, to restrict the formation of viable grass seeds in annual legume based pastures, has been developed over the last decade in Australia (Jones et al., 1984; Dickinson and Stephenson, 1990; Dellow and McDonald, 1991).

Pasture or spray topping restricts the seed formation of grass weed species, thereby reducing their incidence in subsequent pastures and crops and problems with awned grass seeds. It also improves the digestibility of the grasses by halting the plant development and fixing the protein at a high level. The method involves the application of low rates of glyphosate or paraquat to a pasture that has been heavily grazed during early spring and stock removal for 2–3 weeks for the grasses to head evenly. This method is effective in controlling seed set in annual grasses such as annual ryegrass, bromegrass, hare barley, and rattail fescue and broadleaf species such as capeweed and distaff thistle.

Correct timing of application is very important and varies according to the species and the herbicide used. Glyphosate, a translocated herbicide, moves within the plant and is applied earlier than paraquat, which is a contact herbicide. The growth patterns of grass species differ, and they may not reach the optimum stage for spraying at the same time. Therefore, it may not be possible to manage a mixture of grasses with one application. In this situation, it is important to identify the major grass weed problem and spray at the ideal growth stage of that grass for best results. Alternatively, spray at the ideal growth stage of the earliest maturing grass and control the regrowth by heavy grazing or respraying. If paraquat is used on mixed grass swards or where maturity is uneven, the use of a higher rate is likely to improve the control. Care has to be taken when topping pastures as these herbicides can damage pasture legumes and reduce their seed set, especially if applied while legumes are flowering (Dickinson and Stephenson, 1990).

Spray grazing, a common weed management strategy in Australia, involves the application of a sublethal rate of 2,4-D to broadleaf dominant annual legume based pastures, followed by heavy grazing, preferably with sheep at 8–10 times the normal stocking rate, 7–10 days after application. The spray should be applied when the weeds are actively growing in autumn, 6–8 weeks after the break of the season. Application of 2,4-D in combination with heavy grazing is recommended for the control of African mustard, blessed milkthistle, capeweed, distaff thistle, Italian thistle, mustard, Patterson's curse, puncturevine, shepherd's-purse, turnipweed, wild mustard, and wild radish. Care has to be taken in using this method as the growth of established legumes may be retarded, and increased quantities of poisonous plants may be eaten by stock, resulting in death (Dellow and McDonald, 1991).

It must be emphasized that spray topping and spray grazing are practices used in Australia. These practices are not commonly practiced in the United States and advice must be obtained from the state Cooperative Extension Service concerning the herbicide label restrictions prior to using them in these weed management systems.

IV. WEED MANAGEMENT SYSTEMS FOR WARM-SEASON PERENNIAL FORAGE CROPS

Warm-season forage crops are the main forage species found in the tropical and subtropical regions of the world. Most commonly grown warm season grass species are alemangrass, bahiagrass, bermudagrass, blue panicgrass, buffelgrass, creeping signalgrass, dallisgrass, gambagrass, green panicgrass, guinea grass, jaraguagrass, kikuyugrass, limpograss, makarigrass, molassesgrass, Nadi bluegrass, napiergrass, pangolagrass, paragrass, rhodesgrass, setaria grass, signalgrass, speargrass and weeping lovegrass. The most commonly grown warm-season legume species are alfalfa, alyce clover, American jointvetch, calopo, Caribbean stylo carpon desmodium, centro, greenleaf desmodium, hetero desmodium, ipil-ipil, kudzu, lablab bean, lotononis, perennial soybean, phasey bean, shrubby stylo, silverleaf desmodium, siratro, stylo and Townsville stylo.

Except for perennial forage grasses managed for hay, this forage group probably receives less effort for weed management than all other forage systems. It is common to observe many species of weeds in the summer grazing systems, and in most instances their presence can be attributed to mismanagement of the forage species during a stress period and lack of interest in managing the weeds. Most broadleaf annual weeds can be effectively managed for just a few dollars invested in a timely application of a herbicide. However, the weed problem is, generally, of little concern until it becomes obvious during the flowering stage and it is too late to receive a benefit from controlling the annual weed at this stage of development. Annual weeds should be managed early in their growth stage and not when they are mature.

A. Weed Management During Establishment of Seeded Grasses

There are no herbicides labeled for use at the time of seeding the warm-season forage species since there are very few chemicals that will control germinating grass and broadleaf weeds without damaging the germinating forage seedling. There have been several reports of success in the application of certain herbicides at this time, but the risk of crop damage is too high to warrant it. After germination of the forage species and during the establishment period, 2,4-D can be applied to seedling forage grasses that are past the two-leaf stage of maturity to control the broadleaf weed seedlings. The rate of 2,4-D should be as low as possible to minimize damage to the grass seedlings but high enough to control the broadleaf weeds.

Annual grasses such as crabgrass and goosegrass are the most troublesome weeds during the establishment of warm-season grass species. The forage stands on many hectares seeded to warm-season grass species are lost to competition with warm-season weed species. The only weed management system that can give the forage species some relief from competition with the weeds is controlled grazing to reduce the loss of the forage species caused by shading.

B. Weed Management During Establishment of Vegetatively Propagated Grasses

Weed management during the planting and establishment of vegetatively propagated forage grasses is simplified by the difference between vegetative and seed propagation. More bermudagrass is propagated by planting sprigs than by seeding because of establishment is better than in seeding and because only common bermudagrass can be propagated by seeding. The poor seeding habits of the improved bermudagrass cultivars make it mandatory that they be established by vegetative propagation. Methods and equipment for sprigging the improved cultivars of bermudagrass have been developed, making it possible to establish them from sprigs at no greater cost than planting common bermudagrass seed. Coastal bermudagrass is one of the principal hay and pasture grasses in the subtropical and tropical regions of the world (Burton, 1954). Weed management has been found to be especially important during the establishment of Coastal bermudagrass.

Following the proper weed management practices prior to preparing the area for sprigging and the development of the proper soil fertility levels will assist in good crop establishment. Additionally, several herbicides can be used on the sprigged area in a preemergence application. 2,4-D can be applied immediately after the sprigging operation to give some relief from troublesome weeds that could eventually result in loss of the crop. Some states have emergency-use permits for the use of norflurazon at sprigging. Norflurazon applications will give excellent control of annual grasses and some annual broadleaf weeds for periods of up to 2 months after sprigging. The preemergence application of 2,4-D is, generally, not an effective treatment; however, some weed management is achieved through this treatment on sandy soils. Follow-up treatments of 2,4-D applied postemergence for the control of small broadleaf weeds that escape the preemergence application of diuron result in a much more efficient weed management system.

C. Weed Management in Established Stands of Warm-Season Forage Grasses

Diuron applications were labeled for use during the year that Coastal bermudagrass is planted. Until recently, simazine was registered for use in established Coastal bermudagrass pastures and hay fields. However, reregistration was not pursued by the registrants after expiration, thereby eliminating the use of simazine and diuron in the United States for managing weeds in pastures. Under nonirrigated conditions, complete ground cover of the planted area with Coastal bermudagrass is seldom achieved during the first growing season after sprigging or seeding. It is common to enter the second growing season with as little as 50% ground covered by the Coastal bermudagrass. Under high-fertility regimens, the open areas are rapidly invaded by large crabgrass, resulting in loss of the Coastal bermudagrass stand. No herbicide currently registered for use in established stands of Coastal bermudagrass will effectively control large crabgrass. In most instances, the pasture manager must resort to controlled grazing and management of the fertilizer applications to reduce the invasion by the warm-season grass species. Although norflurazon is not presently labeled for use in forage crops, some states have obtained emergency-use permits for its application in established bermudagrass. It has been determined to be very effective in controlling annual grass weeds in Coastal bermudagrass pastures and hayfields when applied before emergence in the early spring. Norflurazon has a long residual period in the soil and will give season-long control of crabgrass. This herbicide will fill a major void in the weed-management strategies for sprigged bermudagrass and should be registered for use in forage crops in the near future (Smith and Martin, 1992).

Broadleaf weeds can be managed in the established-grass pastures by timely postemergence applications of 2,4-D, dicamba, triclopyr, metsulfuron, or picloram. Also, glyphosate can be applied through a wiper applicator if there is adequate difference in height of the weed over the forage species. It is important to apply these herbicides during the early growth stages of the annual weeds for maximum uptake. Quite often it would appear to be advantageous to apply the herbicide in combination with a liquid nitrogen fertilizer formulation; generally, the only benefit is the saving of time and cost when applying the fertilizer and herbicide together. However, research has indicated that a synergistic action is produced by the liquid nitrogen and several ester formulations of the phenoxyalkanoic acids, metsulfuron, and dicamba when applied as a tank mixture for the control of young broadleaf weeds. The mixture results in more rapid desiccation of the foliage of the broadleaf weeds treated with the herbicides in liquid fertilizer than in water.

The rapid desiccation is an advantage when managing annual broadleaf weeds. However, it is not desirable when managing perennial weeds in the pasture ecosystem. These herbicides are dependent on the same mechanisms of transport as naturally occurring food materials (i.e., carbohydrates). Therefore, restrictions to food transport in plants would similarly influence the transport of the acid herbicides in the plant system. Research has indicated that the rapid wilting of horsenettle leaves followed treatment with N solution of 2,4,5-T and ultimately restriction of herbicide transport to the stems and roots of the plant (Smith and Calvert, 1980). The restriction of the herbicide movement from the treated leaves to storage organs is due to an inhibition of the enzymatic hydrolysis of the ester formulation. Hydrolysis of the ester is necessary for continued transport of the acid to the storage organ. Therefore, application of these herbicides in liquid-N diluent for the management of perennial broadleaf weeds is not effective in managing the perennating organs and will not result in a reduction in weed populations in subsequent months and growing seasons.

Several special weed problems that occur the southeastern United States in the warm-season perennial grass forage systems can be managed with specific herbicides. Bahiagrass is a warm-season perennial grass commonly utilized for grazing and along roadsides. It grows well under low fertility regimens and on sites that commonly undergo drought conditions. As a result, bahiagrass is a strong competitor with other forage grasses and readily invades the bermudagrass hayfields. As a result of growth habit, the majority of the foliage lies close to the ground and below the mower setting. Therefore, bahiagrass reduces the hay production by the bermudagrass hayfield. Postemergence applications of metsulfuron will effectively manage the bahiagrass in bermudagrass hayfields without damage to the bermudagrass (Hanna et al., 1989). Often pasture managers delay the management of the bahiagrass until its invasion results in its constituting more than 50% of the hayfield. Management at this time results in large voids in the hayfield that will require several months for the bermudagrass to fill. Often the void will be filled by large crabgrass, resulting in a problem equal to large crabgrass invasion during bermudagrass establishment. Similarly, there is no herbicide labeled for use in controlling grass weeds in established hayfields. When norflurazon is labeled for use in established stands of bermudagrass or in states that have an emergency-use permit, its application will assist in the management of annual grass invasion into the voids resulting from the management of bahiagrass. Mixing metsulfuron in N-fertilizer solutions is not recommended as reduced bahiagrass control results from the rapid burning of the leaves, which restricts the translocation of the herbicide.

Smutgrass has become a troublesome weed in the pasture ecosystem in the subtropical areas of the southeastern United States. Some states have obtained emergency-use registration for hexazinone for the management of smutgrass in bermudagrass and bahiagrass pastures. Some damage to the pasture grass will result. However, smutgrass can be effectively controlled by this management practice. Tebuthiuron pellets will also control smutgrass in bermudagrass pastures. This herbicide has a long residual life in the soil and can cause problems if land use changes within several years. It is important to utilize the pellet formulation of tebuthiuron in Coastal bermudagrass fields since treatment in an aqueous diluent will damage the bermudagrass (Smith, 1982).

Cool-season weeds will result in reduced hay quality at the first harvest of the warm-season grass and should be controlled during early spring. Treatment with paraquat during the dormant period for the warm-season forage crop is an effective and economical method for managing most cool-season weeds with no damage to the forage crop. If burning is an important management tool, paraquat application 1 week prior to burning will increase the dry fuel available for the burn by desiccating the green cool-season weeds.

D. Weed Management in Alfalfa

Seedling legumes are particularly vulnerable to weed competition. They grow slowly and provide the opportunity for rapidly developing weeds to overtop and severely shade them. In several greenhouse experiments we have shown that broadleaf weeds grow up to five times more rapidly than legume seedlings and that weed seedlings from seed of a similar size to the legume seed will survive drought conditions better than legume seedlings. Because of the high cost of planting forage legumes, it is very important to take the necessary precautions for weed management. This must include the necessary management strategies before seedbed preparation as well as during the seeding process. The extra expenditure, at this point, will be appreciated in later months. There are several herbicides registered for use on small-seeded forage legume species at seeding. Benefin and EPTC are registered for application followed by light mechanical incorporation prior to seeding the legumes. These herbicides must be incorporated for maximum effectiveness (Fawcett and Harvey, 1978). Both herbicides effectively control grass weeds and EPTC is more effective in the control of certain broadleaf weeds than benefin. If weeds become a problem after legume germination and during establishment, 2,4-DB can be used for control of broadleaf weeds and pronamide can be applied for control of cool-season grass species. Bromoxynil is effective in the control of wild radish during the establishment year. Pronamide and bromoxynil must be applied during cool temperatures ($<18°C$) for maximum effectiveness.

The management of weeds in established alfalfa stands is as important as management during establishment because of the expense of renovation and reestablishment. In the southeastern United States, we have found that weed management in alfalfa is more important than in most forage crops. First, the quality of hay can be drastically reduced by the presence of a few weeds at haying time. Also, the cool-season weeds will reduce the legume stand, which will not recover, and the open stand allows warm-season weeds to develop in this field. This sequence is a self-perpetuating path of destruction for the alfalfa stand. The importance of stand loss is realized when it becomes necessary to reestablish the alfalfa several years prematurely because of mismanagement of the weeds. When relying on herbicides for the management of weeds in alfalfa, it is very important to be familiar with the label restrictions on the use of these chemicals. Most herbicides registered for use in forage crops have grazing and haying restrictions and it is important that these restrictions be understood.

Cool-season broadleaf weeds such as Carolina geranium, common chickweed, henbit, and shepherd's-purse and the grass weeds such as annual bluegrass, Italian ryegrass, and little barley are easily managed with herbicides if the applications are timely. Postemergence applications of metribuzin, hexazinone, 2,4-DB, bromoxynil and paraquat will adequately manage the cool-season broadleaf species and postemergence applications of pronamide or paraquat are effective in the control of the cool-season grass weeds in alfalfa (Dutt et al., 1979; Smith, 1991b). It is important to apply these herbicides during the early growth stages of the weeds for excellent control.

Paraquat is registered for application to established stands of dormant alfalfa. In some regions of the United States the alfalfa does not become totally dormant during the winter months. Since paraquat is a nonselective contact herbicide it will result in desiccation of the leaves that it contacts (Smith, 1991b). The leaves remaining on the alfalfa plants through the winter will become partially to totally desiccated by the paraquat applications in the early spring, giving the appearance of crop damage. However, results from experiments conducted in Georgia indicate that alfalfa hay production from plots treated with paraquat in March is not adversely affected by the treatment when compared to that of the nontreated check (Smith, 1991b). Probably this is due to the normal loss of these leaves with the onset of regrowth in the early spring and very few of these older leaves remain on the plant at haying regardless of treatment. Additionally, paraquat is registered for application to alfalfa after harvest throughout the United States. This treatment is effective in the management of summer annuals.

Applications of sethoxydim and paraquat will effectively manage the warm-season annual grasses. It is important to apply paraquat as soon as possible after removal of the forage for hay. Delaying the paraquat application and allowing regrowth will result in desiccation of the leaves on the alfalfa plants and ultimately will reduce the yield at the next harvest for hay. This yield reduction is primarily due to loss of leaves and not plants, and yield reduction is not observed in subsequent harvests (Smith, 1991b). Applications of sethoxydim, during the growing season, do not damage the alfalfa plants and sethoxydim is more effective in managing the perennial warm-season grasses such as bermudagrass than applications of paraquat.

Dodder species are annual weeds that are troublesome in both seedling and mature stands of alfalfa, clovers, and sericea lespedeza (Hansen, 1923). As with all weeds, the common name for dodder varies considerably across its range of infestation. In addition to dodder, it is also known as devil's-gut, devils-hair, devil's-ringlet, goldthread, hailweed, hairweed, hellbind, love vine, pull-down and strangleweed (Dawson et al., 1969). It is a troublesome weed in Europe and in certain states in the United States. Although it is a noxious weed, dodder is not as troublesome in the United States as it is in Europe. In parts of Europe, especially certain regions of Germany, seed production for some legumes has been halted by dodder infestations (Agrios, 1978). Dodder seed is similar in size to seeds of the small-seeded legumes and is very hard to separate from the seed of the host plant (Peters and Linscott, 1988). Dawson (1966) found that only 10% of the planted dodder seed germinated during the initial 2 years. Thus once a field is contaminated with dodder, seed will germinate for many years and continue to reinfest the field. Therefore, dodder can be devastating in regions of legume seed production.

The parasitic weed seed germinates, forming a slender yellowish shoot that resembles a toothpick. This leafless and almost rootless stem rotates until it comes in contact with a host plant. Although germination utilizes all of the food reserves in the seed, if the stem does not find a suitable host, it will lie dormant on the soil for 4 or 5 weeks before dying. Upon contacting a suitable host plant, the tiny stem immediately begins to climb by encircling the host and producing haustoria that penetrate the cells of the host plant. The base of the dodder plant then shrivels and dries, losing its connection with the soil, and the dodder plant lives entirely from the nutrients extracted through the haustoria from the host plant (Dawson et al., 1969; Yuncker, 1932). In addition to draining nutrients from its host, dodder causes an increase in toxic compounds in the host plant. Dodder plants grow rapidly and often produce dense mats that cover the alfalfa plants and reduce the amount of sunlight reaching the alfalfa.

For management purposes, a knowledge of the dodder species infesting a crop is frequently of practical importance. For example, clover dodder rarely seeds in the United States and is only of concern during the first season after sowing. Field dodder, largeseed alfalfa dodder, and smallseed alfalfa dodder are the most troublesome in alfalfa production. Large-seeded and small-seeded alfalfa dodder species are troublesome in the western United States but are not a major problem in the eastern United States (Dawson et al., 1969).

Once dodder has been introduced into an area, a well-planned system of attack should be formulated and followed. During the first season of infestation, dodder usually grows in small scattered areas. Such infested areas should be either mowed and removed before dodder matures seed or burned in the field. If seed has formed, burning is the only remedy, since it kills not only the plants but also any seeds of dodder which may be lying on the soil surface. If the infestation is so great that patch treatment is impracticable, the crop may be either plowed under or utilized for hay and the aftermath closely grazed to prevent seed formation. Sheep are especially useful in grazing dodder. If dodder seed has formed and the crop has been very seriously damaged, the safest method is to cut the crop, allow it to dry, and burn it in the field. Dodder does not attack cereals; therefore, oats, wheat, etc., should be incorporated in the rotation following the infestation of the soil with dodder seeds (Dawson et al., 1969).

Pronamide is registered for use in alfalfa and is effective in the management of dodder. It must be moved into the soil by either rainfall or irrigation before the germination of dodder. Preferably, pronamide should be irrigated into the soil within 1 to 3 days after its application, but irrigation can be delayed up to 2 weeks if necessary, provided that it is in the germinating zone at the time of dodder germination. Glyphosate has been found to control dodder effectively when it is attached to established alfalfa (Dawson, 1989). Although the low rates of glyphosate moderately suppress alfalfa growth, the suppression is much less than would result from the presence of dodder (Dawson, 1990). Glyphosate is not registered for this use.

APPENDIX A

List of Weeds Discussed in Chapter

Common name	Scientific name	BAYER code
African mustard	*Brassica tournefortii* Gouan	BRSTO
American falsepennyroyal	*Hedeoma pulegiodes* (L.) Pers.	HEDPU
Annual bluegrass	*Poa annua* L.	POAAN
Annual fleabane	*Erigeron annuus* (L.) Pers.	ERIAN
Absinth wormwood	*Artemisia absinthium* L.	ARTAB
Bahiagrass	*Paspalum notatum* Fluegge	PASNO
Barley	*Hordeum vulgar* L.	-
Barnyardgrass	*Echinochloa crus-galli* (L.) Beauv.	ECHCG
Hairy beggarticks	*Bidens pilosa* L.	BIDPI
Bentgrasses	*Agrostis spp.*	-
Bermudagrass	*Cynodon dactylon* (L.) Pers.	CYNDA
Bitter sneezeweed	*Helenium amarum* (Raf.) H. Rock	HENAM
Blackgrass	*Alopecurus myosuroides* Huds.	ALOMY
Black medic	*Medicago lupulina* L.	MEDLU
Black mustard	*Brassica nigra* (L.) W.J.D. Koch	BRSNI
Black roly-poly	*Sclerolaena muricata* (Moq.) Domin.	-
Blessed milkthistle	*Silybum murianum* (L.) Gaertn.	SLYMA
Bluegrasses	*Poa spp.*	-
Blueweed	*Echium vulgare* L.	EHIVU
Bracken fern	*Pteridium aquilinum* (L.) Kuhn	PTEAQ
Brigalow	*Acacia harpophylla* (F.r.M.)	-
Broadleaf filaree	*Erodium botrys* (Cav.) Bertol.	EROBO
Broadleaf dock	*Rumex obtusifolius* L.	RUMOB
Broadleaf plantain	*Plantago major* L.	PLAMA
Brome	*Bromus spp.*	-
Buckhorn plantain	*Plantago lanceolata* L.	PLALA
Bulbous buttercup	*Ranunculus bulbosus* L.	RANBU
Bull thistle	*Cirsium vulgare* (Savi) Tenore	CIRVU
Buttercup oxalis	*Oxalis pes-caprae* L.	OXAPC
Buttercup	*Ranunculus spp.*	-
Californian burr	*Xanthium orientale* L.	-

(*continued*)

Appendix A Continued

Common name	Scientific name	BAYER code
Canada goldenrod	*Solidago canadensis* L.	SOOCA
Canada thistle	*Cirsium arvense* (L.) Scop.	CIRAR
Cape tulips	*Homeria spp.*	-
Capeweed	*Arctotheca calendula* (L.) Levyns	-
Carolina geranium	*Geranium carolinianum* L.	GERCA
Catchweed	*Asperugo procumbens* L.	ASGPR
Clammy groundcherry	*Physalis heterophylla* Nees	PHYHE
Cluster dock	*Rumex conglomeratutus* Murr.	RUMCO
Cocklebur	*Xanthium spp.*	-
Colonial bentgrass	*Agrostis tenuis* Sibth.	AGSTE
Common chickweed	*Stellaria media* (L.) Vill.	STEME
Common eveningprimrose	*Oenothera biennis* L.	OEOBI
Common groundsel	*Senecio vulgaris* L.	SENVU
Common lambsquarters	*Chenopodium album* L.	CHEAL
Common milkweed	*Asclepias syriaca* L.	ASCSY
Common ragweed	*Ambrosia artemisiifolia* L.	AMBEL
Common speedwell	*Veronica officinalis* L.	VEROF
Common St. Johnswort	*Hypericum perforatum*L.	HYPPE
Common velvetgrass	*Holcus lanatus* L.	HOLLA
Common yarrow	*Achillea millefolium* L.	ACHMI
Cogongrass	*Imperata cylindrica* (L.) Beauv.	IMPCY
Corn gromwell	*Lithospomum arvense* L.	LITAR
Corn spurry	*Spergula arvensis* L.	SPRAR
Crabgrass	*Digitaria spp.*	-
Creeping buttercup	*Ranunculus repens* L.	RANRE
Croftonweed	*Eupatorium adenophorum* Spreng	EUPAD
Cudweed	*Gnaphalium spp.*	-
Curly dock	*Rumex crispus* L.	RUMCR
Dandelion	*Taraxacum officinale* Weber in Wiggers	TAROF
Datura	*Datura spp.*	-
Distaff thistle	*Carthamus lanatus* L.	CAULA
Dock	*Rumex spp.*	-
Dodder	*Cuscuta spp.*	-
Dogfennel	*Eupatorium capillifolium* (Lam.) Small	EUPCP
Downy brome	*Bromus tectorum* L.	BROTE
Eastern baccharis	*Baccharis holimifolia* L.	BACHA
Eastern brackenfern	*Pteridium aquilinum* var. Latiusculum (Desv.) Underw. ex Heller	PTEAL
English daisy	*Bellis perennis* L.	BELPE

Appendix A Continued

Common name	Scientific name	BAYER code
European heliotrope	*Heliotropium europaeum* L.	HEOEU
European blackberry	*Rubus fruticosus* L.	RUBFR
Fall panicum	*Panicum dichotomiflorum* Michx.	PANDI
Fennel	*Foeniculum vulgare* Mill.	FOEVU
Fiddleleaf dock	*Rumex pulcher* L.	RUMPU
Field dodder	*Cuscuta campestris* Yuncker	CVCCA
Field horsetail	*Equisetum arvense* L.	EQUAR
Field pepperweed	*Lepidium campestre* (L.) R.Br.	LEPCA
Field sandbur	*Cenchrus incertus* M.A. Curtis	CCHIN
Field violet	*Viola arvensis* Murr.	VIOAR
Filaree	*Erodium spp.*	-
Foxtail barley	*Hordeum jubatum* L.	HORJU
German velvetgrass	*Holcus mollis* L.	HOLMO
Giant foxtail	*Setaria faberi* Herrm.	SETFA
Goatsrue	*Galega officinalis* L.	GAGOF
Goosegrass	*Eleusine indica* (L.) Gaertn.	ELEIN
Gorse	*Ulex europaeus* L.	ULEEU
Green foxtail	*Setaria viridis* (L.) Beauv.	SETVI
Green sorrel	*Rumex acetosa* L.	RUMAC
Hairgrass	*Deschampsia spp.*	-
Hardhack	*Spiraea tomentosa* L.	SPVTO
Hare barley	*Hordeum leporinum* Link	HORLE
Hawthorn	*Crataegus spp.*	-
Hay-scented fern	*Dennstaedtia punctilobula* (Michx.) Moore	DNNPU
Heal-all	*Prunella vulgaris* L.	PRUVU
Hedge mustard	*Sisymbrium officinale* (L.) Scop.	SSYOF
Henbit	*Lamium amplexicaule* L.	LAMAM
Highbush blackberry	*Rubus argutus* Link	RUBAG
Horsenettle	*Solanum carolinense* L.	SOLCA
Horsetail	*Equisetum spp.*	-
Horseweed	*Conyza canadensis* (L.) Cronq.	ERICA
Indian mustard	*Brassica juncea* (L.) Czern. & Coss.	BRSJU
Indian sweetclover	*Melilotus indica* (L.) All.	MEUIN
Indian tobacco	*Lobelia inflata* L.	LOBIN
Italian ryegrass	*Lolium multiflorum* Lam.	LOLMU
Italian thistle	*Carduus pycnocaphalus* L.	CRUPY
Japanese brome	*Bromus japonicus* Thunb. ex Murr.	BROJA
Johnsongrass	*Sorghum halpense* (L.) Pers.	SORHA
Lady's thumb	*Polygonum persicaria* L.	POLPE
Lanceleaf sage	*Salvia reflexa* Hornem.	SALRE

(*continued*)

Appendix A Continued

Common name	Scientific name	BAYER code
Large crabgrass	*Digitaria sanguinalis* (L.) Scop.	DIGSA
Largeleaf lantana	*Lantana camara* L.	LANCA
Largeseed dodder	*Cuscuta indecora* Choisy	CVCIN
Little barley	*Hordeum pusillum* Nutt.	HORPU
Madrid brome	*Bromus madritensis* L.	-
Marsh thistle	*Cirsium palustre* (L.) Scop.	CIRPA
Mayweed chamomile	*Anthemis cotula* L.	ANTCO
Meadowsweet	*Spiraea latifolia* (Ait.) Borkh.	SPVLF
Mediterranean barley	*Hordeum hystrix* Roth	HORMG
Mugwort	*Artemisia vulgaris* L.	ARTVU
Musk thistle	*Carduus nutans* L.	CRUNU
Noogoora burr	*Xanthium pungens* Wallr.	-
Oniongrass	*Romulea rosea* (L.) Eckl. var. australis (Ewart) de Vos	-
Onionweed	*Asphodelus fistulosus* L.	ASHFI
Paterson's curse	*Echium plantagineum*	-
Pearl gromwell	*Lithospermum officinale* L.	LITOF
Pennyroyal	*Hedeoma spp.*	-
Peppercress	*Lepidium spp.*	-
Peppermint	*Mentha piperita* L.	MENPI
Philadelphia fleabane	*Erigeron philadelphicus* L.	ERIPH
Pigweed	*Amaranthus spp.*	-
Plumless thistle	*Carduus acanthides* L.	CRUAC
Poa tussock	*Poa* labillardieri *Steudel* var. labillardieri	-
Poppies	*Papaver spp.*	-
Poverty oatgrass	*Danthonia spicata* (L.) Beauv. ex R. & S.	DANSP
Pricklypear	*Opuntia spp.*	-
Prostrate knotweed	*Polygonum aviculare* L.	POLAV
Puncturevine	*Tribulus terrestris* L.	TRBTE
Purple nutsedge	*Cyperus rotundus* L.	CYPRO
Quackgrass	*Elytrigia repens* (L.) Nevski	AGRRE
Ragweed parthenium	*Parthenium hysterophorus* L.	PTNHY
Rattail fescue	*Vulpia myuros* (L.) K.C. Gmel	VLPMY
Redroot pigweed	*Amaranthus retroflexus* L.	AMARE
Red sorrel	*Rumex acetosella* L.	RUMMA
Rescuegrass	*Bromus cartharticus* Vahl	BROCA
Ripgut brome	*Bromus diandrus* Roth	BRODI
Rush skeletonweed	*Chondrilla juncea* L.	CHOJU
Rushes	*Juncus spp.*	-
Russian thistle	*Salsola iberica* Sennen & Pau	SASKR

Appendix A Continued

Common name	Scientific name	BAYER code
Sagebrush	*Artemisia spp.*	-
Sandbur	*Cenchrus spp.*	-
Sand rocket	*Diplotaxis muralis* (L.) DC.	DIPMU
Sensitive plant	*Mimosa pudica* L.	MIMPU
Serrated tussock	*Nassella trichotoma* (Nees) Hack.	STDTR
Shepherd's-purse	*Capsella bursa-pastoris* (L.) Medicus	CAPBP
Slenderflower thistle	*Carduus tenuiflorus* W. Curtis	CRUTE
Slender rush	*Juncus tenuis* Willd.	IUNTE
Smallflower buttercup	*Ranunculus abortivus* L.	RANAB
Smallseed alfalfa dodder	*Cuscuta appoximata* Bab. var. urceolata (Ktze.) Yuncker	CVCAV
Smutgrass	*Sporobbolus indicus* (L.) R.Br.	SPZIN
Soft brome	*Bromus mollis* L.	BROMO
Soft rush	*Juncus effusus* L.	IUNEF
Speargrass	*Stipa spp.*	-
Spiny cocklebur	*Xanthium spinosum* L.	XANSP
Spiny emex	*Emex spinosa* (L.) Campd.	EMESP
Spiny sowthistle	*Sonchus asper* (L.) Hill	SONAS
Squirreltail fescue	*Vulpia bromoides* (L.) S.F. Gray	VLPBR
Sticky groundsel	*Senecio viscosus* L.	SENVI
Stinging nettle	*Urtica dioica* L.	URTDI
Swamp milkweed	*Asclepias incarnata* L.	ASCIN
Sweetbriar rose	*Rosa rubiginosa* L.	ROSRB
Sweetfern	*Comptonia peregrina* (L.) L. Coult	COVPE
Swinecress	*Coronopus didymus* (L.) Sm.	COPDI
Tall buttercup	*Ranunculus acris* L.	RANAC
Tall fescue	*Festuca arundinacea* Schreb.	FESAR
Tall thistle	*Cirsium altissimum* (L.) Spreng.	CIRAL
Tansy ragwort	*Senecio jacobaea* L.	SENJA
Thorn apple	*Crataegus spp.*	-
Three-cornered jack	*Emex australis* Steinheil	EMEAU
Turnipweed	*Rapistrum rugosum* (L.) All.	RASRU
Virginia pepperweed	*Lepidium virginicum* L.	LEPVI
Wall barley	*Hordeum glaucum* Steud	HORMC
Wall rocket	*Diplotaxis tenuifolia* (L.) DC.	DIPTE
Western ragweed	*Ambrosia psilostachya* DC.	AMBPS
Western salsify	*Tragopogon dubius* Scop.	TRODM
White horehound	*Marrubium vulgare* L.	MAQVU
Whitestem filaree	*Erodium moschatum* (L.) L'Her. ex Ait.	EROMO

(*continued*)

Appendix A Continued

Common name	Scientific name	BAYER code
Wild artichoke	*Cynara cardunculus* L.	-
Wild carrot	*Daucus carota* L.	DAUCA
Wild garlic	*Allium vineale* L.	ALLVI
Wild marigold	*Tagetes minuta* L.	TAGMI
Wild mustard	*Brassica kaber* (DC.) L.C.Wheeler	SINAR
Wild oats	*Avena fatua* L.	AVEFA
Wild radish	*Raphanus raphanistrum* L.	RAPRA
Wooly plantain	*Plantago patagonica* Jacq.	PLAPR
Yellow foxtail	*Setaria glauce* (L.) Beauv.	SETLU
Yellow hawkweed	*Hieracium pratense* Tausch	HIECA
Yellow nutsedge	*Cyperus esculentus* L.	CYPES
Yellow rocket	*Barbarea vulgaris* R.Br.	BARVU
Yellow toadflax	*Linaria vulgaris* Mill.	LINVU

APPENDIX B

List of Pasture and Hay Crop Species Discussed in Chapter

Common name	Scientific name
Alemangrass	*Echinochloa polystachya* (H.B.K.) Hitchc.
Alfalfa	*Medicago sativa* L.
Alyce clover	*Alysicarpus vaginalis* (L.) DC.
American jointvetch	*Aeschynomene americana* L.
Bahiagrass	*Paspalum notatum* Flugge var. saurae Parodi
Barley	*Hordeum vulgare* L.
Barrel medic	*Medicago truncatula* L.
Bermudagrass	*Cynodon dactylon* (L.) Pers.
Blue panicgrass	*Panicum antidotale* Retz.
Bluegrass	*Poa spp.*
Bufflegrass	*Penniesetum ciliare* (L.) Link
Burr medic	*Medicago polymorpha* L.
Calopo	*Calopogonium mucunoides* L.
Canarygrass	*Phalaris canariensis* L.
Caribbean stylo	*Stylosanthes hamata* L.
Carpon desmodium	*Desmodium heterocarpon* DC.
Centro	*Centrosema pubescens* Benth.
Creeping signalgrass	*Brachiaria humidicola* (Rendle) Schweickt
Dallisgrass	*Paspalum dilatatum* Poir.
Gambagrass	*Andropogon gayanus* Kanth.
Green panicgrass	*Panicum maximum* Jacq. var. trichoglume Eyles
Greenleaf desmodium	*Desmodium intortum* (Mill) Urb.
Guineagrass	*Panicum maximum* Jacq.

Appendix B Continued

Common name	Scientific name
Hetero desmodium	*Desmodium heterophyllum* DC.
Ipil-ipil	*Leucaena leucocephala* L.
Jaraguagrass	*Hyparrhenia rufa* (Nees) Stapf
Kikuyugrass	*Pennisetum clandestinum* Hockst. ex Chiov.
Lablab bean	*Lablab purpureus* L.
Sericea lespedeza	*Lespedeza cuneata* (Dumont) G. Don
Limpograss	*Hemarthria altissima* (Poir.) Stapf & Hubbard
Lotononis	*Lotononis bainesii* Baker
Makarigrass	*Panicum coloratum*L.
Molassesgrass	*Melinis minutiflora* Beauv.
Murex medic	*Medicago murex* L.
Nadi bluegrass	*Dichanthium caricosum* (L.) A. Camus
Napiergrass	*Pennisetum purpureum* Schumach.
Oats	*Avena sativa* L.
Orchardgrass	*Dactylis glomerata* L.
Pangola digitgrass	*Digitaria decumbens* Stent.
Paragrass	*Brachiaria mutica* (Forsk.) Stapf
Pea	*Pisum sativum* L.
Perennial soybean	*Glycine wightii* (R. Grah. ex Wight & Arn.) Verdc.
Perennial ryegrass	*Lolium perenne* L.
Phasey bean	*Macroptilium lathyroides* (L.) Urb.
Red clover	*Trifolium pratense* L.
Rhodesgrass	*Chloris gayana* Kunth
Serradella	*Ornithopus compressus* L.
Setaria grass	*Setaria anceps* Stapf ex Massey
Shrubby stylo	*Stylosanthes scabra* Vog.
Signalgrass	*Brachiaria decumbens* Stapf
Silverleaf desmodium	*Desmodium uncinatum* (Jacq.) DC.
Siratro	*Macroptilium atropurpureum* (DC.) Urb.
Smooth bromegrass	*Bromus inermis* Leyss.
Speargrass	*Heteropogon contortus* (L.) Beauv. ex Roem. & Schult.
Strand medic	*Medicago littoralis* L.
Strawberry clover	*Trifolium fragiferum* L.
Stylo	*Stylosanthes guianesis* (Aubl.) Scv.
Subterranean clover	*Trifolium subterraneam* L.
Tall fescue	*Festuca arundinacea* Schreb.
Timothy	*Phleum pratense* L.
Townsville stylo	*Stylosanthes humilis* H.B.K.
Tropical kudzu	*Pueraria phaseoloides* (Roxb.)
Weeping lovegrass	*Eragrostis curvula* (Schrad.) Nees
Wheat	*Triticum aestivum* L.
White clover	*Trifolium repens* L.

APPENDIX C

Common Weeds of the Pasture Ecosystems with Alleged Allelopathic Potential

Common name	Scientific name
Absinth wormwood	*Artemisia absinthium* L.
Barnyardgrass	*Echinochloa crus-galli* (L.) Beauv.
Bermudagrass	*Cynodon dactylon* (L.)Pers.
Bitter sneezeweed	*Helenium amarum* (Raf.) H.Rock
Canada thistle	*Cirsium arvense* (L.)Scop.
Common chickweed	*Stellaria media* (L.)Vill.
Common evening primrose	*Oenothera biennis* L.
Common milkweed	*Asclepias syriaca* L.
Common ragweed	*Ambrosia artemisiifolia* L.
Dogfennel	*Eupatorium capillifolium* (Lam.) Small
Downy brome	*Bromus tectorum* L.
Fall panicum	*Panicum dichotomiflorum* Michx.
Field sandbur	*Cenchrus incertus* M.A.Curtis
Giant foxtail	*Setaria faberi* Herrm.
Goosegrass	*Eleusine indica* (L.) Gaertn.
Green foxtail	*Setaria viridis* (L.) Beauv.
Hairy beggarticks	*Bidens pilosa* L.
Italian ryegrass	*Lolium multiflorum* Lam.
Japanese brome	*Bromus japonicus* Thunb. ex Murr.
Johnsongrass	*Sorghum halepense* (L.) Pers.
Large crabgrass	*Digitaria sanguinalis* (L.) Scop.
Mayweed chamomile	*Anthemis cotula* L.
Purple nutsedge	*Cyperus rotundus* L.
Quackgrass	*Agropyron repens* (L.) Beauv.
Redroot pigweed	*Amaranthus retroflexus* L.
Russian thistle	*Salsola iberica* Sennen & Pau
Tall thistle	*Cirsium altissimum* (L.) Spreng.
Virginia pepperweed	*Lepidium virginicum* L.
Western ragweed	*Ambrosia psilostachya* DC.
Wild garlic	*Alium vineale* L.
Wild marigold	*Tagetes minuta* L.
Wild oat	*Avena fatua* L.
Woolly plantain	*Plantago patagonica* Jacq.
Yellow foxtail	*Setaria glauca* (L.) Beauv.
Yellow nutsedge	*Cyperus esculentus* L.

Source: Smith (1991a).

APPENDIX D

List of Herbicides Discussed in Chapter

Common name	Chemical name
Benefin	*N*-Butyl-*N*-ethyl-2,6-dinitro-4-(trifluoromethyl)benzenamine
Bromoxynil	3,5-Dibromo-4-hydroxybenzonitrile
Dicamba	3,6-Dichloro-2-methoxybenzoic acid
Diuron	*N'*-(3,4-dichlorophenyl)-*N*,*N*-dimethylurea
EPTC	*S*-Ethyl dipropyl carbamothioate
Glyphosate	*N*-(Phosphonomethyl)glycine
Hexazinone	3-Cyclohexyl-6-(dimethylamino)-1-methyl-1,3,5-triazine-2,4(1*H*, 3*H*)-dione
Metribuzin	4-Amino-6-(1,1-dimethylethyl)-3-(methylthio)-1,2,4-triazin-5(4*H*)-one
Metsulfuron	2-[[[[(4-Methoxy-6-methyl-1,3,5-triazin-2-yl)amino]carbonyl] amino]sulfonyl]benzoic acid
Norflurazon	4-Chloro-5-(methylamino)-2-(3-trifluoromethyl)phenyl)-3(2*H*) pyridazinone
Paraquat	1,1'-Dimethyl-4,4'-bipyridinium ion
Picloram	4-Amino-3,5,6-trichloro-2-pyridinecarboxylic acid
Pronamide	3,5-Dichloro(*N*-1,1-dimethyl-2-propynyl)benzamide
Sethoxydim	2-[1-(Ethoxyimino)butyl]-5-[2-(ethylthio)propyl]-3-hydroxy-2-pyridinesulfonamide
Simazine	6-Chloro-*N*,*N'*-diethyl-1,3,5-trizine-2,4-diamine
2,4-D	(2,4-Dichlorophenoxy)acetic acid
2,4-DB	4-(2,4-Dichlorophenoxy)butanoic acid
2,4,5-T	(2,4,5-Trichlorophenoxy)acetic acid

REFERENCES

Agrios, G. N. (1978). *Plant Pathology*, Academic Press, New York.

Anderson, W. P. (1983). *Weed Science: Principles,* West Publishing Company, St. Paul, Minnesota.

Arnold, G. W. (1981). Pasture production and management, in *Sheep Nutrition* (G. J. Tomes and I. J. Fairnie, eds.), Western Australian Institute of Technolgy, pp. 69–90.

Australian Wool Corporation. (1980). The economic impact of pasture weeds, *Pests and Diseases on the Australian Wool Industry*, Australian Wool Corporation, Melbourne, p. 204.

Bacon, C. W., Porter, J. K., Robbins, J. D. and Luttrell, E. S. (1977). *Epichloe typhina* from toxic tall fescue grasses, *Appl. Environ. Microbiol. 35*: 576–581.

Burton, G. W. (1954). Costal Bermuda grass, *Georgia Agric. Exp. Stn. Bul. NS2:* 12.

Campbell, M. H., Holst P.J., Auld, B. A., and Medd, R. W. (1979). "Control of three pasture weeds using goats," *Proceedings of the 7th Asian-Pacific Weed Science Society Conference*, Sydney, pp. 201–205.

Carpenter, J. A., Boyce, K. G., Cameron, D. G., Collins, W. J., Read, J. N., Ross. B., and Williams, A. (1990). Pastures, in *The Manual of Australian Agriculture* (R. L. Reid, ed.), 5th Edition, Butterworths, Sydney, pp. 228–280.

Daly, G. T. (1990). The grasslands of New Zealand, in *Pastures: Their Ecology and Management* (R. H. M. Langer, ed.), Oxford University Press, Oxford, pp. 1–38.

Dawson, J. H. (1966). Response of dodder to shade, *Weed Sci. 32:* 841–844.

Dawson, J. H. (1989). Established forage alfalfa *(Medicago sativa)* tolerates glyphosate and SC-0224 applied to control dodder *(Cuscuta spp.), Weed Tech. 3:* 560–565.

Dawson, J. H. (1990). Newly seeded alfalfa *(Medicago sativa)* tolerates glyphosate and SC-0224 at doses that control dodder *(Cuscuta spp.), Weed Tech. 4:* 876–879.

Dawson, J. H., Lee, W. O., and Timmons, F. L. (1969). Controlling dodder in alfalfa, *USDA Farmers Bull. 2211* U.S. Govt. Printing Office Washington, D.C.

Dellow, J. J. and McDonald, W. (1991). Weed control in lucerne and pastures, *New South Wales Dept. of Agric. and Fisheries Agdex 100/682:* 46.

Dickinson, J. and Stephenson, D. (1990). Topping pastures with herbicides, *South Australia Dept. of Agric. Agdex 130/640:* 4.

Dietl, W. (1982). Weeds of pastures and meadows in the European Alps, in *Biology and Ecology of Weeds* (W Holzner and M. Numata, eds.), Dr. W. Junk Publishers, The Hague, pp. 375–385.

Dutt, T. E., Harvey, R. G., Fawcett, R. S., Jorgenson, N.A., Larsen, H. J., and Schlough, D. A. (1979). Forge quality and animal performance as influenced by quack grass control in alfalfa with pronamide, *Weed Sci. 27:* 127–132.

Fawcet, R. J. and Harvey, R. G. (1978). Field comparison of seven dinitroaniline herbicides for alfalfa seedling establishment, *Weed Sci. 26*: 123–127.

Field, R. J. and Daly, G. T. (1990). Weed biology and management, *Pastures: Their Ecology and Management* (R. H. M. Langer, ed.), Oxford University Press, Oxford, pp. 409–447.

Gillespie, D. J. and Nicholas, D. A. (1991). Pasture legume recommendations for sowing in 1991, *Western Australian Dept. of Agric. Bul. 4204:* 17.

Gorski, T., Gorska, K., and Rybicki, J. (1978). Studies on the germination of seeds under leaf canopy, *Flora 167:* 289–299.

Haggar, R. J., Standell, C. J., and Birnie, J. E. (1985). Occurrence, impact and control of weeds in newly sown leys, in *Weeds, Pests and Diseases of Grassland and Herbage Legumes,* Monograph No. 29 (J. S. Brockman, ed.), British Crop Protection Council, Occasional Symposium No. 18, British Grassland Society, pp. 11–19.

Hamner, C. L. and Tukey, H. B. (1944). Herbicidal action of 2,4-dichlorophenoxyacetic acid, *Science 100:* 154–156.

Hanna, W. W., Swann, C. W., Schroeder, J., and Utley, P. R. (1989). Sulfometuron for eliminating Bahia grass *(Paspalum notatum)* from centipede *(Eremochloa ophiuroides)* and Bermuda grass *(Cynondon dactylon), Weed Tech. 3:* 509–512.

Hanson, A. A. (1923). Dodder, *Farmers Bul. 161.* USDA Washington D. C.

Hardine, A. R. (1985). Dispersal and establishment of slender thistle *(Carduus pycnocephalus* L.) as affected by ground cover, *Aust. J. Agric. Res. 36:* 791–797.

Harper, J. L. (1977). *Population Biology of Plants,* Academic Press, London.

Harris, W. (1990). Pasture as an ecosystem, in *Pastures: Their Ecology and Management* (R. H. M. Langer, ed.), Oxford University Press, Oxford, pp. 75–131.

Holgate, G. L. (1983). Goats for shrub-weed control, in *May Seminar,* Mohair Producers Assoc. N.Z., Queenstown.

Holst, P. J. (1980). Goats for weed control, *Proc. Postgrad. Comm. Vet. Sci.,* Univeristy of Sydney *52:* 187–189.

Holst, P. J. and Campbell, M. H. (1987). The role of goats in the control of weeds of pastures, in *Temperate Pastures: Their Production, Use and Management* (J. R. Wheeler, C. J. Pearson, and G. E. Robards, eds.), CSIRO, Australia, pp. 262–263.

Hoveland C. S., Haaland, R. L., King, C. C., Anthony, W. B., Clar, E. M., McGuire, J. A., Smith, A., Grimes, H. W., and Holliman, J. L. (1980). Association of *Epichloe typhina* fungus and steer performance on tall fescue pastures, *Agron. J. 72:* 1064–1065.

Intaka Press, (1987). *Australian Weed Control Handbook,* 8th Edition., Melbourne.

Johnson, J. T. (1990). Pastures in Georgia, *Coop. Ext. Service/The Univ. of Georgia College of Agric., Athens, Bull. 573:* 49.

Jones, S.M., Blowes, W. M., England, P., and Fraser, P. K. (1984). Pasture topping using roundup herbicide, *Aust. Weeds 3:* 150–151.

Klingman, D. L. (1956). Weed control in pastures in the North Central Region, *Weeds 4:* 369–375.

Klingman, D. L. and McCarthy, M. K. (1958). Interrelations of methods of weed control and pasture management at Lincoln, Nebraska, *USDA Tech. Bull. 1180:* 49.

Klingman, G. C., Ashton, F. M. and Noordhoff, L. J. (1982). *Weed Science: Principles and Practices,* 2nd Edition, John Wiley & Sons, New York.

Knight, J. (1965). Some observations on the feeding habits of goats in the South Beringo district of Kenya, *East African Agric. For. J. 30:* 182–188.

Lambert, M. G., Clark, D. A., and Rolston, M. P. (1981). "Use of goats for coarse weed control in hill country," *Proceedings of the 33rd Ruakura Farm Conference,* pp. 167–171.

Lamp, C. and Collet, F. (1989). *Field Guide to Weeds in Australia,* 3rd Edition, Intaka Press, Melbourne.

Langer, R. H. M. (1990). Pasture plants, *Pastures: Their Ecology and Management* (R. H. M. Langer, ed.), Oxford University Press, Oxford, pp. 39–74.

Linscott, Dean L., and Vaughan, R. H. (1990). Fenoxaprop for annual foxtail *(Setaria sp.)* control in seedling perennial forages, *Weed Tech. 4:* 560–564.

Lorenzi, H. J. and Jeffery, L. S. (1987). *Weeds of the United States and Their Control,* Van Nostrand Reinhold Company, Inc., New York.

Madin, R. W. and Bowran, D. G. (1989). "Weed control in annual pastures of Western Australia," *Proceedings of the Workshop on Grass Control in Pastures for the Wheat/Sheep Zone* (I. Malling and R. Madin, eds.), Department of Agriculture, Western Australia, pp. 2–7.

Malik, N. (1990). Weed control during establishment and yield response of timothy *(Phleum pratense), Weed Tech. 4:* 598–605.

Matthews, L. J. (1977). Integrated control of weeds in pastures, in *Integrated Control of Weeds* (J. D. Fryer and S. Matsunaka, eds.), University of Tokyo Press, Tokyo, pp. 90–119.

Matthews, L. J. (1982). Pasture weeds in New Zealand, in *Biology and Ecology of Weeds* (W. Holzner and M. Numata, eds.), Dr. W. Junk Publishers, The Hague, pp. 387–394.

Medd, R. W., Kemp, D. R., and Auld, B. A. (1987). Management of weeds in perennial pastures, in *Temperate Pastures: Their Production, Use and Management* (J. R. Wheeler, C. J. Pearson and G. E. Robards, eds.), CSIRO, Australia, pp. 253–261.

Michael, P. W. (1970). Weeds of grasslands, in *Australian Grasslands* (R. M. Moore, ed.), Australian National University Press, Canberra, pp. 349–360.

Mullahey, J. J., Nee, M., Wunderlin, R., and DeLaney, K. R. (1993). Tropical soda apple *(Solanum viarum)* a new weed threat in subtropical regions. *Weed Tech. 7:* 783–786.

National Academy of Sciences. (1968). Weed Control, *Principles of Plant and Animal Pest Control,* Vol. 2, Washington, D. C.

Nemoto, M. (1982). Weeds of pastures and meadows in Japan, in *Biology and Ecology of Weeds* (W. Holzner and M. Numata, eds.), Dr. W. Junk Publishers, The Hague, pp. 395–401.

Pederson, J. F., McGuire, J. A., Schmidt, S. P., King, C. C., Hoveland, C. S., and Smith, L. A. (1986). Steer performance as affected by tall fescue cultivar and level of *Acremonium coenophialum, N.Z. J. Exp. Agric. 14:* 307–312.

Peters, E. J., and Linscott, D. L. (1988). Weeds and weed control, in *Alfalfa and Alfalfa Improvement Agronomy*, Monograph 29, (A. A. Hanson, ed.), ASA, Madison, Wisconsin.

Peters, E. J. and Lowance, S. A. (1969). Gains from timothy forage from goldenrod control with 2,4-D, 2,4-DB, and Picloram, *Weed Sci. 17:* 473–474.

Peters, T. J., Moomaw, R. S., and Martin, A. R. (1989). Herbicides for postemergence control of annual grass weeds in seedling forage grasses, *Weed Sci. 37:* 375–379.

Phung, H. T. and Popay, A. I. (1981). "Effect of Pasture Cover on the Germination of Certain Weeds," *Proceedings of the 34th New Zealand Weed and Pest Control Conference,* pp. 111–113.

Putnum, A. R. and Weston, L. A. (1986). Adverse impacts of allelopathy in agricultural systems, *The Science of Allelopathy* (A. R. Putnam and C. S. Tang, eds.), John Wiley & Sons, New York, pp 43–56.

Radosevich, S. R. and Holt, J. S. (1987). *Weed Ecology,* Wiley Interscience, New York.

Rice, E. L. (1984). *Allelopathy,* Academic Press, Orlando, Florida.

Ross, M. A. and Lembi C. A. (1985). *Applied Weed Science,* Macmillan Publishing Co., New York.

Sampson, K. (1933). The systemic infection of grasses by *Epichole typhina* (Pers.), *Bul. Trans. Br. Myco. Soc. 18:* 30–47.

Schroeder, D. (1983). Biological control of weeds, in *Recent Advances in Weed Research,* (W. W. Fletcher, ed.), Commonwealth Agric. Bureaux, Slough, pp. 41–78.

Sheaffer, C. C. and Marten, G. C. (1986). Effect of mefluidide on cool season perennial grass forage yield and quality, *Agron. J. 78:* 75–79.

Silvertown, J. (1980). Leaf-canopy induced seed dormancy in a grassland flora, *New Phytologist 85:* 109–118.

Smith, A. E. (1977). "Weed Control in Agronomic Crops," *Proc. Agr. Chem. Workshop,* University of Georgia, Tifton.

Smith, A. E. (1982). Chemical control of smutgrass *(Sporobolus poiretii), Weed Sci. 30:* 231–234.

Smith, A. E. (1989). Herbicides for killing tall fescue *(Festuca arundinacea)* infected with fescue endophyte *(Acremonium coenophialum), Weed Tech. 3:* 485–489.

Smith, A. E. (1990). Potential allelopathic influence of certain pasture weeds, *Crop Protection 9:* 410–414.

Smith, A. E. (1991a). The potential importance of allelopathy in the pasture ecosystem, a review, *Adv 1:* 27–37.

Smith, A. E. (1991b). Paraquat for managing weeds in alfalfa *(Medicago sativa), Weed Tech. 5:* 181–184.

Smith, A. E. and Calvert, G. V. (1980). Factors influencing the control of horsenettle in perennial pastures, *Georgia Agric. Exp. Stn. Res. Bul. 255:* 15.

Smith, A. E. and Martin, L. D. (1992). Crabgrass control in coastal Bermuda grass, *Agron. J. 84:* 786–788.

Stachon, W. J. and Zimdahl, R. L. (1980). Allelopathic activity of Canada thistle *(Cirsium arvense)* in Colorado, *Weed Sci. 28:* 83.

Stuedmann, J. A. and Hoveland, C. S. (1988). Fescue endophyte: history and impact on animal agriculture, *J. Prod. Agric. 1:* 39–44.

Stuedemann, J. A., Wilkinson, S. R., Belesky, D. P, Hoveland, C. S., Devine, O. J. Thompson, F. N., McCampbell, H. C., Townsend, W. E., and Ciordia, H. (1986). Effect of level of fungus and nitrogen fertilization rate of KY-31 tall fescue on steer performance, *J. An. Sci. 63:* 290–291.

Wapshere, A. J. (1982). Biological control of weeds, in *Biology and Ecology of Weeds* (W. Holzner and M. Numata, eds.), Dr. W. Junk Publishers, The Hague, pp. 47–56.

Wells, G. J. (1974). The biology of *Poa Annua* and its significance in grassland, *Herb. Abstr. 44:* 385–391.

Wells, M. J. and Stirton, C. H. (1982). Weed problems of South African pastures, in *Biology and Ecology of Weeds,* (W. Holzner and M. Numata, eds.), Dr. W. Junk Publishers, The Hague, pp. 429–434.

Wilson, P. N. (1957). Studies on the browsing and reproductive behavior of the East African dwarf goat, *East African Agric. For. J. 23:* 138–147.

Yuncker, T. G. (1932). The genus *Cuscuta, Mem. Tor. Bot. Club 18:* 113–300.

13

Weed Management Systems for Rangeland

Rodney W. Bovey *
U.S. Department of Agriculture, College Station, Texas

I. INTRODUCTION

Rangelands of approximately 0.4 billion hectares are an essential resource in the United States (Thomas and Ronniger, 1965). Most wildlife and 75% of all domestic livestock depend upon rangelands for survival (Williams et al., 1968). Weeds and brush infestations cause incalculable economic losses by reducing rangeland forage production, water yield, wildlife habitat, and recreational use. Cost of handling livestock, livestock death and injury losses, and human allergies are greatly increased by dense stands of poisonous, thorny, or allergenic pollen producing plants. Estimates of total area infested vary, but weeds and brush are a problem on nearly all grazing land (National Academy, 1968). Weed and brush problems have been created by introduction of exotic plant species, grazing animals, reduction of fire, transport of weed seed by grazing animals, climatic fluctuations, cultivation and subsequent abandonment, local denudation and increased commerce (Vallentine, 1989), and many other factors. More recently today's higher CO_2 levels have been suggested to favor C3

**Current affiliation*: Adjunct Professor, Department of Rangeland Ecology and Management, Texas A&M University, College Station, Texas (shrublands) rather than C4 (grasslands) plants and may partially explain why native woody plants dominate (Mayeux et al., 1991).

Once established, weeds and brush do not disappear even with good or reduced grazing management and eventually become dominant. Removing weeds and brush can only be accomplished by use of an effective control method such as application of chemicals, mechanical means, fire or biological treatments, or a combination of these treatments.

Considering the vast acreages of rangeland infested with weeds, the economic losses, and the perceived need for treatment, philosophies and attitudes have changed in the past decade as to how to manage weeds and brush. On state and federal lands, tactics to control vast acreages of noxious weeds may be extremely costly and difficult, and containing their spread may be all that can be accomplished until further technology is developed. On private lands, indiscriminate brush removal for the sole purpose of increasing herbaceous forage for livestock may have dubious economic justification (Hanselka, 1991). Hanselka (1991) indicated that 10 factors must be considered in managing weeds and brush. The first is to determine whether a weed and brush problem exists. Additional factors concern managerial goals, ranch enterprises, effect of outside forces, type of weeds and brush, available technology, economics, environmental concerns, aesthetics, and weather. The end result is to conserve soil and water while increasing plant and livestock production at minimal cost.

Another consideration in weed and brush control is wildlife management. The aesthetic value of brush and importance of wildlife habitat provided by weeds and brush must also be considered before removal (McGinty, 1991). Endangered species and groundwater contamination may also limit weed and brush control, particularly herbicide use. Therefore, wholesale eradication of the weeds and brush may not be the best approach for the goals of the ranching enterprise. Weed and brush management should fit the economics and goals of the ranch.

II. MANAGEMENT PRINCIPLES FOR WEEDS AND BRUSH

Rangeland vegetation may be characterized by highly diverse mixtures of forage and weed species. The forage component may consist of one or more desirable forage grasses intermixed with forbs and browse.

Woody and herbaceous weeds are vigorous competitors and may dominate rangelands, sometimes as a result of overuse by domestic livestock or other mismanagement. Selective weed control is required for range improvement, especially where these undesirable weeds constitute long-lasting disclimax communities. Maintenance of desirable vegetation discourages reinvasion of weeds and prolongs treatment life.

Weed problems may also exist on abandoned fields that once were in agronomic crops. Soil disturbance followed by poor management has resulted in severe soil erosion and loss of fertility. These areas could be improved by controlling the weeds and brush, by providing adequate soil fertility, and by reestablishing a desirable vegetation. Other reasons for encroachment of weeds and brush on rangelands include spread of seeds and plants by livestock and wild animals, poor grazing management, reduction in rangeland fires, and reduced competition from desirable vegetation. Periodic drought also contributes to weed and brush invasion by reducing competing vegetation cover. Plant diseases, insects, and rodents also may play a significant role in reducing forage cover.

Pastures and rangelands, whether grazed or protected, require some type of management to maintain the desired vegetation. On grazed lands, livestock select the more palatable species, leaving less palatable weeds and brush to spread and to multiply. If grazing is intense, the process is accelerated. After a few years, especially in warm climates, control measures must be employed or the weed problem intensifies. Some areas can be treated by mechanical means with satisfactory results. However, on areas that are too steep, wet, rocky, or subject to wind and water erosion, the only practical means of control may be aerial application or individual plant treatment with herbicides, fire, or biological control.

Many weeds and brush also provide considerable browse, forage, and protection for wildlife and livestock. Land managers need to identify desirable species and adjust control programs to retain them. For example, some oaks provide excellent browse and protection for deer and goats but are of little value for cattle. Retention of some brush may be required to maintain wildlife habitat. On areas subject to wind and water erosion, such as sand dunes, weeds and brush provide necessary cover until more desirable forage is established.

III. METHODS FOR MANAGING WEEDS AND BRUSH

The first step in controlling a problem species is to identify it correctly. Plant species differ greatly in their response to fire, mechanical, chemical, or biological control measures. Selecting the proper method depends on the species, cost, type of soil and terrain, rainfall patterns, existing forage species, proximity to cropland, management objectives, and other factors.

A. Chemicals Used

Herbicides are an important means of weed and brush control on rangelands. Compared to mechanical practices, use of herbicides usually is less expensive, less damaging to the environment, and often more effective. Herbicide sprays, however, are subject to drift that may damage susceptible crops or valuable vegetation on nearby areas if improperly applied. A variety of herbicides and herbicide combinations are available commercially. The user must understand the properties and the effects of herbicides to use them safely and effectively. Herbicides commonly used for weed and brush control include 2,4-D, clopyralid, dicamba, bromacil, picloram, hexazinone, triclopyr, amitrole, and tebuthiuron. Combinations of herbicides such as 2,4-D + triclopyr, triclopyr + picloram, triclopyr + clopyralid, and 2,4-D + picloram or dicamba also have been used (Bovey et al., 1984; Klingman et al., 1983).

Characteristics of Herbicides

Benzoics. Dicamba, a selective translocated herbicide, controls many broadleaf weeds in pastures and cropland and some woody plants. Similar to the phenoxy herbicides in activity and use, it is absorbed through roots as well as through foliage. Dicamba can be applied by either ground or aerial sprays, depending on the weeds to be controlled and their proximity to susceptible crops. Dicamba can be applied in mixtures with 2,4-D to broaden the spectrum of weeds controlled.

Dicamba has low toxicity to wildlife, fish, livestock, and humans. It rapidly degrades and does not accumulate in the environment. Dicamba has a low corrosion hazard to spray equipment. It is formulated as the dimethylamine salt and is sprayed in a water carrier.

Bipyridyliums. Diquat and paraquat are contact herbicides used against some weeds and brush but are sometimes used as selective herbicides. Woody or perennial weeds usually will resprout from foliar sprays of diquat and paraquat.

Paraquat is water soluble and is inactivated by soil contact. It is registered for suppression of existing sod and emerged undesirable weeds and grasses to permit pasture and range seeding. Paraquat can be used to control winter weeds in dormant warm-season grasses. Protective clothing and respirators should be used during application because paraquat is highly toxic when ingested.

Phenoxy Herbicides. Phenoxy herbicides such as 2,4-D, 2,4-DB, dichlorprop, and MCPA have been used for more than 40 years and are effective for control of many weed and brush species. The phenoxy compounds are marketed as liquid concentrates of salts or esters. The ester formulations often are more effective as foliar sprays on trees and brush than are the salts. Esters are classified as high volatile or low volatile, depending on how readily they vaporize. Methyl, ethyl, isopropyl, and butyl esters of 2,4-D are examples of 2,4-D formulations that are highly volatile when sprayed on hot days. Such phenoxy compounds are identified on the herbicide container or package label. Examples of low volatile phenoxy esters include the propylene glycol butyl ether, butoxyethanol, and isooctyl esters. Low volatile esters or salt formulations should be used in areas where sensitive crops or other desirable vegetation is grown. The tendency for a herbicide to vaporize (volatilize) is important because it may be carried as fumes from the target area and may damage nearby valuable vegetation, especially under windy conditions on hot days.

Oil soluble formulations, usually esters, applied in kerosene or diesel oil will penetrate the bark of woody plants and can be used as basal or foliar sprays on individual plants. The phenoxy herbicides, however, are more commonly broadcast to large areas containing dense stands of weeds.

Sulfonylureas. Chlorsulfuron, metsulfuron, and sulfometuron have activity on a broad spectrum of weeds. Metsulfuron has a label use for common broomweed, bitter sneezeweed, woolly croton, and perennial broomweed (broom snakeweed) on rangelands. The compounds are effective at extremely low rates and can be mixed with 2,4-D and other herbicides to control western ragweed.

Triazines. Atrazine applied at rates of 0.9 to 1.1 kg/ha in the fall controls downy brome and other weeds. Perennial grasses can be seeded the following fall and will establish under reduced weed competition. Atrazine also can be used to renovate existing stands of certain perennial range grasses in the western region of the United States (Bovey et al., 1984).

Hexazinone is a triazine compound recommended for forestry site preparation and for pine release where loblolly, longleaf, shortleaf, slash, and Virginia pines are grown. The pelletized product can be spread in a grid pattern for hardwood brush control. It also shows promise for total vegetation control, including control of perennial grasses. It is highly water soluble, and a liquid formulation has federal registration on rangeland for controlling individual woody plants.

Ureas and Uracils. Ureas and uracils are compounds that include bromacil and tebuthiuron. Ureas and uracil-type herbicides may be selective at low rates and nonselective at high rates. They usually are formulated as wettable powders for water sprays or as pellets for dry application.

Bromacil (a uracil) will control a variety of woody species. If rates above 5.6 kg/ha are used, it also will kill many desirable grasses and forbs on grazing lands.

Tebuthiuron excels in controlling a variety of undesirable woody plants. Tebuthiuron is formulated as pellets containing 20% active ingredient; it is commercially available for rangelands and noncrop areas.

The ureas and uracil herbicides are absorbed primarily through the roots of plants. They may be applied in spring or fall when weeds and brush are actively growing and when rainfall is adequate to leach them into the soil.

Fall, winter, and early spring applications of hexazinone and tebuthiuron can be timed to reduce injury to forage plants and to eliminate drift hazards. These compounds may kill trees at a considerable distance from the point of application, when the root system extends into the treated area. The compounds are nonvolatile and do not corrode equipment.

Most urea and uracil herbicides can be injurious to some forage species when applied broadcast, especially as sprays. Applying herbicides as pellets or spheres in grids or bands versus broadcast treatment reduces exposure to most desirable forage plants. Most of the ureas and uracils persist in the soil for several months at rates used for brush control. They are low in toxicity to warm-blooded animals.

Other Organic Herbicides. Amitrole is effective against poison ivy and poison oak. If amitrole is accidentally sprayed on desirable plants, it is less likely to cause severe injury than some other herbicides.

Amitrole is available as a powder containing 50% active ingredient (ai) or as a liquid formulation. Another formulation, Amitrole-T, contains equal amounts of ammonium thiocyanate in addition to amitrole in liquid form. Amitrole is absorbed through the roots and foliage of plants but cannot be used where there is any possibility of residues on food or feed crops.

Clopyralid, an auxin-type selective herbicide related to picloram, is used for control of annual and perennial weeds and woody plants in crops and rangeland. It is highly effective against species of Polygonaceae, Compositae, and Leguminosae families but has little activity against grasses, crucifers, and many woody plants. It is highly phytotoxic to honey mesquite (Bovey et al., 1984).

Fosamine is a liquid herbicide for control of woody vegetation. Response to late summer or early fall foliage sprays is not observed until the following spring, when budbreak is prevented in some species in certain areas.

Glyphosate, a nonselective herbicide, is effective against both grasses and broadleaf plants. It is readily translocated from leaf and stem tissue to roots, killing a high percentage of many weeds. Glyphosate is inactivated by contact with the soil and does not injury newly seeded forages planted in treated soil. It is sprayed in a water carrier and is not corrosive to equipment. Glyphosate is registered for use for noncrop and pretill weed control and as a directed spray in orchards, plantations, Christmas tree farms, and other types of cropland. Broadcast sprays over woody species will damage desirable forage plants, so applications should be made to individual plants on noncrop areas or areas to be renovated. Glyphosate has low toxicity to mammalian species.

Imazapyr controls a wide variety of annual and perennial weeds, deciduous trees, vines, and brambles in noncropland situations. Research is being conducted to develop a granular product for use in site preparation and pine release in forestry. Imazapyr is available as an oil soluble formulation for cut stump, basal bark, and frill application.

Picloram, a selective translocated herbicide, effectively controls many weed and brush species on grasslands. It can be applied to the soil, to foliage, or as an injection/cut surface treatment. Picloram can be applied in spring or fall and foliar sprays are absorbed by both foliage and roots. Most perennial grasses are resistant. Its high activity at moderate rates against many woody plants makes it desirable for brush control.

Care must be taken to prevent drift of picloram to desirable plants. Picloram is relatively persistent in soil, especially in cool climates. Since it is water soluble, care must be taken to limit its movement into water sources to cropped areas. Picloram has low mammalian toxicity. It is only slightly corrosive to spray equipment.

Triclopyr is a selective postemergence herbicide for use on rights-of-way, industrial locations, and forestry sites. Triclopyr has been approved for application on honey mesquite as foliar sprays or basal stem treatments in oil carrier. It also can be used to kill trees and brush by injection/cut surface treatments. Mixtures of triclopyr + clopyralid (1:1) readily control honey mesquite and 1:1 mixtures of triclopyr + picloram have been used on mixed brush species in the southwest United States including honey mesquite. Triclopyr is readily translocated in plants and is reportedly more effective than 2,4,5-T on some species. It is moderately toxic to warm-blooded animals. It degrades rapidly in soil. Ester and amine formulations are available.

Oils. Diesel oil and kerosene are commonly used to control honey mesquite and hiusache. From 0.25 to 4 L of oil is used per tree depending on tree size. The oil is applied around the base of the tree during dry weather when the soil is pulled away from the trunk. Application at this time enables oil to penetrate to the lower buds on the stem.

Oils alone are not very effective herbicides when applied to the foliage of woody plants. Diesel fuel is commonly used as a diluent and carrier for oil soluble herbicides for individual plant treatments. It is also used with and without water as a carrier in aircraft spraying.

Application Methods

Spray applications can be classified according to volume. An ultralow volume signifies a spray volume no greater than 4.7 L/ha. An ultralow volume treatment may be nondiluted herbicide. Low-volume sprays are from 9.4 to 280 L/ha of spray solution. High-volume sprayers apply 280 or more L/ha.

Low-Volume Sprays. Several herbicides control weeds and brush when applied at low volume as broadcast foliar sprayers. Low-volume spraying is usually quicker, easier, and less expensive than high-volume spraying. Low-volume sprays are often applied by aircraft.

Ground equipment is practical for applying low-volume sprays to low-growing weeds and brush on uncleared land or to regrowth on land that has been mechanically cleared. Low-volume sprays applied in swaths up to 15 m wide with either a spray boom with several nozzles or a large, boomless cluster nozzle have produced satisfactory coverage of plant foliage. Spray booms and cluster nozzles on ground sprayers should be mounted to clear the tallest weeds and brush by 1 m. An operating pressure of 207 to 276 kPa (30 to 40 psi) generally is effective. They are most effective under low wind conditions.

Aerial spraying is most efficient for treating large areas or tall and dense stands of brush on rough terrain. Both fixed wing aircraft and helicopters are used and provide excellent coverage in most brush. Sprays applied in tall dense brush may fail to kill understory plants. A second aerial spraying may be necessary a year to two after the first time.

High-Volume Sprays. Control of some woody or resistant species may require complete coverage. High-volume sprays are used to apply herbicides to weeds and brush along roads, rights-of-way, fence rows, and other similar areas or to individual plants on pastures and rangeland. High-volume sprays usually are applied to foliage with power sprayers that can maintain pressures up to 690 kPa. At this pressure, spray is forced spray through foliage to the tops of tall trees. Pressures higher than 690 kPa cause formation of fine spray droplets that may drift.

Individual Plant Treatment

Individual plant treatment is especially useful in controlling hard-to-kill woody species that occur in scattered stands. Individual plant treatments include foliar sprays, basal sprays, cut surface and injections, and soil treatment (Bovey et al., 1984).

Foliar Sprays. Foliar sprays of 2,4-D, dicamba, picloram, or triclopyr usually are applied in spring and summer when plants are actively growing and after the leaves have reached full size. Ester formulations of 2,4-D or triclopyr usually are more effective on woody species than amine formulations and are less likely to be washed off by rainfall soon after application. If complete coverage is necessary, high-volume sprays that wet all foliage, twigs, and terminal stems can be used.

Basal Sprays. Basal sprays are applied to bark at the base of individual plants. Mixtures of 2,4-D ester or triclopyr ester with diesel oil or kerosene carriers or oil alone are commonly applied. Basal sprays can be used to kill brush and trees with main stems up to about 13 cm in diameter any time of the year. Treatment effectiveness may be reduced if the bark or soil is wet. The stem should be sprayed from the soil line to 45 cm above ground. Four liters of spray will treat 50 trees 5 cm in diameter. Compressed air sprayers, knapsack sprayers, or power sprayers also may be used.

Low-volume and streamline basal applications of herbicides for control of brush were introduced in Texas by the State Extension Service in 1987. Since then, many field demonstrations have been established utilizing these two techniques. Most demonstrations include mesquite as the target species although other brush species have been treated.

Low-volume basal solutions contain 25% herbicide formulations and 75% diesel fuel oil. The solution is applied to the lower 30–46 cm of the stem with a fan or hollow-cone nozzle to wet the stem but not to the point of runoff. Less liquid is used per plant than in conventional treatment that allows runoff and puddling at the base of the plant. To be effective, the herbicide solution must be applied completely around the stem and penetrate the bark of the plant.

Streamline basal application utilizes a mixture of 25% herbicide, 10% penetrant, and 65% diesel fuel oil or 25% herbicide and 75% diesel fuel oil. The penetrant that has been tested extensively is d'limonene. It is sold under the product names of Cide-Kick, Cide-Kick II, and Quick Step II. It is a naturally occurring chemical derived from citrus or pine by-products. The mixture is sprayed in a band (7–10 cm wide) with a straight stream nozzle to one or two sides of the stem near ground level or at the line dividing young (smooth) and mature (corky) bark. The material must go completely around the stem. Best results have been obtained on smooth bark stems that are less than 10 cm in diameter.

Streamline basal application and low-volume basal application of herbicides are useful methods for maintenance brush control. They may be used for treating a thin stand of brush, for applying spot treatment, and for controlling brush near desirable plants.

Cut Surface Injection Treatment. Trees larger than 13 cm in diameter often have bark too thick for basal sprays to penetrate. Herbicides can be applied to sapwood of trees through either frills or notches cut into the bark or by mechanical injection.

Frills are cuts made into the sapwood at regular intervals around the trunk and act as cups to hold the herbicide. Frills are made with ax cuts that penetrate the sapwood at least 0.6 cm. Spacing depends on the species, herbicide, and time of year. The frill can be filled with either the same solution used for basal sprays or undiluted herbicide. A notch also can be cut into one or more places at the base of the tree and sprayed with the appropriate herbicide solution.

The same spray equipment used for basal sprays can be used for frills and notches.The solution also can be applied with a properly labeled plastic squeeze bottle or with a small can that has a pouring spout or lip.

Injection treatments are effective for killing woody plants, but the labor requirements make them expensive. Also, to be effective, the correct volume of herbicide solution must be applied in properly spaced cuts. In winter, resistant species may require higher herbicide concentrations or closer spaced cuts with herbicide than in summer. Commercial injection tools are available.

When trees are felled, the freshly cut stump surface should be treated with herbicide to prevent sprouting of certain species. It is more efficient to prevent sprouting than to kill the sprouts. Herbicides mixed with oils are applied directly to the freshly cut stump and thoroughly drenched to ground line.

Soil Treatment

Soil-active herbicide pellets or sprays can be either applied to individual plants, placed in grids or bands, or applied broadcast to the soil to kill plants. Rainfall after herbicide application may be required for satisfactory results. Dry herbicide materials commonly are formulated as extruded pellets 0.13 to 0.38 cm in diameter, spherical pellets from 0.13 to 1.3 cm in diameter, or tablets of varying sizes up to 2.5 cm long. The use of soil active herbicides has become more common since the development of new herbicides effective for brush control (Bovey et al., 1984).

Herbicide pellets can be broadcast or applied in bands in brush-infested areas. The larger spherical pellets and tablets are effective when applied in grids spaced on 1.8 to 2.7 m centers. Plants respond differently to grid applications, depending upon root systems. Gambel oak has a deep root system and does not respond well to grid pattern applications. Because of their shallow, widespread lateral roots, shrub live oak and Utah juniper respond well to a grid pattern application of pellets spaced on 0.9, 1.8, or 2.7 m centers (Bovey et al., 1984). Damage to forage plants is minimized by using grid or band placement. Broom snakeweed control is best achieved with broadcast application. Pellets can be broadcast by aircraft and ground equipment can apply pellets broadcast or spaced in bands 1.8 to 2.7 m apart.

Wipers and Special Techniques

Herbicides can be applied directly to weeds and brush by using specially designed tractor-mounted rope-wick or carpeted applicators. Herbicide solutions are conducted through a porous medium and wiped onto the unwanted plants. Present designs allow treatment of plants not exceeding 1.8 to 2.1 m tall. The ropes or carpets are saturated with herbicide, either undiluted or diluted, in water and wiped onto unwanted plants. Hand-carried wipers for small jobs have been designed. Mayeux and Crane (1983) suggested that a brush roller constructed of household carpet stretched over a 25-cm-diameter by 2-m-long polyvinyl chloride (PVC) cylinder can be used effectively to treat weeds and brush on rangeland. The carpeted roller is tractor-mounted, is saturated with herbicide, and is rotated opposite the direction of travel.

B. Mechanical Methods

Hand-Held Equipment

Hand methods of weed and brush control probably have been practiced since antiquity (Bovey et al., 1984). Although effective, they are slow, costly, and laborious. Hand methods include grubbing, cutting, girdling, and burning.

Grubbing consists of using a grubbing hoe, shovel, or similar tool to expose the root system partially or totally. The operation is difficult, time consuming, and effective if done properly. Axes or saws are effective on woody species that are killed when the top is removed, such as eastern redcedar and Ashe juniper. If species resprout, the cut surfaces can be treated with herbicide.

Girdling, cutting a ring through the bark and cambium layer to prevent movement of nutrients to the roots, is practical in scattered stands. Large non-root-sprouting trees of 15 cm diameter or greater are effectively controlled during summer months. For improved control, herbicides can be applied to the cut ring. Portable chain or power saws and girdlers reduce labor, time, and cost in brush removal but have limited use in dense stands or large areas.

Large Equipment

Bulldozing. Bulldozing, often called dozing, is a widely used method of brush control (Vallentine, 1989). Bulldozing is most commonly practiced in open stands of large trees and brush or on rocky soils where use of other mechanical control methods is limited. It is not desirable in dense brush that sprouts from the roots after top removal. In moderately dense brush, chaining or raking can be done to smooth the land and prepare a seedbed. If raked, brush piles can be left in place for wildlife cover. Follow-up weed and brush control will be necessary within 2 to 5 years after bulldozing to kill resprouts and new seedlings.

The equipment and fuel required for bulldozing are costly. Soil is disturbed, and forage stands and wildlife habitat may be partially or heavily damaged. The soil may be subjected to wind and water erosion, depending on how well the job was done. Under some circumstances, leaving alternate undisturbed strips, particularly on steep or erodible lands or where wildlife is important, may be justified (Scifres, 1980). Much clearing is done with straight dozer blades. Since it is inefficient and often ineffective, many modifications and attachments are available. Ideally, bulldozing removes brush and large trees by pushing or pulling the plants out of the soil. Special attachments to the straight blade include teeth, shearing blades, or U-shaped "stingers" that cut plants below the soil surface and lift out the roots.

Straight blade modification includes the rakelike "stacker" (Rechenthin et al., 1964) with teeth 13 to 36 cm apart mounted on a straight blade or in place of the solid blade. One advantage it has over the straight blade is that the brush rake displaces less soil and reduces the accumulation of soil in brush piled for burning. Should small brush slip between the teeth of the blades, an additional blade or root grubber can be welded across the bottom of the teeth to uproot brush. The stacker is highly maneuverable and prepares a good seedbed. Some brush piles should be left for wildlife, and crushed and broken brush should be left to reduce wind and water erosion. The stacker has been useful for removing thick stands of pricklypear cactus in south Texas.

Large brush rakes and root rakes also are used in south Texas to clean up debris after other mechanical operations such as chaining or root plowing of honey mesquite (Scifres, 1980). Some rakes are towed.

Grubbing. The grubber consists of a U-shaped hydraulically controlled blade with an angle adjustment to sever roots 15 to 30 cm below ground. Wiedemann and Cross (1981) developed a low-energy grubber to control redberry juniper, Ashe juniper, and other small weed trees on rangeland. Mounted on a 65-horsepower crawler tractor, the unit operates at low cost and selectively kills brush with minimal soil disturbance. Foam-filled tires have also allowed use of wheeled farm tractors and wheeled loaders for grubbing small honey mesquite trees where thorns normally would puncture conventional tires (Wiedemann and Cross, 1982).

Chaining. Chaining, a relatively inexpensive means of mechanically clearing brush (Rechenthin et al., 1964), does not remove all brush, and its major effect is temporary. Naval anchor chains are pulled behind two large crawler tractors across brush to be felled or pulled out. The chains may be 30 to 90 m long and weigh 90 to 135 kg/m. Cables, usually 4.3 to 5 cm in diameter, are not as effective as chains.

Many adaptations have been made to chains such as welding on short lengths of angle iron, rails, or disks to make them more efficient (McKenzie et al., 1984). Such modifications included the Ely chain for big sagebrush control, swivels to allow chains to roll, ball-and-chain arrangements for steep terrain, the Dixie-Sager chain for better uprooting of brush and scarification of soil, and the disk chain to reduce draft. Chaining is most effective on large single-stemmed brush and trees growing on sandy, shallow, or moist soils, where many of the plants can be pulled out. Two-way chaining (chaining in opposite directions) increases effectiveness. In some situations, the brush is bent over or sheared off at the ground level before being uprooted, and root sprouting occurs. In dense stands, the chain also may pull over large masses of brush, failing to uproot it.

Chaining can be used to advantage as a pretreatment to raking, root plowing, or goat grazing by knocking down dense brush to make it more accessible. Scifres et al. (1976) found that double chaining followed by raking and stacking mixed brush was more effective than one-way or double chaining. Chaining of brush infested with pricklypear, cholla, and tasajillo may scatter the cactus, making the treatment undesirable if it is not followed by raking. Chaining results in poor seedbed preparation, and residual brush disrupts seeding.

Wiedemann and Cross (1984) developed a disk chain with a triangular pulling configuration that reduced draft requirements by 36% and increased the operating width by 23% compared to the two-tractor diagonal pulling method. The disk-chain, pulled by a single tractor, appears to be cost-effective for preparing seedbeds on rough, log-littered, and root-plowed rangeland.

Railing. Railing is a practice developed for control of pricklypear and tasajillo in south Texas (Rechenthin et al., 1964). Three to nine railroad rails or heavy angle irons are welded or chained together and dragged in tandem in two or three sets behind a tractor. The rails are dragged over the area in one direction, then a second time in the opposite direction, crushing the cactus. Grazing cattle sometimes are used to clean up newly railed areas. Railing is most effective in hot, dry weather. Repeated railing about 1 month after the original treatment removes new sprouts or missed plants.

Chopping. Roller chopping is a quick method of knocking down and crushing brush. Like chaining, it is a temporary measure and kills few resprouting plants. Resprouting will occur soon after treatment unless trees are nonsprouters. Equipment consists of a steel drum or cylinder around which cutter blades are attached. Weight can be added either by filling the roller drum with water or by adding steel or concrete ballast to the frame. Choppers vary in size, depending on the size of brush. A weight of about 700 to 900 kg (1500 to 2000 pounds) per linear foot of cutter blade is suitable for small- and medium-size brush such as catclaw and blackbrush (Rechenthin et al., 1964). Larger choppers are needed for trees with stem diameters of 13 cm or larger.

Chopping is useful only where repeated treatment is planned or where other methods will supplement brush control. Chopping followed with soil-applied herbicides did not improve control over herbicides applied alone (Meyer and Bovey, 1980). Chopping, which removes top growth of brush and makes handling of livestock easier, may be useful where goats are grazed by allowing them to reach the leaves. Chopping is useful for seedbed preparation on log-littered land or where dense brush can be knocked down quickly and burned when dry. When chaining is not practical for tree felling, chopping is usually the most cost-effective method unless trees are large.

Mowing and Shredding. Mowing and shredding are temporary control methods for herbaceous weeds and small brush on pastures and rangeland. Mower types vary, but most consist of sharp rotary blades. Heavy duty shredders can be used on large brush and small trees (8 to 10 cm diameter). Commercially available self-propelled cutters will cut mesquite stumps 35 to 43 m in diameter (Hendon, 1971). Brush should be mowed in spring and early summer. Mowing after July 1 may damage native grasses (Rechenthin et al., 1964). Bermudagrass can be mowed any time during the growing season. In humid areas, the principal benefit is to remove old growth, promote forage regrowth and quality, and reduce selective livestock grazing. Weeds and brush become established in the ungrazed areas. Normally, two to three mowings per year limits the spread of herbaceous or woody perennials in pastures but does not eliminate them.

Root Plowing. Root plowing consists of pulling a horizontal blade through the soil behind a large tractor at depths up to 45 cm to cut roots (Rechenthin et al., 1964). Root plowing was developed originally for land clearing. Blades must be operated at 30 to 40 cm deep in soil to kill large honey mesquite. Root plow blades vary from 1.8 to 4.3 cm wide and fins welded to the top of the blade help lift roots out of the soil. Fisher et al. (1972) indicated finless root plow blades or thin blades gave effective control of honey mesquite and lotebush at lower cost than conventional blades, while allowing better survival of perennial grasses.

Root plowing is useful in dense brush containing species that are resistant to herbicides or other mechanical control methods. Root plowing kills a high percentage of brush and can be done anytime. It is most effective during hot, dry periods but may be detrimental to the forage species at this time. Root plowing in central and northwest Texas is done predominantly during winter months when native forage species are dormant, allowing maximum survival.

If seeding is needed, it should be done during the optimum time for the area. After root plowing, the seedbed can be improved with tillage, such as chaining, roller chopping, or disking. Seeding is done by either aircraft or ground drills. Root plows equipped with grass seeders have had limited success. Root plowing is an effective brush control treatment, but it is expensive, may destroy the grass stand, and may subject the soil to wind and water erosion. Some areas of brush should be left for wildlife habitat.

Disking. Large disk plows or tandem disks will destroy stands of smaller brush as well as the grass stand. Disking is limited to tillable soils. Disking prepares a good seedbed, although compaction by a cultipacker, roller, or other implement may be desired to maximize soil-seed contact. Land clearing disks are an adaptation of standard agricultural implements but weigh from 70 to 450 kg per blade. They are heavy duty implements and are effective in uprooting, chopping, and mulching brush, even in dense stands. Commercial seeding attachments are available, but seeding usually is conducted as a separate operation.

C. Biological Control

With the high cost of herbicides and mechanical means for weed and brush control, biological control is attractive. In nature, biological control is already at work since plant species have natural enemies that feed on them, especially plants in their native habitat. Scientists recognize this natural phenomenon and attempt to restrain enemies of desirable plants and augment their activity against undesirable plants. Schroeder (1983) stated that since a large number of economically important weeds were introduced from other parts of the world, biological control is concerned primarily with importing exotic organisms to control naturalized weeds. Rangelands offer unique possibilities for biological control because vast areas often are infested with one or more undesirable weed and brush species, many of foreign origin. Huffaker (1964) indicates that the primary objective in biological control of weeds is not eradication but use of the plant's natural enemies to lower its density to a noneconomic level.

The procedure in developing biological control consists of determining the suitability of the weed problem for biological control, surveying the weed for natural enemies in both naturalized and native habitats, studying the biological characteristics and host relations of the natural enemies to determine how they may be used to solve the problem, implementing their use in weed control, and evaluating their effectiveness (Harris, 1971). The most common method of implementing biological control has been through the introduction of new natural enemies. Other methods of implementation manipulate or augment already existing natural enemies of weeds (Rosenthal et al., 1984).

DeLoach (1991) indicated that the greatest advantage of augmentation is that it's active only where applied so no damage occurs to plants in untreated areas. Augmentation effects can be ended by discontinuing it. The method is similar to herbicidal control in that the entire affected area must be treated at repeated intervals. Augmentation is unlikely to be economical in areas of low economic return such as rangelands.

Biological weed control dates back to the 1860s, but it was little used until the 1920s. Some success with lantana and spectacular success with pricklypear and St. Johnswort stimulated worldwide biological weed control research. Most biological control agents have been insects, but mites, snails, nematodes, fish, mammals, birds, plant pathogens, and even other plants also can be valuable (Rosenthal et al., 1984).

A lengthy and comprehensive procedure is followed to make certain that the insect or organism, especially of foreign origin, is effective on the target weed and will not injure desirable plants or other organisms before release from quarantine into the field (DeLoach, 1991).

Control of Woody Plants with Insects

The best example of effective control by insects involves pricklypear control in Australia by the Argentine moth borer. Pricklypear were used as ornamentals and hedges around homesteads in the late eighteenth century. By 1925, a total of 24 million hectares was dominated by pricklypear. Mechanical and chemical control were too expensive. Although several insects were introduced, none was as successful as the Argentine moth borer. The larvae feed on cactus by tunneling through the pads and root tissue. Damaged tissue is subjected to secondary attack by bacteria, fungi, and adverse weather (Schroeder, 1983; Rosenthal et al., 1984; Vallentine, 1989). Goeden et al. (1968) reported successful reduction of pricklypear species (*Opuntia littoralis* and *O. oricola*) and their hybrids on Santa Cruz Island off the coast of Santa Barbara, California, using a cochineal insect (*Dactylopius* spp.). The insect was introduced from

Mexico via Hawaii. Attempts to control pricklypear in Texas by biological means have been discouraged because some ranchers use pricklypear for livestock feed during periods of severe drought.

Insects have been used to control lantana, an escaped ornamental (Schroeder, 1983). Lantana, a perennial shrub native to Central America, has become a pest of rangeland and coconut plantations in Hawaii, Fiji, India, and Australia. The lace bug was the most important control agent in Australia (Harley, 1973); however, a complex of organisms may be required to control lantana (Schroeder, 1983).

Gorse, a native of Western Europe and North America, has become a pest in New Zealand, Australia, and the United States (Hawaii, California, and Oregon). It is used extensively as an ornamental hedge but escaped to become a serious pasture weed. The gorse weevil (*Apion ulicis*) prevents seeding in Europe and New Zealand but does not affect established plants. Several other insects have been tested, but none has proved as useful as *A. ulicis*. Although it has been released in Hawaii and on the Pacific coast (United States of America), gorse control has not been outstanding (Holloway, 1964).

DeLoach (1990) indicated that top weed candidates for biological control in the southwestern United States included snakeweed and broomweed, *Baccharis* spp., mesquite, saltcedar, and creosotebush. Promising foreign insects attack these species and control strategies are being considered. Conflicts between beneficial and harmful potential should be resolved (DeLoach, 1990). Other possible biocontrol candidates include whitebrush, tarbush, pricklypear, cholla, tasajillo, huisache, retama, and Macartney rose.

Control of Herbaceous Weeds with Insects

Another example of successful biological weed control is that of St. Johnswort or klamath weed by a leaf-eating beetle, *Chrysolina quadrigemina*, and by the root-boring beetle (*Agrilus hypericic*) (Holloway, 1964). St. Johnswort is an aggressive perennial forb accidentally introduced into the United States from Europe. This poisonous plant occurs mainly in California and the Pacific Northwest. Biological control of St. Johnswort with the *Chrysolina* beetle has generally been excellent (Holloway, 1964; Rosenthal et al., 1984; and Vallentine, 1989).

On western U.S. rangelands, puncture vine, tansy ragwort, musk thistle, rush skeletonweed, and toadflax have been controlled biologically (Julien, 1987; Quimby et al., 1991). Control is sometimes incomplete, and research continues. The USDA-ARS has nine laboratories concerned with developing biological control of at least 16 genera and more than 30 species of rangeland weeds (Quimby et al., 1991). They include snakeweed, creosotebush, tarbush, starthistles, toadflax, gorse, dryer's woad, Scotch thistle, poison hemlock, knapweeds, leafy spurge, saltcedar, St. Johnswort, larkspur, whitetop, field bindweed, cocklebur, and bitter sneezeweed.

Rees and Spencer (1991) indicated that there are several Eurasian insects that attack leafy spurge on U.S. rangelands and an additional eight insects are in various stages of clearance. With biocontrol, the goal is to reduce leafy spurge in North American plant communities to noneconomic levels as found in Eurasia.

Rees (1991) indicated that seed heads of both musk and Canada thistles are attacked by a seed head weevil (*Rhinocyllus conicus*) while the larvae of a second weevil (*Trichosirocalus horridus*) feed on the growing point. Larvae of *Cheilosa corydon* burrow into musk thistle stems and roots. The larvae of the weevil *Ceutorhynchus litura* reduce overwintering survival of Canada thistle by burrowing in the stems and roots. The larvae of the gall-producing fly *Urophora cardui* also burrow into Canada thistle stems and induce large galls. Galls act as a nutrient sink to stress the plant and prevent flowering in some cases.

Rosenthal et al. (1991) are developing biological control measures for yellow starthistle and several knapweeds (diffuse, spotted, squarrose and Russian) using insects but indicate that a wide variety of natural enemies will be necessary to reduce weed populations to tolerable densities. Researchers are also investigating mites, nematodes, and pathogens. Integration of biological control with other range management practices will be necessary to suppress the knapweeds.

Biocontrol with Plant Pathogens

Wilson (1969) described the use of plant pathogens in weed control. The successful control of pricklypear cactus in Australia is attributed to secondary attack by associated microorganisms following the attack of *Cactoblastus cactorum*. *Gloeosporium lunatum* E&E and bacterial soft-rot organisms are considered the primary invaders following *C. cactorum* attacks. Fungi that cause disease in pricklypear also have been described, although their importance has not been fully assessed in relation to the decline of various cactus species.

Tainter (1972) evaluated oak wilt fungus as a selective silvicide in northern Arkansas. Post oak, white oak, blackjack oak, southern red oak, and black oak were inoculated with an Arkansas isolate of the oak wilt fungus in April 1971. Results indicated that oak wilt fungus effectively killed inoculated red oaks and prevented sprouting from the cut stump. The oak wilt fungus is reported to be nonhazardous to tree species other than those in the red oak group and may be useful as a selective silvicide since it does not spread by root grafts to untreated trees.

Van Arsdale and Halliwell (1970) reported that live oak decline is a complex of diseases that affect other oaks as well as live oak. A fungus (*Cephalosporium* spp.) has been identified as the major component of the disease complex. It causes a vascular wilt that is apparently root graft–transmitted. Since live oak often is considered a desirable shade tree and produces acorns for wildlife, it is doubtful that live oak decline can be used as a means of biocontrol for oaks.

Templeton (1982) reported that rush skeletonweed has been successfully controlled in Australia by a rust fungus (*Puccinia chondrillina*) from the Mediterranean area where the weed originated. *P. chondrillina* was released on rush skeletonweed populations in the western United States in 1978.

Quimby et al. (1991) indicated northern joint vetch in Arkansas rice crops and stranglervine in Florida citrus orchards have been controlled with native pathogens. Major targets for biological control using plant pathogens are leafy spurge, musk thistle, the knapweeds, and the starthistles. A leaf-spotting fungus (*Alternaria* sp.) collected in North Dakota and Nebraska is being evaluated on leafy spurge. *Puccina* sp., rust disease, is being investigated on musk thistle, starthistles, and knapweeds (Quimby et al., 1991).

Selective Grazing

Selective grazing is a wide-scale practice to control undesirable weeds and brush with cattle, sheep, goats, horses, poultry, and some wildlife species. In some cases, weeds may be so unpalatable, poisonous, or vigorous that grazing management is not practical.

In central Texas, California, Arizona, and other areas, goats use browse as a large part of their diet. Goats consume woody plants, such as live oak, blackjack oak, sumac, greenbriar, hawthorn, American beautyberry, yaupon, sweetgum, and retama. Brush and trees can be cut, chained, roller chopped, shredded, or bulldozed down when too tall for animals to reach the foliage.

Goats are an effective means of brush management by grazing one goat per hectare on a year-long basis (Vallentine, 1989). For short-term grazing, a 30-day period, 12 to 20 animals per hectare, is required, depending upon kind and density of brush (Rechenthin et al., 1964). The short-term approach, with high stocking rates, controls most brush with less damage to other forage plants. Three years usually is required for this system to reduce the brush to a manageable level. A lighter stocking rate can then be used to maintain brush populations.

Warren et al. (1984) reported that Spanish goats grazing on mixed brush in the South Texas Plains used shrubs as food during autumn, winter, and summer. The goats browsed blackbrush acacia, *Condalia* spp., guajillo, guayacan, and Berlandier wolfberry, suggesting their potential for use in conjunction with other brush control practices to increase the efficiency of forage use on mixed-brush sites.

Goats were superior to cattle and sheep for brush control in New England (Wood, 1987). Goats destroyed small trees and saplings by debarking them. As brush was destroyed, grass increased. In Canterbury, New Zealand, gorse can be a year-round maintenance feed for goats (Radcliffe, 1986). Angora goats were used to control gorse in Australia in combination with sheep (Harradine and Jones, 1985). Goat grazing is not a simple operation since special goat-proof fences, management, and protection from adverse weather and predators are required. In some areas green foliage is not available year round so animals must be sold or moved. Predators, such as bobcats, coyotes, and occasionally dogs, may injure or kill the goats, resulting in severe economic loss.

In Montana, 40% to 50% of sheep's diet consisted of leafy spurge in heavily infested pastures (Landgraf et al., 1984). Ewe weight gain in leafy spurge pastures was not different from weight gain in pastures free of leafy spurge. Sheep consumed leafy spurge without apparent harmful effects. Therefore, sheep can be considered an effective biological control agent of leafy spurge.

Fay (1991) reported that some research showed that sheep refused to eat leafy spurge and suffered deleterious effects when they did eat it. At the present time, sheep are being used to control leafy spurge in Montana because of high herbicide costs and lack of success of other biological control methods. Problems occurred because there are not enough sheep to control the weed adequately to prevent seed production without overgrazing infested areas. Sheep loss to predators and containment of sheep on infested areas were also problems (Fay, 1991).

Fertilizer and Plant Competition to Control Weeds and Brush

Herbaceous weeds sometimes can be effectively eliminated or reduced by encouraging growth of competing desirable vegetation. For example, broomsedge can be controlled on unproductive pastures in Virginia with annual nitrogen fertilizer applications for 5 years. Neither burning nor mowing controls the weed, which may make up 70% of the vegetation (Semple, 1970). Annual nitrogen fertilization after the first year maintained superior stands of five warm-season prairie grasses and reduced weed invasion at 12 locations in Nebraska (Warnes and Newell, 1969).

Johnston and Peake (1960) seeded crested wheatgrass on a 12 hectare field infested with leafy spurge. While excellent stands of crested wheatgrass were established, they did not control leafy spurge.

Control of woody plants by fertilization has not been successful because fertilization may encourage rapid growth of dominant brush. In some situations plants are used as browse and fertilization may increase production and palatability, but treatment cost may not be economical on rangelands (Bovey et al., 1984).

A vigorous, dense stand of grass and broadleaf species may discourage weed and brush encroachment, especially as seedlings, into rangelands. Once woody plants become established and dominant, however, they usually cannot be controlled or eliminated by other competitive plants.

D. Control with Fire

Fire is an ancient method of suppressing weedy vegetation. In the past, it occurred naturally in many areas. Whether man-made or natural, however, fire has been a strong influence on the vegetation type of an area. There is still great diversity of opinion and inconclusive data about prescribed burning as a valid weed control method. It has been used with great benefit in some grassland situations but uncontrolled can have disastrous effects. Burning is restricted by local or state air pollution laws in some locations.

Scifres (1980), Vallentine (1989), and Kozlowski and Ahlgren (1974) have reviewed range improvement by burning. Wright and Bailey (1982) have evaluated fire for rangelands and forests in the United States and southern Canada. Fire has been used to control sagebrush in the western United States (Pechanec et al., 1965),. convert chaparral to grasslands in California (Bentley, 1967), control pinyon-juniper communities in the Southwest (Vallentine, 1989), and control mixed brush in South Texas (Box and White, 1969).

Fire can be used on rocky or rough terrain, at moderate cost, with fast coverage versus costly alternative methods (Scifres, 1980; Vallentine, 1989). It can also be used to maintain or extend the treatment life of initial brush control methods (Scifres, 1980). Prescribed burning may increase palatability and forage production, kill or suppress undesirable plants, improve grazing use and distribution, and release plant nutrients for wildlife and livestock and plant growth. Much information has been accumulated on fire effects on desirable and undesirable vegetation and the reader is referred to the literature for specific vegetation types and plant responses. Britton and Wright (1983) indicated that historically, fire was a major factor that kept shrubs suppressed in grasslands. For fire to be effective, however, adequate vegetative fuel must be produced or other methods of weed and brush control must be considered. Wright and Bailey (1982) indicated that with an adequate fuel load, burroweed, broom snakeweed, creosotebush, young mesquite trees, and cacti can be suppressed, and tobosagrass, big sacaton, alkali sacaton, and mixed grass ranges can be managed. Black grama grasslands appear too delicate to manage with fire (Wright and Bailey, 1982). In Arizona, Cox (1988) found that burning or mowing big sacaton at any season removed green foliage available to livestock and reduced the amount of green forage that may accumulate for at least two summer growing seasons.

Big sagebrush (a nonsprouting species) was killed in Wyoming when a uniform fire spread occurred (Smith et al., 1985). Green rabbitbrush and horsebrush resprouted. Greasewood was sometimes killed as a result of a high fuel load. Aspen, snowberry, and serviceberry resprouted while bitterbrush did not (Smith et al., 1985). Wambolt and Payne (1986) found that sagebrush canopy was most effectively reduced by burning and that burning provided the most production from dominant forage species when compared to other methods.

Saw palmetto, a low growing shrub dominating Florida rangeland, is only temporarily controlled by fire (Tanner et al., 1986). Burning of saw palmetto reduced carbohydrate content and increased mortality if subsequent chemical or mechanical treatment was applied.

Conversion of chaparral to grasslands is done to improve fire control, to increase water yield, and to develop new forage for livestock and deer (Bentley 1967). The control method depends upon chaparral type, terrain, resistance of brush species to herbicides, and need for revegetation. Herbicides are preferred where practical for control of the chaparral. In California, Bentley (1967) indicated that burning standing chaparral seldom removed the brush satisfactorily. Chaparral should be prepared to permit safe burning during weather unfavorable for ignition, buildup, and spread of fire to natural fuels. Chaparral will burn readily if crushed and compacted before broadcast burning. A bulldozer blade or roller chopper will compact the brush. Spraying chaparral with herbicides for fuel preparation is recommended only in light or medium density brush.

In pinyon-juniper communities, a uniform burn has been an effective, economical method of controlling nonsprouting junipers (Vallentine, 1989). Uniform burn depends upon the presence of enough juniper and understory plants to carry the fire. Barney and Frischknecht (1974) studied the vegetation changes following a fire of the pinyon-juniper type in Utah. Woodlands were well developed 85 to 90 years after the fire. Weedy annuals reached peak levels 3 to 4 years after fire. Intermediate successional stages followed a general pattern of perennial grasses, perennial grasses-shrubs (sagebrush), and perennial grasses-shrubs-trees.

Postsettlement invasion of trees and shrubs on a tallgrass prairie in the Kansas Flint Hills was assessed by using aerial photography, General Land Office survey data, and field observations (Bragg and Hulbert, 1976). On unburned sites, aerial photographs showed that combined tree and shrub cover increased 34% from 1937 to 1969; section-line data showed that tree cover alone increased 24% from 1856 to 1969. Data from two sites suggested that herbicide spraying only slowed invasion rate. Percentage of woody plant cover varied with soil site but increased most rapidly on lowland soils. The authors concluded that on the Flint Hills tallgrass prairie, regular burning (every 1 to 5 years) has restricted woody plant invasion to natural presettlement amounts, although rate of woody plant invasion varies with soil type and topography.

On Oklahoma rangeland, burning had a negative effect on woody plant control (Elwell et al., 1974). Burning alone contributed little to control of larger trees (post oak and blackjack oak).Fire increased root sprouting of blackjack oak, sumac, and dogwood. Where 2,4,5-T was used to control woody plants, herbaceous vegetation increased and provided more fuel for fire. An increase in stands of decreaser (highly palatable to livestock) grass species occurred only in herbicide-treated plots. Burning either alone or combined with 2,4,5-T resulted in slight increases in grasses. Elwell et al. (1970) indicated that burning controlled small eastern redcedar trees but had little effect on post oak and blackjack oak. Engle and Stritzke (1991) developed a self-contained backpack propane burner to ignite larger eastern redcedar trees not controlled by initial prescribed burning. Reburning of larger trees on 32 hectare pastures ranged from 21 minutes to 9.5 hours at a cost comparable to that of chemical or mechanical methods.

Wright et al. (1976) found honey mesquite difficult to kill with fire on the High Plains and Rolling Plains of Texas. On upland sites in the Rolling Plains, 27% of the mesquite trees were killed by a single fire, but repeated fires at 5- and 10-year intervals killed 50% of the older trees. Seedlings 1 1/2 years of age were easily killed with fire but plants 3 1/2 years of age were tolerant of fires.

In an Ashe juniper community, a minimum of 1200 kg/ha of fine fuel was needed to carry a fire to kill juniper seedlings and burn piles of bulldozed juniper (Wink and Wright, 1973). Grasses recovered quickly and soil erosion was minimal when burning was conducted during a wet winter and spring. During a dry winter and spring, however, burning increased drought stress, reduced herbaceous plant yield, and exposed soil to wind and water erosion.

In South Texas, Scifres (1975) found best control of Macartney rose occurred where individual plants were treated with 2,4,5-T + picloram (1:1) and followed by a prescribed burn 18 months later. In British Columbia, Wikeem and Strang (1983) indicated that prescribed burning is being used increasingly for rangeland management, but fire ecological research is needed for intelligent use. Whisenant (1989) showed that frequent fires (<5 year apart) in cheatgrass ranges in Idaho retarded normal vegetation replacement sequences leading to vegetation that resembled that of less frequently burned areas. Reducing frequency and size of fires should be the primary management objective.

White and Hanselka (1991) indicated that prescribed burning is a viable improvement practice for most Texas rangelands. When integrated with other practices, fire can be used to maintain a desired vegetation composition. Landers (1991) indicated a detailed plan is required for a successful burn, including adequate fuel, appropriate fire control, interpersonal communications, means for emergency calls, proper burning procedure, and a follow-up plan.

E. Combinations of Control Methods

Selection of combinations of methods for weed and brush control is based on maximum effectiveness with lowest possible cost using good conservation practices. Use of two or more control methods is referred to as an integrated weed or brush management system. Rechenthin et al. (1964) reported using goats for successful control of a large number of brush species in Texas. If brush is too large, it can be chopped, chained, or bulldozed down within reach of the goats.

Davis et al. (1974) reported that Gambel oak, an important component of several million hectares in foothill rangelands of Arizona, Colorado, New Mexico, and Utah, can be controlled with a mechanical treatment followed by goat grazing. The brush is first treated by roller chopping, bulldozing, or undercutting to allow the animals full access to all foliage. Repeated annual defoliation by goats during late June and late August gave best brush control.

Bentley (1976) provided a list of brush removal combinations for chamise-chaparral and mixed chaparral in California along with costs associated with each operation. For chaparral, crushing and burning is a common practice, followed by herbicide spraying 1 or 2 years after top removal.

Scifres (1975) proposed that undisturbed dense stands of Macartney rose can be controlled in South Texas by mechanical treatment followed by herbicide treatment 2 years later. Prescribed burning or herbicide treatment can be used for maintenance.

Scifres et al. (1983) concluded that herbicide-fire combinations improved rangeland supporting dense stands of running mesquite or whitebrush. The herbicide reduces the brush cover and releases herbaceous vegetation either for livestock or for fuel for burning. The burn expedites forage production, improves botanical composition of herbage, and suppresses brush. A burn at 3- to 5-year intervals was suggested to maintain desirable species. Wink and Wright (1973) used prescribed burning to reduce woody debris, suppress missed and newly established Ashe juniper plants, and enhance forage production after tree dozing. Ueckert et al. (1988) found that controlling pricklypear cactus after fires in lighter fine fuel vegetation such as sideoats grama and buffalo grass was often not satisfactory. Fire followed by aerial sprays of picloram at 0.14 kg/ha reduced pricklypear cover by an average of 98% and represents an economically significant level of control.

Mayeux and Hamilton (1983) found that burning prior to application of tebuthiuron synergistically enhanced the control of common goldenweed compared to each treatment applied alone. Buffelgrass yield was increased as much as threefold with the burn-herbicide concentration.

Johnston and Peake (1960) investigated the effect of selective grazing by sheep and crested wheatgrass competition on leafy spurge control. While excellent crested wheatgrass stands were established, the competition did not control leafy spurge. Twelve years after seeding, ewes were placed on the pasture from May to September annually for 5 years. At the conclusion of the study, basal leaf area of the leafy spurge was reduced 98% while the basal areas of crested wheatgrass increased 20%.

Chaining is an inexpensive method to knock down brush after herbicide treatment to hasten decomposition or to facilitate burning (Bovey et al., 1984).

Burning may be desirable to remove understory woody plants partially killed by foliar herbicide application. Individual plant treatments, such as grubbing, cutting, or herbicide application, may be used for maintenance control of plants missed by mechanical or other control methods.

Treatment combinations are selected on the basis of availability, weed species, cost, and effectiveness. The integrated approach to weed and brush control is recognized as an important area of research and field use.

F. When to Manage Weeds and Brush

The density and type of weed infestation warranting control are important and require serious thought and calculation (Bovey et al., 1984). The decision to initiate weed and brush control depends upon rainfall, terrain, vegetative composition, the livestock production system, control techniques, and cost.

At some level, weeds and brush can become severe enough to limit forage production economics, and land managers must choose the most economical strategy of weed and brush management to continue livestock and wildlife production. Once efforts are made to control heavy infestations, continued efforts are necessary to control resprouts and scattered plants. Good management involves recognizing potential weed problems and controlling them promptly before they become expensive.

Weeds and brush left in strips maintain wildlife habitat and prevent wind and water erosion. Proper use and timing of herbicides, fire, or mechanical methods can control unwanted plants and enhance forage production. Improper use of chemicals or mechanical treatments such as bulldozing or root plowing also can destroy desirable vegetation. Where poor stands of low quality forages occur, revegetation may be warranted. Land managers must select weed control treatments that adequately return their investment.

G. Maintenance Control

Once an effective weed and brush control method has been used, a maintenance program is required to prevent reinfestation (Bovey et al., 1984; Klingman et al., 1983). This commonly involves mowing or applying 2,4-D or other herbicides during spring or summer to control herbaceous annual weeds. Woody plants and perennial herbaceous weeds sometimes can be managed by mowing or using herbicides, but often, because of their persistent nature and rapid recovery, individual plant treatment may be necessary. Depending on the species, sprays or pellets of the appropriate herbicides are used. Hand or power grubbers also are used as well as suppression by fire and animals.

H. Forage Production and Revegetation

Some depleted ranges and pastures can be restored by improved management of existing natural vegetation. Improved management, particularly better control of grazing, may be necessary to restore forage production. When undesirable vegetation limits forage production, removing weeds restores grazing potential. Where residual desirable native vegetation is lacking, revegetation may be necessary.

IV. FUTURE WEED AND BRUSH MANAGEMENT

The need to manage and manipulate weeds and brush on rangelands will not diminish. Range weeds will continue to dominate and will not be replaced with desirable vegetation without human interference. One only has to evaluate the immense hectares infested with honey mesquite, junipers, chaparral, prickly-pear cactus, halogeton, downy brome, medusahead, leafy spurge, knapweeds, yellow starthistle, and many other plants to recognize their persistent nature, spread, and dominance. Future research may elucidate reasons for range weed aggressiveness and will be invaluable in management decisions, but may not lead to their containment. In other situations, weeds and brush may have some value in wildlife habitat and commercial ventures. However, weeds and brush with no redeeming qualities are harmful to the environment and must be contained. One control method may be useful in one situation, but may have to be part of an integrated approach in other situations to be effective, economical, and environmentally sound.

Insects and plant pathogens are gaining attention for weed control on U.S. rangelands because of vast areas infested. The USDA-ARS and several state experiment stations have programs and foreign cooperators in search of organisms that mainly attack introduced weeds.

Allelopathy may be defined as biochemical interaction between plants (Anderson, 1983). Plants may compete with one another by releasing growth inhibiting chemicals into the environment from roots or leachate from dead, decaying vegetation, or plants may be affected by secondary compounds from microorganisms. Rice (1984) showed evidence of allelopathy in the natural environment and numerous chemical groups have been isolated from terrestrial and aquatic plants (Putnam, 1984). Examples of allelopathy for use in weed control are limited because presently used synthetic herbicides are more effective and less costly.

Weed and brush management may be enhanced by intensive short duration grazing systems in which deer and cattle use woody plants more efficiently. Leaving brush in strips or blocks for wildlife cover is advocated where certain stocking ratios of livestock and deer are desired. Forthcoming research will be concentrated on developing the best stocking ratios of livestock and wildlife species. Exotic game animals also may have a place in weed and brush management systems. Goats will continue to be important in brush management.

The search will continue for new herbicides and formulations that are safe, effective, and economical. Development has been slowed by high production costs and application and legal restrictions. New herbicides such as the sulfonylureas show promise for weed and brush control on grazing lands. These herbicides are desirable since lower amounts in the environment reduce the risk of residues in forages, livestock, wildlife, water sources, or soil.

Methods and techniques to enhance efficacy, safety, and economics of existing herbicides may result from new formulations, herbicides, and mixtures and improved placement. New herbicides are needed for establishment in grasses and forbs on grazing lands.

Improved delivery systems are required to apply new herbicides at extremely low rates per hectare. Aircraft can now apply small amounts of pelleted herbicides (Bouse et al., 1982). Low-volume application from ground sprayers with controlled droplet atomizers and electrostatic sprayers is now feasible (Bode and Butler, 1983). Electronic sensors, monitors, and control systems assure more precise and less costly herbicide application. Rope-wick applicators and wiping devices assure better placement and reduce herbicide used. These and future developments will cut costs and will improve herbicide efficacy and safety.

Modifications to chains make them more effective for brush control. Such chains include the Ely chain for big sagebrush control, swivels to allow chains to roll, ball and chain arrangements for steep terrain, and the Dixie-Sager chain for better uprooting of brush and scarification of soil.

Wiedemann and Cross (1984) developed a new disk chain with a triangular pulling configuration that reduced draft requirements by 36% and increased the operating width by 23% when compared to the two-tractor diagonal pulling method. The disk chain, pulled by a single tractor, appears to be a cost-effective method to prepare seedbeds on rough, log-littered, and root-plowed rangeland for grass establishment and control of small brush.

The low-energy grubber is one of the most promising mechanical concepts for effective economical control of sparse to moderate stands of brush (Wiedemann, 1982).

Fire (prescribed burning) will continue to be important, especially in combination with other practices. Proper burning is effective for weed control and rejuvenation of forage species, however, skill and knowledge of planning a burn and fire behavior is essential. Future research will emphasize combinations of treatments for economical and effective weed control, as discussed in the Combination of Control Methods section.

V. SUMMARY

Weeds and brush infest most grazing land in the United States and are among the most serious barriers to economic livestock production. Weeds and brush reduce the quantity and quality of desirable forages, harbor insect vectors and predators, make handling of livestock difficult, reduce land values, influence recreation opportunities, and increase production costs. Many unpalatable, mechanically injurious, and poisonous plants reduce grazing.

Several herbicides are available for managing weeds and brush. Herbicides are applied to foliage (sprays) or roots (sprays or pellets), depending on the species and herbicide. Sprays can be applied to weeds and brush with hand-carried, ground, or aircraft equipment. Selection of application equipment depends on weed and brush density, species, and area. Rough terrain and tall vegetation may limit the use of ground or hand-carried equipment. Where stands are scattered, individual plants can be treated with foliar sprays; basal stem, cut surface, and injection treatment; or soil application. Special wiping devices also are used to treat individual plants in small or scattered stands. The low energy grubber is especially effective in removing small, scattered trees.

Mechanical control methods for woody plants include hand-held saw, axes, girdling, and grubbing equipment. Hand methods are costly, laborious, and useful only for small areas. Practices requiring large equipment include bulldozing, chaining, railing, chopping, mowing, grubbing, root plowing, and disking. Selection of method is determined by available equipment, species and density, type of land, and terrain.

Biological control is possible with selective grazing by cattle, sheep, goats, poultry, and certain wildlife species. Insects and plant pathogens have potential to control certain species.

Fire has been a natural event on many rangelands in the past and can be used alone or with other methods on some grassland situations for weed and brush control and maintenance of desirable species, provided there is adequate fuel for burning.

Deciding when to control weeds or brush sometimes is difficult because many factors must be considered in addition to cost. Recognizing a potentially important weed and controlling it before it becomes a serious pest is the best approach.

Integrated weed and brush management systems will become more important in the future. More effective and safer new herbicides, herbicide formulations, and herbicide mixtures will be developed. Improvements in mechanical equipment will continue. Biological control methods, including the use of grazing animals, insects, pathogens, allelopathy, and other agents, will gain greater attention. Fire will continue to be an important management tool in grazing lands and forests, especially when combined with other practices. Future weed and brush control methods will be more effective, less damaging to the environment, and more cost effective.

Appendix A List of Plants Mentioned in Chapter

Common name	Scientific nomenclature	BAYER code
Acacia, blackbrush	*Acacia rigidula* Benth.	ACARI
Aspen, quaking	*Populus tremuloides* Michx.	POPTM
Baccharis	*Baccharis* spp.	
Beautyberry, American	*Callicarpa americana* L.	
Bitterbrush, antelope	*Purshia tridentata* (Pursh.) DC.	
Brome, downy	*Bromus tectorum* L.	BROTE
Broomsedge	*Andropogon virginicus* L.	ANOVI
Broomweed, common	*Gutierrezia drancunculoides* (DC.) Blake	GUEDR
Burroweed	*Haplopappus tenuisectus* (Greene) Blake ex Bensen	HAPTE
Creosotebush	*Larrea tridentata* (Sesse & Moc. ex DC.) Coville	LARTR
Croton, wooly	*Croton capitatus* Michx.	CVNCP
Dogwood	*Cornus* spp.	
Dryer's wood	*Isatix tinctoria* L.	ISATI
Gorse	*Ulex europaeus* L.	ULEEU
Grama, black	*Bouteloua eriopoda* (Torr.) Torr.	
Greasewood	*Sarcobatus* vermiculatus (Hook.) Torr.	SAYVE
Guajillo	*Acacia berlandieri* Benth.	ACABE
Guayacan	*Porlieria angustifolia* (Engelm.) Gray	POKAN
Hawthorn	*Crataegus* spp.	
Horsebrush, spineless	*Tetradymia canescens* DC.	TEYCA
Hiusache	*Acacia farnesiana* (L.) Willd.	ACAFA
Jointvetch, northern	*Aeschynomene virginica* (L.) B.S.P.	AESVI
Juniper, Ashe	*Juniperus ashei* Buchh.	IUPAS
Juniper, Utah	*Juniperus ostersperma* (Torr.) Little	
Lantana	*Lantana camara* L.	LANCA
Mesquite, honey	*Prosopis glandulosa* Torr.	PRCJG
Oak, blackjack	*Quercus marilandica* Muenchh.	QUEMR
Oak, Gambel	*Quercus gambelii* Nutt.	QUEGA
Oak, live	Quercus virginiana Mill.	QUEVI
Oak, post	*Quercus stellata* Wangenh.	QUESL
Oak, scrub live	*Quercus germinata* Small	QUEVM
Oak, souther red	*Quercus falcata* Michx.	QUEFC
Palmetto, saw	*Serenoa repens* (Bartr.) Small	SERRE
Pine, loblolly	*Pinus taeda* L.	
Pine, longleaf	*Pinus palustris* Mill.	
Pine, shortleaf	*Pinus echinata* Mill.	
Pine, slash	*Pinus elliottii* Engelm.	
Pine, Virginia	*Pinus virginiana* Mill.	
Pricklypear, cholla	*Opuntia* spp.	
Puncturevine	*Tribulus terrestris* L.	TRBTE
Rabbitbrush, green	*Chrysothamnus viscidiflorus* (Hook.) Nutt.	CYTVI

Appendix A Continued

Common name	Scientific nomenclature	BAYER code
Ragweed, western	*Ambrosia psilostachya* DC.	AMBPS
Ragwort, tansy	*Senecio jacobaea* L.	SENJA
Redcedar, eastern	*Juniperus virginiana* L.	IUPVI
Retama	*Parkinsonia aculeata* L.	PAKAC
Rose, Macartney	*Rosa bracteata* J.C. Wendl.	ROSBC
Sacaton, alkali	*Sporobolus airoides* (Torr.) Torr.	
Sacaton, big	*Sporobolus wrightii* Scribn.	
Sagebrush, big	*Artemisia tridentata* Nutt.	ARTTR
St. Johnswort	*Hypericum perforatum* L.	HYPPE
Saltcedar	*Tamarix* spp.	
Serviceberry	*Amelanchier* spp.	
Skeletonweed, rush	*Chrondrilla junea* L.	CHOJU
Snakeweed, broom	*Gutierrezia sarothrae* (Pursh) Brutt. & Rusby	GUESA
Sneezeweed, bitter	*Helenium amarum* (Raf.) H. Rock	HENAM
Snowberry	*Symphoricarpos oreophilus* (Gray)	
Starthistle, yellow	*Centaurea solstitialis* L.	CENSO
Stranglervine	*Morrenia odorata* (H. & A.) Lindl.	MONOD
Sumac	*Rhus* spp.	
Sweetgum	*Liquidambar styraciflua* L.	LIQST
Tarbush, American	*Flourensia cernua* DC.	
Tasajillo	*Opuntia leptocaulis* DC.	OPULE
Thistle, musk	*Carduus nutans* L.	CRUNU
Thistle, Scotch	*Onopordum acanthium* L.	ONRAC
Toadflax, Dalmation	*Linaria genistifolia* spp. *dalmatica* (L.) Maire & Petitmengin	LINDA
Toadflax, yellow	*Linaria vulgaris* Mill.	LINVU
Tobosagrass	*Hilaria mutica* (Buckl.) Benth.	HILMU
Wheatgrass, crested	*Agropyron cristatum* L.	
Whitebrush, Texas	*Aloysia gratissima* (Gilles & Hook.) Troncoso	ALYLY
Wolfberry, Berlandier	*Lycium berlandieri* Dunal	LYUBE
Yaupon	*Ilex vomitoria* Ait.	ILEVO

Appendix B List of Herbicides Mentioned in Chapter

Common name	Chemical nomenclature
Amitrole	1*H*-1,2,4-Triazol-3-amine
Atrazine	6-Chloro-*N*-ethyl-*N*′-(1-methylethyl)-1,3-5-triazine-2,4-diamine
Bromacil	5-Bromo-6-methyl-3-(1-methylpropyl)-2,4(1*H*,3*H*)pyrimidine-dione

continued

Appendix B Continued

Common name	Chemical nomenclature
Chlorosulfuron	2-Chloro-*N*-[[(4-methyoxy-6-methyl-1,3,5-triazin-2-yl)amino]-carbonyl]benzenesulfonamide
Clopyralid	3,6-Dichloro-2-pyridinecarboxylic acid
Dicamba	3,6-Dichloro-2-methoxybenzoic acid
2,4-D	(2,4-Dichlorophenoxy)acetic acid
2,4-DB	4-(2,4-Dichlorophenoxy)butanoic acid
Dichlorprop	(±)2(2,4-Dichlorophenoxy)propanoic acid
Diquat	6,7-Dihydrodipryrido[1,2-α:2′,1′-*c*]pyrazinediium ion
Fosamine	Ethyl hydrogen (aminocarbonyl)phosphonate
Glyphosphate	*N*-(Phosphonomethyl)glycine
Hexazinone	3-Cyclohexyl-6-(dimethyl-amino)-1-methyl-1,3,5-triazine-2,4-(1*H*,3*H*)-dione
Imazapyr	(±)-2-[4,5-Dihydro-4-methyl-4-(1-methylethyl)-5-oxo-1*H*-imidazol-1-yl]-3-pyridinecarboxylic acid
MCPA	(4-Chloro-2-methylphenoxy)acetic acid
Metsulfuron	2-[[[[(4-Methoxy-6-methyl-1,3,5-triazin-2-yl)amino]carbonyl]-amino]sulfonyl]benzoic acid
Paraquat	1,1′-Dimethyl-4-4′-bipyridinium ion
Picloram	4-Amino-3,5,6-trichloro-2-pyridinecarboxylic acid
Sulfometuron	2-[[[[(4,6-Dimethyl-2-pyrimidinyl)amino]carbonyl]amino]sulfonyl]benzoic acid
2,4,5-T	(2,4,5-Trichlorophenoxy)acetic acid
Tebuthiuron	*N*-[5-(1,1-Dimethylethyl)-1,3,4-thiadiazol-2-yl]-*N*-*N*′-Dimethyl-urea
Triclopyr	[(3,5,6-Trichloro-2-pyridinyl)oxy]acetic acid

REFERENCES

Anderson, W. P. (1983). *Weed Science Principles*, 2nd Edition, West Publishing Co., New York, p. 34.

Barney, M. A. and Frischknecht, N. C. (1974). Vegetation changes following fire in the pinyon-juniper type of west central Utah, *J. Range Manage. 27*: 91.

Bentley, J. R. (1967). *Conversion of Chaparral Areas to Grassland: Techniques Used in California*, USDA, Forest Service Agricultural Handbook No. 328, pp. 1–35.

Bode, L. E. and Butler, B. J. (1983). "New techniques and equipment for ground application of herbicides in the U.S.A." Proceedings of the 10th International Congress on Plant Protection, Brighton, England, pp. 478–485.

Bouse, L. F., Carlton, J. B., and Brusse, J. C. (1982). Dry materials metering system for aircraft, *Trans. ASAE 28*: 316.

Bovey, R. W., Wiese, A. F., Evans, R. A., Morton, H. L., and Alley, H. P. (1984). Control of weeds and woody plants on rangelands, USDA and University of Minnesota, AB-BU-2344.

Box, T. W. and White, R. S. (1969). Fall and winter burning of south Texas brush ranges, *J. Range Manage.* *22*: 373.

Bragg, T. B. and Hulbert, L. C. (1976). Woody plant invasion of unburned Kansas bluestem prairie, *J. Range Manage.* *29*: 19.

Britton, C. M. and Wright, H. A. (1983). "Brush management with fire" (K. C. McDaniel. ed.), *Proceedings of the Brush Management Symposium,* Society of Range Management, Albuquerque, New Mexico, pp. 61–68.

Cox, J. R. (1988). Seasonal burning and moving impacts on *Sporobolus wrightii* grasslands, *J. Range Manage.* 41: 12.

Davis, G. G., Bartel, T. E., and Cook, C. W. (1974). Control of Gambel oak sprouts by goats, *J. Range Manage.28*: 216.

DeLoach, J. C. (1990). Terrestrial weeds, classical biological control in the Southern United States (D.H. Habeck, F. D. Bennet, and J. H. Frank, eds.), *South Coop Series Bull.* No. 355, IFAS Editorial, University of Florida, Gainesville, p. 157.

DeLoach, J. C. (1991). Past success and current prospects in biological control of weeds in the United States and Canada, *Natural Areas J. 11*: 129

Elwell, H. M., McMurphy, W. E., and Santelmann, P. W. (1970). Burning and 2,4,5-T on post- and blackjack oak rangeland in Oklahoma, *Oklahoma Agric. Exp. Stn. Bull.* B-675: 1–11.

Elwell, H. M. and Santelmann, P. W., Stritzke, J. F., and Greer, H. (1974). Brush control research in Oklahoma, *Oklahoma Agric. Exp. Stn. Bull* B-712: 1–42.

Engle, D. M. and Stritzke, J. F. (1991). Igniting crowns of partially scorched juniper: 1983-1991 Range Research Highlights, Oklahoma, *Coop Ext. Serv., Cir.* E-905: 15–16.

Fay, P. K. (1991). Controlling leafy spurge with grazing animals, in *Noxious Range Weeds* (L. F. James, J. O. Evans, M. H. Ralphs, and R. D. Child, eds.), Westview Press, Boulder, Colorado, p. 193.

Fisher, C. E., Wiedemann, H. T., Walter, J. P., Meadors, C. H., Brock, J. H., and Cross, B. T. (1972). Brush control research on rangeland, *Tex. Agric. Exp. Stn.* MP-1043: 1–18.

Goeden, R. D., Fleschner, C. A., and Richer, D. W. (1968). Insects controlling pricklypear cactus, *California Agric.* *22*: 8.

Hanselka, C. W. (1991). "Brush management decisions: ten factors to consider," Proc. Brush Manage. Symp, Giddings, Texas, Tex. Agric. Ext. Serv., College Station, Texas, pp. 16–22.

Harley, K. L. S. (1973). "Biological control of *Lantana* in Australia," Preceedings of the 3rd International Symposium on the Biological Control of Weeds, Montpelier, Canada, pp. 23–29.

Harradine, A. R. and Jones, A. Z. (1985). Control of gorse regrowth by Angora goats in the Tasmanian Midlands, *Aust. J. Exp. Agric., 25*: 550.

Harris, P. (1971). Current approaches to biological control of weeds, *Tech. Commun.*, Commonwealth Institute of Biological Control No. 4, pp. 67–76.

Hendon, E. P. (1971). Kershaw brush cutter, in *Noxious Brush and Weed Control Research Highlights - 1971, Tex. Tech. Univ. Spec. Rep. 51*: 28.

Holloway, J. K. (1964). Projects in biological control of weeds, in *Biological Control of Insects, Pests and Weeds* (P. DeBach, ed.), Reinhold, New York, p. 650.

Huffaker, C. B. (1964). Fundamentals of biological weed control, in *Biological Control of Insects, Pests and Weeds* (P. DeBack, ed.), Reinhold, New York, pp. 631-649.

Johnston, A. and Peake, R. W. (1960). Effect of selective grazing by sheep on the control of leafy spurge (*Euphorbia esula* L.), *J. Range Manage. 13*: 192.

Julien, M. H. (1987). Biological control of weeds: a world catalogue of agents and their target weeds, 2nd Edition, Commonwealth Agric. Bureau Farnham Royal, Slough, UK, p. 144.

Klingman, D. L, Bovey, R. W., Knake, E. L., Lange, A. H., Meade, J. A., Skrock, W. A., Stewart, R. E., and Wyse, D. L. (1983). Systemic herbicides for weed control-phenoxy herbicides, dicamba, picloram, amitrole and glyphosate, USDA and University of Minnesota, AD-BU-2881.

Kozlowski, T. T. and Ahlgern, C. E., eds., (1974). Fire and Ecosystems, Academic Press, New York.

Landers, R. Q., Jr. (1991). Planning a prescribed burn, *Tex. Agric. Ext. Serv.* L-2461: 1-5.

Landgraf, B. K., Fay, P. K., and Havstad, K. M. (1984). Utilization of leafy spurge (*Euphorbia esula*) by sheep, *Weed Sci.* 32: 348.

Mayeux, H. S., Jr. and Crane, R. A. (1983). The brush roller - an experimental herbicide applicator with potential for range and brush control, *Rangelands* 5: 53.

Mayeux, H. S., Jr. and Hamilton, W. T. (1983). Response of Common goldenweed (*Iscoma coronopifolia*) and buffelgrass (*Cenchrus ciliaris*) to fire and soil-applied herbicides, *Weed Sci.* 31: 355.

Mayeux, H.S., Johnson, H. B., and Polley, H. W. (1991). Global Change and vegetation dynamics, in *Noxious Range Weeds* (L. F. James, J. O. Evans, M. H. Ralphs and R. D. Childs, eds.), Westview Press, Boulder, Colorado, pp. 62-74.

McGinty, A. (1991). "Rangeland weed and brush management technologies past and present," Proc. Brush Manage. Symp., Giddings, Texas, Agric. Ext. Serv., College Station, Texas, pp. 31-38.

McKenzie, D., Jensen, F. R., Johnsen, T. N., Jr., and Young, J. A. (1984). Chains for mechanical brush control, *Rangelands*, *6*: 122.

Meyer, R. E. and Bovey, R. W. (1980). Control of live oak (*Quercus virginiana*) and understory vegetation with soil-applied herbicides, *Weed Sci. 28*: 51.

National Academy of Science Publication. (1968). Brush and weed control on range and pastures, in *Principles of Plant Animal Pest Control*, Vol. 2, *Weed Control*, Washington, D.C.

Pechanec, J. F., Plummer, A. P., Robertson, J. H., and Hull, A. C. (1965). Sagebrush control on rangelands, USDA Handbook No. 277, pp. 1-40.

Putnam, A. R. (1988). Allelochemicals from plants as herbicides, *Weed Tech.* 2: 510-518.

Quimby, P. C., Jr., Bruckart, W. L., DeLoach, C. J., Knutson, L., and Ralphs, M. H. (1991). Biological control of rangeland weeds, in *Noxious Range Weeds* (L. F. James, J. O. Evans, M. H. Ralphs, R. D. Child, eds.), Westview Press, Boulder, Colorado, p. 84.

Radcliffe, J. E. (1986). Gorse - a resource for goats? *New Zealand J. Exp. Agric. 14*: 399.

Recenthin, C. A., Bell, H. M., and Pederson, R. J. (1964). Grassland restoration. Part II. Brush control, USDA, Soil Conservation Service, Temple, Texas, pp. 1–39.

Rees, N. E. (1991). Biological control of thistles, in *Noxious Range Weeds*, Westview Press, Boulder, Colorado, p. 264.

Rees, N. E. and Spencer, N. R. (1991). Biological control of leafy spurge, in *Noxious Range Weeds* (L. F. James, J. O. Evans, M. H. Ralphs, and R. D. Child, eds.), Westview Press, Boulder, Colorado, p. 182.

Rice, E. L. (1984). *Allelopathy* 2nd Edition Academic Press, New York, pp. 1–422.

Rosenthal, S. S., Campobasso, G., Fornasari, L., Sobhian, R., and Turner, C. E. (1991). Biological control of *Centaurea* spp., in *Noxious Range Weeds* (L. F. James, J. O., M. H. Ralphs, and R. D. Child, eds.), Westview Press, Boulder, Colorado, p. 292.

Rosenthal, S. S., Maddox, D. M., and Brunetti, K. (1984). Biological methods of weed control, Monograph No. 1, California Weed Conference, Thomson Publishers, Fresno, California pp. 1–88.

Schroeder, D. (1983). Biological control of weeds, in *Recent Advances in Weed Research* (W. W. Fletcher, ed.), Common Wealth Agric, Bureaux, Slough, England, pp. 41–78.

Scifres, C. J. (1975). Systems for improving Macartney rose infested coastal prairie rangeland, *Tex. Agric. Exp. Stn.* MP-1225: 1–12.

Scifres, C. J. (1980). Mechanical methods of brush management, *Brush Management Principles and Practices for Texas and the Southeast*, Texas A&M University Press, College Station, Texas, pp. 263–275.

Scifres, C. J., Mutz, J. L., and Durham, G. P. (1976). Range improvements following chaining of south Texas mixed brush, *J. Range Manage. 29*: 418.

Scifres, C. J., Mutz, J. L., Rasmussen, B. A., and Smith, R. P. (1983), Integrated brush management systems (IBMS): concepts and potential technologies for running mesquite and whitebrush, *Tex. Agric. Exp. Stn.*, B-145: 1–13.

Semple, A. T. (1970). Control of undesirable vegetation, other pests and diseases: eradication and control of weeds and brush, in *Grassland Improvement*, Chemical Rubber Co. Press, Cleveland, Ohio, p. 213.

Smith, M. A., Dodd, J. L., and Rodgers, J. D. (1985). Prescribed burning on Wyoming rangeland, *Agric. Ext. Serv., Univ. Wyoming* B-810: 2–10.

Tainter, F. H. (1972). Oak-wilt, a selective silvicide in Arkansas? *Arkansas Farm Res.*, University of Arkansas, Fayettville, pp. 8–9.

Tanner, G. W., Kalmbacher, R. S., and Prevatt, J. W. (1986). Saw-palmetto control in Florida, *Agric. Ext. Serv., Univ. Florida*, Cir. 668, pp. 4–5.

Templeton, G. E. (1982). State of weed control with plant pathogens, in *Biological Control of Weeds with Plant Pathogens* (R. Charudattan and H. L. Walker, eds.), John Wiley & Sons, New York, p. 38.

Thomas, G. W. and Ronninger, T. S. (1965). Rangelands - our billion acre resources, *Agric. Sci. Rev. 3*: 11.

Ueckert, D. N., Petersen, J. L., Potter, R. L., Whipple, J. D., and Wagner, M. W. (1988). Managing pricklypear with herbicides and fire, *Sheep & Goat, Wool & Mohair, 1988, Texas Agric. Exp. Stn.* Prop. Rept. -4570, pp. 10–15.

Vallentine, J. F. (1989). Range Development and Improvements, 3rd Edition, Brigham Young University Press, Provo, Utah, pp. 11, 168.

Van Arsdale, E. P. and Halliwell, R. W (1970). Progress in research in live oak decline, *Plant Dis. Rep.*, *54*: 669.

Wambolt, C. L. and Payne, G. F. (1986). An 18-year comparison of control methods for Wyoming big sagebrush in Southwestern Montana, *J. Range Manage.39*: 314.

Warnes, D. D. and Newell, L. C. (1969). Establishment and yield responses of warm-season grass strains to fertilization, *J. Range Manage.* *22*: 235.

Warren, L. E., Ueckert, D. N., Shelton, M., and Chamrad, A. D.(1984). Spanish goat diets on mixed-brush rangeland in the South Texas Plains, *J. Range Manage.* *37*: 340.

Whisenant, S. G. (1989). "Changing fire frequencies on Idaho's Snake River Plains: ecological and managment implications," Proceedings - Symposium on Cheatgrass Invasion, Shrub Die-Off, and Other Aspects of Shrub Biology and Management, pp. 4–10.

White, L. D. and Hanselka, C. W. (1991). Prescribed range burning in Texas, *Texas Agric. Ext. Serv.* B-1310: 1–8.

Wiedemann, H. T. (1982). "New developments in mechanical brush control," Proceedings of the International Rancher Roundup, Del Rio, Texas, pp. 181–187.

Wiedemann, H. T. and Cross, B. T. (1981). Low-energy grubbing for control of junipers, *J. Range Manage.* *34*: 235.

Wiedemann, H. T. and Cross, B. T. (1982). Performance of front-mounted grubbers on rubber-tired equipment, in *Brush Management and Range Improvement Research, 1980-1981*, CRP 3968-4014, Tex. Agric. Exp. Stn., PR-3982, College Station, Texas, pp. 49–53.

Wiedemann, H. T. and Cross, B. T. (1984). Influence of pulling configuration on draft of disk-chains, *Trans. ASAE* *28*: 79.

Wikeem, B. M. and Strang, R. M. (1983). Prescribed burning on B.C. rangelands: the state of art, *J. Range Manage.*, *36*: 3.

Williams, R. E., Allred, B. W., Denio, R. M., and Paulsen, H. A., Jr. (1968). Conservation, development and use of the world's rangelands, *J. Range Manage.* *21*: 355.

Wilson, C. L. (1969). Use of plant pathogens in weed control, *Annu. Rev. Phytopathol.* *7*: 411.

Wink, R. L. and Wright, H. A. (1973). Effects of fire on an Ashe juniper community, *J. Range Manage.* *23*: 376.

Wood, G. M. (1987). Animals for biological brush control, *Agron. J.*, *79*: 319.

Wright, H. A. and Bailey, A. W. (1982). *Fire Ecology: United States and Southern Canada*, John Wiley & Sons, New York.

Wright, H. A., Bunting, S. C., and Neuenschwander, L. F. (1976). Effect of fire on honey mesquite, *J. Range Manage.* *29*: 467.

14

Weed Management Systems for Horticultural Crops

Reid J. Smeda
U.S. Department of Agriculture, Stoneville, Mississippi

Leslie A. Weston
University of Kentucky, Lexington, Kentucky

I. INTRODUCTION

The development and implementation of successful weed management systems for horticultural crops depend upon the consideration of many factors relevant to both crops and weeds. Some of these factors include growth cycle (annual, biennial, perennial), growth habit (prostrate, vining, tree), habitat (well-drained soils [e.g., strawberries]; acidic soils [e.g., blueberries]; high organic matter soils [e.g., onions, celery]), and climate (cool-season [e.g., broccoli], warm-season [e.g., tomatoes], long growing season [e.g., citrus]). See Chapter 2 for a detailed discussion. Because all of these factors together define a unique production system for any one crop, and because weed management is only one aspect of overall crop production, weed management practices should be designed to minimize interference with other aspects of crop production. For example, broadcast herbicide applications to some fruit crops during flowering may disrupt insect pollination and reduce fruit set and yields.

Identification and distribution of problem weeds in a cropping system are among the first priorities in determining which weed control program should be used. A high incidence of established perennial weeds necessitates immediate control measures to prevent losses in crop production, especially for perennial crops. In this situation, chemical control measures may be more effective than cultivation practices, since many perennial weeds can be spread by cultivation. Although herbicides can control many weed species, crop safety and the spectrum of weeds controlled vary among herbicides. In addition, some herbicides are registered for application prior to weed emergence, while others are used on emerged weeds. Therefore, proper identification of problem weeds is important in determining the optimal, cost-effective control measure.

Before addressing weed management practices for various horticultural crops, one should consider the problems associated with weed control which are unique to the horticultural industry. Collectively, there is a large potential for herbicide development in horticultural crops, but the acreage of any one crop is small compared to that of many agronomic crops. In addition, horticultural crops are often high in value and many companies remain concerned about liability issues. Therefore, commercial development of many herbicides for use in horticultural commodities is of limited interest to chemical companies and is reflected in the relatively small number of registered compounds. With the ongoing process of reregistration, availability of registered herbicides may be further diminished.

Soil persistence of selected herbicides can also be a problem for horticultural producers. Many vegetable crops are rotated with other crops, often in the same growing season, necessitating herbicides with minimal residual carryover. Application of labeled rates of imazethapyr or imazaquin on soybeans can severely injure some successive vegetable crops such as celery, cole crops, and tomatoes, for up to 2 years (Wiley & Sons, 1993). For perennial crops such as tree fruits, grapes, and cane fruits, long-term (3–30+ years) weed management practices are required. In this situation, many growers have repeatedly applied the same herbicide, thus allowing tolerant weed species to proliferate by shifting the weed spectrum. For many other crops, the lack of diverse herbicide choices creates a dependence upon specific compounds, often shifting the weed spectrum to species taxonomically related to the crop. One example is the difficulty encountered in controlling nightshade (*Solanum* species) in tomatoes, because both are in the Solanaceae family and respond similarly to many herbicides (Gorski and Wertz, 1987). Repeated application of herbicides with similar modes of action (e.g., photosynthetic inhibitors) can also select for herbicide resistant species. In fact, the first reported observation of herbicide resistance occurred in an apple orchard receiving annual applications of triazine herbicides over many years (Ryan, 1970). Paraquat resistant weeds have also been reported in grape vineyards in Japan (Kato and Okuda, 1983).

Another problem associated with weed management in horticultural crops is the inability of the crop to compete with surrounding weeds. Although crop tolerance to weed competition varies, some growers are obligated to maintain near-weed-free conditions for other reasons. These growers typically operate U-pick farms, which depend upon consumers to harvest the crops, or in other instances farms which depend upon hired labor. For aesthetic reasons as well as the desire to satisfy consumers and retain hired workers, growers often strive for weed-free crops. Aside from harvesting, proper weed management is also required to ensure uniform application and optimal activity of pesticides which protect crops from insects and diseases.

The decision of when to implement weed management practices is also influenced by specific crop and weed problems. For strawberries in the Midwest or Northeast, winter annual and biennial weeds can often interfere with fruit set and harvest, while summer annuals can reduce runner production and setting, reducing yields the following year. In established tree fruits, some vegetation around the base of trees will not detrimentally affect tree growth throughout the growing season. However, over the winter months, this vegetation may harbor rodents which can feed on the tree roots or even girdle and kill young trees.

The purpose of this chapter is to summarize the weed management practices recommended for horticultural crops. This will be done in different sections: 1) vegetables (addressed collectively by weed management technique); 2) nursery and landscape plantings (ornamentals); 3) small fruits (strawberries, cranberries, grapes, blueberries, and cane fruits); 4) tree fruits, and 5) nut crops. While many weed management systems emphasize the use of herbicides, nonchemical or alternatives practices will be described. Because the rates of herbicides which are labeled for use vary from crop to crop and even from one geographical area to another, specific mention of usage rates will be avoided. In considering the optimum weed management system for a crop, it is important to emphasize to growers that weed control techniques should be integrated to provide long-term effective management.

II. VEGETABLES

Identification and discussion of specific weed management techniques for vegetables are very difficult considering the botanical diversity of these crops. By one account, there are over 100 different kinds of vegetables, which are classified into 17 families. The grouping of vegetables based on weed management strategies can use numerous classifications, but these classifications are often too general to describe the weed management techniques widely practiced among growers for any crop adequately. Classifications include life cycles, such as annuals, biennials, or perennials; climatic conditions for production, such as warm or cool season crops; or the portion of the plant consumed (e.g., root [carrot], stem [asparagus], petiole [rhubarb], leaf [lettuce], flower [broccoli], fruit [cucumber], seed [lima bean]) (Pierce, 1987).

As with other cropping systems, herbicides are one of the main tools of choice for weed management in vegetables. However, regulatory constraints imposed at the federal (EPA), state, and local levels as well as public concern with pesticide residues in vegetables (e.g., aldicarb in watermelons) have resulted in the loss of herbicide labels for many vegetables. Also, the cost of conducting research to reregister a herbicide is often not warranted because of the limited vegetable acreage. In fact, some herb and vegetable crops (e.g., leeks and parsley) are not currently covered by any herbicide label (Majek, 1994). Vegetable growers must now consider alternative weed control strategies to maintain productivity and crop quality. Essentially, weed management strategies should consider the cost of the potential strategy and the benefits obtained in increased crop production and/or quality. In the following sections, cultural, biological, and chemical approaches to weed management in vegetables will be considered. Although individual crops will not be covered in separate sections, examples will be used where appropriate.

A. Cultivation

Before herbicides were developed, many growers utilized only manual or mechanical cultivation for weed control. Although growers with small acreages often depend primarily upon cultivation, cultivation alone is not practical today for growers with larger acreages. The practice of planting vegetables in rows facilitates the use of cultivation equipment. Proponents of cultivation state that it is economical, improves soil aeration, increases rainfall penetration, and may raise soil temperatures by minimizing the interference of sunlight with surface residues. However, cultivation also has negative consequences such as rendering soil more susceptible to erosion and causing damage to crop roots. Many vegetables have shallow feeder roots. Weed management via cultivation is best carried out by shallow cultivation early in the season, with later cultivation primarily restricted to areas between the rows. For curcurbit crops (cucumbers, cantaloupe, pumpkin, watermelon), cultivation is only possible before plants begin to vine extensively. Although the intent of cultivation is to destroy weeds, disturbance of the soil surface actually stimulates the germination of additional weeds. Cultivation is most effective on small seedlings of perennials and annuals. In vegetables such as potatoes, asparagus, and sweet corn, growers can effectively bury small weeds in the plant row with minimal negative effects on the larger crop.

Proper soil moisture is essential to the success of cultivation. If soils are too moist, tillage equipment can compact the soil and weeds may become reestablished. If the soil remains too wet to work, weeds can become too large to be controlled adequately by cultivation. In very dry conditions, cultivation may result in soil clodding and lack of weed control. Hand weeding was widely used in the past and is effective in removing weeds within the crop row. The time involved and increasing expense for labor make hand weeding cost prohibitive, but it is still used in production of high value crops. Migrant or local labor crews in fresh market tomatoes, cucumbers, or melons often rogue weeds by hand.

B. Mowing

Unlike many fruits and ornamental or nursery crops, vegetables are predominantly annuals or biennials. Therefore, mowing is not used for weed management in vegetables, except perhaps between rows of perennial crops such as artichoke, asparagus, or rhubarb (Foster, 1994). Mowing is also practiced in asparagus plantings in late fall to remove ferns and prevent interference with developing spears the following spring.

C. Mulches

Mulches are widely used in production practices of certain vegetables as well as nursery and ornamental crops. Mulches for vegetables can be organic (plant residues, paper) or inorganic (plastic). Commonly cited benefits of mulches include erosion control, weed suppression, moisture retention, and moderation of soil temperatures. Soil temperatures are generally increased 1° to 3°C for black and 3° to 5°C for clear plastic (Ashley, 1992). This rise in temperature provides early season growth stimulation of crops which respond to elevated soil temperatures (cucumbers, muskmelon, summer squash, watermelon, pepper, tomatoes, and sweet corn). Only opaque (brown or black) plastics provide any measure of weed control, primarily by restricting water and light penetration to the soil surface. Very few herbicides can be applied beneath clear plastic, although oxyfluorfen provides effective weed control for transplanted curcurbit crops (Bellinder et al., 1993). Plastic mulches should not be used where creeping perennials are present (e.g., nutsedge), since these weeds can puncture plastic, providing light to stimulate germination of additional weeds.

Although organic mulches are commonly used between rows of strawberries and by home gardeners, this strategy traditionally has not been exploited by growers with larger acreage. The use of organic mulch can be expensive and labor intensive. However, since the late 1970s, numerous reports indicate that residues of fall-seeded small grains provide a measurable level of weed control in successive vegetables (Phatak, 1992; Putnam, 1990). Cover crops are generally sown in the late summer or fall and killed by frost, mowing, or with herbicides the following spring, prior to planting a designated vegetable. Residues from the cover crop would suppress weeds by their physical presence on the soil surface, and possibly through the transient release of allelochemicals. For example, up to 90% weed suppression was noted 60 days after the termination of winter rye cover crop growth (Putnam et al., 1983). Since many vegetable crops mature in 60–90 days, this technique may offer potential for commercial producers. Throughout the United States, transplanted tomatoes have been successfully grown in rye and hairy vetch residues (Abdul-Baki and Teasdale, 1993; Masiunas et al., 1994). Many researchers have found that production of direct-seeded larger-seeded vegetables (pumpkins, snap beans, peas, cucumbers, etc.) in cover crop residues was successful, whereas lack of establishment or injury occurred in lettuce, radish, and tomato (Putnam and DeFrank, 1983). Therefore, the use of cover crops may be limited to transplant crops (tomatoes, peppers, melons, cucumbers, and squash) or direct seeding of larger-seeded vegetables.

In addition to planting vegetables into cover crops, intercropping or relay cropping of vegetables has also been investigated. This practice involves the simultaneous production of two or more crops on the same piece of land. Although this practice was common in the United States prior to 1940, synthetic chemicals and mechanization favored monocrop agriculture. Intercropping is common in various parts of the world (Horwith, 1985), especially lesser developed countries. Intercropping can be practiced by planting two crops into an area where at least one crop of interest will benefit (e.g., a legume to provide nitrogen for other crops), and yields will only be taken for the main crop. Nicolson and Wien (1983) found that certain turfgrasses and clovers provided either weed suppression or nitrogen for intercropped sweet corn or cabbage. Alternatively, a grower may plant two crops into an area with the intent of harvesting both crops. Putnam (1990) reported success with intercropping peas and sweet corn as well as broccoli and sweet corn. In general, weed suppression by an intercrop system surpasses that where the crops are grown in monoculture (Altieri et al., 1983). However, since each crop in an intercrop system is competing for available resources, intercropping may limit production of optimal yields as a result of competition for nutrients or perhaps water. Intercropping may also limit a grower's ability to apply necessary pesticides for control of insects and/or diseases. Where land availability is limited, growers might consider intercropping vegetables as a means of suppressing weeds.

D. Stale Seedbed

Utilization of a stale seedbed is common among growers of agronomic crops who desire to reduce tillage. However, this technique may also be beneficial for vegetable growers who have a limited selection of herbicides labeled for use. In essence, soil is prepared for crop seeding some length of time prior to planting. A majority of weeds will germinate within 2 weeks of soil disturbance under favorable conditions. Growers can allow weeds to germinate and then kill these weeds with a broadcast application of a nonselective herbicide (paraquat or glyphosate). The vegetable can then be seeded, preferably with minimal soil disturbance. Asparagus production commonly utilizes a stale-seedbed technique. Crown establishment occurs by seeding or planting young crowns in furrows. The slow germination of seeds (2 to 3 weeks) or time for shoots to emerge from crowns in early spring is ideal for application of burndown herbicides. For some perennial crops including asparagus, residual herbicides can be applied together with burndown herbicides. The limitation of this system is that cultivation is discouraged early in the season, since soil disturbance will trigger another flush of weeds. Growers who depend on cultivation may not want to risk increased weed problems by using this method, considering the relatively small savings in fuel and labor compared to the potential losses in yields for high-value crops. Specialized seeders for planting into reduced tillage systems have been designed for some vegetable crops; however, these systems are more successful for larger seeded vegetables.

E. Herbicides

As in many cropping systems, herbicides are primarily used for weed management in vegetables (Table 1). However, for reasons stated earlier, the number of compounds registered for use in vegetables is limited. Many vegetables are small-seeded, are slow-growing early in the season, or are weak competitors with weeds. The length of a critical weed-free period without a significant loss in crop yield depends upon the vegetable. For many vegetables, adequate weed control is needed beyond the time that competition alone would reduce yields, in order to optimize both the application of other pesticides (fungicides, insecticides) and crop harvest. The majority of herbicides labeled for use in vegetables are applied prior to weed emergence. For postemergence weed management, selective grass herbicides are widely used since most vegetables are dicots. Very few broadleaf herbicides are registered for use in vegetables, with the margin of safety between effective weed management and crop injury quite narrow. Some vegetable growers rotate land with agronomic crops such as soybeans and corn. This can be a very effective means of reducing populations of problem weeds that flourish in a vegetable monoculture. Many herbicides labeled for use in these agronomic crops, however, have rotational crop restrictions. Before using a particular herbicide, consult the product label to determine when planting can be conducted safely. This may also be important for growers who produce multiple vegetable crops in one season on the same land.

Because the outlook for new vegetable herbicides remains bleak, growers must develop new approaches to weed control with existing herbicides. One

Table 1 Herbicides Registered for Use on Vegetables[a]

Common name	Trade name[b]	Crop designation[e]
Alachlor	Lasso[d]	BE[c],CR
Ametryn	Evik	CR
Atrazine	Aatrex Nine-O[d]	CR
Benefin	Balan[d]	LT
Bensulide	Prefar[d]	CA,CO,CU,LT,MU,ON, PP,PU, SQ,TM,WA
Bentazon	Basagran[d]	BE,CR,PE
Bromoxynil	Buctril	CR,GA,ON
Butylate	Sutan	CR
Clomazone	Command	PE,PP,PU
Cyanazine	Bladex	CR
Cycloate	Ro-Neet	SP,TB
DCPA	Dacthal[d]	BE,CO,CU,EG,GA,GR,HO,MU, ON,PE,PP,PT,PU,RA,SQ,SW, TM,WA
Dicamba	Banvel[d]	AS,CR
Diglycolamine	Clarity	CR
Diquat	Diquat H/A	PT
Diuron	Karmex[d]	AR,AS
EPTC	Eptam[d]	BE,CR,PT,SW
Ethafluralin	Curbit[d]	BE,CU,MU,PE,PU,SQ,WA
Fluazifop-P	Fusilade DX	AS,CA,EN,GA,ON,ON,PP,RH, SW
Glyphosate	Roundup[d]	AS,BE,CA,CE,CO,CR,CU,EN, GA,GR,HO,LE,LT,MU,OK, ON,PE,PP,PT,PU,RA,SP,SQ, SW,TB,TM,WA
Imazethapyr	Pursuit	BE,PE
Linuron	Lorox[c]	AS,CA,CE,CR,PA,PT
MCPA	MCPA Amine[d]	PE
MCPB	Thistrol	PE
Metolachlor	Dual	BE,CH,CR,PE,PT
Metribuzin	Sencor[d]	AS,CA,PT,TM
Monocarbamide dihydrogensulfate	Enquik	CE,CO,CU,EG,EN,GA,GR,LT, MU,ON,PP,SP,TM,WA
Napropamide	Devrinol	AR,AS,CO,EG,PP,TM
Naptalam	Alanap	CU,MU,WA
Nicosulfuron	Accent	CR
Norflurazon	Zorial[d]	AS
Oxyfluorfen	Goal[d]	AR,CH,CO,GA,HO,ON

continued

Table 1 Continued

Common name	Trade name[b]	Crop designation[e]
Paraquat	Gramoxone Extra[d]	AS,BE,CA,CO,CP,CR,CU,EG, EN,GR,LT,ON,PA,PE,PP,PT, PU,RA,SP,SQ,TB,TM,WA
Pebulate	Tillam	TM
Pendimethalin	Prowl[d]	BE,CR,GA,ON,PT
Phenmedipham	Spin-Aid	SP,TB
Prometryn	Caparol[d]	CE
Pronamide	Kerb	EN,LT,PE,RH
Propachlor	Ramrod	CR,PE
Pyrazon	Pyramin	TB
Pyridate	Lentagran[d]	CO
Sethoxydim	Poast[d]	AR,AS,BE,CE,CO,CP,CR,CU, EG,GR,LE,LT,PE,PP,PT,PU, RH,SP,SQ,TM,WA
Simazine	Princep[d]	AR,AS,CR
Stoddard Solvent		CA,PA
2,4-D amine	Weedar 64[d]	AS,CR
Terbacil	Sinbar	AS
Triallate	Far-Go	LE,PU
Trifluralin	Treflan[d]	AS,BE,CA,CE,CO,CP,CU,GR, OK,PE,PP,PT,TM,WA

[a]Note that some crops within designated groups (e.g., cole crops) may not be covered under a herbicide listed for use on that group. Consult herbicide labels to determine geographical restrictions and specific recommendations for use of each herbicide.
[b]Trade names given are for U.S.A. labels.
[c]Some crop designations include more than one vegetable. The list of crops shown is not all-inclusive, but covers the major vegetables.
[d]More than one trade name and/or manufacturer is represented for this active ingredient.
[e]AR, artichoke; AS, asparagus, BE, beans (dry, great northern, kidney, lima, navy, pink, pinto, red, snap, white); CA, carrots; CE, celery; CH, chickpeas; CO, cole crop (broccoli, brussels sprouts, cabbage, cauliflower, Chinese cabbage, turnips); CP, cantaloupe; CR, corn (sweet corn, popcorn); CU, cucumbers; EG, eggplant; EN, endive; GA, garlic; GR, greens (collards, kale, mustard, turnip greens); HO, horseradish; LE, lentils; LT, lettuce; MU, muskmelon; OK, okra; ON, onions; PA, parsnips; PE, peas; PP, peppers, PT, potatoes; PU, pumpkins; RA, radish; RH, rhubarb; SP, spinach; SQ, squash; SW, sweet potatoes; TB, table beets; TM, tomatoes; WA, watermelon.

such option is the use of a hooded sprayer. This sprayer has nozzles enclosed in a plastic shield which allows the application of nonselective herbicides between crop rows.

The latest technological advancements in weed control involve genetic alteration of the vegetable to exhibit resistance to herbicides with improved spectrums of weed control. Currently, only vegetables with large acreages are targeted for this technology (e.g., glyphosate resistance in tomatoes), but this may be expanded eventually to other minor crops.

F. Miscellaneous

Recent articles in popular farm magazines (Leidner, 1994) have advocated the application of hot water as a directed method of weed management between crop rows. Water heated to 99°C was applied over the top of weeds in a volume up to 4680 liters per hectare, with one application apparently effective on annual weeds and two applications exhibiting good control of perennial weeds such as johnsongrass. A number of application units are being built and this method may be used more extensively in the future.

Flame cultivation is also becoming a more attractive method of weed management. This technique was previously used in the 1950s and 1960s in cotton, but the rising cost of propane and butane fuel made it too costly compared to application of herbicides. In short, a series of burners (two burners per row) are directed at weeds between row crops with a band of water applied at the base of the crop for protection. Since plant stems are less sensitive to flames than foliage, flame cultivation is only effective when used in row crops which are taller than weeds. For cotton, flame cultivation was initiated when plants were 25 to 30 cm in height (Parker et al., 1965). With increasing restrictions on pesticide applications and soil erosion, flame cultivation may be more widely used in the future.

G. Summary

Weed management systems in vegetables are highly variable, depending upon the vegetable, time of year grown, and geographical location. The small seed size and growth habit of many vegetables make effective weed control essential for optimal crop yields. With relatively few herbicides labeled for use in vegetables, growers often depend heavily upon mechanical cultivation and hand labor. Vegetables are frequently produced in multiple crop systems, with several crops produced per growing season. This complicates the use of herbicides with residual activity since successive crops may be injured. Organic and inorganic mulches play a major role in weed management of some crops, but these techniques increase production costs and may require specialized planting equipment. Some growers are trying stale seedbed or reduced tillage practices, but the spectrum of herbicides available for use often does not compensate for the loss in weed control afforded by mechanical cultivation. With high value cash crops like vegetables, growers hesitate to experiment with new weed management techniques that may not prove effective.

III. NURSERY AND LANDSCAPE ORNAMENTALS

This section summarizes many recommended weed management practices for use in nursery crops and landscape ornamentals. We discuss concepts of weed management which are commonly used in the United States nursery and landscape maintenance industries. Specific recommendations for weed management are not provided; rather examples of commonly used systems are described. Both chemical and nonchemical weed management strategies will be outlined. Nonchemical controls include hand weeding and other physical means of weed control such as the use of mulches for weed suppression. Chemical control measures will describe the use of pre- and post-emergence herbicides for weed control as well as use of soil fumigants.

The United States has a diverse nursery production industry which has been influenced by existing environment, water availability, soils, labor, and markets. Nursery production in the United States is widespread with concentrated areas of production found in Alabama, Florida, Georgia, Ohio, Tennessee, Texas, Oregon, and California. Across the country, an assortment of high value ornamentals are grown and marketed in size ranging from a few cells to large caliper trees. Propagation techniques, production systems and plant materials produced vary with the firm and region of the country. Plants may be grown exclusively in the field, where the product is harvested, balled and burlapped, or barerooted in one or more production phases. In contrast, production may occur solely in containers. In either situation, production practices are extremely diverse, with both hand and mechanical labor used to develop the final product for residential or commercial landscapes (McNiel and Weston, 1992).

Professionally developed landscapes are an artistic blending of hardscape and vegetation elements. During the past two decades, quality of both components of American landscapes has improved as increased emphasis has been placed on the form, texture, color, and maintenance requirements of plant material. The increased use of high-quality nursery stock has placed great demands on plant suppliers and trained personnel who can assemble and adequately maintain the sites.

Weed management is perhaps the most serious cultural problem confronting the nursery production and landscape management industries. Full season weed management is often necessary for producers of high value ornamentals to achieve the desired plant quality and size. Weed competition decreases the fresh and dry weights and quality of nursery crops, although the degree of reduction varies with weed and ornamental species and/or the level or length of weed infestation. Weed seed contamination can easily occur in container production systems. The weed seed could be present in the container media itself or transported in from other nurseries on nursery stock. Weed seed can also be wind blown from adjacent areas. Most weed seed contamination appears to be associated with movement of water, either irrigation water or rainfall splashing seed into containers (Skroch, 1990). Weeds increase container crop costs by slowing plant growth and possibly through production of allelopathic compounds (Whitwell et al., 1990). Weed competition may also increase the need for irrigation and fertilization in the container. Weedy containers are unattractive to consumers and some weeds, once established, are nearly impossible to remove by hand.

Effective weed management in the nursery or landscape begins with an understanding of weed biology (Whitwell et al., 1990). An integrated program for weed management should be developed by landscapers and nursery managers. Most successful weed management strategies will utilize preventative practices, mulching, cultivation, and herbicides. Initial weed control efforts should be primarily preventative in nature. Weed seed sources should be eliminated by vegetation removal prior to establishment of ornamental beds or container nurseries. Perennial weeds must be eliminated, or successive plants may become infested. All areas surrounding production sites should be mowed and maintained in a clean, organized manner. All mulches, media, containers, and plant stock introduced into the production system or landscape should be weed-free. If ornamentals are maintained in healthy condition, they will likely be more competitive with interfering weeds. Maintaining relatively weed-free plantings also minimizes the infestation of areas where nursery plants and ornamentals are ultimately established (homes, businesses, etc.). The following strategies are commonly utilized to manage weed populations in the landscape or nursery.

A. Cultivation

Nonchemical strategies can be implemented to achieve efficacious weed management in the landscape or nursery. Cost of efficacious control can be very high, with hand weeding being most costly and highly labor intensive. It has been estimated that 625 total hours of manual labor would be required for weed management to produce 0.4 ha of marketable plants grown in number 1 containers (Carpenter, 1983; Fretz, 1972). In some cases, hand weeding is the only viable alternative for effective weed management. Hand weeding itself cannot economically control weeds because of the intense labor requirements. It has been estimated that hand weeding costs would be approximately $9880 per hectare per year of container ornamentals (Whitwell et al., 1990). With certain plant materials, especially expensive or newly established plant materials, the use of herbicides is undesirable and risky. With increased emphasis on environmental quality, it may be preferable to reduce herbicide usage, especially in areas where human exposure is likely. Hand or mechanical weed removal may then be used repeatedly to maintain a weed-free site. Unfortunately, damage to ornamentals may occur as a result of weed removal by hand in areas in close proximity to established ornamental beds.

Other physical weed management measures are commonly utilized to reduce weed populations nonchemically. A number of mechanical control methods used in the landscape or nursery include rototilling, cultivation, hoeing, disking, mulching, and establishment of competitive crops to minimize weed populations (Whitwell, et al., 1990). Care must also be taken to select control measures which minimize potentially deleterious effects encountered during mechanical cultivation, including damage of shallow-rooted ornamentals or the spread of perennial weed propagules.

B. Cover Crops

The establishment of cover or row crops in nurseries has long been utilized to improve weed management in open areas surrounding ornamentals. This technique is also successfully used in vineyards and orchards across the United States. The cover crop selected will depend on soil type, environmental conditions, and crops under protection. The use of allelopathic or weed suppressive covers can provide additional long-term weed suppression. The cover crop can be maintained as a living mulch and repeatedly mowed to minimize competition or the cover can be suppressed chemically with postemergence herbicides. Wheat, oats, rye, and hairy vetch are often used as winter annual cover crops. Unfortunately, difficulty in cover crop establishment or overwintering ability can be encountered in northern temperate areas. Certain covers can also be hard to suppress by using labeled postemergence herbicides. Alternatively, an annual cover crop can be established and allowed to senesce naturally or be killed back by exposure to frost. Summer annual cover crops include sorghum and sorghum–sudangrass hybrids.

C. Mowing

Mowing is used to prevent the formation and spread of weed seed into cultivated areas. Mowing has been shown to be most effective when used on the perimeter of cultivated areas with taller grass weeds (Whitwell et al., 1990). Species which flower at a level lower than the mowing blade are not effectively controlled in this manner. The proper timing of mowing operations can also improve suppression of such perennial weeds as johnsongrass. Long-term mowing tends to favor the establishment of perennial low-growing grasses, which are very competitive for water and nutrients.

D. Mulches

Mulching is a physical control measure used to prevent weed emergence in the landscape or nursery. Mulching may result in the elimination of light and oxygen or gas exchange required for weed seed germination. Like all other components of the landscape, mulch types have diverse appearances and physical properties. A *mulch* is defined as any material placed on the soil to cover and protect it (Harris, 1983). Mulch types commonly marketed today include natural organic, natural inorganic, and synthetic mulches. The thickness or depth of a mulch necessary to suppress weed growth adequately depends on the mulch type and the weed pressure. Weeds such as field bindweed and nutsedge often have sufficient root reserves to enable them to penetrate thick mulches (> 15 cm).

In local environments across the country, there will usually be a natural organic mulch which is a by-product of a living crop previously in production (Robinson, 1988). Common natural organic mulches include composted yard wastes which exhibit fine particle size and changing characteristics during the decomposition process. Pine straw is common in the southern United States and exhibits fine particle size and slight to moderate stability over time. Preferred materials in the nursery and landscape include wood chips and hardwood or softwood bark chips or nuggets. Bark chips have moderate particle size (0.6 to 1.3 cm) and moderate to good stability, while bark nuggets have larger particle sizes (1.3 to 7.6 cm) and excellent stability over time. These materials can be used in nurseries and landscape beds containing herbaceous or woody ornamentals.

Natural inorganic mulches are generally more expensive and less widely used in the nursery and landscape. Natural inorganic mulches are stable over time and include such materials as sand, which has fine particle size, and gravel or pebbles, which have moderate particle size. Shale and other rock products offer a diversity in color, particle size, and shape. Most of the products are used in public and commercial plantings.

Synthetic mulches are manufactured materials which are often termed *geotextiles* since they are manufactured from petroleum or petroleum by-products (Appleton et al., 1990). Synthetic mulches vary in appearance and stability depending upon the composition and exposure to ultraviolet (UV) light or other environmental extremes such as temperature. These materials include solid films, often polyethylene, which are nonpermeable if not perforated. Composite mulches made up of two or more nonpermeable materials are also used. Woven fabrics made of synthetic materials are also available and tend to be water permeable. Nonwoven fabrics are becoming more popular in the landscape industry beneath an organic or inorganic mulch for weed suppression or soil stabilization on sloped sites. Nonwoven fabrics also tend to be water permeable. Synthetic mulches are often used in container nurseries or smaller or specialized landscape sites. They are relatively expensive but vary in price.

Mulches which are placed in the nursery or landscape become barriers. Several phenomena occur at the interfaces which develop both above and below the mulch as it contacts the adjacent material. The influence of this interface on the movement of air, moisture, plant roots, and soil organisms will determine whether mulch usage will be beneficial (Harris, 1983). As pore size between two materials becomes more divergent at the interface (i.e., soil and mulch), water movement across the interface is impeded. One of the most beneficial aspects of mulch usage is the retention of moisture below the landscaped surface. Pore space also influences the water holding capacity inherent in each mulch. Mulches with larger pores retain less moisture than mulches with smaller pores (Gartner, 1978).

Mulches which are or become impervious over time create a landscape barrier in which no movement occurs across the interface. Not only is vertical water movement impeded, impervious mulches reduce air, root, and organismal movement and subterranean perennating organs. Clear polyethylene mulches will also markedly increase soil temperatures. Many of these influences are detrimental to landscape ornamentals. However, these characteristics may contribute to enhanced weed suppression in the nursery or landscape.

Mulches have been investigated for stability and long-term weed suppression with varying results (Appleton et al., 1990; Derr and Appleton, 1989; Martin et al., 1991; Powell et al., 1989; Whitcomb, 1980). Generally, mulches with larger particle size are more stable and provide prolonged weed control in the landscape. Mulches with finer texture tend to degrade more quickly and provide less weed suppression over a longer time frame. Stability is dependent upon mulch type, particle size, and environmental conditions. Weed suppression is provided through the physical effects of the mulching material which prevent light, oxygen, and perhaps moisture from reaching the nondormant weed seed. It has been shown that depth of the organic mulching material strongly impacts weed germination and growth. Weed growth was reduced by greater than 90% when shredded wood bark mulch was placed to a depth of 15 to 25 cm around ornamentals (Rakow, 1992). At less than a 15 cm depth, weed suppression was significantly reduced later in the same growing season. However, annual weeds tend to germinate over time on mulching media which provide adequate moisture and temperatures for germination. Synthetic or organic mulches of finer texture and smaller pore space retain higher moisture and create ideal sites for weed growth (Billeaud and Zajicek, 1989). Over time, considerable weed numbers can become established in upper surfaces of porous mulches. Perennial weeds can also be problematic and generally survive from year to year by developing extensive root systems which serve as storage organs. Spring growth from perennial weeds is supported by the storage organs and thus can produce vegetation which protrudes through mulches more easily than an emerging seedling.

Limited information is available on natural or synthetic mulch degradation, accumulation, and incorporation into the soil and the impact of these factors on ornamental and weed growth. Weed suppression associated with synthetic fabrics covered with hardwood mulch was investigated over a 3 year period (McNiel and Weston, 1992). Several fabric treatments provided greater weed suppression than did other fabric treatments or shredded hardwood mulch itself (Table 2). Over the course of the study, Reemay T3301 and Dalen's Weed X produced greater overall weed suppression and reduced growth of weed seeds germinating under the mulch fabric when compared to other treatments. Derr (1991) also reported greatest weed suppression associated with Weed X in a comparative study evaluating synthetic landscape fabrics.

Generally, organic hardwood mulch placed over synthetic fabrics proved to be a good medium for the germination and growth of weed seedlings. Over one growing season, hardwood mulch alone became overgrown with invasive annual and perennial weed species. Synthetic fabrics provided some additional weed suppression, but substantial weed growth occurred above the fabric in the bark mulch in each treatment. Greatest weed densities were associated with bare ground treatments (McNiel and Weston, 1992).

The physical properties of synthetic mulches vary considerably and may result in observed differences in ability to suppress weeds. Differences in fabric thickness, tensile strength, unit weight, and air permeability are shown in Table 3. In general, fabric unit weight and thickness are not well correlated with ability to suppress weeds. However, fabrics with reduced air permeability appear to provide greatest weed suppression, and these include Reemay T3301, Dalen's Weed X, and DeWit's Pro 5. These fabrics were also most effective

Table 2 The Average Percentage of Ground Cover (Weeds) Within Plots, 7 and 9 Months After Fall Establishment of Mulching Materials in Lexington, Kentucky, 1991

Mulching treatment	July 18[a] % Cover	September 17[a] % Cover
Reemay T 534	56.7 b	91.7 ab
Reemay T 3201	48.3 bc	83.3 c
Reemay T 3301	33.3 d	75.0 d
Dewit's Pro 5	55.0 b	90.0 abc
Weed X Dalen	36.7 cd	75.0 d
Blunk's Fabric	60.0 b	93.3 a
Weed Block	60.0 b	85.0 bc
BareGround	95.7 a	95.0 a
Bark Mulch Only	55.0 b	91.7 ab
LSD 0.05	11.90	6.84

[a]Numbers followed by different letters are significantly different at the 0.05 level.

able 3 Physical Characteristics of Selected Mulching Materials Under Evaluation in Lexington, entucky, 1991

ulching materials	Unit weight (ox/yd^2)	Thickness (mils)	Tensile strength (lb md)	Air permeability (cfm/ft^2)
eemay T 534	1.9	11.2	57	452
eemay T3201	1.9	10.2	74	327
eemay T3301	2.9	12.3	120	149
ewit's Pro5 (woven slit film)	4.5	38.6	119	64
'eed X Dalen (perforated composite film)	1.5	6.0	31	23
unk's Fabric (needled nonwoven polypropylene)	2.8	21.3	62	339
eed Block (perforated polyethylene film)	1.7	22.2	13	376

in suppressing root establishment of weeds underneath the fabric itself (Table 4). Although fabrics provided some level of weed suppression over spring and summer months, the expense of the fabric itself probably does not warrant use for weed control purposes alone (Derr, 1991; McNiel and Weston, 1992). A combination of herbicide plus mulch or landscape fabric provides most effec-

Table 4 Root Penetration Through Selected Mulching Materials in the spring of 1992, 18 Months After Experimental Initiation

Mulching treatment	Root penetration rating[a,b]
Reemay T 534	2.83 c
Reemay T3201	2.67 c
Reemay T3301	1.83 d
Dewit's Pro5	3.67 b
Weed X Dalen	1.08 e
Blunk's Fabric	3.17 bc
Weed Block	4.67 a
LSD 0.05	0.64

[a]0, no root penetration; 5, extensive root penetration.
[b]Numbers followed by different letters are significantly different at the 0.05 level.

tive long-term control of both annual and perennial weeds. In container production of ornamentals, black plastic is commonly used beneath containers to prevent weed growth. Other synthetic mulches are also used.

E. Herbicides

Herbicides are widely used for weed management in the home and commercial landscape maintenance industries. Increased urban acreage is treated by lawn care and tree maintenance (Ball, 1987). Producers of field- and container-grown ornamentals in the United States also utilize herbicides for effective weed control. However, up-to-date information on the use of herbicides in nursery crop production is limited. Several excellent extension bulletins on weed management in ornamental crops are available from land grant universities across the southern United States. Achieving satisfactory weed control, especially in the South, a major production area for ornamentals, can be difficult since the growing season is longer and warmer soil temperatures may hasten herbicide dissipation (Wehtje and Gilliam, 1991).

Herbicides are one way to reduce weed numbers effectively in field nurseries or in container production of woody ornamentals. If used properly, herbicides are often less expensive and more convenient than hand weeding or cultivation. In addition, injury to the crop from close cultivation or hand weeding is minimized (Welker, 1984). Herbicide combinations as well as multiple herbicide applications are often used in field situations (Gilliam et al., 1989). However, the timing of application is sometimes more important than the specific herbicides used (Wehtje and Gilliam, 1991).

An increasingly wide selection of chemical products is available for weed management in turf and ornamentals. Most major chemical companies have developed diverse product lines for this expanding market. However, herbicide tolerance varies with ornamental species and within cultivars of that species. New products must be evaluated for potential damage to a diverse group of ornamentals. This presents a formidable problem, given the extensive selection of trees, shrubs, annuals, and perennials now available. Field trials are conducted by both the chemical industry and university research programs to evaluate ornamental tolerance to newly developed products. It is best to plan a crop planting scheme with weed control procedures in mind to make weed management easier in the future. For example, do not plant species together that differ in tolerance to potential herbicides to be applied. When incompatibility exists among species, weed management may not be possible. Certain weeds are not chemically controlled by any labeled products which are safe for ornamentals. In this case, it is best not to plant a particular ornamental in an area which is infested by an uncontrollable weed (e.g., do not plant liriopes in an area infested with yellow nutsedge because the products labeled for use in liriope do not control yellow nutsedge).

Effective weed control for field-grown ornamentals requires planning and management skills. Weed pressures will be greatest during the first years of establishment. After several years of intensive management, pressures may be reduced (Whitwell et al., 1990). However, the repeated use of selective herbicides for weed control without herbicide rotation can allow the buildup of undesirable or difficult to control weeds over time. A combination of timely cultivation along with the use of several herbicides or combinations of herbicides is often essential. Multiple applications of herbicides are often required for full season weed control. Timing of herbicide application is also critical to achieve maximum weed control. Preemergence herbicides should be applied when ground is freshly tilled or cultivated and weeds have not yet germinated. Herbicides should be incorporated or watered in for maximum performance. Postemergence herbicides may be applied to vegetation that is already established. Postemergence products are most effective if applied in dry conditions with surfactants and without rainfall during a 24-hour period following application. A combination of pre- and postemergence herbicides may be most effective for weed management in the nursery or landscape.

Many soil-applied preemergence herbicides are now labeled for use in selected herbaceous and woody ornamentals by certified applicators. These herbicides are active in the soil from a period of 4 weeks to 4 months after application. The chemicals available control seedling weeds through many different modes of action, including inhibition of carotenoid biosynthesis, photosynthesis, cell division, or amino acid biosynthesis. Commonly used preemergence herbicides are listed in Table 5. Many of the recently developed products have proved useful because of extended soil persistence and activity, resulting in increased seasonal control of annual or perennial weeds. Other products or product combinations provide improved control of a broader spectrum of annual or perennial weeds. Most products exhibit considerable selectivity in the weeds controlled and tolerance of ornamentals to them. Nonselective soil-applied herbicides include only a few products; the fumigants methyl bromide and metham are most common.

Postemergence herbicides are also labeled for use in ornamentals and landscapes. Many of these products are selective grass herbicides only. A table of commonly used postemergence herbicides in nursery and landscape settings is included (Table 6). Several products are translocated and act as systemic inhibitors. Other products exhibit contact activity only and are not translocated throughout the plant. Some are selective while others are nonselective. Glyphosate is probably the most versatile postemergence herbicide available for controlling perennial weeds in ornamentals. The difference in height between ornamentals and most emerging weeds allows glyphosate to be postdirected with minimal injury to ornamentals. It is also effective in controlling cover crops which may be used between rows of ornamentals for weed suppression.

Table 5 Preemergence Herbicides Labeled for Use in Nursery and Landscape Plantings[a]

Common name	Trade name[b]	Active ingredient (kg per hectare)
Prodiamine	Barricade[c]	0.7 to 0.8
Dazomet	Basamid	388
Bensulide	Betasan[c]	8.4 to 14
Methyl bromide	Brom-O-Gas[c]	268 to 504
Dichlobenil	Casoron[c]	4.5 to 6.7
DCPA	Dacthal[c]	6.7 to 16.8
Simazine + metolachlor	Derby	3.4 to 5.6
Napropamide	Devrinol	4.5 to 6.7
EPTC	Eptam[c]	5.6 to 6.7
Isoxaben	Gallery	0.6 to 1.1
Oxyfluorfen	Goal[c]	0.3 to 2.2
Imazaquin	Image[c]	0.4 to 0.6
Pronamide	Kerb	1.1 to 2.2
Alachlor	Lasso[c]	4.5
Oxyfluorfen + pendimethalin	OH2 Ornamental Herbicide II	2.2 + 1.1
Pendimethalin	Pendulum[c]	2.2 to 3.4
Metolachlor	Pennant[c]	2.2 to 4.5
Norflurazon	Predict[c]	2.7
Simazine	Princep[c]	2.2 to 3.4
Oxadiazon	Ronstar	2.2 to 4.5
Oxyfluorfen + oryzalin	Rout	2.2 + 1.1
Isoxaben + orzyalin	Snapshot DF	2.2 to 4.5
Isoxaben + trifluralin	Snapshot TG	2.6 to 5.2
Oryzalin	Surflan	2.2 to 4.5
Trifluralin	Treflan[c]	0.6 to 1.1
Metham	Vapam[c]	267 to 356
Benefin + oryzalin	XL	5.6 to 6.7

[a]Timing of application, usage rates, locations for use, and other restrictions should be determined from product labels prior to application.
[b]Trade names given are for U.S.A. labels.
[c]More than one trade name and/or manufacturer is represented for this active ingredient.

Products are available in many different formulations including liquids, emulsifiable concentrates, flowables, dry flowables, and granulars. Liquid and flowable formulations are most common in applications for larger sites. Granules are often preferred for application to container ornamentals or for use by homeowners. All herbicides should be considered as potentially dangerous

Table 6 Postemergence Herbicides Labeled for Use in Nursery and Landscape Plantings[a]

Common name	Trade name[b]	Active ingredient (kg per hectare)
Fenoxaprop	Acclaim[c]	0.1 to 0.3
Bentazon	Basagran T/O[c]	0.8 to 1.1
Clethodim	Prism[c]	0.06 to 0.3
Triclopyr	Redeem[c]	0.6 to 1.1
Diquat	Reward[c]	0.6 to 1.1
Glyphosate	Roundup[c]	1.1 to 5.6
Potassium salts of fatty acids	Sharpshooter	—
Clopyralid	Stinger[c]	0.1 to 0.3
Sethoxydim	Vantage[c]	0.3 to 0.6

[a]Timing of application, usage rates, locations for use, and other restrictions should be determined from product labels prior to application.
[b]Trade names given are U.S.A. labels.
[c]More than one trade name and/or manufacturer are represented for this active ingredient.

to the applicator. The herbicide labels and material safety data sheets pertaining to the herbicide should be read thoroughly before application.

Most commonly, weed management systems for field-grown ornamentals initially utilize thorough cultivation to provide a smooth weed-free planting site. Postemergence herbicides, such as glyphosate, can be applied to treat problem areas or perennials before cultivation. Following cultivation, application of a soil-applied preemergence herbicide or herbicide combination provides annual weed control. Perennial weed control may be accomplished chemically or culturally. Cultivation in late summer and/or fall will generally be performed again to minimize weed populations. Grassy or weedy areas surrounding field production sites should be mowed repeatedly throughout the growing season to minimize sources of contaminating weed seed.

Production systems for container ornamentals will generally involve the creation of a weed-free, well-drained site for containers. Gravel, concrete, or a surface covered with landscape fabric is commonly used. After plants are established, both pre- and postemergence herbicides are applied repeatedly throughout the year for weed management. Hand pulling of escaped weeds is undesirable but may be unavoidable.

Herbicide injury in ornamentals occurs all too often in the landscape or nursery. This causes serious problems for landscape or nursery managers because of the high value of the crops under production and the visual damage which may result in the landscape. Misapplication of herbicides is the most common cause of injury, with direct spray contact to nontarget plants causing most injury (Whitwell et al., 1990). The application of nonselective postemergence herbicides such as paraquat or glyphosate can result in serious long-term injury to valuable species. Applications of these herbicides should be made with low pressure and large droplets on a calm day. Glyphosate should be applied using low volumes of solution to provide more effective weed control. Injury from paraquat will appear within 1 day after application, while glyphosate injury may be delayed and even be indicated by stunted growth several months after application. Application to ornamentals should be made using shielded sprayers and care must be taken to prevent contact with foliage. Glyphosate applications are also common in dormant ornamentals (e.g. shrubs and trees in late fall or very early spring) which are not actively growing at the time of application.

Herbicide injury from soil-applied chemicals can also occur in newly established ornamentals or actively growing plants. This injury is often temporary in nature, but can result in serious growth inhibition to newly established plants. Many plants outgrow these injury symptoms. Herbicides which contain oxyfluorfen or occasionally isoxaben are more likely to cause this injury. Injury may result from more persistent herbicides which are applied to surrounding areas for weed control in turf, agronomic crops, or complete vegetative control under pavement (Whitwell et al., 1990). These products may cause severe stunting or death. In some cases, activated charcoal may be utilized to absorb the herbicide and minimize injury. In other cases, time is required to allow complete herbicide dissipation.

Herbicide injury symptoms vary depending on the plant species and the herbicide applied. Injury can range from chlorosis and/or bleaching caused by photosynthetic or carotenoid biosynthesis inhibitors to root stunting caused by inhibitors of cell division. Herbicides with plant growth regulatory activity may cause twisting and elongation of foliage and/or epinastic appearance, while membrane disrupters may cause severe necrosis to affected foliage. It is often difficult to diagnose herbicide injury clearly because environmental conditions and insect or disease infestation may result in similar symptoms. Soil testing for residual herbicide use or use of a bioassay to document sensitivity of particular ornamental species may aid in determination of herbicide injury.

F. Summary

Weed management systems for ornamentals in the landscape or nursery are diverse. Generally, a combination of mechanical and chemical strategies is utilized for weed suppression in containers or field-grown stock. The use of a diverse group of mulching materials is common in the landscape for aesthetic purposes as well as weed suppression. Herbicides often provide cost effective weed suppression in ornamentals. However, care must be exercised to select an herbicide which can be used safely on the ornamental areas established. Herbicide injury to ornamentals can result from misapplication or lack of crop tolerance.

IV. SMALL FRUITS

A. Strawberries

Strawberries are widely grown throughout the United States and the life of each planting ranges from 1 to 3 years. Effective weed control is required since strawberries are low-growing and, therefore, poor competitors with weeds. Weed control practices vary for this crop, depending on geographical location and whether the plantings are new or established.

Cultivation

Mechanical cultivators (e.g., power tillers) are routinely used for new plantings to eliminate weeds between plant rows. As the growing season progresses, the width of cultivation between rows is narrowed to permit the establishment of runner plants. Cultivation both removes weeds and provides loose soil near the matted plants where runner plants can more easily set down roots. Weeds emerging between plants but within the row must be removed by hoeing or by hand.

In established plantings, mechanical cultivation is also effective in removing weeds between matted rows, with hand labor used to remove weeds within the row. Within densely planted rows, strawberries can be quite efficient at eliminating many weeds. Although hand labor is expensive, the high cash value of strawberries continues to make it a viable option for weed control. For strawberries grown under black polyethylene, hand weeding is typically limited to the planting slit. In some regions of the country, strawberries are covered in the fall with straw mulch for winter protection and the mulch removed the following spring to areas between the rows. This method affords some suppression of annual weeds but precludes the use of mechanical cultivation. In addition, the contamination of straw with grain seeds or weed seeds can introduce new populations of unwanted species.

Herbicides

The life expectancy of a strawberry planting is dependent upon production practices and often on disease and weed pressure. Perennial weeds pose one of the major threats to maintaining long-term plantings and must be removed prior to planting strawberries. A list of herbicides labeled for use in strawberries is found in Table 7. The stage of plant development is critical for use of many herbicides labeled for strawberries (Hemphill, 1980).

Table 7 Herbicides Labeled for Use in Strawberries[a]

Common name	Trade name[b]	Application
Dazomet	Basamid	Preplant soil sterilant
DCPA	Dacthal[c]	Preemergence
Fluazifop-P	Fusilade DX	Postemergence
Napropamide	Devrinol	Preemergence
Paraquat	Gramoxone Extra[c]	Postemergence
Sethoxydim	Poast[c]	Postemergence
Simazine	Princep[c]	Preemergence
2,4-D	Weedar 64[c]	Postemergence

[a]Timing, usage rates recommended, locations for use, and other restrictions should be determined from product labels prior to application.
[b]Trade names give are for U.S.A. labels.
[c]More than one trade name and/or manufacturer is represented for this active ingredient.

Soil fumigation prior to planting is a common practice in some areas of the country, with metham or a mixture of methyl bromide and chloropicrin commonly used. These fumigants are effective in killing weed seeds and vegetative structures of some perennial weeds. After the injection of fumigants, the soil is irrigated, cultivated or rolled, or covered with tarpaulin or plastic to minimize volatilization. Strawberries can be planted without threat of injury within 3 days of removing the soil covers.

Loss of registration for diphenamid, chloroxuron, and terbacil leaves napropamide and DCPA as the only preemergence herbicides labeled for use on new plantings. In the states of Oregon and Washington, simazine is registered for use, and application is recommended when strawberries are dormant.

Growers should exercise caution when considering the use of preemergence herbicides for strawberries. There is a wide spectrum of varietal tolerance, and growers should consult research bulletins to prevent herbicide use on sensitive varieties. Also, the soil type can influence herbicide availability, and rates should be minimized on soils low in organic matter and clay content. For new plantings, napropamide rates should be carefully selected since this herbicide can inhibit rooting of runner plants during the growing season.

Numerous postemergence herbicides are labeled in strawberries, but there are time restrictions when applications can be made. Fluazifop-P cannot be applied to bearing plants and sethoxydim cannot be applied within 7 days of harvest. Application of 2,4-D is restricted to periods when strawberries are dormant. This is during late fall (after several hard freezes) or renovation after harvest. Paraquat, a nonselective postemergence herbicide, can be applied as a directed spray between rows. Shields should be used to prevent contact with strawberry plants. Some growers have also used glyphosate, another nonselective postemergence herbicide, to control problem weeds in strawberries. This herbicide is typically applied to weeds taller than strawberry plants using a ropewick applicator. Since glyphosate is translocated readily throughout plants, caution should be used to prevent all contact with actively growing crop tissue.

Miscellaneous

In the past, geese have been employed to control various annual and perennial grasses in strawberry plantings (Doll, 1980). However, they do not generally control broadleaf weeds and other management techniques must be used.

More recently, cereal grains have been fall planted in the Midwest with applications of selective grass herbicides early the following spring to control cereal grain growth (Smeda and Putnam, 1988). Winter annual weeds were suppressed by overwintering cover crops, as were some summer annual weeds following the desiccation of cover crops. This technique may provide an alternative to the application of preemergence herbicides during dormant periods.

Summary

Prior to planting, remove annual and perennial weeds utilizing a combination of tillage and nonselective herbicides. Soil fumigants can also be applied to reduce the population of weed seeds in the soil as well as the incidence of soilborne pathogens, insects, and nematodes. At planting, a preemergence herbicide can be applied, but users should be aware of varietal sensitivity and soil type. Cultivation, hand labor, and the use of postemergence herbicides should minimize weed competition during mother plant establishment and runner plant production. Once plants are established, growers can use a number of different herbicides at various times of the year as well as continue modified cultivation, hand labor, and/or use of mulches between strawberry rows.

B. Cranberries

The unique environment reserved for cranberry production (acidic bogs) and the cultural practices for production (e.g., flooding for insect control, frost protection, and protection against winter injury) presents some challenging weed control practices for this crop. Because cranberries grow as intertwining vines, certain cultural practices and herbicide use are the primary weed management practices.

Cultural

As in other perennial crops, prevention of perennial weed establishment prior to planting and efficient vegetation control in areas surrounding the bogs can reduce the potential for perennial weed establishment. By manipulation of soil moisture levels, the severity of some weeds can be manipulated. Under drier conditions, control of rice cutgrass and arrowhead is improved. However, increases in soil moisture reduce infestations of common ragweed and tall goldenrod (Mahr et al., 1990; Demoranville, 1984, 1986). The practice of flooding distributes weeds that reproduce by seed, requiring additional weed management practices.

Various species of ferns are common weeds in cranberries. The use of ferrous sulfate as a soil acidifying agent is reported to offer some control of ferns. However, it is difficult to distribute ferrous sulfate evenly to control the fern species with minimal damage to the cranberry vines (Demoranville, 1984).

One of the most effective weed suppression agents is a healthy cranberry vine. Numerous weeds found in cranberry bogs are established in open areas. Promoting the vigorous growth of vines prevents many weeds from becoming established.

Herbicides

A limited number of herbicides are registered for use in cranberries (Table 8), with the majority applied prior to weed emergence. The use of preemergence herbicides is influenced by various factors. Since cranberry vines are intertwined, placement of the preemergence herbicides is possible only by irrigation or timely rains. The acidic bogs used for cranberry production are highly variable in organic matter content, which influences the activity of preemergence herbicides. Cranberries have shallow, fine, fibrous root systems which can be damaged if herbicide rates are not carefully controlled (Cain and Eck, 1966).

Table 8 Herbicides Labeled for Use in Cranberries[a]

Common name	Trade name	Application[a,b]
Dazomet	Basamid	Preplant soil sterilant
Dichlobenil	Casoran[c]	Preemergence
Fluazifop-P	Fusilade DX	Postemergence
Glyphosate	Roundup[c]	Postemergence
Napropamide	Devrinol	Preemergence
Norflurazon	Zorial[c]	Preemergence
Sethoxydim	Poast[c]	Postemergence
Sulfosate	Touchdown	Postemergence
2,4-D	Weedar 64[c]	Pre- and postemergence

[a]Timing, usage rates recommended, locations for use, and other restrictions should be determined from product labels prior to application.
[b]Herbicide application refers to the stage of weed development.
[c]More than one trade name and/or manufacturer is represented for this active ingredient.

In the spring, prior to weed emergence, several preemergence herbicides can be broadcast. Some cultivars are injured by norflurazon. For new plantings, metham, a soil fumigant, can be applied to kill all existing vegetation and viable weed seeds.

During the growing season, postemergence herbicides can be applied through rope-wick applicators or some other equipment which places the herbicide only on target weeds and not cranberry vines. For nonbearing vines, selective grass herbicides can be applied. Attempts are being made to label sethoxydim for use on bearing vines.

After harvest and prior to winter flooding, preemergence herbicides can be applied again. Restrictions apply regarding the interval between herbicide application and winter flooding for proper activation.

Summary

The unique environment and growth habit of cranberries have created a heavy dependence upon herbicides for effective weed control. Eradication of perennial weeds is necessary prior to planting. Once cranberries are planted, growers apply preemergence herbicides and spot treat or wipe on nonselective herbicides as needed. In the past, cultural techniques which controlled soil moisture were used, but now most growers depend upon herbicides. In some instances, hand labor is used to remove problem weeds.

C. Grapes

Grapes are adapted to grow throughout much of the United States as well as other parts of the world. Commercial production in the United States is concentrated in the Napa, San Joaquin, and Coachella valleys and coastal regions of California, as well as small areas in Michigan, Oregon, and parts of New England. Vines can survive more than 80 years; however, commercial producers maintain vineyards for approximately 25 years. Typically, vines are trained on a fence or other support in rows up to 3.6 meters apart. Although well established vines can shade out some weeds, vegetation management, especially in the grape row, is important to minimize weed competition.

Cultivation

Prior to the development of herbicides, mechanical cultivation was frequently carried out throughout the growing season to minimize weed competition for available nutrients and water. Within the vine row, specialized equipment such as the grape hoe reduced hand labor costs for weed removal. However, results from long-term practices of clean cultivation indicate that losses in soil organic matter and increases in soil compaction, along with physical losses in soil due to erosion (especially on sloped soils), have contributed to yield declines (Pool et al., 1990). Trash cultivation is practiced in some areas. In this system, hay or straw is broadcasted between vine rows and partially incorporated into the soil with a disk or other implement (Fleming and Alderfer, 1956). The objective is to control losses in soil organic matter as well as minimize soil and water runoff. In spite of the negative effects of clean cultivation between the vine rows, it remains a common practice.

Grass Sods and Mulches

An alternative to clean cultivation is maintenance of a grass sod strip between vine rows. With the development of heavy, mechanized harvesters, a sod middle allows grape harvesting during periods when soils are wet. Sod middles also minimize water induced soil erosion, a severe problem on sloped soils. Under conditions of adequate soil moisture, competition for water by grasses does not limit vine growth. Vineyard productivity in the short term is lower in managed sods than in vineyards with cultivated row middles; productivity of these systems is much lower compared to that of vineyards where herbicides are predominantly used.

Maintenance of soil properties and promotion of vine growth can be achieved by mulching with straw from cereal grains or growing cover crops (Shoemaker, 1978). The expense of straw mulches compared to herbicides has minimized the use of this practice. Small cereal grains and legumes can be sown in the fall or early spring and cultivated into the soil or killed with herbicides prior to vine growth in the spring. Although the cost of seed for some cover crops is relatively high, benefits include improvement in soil structure and organic matter, reduction of soil erosion, and conservation of soil moisture. Overall, the integration of cover crops into vineyard weed management practices requires more intensive management than cultivation or application of herbicides alone.

Herbicides

The availability of both broad-spectrum and selective herbicides has led to their widespread use in both vine rows and row middles. The herbicide labeled and the recommended use rate depend on whether the vineyard is new or established. Numerous preemergence and postemergence herbicides can be used (Table 9), with specific restrictions found on the product label. The use of herbicides, especially in row middles, simplifies weed control and reduces cost. Also, certain negative effects from cultivation and sod strips are prevented. However, overdependence on herbicides has led to increased erosion and the selection of herbicide resistant weeds. Weed resistance to paraquat has been documented in Japanese vineyards (Kato and Okuda, 1983).

Recent developments in technology have resulted in automated herbicide application equipment. Basically, energized photocells which scan the ground

Table 9 Herbicides Labeled for Use in Grape Vineyards[a]

Common name	Trade name	Application[b]
Dazomet	Basamid	Preplant soil sterilant
Dichlobenil	Casoran[d]	Preemergence
Diuron	Karmex[d]	Preemergence
Fluazifop-P	Fusilade DX	Postemergence[c]
Glufosinate	Ignite	Postemergence
Glyphosate	Roundup[d]	Postemergence
Isoxaben	Gallery	Preemergence
Metolachlor	Dual[d]	Preemergence
Napropamide	Devrinol	Preemergence
Oryzalin	Surflan	Preemergence
Oxyfluorfen	Goal[d]	Preemergence
Paraquat	Gramoxone Extra[d]	Postemergence
Pendimethalin	Prowl[d]	Preemergence
Pronamide	Kerb	Preemergence
Sethoxydim	Poast[d]	Postemergence
Simazine	Princep[d]	Preemergence
Sulfosate	Touchdown	Postemergence
Trifluralin	Treflan[d]	Preemergence

[a]Timing, usage rates recommended, locations for use, and other restrictions should be determined from product labels prior to application.
[b]Herbicide application refers to the stage of weed development.
[c]Under Experimental Use Permit; registration pending.
[d]More than one trade name and/or manufacturer is represented for this active ingredient.

below crop canopies or between crop rows detect weeds by sensing chlorophyll. This triggers the sprayer to apply a small volume of herbicide directly onto weed foliage (Mortenson et al., 1995). These systems have been designed for vineyards and orchard crops in the United States, but applications in row crops are being pursued.

Summary

As in other perennial cropping systems, the removal of perennial weeds is crucial prior to planting grape vines. Numerous preemergence herbicides can be applied in the vine row to minimize weed competition during plant establishment, with additional herbicides available in established plantings. Alternatively, cover crops can be sown in vine rows to protect the soil when vine growth has stopped, and cover crops desiccated with herbicides prior to the onset of new vine growth. Row middles can be managed with herbicides alone or in conjunction with cultivation. Another option is the application of mulches or the use of cover crop residues. Establishment of mowed sod in row middles is also possible, especially on highly erodible soils. Fleming and Alderfer (1956) have recommended the rotation of grass sods with cultivation in the row middle. An effective, long-term management system for grapes should utilize a number of the approaches outlined.

D. Blueberries

Weed management practices for blueberries depend upon both the species grown and the age of the planting. There are three major species of blueberries cultured, and their optimum habitat is temperature dependent. Throughout eastern Canada and the northeastern region of the United States low bush blueberries are predominantly grown. Typically, plantings are grown as solid stands. In Atlantic coast states down through Florida, as well as midwestern states including Michigan and Indiana, highbush blueberries are grown. In the southeastern United Sates, rabbiteye blueberrries are primarily grown. Both rabbiteye and highbush plantings are established in rows with sufficient space between them to allow access by equipment.

Weed management practices involve pruning (mowing or burning) or use of herbicides for lowbush blueberries, because plantings result from naturally occurring stands and, therefore, are inaccessible to mechanical cultivation. Both highbush and rabbiteye blueberries are managed by cultivation, use of herbicides, and mowing.

Cultivation

Highbush plantings are often mechanically cultivated in plant rows to minimize weed competition as well as minimize surface litter which can harbor various diseases such as mummyberry (Eck, 1988). However, cultivation also stimulates plant growth late in the growing season, which can result in more winter injury because of insufficient hardening-off. Only shallow cultivation is recommended because blueberries have shallow root systems. Hand pulling can also be used effectively in small areas with concentrated weed infestations.

Mulches

Sawdust, pinebark, and composted sludge have been used to create 10 to 15 cm thick mulches under highbush blueberries. This practice eliminates most annual weed pressure because weed seeds don't have the reserve to penetrate thick mulches. These mulches also benefit blueberries by stabilizing soil moisture fluctuation, minimizing soil erosion, and producing acidic soil conditions. Perennial weeds with vigorous underground reproductive systems often grow through mulches and other management practices for these weeds are required prior to the use of mulches (Eck, 1988).

Planting annual cover crops in the fall is a former practice used to harden off blueberry plants and minimize cold injury (Eck, 1988). Cover crops also increase soil organic matter and reduce soil erosion as well as the emergence and growth of winter annual weeds. The ease of herbicide application and cost of planting and maintaining cover crops have limited the use of this management technique.

In some areas, permanent turfgrass sods such as Kentucky bluegrass are managed between blueberry rows with bare ground areas maintained within rows. These sods minimize soil erosion and prevent the invasion of highly competitive annual weeds. In dry years or without proper sod management, however, grass sods can compete with blueberry bushes for water and nutrients (Eck, 1988).

Mowing

Mowing is a vital part of weed management in lowbush blueberries but is not used in highbush plantings except for grass sod strips between plant rows. Mowing is used to control tall weeds such as brackenfern, sweet fern, poplar, dogbane, and other species. Repeated mowings are necessary to exhaust plant reserves and kill these weeds. In some instances, flail mowing is used to cut all vegetation, including blueberries, to ground level. This can reduce the incidence of weeds which form crowns at or above ground level. Flail mowing is typically only recommended for lowbush plantings (Yarborough and DeGomez, 1990).

Miscellaneous

Certain weed management techniques are unique to lowbush plantings. Filling in bare spots with new blueberry plants can minimize the introduction of annual weeds. It is important to use clean straw for mulching to minimize the potential of introducing new weeds from infested straw. This practice can effectively control coniferous trees and some weeds spread by seed. In some instances, lowbush plants are pruned by burning entire fields using straw or oil burners (Smagula and Yarborough, 1990). However, the elimination of ground litter can promote the invasion of early successional stage weed species. Therefore, fire pruning may alter the weed species present and require additional weed management practices. Since blueberries flourish in soils with a pH between 4.3 and 4.8, maintaining an optimum pH can discourage the growth of many weeds, especially grasses.

Herbicides

Numerous herbicides are registered for use in lowbush, highbush, and rabbiteye blueberry plantings. Since herbicide use varies with the type of blueberry and soil conditions, growers should determine specific label restrictions prior to use of these compounds.

Preemergence herbicides are commonly applied in blueberry plantings for season-long control of many weeds. Blueberries are grown on both low and high organic matter soils; therefore, the usage rate of these compounds is dependent upon organic matter content. Herbicides commonly used are listed in Table 10. Some of these materials should be used only on established plantings or nonbearing bushes. In lowbush blueberry plantings, continual mowing of plants precludes the use of certain preemergence herbicides, except during periods after burning, but prior to emergence of new shoots.

Postemergence herbicides are primarily effective as spot applications for escaped weeds or those not controlled by preemergence herbicides. Registered compounds include fluazifop-P and sethoxydim, grass-selective compounds, and the nonselective herbicides glyphosate and paraquat. Glyphosate should be applied directly on weeds through rope-wick or roller equipment, or as spot applications. Paraquat can be applied at the base of blueberry bushes as a coarse spray, but care is needed to prevent plant contact.

Summary

Perennial weed and brush control is a prerequisite before establishing highbush, lowbush, and rabbiteye blueberry plantings. A combination of cultivation,

Table 10 Herbicides Labeled for Use in Blueberries[a]

Common name	Trade name	Application[b]
Dazomet	Basamid	Preplant soil sterilant
Dichlobenil	Casoron[d]	Preemergence
Diuron	Karmex[d]	Preemergence
Fluazifop-P	Fusilade DX	Postemergence[c]
Isoxaben	Gallery	Preemergence
Glyphosate	Roundup[d]	Postemergence
Hexazinone	Velpar	Preemergence
Napropamide	Devrinol	Preemergence
Norflurazon	Zorial[d]	Preemergence
Oryzalin	Surflan	Preemergence
Paraquat	Gramoxone Extra[d]	Postemergence
Pronamide	Kerb	Preemergence
Sethoxydim	Poast[d]	Postemergence
Simazine	Princep[d]	Preemergence
Sulfosate	Touchdown	Postemergence
Terbacil	Sinbar	Preemergence

[a]Timing, usage rates recommended, locations for use, and other restrictions should be determined from product labels prior to application.
[b]Herbicide application refers to the stage of weed development.
[c]Apply to emerged weeds.
[d]More than one trade name and/or manufacturer is represented for this active ingredient.

mulches, or specified preemergence herbicides can be used in the planting row during plant establishment. Alternatively, glyphosate can be applied through a wiper or paraquat directed at the base of plants. Between crop rows, areas can be clean cultivated, treated with herbicides, or established with a grass sod.

Once crops are established, additional preemergence herbicides can be used for lowbush plantings. Compounds are broadcast following fire pruning or mowing. In rabbiteye or highbush plantings, preemergence herbicides are applied in plant rows or substituted for by mulches. Broad spectrum herbicides, mowed sods, or cultivation is used between plant rows. In highbush and rabbiteye plantings where cultivation is used, annual weeds are the greatest problem. In lowbush plantings, many of the major weeds are perennials. The weed management practices established for blueberries must consider the type of blueberry plantings and the most prevalent weed species.

E. Cane Fruits

Bush berry fruits (blackberry, boysenberry, currants, dewberry, elderberry, huckleberry, gooseberry, loganberry, raspberry) are widely grown throughout the United States, with individual fruits grown in specific geographical regions because of climatic requirements. Canes are generally grown in rows with bare soil or grass sods maintained between rows. Weed management is important for these crops, especially for the brambles (blackberries, boysenberries, dewberries, huckleberries, loganberries, and raspberries), because fruit production is dependent upon new canes which develop from the root suckers of parent canes (Lawson and Wiseman, 1976). The development of new canes depends upon the overall vigor of parent canes.

Cultivation and Mowing

Mechanical cultivation is commonly practiced between plant rows. Since cane fruits have shallow root systems, cultivation should be shallow (5 to 7.5 cm). In new plantings, cultivation can be practiced between plants within a row, but rows quickly fill in to make this method impractical. Ironically, pruning the roots of brambles in established plantings encourages sucker growth and will boost new cane production. In established plantings, cultivation should cease prior to harvest.

In some areas, mowed grass sods are maintained between plant rows to keep out undesirable weeds and minimize soil erosion. Grass species with aggressive growth habits should not be used, to prevent their spread into plant rows. Cane fruits will grow on a wide range of well-drained soils and generally require a lot of moisture. Therefore, it's paramount to control the vigor of grass sods to conserve soil moisture.

Cultural

The longevity of bush berry fruits necessitates the eradication of perennial weeds prior to crop establishment. Once established, these fruits, especially brambles, can form an extensive canopy and shade out many annual weeds (Lawson and Wiseman, 1976). The ability to exclude weeds within the plant row depends on optimizing vegetative growth of crop plants.

Small cereal grains and legumes can be sown as cover crops between plant rows in the fall to prevent establishment of winter annual weeds and to boost soil structure and organic matter. For cover crops which survive the winter, cultivation or herbicides should be used the following spring, prior to cane fruit growth, to prevent competition.

Herbicides

Before applying herbicides, product labels should be read carefully. The genetic diversity exhibited among the many species classified as cane fruits indicates that application of one herbicide registered for use on cane fruits may be safe for one species, but lethal for another.

Herbicide programs for cane fruits can generally be broken down into two categories, those for new or established plantings. For new plantings, a preemergence herbicide is generally not recommended until the fall if cane fruits were planted in the spring. Under some situations, napropamide can be used under new plantings. Fluazifop-P and sethoxydim can also be applied for selective removal of problem grass weeds.

Once established, cane fruits typically tolerate higher rates of herbicides. Additional compounds such as dichlobenil, diuron, norflurazon, and terbacil can be used. Glyphosate, paraquat, and sulfosate can be used for broad-spectrum weed control, but contact with actively growing crop tissue should be prevented. A list of herbicides registered for use in cane fruits is given in Table 11.

Summary

For successful control of weeds in cane fruits eliminate perennial weeds the year before planting using cultivation and herbicides. Napropamide can be applied broadcast before planting or shallow cultivation can be used throughout the first growing season. During the fall and early spring, preemergence herbicides can be applied in the plant row and cultivation, herbicides, or a sod cover crop used between rows. In the fall, a cereal grain or legume cover crop can be sown for winter cover and controlled the following spring. Weed management systems should comprise numerous techniques used interchangeably. Herbicides used within plant rows should be rotated to prevent the buildup of tolerant species.

V. TREE FRUITS

Vegetation management in tree fruit crops has evolved from no control or grazing by animals, to the use of selective herbicides, mechanical cultivation, mulching, or managed grass sods. Each management system has both beneficial and detrimental effects on both the soil and crop. The optimum system for

ble 11 Herbicides Labeled for Use in Cane Fruits[a]

mmon name	Trade name[b]	Application	Crop[c]
zomet	Basamid	Preplant soil sterilant	BL,CU,EL,GO,RA
:hlobenil	Casoron[d]	Preemergence	BL,RA
ıron	Karmex[d]	Preemergence	BL,BO,DE,GO,LO,RA
ːazifop-P	Fusilade DX	Postemergence	BL,BO,CU,DE,EL, GO,HU,LO,RA
/phosate	Roundup[d]	Postemergence	BL,BO,CU,DE,EL,GO, HU,LO,RA
xaben	Gallery	Preemergence	BL,BO,CU,DE,EL,GO, LO,RA
propamide	Devrinol	Preemergence	BL,BO,LO,RA
rflurazon	Zorial[d]	Preemergence	BL,RA
/zalin	Surflan	Preemergence	BL,BO,CU,DE,EL,GO, LO,RA
·aquat	Gramoxone Extra[d]	Postemergence	BL,BO,RA
›namide	Kerb	Preemergence	BL,BO,RA
hoxydim	Poast[d]	Postemergence	BL,BO,LO,RA
ıazine	Princep[d]	Preemergence	BL,BO,LO,RA
fosate	Touchdown	Postemergence	BL,BO,DE,EL,GO,LO,
·bacil	Sinbar	Preemergence	BL,BO,DE,LO,RA

ning, usage rates recommended, locations for use, and other restrictions should be determined from product ːls prior to application.
ade names given are for U.S.A. labels.
., blackberry; BO, boysenberry; CU, currants; DE, dewberry; EL, elderberry; GO, gooseberry; HU, kleberry; LO, loganberry; RA, raspberry.
ɔre than one trade name and/or manufacturer is represented for this active ingredient.

a grower depends upon many factors and thus may be unique for each crop and location.

Design and implementation of weed management systems for tree fruits require the consideration of many factors unique to these crops. Tree fruits, including pome and stone fruits, citrus and nut trees, are maintained long-term on the same site, and serious perennial weed problems are often encountered. Small trees are influenced by annual weeds, but as trees mature, canopy shading of the soil surface suppresses competition from many annuals. Mechanical cultivation is used in many situations, but as trees mature, cultivation equipment is more difficult to maneuver and physical damage to trees is common on high density plantings. Another factor somewhat unique to tree fruits is the influence of the vegetation beneath the tree on insects and rodents. Many insects associated with tree fruits, both beneficial and destructive, overwinter in the soil or plant residues beneath the tree. Few studies have documented the influence of any one plant species on specific insects, although mite damage to apple trees is observed frequently in warm, dry periods with little or no vegetation under the tree. Higher soil surface temperatures in cultivated citrus groves are associated with greater infestations of urst mites compared to those of comparable groves with cover crops (Tucker and Singh, 1983). When trees are dormant, rodent damage to tree trunks and even exposed roots is usually more severe where adequate plant cover exists. Also, poisonous snakes have been reported to hide in tall grasses surrounding nut trees in some areas of Northern California (Lange, 1989).

Herbicide use in tree fruits is also influenced by location of tree foliage and roots. A majority of the roots of tree fruits lie below the zone inhabited by roots of many weed species. This physical separation is the basis for frequent use of residual herbicides which remain in the upper soil profile. In addition, tree foliage is physically separate from many unwanted weeds, making the directed application of nonselective postemergence herbicides a popular means to control weeds.

A. Citrus

Over 95% of the citrus (lemon, lime, grapefruit, orange, tangelo, tangerine) production in the United States takes place in three states, Florida, California, and Texas (Mersie and Singh, 1989). Because citrus trees are grown in subtropical to tropical regions with long growing seasons, weed pressure is heavy and weed management must be timely to minimize yield losses (Mersie and Singh, 1989). Many of the major weeds encountered consist of perennial grasses and vines, particularly in Florida. Weed management practices consist of mechanical cultivation, mowing of grass sods, and use of cover crops, mulches, and herbicides.

Cultivation

Prior to the development of selective herbicides, mechanical cultivation was widely practiced. Tillage is simple to perform and results in improved water penetration into soils prone to surface crust formation. Tillage also can prepare soil for cover crop establishment or incorporation of plant residues. However, tillage can negatively influence citrus production because of rising labor, equipment, and fuel costs. In addition, cultivating sites infested with perennials can spread plant parts, resulting in redistribution of perennial weeds. For citrus, many feeder roots are close to the surface and cultivation often prunes roots, ultimately reducing plant productivity (Tucker and Singh, 1983). On sloped sandy soils, soil erosion via surface water runoff can be severe where cultivation is practiced.

Mowing

On sites susceptible to erosion, a strip of grass or mixed plant species between tree rows is often maintained by mowing. These strips can also be maintained by chemical mowing using infrequent application of herbicides at sublethal doses (Tucker, 1989). This practice has both advantages and disadvantages. Mowing is simple to perform; however, vegetative propagules or weed seeds can be spread by mowing from the vegetative strip to the tree row. Therefore, mowing must be performed before seed formation of plants in the target area. Long-term mowing often shifts the species in row middles from tall growing to short perennial grasses such as bermudagrass. These perennial grasses are very competitive for available nitrogen and water. Air temperatures are generally 1–2°C cooler in groves with mowed sod middles than groves maintained weed-free with herbicides (Mersie and Singh, 1989). Because cold temperatures, especially freezing, can severely damage citrus fruit and trees, this temperature difference may be critical in certain locales.

Cover Crops and Mulches

Some citrus growers raise cover crops in the tree row. Cover crops are used to conserve soil moisture, increase soil organic matter, stabilize soil pH, improve soil nutrient status, reduce soil erosion, and lower weed competition. However, cover crops can also compete with trees, especially for soil moisture. In addition, without proper management, cover crop stands progressively become infested with weeds, resulting in a predominant stand of highly competitive perennial grasses. Therefore, the type of cover crop selected and proper management to minimize competition with the trees are important factors to consider.

Mulches consisting of plant residues can also be used under trees to manage weeds and provide additional benefits as mentioned for cover crops. The use of mulches, however, is often limited by cost and availability. In warm climates, mulches also break down quickly and require frequent replacement. The more intensive and timely management practices required to interplant cover crops in citrus groves or apply and maintain mulches have resulted in the use of periodic applications of herbicides by many citrus growers.

Herbicides

A long growing season, the severity of weed pressure, and the spatial separation between weed and crop provide the basis for widespread use of relatively inexpensive, soil-active herbicides. These herbicides remain active in the upper soil profile of many soils for months. Application of herbicides is traditionally done by spray booms, but increasingly growers are injecting herbicides through portals and applying the chemicals with irrigation (chemigation). Many different preemergence soil applied herbicides are labeled for use.

Selection of the appropriate herbicide for use is determined by tree age, soil type, and target weed species. Because of the relatively long life of citrus groves, mixed stands of mature and immature trees result from replacement of older, unproductive trees. This can be a problem because many soil applied herbicides registered for citrus cannot be used on immature trees. Herbicides registered for use in citrus are listed in Table 12.

A potential substitute for synthetic herbicides is the use of the mycoherbicide Devine. The commercial formulation consists of spores of the fungus *Phytophthora palmivora*, which selectively attacks the perennial milkweed or stranglervine, a weed problem that has caused severe problems in Florida (Tucker and Singh, 1983). Unfortunately, this herbicide is species specific, requiring additional herbicides for overall weed management. However, the success of Devine should encourage the development of other mycoherbicides which may have application for citrus and other crops.

Summary

Citrus production occurs predominantly in areas with long growing seasons where weed pressures are severe. Trees are established and maintained in rows to allow access by equipment for pesticide or fertilizer application and soil or ground cover maintenance. The season-long efficacy of numerous, economical soil-applied herbicides and postemergence herbicides makes herbicides the most widely utilized weed management tool in citrus. This is especially true under trees versus strips between tree rows. During the year of tree establishment, cultivation is also practiced since the use of many soil-applied herbicides is restricted. Sod strips or bare soil is maintained between tree rows. Because of cost and more intensive management, only a minority of growers utilize cover crops or mulches beneath trees. Long-term use of herbicides will shift weed species to those which tolerate or exhibit resistance to these herbicides. Therefore, growers should alternate herbicide chemicals and consider the use of cover crops and/or mulches.

B. Pome and Stone Fruits

Weed control in pome (apple and pear) and stone (peach, cherry, apricot, prune, nectarine, and plum) fruit orchards is a significant component of the overall fruit production system, and the methods utilized have a direct and indirect effect on other components of the system. Because weed management practices vary with the type of pome and stone fruit, tree age, geographical location, weed populations, and soil conditions, it is imperative that growers adopt management practices suitable for their location and crop.

Table 12 Herbicides (Common Names) Labeled for Use in Citrus[a]

Herbicide	Crop						
	Grapefruit	Kumquat	Lemon	Lime	Orange	Tangelo	Tangerine
Ametryn[b]	x				x		
Bromacil[c]	x		x		x		
Diuron[c]	x		x		x	x	x
DSMA[b]	x		x	x	x		x
EPTC[c]	x		x		x		x
Fluazifop-P[b]	x		x	x	x	x	x
Glufosinate[bd]	x	x	x		x		
Glyphosate[b]	x	x	x	x	x	x	x
Isoxaben[c]	x		x		x		
Metolachlor[c]	x	x	x	x	x	x	x
MSMA[b]	x		x	x	x		x
Napropamide[c]	x		x		x		x
Norflurazon[c]	x		x	x	x		x
Oryzalin[c]	x		x		x		
Oxyfluorfen[c]	x	x	x	x	x	x	x
Paraquat[b]	x	x	x	x	x	x	x
Pendimethalin[c]	x	x	x	x	x	x	x
Sethoxydim[b]	x		x	x	x	x	x
Simazine[c]	x		x		x		
Sulfosate[b]	x		x	x	x	x	x
Trifluralin[c]	x		x		x	x	x

[a]Timing, usage rates recommended, locations for use, and other restrictions should be determined from product labels prior to application.
[b]Apply to emerged weeds.
[c]Apply prior to weed emergence.
[d]Registration pending.

Cultivation

Prior to the development of herbicides with residual soil activity, various forms of cultivation were used primarily in orchards to minimize weed interference. Tillage is effective in eliminating annual weeds, but control of perennial weeds is temporary and vegetative propagules are typically spread, ultimately increasing the severity of perennials. In addition, cultivation around the tree destroys small feeder roots in the upper soil profile and may damage the trunk or lower limbs, creating a wound site for the entrance of detrimental insects and/or pathogens. Because pome and stone fruits are frequently grown on sloped soils to encourage drainage of cold air (to minimize frost damage), tillage in these situations also promotes soil erosion. Increasing equipment, fuel, and labor costs have also contributed to shifting from tillage to more cost effective weed control methods.

Mowing

The predominant method of managing row middles for orchards is mowing native vegetation or established grass species. This practice reduces soil erosion and allows greater flexibility for equipment movement through orchards under wet conditions. However, the plants which compose the mowed strip can negatively influence production. In general, broadleaf plants are less desirable in orchards than grass species because they are often hosts for damaging insects or pathogens. Buckhorn plantain, a common weed problem in California apple orchards, is a principal host for the rosy apple aphid (Sieckert et al., 1989).

Proper maintenance of vegetation between tree rows is also important in orchards. In general, humidity within an orchard is greater when vegetation is mowed infrequently. Humidity can affect how quickly tree foliage dries after a dew or rain, ultimately influencing the severity of foliar pathogens such as scab and brown rot. Therefore, vegetative strips should be mowed frequently (Savage, 1988). This is also beneficial in limiting the competition of vegetation with trees for available moisture during dry periods. Mowing favors the establishment of low-growing species, both perennial broadleaves and grasses. However, some highly competitive weeds such as bermudagrass and quackgrass can become the dominant species. Growers may need to monitor the species present in the vegetative strip and reseed desirable species if necessary. The type of perennial grass recommended for row middles is dependent on the geographical region. The proximity of the vegetative strip to the tree is also important, because of the competition for available nutrients, soil, space, and water (Welker and Glenn, 1989).

Cover Crops

Numerous studies have shown that permanent grass sods under pome and stone fruit trees reduce productivity significantly compared to tillage or treatment with herbicides (Hogue and Neilson, 1987; Jordan and Jordan, 1985). However, the negative aspects of cultivation (previously discussed) render this management practice undesirable as a primary means of weed control.

Some studies indicate that companion cropping small cereal grains or legumes with trees or using cover crop residues as mulches has afforded varying levels of weed suppression (Putnam et al., 1983) and improved soil composition, soil structure, and nutrient availability (Hogue and Nielson, 1987; Jordan and Jordan, 1985). However, just as there are positive attributes related to the use of cover crops, there are potentially negative ones. Rodents frequently inhabit vegetated areas, and the incidence of tree girdling is significantly greater where vegetation is present around the base of the tree. Cover crop residues, especially legumes, can release available nitrogen, which deters pome and stone fruit trees from hardening off, rendering trees more susceptible to cold injury during the late fall and winter. Periods of cover crop and tree growth must not directly coincide or there will be competition for available water and nutrients. In addition, flowering vegetation beneath trees can attract pollinating insects such as bees. If insecticides are simultaneously being used to protect developing fruit on the trees, these beneficial insects may be destroyed, impairing pollination of future tree fruit crops. Considering all factors, properly managed cover crops within the tree row may be the best system, but the requirement for more intense management and the added expense lead many growers to the use of herbicide strips in the tree row.

One of the recent suggestions for orchard floor management in newly established orchards is the killed sod technique (Glenn and Welker, 1989; Welker and Glenn, 1990). In short, a living grass sod is established more than 1 year prior to planting trees, then desiccated with glyphosate at the time of planting. By maintaining this killed sod with periodic applications of herbicides, peach tree growth during the first three growing seasons was greater than that of a herbicide strip system. Improvements in soil structure and water infiltration contributed to greater tree growth. The killed sod system developed for peaches may be beneficial for other tree fruits.

Herbicides

The availability of preemergence herbicides as well as broad spectrum postemergence herbicides and their effectiveness in cutting weed control costs, have led to the widespread use of herbicides in maintaining a vegetation-free strip beneath the fruit trees. Numerous herbicides provide long-term control of weeds arising from seeds and postemergence herbicides control both annual and perennial weeds (Table 13). Herbicides are more economical than tillage and cover crops for weed management, and prevent the physical damage to trees that results from tillage. Herbicide use also requires less overall management than cover crops.

Table 13 Herbicides (Common Names) Labeled for Use in Pome and Stone Fruits[a]

	Crop							
Herbicide	Apple	Pear	Apricot	Cherry	Nectarine	Peach	Plum	Prune
Dazomet[b]	x	x	x	x	x	x	x	x
Dichlobenil[c]	x	x		x	x	x	x	x
Diuron[c]	x	x			x	x		
Fluazifop-P[d]	x	x	x	x	x	x	x	x
Glyphosate[c]	x	x	x	x	x	x	x	x
Glufosinate[d]	x	x	x	x	x	x	x	x
Isoxaben[c]	x	x	x	x	x	x	x	x
MSMA[d]	x	x		x		x	x	x
Napropamide[c]	x	x	x	x	x	x	x	x
Norflurazon[c]	x	x	x	x	x	x	x	x
Oryzalin[c]	x	x	x	x	x	x	x	x
Oxyfluorfen[c]	x	x	x	x	x	x	x	x
Paraquat[d]	x	x	x	x	x	x	x	x
Pendimethalin[c]	x	x	x	x	x	x	x	x
Pronamide[c]	x	x		x	x	x	x	x
Sethoxydim[d]	x	x	x	x	x	x	x	x
Simazine[c]	x	x		x		x	x	
Terbacil[c]	x					x		
Trifluralin[c]			x		x	x	x	x
2,4-D[d]	x	x	x	x	x	x	x	x

[a]Timing, usage rates recommended, locations for use, and other restrictions should be determined from product labels prior to application.
[b]Apply preplant; soil sterilant.
[c]Apply prior to weed emergence.
[d]Apply to emerged weeds.

As in all systems, proper use of herbicides is important in long-term weed management. Some herbicides (e.g., glyphosate and 2,4-D) should not contact tree foliage or severe injury may result. Also, many preemergence herbicides should only be used on established plantings. Many growers find one or a few herbicides which achieve a desired level of weed control and continually use those products for many years. However, this practice has shifted the weed species present in the tree row to those exhibiting a natural level of tolerance (Skroch et al., 1975), or, as is more frequently reported, species have developed resistance to compounds which previously were effective.

Summary

Maintenance of pome and stone fruits in the same location over a long period requires eradication of perennial weeds prior to planting. This is typically done through the use of nonselective herbicides in combination with tillage. An alternative is the growth of a grass sod for 1 to 2 years prior to planting and killing the sod with herbicides prior to planting. There are numerous factors to consider before implementing a weed management system in new or established orchards. Cultivation, mowing, cover crops, and herbicides are used to manage weeds in orchards, and there are advantages and disadvantages for each method. Although most growers prefer the most simple and cost-effective method available, integrating various methods is a sound approach for maintaining long-term production and optimal stewardship of our natural resources.

VI. NUT CROPS

Many of the weed management practices employed by citrus growers are utilized in nut crop orchards, although the geographic locations and weed species encountered may be different. Since many nut trees are grown in areas with long growing seasons, weed control is necessary over a longer period than in northern temperate climates. A popular practice is mowing native vegetation or grasses in row middles or tillage of row middles, and chemical management of strips beneath trees. Mowing row middles is typically practiced on sloped land susceptible to soil erosion. Many growers who specialize in nut crop production typically employ trickle irrigation to prevent competition for water between trees and sod.

The popularity of vegetation-free strips beneath trees is encouraged by factors in addition to easy management and minimal interference with tree growth. In one California study, the incidence of ant damage to nutmeats was higher in orchards where vegetation was maintained beneath almond trees (Klonsky and Elmore, 1989). Also, many different nuts are mechanically harvested, and the presence of excessive vegetation beneath trees decreases the efficiency of nut harvest (Lange, 1989). Nut trees typically have deep root systems, allowing the use of many herbicides with persistent soil activity. One exception are almonds, which are typically grafted on peach rootstocks, and have shallow feeder roots (Lange, 1989).

Effective weed control is more critical to newly planted trees (up to about 4 years old) than to established trees, since many weeds are shaded out as trees grow larger. However, vining perennials such as poison ivy, trumpet creeper, and field bindweed can reduce crop yields or harvesting efficiency in even large trees.

Numerous herbicides, both preemergence and postemergence, are registered for use in nut crops (Table 14). In situations where perennial weeds are

Table 14 Herbicides Labeled for Use in Nut Crops[a]

Common name	Trade name[b]	Crop[c]
Dazomet[e]	Basamid	FI,WA
Atrazine[f]	Aatrex-Nine-O[d]	MA
Dichlobenil[f]	Casoron[d]	AL,FI,WA
Diuron[f]	Karmex[d]	MA,PE,WA
EPTC[f]	Eptam[d]	AL,WA
Isoxaben[f]	Gallery	AL,FI,MA,PE,PI,WA
Glyphosate[g]	Roundup[d]	AL,FI,MA,PE,PI,WA
Fluazifop-P[g]	Fusilade DX	AL,FI,MA,PE,PI,WA
MSMA[g]	MSMA[d]	AL,WA
Napropamide[f]	Devrinol	AL,CH,FI,HA,HI,PE, PI,WA
Norflurazon[f]	Zorial[d]	AL,FI,PE,WA
Oryzalin[f]	Surflan	AL,FI,MA,PE,PI,WA
Oxyfluorfen[fg]	Goal[d]	AL,CA,CH,FI,HI,MA, PE,PI,WA
Paraquat[g]	Gramoxone Extra[d]	AL,FI,MA,PE,PI,WA
Pendimethalin[f]	Prowl[d]	AL,PI,WA
Sethoxydim[g]	Poast[d]	AL,FI,MA,PE,PI,WA
Simazine[f]	Princep[d]	AL,FI,MA,PE,WA
Sulfosate[g]	Touchdown	AL,FI,MA,PE,PI,WA
Terbacil[f]	Sinbar	PE
Trifluralin[f]	Treflan[d]	AL,PE,WA

[a]Timing, usage rates recommended, locations for use, and other restrictions should be determined from product labels prior to application.

[b]Trade names given are for U.S.A. labels.

[c]AL, almond; CA, cashew; CH, chestnut; FI, filbert; HA, hazelnut; HI, hickory nut; MA, macademia nut; PE, pecan; PI, pistachio; WA, walnut (label restrictions are made between black and English walnuts).

[d]More than one trade name and/or manufacturer is represented for this active ingredient.

[e]Apply preplant; soil sterilant.

[f]Apply prior to weed emergence.

[g]Apply to emerged weeds.

present, one or more applications of registered postemergence herbicides such as glyphosate are necessary prior to the use of longer-lived preemergence herbicides. Growers are encouraged to diversify or rotate herbicide chemistryl for weed control.

Development of weed management practices for nut crops is dependent on soil and climatic conditions, and the weeds present. Vegetation-free areas beneath trees are typically maintained to facilitate tree growth and harvesting efficiency.This is mainly accomplished with herbicides. Vegetation between tree rows is usually mowed or cultivated.

VII. CONCLUSION

Successful production of horticultural crops requires effective weed management practices. The wide diversity of crops grown, unique cultural practices, and varying environmental conditions create a multitude of weed control problems and methods for management. Each method will optimally control only certain weeds, while encouraging others. Although herbicides are the major method for weed management, the most prudent approach to weed control in horticultural crops is to employ a combination of methods and to be flexible as weed populations shift, herbicide registrations change, and perennial crops age.

APPENDIX

List of Weeds Mentioned in Chapter

Common name	Scientific binomial	BAYER code
Arrowhead	*Sagittaria sagittifolia* L.	SAGSA
Balsam poplar	*Populus balsamifera* L.	POPBA
Bermudagrass	*Cynodon dactylon* (L.) Pers.	CYNDA
Brackenfern	*Pteridium aquilinum* (L.) Kuhn	PTEAQ
Buckhorn plaintain	*Plantago lanceolata* L.	PLALA
Common ragweed	*Ambrosia artemisiifolia* L.	AMBEL
Field bindweed	*Convolvulus arvensis* L.	CONAR
Hemp dogbane	*Apocynum cannabinum* L.	APPCA
Johnsongrass	*Sorghum halepense* (L.) Pers.	SORHA
Kentucky bluegrass	*Poa pratensis* L.	POAPR
Poison ivy	*Rhus radicans* L.	TOXRA
Rice cutgrass	*Leersia oryzoides* (L.) Sw.	LEROR
Quackgrass	*Agropyron repens* (L.) repens	AGRRE
Stranglervine	*Morrenia odorata* (H. and A.) Lindl.	MONOD
Tall goldenrod	*Solidago altissima* L.	SOOAL
Trumpetcreeper	*Campsis radicans* (L.) Seem. ex Bureau	CMIRA
Sweetfern	*Comptonia peregrina* (L) L. Coult.	COVPE
Yellow nutsedge	*Cyperus esculentus* L.	CYPES

REFERENCES

Abdul-Baki, A. A. and Teasdale, J. R. (1993). A no-tillage tomato production system using hairy vetch and subterranean clover mulches, *HortScience 28*: 106–108.

Altieri, M. A., Letourneau, D. K., and Davis, J. R. (1983). Developing sustainable agroecosystems, *Bioscience, 33*: 45–49.

Appleton, B. L. and Derr, J. F. (1988). The wide world of geotextiles, *Grounds Maintenance 23*(4): 42–48.

Appleton, B. L., Derr, J. F., and Ross, B. B. (1990). The effect of various landscape weed control measures on soil moisture and temperature, and tree root growth, *J. Arboriculture 16*(10): 264-268.

Ashley, R. A., ed. *New England Vegetable Management Guide*, University of Connecticut, Storrs, Connecticut, pp. 23–24.

Ball, J. (1987). Efficient monitoring for urban IPM program, *J. Arboric. 13*(7): 174–177.

Bellinder, R. R., Binning, L. K., Yourstone, K. S., Bonanno, A. R., Gorski, S. F., Majek, B. A., Neary, P. E., Baron, J. J., Holmdal, J., and Wallace, R. W. (1993). Oxyfluorfen under clear polyethylene film controlled weeds in transplanted curcubits, *Weed Tech. 7*: 585-593.

Billeaud, L. A. and Zajicek, J. M. (1989). Mulching for weed control, *Grounds Maintenance 24*(2): 16, 20–107.

Cain, J. C. and Eck, P. (1966). Blueberry and cranberry, in *Fruit Nutrition* (N. F. Childers, ed.), Somerset Press, Inc., Somerville, New Jersey, pp. 101–129.

Carpenter, P. L. (1983). Chemical weed control in container-grown nursery stock, *HortScience 8*(5): 485-486.

Cultural Practices. (1988). *Blueberry Science* (P. Eck, ed.), Rutgers University Press, New Brunswick, New Jersey, pp. 135–169.

Demoranville, I. E. (1984). *Weeds of Massachusetts Cranberry Bogs,* Part I, University of Massachusetts Cooperative Extension Service, Amherst, Massachusetts, #SP–148.

Demoranville, I. E. (1986). *Weeds of Massachusetts Cranberry Bog,* Part II, University of Massachusetts Cooperative Extension Service, Amherst, Massachusetts, #SP–149.

Derr, J. F. (1991). Research: landscape fabrics for weed control in commercial and home landscapes, *Vir. Nur. Assoc. Newsletter 61*(5): 48.

Derr, J. R., and Appleton, B. L. (1989). Weed control with landscape fabrics, *J. Environ. Hortic. 7*(4): 129-133.

Doll, C. C. (1980). Geese for weeding strawberries, in *The Strawberry* (N. F. Childers, ed.), Horticultural Publications, Gainesville, Florida, pp. 119–120.

Eck, P. ed. (1988). *Blueberry Science*, Rutgers University Press, New Brunswick, New Jersey, pp. 3–9.

Fleming, H. K. and Alderfer, R. B. (1956). *Cultivation and concord grape production, Penn State Univ. Agr. Exp. Stat. Bull.* #616.

R. Foster, ed. (1994). *Midwest Vegetable Production Guide for Commercial Growers* Purdue University Cooperative Extension Service, West Lafayette, Indiana, p. 19.

Fretz, T. A. (1972). Weed competition in container-grown Japanese holly, *HortScience 7*(5): 485-486.

Gartner, J. B. (1978). Using bark and wood chips as a mulch for shrubs and evergreens, *Am. Nurseryman, 147*(10): 9, 53–55.

Gilliam, C. H., Wehtje, G., Eason, J. E., Hicks, T. V., and Fare, D. C. (1989). Weed control with Gallery and other herbicides in field-grown nursery crops, *J. Environ. Hort. 7*(2): 69–72.

Glenn, D. M. and Welker, W. V. (1989). Cultural practices for enhanced growth of young peach trees, *Am. J. Alternative Agric. 4*: 8–11.

Gorski, S. F. and Wertz, M. K. (1987). Tomato (*Lycopersicon esculentum*) and eastern black nightshade (*Solanum ptycanthum*) tolerance to acifluorfen, *Weed Tech. 1*: 278–281.

Harris, R. W. (1983). Soil Management, in *Arboriculture: Care of Trees, Shrubs, Vines, in the Landscape*, Prentice-Hall Press, Englewood Cliff, New Jersey, pp. 356–361.

Hemphill, D. D. (1980). Weed control in strawberries, in *The Strawberry* (N. F. Childers, ed.), Horticultural Publications, Gainesville, Florida, pp. 309–317.

Hogue, E. J. and Neilsen, G. H. (1987). Orchard floor vegetation management, *Hort. Rev. 9*: 377–430.

Horwith, B. (1985). A role for intercropping in modern agriculture, *BioScience 35*: 286–291.

Jordan, L. S. and Jordan, J. L. (1985). Tillage reduction in orchards with cover crops and herbicides, in *Weed Control in Limited Tillage Systems* (A. F. Wiese, ed.), Weed Science Society of America, Champaign, Illinois, pp. 127–139.

Kato, A. and Okuda, Y. (1983). Paraquat resistance in *Erigeron canadensis* L., *Weed Res.* (Japan) *28*: 54–56.

Klonsky, K. and Elmore, C. L. (1989). Almond orchard floor management costs, *California Agric. 43*: 25–26.

Lange, A. H. (1989). Horticultural crops, in *Principles of Weed Control in California*, Thomson Publishers, Fresno, California, pp. 371–376.

Lawson, H. M. and Wiseman, J. S. (1976). Weed competition in spring-planted raspberries, *Weed Res. 16*: 155–162.

Leidner, J. (1994). Hot water kills weeds, *Progressive Farmer*, July, p. 24.

Mahr, D. L., Jeffers, S. N., Stang, E. J., and Roper, T. R. (1990). *Cranberry Pest Management in Wisconsin*, Univ. of Wis. Coop. Ext. Serv., Madison, Wi #A-3276.

Majek, B. A. (1994). Vegetables, *Weed Control Manual*, Meister Publishing Co., Willoughby, Ohio, pp. 178–222.

Martin, C. A., Ponder, H. G., and Gilliam, C. H. (1991). Evaluation of landscape fabrics in suppressing growth of weed species, *J. Environ. Hort. 9*(1): 38–40.

Masiunas, J. B., Weston, L. A., and Weller, S. C. (1995). The impact of rye cover crops on weed populations in a tomato cropping system, *Weed Sci.*, In press.

McNiel, R. E. (1993). "Nurseries across the United States: Principles, practices, and technologies," *Proceedings of the Southern Nurserymen's Association*. Joint International Symposium Efficiencies of Producing and Marketing Landscape Plants, Boskoop, Netherlands, pp. 20–25.

McNiel, R. E. and Weston, L. A. (1992). Landscape fabrics as weed blockers: differences in weed cover suppression, *Proceedings S. Nurserymen's Assoc.* 37:301–304.

Mersie, W. and Singh, M. (1989). Benefits and problems of chemical weed control in citrus, *Rev. Weed Sci. 4*: 59–70.

Mortensen, D. A., Meyer, G. E., and VonBargen, K. (1995). Plant sensing and intermittent herbicide application, *Rev. Wed Sci. 7*: In press.

Nicholson, A. G. and Wien, H. C. (1983). Screening of turfgrasses and clovers for use as living mulches in sweet corn and cabbage, *J. Am. Soc. Hortic. Sci. 108*: 1071-1076.

Parker, R. E., Holstan, J. T., Jr., and Fulgham, F. E. (1965). Flame cultivation equipment and techniques, *USDA Res. Rept. #86*.

Phatak, S. C. (1992). An integrated sustainable vegetable production system, *HortScience 27*: 738-741.

Pierce, L. C., ed. (1987). *Vegetables: Characteristics, Production and Marketing* Wiley & Sons, Inc., New York, pp. 163-172.

Pool, R. M., Dunst, R. M., and Lakso, A. N. (1990). Comparison of sod, mulch, cultivation, and herbicide floor management practices for grape production in non-irrigated vineyards, *J. Am. Soc. Hortic. Sci. 115*: 872-877.

Powell, M. A., Skroch, W. A. and Bilderback, T. E. (1989). Landscape mulch evaluation: A three-year study, *Proceedings S. Nurserymen's Assoc. 34*: 274-278.

Putnam, A. R. (1990). Vegetable weed control with minimal herbicide inputs, *HortScience 25*: 155-159.

Putnam, A. R. and DeFrank, J. (1983). Use of phytotoxic plant residues for selective weed control, *Crop Protection 2*: 173-181.

Putnam, A. R., DeFrank, J., and Barnes, J. P. (1983). Exploitation of allelopathy for weed control in annual and perennial cropping systems, *J. Chem. Ecol. 9*: 1001-1010.

Rakow, D. (1992). Mulching: Benefits backed by survey, *Arbor Age, 12*(9):22-29.

Robinson, D. W. (1988). Mulches and herbicides in ornamental plantings, *HortScience 23*(3): 547-552.

Ryan, G. F. (1970). Resistance of common groundsel to simazine and atrazine, *Weed Sci., 18*: 614-616.

Savage, E. E. (1988). Do peaches need cultivation? In *The Peach* (N. F. Childers and W. B. Sherman, eds.), Horticultural Publications, Gainesville, Florida, pp. 629-632.

J. S. Shoemaker, ed. (1978). Grapes, in *Small Fruit Culture*, AVI Publishing Co., Inc., Westport, Connecticut, pp. 46-50.

Sieckert, E. E., Elmore, C. L., and Lange, A. H. (1989). Horticultural crops, in *Principles of Weed Control in California*, Thomson Publishers, Fresno, California, pp. 393-400.

Skroch, W. (1990). Short course notes, *Ornamentals*, North Carolina State University, Raleigh.

Skroch, W. A., Sheets, T. J., and Monaco, T. J. (1975). Weed populations and herbicide residues in apple orchards after five years, *Weed Sci. 23*: 53-57.

Smagula, J. M. and Yarborough, D. E. (1990). Changes in the lowbush blueberry industry, *Fruit Varieties J. 44*: 72-77.

Smeda, R. J. and Putnam, A. R. (1988). Cover crop suppression of weeds and influence on strawberry yields, *HortScience, 23*: 132-134.

Tucker, D. P. H. (1989). Planning, conduct and evaluation of citrus weed control programs, *Citrus and Vegetable Magazine*, May, p. 18.

Tucker, D. P. H. and Singh, M. (1983). Section XII-Florida citrus weed management, in *Florida Citrus Integrated Pest and Crop Management Handbook*, University of Florida Cooperative Extension Service, #SP-14, pp. 1-31.

Wehtje, G. and Gilliam, C. H. (1991). Weed control in field-grown holly, *J. Environ. Hortic. 9*(1): 29-32.

Welker, W. V. (1984). The effects of oryzalin alone and in combination with diuron and simazine on young peach trees, *HortScience 19*(6): 824-826.

Welker, W. V. and Glenn, D. M. (1989). Sod proximity influences the growth and yield of young peach trees, *J. Am. Soc. Hortic. Sci. 114*: 856-859.

Welker, W. V. and Glenn, D. M. (1990). Peach tree growth as influenced by grass species used in a killed-sod planting system, *HortScience 25*: 514-515.

Whitcomb, C. E. (1980). Effects of black plastic and mulches on growth of landscape plants, *J. Arboriculture, 6(1)*: 10-12.

Whitwell, T., Kelly, J., and George, B. (1990). *Weed Management in Ornamental Plantings*. Clemson University Cooperative Extension Service, Clemson, South Carolina, #EB137.

Wiley and Sons, Inc., (1993). *Crop Protection Chemicals Reference*, New York, p. 126-130.

Yarborough, D. and DeGomez, T. (1990). *Weed Management in Lowbush Blueberry Fields*, Univ. of Maine Cooperative Extension Service, Orono, #236.

RELATED REFERENCES

Ashton, F. M. and Monaco, T. J. (1991). *Weed Science, Principles, and Practices*, Wiley & Sons, Inc., New York.

Haynes, R. J. (1980). Influence of soil management practice on the orchard agro-ecosystem, *Agro-Ecosystems 6*: 3-32.

Rice, R. P., Jr. (1992). *Nursery and Landscape Weed Control Manual*, Thomson Publishers, Fresno, California.

Shoemaker, J. S. (1978). *Small Fruit Culture*, AVI Publishing Co., Inc., Westport, Connecticut.

Standifer, L. C. and Beste, C. E. (1985). Weed control methods for vegetable production with limited tillage, *Weed Control in Limited Tillage Systems* (A. F. Wiese, ed.), Weed Science Society of America, Champaign, Illinois, pp. 93-100.

15

Weed Management Systems for Turfgrass

S. Wayne Bingham
Virginia Polytechnic Institute and State University, Blacksburg, Virginia

William J. Chism
Rhône-Poulenc Ag Company, Research Triangle Park, North Carolina

Prasanta C. Bhowmik
University of Massachusetts, Amherst, Massachusetts

I. INTRODUCTION

A. General Comments

Over 12.2 million hectares of turfgrass are managed in America (Table 1). These turfgrass areas contribute to our lives through a variety of functional, aesthetic, and recreational roles. Turfgrass provides soil stabilization, dust, and erosion control; clean, cool air; a recreational surface; and ornamentation to businesses, homes, and parks. Turfgrass management consists of all activities that influence the growth and maintenance of the grass. Proper management must consider and respond to biological conditions such as turf type, fertility, wear, pest problems, and environmental conditions such as light, rain, drought, shade, temperature fluctuations, snow, and ice. Weeds reduce the erosion control and cooling effects of turf, the visual aesthetics of parks and gardens, and the playability of golf courses and athletic fields. Because of their adaptability weeds quickly fill in open areas left by soil fertility or pH problems, insect or disease problems, shady or droughty areas, and newly seeded turf. Many turfgrass management strategies have been designed to maximize turf growth

Table 1 Hectares of Maintained Turfgrass in the United States in 1988

Category	Hectares millions
Home lawns	2.6
Golf courses	0.6
Sod production	0.4
Commercial and public grounds	3.3
Roadsides	4.1
Cemeteries	0.2
Other	1.0
Total	12.2

Source: Duble (1989).

and minimize the influence of weedy species. But, despite prescription fertilizer, optimized mowing, adapted turf types, disease and insect control programs, and pre- and postemergence herbicide programs, weeds continue to be a problem in turfgrass.

B. Characteristics of Turf Weeds

A plant becomes undesirable in turf when it disrupts the uniformity of the surface because of differences in density, color, texture, leaf blade size and shape, growth habit, smoothness, rooting, or wear resistance. Turf weeds typically exhibit one or more of the following characteristics: low growth (mowing avoidance), rapid seed production, flowers set low to the ground, tolerance of extreme environmental condition (e.g., compacted soil [knotweed], wet soil [nutsedge], or low light [ground ivy]), meristematic region low to the ground or below ground, adaptation to cycles of other weeds (e.g., annual bluegrass alternating with goosegrass), or high rate of seed germination. The presence of weeds may also be indicative of other problems inherent at that site, e.g., crabgrass species in thin turf, knotweed in compacted soil, or ground ivy in shady areas.

II. WEED BIOLOGY

Climatic, edaphic (soil), and biotic factors that characterize an environment provide an appropriate and stable matrix for the turfgrass within which the weeds determine their invasion, abundance, and distribution. Persistence is a measure of the adaptive potential of a weed species that enables it to grow in a turfgrass environment. However, all weeds will not necessarily pose a serious threat to turfgrass. The different turfgrass environments (low maintenance or high maintenance areas) will determine the invasion and persistence of weed species and thus particular turfgrass-weed associations.

Turfgrass species have a specific zone of adaptation. Similarly weed species have specific adaptation. For example, yellow nutsedge is adapted to temperate regions, while purple nutsedge is adapted to tropical regions.

The important climatic factors of the environment that relate to persistence of weeds are light, temperature, water, wind, and humidity. Light intensity, quality, and duration are important in determining the growth, reproduction, and distribution of weeds. The evolution of various ecotypes of weeds within a weed species is partly a response to photoperiod. Tolerance to shading is a major adaptation that enables weeds to persist in a turfgrass habitat. For example, ground ivy is widely adapted to shady areas or low light incidence areas.

Temperature of air and soil, and length of frost free periods are major limits to the distribution of weeds. Soil temperature as it relates to seed germination and low-temperature effects of soil on seed dormancy during temperate zone winters are critical factors in the adaptation and persistence of weeds. Weed persistence in a given environment is heavily influenced by dormancy of weed seeds.

Several factors influence the growth of each species, including climate, soil, and cultural or maintenance practices. Cultural practices applied to the turfgrass during routine maintenance exerts the greatest influence on weed persistence. For example, overirrigation promotes germination of annual bluegrass or crabgrass that thrive in a moist environment. The timing and frequency of aerification, mowing, or vertical mowing can influence the germination of some weed species. Soil fertility can affect the growth of certain weed species. Annual bluegrass responds positively to high levels of soil phosphorus.

Soil factors influencing weed persistence include soil water, aeration, temperature, pH, and fertility level. Prostrate knotweed tolerates both acid and alkaline soil conditions, while it also thrives under compact soil conditions. Red sorrel, on the other hand, is abundant in acid soils.

Annual weeds are prolific seed producers, and seed is their primary mechanism for survival. Perennial weeds possess the additional mechanisms of buds, bulbs, rootstocks, rhizomes, and tubers that favor vegetative propagation. Many weeds such as crabgrass are low growing and survive repeated mowings. Other mechanisms include effective means of dissemination of seeds, ability of seeds and vegetative propagules to resist detrimental factors of the environment, and dormancy of seeds and propagules that remain in the soil. Vegetative propagules include rhizomes, tubers, bulbs, and rootstocks.

Weeds exhibit the greatest influence on turfgrass through direct competition for light, moisture, and nutrients. The degree of competition by various weed species is related to turfgrass species, density of turfgrass stand, as well as cutting height. Weeds generally prevail under a thin and weak stand of any turfgrass species. Weeds also greatly reduce the aesthetic value of turfgrass. Under stress conditions, turfgrass may lose its quality and turfgrass cover, resulting in a thin strand of turfgrass or even bare ground. Weeds often occupy these areas more rapidly than any desired turfgrass species.

It is essential to identify the weed species present before a timely and economical weed management program can be adopted. It is a good idea to identify problem weed species as grass species or broadleaf species. All grass weeds have relatively narrow leaves with veins running parallel to each other (Figure 1). Most grasses have fibrous root systems with all stems meeting at the crown. Nongrass monocotyledonous plants such as wild garlic and nutsedge are often confused with grasses. Broadleaf weeds have leaves with netlike venation (Figure 2). Most have a tap root system with many branches; however, a few, such as the plantains, have fibrous roots.

Some of the weeds commonly found in turfgrass can be grouped into three categories based on their life cycle: annuals, perennials, and biennials.*

A. Annual Grass Weeds

Annual weeds are those that live for one growing season only. Annual bluegrass (Figure 3) is a winter annual; however, it germinates both in spring and in early fall. As one of the worst turf weeds in golf courses, it grows profusely and flowers at any mowing height. Annual bluegrass has shallow, fibrous roots and flattened stems. Soft apple-green leaves are folded. Its membranous ligule is larger than in Kentucky bluegrass. Blades are parallel, with boat shaped tips. The greenish white panicle produces many seeds. Hot, dry weather causes it to brown out and die, resulting in brown spots during the summer.

Barnyardgrass (Figure 4) is a coarse-bladed summer annual grass that produces leaves that are rolled in the bud. Blades are flat with sparse hair on edges and thick midrib. No ligules or auricles are present. Sheaths are flat and compressed with barbed spikelets and awns. It usually germinates later than crabgrass in spring.

Large crabgrass (Figure 5) is also known as hairy crabgrass. It is a summer annual that is widespread and has long stems often rooting at the basal nodes. Yellow-green leaves are rolled in the bud with hairy edges. Blades are flat with a sharp point, broad collar, membranous ligule, and hairy sheath. It reproduces by seed. The seed head, composed of 3-10 fingerlike racemes or spikes, may appear purple to tan in color. Smooth crabgrass is very similar, without hairs. Crabgrass is very competitive and withstands dry weather conditions better than most turfgrass species, because of its physiological makeup

*Weed illustrations excerted from *Selected Weeds of the U.S.*, *Weeds of the North Central States*, and *Ontario Weeds*.

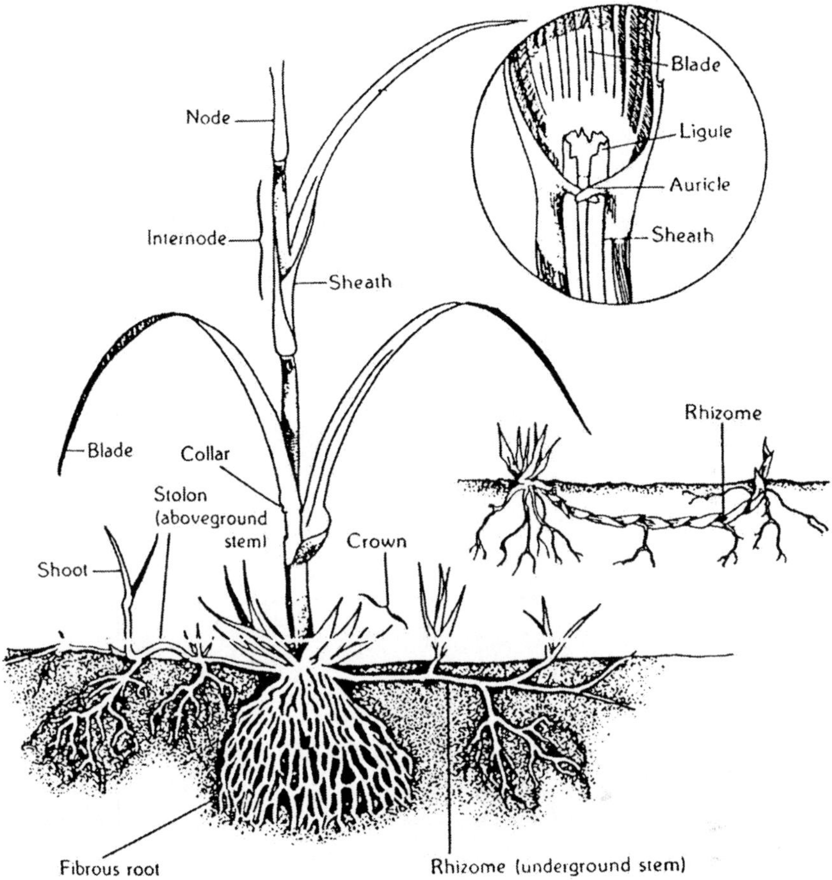

Figure 1 Major vegetative parts of grass weeds.

(C_4, or efficient species). Light is required for seed germination and crabgrass is abundant in thin stands of turfgrass.

Goosegrass (Figure 6) is also known as silver crabgrass, crowfoot grass, wire grass, and yard grass. It is a summer annual, heavily infesting the Midwest, East, and South, especially in compacted soil. This annual forms a tiller of flat stems, usually silvery at the center. Leaves are folded in the bud with overlapping sheaths. It has a membranous ligule and sparse hairs on the collar. The seed head forms 2-10 (mostly 4 to 5) fingerlike spikes much broader than those of crabgrass. It germinates 3 to 6 weeks later in the season than crabgrass.

Yellow foxtail (Figure 7) is a summer annual, usually germinating between crabgrass and goosegrass in the spring, with an erect round stem. Leaves are

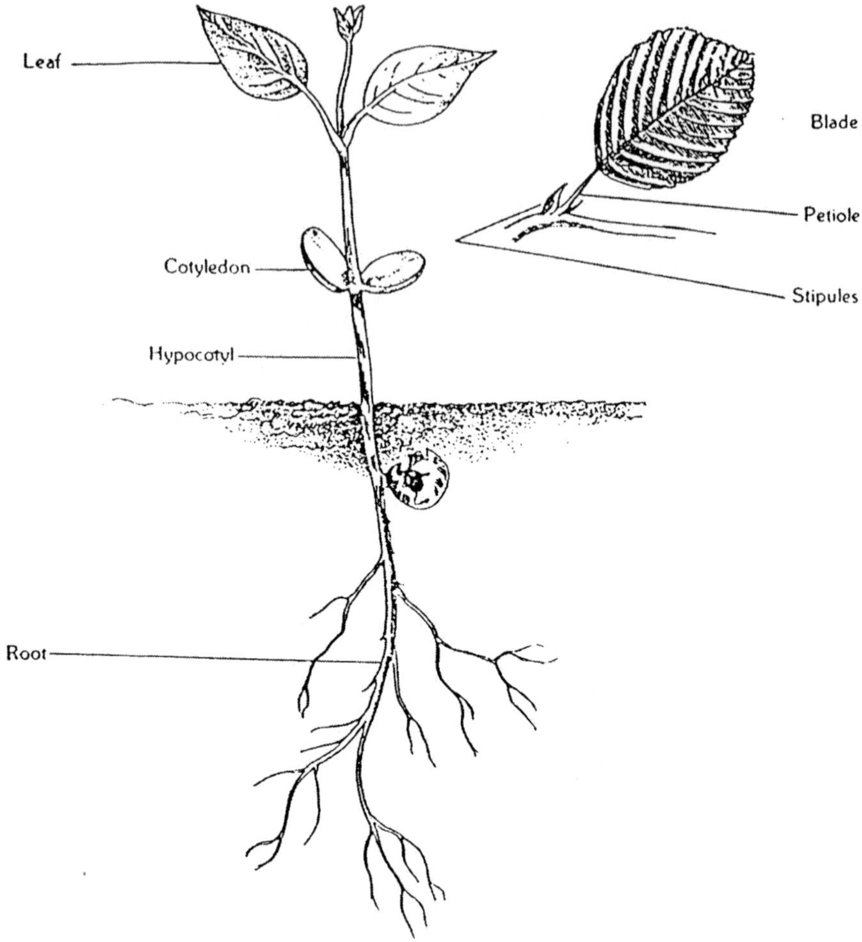

Figure 2 Major vegetative parts of broadleaf weeds.

rolled in the bud. Blades are flat with long sparse hairs and prominent midrib. The collar is narrow and a continuous ligule is a ring of hair. The seed head is a bristly spikelike panicle. Green, giant, and bristly foxtail are species of the same genus.

B. Annual Broadleaf Weeds

Carpetweed (Figure 8) is a summer annual that spreads rapidly during hot weather. The branched stems form prostrate mats. Leaves are slender and radiate from the nodes. Flowers are small with greenish white petals and arise from the nodes.

Figure 3 Annual bluegrass (*Poa annua* L.) POAAN.

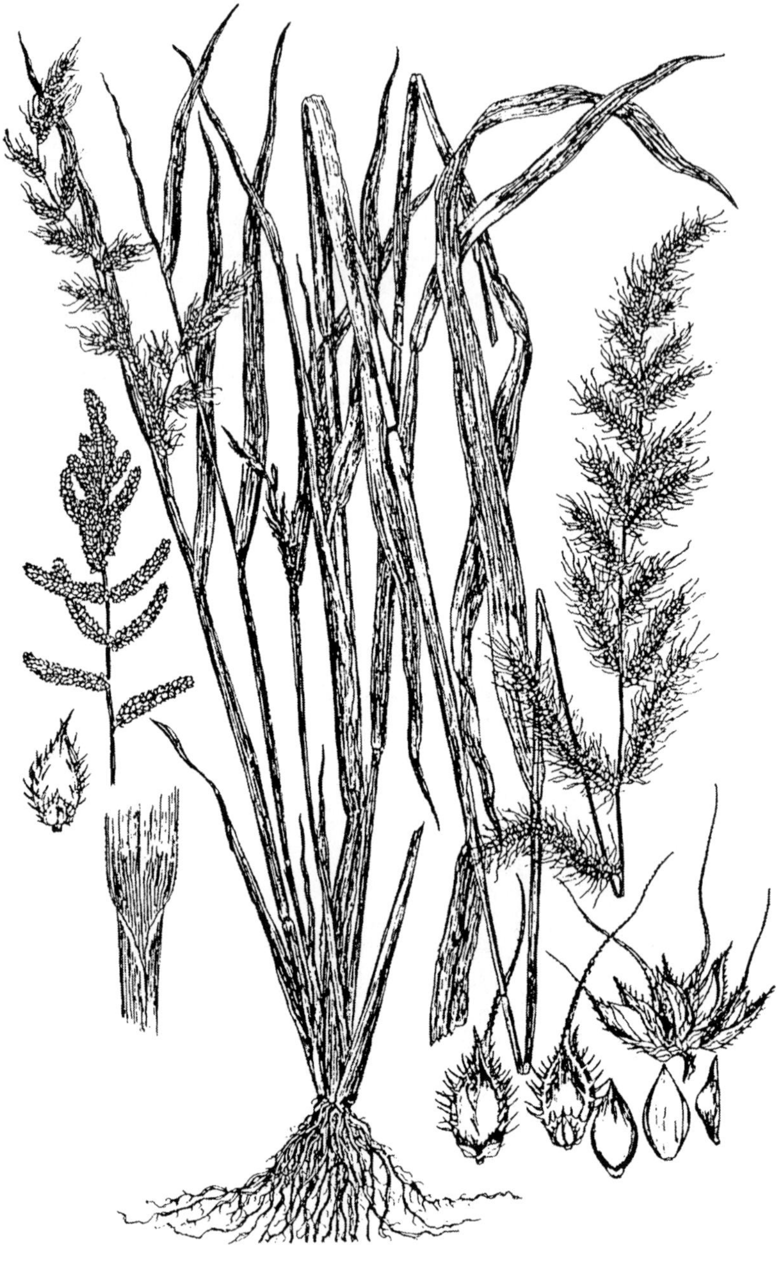

Figure 4 Barnyardgrass (*Echinochloa crusgalli* (L.) Beauv.) ECHCG.

Figure 5 Large crabgrass (*Digitaria sanguinalis* (L.) Scop.) DIGSA.

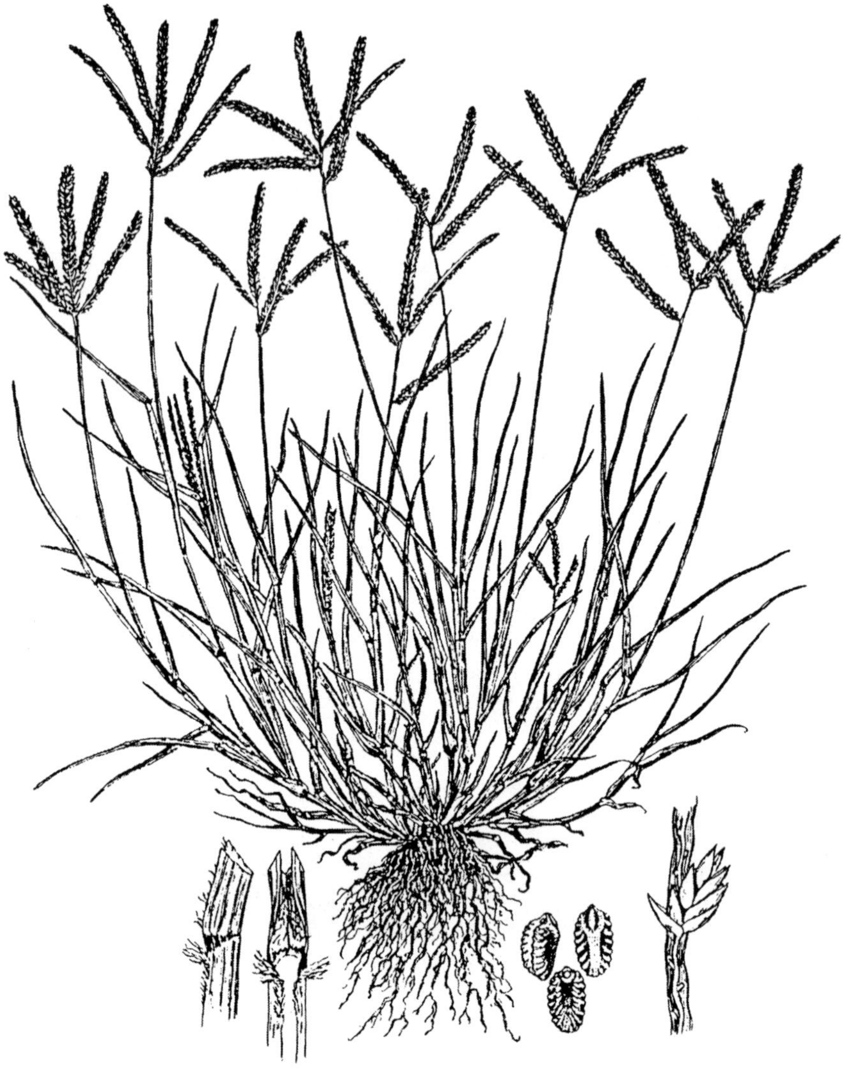

Figure 6 Goosegrass (*Eleusine indica* (L.) Gaertn.) ELEIN.

Common chickweed (Figure 9) is a winter annual with low, slender, branched creeping stems with rows of hair, and potentially with roots at the nodes. Leaves are opposite, yellowish green, oval, or elliptic shaped with pointed tip. White five petaled flowers are small but conspicuous. It grows best during cool, wet weather. It rarely tolerates hot, dry conditions. It is sometimes used as an indicator plant for moisture stress.

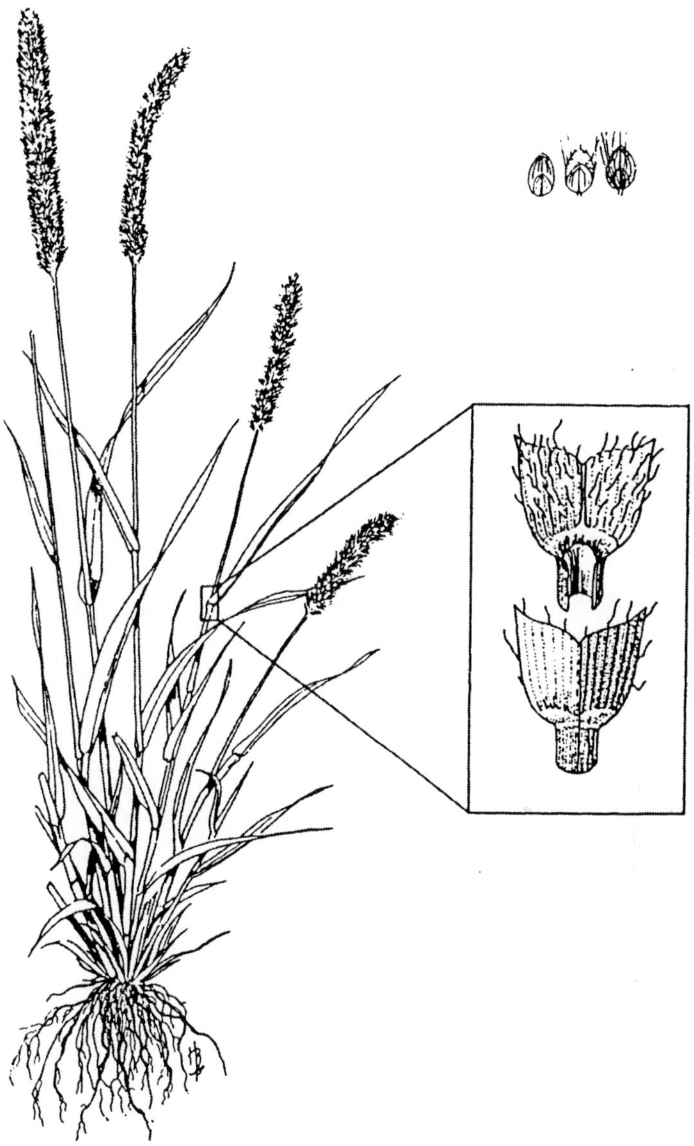

Figure 7 Yellow foxtail (*Setaria glauca* (L.) Beauv.) SETLU.

Henbit (Figure 10) is a winter annual that grows mainly as an upright plant; however, this winter annual may root at stem nodes and also reproduce by seed. Because it is a member of the mint family, the stems are square and are branched near the ground. Leaves are deeply veined, hairy, coarsely

Figure 8 Carpetweed (*Mollugo verticillata* L.) MOLVE.

toothed, and rounded. Flowers are tubular and pink to purplish. It dominates in early spring.

Prostrate knotweed (Figure 11) is a summer annual that germinates quickly in the early spring and is often mistaken as crabgrass seedlings. Wiry stems are prostrate and branching, quickly forming a mat. Bluish green leaves are small and narrow at the base with a rounded tip. The small pale green flowers with pink tips produce dark, three-cornered seeds. It is known to prosper in hard,

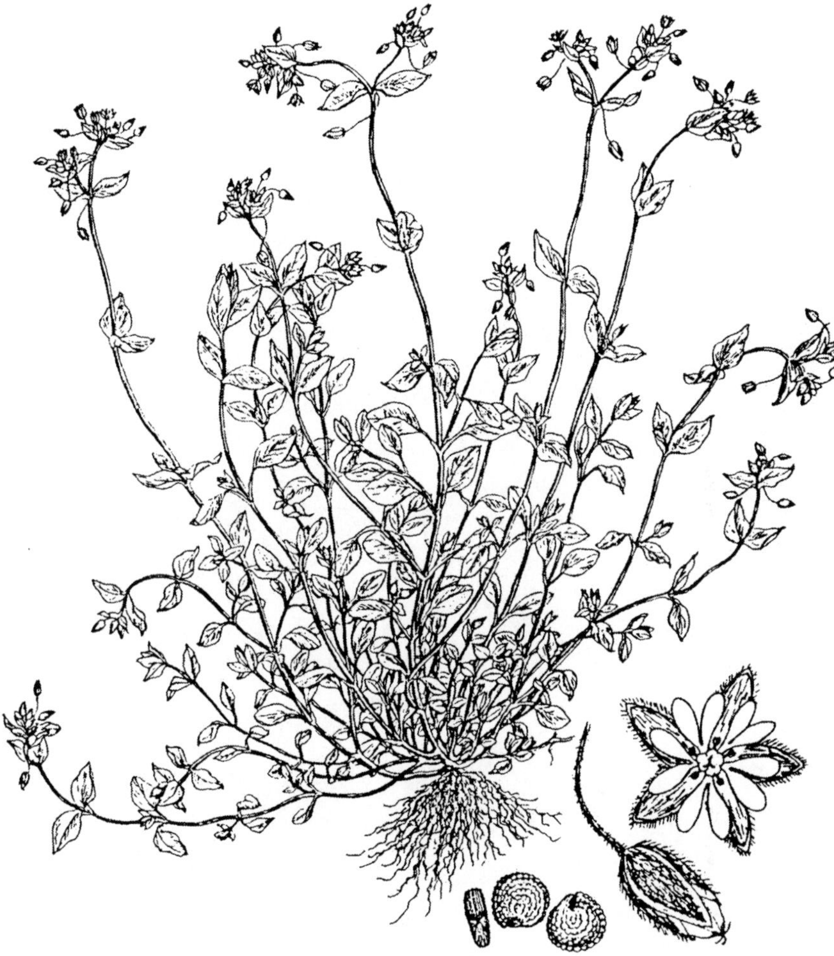

Figure 9 Common chickweed (*Stellaria media* (L.) Vill.) STEME.

compacted soil and is sometimes used as an indicator plant for compact soil. It may have a single deep root in hard soil and reproduces by seed.

Common purslane (Figure 12), a summer annual, has the ability to withstand drought with its thick, fleshy leaves and stems. The purple stems are prostrate, bearing fleshy, shiny, green leaves. Tiny yellow flowers produce an abundance of dark, rounded seed. Mats which actually smother turf can be formed.

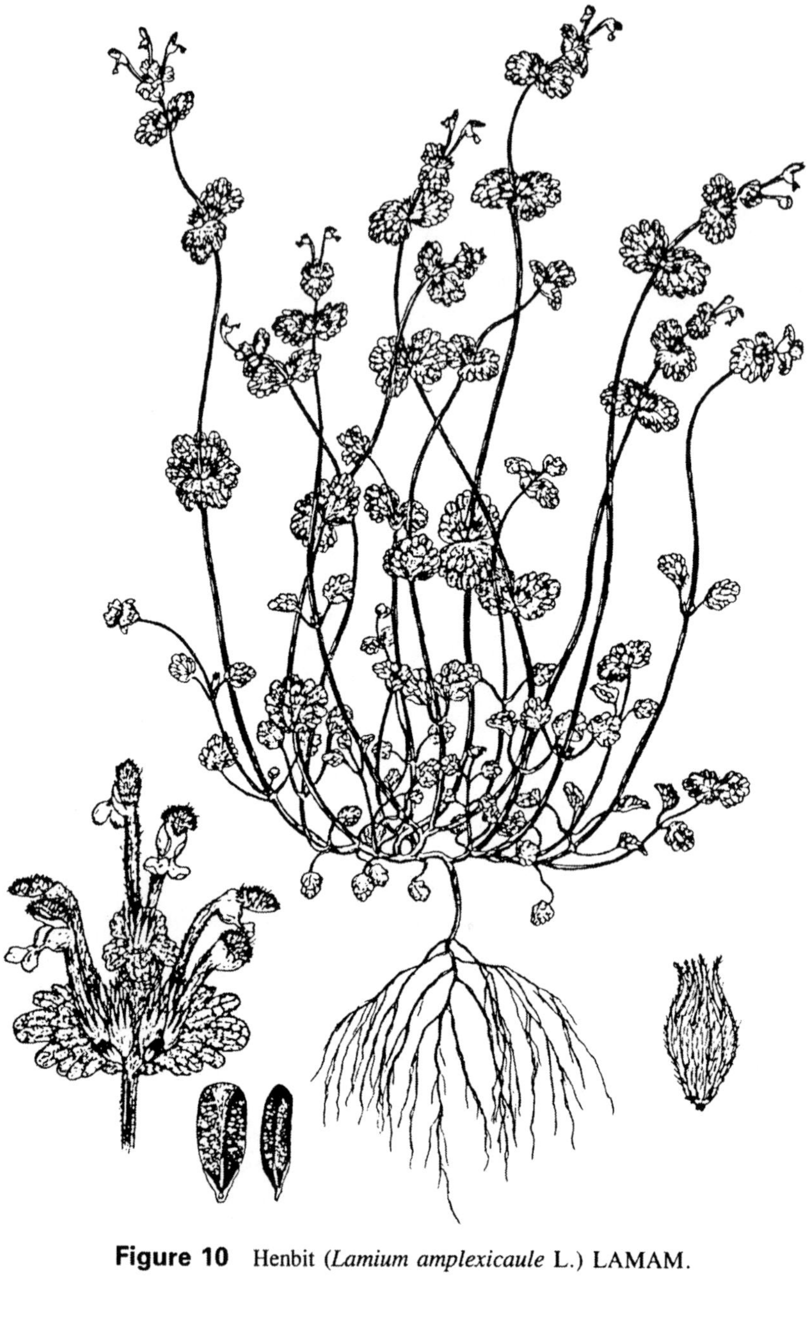

Figure 10 Henbit (*Lamium amplexicaule* L.) LAMAM.

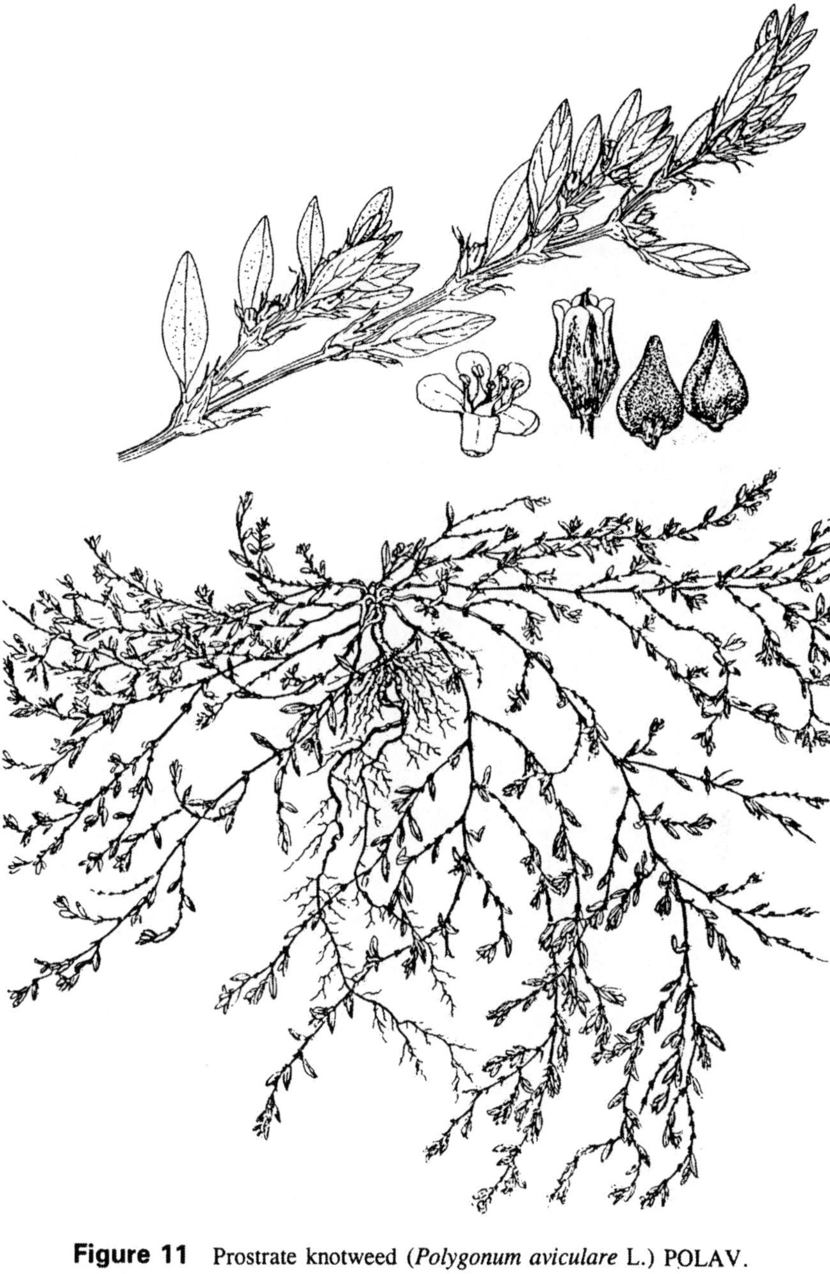

Figure 11 Prostrate knotweed (*Polygonum aviculare* L.) POLAV.

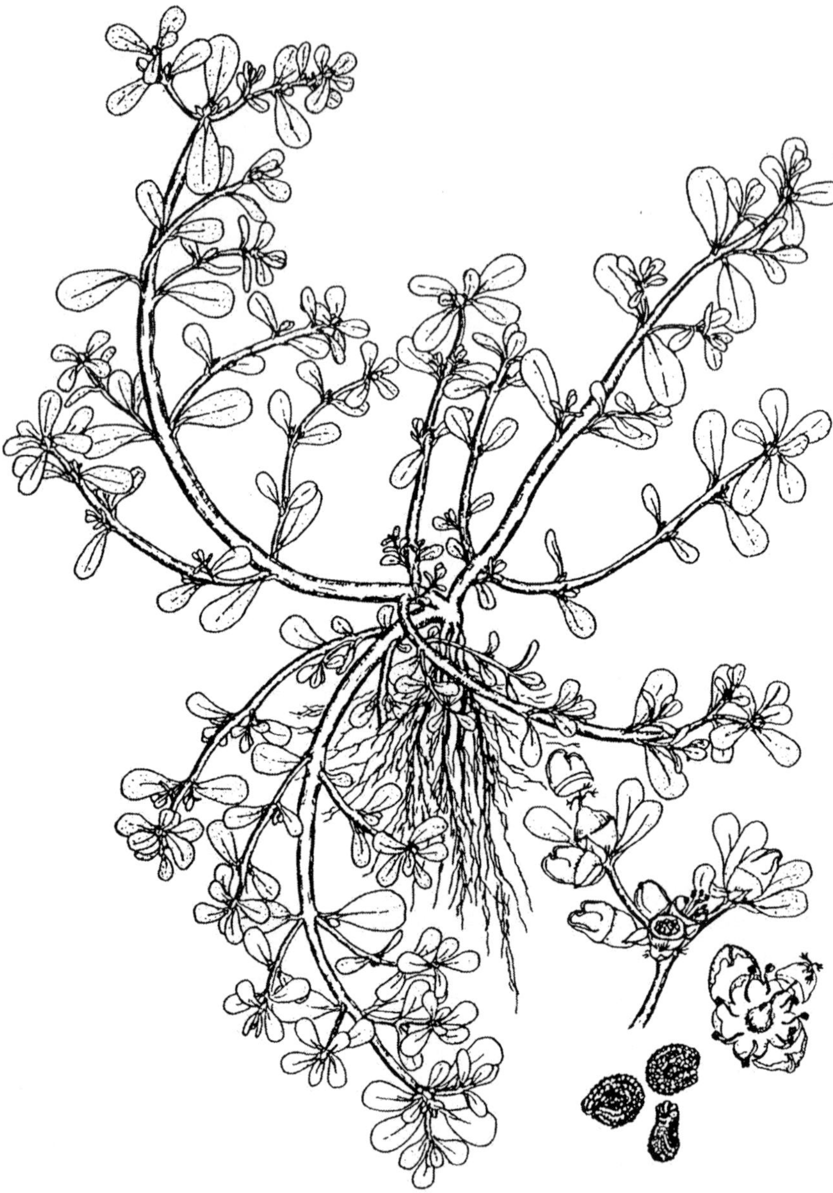

Figure 12 Common purslane (*Portulaca oleracea* L.) POROL.

Shepherd's-purse (Figure 13) is a winter annual, reproduces by seed, and is a member of the mustard family. It grows as a rosette and forms a slender taproot. Leaves alternate and are variously toothed or lobed, similar to dandelion. If allowed to grow, erect stems produce arrow-shaped leaves. Flowers are tiny and white with four petals. Heart-shaped or triangular seed capsules are obvious.

Prostrate spurge (Figure 14) is a summer annual, germinating in late spring and early summer. Stems are often reddish and filled with milky sap. Leaves are green to reddish green, opposite, and oblong, often with a red blotch. Flowers are tiny and inconspicuous. It forms a "mat" which choke desirable grasses and can easily be confused with spotted spurge in turfgrass.

C. Perennial Grass Weeds

Perennial weeds are those that produce vegetative propagules that allow them to live for more than 2 years. Nimblewill (Figure 15) is a perennial reproducing by rootstocks and sometimes seeds. Leaves are rolled in the bud and have compressed sheaths with white edges. It has a membranous ligule and a medium collar with hairs at the edges. The short, veined blades are flat and pointed. The seed head is a spiked panicle containing awned seeds. Stolons produce dense patches, which turn brown with frost. It is often mistaken for bentgrass or bermudagrass.

Quackgrass (Figure 16) is also known as couchgrass, witchgrass, and wheatgrass. It is a blue-green bladed perennial that reproduces by seeds and aggressive rhizomes. The erect stems have leaves rolled in the bud and hairy sheaths. Flat bush blades are rough on top and smooth below. It has a short membranous ligule and small auricles. The spiked seed head resembles ryegrass. This very strong plant is difficult to control because it primarily reproduces by rhizomes.

Yellow nutsedge (Figure 17) is a low-growing perennial found throughout the northern parts of the United States, particularly in turfgrass under high moisture conditions. The plants are yellow green with triangular stems bearing three-ranked leaves. Yellow nutsedge has yellowish scales at the base of the stem and tubers (nutlets). It has a fibrous root system. Tubers have the reproductive potential for new plants. The flowers are yellowish brown and three-ranked in a paniclelike arrangement.

D. Perennial Broadleaf Weeds

Black medic (Figure 18) is a perennial (sometimes a biennial) often confused with white clover because its leaves have three wedged-shaped leaflets. It is a legume with slender, branching stems which are dark green and reach 1–2 feet.

Figure 13 Shepherd's-purse (*Capsella bursa-pastoris* (L.) Medicus) CAPBP.

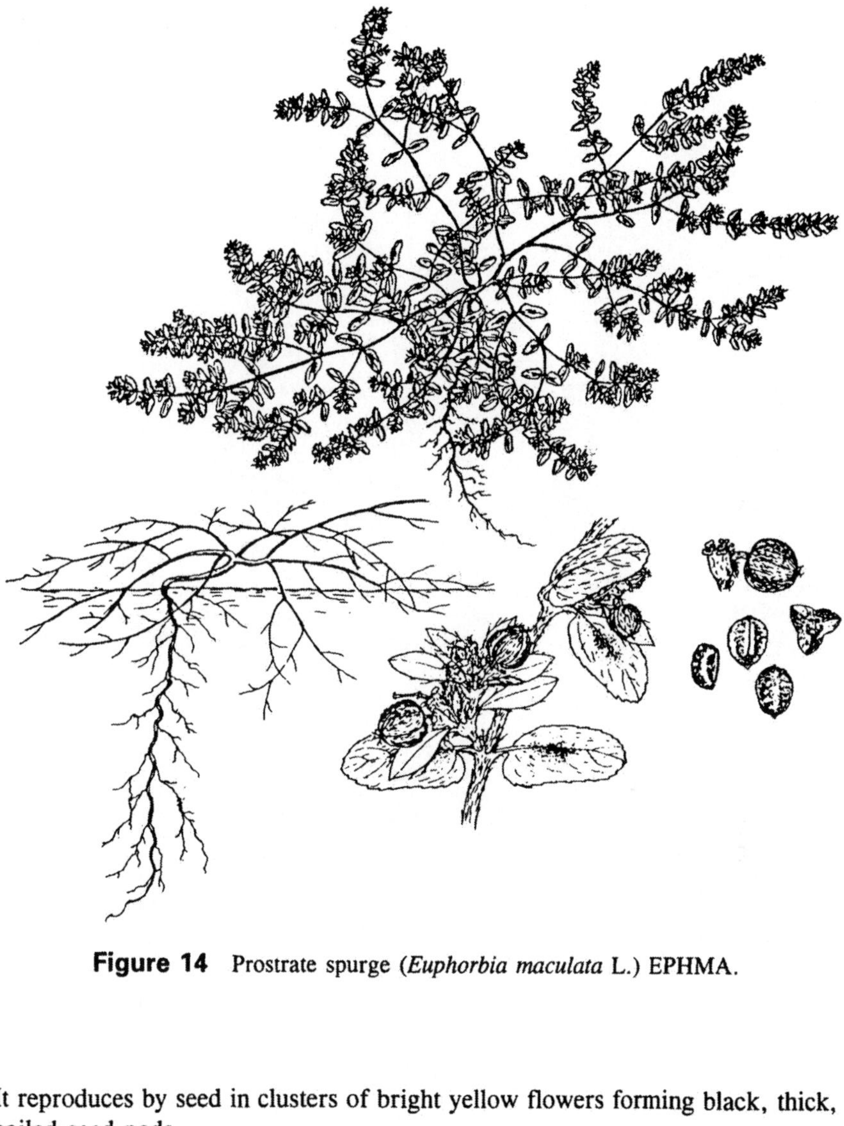

Figure 14 Prostrate spurge (*Euphorbia maculata* L.) EPHMA.

It reproduces by seed in clusters of bright yellow flowers forming black, thick, coiled seed pods.

Mouseear chickweed (Figure 19) is a low growing perennial which reproduces by seed and creeping, hairy stems. Dark green leaves are oblong, opposite on the stem, and very hairy. Small white flowers contain five notched petals similar to those of common chickweed. It is a widespread weed which persists throughout the summer.

Figure 15 Nimblewill (*Mulenbergia schreberi* J.F. Gmel.) MUHSC.

Figure 16 Quackgrass (*Elytrigia repens* (L.) Nevski) AGRRE.

Figure 17 Yellow nutsedge (*Cyperus esculentus* L.) CYPES.

Curly dock (Figure 20) is a hardy perennial reproducing by seed. Rosettes contain long, narrow, crinkle edged leaves which are shiny in spring and purplish later in the season. Green flowers are seldom seen in mowed turf; however, seed is usually produced in fields, waste areas, and occasionally mowed rights-of-way.

Figure 18 Black medic (*Medicago lupulina* L.) MEDLU.

Dandelion (Figure 21) is a perennial with a thick, long taproot and reproduces by seed. Leafless stems are very short, arising from a rosette bearing oblong, lobed, and narrow leaves. Flowers are large, yellow, and mature into round puffballs full of seeds. Winds carry the seeds several miles. Seed germinate mainly in the fall; however, some seedlings emerge anytime during the growing season in favorable moisture conditions. New rosettes form from the root in the fall.

Ground ivy (Figure 22) is a perennial broadleaf weed reproducing by seeds and creeping stems sometimes called creeping Charlie. This perennial of the mint family has square stems and forms dense patches in the sun or shade. The

Figure 19 Mouseear chickweed (*Cerastium vulgatum* L.) CERVU.

creeping stems root at the nodes. Leaves are opposite, round with scalloped edges, and rough on the upper side. Plants flower in the spring and have a trumpet shaped purplish flower.

Orange hawkweed (Figure 23) is a perennial sometimes called orange paintbrush or devil's paintbrush. This plant forms a rosette and reproduces by seeds and runners, forming patches which smother desirable turf. Long, narrow leaves are hairy and rounded at the tip. Runners are also hairy and set new

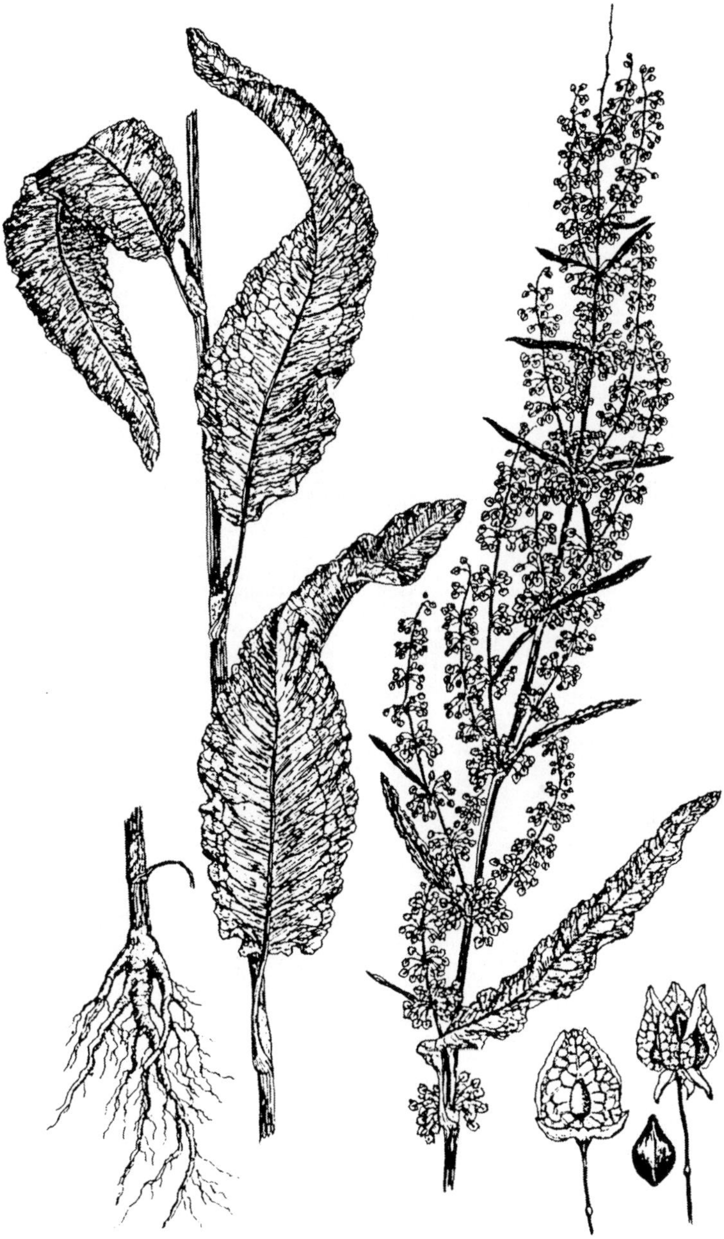

Figure 20 Curly dock (*Rumex crispus* L.) RUMCR.

Figure 21 Dandelion (*Taraxacum officinale* Weber in Wiggers) TAROF.

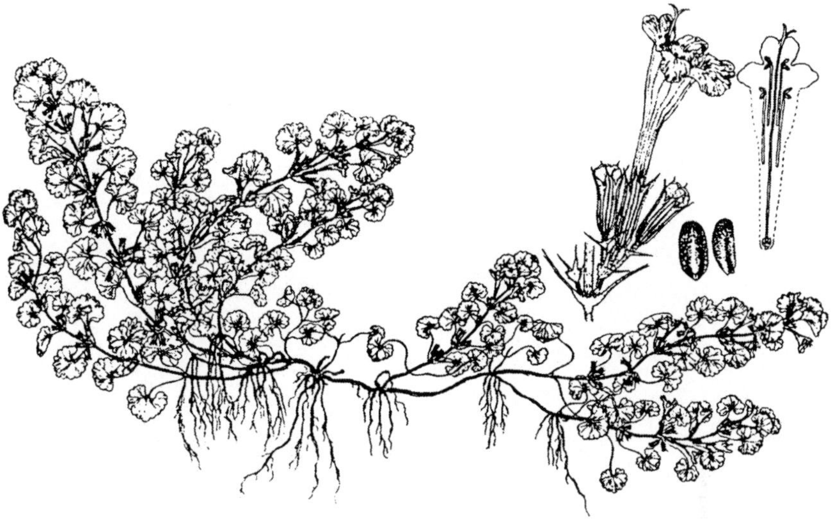

Figure 22 Ground ivy (*Glechoma hederacea* L.) GLEHE.

plants during summer. Flowers are orange-red and puffballs are about 1.9 cm in diameter. Yellow hawkweed is also common in turfgrass. These two species are known to prosper in low maintenance turfgrass.

Healall (Figure 24) is a perennial reproducing by seeds and short rootstocks and a member of the mint family. Oblong, opposite leaves with hairs taper to a rounded tip. Stems are square and prostrate in mowed turf. Flowers are pale violet to deep purple, are tubular, and form clusters at branch tips. It forms thick patches in mowed grass.

Yellow woodsorrel (Figure 25) is an annual with hairy stems weakly branched at the base. Leaves are heart shaped, pale green to purple and bitter to the taste because of the presence of oxalic acid. Creeping oxalis spreads by rhizomes. Bright yellow flowers have five petals. Mature seed pods will scatter seeds about 2 meters when touched. It can be easily mistaken for a clover. Yellow woodsorrel occurs extensively throughout the United States.

Broadleaf plantain (Figure 26) is a perennial forming a rosette and reproducing by seeds. Leaves are rounded, large, and deeply veined with wavy edges and purplish petioles. Stems support small flowers and the seed head. The fibrous, white root system is extensive.

Buckhorn plantain (Figure 27) is a perennial with long lancelike leaves and bulletlike seed heads on slender stalks. It has extensive fibrous roots. It reproduces by seed and forms new rosettes at the crown in the fall. Leaves are long,

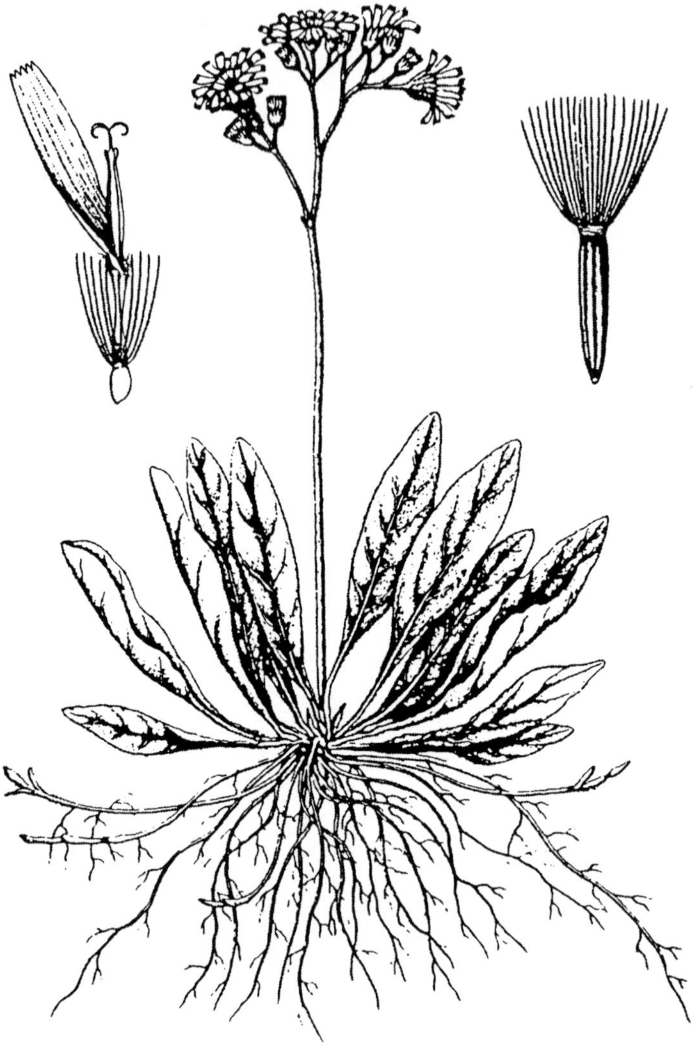

Figure 23 Orange hawkweed (*Hieracium aurantiacum* L.) HIEAU.

narrow, and pointed with deep veins. Leafless stalks from the rosette produce flowers in heads several times during the growing season to produce an abundance of seed.

Creeping bellflower (Figure 28) is a perennial, spreading throughout the turfgrass by short stolons. The stems are generally erect, reaching heights of 1 meter and containing milky juice. The basal leaves have a long petiole and

Figure 24 Healall (*Prunella vulgaris* L.) PRUVU.

are heart-shaped with minimal separate margins. The upper leaves are smaller and directly attached to the stem. The flowers are bell-shaped with five purple teeth.

Poison ivy (Figure 29) is a perennial reproducing by seeds and creeping rootstocks, often appearing at edges of turf areas. Stems reproduce through aerial rootlets and also run horizontally underground, producing new plants at the nodes. Alternate leaves are composed of three glossy, generally lobed leaf-

Figure 25 Yellow woodsorrel (*Oxalis stricta* L.) OXAST.

lets. Tiny green flowers are produced on stalks between the leaf stem and branch. Attractive gray berries and leaves cause an irritating rash on the skin.

Red sorrel (Figure 30) is a perennial known for its extensive rhizome system that also reproduces by seed. Low stems produce many creeping roots and alternate leaves are arrow-shaped. Male plants bear very small yellow-green flowers, and female, reddish-brown. It is known for its persistence in acid soils and is an indicator plant for acid soils whose dense patches tend to choke out desirable turf.

Canada thistle (Figure 31) is a perennial weed infesting turf. It reproduces by seed and creeping rhizomes. Leaves are oblong, usually lobed irregularly,

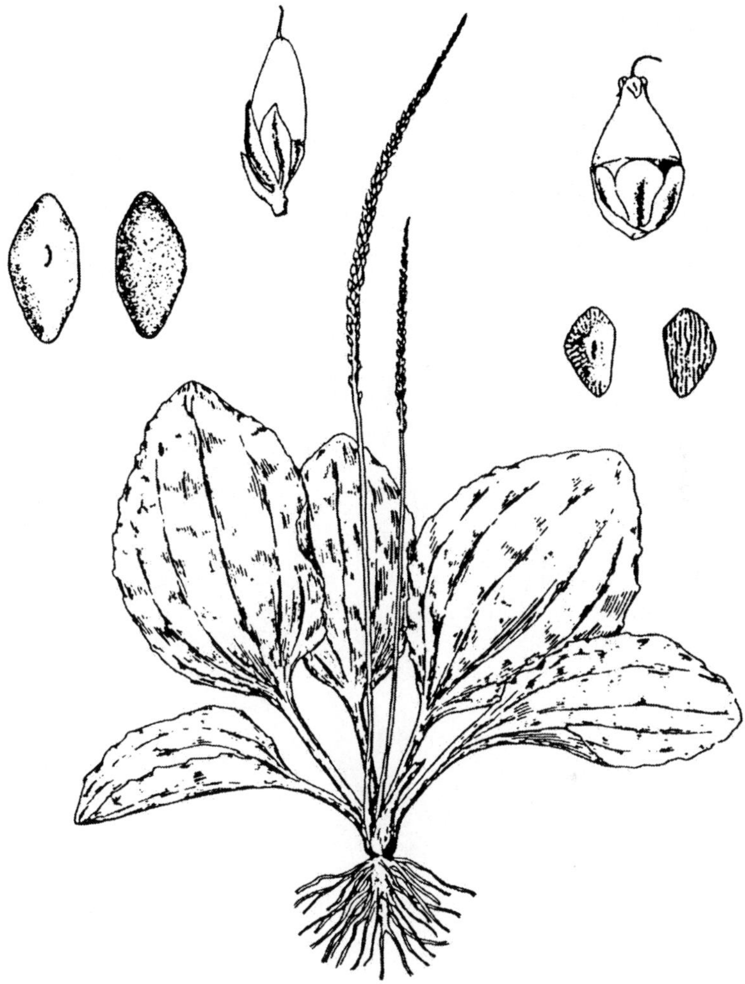

Figure 26 Broadleaf plantain (*Plantago major* L.) PLAMA.

dark green, and spiny-toothed on the margins. They are smooth on top and hairy underneath when mature. Flowers formed in terminal clusters are rose-purple, lavender, or white.

White clover (Figure 32) is a dark green perennial legume which is used in some lawn seed mixes. Leaves contain three leaflets which display white water marks. Creeping stems are very aggressive and root at the nodes. White flowers sometimes seen with a tinge of pink are a favorite of bees and other insects. It is known for its abundance during cool and wet springs.

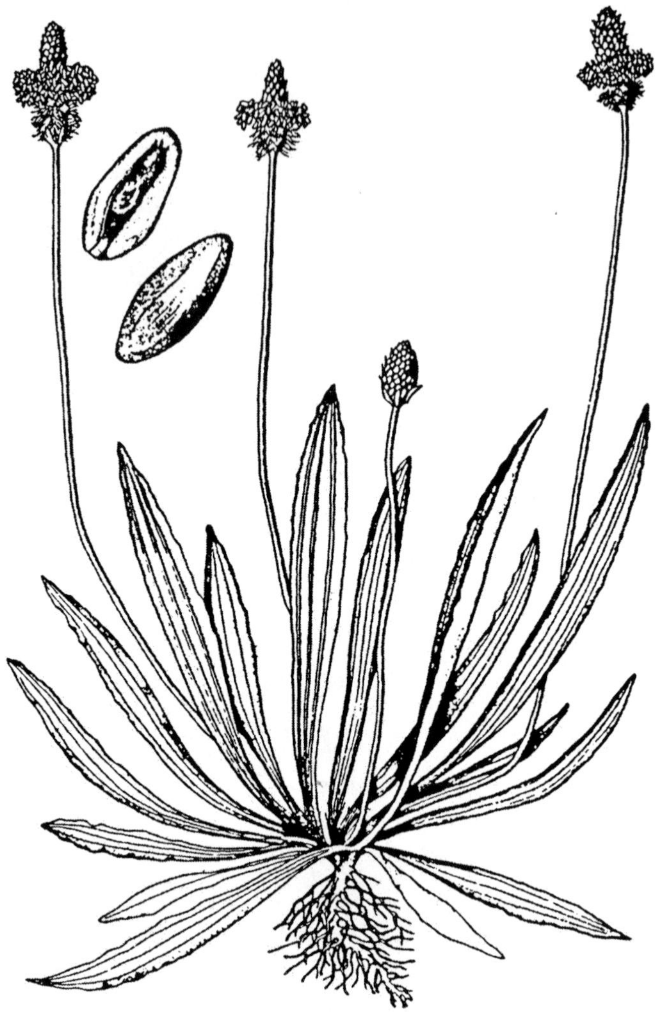

Figure 27 Buckhorn plantain (*Plantago lanceolata* L.) PLALA.

Common yarrow (Figure 33) is a fernlike perennial reproducing by seeds and rhizomes. Erect stems are sometimes branched; however, under close mowing, it forms a dense mat and reproduces by rhizomes as a very aggressive weed in turf. Alternate leaves are delicate and hairy. Petals are white with pale yellow centers in cluster of flowers.

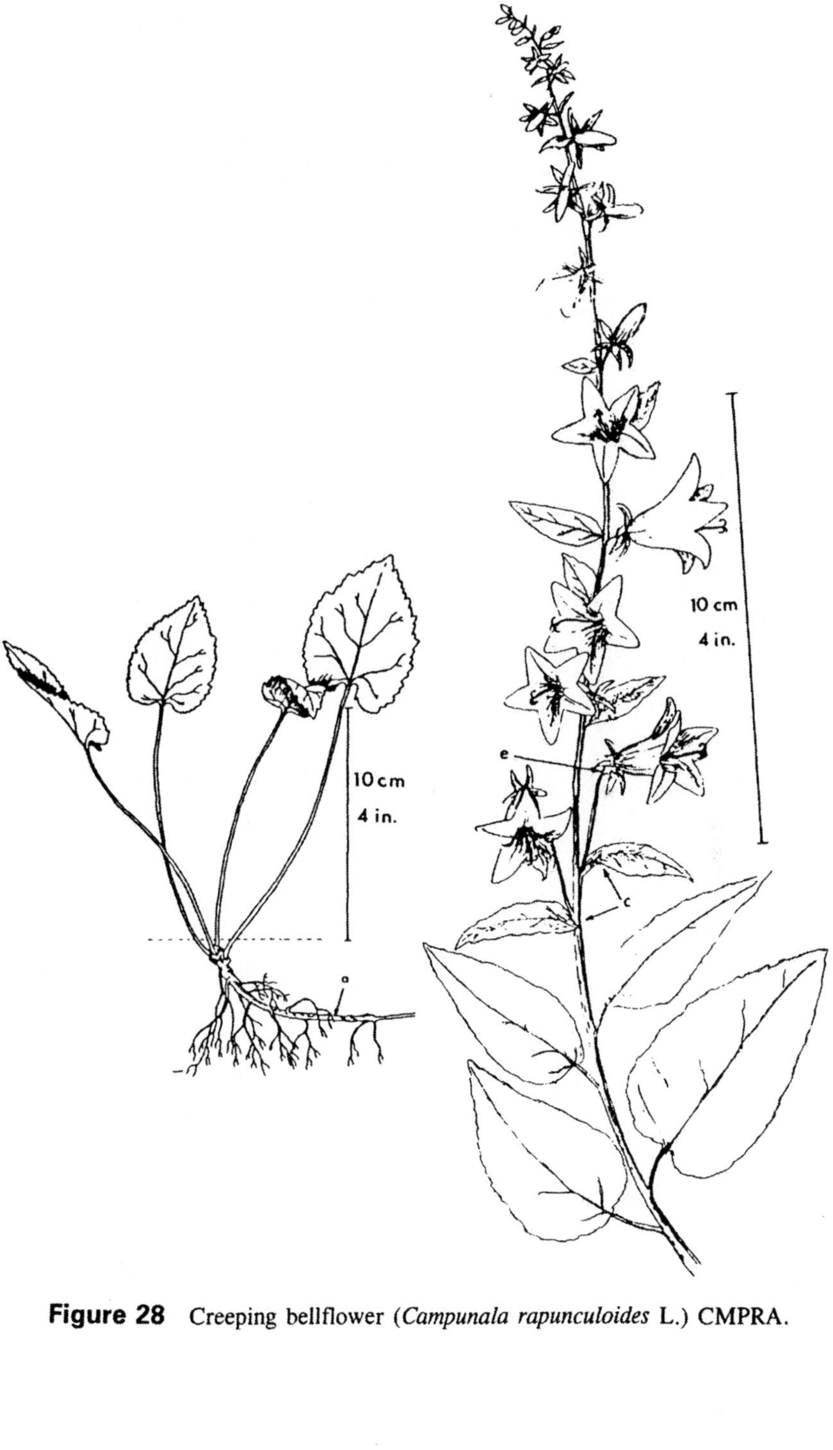

Figure 28 Creeping bellflower (*Campunala rapunculoides* L.) CMPRA.

Figure 29 Poison ivy (*Toxicodendron radicans* (L.) Ktze.) TOXRA.

Figure 30 Red sorrel (*Rumex acetosella* L.) RUMAA.

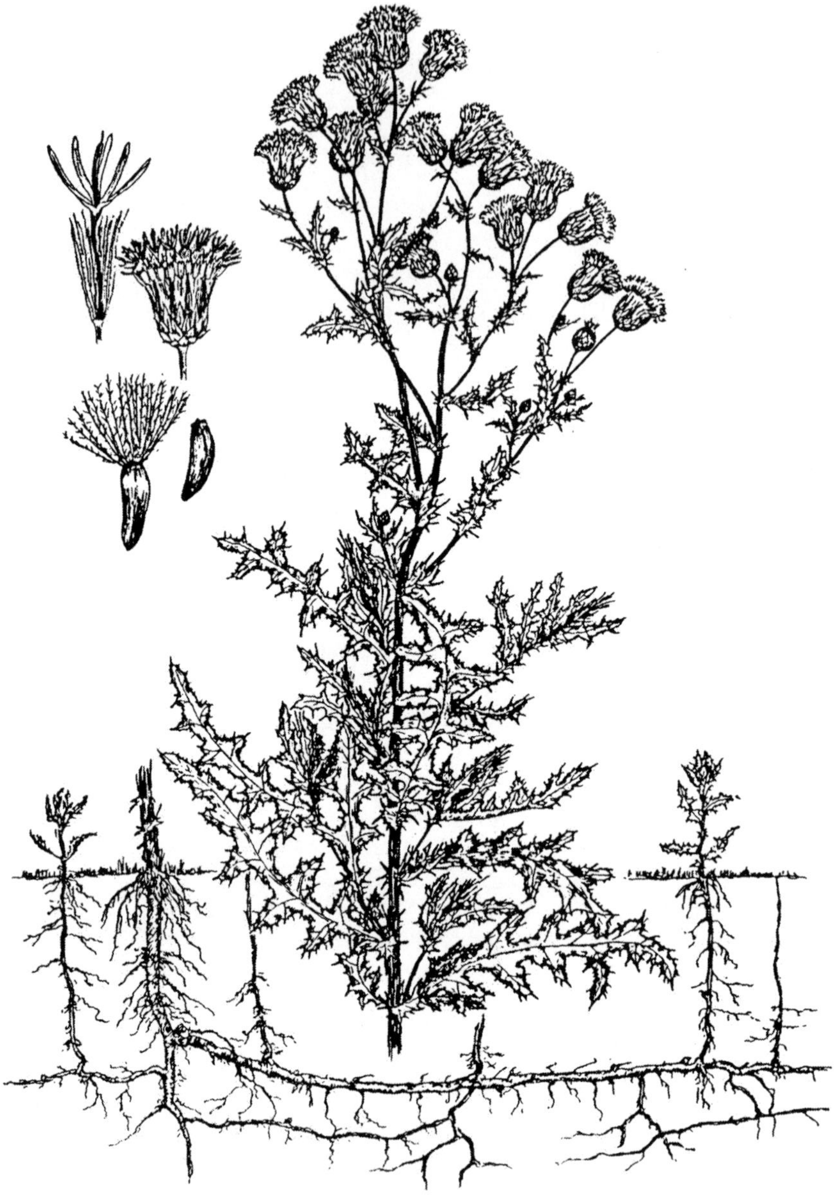

Figure 31 Canada thistle (*Cirsium arvense* (L.) Scop.) CIRAR.

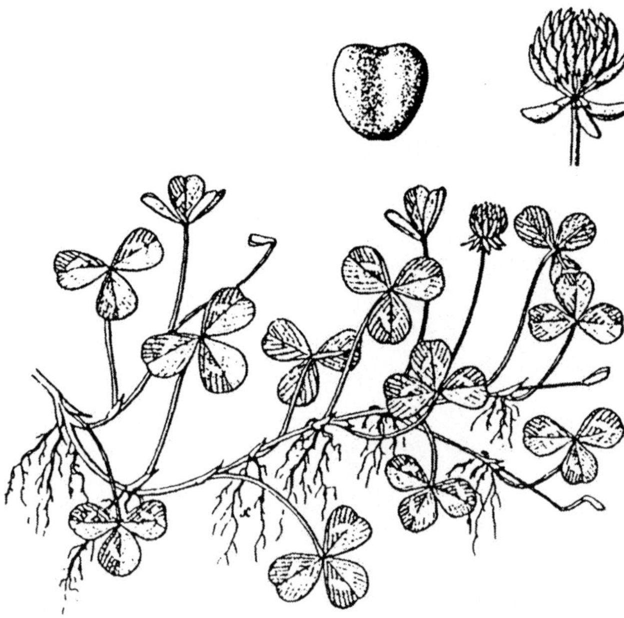

Figure 32 White clover (*Trifolium repens* L.) TRFRE.

E. Biennials

Biennial weeds are those plants that require two growing seasons to complete their life cycles. Common mallow (Figure 34) is a biennial broadleaf weed with widely spreading stems and a deep taproot. Stems do not root at the nodes. Leaves are alternate, deeply veined, round, and slightly lobed. Pinkish white flowers emerge from the petiole-stem junction. Seeds resemble a wheel of cheese. It is known to spread into lawns from gardens and bed areas.

Wild carrot (Figure 35), also known as Queen Anne's lace, is a biennial plant that reproduces by seed. Erect stems and branches arise from a taproot and rosette the first year. Leaves are lacy, resemble those of a carrot, are lobed, and are somewhat hairy. Compound white flowers form in umbrellalike clusters which become concave as seeds mature. It is a member of the parsley family and often misidentified as yarrow.

III. MANAGEMENT STRATEGIES AND CULTURAL WEED CONTROL

A web with each strand representing a management practice is used to illustrate the interrelationships involved during the establishment and maintenance

Figure 33 Common yarrow (*Achillea millefolium* L.) ACHMI.

Figure 34 Common mallow (*Malva neglecta* Wallr.) MALNE.

of high quality turfgrass (Figure 36). Sections of the web may be represented by the propagation medium, nutrient requirement, cultural practices, pest management, turfgrass and cultivar selection, and economic aspects. The pest management section includes weed, insect, and disease control as well as application equipment. Weed control is impacted by each of the other strands

Figure 35 Wild carrot (*Daucus carota* L.) DAUCA.

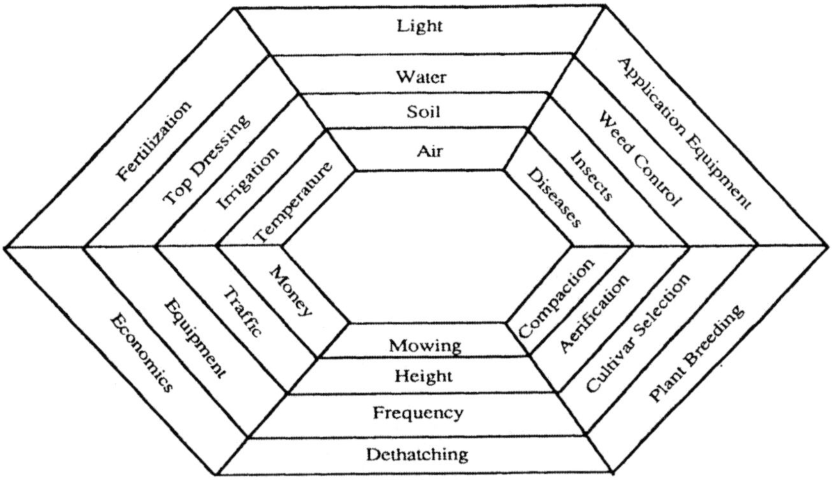

Figure 36 A web is used to illustrate the influence of each management practice or growth factor on the effectiveness of all the other practices.

represented in the web, and the principle is generally related to the turfgrass cover providing competition for emergence and growth of weeds.

A. Turfgrass Species Selection

Weed problems usually develop during periods of poor turfgrass growth and density. Rapid establishment is desired to allow less invasion with weeds (see Table 2). The rapid germination of perennial ryegrass and tall fescue provides cover to reduce weed invasion, while the slow germination and growth of Kentucky bluegrass require more intensive use of herbicides for initial maintenance. The establishment rate for cool season turfgrasses range from about 7 days for perennial ryegrasses and tall fescue to 20 days for Kentucky bluegrass. In warm season turfgrasses, the range is from 7 days for bermudagrass to more than 3 months for zoysiagrass. Bermudagrass germinates rapidly, produces a dense cover, and because of its heat and drought tolerance provides an excellent warm season turfgrass. It produces dense stolon growth and thus can be maintained at lower mowing heights while still competing well for space with weeds.

After establishment, the range of turfgrass density varies considerably with species (see Table 3). In cool season turfgrasses, creeping bentgrass has a high density, perennial ryegrass and tall fescue are relatively low in density. Bermudagrass and zoysiagrass become very dense, where as bahiagrass is a low density turfgrass. Thus, rapid establishment of a dense cover is quite impor-

Table 2 Establishment Vigor of Popular Turfgrasses

ESTABLISHMENT RATE	Cool season	Warm season
FAST	Perennial ryegrass	Bermudagrass
	Tall fescue	Saint Augustinegrass
	Fine fescues	Bahiagrass
	Creeping bentgrass	Centipedegrass
	Colonial bentgrass	Carpetgrass
SLOW	Kentucky bluegrass	Zoysiagrass

Source: A. J. Turgeon (1980).

tant for reduced weed contamination in new turfgrass sites. Check with your local extension agent for specific turfgrass recommendations for your area.

Turfgrass seed selection is important to reduce the spread of many weeds. The seed label contains the percentage of pure seed, percentage germination, and other information such as the presence of weed seed. In many instances, the soil contains a supply of common lawn weed seed. In a few instances, weed seed is introduced into the area with the turfgrass seed. Serious weed introduction with turfgrass seed may involve annual bluegrass with Kentucky bluegrass, bermudagrass, or bentgrass seed; tall fescue in Kentucky bluegrass or perennial ryegrass; goosegrass in bermudagrass, Kentucky bluegrass, or bentgrass; and bentgrass in Kentucky bluegrass or tall fescue.

B. Mowing Height and Frequency

Mowing turfgrass is generally done to maintain top growth within specific limits and involves clipping off a portion of the shoot to prevent growth beyond the limits. The removal of leaf tissue reduces carbohydrate synthesis and storage in the roots. A reduced root growth rate is associated with the lower carbohy-

Table 3 Shoot Density of Popular Turfgrasses

SHOOT DENSITY	Cool season	Warm season
HIGH	Creeping bentgrass	Bermudagrass
	Colonial bentgrass	Zoysiagrass
	Fine fescues	St. Augustinegrass
	Kentucky bluegrass	Centipedegrass
	Perennial ryegrass	Carpetgrass
LOW	Tall fescue	Bahiagrass

Source: A. J. Turgeon (1980).

drate production. Removal of large portions of the leaf tissue stimulates translocation of carbohydrates to shoots to provide for replenishing the plant's photosynthesis apparatus. In that respect, mowing is harmful to the plant and the larger amount of leaf tissue removed each time, the more drastic the detrimental effect.

The mowing requirements vary considerably among the various turfgrasses and cultivars when maintaining shoot densities necessary to reduce weed invasion. Bermudagrasses provide quite dense shoots at low mowing heights while tall fescue deteriorates, leaving voids for weed emergence under similar heights. Juska and Hanson (1961) demonstrated that Kentucky bluegrass mowed at 2.5 cm weekly provided 14% fewer roots and 41% fewer rhizomes than the same turfgrass mowed at 5 cm height. A 5 cm mowing height allowed the development of 6% more leaves per tiller, and 51% less thatch was observed at 2.5 cm mowing height (Sherman et al., 1980). The frequency of mowing was especially important to root and rhizome growth. Crider (1955) showed that removing 60% of the shoots at a mowing leads to cessation of root growth within several days while removing 40% or less of the foliage was not detrimental to root growth in Kentucky bluegrass. Juska and Hanson (1961) showed that frequent mowing of 2.5 cm tall Kentucky bluegrass reduced plant growth more than the same frequency of 5 cm tall turfgrass. In general, for high quality the appearance of each turfgrass species requires that specific limits be maintained in heights (Table 4). The amount of foliage removed at any mowing may be 30% to 40% of the total to produce the lowest detrimental effect on plant growth.

Intensively maintained Kentucky bluegrass and Kentucky 31 tall fescue at 5 cm mowing height prevent crabgrass emergence and growth almost completely, while under 2.5 cm heights, a preemergence herbicide will be required to maintain a crabgrass free lawn (Table 5). At a 2.5 cm height, Kentucky bluegrass is generally overrun with annual grasses during the summer, requiring intensive use of both annual grass and broadleaf herbicides. Below the mowing height acceptable tolerance range, the turfgrasses become thin during high stress periods and are poor competitors with weeds.

Generally, turfgrasses maintained at lower mowing heights (Table 4) require greater mowing frequency to produce a high density cover for reduced weed invasion. Thus, Kentucky bluegrass mowed at 3 to 4 cm will require mowing at 5 day intervals during active growth periods. However, at 6.2 cm mowing height, the frequency may be every 7 days. The frequency is also influenced by nitrogen fertilization.

Mowing equipment may transport weed seed from infested areas to uncontaminated sites. This is particularly important with high seed producing weeds over extended periods. This type of weed spread is particularly evident with annual bluegrass and goosegrass in golf courses and lawn care programs. The

Table 4 Recommended Turfgrass Mowing Heights Under General Agronomic Practices for Cool and Warm Season Species

Climatic type	Turf	Mowing height	Mowing frequency[a]
Cool season		cm	days
	Creeping bentgrass	0.6–1.2	3–4
	Perennial ryegrass	3.8–6.2	5–7
	Kentucky bluegrass	3.8–6.2	5–7
	Red fescue	3.8–6.2	5–7
	Tall fescue (KY 31)	5.0–10.0	7–14
	(turf types)	3.8–7.5	5–7
Warm season			
	Bermudagrasses		
	common and	1.2–2.5	3–4
	medium texture	3.8	4–7
	hybrid & fine texture	1.2–2.5	3–4
	Carpetgrass	1.2–5.0	5–8
	Centipedegrass	3.8–6.2	7–14
	Saint Augustinegrass	3.8–6.2	5–10
		6.2–7.5	10–14
	Zoysiagrass	1.2–2.5	3–5
		2.5–5.0	5–7

[a]Mowing frequency should be adjusted so that no more than one-third of the leaf area is removed per mowing.

prevention of weed seed spread through equipment requires cleaning of greens mowers or mowing equipment between site uses.

C. Aerification and Verticutting

Aeration of soils allows improved root growth and plant response to plant nutrients while releasing toxic gases produced in hydrophobic soils. Aerification reduces thatch and provides for better infiltration of water. Vertical mowing is used to break up the cores removed during aerification and also to remove accumulated thatch on a broader scale than just aeration. The timing of these management practices is quite important to maintain good quality turfgrass. They may cause less disturbance in turfgrass quality when growth is vigorous under favorable environmental conditions. These practices also open up the turfgrass for weed invasion and must be performed at the time when weed seed germination is low. Aerification and vertical mowing at the time crabgrass and goosegrass are at peak germination allow serious infestations. Fall aerification and vertical mowing to remove thatch also allow more broadleaf weed emer-

Table 5 Crabgrass Cover of Kentucky 31 Tall Fescue and Kentucky Bluegrass Turfgrass as Affected by 3 Years of Nitrogen Timing and Mowing Height

	Crabgrass cover Oct. 1975 Mowing height, cm		
Nitrogen timing[a]	2.5	5.0 %	7.5
Early Oct. and early Dec	52	23	0
Early Sept. and early Dec.	40	10	0
Early Mar. and early Dec.	63	23	3
Early Mar. and early May	67	22	0
Early May and mid-July	60	27	2
Early Sept. and early Oct.	35	12	0

Source: Hall (1980).
[a]Sod was seeded on 3 October 1972. Nitrogen rate was 146 kg N/ha, one-half applied at each timing period. Crabgrass cover was determined by visual estimate.

gence, especially of perennial weeds that become vegetative during the fall and early winter.

D. Fertilization and Weeds

The level and season of nitrogen fertilization have a direct effect on the mowing frequency and shoot density of turfgrasses. With soluble nitrogen applications, growth increases sharply within 2 or 3 days, reaches a peak in about a week, and slows to a constant rate in 1 to 2 months. The increased rate of growth from nitrogen applications is also dependent upon the season and the type of turfgrass. Warm season turfgrass should not be overstimulated by late summer and fall nitrogen. Similarly, cool season turfgrasses should not be overstimulated by early spring nitrogen. Nitrogen applications in the fall prolong the dark green color and extend the usefulness of turfgrass for golf course fairways and athletic fields. Turfgrass density and recuperative capacity are enhanced by late fall fertilization. The enhanced dark green color and density of cool season turfgrasses may even carry throughout the winter with late fall fertilization. Keeping spring nitrogen application at a very low level, just to maintain adequate color, reduces the rate of shoot growth of cool season grasses and prevents carbohydrate utilization from the roots. Late fall fertilization favors root growth and carbohydrate accumulation. Thus, fall fertilization reduces the need for more frequent mowing during spring growth of cool season turfgrass and with more favorable shoot density reduces the need for excessive use of herbicides (see Table 5). Fall fertilization improves turfgrass density without the excessive shoot growth associated with the spring nitrogen application.

E. Irrigation and Weeds

Turfgrass species differ in their requirements for irrigation, and depth of root development is a major factor related to water stress. For example, a deep rooted tall fescue is much more tolerant of drought injury than a shallow-rooted Kentucky bluegrass. Cool season turfgrasses require more water than warm season species to maintain similar quality fairways or lawns. Generally, cool season turfgrasses utilize about 0.8 cm of water per day during high water use periods compared to 0.6 cm per day for warm season turfgrasses. Generally under drought stress the turfgrass becomes less dense, allows more light to penetrate to the soil surface, and allows space for weed invasion into the canopy. All turfgrasses require supplemental irrigation during deep drought periods. The management practices that encourage deep root development enhance turfgrass quality with few requirements for irrigation; however, water use should be considered essential to prevent loss of density and cover to discourage weed invasion.

F. Influence of Disease and Insect Control on Weeds

As with weed control, insects and diseases are minimized by altering management practices to enhance a healthy turfgrass. Nitrogen fertilization and irrigation are important practices which influence disease severity. Even with the best management practices, there are environmental conditions over extended periods that become favorable for disease development requiring fungicide programs. The relationship to weed invasion is related to the maintenance of turfgrass cover and density to prevent weed germination and establishment in open areas. Insect damage to turfgrass shoots and/or roots may allow less density and more weeds.

IV. MANAGEMENT STRATEGIES IN CHEMICAL WEED CONTROL

A. Application Timings

Preemergence herbicides are typically timed to occur prior to weed emergence when the turfgrass is growing and well established. Most preemergence compounds are primarily active on grass species (Bhowmik and Bingham, 1990). One herbicide that does not follow this use pattern is siduron, which is registered for use on newly seeded turfgrass. Siduron is applied preemergence to the weeds and to the turfgrass.

Sequential applications are made where the margin of safety between turfgrass and weeds is narrow or repeated germination of weed seed is a problem. Sequential applications, e.g., by organic arsenicals or preemergence herbicides, are commonly necessary for full season weed control.

Herbicide and fertilizer mixtures are also applied commonly for turfgrass weed control. When properly utilized, herbicide and fertilizer mixtures can produce a beautiful turf while reducing the number of trips across the turf. They require additional care in determining the optimum time of application to ensure that fertilization and weed control requirements can be met at one timing.

Postemergence applications on established turfgrass are timed to occur after the weed species has emerged. Adequate coverage of the weed species is critical for optimum weed control. And timing of the first mowing after postemergence treatment is critical to prevent removing the treated weed leaves before herbicide translocation can occur.

Pre- and postemergence applications are treatment regimens designed for the control of individual problem weeds or more commonly for control of several weed species. Pre- and postemergence treatment for an individual problem weed such as crabgrass might consist of a preemergence application of material to control germinating weed seeds followed by a postemergence application of another product to control crabgrass that escaped the preemergence application. When used for the control of multiple weed species the preemergence material might be chosen to control grass species while the postemergence application would control broadleaf weeds.

Wetting agents and adjuvants are most commonly used on turfgrass with plant growth regulators. However, for some compounds such as the imidazolinones and sulphonylureas the use of wetting agents is specifically recommended for maximum effectiveness. When properly used these spray additives increase the uptake of these compounds without reducing the selectivity between desirable and undesirable species.

B. Establishing a Chemical Program

If the individual has done everything to optimize the environment for turf growth and weeds are still a problem then herbicide usage may be necessary. The first step in designing a chemical weed management program is selecting the level of management intensity. This decision involves the quality of the turf, the use and season of usage of the turf, and the resources available for weed management. The level of management will influence the compounds, methods, and timing of application available to the individual. High input weed management systems may require modification of cultural practices as well as multiple applications of preemergence and postemergence herbicides for acceptable levels of control, whereas a program in a less demanding site may require only moderate levels of herbicide application.

Weed management strategies should be based on long term evaluation of the weed problem. Minor weed problems may respond to cultural control or spot application of herbicides. Heavy weed infestations may require changes in cultural techniques as well as herbicide applications. In some cases control of a minor weed infestation in the current year may reduce the weed population in subsequent years. An alternative on this theme is the control of fall weeds to help in the control of summer weeds. A common example in temperate regions is the control of annual bluegrass (a winter annual) to help control goosegrass (a summer annual) the next year. In this example, annual bluegrass dies as temperature increases in the spring and leaves areas for goosegrass to invade in the summer, with the cycle repeating in the fall.

C. Herbicide Selection

If a high level of weed management is desired for a specific site then an herbicide should be selected. Even if cultural control methods will be tried first, an herbicide that can be used in conjunction with that cultural program should be selected to allow an alternative in case the weeds are not controlled. Herbicide selection should consider the biological characteristics of the weed (time of emergence, life span, invasiveness), available application equipment, time of application, level of management desired, and cost of the alternative herbicides.

Selecting a specific herbicide requires consideration of several possibilities. Will the application require a selective compound? Will the best weed control occur with a preemergence or a postemergence application? Will adequate control require a combination of both pre- and postemergence applications? Does the area have a single weed problem or are there several weeds present? Mixed stands of weeds often require a mixture of herbicides for best control. Proper weed identification is critical to the design of a weed management strategy (see the section on weed identification). A listing of herbicides and their ability to manage specific weeds is given at the end of the chapter (Tables 8 and 9). In addition to weed control an equal consideration is the relative tolerance of the turf to the selected herbicide, as shown in Table 10.

Another consideration is the regulatory aspects of the compound selected. General use herbicides allow usage by all types of turf care personnel. This includes professional turf specialists as well as homeowners. Professional use is listed on products which require that the user be an employee of a professional nursery, turf, or landscape company. Restricted use pesticides require that the applicator or immediate supervisor be licensed by the state to use that type of pesticide. Restricted use pesticides generally pose additional hazards to the applicator or the environment.

One final consideration prior to selecting an herbicide is a consideration of the environmental impact of the compound to be applied. Knowledge of an herbicide's longevity, potential for drift, and leaching will all help the individual to make an informed choice. Herbicides with long soil residual activity could lead to problems with fall overseeding. Spray drift of phenoxy compounds is a well documented problem. The soil movement of dicamba can seriously injure expensive ornamental plantings if care is not used in its application. And finally, if the turf site is located in an environmentally sensitive area the only weed control option may be cultural control methods.

V. MANAGEMENT STRATEGIES IN INTEGRATED PEST MANAGEMENT

Any turfgrass maintenance program lends itself to the incorporation of integrated pest management strategies. Weed management is no exception. Practices such as proper mowing height, watering, turfgrass species selection, fertilization, and thatch control can be included in any routine weed management program.

To develop a weed management program, one must identify the present weed problem. At present there is no method by which one can predict the cooccurrence or invasion of any weed species. Therefore, scouting information in the fall and spring is essential for developing a weed program. A well planned program can use the scouting form (Table 6) to help provide the necessary information to diagnose the problem.

The following checklist should be useful in selecting and implementing a weed management problem.

1. Problem diagnosis: Information on weeds, turfgrass species, soil type, etc., can be obtained by scouts (Table 6).
2. Selection of a program: Method of control to be used, cost effectiveness, custom application versus owner application.
3. Execution of the program: The selected method must be properly executed in relation to timing of application, equipment, etc.
4. Evaluate the results: Was the program successful or should it be modified to provide better results?

VI. ATHLETIC FIELD AND GOLF COURSE WEED MANAGEMENT

Athletic fields and golf courses are more intensively managed and used than home lawns. The wear and tear that equipment and athletes inflict on the turfgrasses must be carefully managed to provide acceptable quality playing surfaces. Weed management strategies must consider the intensity of play and the relationship to periods of active turfgrass growth. Spectator fields with one or

Table 6 Lawn Analysis Form

			RECOMMENDATIONS
Homeowner	Turf Builder		
Date	October 27, 1993		
Scout	Weed Doctor		
MAJOR TURFGRASS SPECIES PRESENT			Kentucky bluegrass
Minor species present			
LAWN ESTABLISHMENT			
Age	4 years old		
Seeded	X		
Sodded			
THATCH ACCUMATION			
<1 cm	X		
1.25–2.5 cm			
<5.0 cm			
DENSITY OF TURFGRASS			
Thick			
Moderate	X		
Thin			
Sparse			
COLOR OF TURFGRASS			
Dark green			
Light green	X		
Yellow green			
Dormant turf			
SOIL CHARACTERISTICS			
Texture	Silt loam		
Depth	7.5 cm top soil		
Organic matter	4% O.M.		
Condition			
WEEDS PRESENT			
Annuals			
Grass	Large crabgrass, yellow foxtail (80%)		
Broadleaf	Common chickweed (10%)		
Perennials			
Broadleaf	Dandelion, violet in patches		
Grass	None		
CULTURAL PRACTICES			
Mowing	3.75 cm	Cut at 6.2 cm	
Watering		Deep watering	
Fertilization		1 kg N/100 m^2, apply in the Fall.	
PAST HISTORY ON WEED CONTROL, METHODS?			2,4-D (Fall 1991)
COMMENTS			

two games each week are easily maintained during periods of active turfgrass growth. However, with daily use during periods of slow growth or even dormancy, weeds compete with the thinning turfgrass to become challenging to the manager (Bingham, 1992). The same principles discussed for cultural management strategies apply in golf courses and athletic fields. However, a compromise between the golfer's need to achieve accurate shot control from the fairway or tee and the agronomic requirements for quality turfgrass performance is a decision confronting the manager (Beard, 1982).

A. Turfgrass Species Selection

The species and cultivar selection process becomes more critical as mowing height is reduced to allow accurate control of the ball in golf course play. The stress of mowing at the low extremes may be offset partially by cultivar selection, intensive nutrition, irrigation, and frequent grooming techniques. As wear, tear, and compaction becomes more intense, the turfgrass thins. Weeds quickly encroach into heavily used fields and golf courses, requiring use of herbicides to maintain acceptable turfgrass quality.

The mowing height preferred for greens, tees, and fairways (as well as baseball fields) is generally the lowest that will allow continued acceptable turfgrass quality for the area (Table 7). For example, for bluegrass fairways or baseball fields, mowing heights are frequently 1.8 cm using selected varieties that will persist under a 2 day mowing frequency. Under these circumstances, a preemergence herbicide must be used to prevent competition from annual grasses and postemergence herbicides are used to control broadleaf weeds.

Table 7 Mowing Height and Frequency on Golf Course Fairways and Greens[a]

Turfgrass type	Mowing height, cm	Mowing frequency, days
Kentucky bluegrass	1.9–2.5	2–3 1/2
Bentgrasses	0.3–0.6 (green)	1–2
	1.2–1.9	2–3 1/2
Bermudagrasses	1.2–1.9 (fairway)	2–3 1/2
	0.3–0.6 (green)	1–2
Annual bluegrass	1.2–2.5	2–3 1/2
Zoysiagrass	1.2–2.5	2–3 1/2

[a]The needs of golfers will necessitate compromises in the desirable agronomic practices for a high-quality turfgrass.

B. Weed Management on Established Turfgrass

Winter Weeds

A number of annual and perennial weeds encroach into turfgrass during fall and winter. Winter annuals germinate and form rosettes or creeping vegetation in late summer, persist as young plants through the winter, and mature, setting seed, during spring. Some perennials return to vegetative status again in late summer after flower stalks die back, persist in a more juvenile status through the winter, and are a year-long problem. Winter weeds are common to all turfgrass areas; they include dandelion, common chickweed, mouse-ear chickweed, buckhorn plantain, broadleaf plantain, white clover, henbit, shepherd's-purse, black medic, corn speedwell, wild violets, and wild garlic. Use broadleaf herbicides on young weeds while they are still in a vegetative stage in late fall to achieve successful control (Bingham, 1995). The timing also allows cool season turfgrass to become a dense winter cover.

A second choice for application timing is spring; however, winter weeds are flowering, maturing, and setting seed at this time. The spring timing allows weeds to compete with turfgrass for extended periods and produce seed for reinfestation in the fall.

Summer Weeds

Summer annuals germinate in the late winter or spring, develop to maturity in summer, seed and die in early fall or by arrival of a killing frost. Broadleaf herbicides are used but with different timing. Knotweed, spotted spurge, carpetweed, and yellow woodsorrel, for example, are effectively controlled with spring treatment. Where both winter and summer weeds are a problem, broadleaf herbicides are used in both seasons. Isoxaben is an effective preemergence broadleaf herbicide used in early fall and/or early spring.

Crabgrass, goosegrass, foxtail, and barnyardgrass are also summer weeds and require early spring preemergence herbicides for effective control. A wide selection of products are available and are effective for various lengths of time. Oxadiazon, bensulide, dithiopyr, and prodiamine are generally applied only once annually, while DCPA, benefin, siduron, and pendimethalin are used a second time for full season annual grass control. The application rate has some effect on the length of effectiveness. Oxadiazon, pendimethalin, and prodiamine are sometimes selected for goosegrass control. Siduron is a particularly good selection for turfgrass areas that will receive additional turfgrass seed (cool season) in the spring. Athletic fields, requiring overseeding or renovation with cool season turfgrass during spring, may be treated with siduron for annual grass control during establishment.

Postemergence annual grass control can be achieved with DSMA, MSMA, fenoxaprop, and dithiopyr. Adequate soil moisture improves effectiveness, and applications before annual grass tillering provides a high level of control. In many instances, tank mixing the postemergence herbicide with a preemergence herbicide extends effectiveness over the season.

The herbicides are selected for the weeds and tolerance of the specific turfgrass. In mixed weed populations, a formulation that contains two or three herbicides may be required (Tables 8 and 9 and pesticide labels).

C. Bare Areas in Athletic Fields

Bare soil areas are maintained in certain athletic fields; glyphosate is used to remove existing turfgrass and weeds. If weeds are small, glyphosate is tank mixed with a preemergence herbicide and after 1 or 2 days the surface soil is lightly raked or stirred to incorporate the material. The highest rate provides the best control and repeating the preemergence application after 6 weeks may be required.

Table 8 Annual Grass Control in Kentucky Bluegrass, Tall Fescue, Perennial Ryegrass, and Common Bermudagrass[a,b]

Herbicide	Crabgrass species	Goose-grass	Annual bluegrass	Foxtail species	Sandbur
Preemergence					
Benefin	S	I	S	S	I
Bensulide	S	R	S	S	-
DCPA	S	I	S–I	S	I
Dithiopyr	S	S–I	-	S	-
Ethofumesate	-	-	S	-	-
Oryzalin	S	I	S–I	S	S–I
Oxadiazon	S	S	S–I	S	I
Pendimethalin	S	S–I	S–I	S	S–I
Siduron	S	R	R	S	-
Benefin + oryzalin	S	S–I	S	S	S–I
Benefin + trifluralin	S	S–I	S	S	S–I
Postemergence					
Dithiopyr	S	I	S–I	S	-
DSMA, MSMA	S	I–R	R	S	S–I
Fenoxaprop	S	S–I	-	S	-

[a]Relative effectiveness of commonly used herbicides for selected weeds is indicated as follows: S, weed susceptible; I, intermediate, good control at times with high rates, sometimes poor, may require more than one treatment; R, resistant weeds in most instances.
[b]Lower label rates or heavy weed pressure may require additional applications. See individual labels for specific recommendations.

Table 9 Postemergence Broadleaf Weed Control Showing Herbicides and Active Ingredient per ha in Kentucky Bluegrass, Tall Fescue, Perennial Ryegrass, and Common Bermudagrass[a]

Weed	Classification	2,4-D 1.5–2.0[b]	Dicamba 0.33–0.5	Mecoprop 1.5–2.0	2,4-D + Dicamba 1.0+0.33	2,4-D+ Dichlorprop 1.25+1.25	2,4-D+ Triclopyr 1.0+0.5	Preferred time to treat
Bedstraw	A	R	S	I	S	—	—	April & May
Bindweed	P	S	S	S–I	S	S	S	May & June
Bittercress	WA or B	S	S	I	S	S	S	Oct. & Nov.
Black medic	A,B, & P	I	S	I	S	S	S	April & May
Buttercup	WA,B, & P	S–I	S	I	S	I	I	Oct. & Nov.
Buttonweed,								
Virginia	P	R	R	R	I	I	R	May & repeat
Carpetweed	SA	S	S	I	S	S	S	May & June
Chickweed,								
Common	WA	R	S	S–I	S	S	S	Oct. & Nov.
Mouseear	P	I–R	S	S–I	S	S	S	Oct. & Nov.
Chicory	P	S	S	S	S	S	S	Oct. & Nov.
Cinquefoil,								
Common	A	S	S	S	S	S	S	May & June
Clover,								
Crimson	SA	S	S	S	S	S	S	May & June
Hop	SA	I	S	S	S	S	S	April & May
White	P	I	S	S	S	S	S	Oct. & Nov.
Geranium,								
carolina	WA & B	S	S	S–I	S	S	S	Oct. & Nov.
Daisy, oxeye	P	I	I	I	I	—	—	Oct. & Nov. or May
Dandelion	P	S	S	S	S	S	S–I	Oct. & Nov.
Dock	P	I	S	I–R	S	I	I	Feb.–April
Dog fennel	P	I	S	I	S	—	—	Oct. & Nov. or April
Garlic, wild	P	I	I	R	S–I	I	—	Oct.–Nov. & Feb.–March

Ground ivy	P	I–R	S–I	I	S–I	I	S	April–May
Hawkweed	P	S–I	S–I	R	S	—	S	Aug. & Sept.
Healall	P	S	S–I	R	S	S	—	Oct. & Nov.
Henbit	WA	I	S	I	S	S	S	Oct. & Nov.
Honeysuckle	P	S–I	S	I	S	S	S	May & June
Horsenettle	P	I–R	I	R	I	—	I	May & June
Knapweed, spotted	B	I	S	I	S	I	S	Oct. & Nov.
Knawel	WA	R	S	I	S	S	—	Oct. & Nov.
Knotweed, prostrate	SA	R	S	I	S	I	—	March–April
Lambsquarters	SA	S	S	S	S	S	S	April & May
Lespedeza	SA	I–R	S	S	S	S	—	April & May
Mugwort	P	I	S—I	I–R	S–I	I	—	March
Mustards	WA & B	S	S	I	S	S	—	Oct. & Nov.
Onion, wild	P	I	I	R	S–I	I	—	Oct.–Nov. & Feb.–March
Ornamental plants	P	S–I	S	S–I	S	S	S	Most likely to injure April to June
Oxalis	A,P	I–R	R	I–R	I	S	I	April–May
Pennycress, field	A	S	S	I	S	—	—	Oct. & Nov.
Pepperweed, Virginia	WA or B	S	S	I	S	S	—	Oct. & Nov.
Pigweed	SA	S	S	S	S	S	S	April & May
Plantains	P	S	I–R	I	S	S	S	Oct. & Nov.
Poison ivy	P	I	S–I	R	S–I	I	S–I	June
Poorjoe (Diodea)	A	S–I	—	I	S	—	—	May & June
Prostrate spurge	SA	I	S	I	S	S–I	S	April–May
Purslane	SA	I	S	R	S	I	S–I	May & June
Red sorrel	P	R	S	S	S	I	S–I	Oct. & Nov.

Table 9 Continued

Weed	Classification	2,4-D 1.5–2.0[b]	Dicamba 0.33–0.5	Mecoprop 1.5–2.0	2,4-D + Dicamba 1.0+0.33	2,4-D+ Dichlorprop 1.25+1.25	2,4-D+ Triclopyr 1.0+0.5	Preferred time to treat
Shepherd's purse	WA	S	S	S–I	S	S	S	Oct. & Nov.
Smartweed	SA	I–R	S	I–R	S	I	—	April & May
Sowthistle	WA	S	S	I	S	S	—	Oct. & Nov.
Speedwell	SA or WA	R	R	R	I–R	I	I	April
Spotted spurge	SA	I–R	S–I	S–I	S–I	S –I	S	May & June
Star of Bethlehem	P	R	I–R	R	I–R	R	—	April
Thistle,								
Bull	B	S–I	S	I	S	S–I	S–I	Oct. & Nov.
Canada	P	I	S	I	S	I	I–R	Oct. & Nov.
Curl	B or WA	S	S	I	S	S	S	April
Musk	B	S	S	I	S	S	S	April
Violet	P	I–R	I	I–R	I	I	I	April
Wild carrot	B	S	S	I	S	S	S	Oct. & Nov.
Wild strawberry	P	R	S–I	R	S–I	I	I	Oct. & Nov.
Woodsorrel,								
yellow	A	I	R	I–R	I	S	I	April & May
Yarrow	P	I	S	I–R	S	I	I	Oct. & Nov.
Yellow rocket	B or P	S–I	S–I	I	S	S	S	Oct. & Nov.

[a]Relative effectiveness of commonly used herbicides for selected weeds is indicated as follows: S, weed susceptible; I, intermediate, good control at times with high rates, sometimes poor, may require more than one treatment; R, resistant weeds in most instances.

[b]Lower label rates or heavy weed pressure may require additional applications. See individual labels for specific recommendations.

D. General Principles and Strategies

Apply preemergence herbicides before weed seed germination. They are not effective after weed emergence.

Lightly irrigate after preemergence herbicide application to place it in an active site in the soil surface.

Apply postemergence broadleaf herbicides to weeds that are actively growing with juvenile tissue for uptake. Weed photosynthesis and carbohydrate translocation to roots will assist in control of perennial plants (temperature range 16 to 27°C and adequate soil moisture are best).

Ample leaf area for absorption is needed for broadleaf herbicides and 24 hours without rainfall or irrigation.

Select the herbicide for the weed and turfgrass on your site.

Follow good management practices for growing the desired turfgrass. The turfgrass placed under stress from low mowing height or other management stress will require high levels of herbicides for adequate weed control.

VII. HOME LAWN WEED MANAGEMENT

Home lawns have a set of considerations differing substantially from those of other use areas. The typical homeowner would like lush growth, resistance to wear, shade and drought tolerance, and few management decisions. The problems typically faced by home lawns are turfgrass cultivars that may not be adapted to their particular conditions, incorrect mowing practices, wear patterns due to children and pets, close proximity to ornamentals, and a less than ideal level of knowledge of turfgrass cultural practices. Cultural practices such as aerification, multiple fertilizer applications, proper mowing height, and timely pest and disease control are erratically applied under these conditions. While the level of care for the turfgrass may not be optimum, the level of expectations typically is very high. The best long term remedy for weed problems in home lawns is healthy, vigorously growing turf.

A. Renovation or Establishment

Weed management during turfgrass renovation or establishment requires patience and optimism. Because of the lack of herbicides available for use at this growth stage any management strategy must maximize cultural control methods. Vigorous turf establishment and a high density turfgrass species will reduce the overall weed management needs (Tables 2 and 3). The previous weed history and turf growth patterns will also suggest the level of weed management that will be necessary. The one herbicide registered for use on cool season turf at seeding is siduron. Siduron is applied pre emergence to the weeds and to the turfgrass at 5.6 to 6.7 kg ai/ha.

The best management strategy is to ensure optimum turf growth. Establishing a high quality turfgrass must include proper seedbed preparation, adequate moisture, proper mowing height and frequency for the turf species (Table 4), and pest and disease management. One weed management tool is controlling the existing weeds before planting the new turfgrass. Glyphosate does an exceptional job of controlling weeds and turf prior to reseeding. Care must be taken not only to follow the label recommendations but to prevent drift to desirable plants near the lawn. Another strategy in weed management is to apply a material such as DCPA after the turf has emerged and is greening up with 2.5 to 5.0 cm of height.

B. Established Home Lawns

Weed management begins with proper turf management. Section III in this chapter covers the cultural and management practices that will produce a healthy, vigorous turf. When management strategies have allowed weeds to survive, then a herbicide may be necessary. A wide variety of herbicides are registered for the control of weeds in turfgrass. Proper herbicide selection begins with weed identification and assessment of the problem, evaluating the potential herbicide options (Table 8 or 9), determining the turfgrass tolerance to the potential herbicides (Table 10), and considering the impact on adjacent ornamentals or a fall overseeding. The applicator must remember to evaluate the result (Figure 37) and decide whether the material was effective.

C. Bare Soil Areas in Home Lawns

Bare soil areas may be desirable around ornamentals, walks, or areas adjacent to roadways. Several herbicides can replace or supplement mechanical removal. Postemergence materials may be used alone or in combination with preemergence herbicides to provide long term weed management. Care must be taken that these materials are safe around the turfgrass and ornamentals common for homes. For this reason long residual and highly soil mobile herbicides are generally not recommended.

VIII. SUMMARY

Proper weed management in turfgrass begins with producing the most vigorous turf possible under the available conditions. If weeds have become a problem the solution must begin with evaluating current turfgrass management practices. In many circumstances, even with the best management practices, herbicides will be necessary to provide proper weed management. This chapter has provided the background for designing a weed management program. Weed management in turfgrass has more management tools than almost any other crop. The trick lies in selecting the proper combination of available tools to provide the desired level of weed management. Figure 37 describes the four

Table 10 Turfgrass Tolerance to Herbicides[a]

Herbicide	Bluegrass	Bermuda-grass	Tall fescue	Fairway bentgrass	Fine fescue	Zoysia-grass
Preemergence						
Benefin	T	T	T	I	S	T
Bensulide	T	T	T	T	T	T
DCPA	T	T	T	T	T	T
Dithiopyr	T	T	T	T	T	T
Ethofumsate	–	–	–	T	–	–
Oryzalin	S	I	S	S	S	T
Oxadiazon	T	T	T	I	I	T
Pendimethalin	T	T	T	I	T	T
Prodiamine	T	T	T	–	–	T
Siduron	T	I	T	I	T	–
Benefin + oryzalin	S	T	T	S	S	T
Benefin + trifluralin	T	T	T	I	I	T
Bensulide + oxadiazon	T	T	T	I-T	–	T
Postemergence						
Dithiopyr	T	T	T	–	T	T
DSMA, MSMA	T	T	T	I	I	I-T
Fenoxaprop	T	I	T	S	T	T

[a]Relative turfgrass tolerances to commonly used herbicides are indicated as follows: T, turfgrass is tolerant to normal use rates; I, turfgrass is injured slightly to moderate, good recovery expected within a week; S, turfgrass injury may be moderate to susceptible, recovery may occur over long time periods.

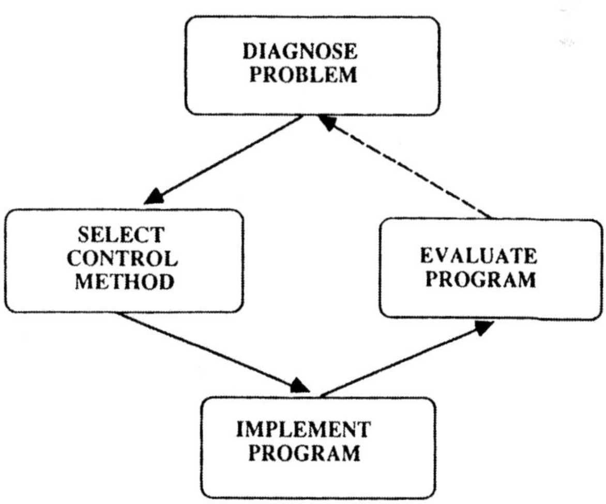

Figure 37 Selecting and evaluating a weed management strategy.

steps in selecting a weed management strategy. The turfgrass manager must diagnose the problem accurately. Select the best combination of cultural methods and herbicides (where necessary). Properly carry out the weed management program selected. And, finally, evaluate the results of the methods used. Unfortunately, the final step is the one most commonly neglected in the process. With a thorough understanding of the different types of weed management tools, the turfgrass manager can effectively and economically control a wide range of weed species.

APPENDIX A

List of Herbicides Mentioned in Chapter

Common name	Chemical name
Benefin	*N*-Butyl-*N*-ethyl-2,6-dinitro-4-(trifluoromethyl)benzenamine
Bensulide	*O,O*-bis(1-Methylethyl) *S*-[2-[(phenylsulfonyl)amino]ethyl] phosphorodithioate
2,4-D	(2,4-Dichlorophenoxy)acetic acid
DCPA	Dimethyl 2,3,5,6-tetrachloro-1,4-benzenedicarboxylate
Dicamba	3,6-Dichloro-2-methoxybenzoic acid
Dichlorprop	(±)-2-(2,4-Dichlorophenoxy)propanoic acid
Dithiopyr	*S,S*-Dimethyl 2-(difluoromethyl)-4-(2-methylpropyl)-6-(trifluoromethyl)-3,5-pyridinedicarbothioate
DSMA	Disodium salt of MAA
Ethofumesate	(±)-2-[4-[(6-Chloro-2-benzoxazolyl)oxy]phenoxy]propanoic acid
Fenoxaprop	(±)-2-[4-[(6-Chloro-2-benzoxazolyl)oxy]phenoxy]propanoic acid
Glyphosate	*N*-(Phosphonomethyl)glycine
MAA	Methylarsonic acid
Mecoprop	(±)-2-(4-Chloro-2-methylphenoxy)propanoic acid
MSMA	Monosodium salt of MAA
Oryzalin	4-(Dipropylamino)-3,5-dinitrobenzenesulfonamide
Oxadiazon	3-[2,4-Dichloro-5-(1-methylethoxy)phenyl]-5-(1,1-dimethylethyl)-1,3,4-oxadiazol-2-(3*H*)-one
Pendimethalin	*N*-(1-Ethylproply)-3,4-dimethyl-2,6-dinitrobenzenamine
Prodiamine	N^3,N^3-di-*n*-Propyl-2,4-dinitro-6-(trifluoromethyl)-*m*-phenylenediamine
Siduron	*N*-(2-Methylcyclohexyl)-*N'*-phenylurea
Triclopyr	[(3,5,6-Trichloro-2-pyridinyl)oxy]acetic acid
Trifluralin	2,6-Dinitro-*N,N*-dipropyl-4-(trifluoromethyl)benzenamine

APPENDIX B

List of Weeds and Crops Mentioned in Chapter

Common name	Scientific name	BAYER Code
Common yarrow	*Achillea millefolium* L.	ACHMI
Creeping bellflower	*Campunala rapunculoides* L.	CMPRA
Shepherd's-purse	*Capsella bursa-pastoris* (L.) Medicus	CAPBP
Mouseear chickweed	*Cerastium vulgatum* L.	CERVU
Canada thistle	*Cirsium arvense* (L.) Scop.	CIRAR
Yellow nutsedge	*Cyperus esculentus* L.	CYPES
Wild carrot	*Daucus carota* L.	DAUCA
Large crabgrass	*Digitaria sanguinalis* (L.) Scop.	DIGSA
Barnyardgrass	*Echinochloa crusgalli* (L.) Beauv.	ECHCG
Goosegrass	*Eleusine indica* (L.) Gaertn.	ELEIN
Quackgrass	*Elytrigia repens* (L.) Nevski	AGRRE
Prostrate spurge	*Euphorbia maculata* L.	EPHMA
Ground ivy	*Glechoma hederacea* L.	GLEHE
Orange hawkweed	*Hieracium aurantiacum* L.	HIEAU
Henbit	*Lamium amplexicaule* L.	LAMAM
Common mallow	*Malva neglecta* Wallr.	MALNE
Black medic	*Medicago lupulina* L.	MEDLU
Carpetweed	*Mollugo verticillata* L.	MOLVE
Nimbleweed	*Mulenbergia schreberi* J.F. Gmel.	MUHSC
Yellow woodsorrel	*Oxalis stricta* L.	OXAST
Broadleaf plantain	*Plantago major* L.	PLAMA
Buckhorn plantain	*Plantago lanceolata* L.	PLALA
Red sorrel	*Rumex acetosella* L.	RUMAA
Annual bluegrass	*Poa annua* L.	POAAN
Prostrate knotweed	*Polygonum aviculare* L.	POLAV
Common purslane	*Portulaca oleracea* L.	POROL
Healall	*Prunella vulgaris* L.	PRUVU
Curly dock	*Rumex crispus* L.	RUMCR
Yellow foxtail	*Setaria glauca* (L.) Beauv.	SETLU
Common chickweed	*Stellaria media* (L.) Vill.	STEME
Dandelion	*Taraxacum officinale* Weber in Wiggers	TAROF
Poison-ivy	*Toxicodendron radicans* (L.) Ktze.	TOXRA
White clover	*Trifolium repens* L.	TRFRE
Bentgrass	*Agrostis stolonifera* L.	AGSST
Bermudagrass	*Cynodon doctylon* (L.) Pers.	CYNDA
Kentucky bluegrass	*Poa Pratensis* L.	POAPR
Perennial ryegrass	*Lolium Perenne* L.	LOLPE
Red fescue	*Festuca rubra* L.	FESRU
Tall fescue	*Festuca arundinacea* Schreb.	FESAR
Centipedegrass	*Eremochloa ophiuroides* (Murro) Hock.	ERLOP
St. Autustinegrass	*Stenotaphrum secundalum* (Walt.) Ktze.	STPSE
Zoysiagrass	*Zoysia japonica* Steud.	ZOYJA

REFERENCES

Anderson, W. P. (1983). *Weed Science Principles*. West Publishing Co., St. Paul, Minnesota.

Bhowmik, P. C. and Bingham, S. W. (1990). Preemergence activity of dinitroamiline herbicides used for weed control in cool-season turfgrasses. *Weed Technol. 4*: 387–393.

Bingham, S. W. (1992). Weed management in athletic fields, *Grounds Maintenance 27*(6): 50–52.

Bingham, S. W. (1995). Weeds, in *Pest Management Guide for Horticultural and Forest Crops*, Virginia Cooperative Extension, Blacksburg, Virginia, pp. 151–162.

Beard, J. A. (1973). *Turfgrass: Science and Culture*, Prentice-Hall, Inc. Englewood Cliffs, New Jersey.

Beard, J. A. (1982). *Turf Management for Golf Courses*, Burgess Publishing Co., Minneapolis.

Beard, J. A. (1992). Turf regrowth, *Grounds Maintenance 27*(5): 36–38.

Chalmers, D. R., Hall III, J. R., and Schmidt, R. E. (1987). Lawn establishment in Virginia. Virginia Cooperative Extension, Blacksburg, Virginia, Pub. 430-009.

Crider, F. J. (1955). Root growth stoppage resulting from defoliation of grasses, U.S. Technical Bulletin #1102.

Duble, R. L. (1989). *Southern Turfgrasses: Their Management and Use*, TexScape, Inc. College Station, Texas.

Hall, J. R., III. (1980). "Effect of cultural factors on tall fescue: Kentucky bluegrass sod quality and botanical composition," Proceedings of the Third International Turfgrass Research Congress, pp. 367–377.

Juska, F. V. and Hanson, A. A. (1961). Effects of interval and height of mowing on growth of Marion and common Kentucky bluegrass (*Poa pratensis L.). Agron. J.* 53: 385–388.

Kenna, M. K., Pinkston, M., Andrews, J., Criswell, R., Taylor, W., Downs, G., Cuperus, Barber, J., Price, R., and Arnold, D. Turfgrass pest management, Pub. No. E-879. Cooperative Extension Service and Division of Agriculture, Oklahoma State University, Stillwater.

Ontario Weeds. Ontario Ministry of Agriculture and Food, Publ. No. 505.

Sherman, R. C., Kinbacher, E. J., Riordan, T. P., and D. H. Steinegger, (1980). Thatch accumulation in Kentucky bluegrass as influenced by cultivar mowing and nitrogen, *Hortic. Sci.* 5: 312–313.

Turgeon, A. J. (1980). *Turfgrass Management*, Reston Publishing Co., Reston, Virginia, pp. 219–235.

Turgeon, A. J. (1991). *Turf Management*, Prentice-Hall, Inc., Englewood Cliffs, New Jersey.

U.S. Department of Agriculture Handbook. 1970 *Selected Weeds of the United States*. USDA Handbook No. 366, Washington, D.C.

Weeds of the North Central States. (1981). North Central Regional Res. Publ. No. 281.

RELATED REFERENCES

Hitchcock, A. S. and Chase, A. (1971). Manual of the grasses of the United States, Vol. 2, Dover Publications, Inc., New York.
Kummer, A. P. (1951). Weed Seedlings, University of Chicago Press, Chicago.
Southern Weed Science Society. *Weed Identification Guide*. Champaign, Illinois.

16

Weed Management Systems for Forest Nurseries and Woodlands

Ken Mc Nabb and David B. South
Auburn University, Auburn, Alabama

Robert J. Mitchell
Joseph W. Jones Ecological Research Center, Newton, Georgia

I. INTRODUCTION

Vegetation management in forest nurseries and woodlots is unique with respect to the array of problems encountered, the variety of tools available to address weed problems and the social and economic context in which vegetation management decisions are made. The purpose of this chapter is to describe the nature and scope of weed management practices as they are used today in forestry. We hope to describe the available tools and the manner in which they can be properly used, and to discuss the consequences of alternatives. The biological and economic considerations needed to develop integrated approaches to vegetation management and important future challenges that will need to be considered in moving the science forward as well as preserving suitable tools for growing trees are discussed.

Plants are not inherently undesirable; they only become unwanted (i.e., weeds) when they hinder a manager from reaching a desired goal; thus weed management implies an objective. Although forests can be managed for a wide variety of objectives, it is beyond the scope of this chapter to review all potential weed management practices that would be used for every conceivable landowner objective. Our discussion will relate principally to forest vegetation management for profitable timber production, even though many of the tools and considerations in using weed management techniques have relevance to many forest management objectives. Production of timber for profit is predominantly an exercise in growing softwoods whether they be in the southeastern, northwestern, or northeastern United States. Although our purpose is not to be parochial with respect to region, many of the examples will come from southeastern silviculture, largely as a result of the breadth of the experiences of the authors.

II. WEED MANAGEMENT IN FOREST NURSERIES

A. Importance of Weed Management

Weed management can be one of the most expensive steps in the production of tree seedlings. At some nurseries, the cost of hand weeding may constitute 25% to 90% of the total production costs (Steven, 1932; Wakeley, 1935; McKeller, 1936; Abbott and Fitch, 1977; Harstela and Tervo, 1983). Nursery managers employing efficient weed management strategies can reduce hand weeding cost to less than 1% of production costs and the total cost of weed management to less than 8%.

In addition to reducing hand weeding costs, efficient weed management is important when using valuable seed. Seed may be valuable because it is difficult to collect or because it is genetically improved and is in short supply. Western white pine seed are worth approximately $0.02 per viable seed. Therefore, a 20% loss during hand weeding due to dislodging seed or pulling up young germinants could result in wasting seed worth $12,000/ha. In some cases, 50% of seedlings have been pulled up during hand weeding (Duthie, 1927).

Even when the total production of seedlings is not affected, the production of plantable seedlings can be reduced by inefficient weed management. An increase in cull percentage (seedlings that do not meet a minimum standard for outplanting) is caused by increased competition for light, moisture, and nutrients. In one case, competition from bermudagrass reduced the number of plantable seedlings by 66% (South et al., 1978). When seedlings are sold for 3 cents each, this could cost the nursery $48,000/ha. For this reason, the real cost of inefficient weed management entails more than just high weed management cost per hectare.

In addition to the economic aspects, weed management can influence the level of pest populations in the nursery. Large weeds change microenvironment humidity and temperature to levels that are more suitable to various fungal diseases. Certain insects may be attracted by the presence of weeds (such as *Lygus* spp. to common groundsel). After weeds are disturbed or hand weeded, the insects may injure young pine seedlings. Rodents are a problem in nurseries that have sufficient weeds to provide cover from predators. An efficient weed management system is used to alter the habitat and eliminate need to use cats or poisonous baits. In the past when weed populations were high, poison baits were often used to reduce rodent populations (Wakeley, 1935).

Efficient weed management systems involve a combination of methods which may include use of fumigation, herbicides, and various nonchemical techniques. Inefficient systems usually rely heavily on just one or two methods of weed management. In addition, systems that control weeds when they are small and vulnerable are more efficient than management systems that are initiated when weeds are mature.

B. Pine Seedbeds

Weed management systems developed for forest nurseries can be classified according to the management techniques employed. In the United States, 71% of the nearly 1.7 billion seedlings planted each year come from bare-root southern pine nurseries (Mangold et al., 1991). For this reason, most of the weed management practices discussed will be relevant to pine seedbeds. The following systems are classified according to the types of chemical techniques used.

Systems with No Fumigation or Herbicides

Prior to 1947, most nursery managers relied on cultural and biological methods of weed management. Nursery managers depended mainly on hand weeding to prevent weeds from overtopping the young pine seedlings. Weed populations were high and the time required for hand weeding often exceeded 3,000 hr/ha/yr (Figure 1). Frequent hand weeding was very important to ensure against seed production. Additionally, frequent removal of small weeds required less time than infrequent removal of large weeds. Weeding at 6 to 8 week intervals involved 30% to 100% more time than weeding at 3 to 4 week intervals (Aldhous, 1972).

Managers used various methods to reduce hand weeding, including sanitation practices, cultivation, cultural treatments, and animals. For example, soil amendments contaminated with weed seed were avoided. Cover crops were used to shade out weeds during the off-years. Toothed harrows were recommended for controlling bermudagrass since rhizomes would be brought to the surface and exposed to freezing temperatures over the winter. Nursery managers could cultivate germinating weeds in the spring and before they went to seed in the fall. Broadcast sowing was done at high densities to shade out weeds, while drill sowing was spaced to facilitate hoeing and mechanical cultivation. At some

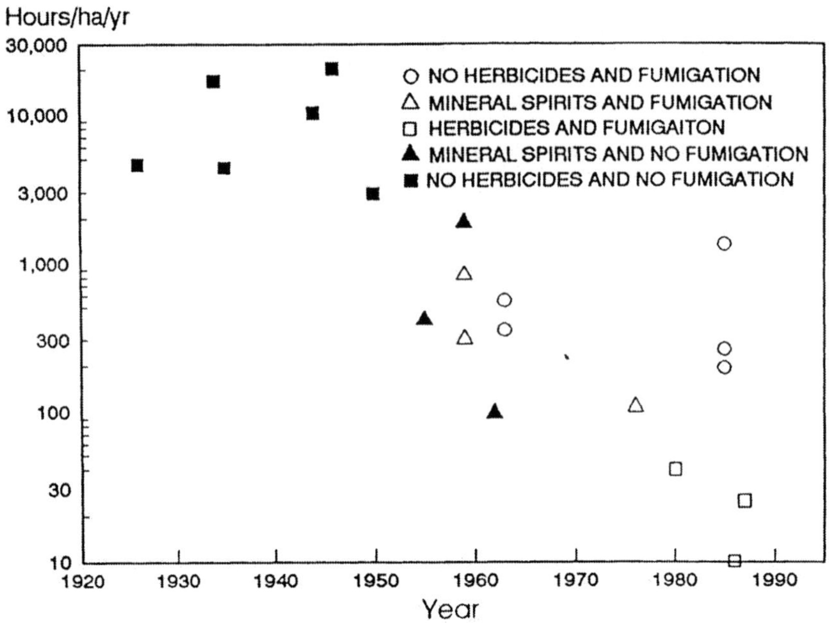

Figure 1 Time required to hand weed 1-year-old seedbeds in forest nurseries from 1930 to 1990 (From McKellar, 1936; MacDonald, 1953; Umland, 1946; Wycoff, 1954; Chapman, 1962; Shoulders et al., 1965; Boyd, 1971; South, Crowley, and Gjerstad, 1976; Boyer and South, 1984; South, 1986; Stringfield, 1987; Garrett et al., 1991.)

nurseries, mechanical weeding reduced the amount of hand weeding by as much as 40%.

Mechanical weeding has been used on a limited basis in about one-fourth of the nurseries in the Northwest (Owston and Abrahamson, 19840. A "blow-lamp" (similar to a flame-thrower) was used in forest nurseries in the early 1920s (Duthie, 1927), and its use prior to seed germination became a common practice for reducing weeding in European nurseries. Flame cultivation along irrigation pipelines has been used in nurseries in Alabama and Arkansas but is no longer feasible because of high petroleum cost.

Biological weed management techniques were applied by early nursery managers. Wakeley (1935) stated, "Geese eat the leaves of nutsedge and thus help exhaust the food stored in the tubers, and can be allowed to feed on infested areas where soil-improving crops are being grown." Chinese weeder geese have been used in nurseries in West Virginia (Woodrum, 1952) and Washington (Dutton, 1983). Hogs were sometimes brought into the nursery during the time between lifting and sowing (Wakeley, 1935). Hogs have an affinity for nutsedge tubers and will dig some distance to find them.

Seed efficiency (number of plantable seedlings produced per viable seed sown) without the use of fumigation and herbicides was often less than 50% (Sutherland and Adams, 1965). Low seed efficiency resulted not only from a loss of young seedlings during the hand weeding process but also from losses due to fungal diseases, nematodes, white grubs, birds, insects, and rodents (present because of the availability of protective weed cover).

Systems with Fumigation and Mineral Spirits

Weed management systems in forest nurseries changed dramatically when mineral spirits (a petroleum product commonly used as a cleaning fluid) were tested on conifer seedlings in Louisiana and Michigan in 1946. The discovery that conifers were tolerant of mineral spirits revolutionized management. Not only did the use of this practice rapidly increase in the southern United States, but it spread quickly throughout the world. By 1954, one-third of the large nurseries in Finland were using mineral spirits (Mikola, 1957). The use of mineral spirits was equated to a "magic wand" that could be waved over the seedbeds to kill most weeds (Cossitt, 1951). When using mineral spirits to manage weeds, the median seed efficiency at southern pine nurseries improved to 65% by 1965 (Abbott and Eliason, 1968). It was initially believed necessary to use 375 to 935 L/ha at 3 or 4 week intervals (Stoeckeler, 1951; Schubert and Adams, 1971). However, it was soon realized that it was much easier to kill young newly emerged weeds with frequent application at smaller doses than to use large infrequent doses to kill mature weeds with thick layers of epicuticular wax. The general practice adopted was to apply 95 L/ha two or three times a week during seedling germination. After the seedlings were older and tougher, the rate could be increased up to 290 L/ha without causing seedling injury. At some nurseries, over 30 tractor trips per year were used in applying mineral spirits. At one nursery in Alabama, a total of 3800 L/ha was applied. This nursery used no fumigation and even with the application of mineral spirits, 412 hr/ha of hand weeding was required in 1955.

In reality, mineral spirits is the inert ingredient and xylene is the active ingredient. In some cases, nursery managers purchased mineral spirits with no aromatic content, and this resulted in no weed control. For good weed control, the aromatic content of mineral spirits should be 10% to 20%.

The application of mineral spirits was common in 1975 (Table 1). However, the oil embargo caused a rapid increase in price and more effective herbicides were being developed. By 1980, less than a third of southern pine nurseries were using mineral spirits. By 1987, only two nurseries reported using it (mainly during emergence, when seedlings are susceptible to injury from diphenylether herbicides).

Table 1 Frequency of Chemicals Used for Weed Management in Southern Forest Nurseries as Determined by Questionnaires

Trade name	Common name	1975 (52)[a], %	1980 (51), %	1987 (39), %
Various	Methyl bromide	77	88	100
Basamid/Mylone	Dazomet	0	0	8[b]
Various	Mineral spirits	90	29	5
Goal	Oxyfluorfen	NA	73	100
Poast	Sethoxydim	NA	NA	85
Cobra	Lactofen	NA	NA	26
Roundup	Glyphosate	NA	29	23
Oust	Sulfometuron methyl	NA	NA	15
Modown	Bifenox	NA	67	10
Fusilade 2000	Fluazifop-butyl	NA	NA	3
Devrinol	Napropamide	NA	18	0
Enide	Diphenamid	46	0	0
Treflan	Trifluralin	29	12	5
Caparol	Prometryn	29	10	5
Eptam/Genep	EPTC	8	2	5
TOK E25	Nitrofen	8	26	NA
Destun	Perfluidone	6	4	NA
Attac	Toxaphene	0	2[a]	NA

[a]Numerals in parentheses indicate number of responses.
[b]Only on a trial basis. NA, not available.

By 1975, most nursery managers (73%) in the southern United States had adopted the system of fumigating with methyl bromide and applying mineral spirits as a selective herbicide (Table 1). The average hand weeding for 36 nurseries that used this combination was 380 hr/ha. The 20 nurseries operated by state and federal governments tended to fumigate less frequently and averaged 545 weeding hr/ha. Industry nurseries (16) fumigated more frequently and averaged 182 hand weeding hr/ha. Although these chemicals were very helpful in reducing the level of weed populations, a variety of weed species persisted (Table 2).

Three fumigants used by nursery managers since 1950 are dazomet, methyl bromide with 2% chloropicrin (MB-2C), and methyl bromide with 33% dichloropicrin (MB-33C). The methyl bromide/chloropicrin (MBC) mixtures require a plastic tarp while dazomet can be applied with or without a plastic cover. Product preference is influenced by efficacy, cost, ease of application, and, to some extent, politics. When using a plastic tarp, the cost of application with dazomet is higher than that of either MB-2C or MB-33C (about twice

Table 2 Percentage of Southern Pine Nurseries with Troublesome Weeds

Weed	1975, %	1980, %	1987, %
Nutsedge	52	62	69
Flathead sedge	16	4	8
Large crabgrass	32	64	51
Goosegrass	10	23	0
Bermudagrass	2	36	13
Crowfootgrass	4	4	5
Common purslane	20	30	8
Dogfennel	20	13	5
Morningglory	16	28	21
Sicklepod	14	23	41
Carpetweed	12	17	5
Florida pusley	10	4	5
Redroot pigweed	12	0	0
Clover	4	6	3
Prostrate spurge	4	2	38
Horseweed	0	0	8

as great per kilogram of active ingredient). However, when comparing non-tarped dazomet (applied by the nursery staff) with tarped MB-33C (applied by a contractor), the per hectare cost is approximately the same (Landis and Campbell, 1989). Some nursery managers prefer to use dazomet since it does not require a tarp. However, the efficacy is less when a tarp is not used and the potential for drift of methyl isothiocyanate and formaldehyde is greater. In some cases, adjacent plants have been injured when dazomet was applied without a tarp (Scholtes, 1989). Early comparison tests found MB-33C produced more plantable seedlings than untarped dazomet (Foster, 1961; Hodges, 1962). In addition, efficacy on weeds is often better with MBC than with dazomet. For these reasons, nursery managers in the southern United States prefer to fumigate with MBC.

Rates commonly used in the southern United States are 390 kg ai/ha for MB-33C and 450 kg ai/ha for MB-2C. A reduction in hand weeding can be obtained with either formulation. However, when fumigating mainly to kill nutsedge tubers, MB-2C is recommended since up to 494 kg/ha of methyl bromide can be applied. Effective control of nutsedge can be obtained by fumigating in the fall or spring when proper soil conditions exist. However, there are many more days in the fall when conditions are proper than in the spring. Because of frequent spring rains, some nursery managers have fumigated when soil conditions were too wet while others have opted to forgo fumigation altogether. Waiting for proper soil conditions in the spring can delay sowing. For these reasons, fall fumigation is recommended when fumigating to suppress nutsedge. When fumigating in September or October, some nursery managers keep the plastic tarp down for 1 to 2 months. This provides a degree of soil solarization and some resistant weeds, such as morning glory, will germinate under the plastic and will be killed by high temperatures.

When fumigating mainly to kill fungi, MB-33C is recommended since 130 kg/ha of chloropicrin will be applied. Although fumigation with MB-2C can reduce weeding times when populations are high, the chance of a reduction is less when weed populations are low (Table 3). This is because the fumigated soil can be contaminated with wind-blown seed or mulches containing weed seed. In addition, some weed species (e.g., morningglory and sicklepod) are not killed by fumigation and germination of these weeds can increase after removal of the tarp.

For southern pines, the frequency of soil fumigation varies with nursery, but most (74%) produce two crops for each fumigation. Those with disease problems (18%) fumigate prior to each seedling crop. The remainder fumigate prior to three or four seedling crops. About 33% fumigate in the fall while 26% fumigate in the spring (the remaining 41% fumigate some areas in the fall and some areas in the spring). The cost of fumigation varies depending on formulation, rate applied, tarp thickness, area treated, and commercial or private application. In 1987, the general price for contract fumigation was about $2470/ha. Some researchers think this is an expensive investment and justified only when multiple pest problems exist (Sutherland and van Eerden, 1980). However, many nursery managers know that when fumigation increases the production of plantable seedlings, the savings in seed cost or the return in seedling sales often exceeds the cost of fumigation. In one case, $2718 worth of seed per hectare was saved by fumigation (Boyd, 1971). In addition, when 2-year-old seedlings sell for 17 cents each, a loss of only 15,000 seedlings/ha (equivalent to about a 1% reduction in plantable seedlings) justifies the cost of fumigation. For various reasons, including weed control, fumigation with MBC can sometimes produce substantial gains in seedling production (Hill, 1955; Boyd, 1971; Tanaka et al., 1986; Campbell and Kelpasas, 1988; Grierson, 1989). Fumigation with MBC increased seedling production by more than 75% at one nursery (Sutherland and Adams, 1965).

Currently, there is concern that man-made sources of methyl bromide can act to deplete the ozone layer. If this fumigant is removed from the market, then use of alternatives such as 100% chloropicrin will increase. Without the use of methyl bromide, some nursery managers may decide to control nutsedge tubers with herbicides like alachlor, atrazine, and bentazon. Unlike methyl bromide, these herbicides have been detected in domestic wells. There is little doubt that use of other pesticides will increase if methyl bromide is withdrawn from the market.

Table 3 Hand Weeding in Forest Nurseries as Affected by Herbicides and Methyl Bromide (2% Chloropicrin)

Nursery	Year	Herbicides and rates, kg ai/ha	Herbicides + fumigant, hr/ha/yr	Herbicides hr/ha/yr	Fumigant hr/ha/yr	No fumigation, no herbicides, hr/ha/yr
Pine Seedbeds						
Ft. Towson	1977	Bifenox 3.4 + napropamide 1.1	2	2	14	12
Magnolia	1976	Bifenox 3.4 + napropamide 1.1	0	3	10	14
Aliceville	1977	Bifenox 3.4 + napropamide 1.1	18	17	31	26
Ft. Towson	1976	Bifenox 3.4 + napropamide 1.1	4	2	47	64
Hauss	1976	Diphenamid 4.5 + prometryn 1.1	—	86	100	131
Claridge	1972	Diphenamid 4.5 + prometryn 1.1	41	71	77	146
Lee	1982	Bifenox 3.4 + oxyfluorfen 2.3	0	3	29	180
Claridge	1973	Diphenamid 4.5 + prometryn 1.1	40	99	47	187
Miller	1976	Diphenamid 4.5 + prometryn 1.1	—	10	142	199
Hardwood Seedbeds						
Capron	1982	Trifluralin 1.1 + oryzalin 2.2	59	141	193	437
Hauss	1977	Trifluralin 1.1 + napropamide 1.7	230	257	576	497
Pinson	1977	Trifluralin 1.1 + napropamide 1.7	183	618	427	1378
Natchez	1982	EPTC 7.8	355	500	1272	2408

Systems with Methyl Bromide Fumigation and Herbicides

Diphenylether Herbicides. The need for testing of herbicides for selectivity in southern pine seedbeds increased after Congress passed the Federal Insecticide, Fungicide, and Rodenticide Act (FIFRA) of 1972. By 1975, only soil fumigants, diphenamid, EPTC, and mineral spirits (Amoco weed killer) could be legally used to manage weeds in pine seed beds. Research at Auburn University provided data in support of registering a number of herbicides for use on pine seedbeds (diphenamid, nitrofen, bifenox, oxyfluorfen, oxdiazon, perfluidone, prometryn, lactofen, sethoxydim, fluazifop-butyl). Although certain herbicides from the triazine, dinitroaniline, thiocarbamate, and substituted amide families showed promise, problems with each have occurred. Some nurseries observed injury from prometryn and napropamide while efficacy has been less than desired for trifluralin, diphenamid, and EPTC. In contrast, pine seedlings have demonstrated excellent tolerance to diphenylether herbicides.

Nitrofen and bifenox were tested in the early 1970s and then adopted as more cost effective alternatives to mineral spirits. By 1980 two-thirds of southern nurseries were using bifenox operationally. After the labeling of oxyfluorfen in 1979, the use of these two chemicals declined and neither is presently sold on the U.S. market.

It was initially believed that a preemergence application of 0.5 kg ai/ha of oxyfluorfen followed by three monthly applications (at the same rate per application) would be sufficient to provide season long management of weeds (South and Gjerstad, 1980). The increase in weed suppression is attributable to two factors. Field observations and subsequent research have shown that the efficiency of herbicide use is increased with light frequent applications. Weekly applications ensure that newly germinated weeds are treated before they have sufficient time to produce a thick epicuticular wax. Several weed species are more susceptible when they are 1 week old than when they are 4 weeks old. Holes in the chemical barrier may result when irrigation, heavy rainfall, or tractor tires disturb the soil surface. Weed seed can germinate and emerge in these voids without contacting the herbicide. With weekly applications, the chemical barrier is reestablished and the number of weeds that escape contact with the barrier is decreased. This is especially important for species like prostrate spurge (Blake and South, 1987) and flathead sedge. Although the cost of application increases, the same amount of herbicide is used per year. Many nursery managers in the South now use frequent applications of low rates of diphenylether herbicides.

At some nurseries, a liquid fertilizer is applied in a tank mix with weekly applications of oxyfluorfen. This "weed and feed" program has an advantage in that the application cost of the fertilizer is eliminated. In some cases, the cost of fertilizer is less in liquid form and the uniformity of fertilizer application is better than with granular fertilizer. The effect of adding ammonium to oxyfluorfen might provide better contact activity for some weeds.

In 1988, lactofen was registered for use on southern pine nurseries. This herbicide has more activity on certain broadleaf weeds than oxyfluorfen. However, it is less effective on certain grasses and sedges. Since newly emerged southern pines have a greater tolerance to lactofen than to oxyfluorfen, a number of nursery managers use lactofen for the first postemergence application.

The sodium salt of fomesafen is currently registered in the state of Georgia as a preemergence application on pine seedbeds. As a salt, this herbicide will leach into the soil, resulting in activity against yellow nutsedge. Since nutsedge is a troublesome weed, this herbicide has been tested in forest nurseries throughout the southern United States. At nurseries with silt loam soils, pine seedlings have been injured with a preemergence application of 0.5 kg ai/ha. Similar injury has not been observed on sandy soils, where the herbicide apparently is leached away from germinating seeds by frequent irrigation. Injury has also been observed on newly emerged pine seedlings when a surfactant or crop oil was used with a postemergence application.

Grass Herbicides. Prior to the introduction of selective grass herbicides, various annual and perennial grasses were often considered major weed pests in forest nurseries (Table 2). Currently, sethoxydim and fluazifop-butyl are registered for use on various tree seedlings and the need to hand weed grasses has been eliminated. The selective nature of these herbicides allows nursery managers to apply these materials even during germination (injury can result when using a crop-oil adjuvant). One advantage of these herbicides is that they work only after the grass seedlings have emerged. This allows the nursery manager to apply the herbicide when and where needed. Spot applications are often used to prevent wasting herbicide. With the use of sethoxydim, the median weeding times in southern pine nurseries is 25 hr/ha/yr. In 1980 (when sethoxydim was not used) the median weeding time was 62 hr/ha/yr (South, 1986).

Use of oxyfluorfen and sethoxydim on soil fumigated with MBC has reduced the median hand weeding times to approximately 25 hr/ha/yr. When using these herbicides, the formulation of methyl bromide apparently does not greatly influence the amount of hand weeding required. Assuming that the total production cost is 2.5 cents per seedling (and that 1.5 million seedlings/ha are produced and hand weeding costs $6/hr) the cost of hand weeding would represent less than 0.5% of the total production cost. When adding the costs of hand weeding, fumigation, herbicides, and application, the percentage would often be less than 8% of the total production cost.

Systems Using Methyl Bromide and No Herbicides

The use of herbicides in southern forest nurseries helps to produce seedlings at a reasonable cost. In the past, a high seedling cost has inhibited private investment in reforestation. With the use of MBC and selective herbicides, the cost of growing, lifting, and packaging 1-year-old seedlings at large nurseries is often less than 3 cents each. At some government operated nurseries that do not use herbicides the cost can exceed 12 cents each.

The use of herbicides sometimes becomes a political issue, as evidenced by the U.S. District Court order of 1984 that banned the use of herbicides on national forest lands in Washington and Oregon. The consequences of ceasing the use of herbicides in a nursery weed management program can be documented by observing the effect this ban had on weed management costs at the J. Herbert Stone Nursery (Figure 2). Even with the use of MB-33C and mechanical cultivation, hand weeding cost in 1-year-old seedbeds after 1985 was five times greater than the total weed management costs prior to the ban. In addition, seed efficiency at the Wind-River Nursery was reduced to the point where 25% more seed was required to produce the same number of plantable seedlings. In contrast, seed efficiency can often be greater than 80% with the use of effective herbicides and methyl bromide fumigation (South, 1991). Ceasing the use of herbicides in forest nurseries not only increases the cost of seedling production, but also reduces the number of seedlings available for reforestation.

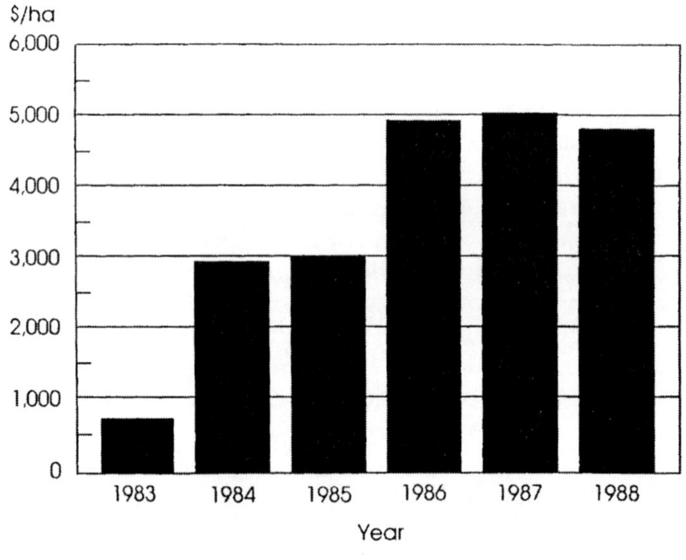

Figure 2 Increase in weed management costs in 1-year-old seedbeds at the J. H. Stone Nursery in Oregon after a ban on herbicide use. (1983 costs include hand weeding and herbicides while other years include only hand weeding costs).

C. Hardwood Seedbeds

Growth of weeds can be rapid in hardwood seedbeds because they are often irrigated and fertilized more often than pine seedbeds. In addition, the alkalinity of hardwood seedbeds is often above pH 6 while pine seedlings are often grown in soil that is below pH 5.5. A higher alkalinity is conducive to greater growth of various nursery weeds (Aldhous, 1972). Weed management systems for hardwood seedbeds often rely on fumigation with MBC. Some nurseries use no preemergence herbicide while others use either EPTC incorporated prior to sowing or a spray of trifluralin applied just after sowing and incorporated with irrigation. The few postemergence herbicides operationally used have no contact activity on broadleaf weeds. As a result of all these factors, weed populations in hardwood seedbeds are often higher than in pine seedbeds. Without using some chemical treatments, hand weeding times for hardwood seedbeds are often greater than 200 hr/ha/yr (Table 3).

Sethoxydim and fluazifop-butyl are currently applied as needed soon after grasses emerge in hardwood seedbeds. Certain herbicides (e.g., napropamide, trifluralin, oryzalin) are also used soon after the hardwoods are established (just after complete emergence) to suppress subsequent emergence of small seeded broadleaf weeds. Broadcast sprays of diphenylether herbicides are not labeled and are not used after emergence because of potential phytotoxicity to developing leaves. A period of several weeks is required to allow leaves of sweetgum to build up enough epicuticular wax to protect against damage from a liquid application of oxyfluorfen (South, 1982). However, some nurseries use granular applications of herbicide mixes that include oxyfluorfen. Incorporation of oxyfluorfen on granular fertilizer has been an effective way to apply it without injuring young hardwood leaves (South, 1983). At some nurseries, mechanical cultivation is used when the distance between drill-rows is 30 cm or more (South, 1987; Barham, 1980). A combination of using fumigation with MBC, herbicides, and mechanical cultivation can result in the lowest hand weeding times in hardwood seedbeds (South, 1988).

III. STAND REGENERATION AND CROP MANAGEMENT

A. Ecological Aspects of Vegetation Management

Nursery produced seedlings are normally out-planted in an environment that differs considerably from the nursery bed. In addition to stresses associated with suboptimal soil moisture and nutrient conditions, competitive pressures are much greater. The establishment and management of forests for timber production are largely a struggle to channel site resources (nutrients, water, and light) to plants that will be harvested at the end of the rotation in a manner that is both economical and environmentally sound. The ability to manage land in this manner is dependent on understanding plant community dynamics on sites with important commercial species.

For the purposes of this section we will consider *competition* as the negative influence on plant performance (growth, survival, etc.) resulting when two or more plants, growing in close proximity to each other, use the same limited resources at the same time. This operational definition suggests that competitive interactions arise largely because crop plants or their neighbors obtain resources at the expense of the others in the neighborhood. This definition discounts the effects of allelopathy, the direct interference of one plant with another through the release of a chemical into the environment. Although the existence of allelopathy is not questioned, for important forest types a large body of literature shows that trees compete with other plants through resource consumption. No work has been published which shows in any demonstrable way that allelopathy influences growth of important trees of the magnitude necessary to influence management decisions.

This definition of competition inherently separates plant response (negative influence on plant performance) from competitive effects (use of the same limiting resources) on resource availability. Although on the surface this distinction may not appear to provide insight into the way that competitive processes operate, this definition provides a new perspective on competitive processes and what or who is a good competitor. For instance, a plant can be a strong competitor by either depleting resource levels or growing rapidly when resources are reduced (Goldberg, 1990). The way plants influence soil resource levels is also overlooked in many cases. Reduction in resources due to uptake by a plant is intuitively understood, but plants can influence the availability of soil borne resources in ways other than through uptake. For instance, plants producing litter with high C:N or lignin:N ratios will reduce the rate of N mineralized in the soil. Surface soil moisture can often be increased during substantial parts of the growing season by shade and increased infiltration (Gill and Marks, 1991). Thus competitive relationships and the manner in which competitive processes unfold in widely different vegetation types and site conditions may be best analyzed by viewing both processes—the influence of the community on resources availability, then the response of the crop to resources supplied by the abiotic environment and modified by the competitive neighborhood.

Considering this perspective of competition, the various competitive situations commonly encountered in a tree crop rotation will be discussed. For illustrative purposes we will begin with a mixed upland loblolly pine hardwood site, typical of those found throughout the southeastern United States. When these areas are cut over, early successional plant communities comprise a large number and variety of herbaceous species and hardwood sprouts. The relative dominance of these plants depends on the types and amounts of the species present before harvest (i.e., understory hardwoods provide the source for sprouts of the next generation) and the practices used to site prepare the area (chopping will encourage sprouting, particularly for species such as sweetgum that can sprout root suckers).

Unless herbaceous weeds are controlled, competition from the herb community constrains pine growth to the greatest extent during the first several years of stand development. Miller et al. (1991) report on a study conducted across 14 sites throughout the Southeast in which four distinct communities were constructed, pine alone, pine plus herbaceous plants, pine plus hardwoods, and pine with both herbaceous and hardwood components. After 5 years, growth response was the greatest where herbaceous plants were managed. Growth responses to eliminating the hardwood component lagged behind as a result of elimination of competition from herbaceous plants, but some evidence suggested that the importance of woody vegetation was increasing as the stands aged through the 5 year period.

Evidence which explains the dynamics of competition in upland pine stands is found in Figure 3. Herbaceous plants quickly invade cutover areas, particularly when some type of mechanical site preparation and burning precedes planting (this site was chopped and burned). After the first year, bare ground averaged approximately 10% in the pine-hardwood-herbaceous community and the pine-herbaceous community. However, in the pine-hardwood and pine alone communities bare ground exceeded 80%. Moreover, as stands develop, herbaceous plants are shaded out and diminish in importance and hardwood sprouts continue to increase in site domination.

Herbaceous plants are strong competitors for soil resources such as soil moisture (Britt et al., 1990) and soil nutrients, particularly N (Smethhurst and Nambiar, 1989). Herbaceous plants, in addition to rapidly building above ground biomass, have the ability to colonize the soil rapidly with a thick mass of roots that severely limit the relatively small root system of a pine seedling to compete effectively in acquiring soil borne resources. However, as pines grow, the root stem extends throughout the upper portion of the soil matrix, allowing for greater uptake of N, and penetrates deeper into the soil, where roots of herbaceous plants are not abundant, thus decreasing competition for soil moisture.

These data (Figure 3) also illustrate a second feature of ecological responses to manipulation of vegetation. Vegetation is not controlled; it is replaced. When one component of the vegetation complex is disturbed, the resources (light water and nutrients) that are liberated (made available to other plants in the community as a result of the absence of the plants removed) increase the growth of those plants positioned to obtain the additional resources. The additional water, light, and/or nutrients increase the growth of the remain-

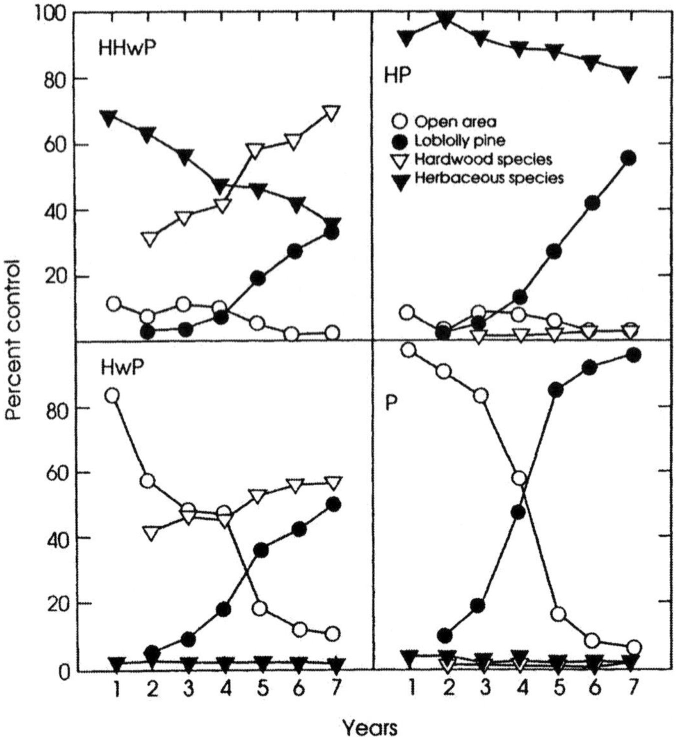

Figure 3 Vegetation dynamics of four communities that have been created through selective herbicide application: HHwP, hardwood herbaceous and pine; HP, herbaceous and pine; HwP, hardwood and pine; P pine alone. (From Wood et al., 1992.)

ing plants until resource levels fall to a new equilibrium, again limiting growth. Further evidence of this is illustrated in Figure 4. These data show that after 4 years, the water potential of pines grown with or without herbaceous weed control was nearly identical, on both a diurnal and a seasonal basis (Green et al., 1991). However, earlier work on this site showed that during a first year drought, water stress was significantly reduced when pines were grown without herbaceous weeds (Miller et al., 1991). Treatments that are marginally effective in controlling competitors will release not only the crop trees but those plants not affected by the treatment as well. Also, treatments that are more effective at eliminating interspecific competitors will accelerate the time in which crop trees begin to compete with each other.

A new stage in stand development can be considered after canopy closure has been achieved (from approximately age 10 to 15 for southern pines) until

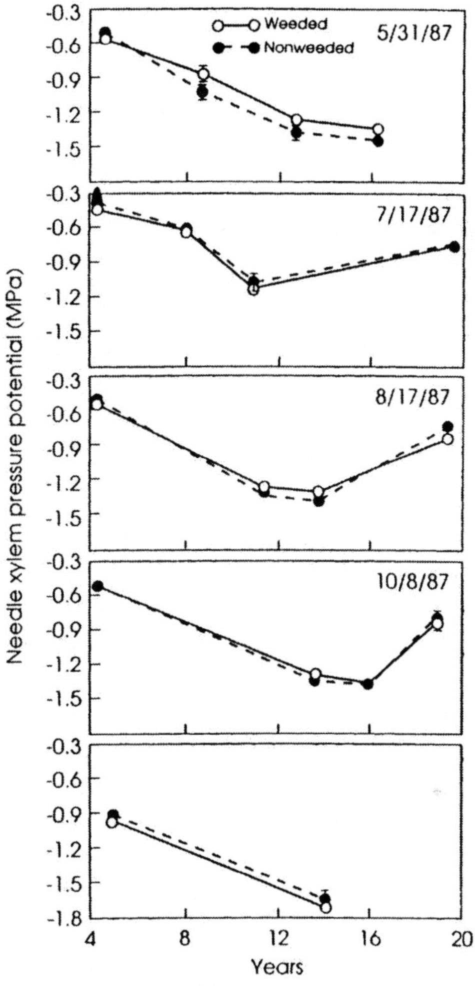

Figure 4 Xylem pressure potential of 4-year-old loblolly pine seedlings that have grown with herbaceous vegetation (nonweeded) or without herbaceous vegetation (weeded) throughout their development after outplanting. Note: large differences were noted between the two treatments during their first year after outplanting. (From Green et al., 1991.)

the next harvest. Shade from the canopy will eventually eliminate herbaceous plants. Also, some evidence suggests that the percentage basal area of dominant and codominant hardwood trees within a pine stand remains relatively constant from age 10 to age 35 (Burkhart and Sprinz, 1984; Burkhart et al.,

1987). Although a relatively constant proportion of basal area between pine and hardwood exists, reduction in pine volume is proportionally more than hardwood basal area would indicate. For example, a stand that has 30% hardwood basal area at age 10 can experience as much as a 50% reduction in pine volume at age 30 (Burkhart et al., 1987). In addition to hardwood competition from dominant and codominant trees limiting pine growth, invading understory hardwood plants become established and cause competition problems for the next stand. The degree to which the understory hardwoods influence the overstory pine is questionable; however, their competitive effects on the next stand are likely to be significant. Historically, few attempts have been directed to controlling vegetation during this preharvest stage even though these competition problems are less expensively and more easily managed during this stage of stand development.

B. Artificial Regeneration

Mechanical Site Preparation

The term *artificial regeneration* in a forestry context refers to the establishment of a stand of trees by planting nursery-produced seedlings or direct sowing of seed. The majority (98%) of artificially regenerated sites in the United States are planted and few are direct sown (Mangold et al., 1991). Artificial regeneration systems are usually capital intensive and costs may vary from $250 to $600 per hectare for stand establishment in the southeastern United States. This includes the cost of site preparation, seedlings, and planting. Although seedling and planting costs are usually around $150/hectare (Belli et al., 1993), site preparation is the most significant expenditure for artificial regeneration. The objectives of site preparation are weed control, amelioration of adverse soil conditions such as poor drainage, and removal of logging slash, which facilitates planting and reduces fire hazard. The principal objective of site preparation, however, is the reduction of competing vegetation.

The general forestry site preparation methods are mechanical and chemical. Mechanical site preparation uses heavy equipment to push down, chop, or remove living and dead plant material on the site. Additional operations are used to manipulate the soil through subsoiling, disking, or bedding.

Roller Drum Chop and Burn: After harvest, an area is rolled, using a large rotating drum partially or completely filled with water. The drum has large knives perpendicular to the direction of the roll and is pulled by a large rubber tire skidder or crawler tractor. Standing trees that are not too large are pushed down and rolled over. After the chopped vegetation has suitably dried, usually for a month or more, the area is burned.

Considerations: Chop and burn operations are relatively cheap and leave a great deal of biomass on the site, contributing to soil stability and nutrient pools. From a vegetative control standpoint, this technique usually results in prolific sprouting and may require a chemical release treatment sometimes before the fifth or sixth year after planting to reduce hardwood competition.

Shear and Pile: A crawler tractor with a shearing blade mounted on the front is used to cut all standing vegetation at or near ground level. After shearing, the same or another tractor pushes the residue into large piles in long lines (windrows) or circular piles. These piles are usually burned. The pushing blades often have tines or teeth on them and are used to uproot stumps and roots for removal from the site.

Considerations: Weed control using this type of mechanical site preparation is usually expensive but effective as hardwood rootstocks are uprooted and removed. If done incorrectly, however, topsoil can be moved into windrows and piles, resulting in a site productivity decline even with reduced weed competition (Morris et al., 1983).

Ripping, Disking, and Bedding: These practices may improve soil conditions such as surface soil compaction and internal drainage. Ripping is deep subsoiling and is most effective if done in the summer when soils are dry (Lowery and Gjerstad, 1991). Forestry disking implements are heavier and stronger than disk harrows used in row-crop production, although similar in design. *Bedding* refers to the use of a bedding plow which forms large continuous furrows in poorly drained areas to improve aeration for seedling roots. The effect of these treatments on crop growth and development can be significant (Morris and Lowery, 1988).

Disking will provide additional weed control beyond shearing or burning through more complete chopping and uprooting of smaller hardwood rootstocks, vines, and sprouts. Bedding provides a "cultivation" effect and an additional level of weed control during stand establishment. However, a serious disadvantage to both disking and bedding is the added cost beyond that of shearing and raking. It is recommended only when soil conditions warrant or potential gains in site productivity are significant.

Chemical Site Preparation

Although the majority of artificially regenerated sites in the Southeast have been mechanically prepared (Lowery and Gjerstad, 1991), the use of herbicides is increasing. The reasons for this increase are 1) new herbicides that are environmentally safer, have lower mammalian toxicity, and are effective at lower doses (Table 4); 2) the increasing number of applicators trained in the proper methods of herbicide prescriptions and safe utilization; and 3) an increase in practical research which has improved the efficiency of chemical application technology and documented its cost effectiveness. When compared to mechanical site preparation, chemical methods have advantages and disadvantages.

Table 4 Herbicides Labeled for Forestry Application

Product	Active ingredient	Formulation[a]	Labeled uses[b] SP	CS	RE	HB
Aatrex 4L	43% atrazine	WSL				X
Aatrex 80W	80% atrazine	WP				X
Aatrex nine-O	90% atrazine	WDG				
Access	17% picloram + 32% triclopyr	EC			X	
Accord	41% glyphosate	WSL	X		X	X
Arsenal	54% imazapyr	WSL	X	X	X	X
Banvel	60% dicamba	WSL	X	X		
Banvel CST	17% dicamba	RTU		X		
Banvel 720	13% dicamba + 24% 2,4-D	WSL	X			
Chopper	4% imazapyr	RTU		X	X	
Escort	60% metsulfuron	DF	X		X	
Fusilade	13% fluazifop	EC				X
Garlon 3A	44% triclopyr	WSL	X	X	X	
Garlon 4	62% triclopyr	EC	X		X	
Hi-Dep	50% 2,4-D	WSL	X	X		
Krenite K	42% fosamine	WSL	X			
Oust	75% sulfometuron	WDG				X
Pathfinder	17% triclopyr	RTU		X	X	
Poast	18% sethoxydim	WSL				X

Pronone MG	10% hexazinone	G	X		X	X
Pronone 10G	10% hexazinone	G	X		X	X
Pronone 25G	25% hexazinone	G	X		X	X
Pronone Power Pellet	75% hexazinone	G	X		X	X
Roundup	41% glyphosate	WSL	X	X	X	X
Super Brush Killer	32% 2,4-D + 32% 2,4-DP + 5% dicamba	EC	X	X	X	
Tordon 101M	10% picloram + 40% 1,4-D	WSL	X	X		
Tordon K	24% picloram	WSL	X			
Tordon 101R or Tordon RTU	5% picloram + 21% 2,4-D	RTU		X		
Velpar	25% hexazinone	WSL	X		X	X
Velpar ULW	75% hexazinone	G	X		X	
Weedone 170	29% 2,4-DP + 30% 2,4-D	EC	X		X	
Weedone CB	12% 2,4-DP + 30% 2,4-D	RTU		X	X	
Weedone 2,4-DP	59% 2,4-DP	EC	X		X	
Weedone LV4	61% 2,4-D	EC	X		X	

[a]DF, dry flowable; EC, emulsifiable concentrate; G, granular; RTU, ready to use; WDG, water dispersable granule; WSL, water soluble liquid; WP, wettable powder.

[b]SP, site preparation; CS, cut surface (injection and stump sprays); RE, release (includes basal sprays and streamlining); HB, herbaceous weed control.

ADVANTAGES

Costs of the more common chemical site preparation applications are in the lower range of the costs for mechanical methods.

Chemical site preparation is more productive in terms of the number of acres that can be prepared in a day.

Chemical site preparation requires lower capital expenditures than mechanical methods.

There is less soil disturbance with chemical methods. Improper use of some mechanical site preparation equipment may result in decreased site productivity through loss of topsoil and/or soil compaction.

Because chemical methods leave the root mat intact, soil erosion is significantly reduced. The effect is an increase in water quality in adjacent streams and other bodies of water when compared to that of mechanical site preparation (Beasley et al., 1986; Neary et al., 1986). Moreover, water infiltration rates, soil organic matter, and nutrient contents remain constant or even increase as a result of chemical site preparation.

Chemical methods can be used in more rugged terrain where machinery may have difficulty in access or operation.

The use of chemical weed control usually results in less resprouting of woody competitors.

DISADVANTAGES

Chemical site preparation requires a high level of technical input. The prescription process is site specific and demands considerable expertise. Application requires trained and certified personnel who are knowledgeable in handling pesticides.

Application timing is more dependent upon weather than mechanical methods.

Chemical applications carry a risk of damaging crops or vegetation in nearby areas which may lead to indemnity or litigation.

The use of herbicides can be controversial in certain areas because of public concern regarding their environmental and human health effects.

Chemical methods cannot correct adverse soil conditions such as poor drainage or soil compaction due to harvesting.

The single objective of chemical site preparation is to control competing vegetation. Properly done, chemical methods will greatly reduce competition to crop trees and enhance plantation productivity. Applications are usually broadcast over the entire area during the growing season after stand harvesting. Some areas are allowed to lay fallow for 1 year after harvest so that additional resprouting of hardwood rootstocks occurs before chemical application. This produces a larger leaf surface area and transpirational pull, resulting in more chemical uptake. As a general rule, chemically prepared sites are burned prior to planting to facilitate planting and obtain additional weed control. Burning should be done 2 to 12 weeks after application with the objective of burning after adequate "browning up" has occurred along with complete translocation of absorbed herbicides.

Tank mixes of two or more chemicals are often used for site preparation treatments. The two most common tank mixes are picloram/triclopyr and imazapyr/glyphosate. The picloram/triclopyr mix is particularly effective for Lower Coastal Plain sites, where waxy cuticled species predominate, although this mix can be used on a variety of locations. The imazapyr/glyphosate tank mix is routinely used on Upper Coastal Plain and Piedmont sites characterized by hardwood species diversity. Both tank mixes will normally produce a browning up sufficient for a good fire. One of the advantages of these tank mixes is that they broaden the spectrum of control. Imazapyr, for example, is relatively weak on black locust (Table 5) but is effective for controlling red maple. Glyphosate, on the other hand, can control black locust but has relatively limited effect on red maple. Moreover, because each compound has a different site of activity in the plant, the effectiveness of one complements that of the other. Therefore, application rates for each compound are lower than they would be if used alone.

Herbaceous Weed Control

Conifer Plantations. *Herbaceous weed control* refers to the control of annual and perennial herbaceous weeds that compete strongly with the crop species during the first year or two after planting. It is during this period that young seedlings are particularly sensitive to competition for soil water (Britt et al., 1990). Herbaceous weeds are aggressive and efficient at acquiring site resources. Established pasture grasses, for example, can significantly reduce crop survival in dry years. Positive responses to herbaceous weed control in plantations are well documented (Creighton et al., 1987; McKee and Wilhite, 1988; Glover et al., 1989). Miller et al. (1991), for example, found that total herbaceous weed control increased loblolly pine volume index at age 5 by 32% to 314% above the check plots (no weed control). The long term effect of this early growth advantage continues to be monitored.

Most herbaceous weed control in forestry is done on pine plantations in the southeastern United States. These applications are made right before or soon after weed emergence in March and April after planting the previous winter. Applications can be made broadcast over the top, in 1 to 2 meter bands over the top of seedling rows, or as spot applications over the top or at the base of individual seedlings. Studies have shown that band application is more cost effective than the other two methods (Creighton et al., 1987).

A second application during the spring of the second growing season can be made for additional weed control and may increase plantation productivity. This application is not usually done operationally as application costs remain the same while crop tree response is relatively low.

Table 5 The Relative Susceptibility and Resistance[a] of Common Forestry Weed Species to Various Herbicides

Atrazine		
Susceptible	-	Broad spectrum control of annual grasses and forbs
Resistant	-	Bermudagrass, broomsedge, johnsongrass, lespedeza, fall panicum, beggarweed
Dicamba		
Susceptible	-	White oak, pines, sumac
Resistant	-	Red maple, ash, hickory, elm, dogwood, persimmon
Diclorprop (2,4-D)		
Susceptible	-	Black locust, blackgum, sumac, multiflora rose sweetgum, sassafras, post oak, blackjack oak, yellow poplar
Resistant	-	Red maple, ash, elm, dogwood, persimmon, willow, kudzu, grasses
Fluazifop-butyl		
Susceptible	-	Annual and perennial grasses including bermudagrass and young johnsongrass
Resistant	-	All broadleaf weeds, nutsedge
Glyphosate		
Susceptible	-	Broad spectrum control of hardwoods, pines, annual and perennial grasses and forbs
Resistant	-	Red maple, black cherry, ash, hickory, dogwood, greenbriar, Virginia creeper, trumpet creeper
Hexazinone		
Susceptible	-	Broad spectrum control of hardwoods including red and white oaks, winged elm, and hawthorn, annual grasses and forbs, white pine
Resistant	-	Yellow poplar, eastern red cedar, sassafras, persimmon, blackgum, pines, hollies, American beautyberry, bermudagrass, white snakeroot, broomsedge, johnsongrass, sicklepod, trumpet creeper, morning glory
Imazapyr		
Susceptible	-	Broad spectrum control of hardwoods including persimmon, sassafras, and red maple, annual and perennial grasses and forbs, longleaf pines, greenbriar, trumpet creeper
Resistant	-	Elm, pines, buckeye, wax myrtle, holly, magnolia, hawthorn, blackberry, tropical croton, broomsedge, and many legumes including locust, redbud, coffeeweed, and sicklepod
Picloram		
Susceptible	-	Broad spectrum control of pines and hardwoods including blackgum and red maple, kudzu
Resistant	-	Ash, eastern red cedar, yellow poplar, privet, sumac, grasses

Table 5 Continued

Sethoxydim		
Susceptible	-	Annual and perennial grass including bermudagrass
Resistant	-	All broadleaf weeds
Sulfometuron methyl		
Susceptible	-	Broad spectrum control of annual grasses and forbs and certain perennials, blackberry
Resistant	-	Bermudagrass, tropical croton, johnsongrass, trumpet creeper, broomsedge
Triclopyr		
Susceptible	-	Broad spectrum control of pines and hardwoods including yellow poplar, sassafras, willow, blackberry
Resistant	-	Black cherry, eastern red cedar, honeysuckle, persimmon, greenbriar, hawthorn, sumac, beautyberry

[a]The terms *susceptible* and *resistant* relate to consistently demonstrated application effectiveness at labeled rates.
[b]Sources: Nelson and Cantrell (1990); Miller and Mitchell (1988).

Hardwood Plantations. The positive response of hardwood seedlings to weed control is well established (Kennedy, 1984; Zutter et al., 1987). Herbaceous weed control in hardwood plantations is more problematic because most available herbicides for herbaceous weed control are toxic to hardwood seedlings. Fluazifop-butyl, a grass herbicide, is the exception to this rule with excellent tolerance by most broadleaved species. Glyphosate, although labeled for use in "silvicultural plantings," has no selectivity and must be used as a directed spray. One of the best ways to provide increased herbaceous weed control in hardwood plantations is to band glyphosate in the fall *before* planting. This is perhaps the best option when the site is occupied by a well established competitor such as bermudagrass or *Lespedeza* spp. which would be difficult to control at normal herbaceous weed control rates applied in the spring.

Mowing has been used for early weed management in hardwood plantations. However, mechanical methods have several disadvantages. Weeds are not killed by mowing and repeated mowings are necessary for control. Seedlings can be damaged when the tractor or mower bumps into them. Finally, weeds adjacent to the seedlings, which are the most competitive, are usually not cut. It is apparent that weed control in hardwood plantations is an area requiring additional research, particularly in terms of selectivity.

C. Natural Regeneration

Natural regeneration systems use the seed fall from trees on site to establish the next crop. Although considerable skill is required and weather related risks are encountered, natural regeneration is a successful and often used stand establishment technique. As with artificial regeneration methods, success is dependent upon providing environmental conditions conducive to seed germination and growth of very young seedlings. Most commercial forestry species are characteristically early successional, requiring clean seedbeds and full sunlight for establishment. This is true of the pines, several other conifer genera, and many commercial hardwood species. Therefore, weed control is an essential component in site preparation for natural regeneration systems.

When a stand is naturally regenerated, a specific number of trees are intentionally left at the time of harvest to serve as seed trees. Since these are normally the best trees, mechanical damage to these individuals must be prevented to reduce potential problems of disease and bark beetle infection. For pines, site preparation usually consists of a summer fire which burns off herbaceous and small hardwood weeds before seed dispersal in the fall. In some instances disking will further weed control as well as provide a better seedbed.

The use of chemicals for preparing seed beds is limited. Injection of large hardwood trees prior to seedfall is often used for weed control in natural regeneration of pine. Applications of the soil active chemicals imazapyr and hexazinone or directed foliar applications of glyphosate can be used in natural regeneration systems; the cost of these applications is normally incompatible with a principal objective of natural regeneration, which is low cost stand establishment.

D. Established Stand Management

The "weed" component in established stands consists of woody species that compete with designated crop trees for site resources. "Release" is the removal of woody competitors from managed stands, either artificially or naturally regenerated. The principal objective of release is to maximize crop productivity by increasing light, nutrient, and moisture resource availability to the crop species. Pine productivity (volume per unit of area) is inversely proportional to the amount of hardwood basal area on the site (Figure 5). This trend has been found across numerous locations and site conditions (Burkhart et al., 1987; Glover and Dickens, 1985).

Most of the release applications in the United States are done in managed pine stands in the southeastern United States. In artificially regenerated sites where mechanical site preparation was used or where chemical site preparation was ineffective, a release may be required in the fourth to sixth year after establishment. At this age, hardwood competitors are reaching the same height as pine or even beginning to overtake them. In these situations aerial-broadcast applications of selective herbicides are the preferred release technique. The more commonly used compounds are imazapyr and hexazinone, although several are

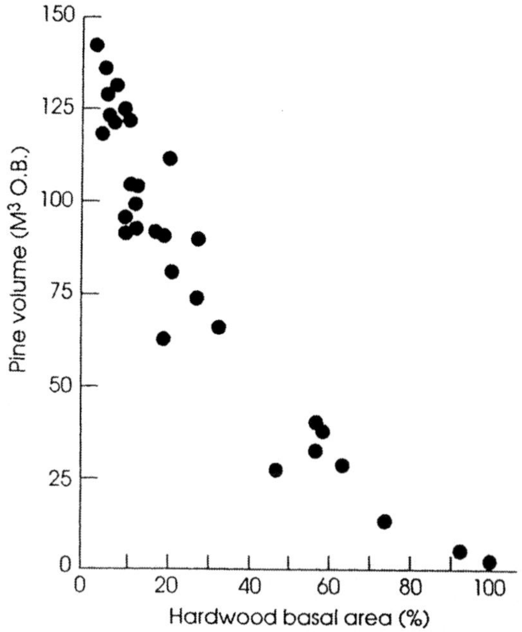

Figure 5 Comparison of pine volume per hectare to percentage hardwood basal area at Fayette, Alabama (age 24).

labeled for release work (Table 4). It is essential, in broadcast release, that chemicals be prescribed and applied skillfully because of the presence of the crop species. Insufficient quantities of an inappropriate chemical can result in wasted money, and an overdose or use of chemicals with poor selectivity will cause significant crop damage.

Pine release is also done using directed chemical applications to individual trees. There are a number of application methods suitable for this purpose, including injection, basal sprays, foliar sprays, soil spots, and streamlining (Miller and Mitchell, 1988). Directed applications have the distinct advantage of minimizing crop-tree damage. Directed applications can skip gaps in the crop tree canopy, and other areas where release is not needed. If specific hardwood species are deemed valuable (e.g., for wildlife), they can be omitted during directed applications. The principal disadvantage of directed stem applications is that costs are directly related to the number of stems treated per hectare. A hectare with 4000 hardwood rootstocks will cost twice as much to treat as a hectare with 2000 hardwood rootstocks.

One of the most potentially beneficial uses of individual tree application for crop release is during the final 5 years of the rotation. At this point both crop and competing species are large. Passing through a stand and injecting noncrop species above 5 cm DBH produces two distinct advantages. First, the crop species may respond with a significant increase in diameter growth in the final years before harvest and at a period when they are most valuable. Second, hardwood control, prior to harvest, will reduce sprouting and root suckering after harvest. This will improve the effectiveness of site preparation. The effectiveness of chemical weed control, late in the rotation, will be enhanced if the stand has been managed under a regular regimen of prescribed burning after age 10.

IV. FORESTRY WEED MANAGEMENT TECHNIQUES

A. Chemical Application Techniques

Selectivity

Selectivity in forestry herbicides is related to 1) the physical barrier provided by the waxy cuticle found on many species of conifers and 2) differential metabolism, whereby species or genera are able to transport, metabolize, or otherwise deactivate a chemical.

The selectivity of the phenoxyalkanoic herbicides (e.g., 2,4-D) for pine is due to the physical barrier of needle cutin. Hardwoods are killed at labeled application rates because of greater uptake. Glyphosate is another foliar absorbed compound which is differentially absorbed because of the cuticle barrier. On the other hand, the grass herbicides sethoxydim and fluazifop butyl have a high degree of selectivity that is due to metabolic differences between broadleaf and grass species.

Both of the soil absorbed triazine compounds atrazine and hexazinone possess some selectivity for conifers. However, conifer mortality is often a problem when hexazinone is used and the application requires careful matching of soil type with application rates. Imazapyr, a compound with both soil and foliar absorption, exhibits a high degree of selectivity between conifers and hardwood species which is possibly related to differential translocation.

The susceptibility of plants to specific chemical applications can be influenced by factors which either differentially promote or inhibit chemical uptake. When using foliar absorbed chemicals for site preparation, the addition of surfactants is highly beneficial. Application effectiveness is also improved by environmental conditions such as adequate soil moisture and high humidity, which tends to hydrate the leaf cuticle, resulting in more herbicide penetration. Older plants tend to have a thicker leaf cuticle compared to younger plants of the same species. Thus, differential leaf penetration is influenced by plant age, use of wetting or sticking agents, and weather at the time of application.

Foliar Applications

The selection of an application technique is influenced by a number of factors, including owner objectives, type and intensity of competition, equipment availability, cost, size of tract, and soil type. Probably most forestry herbicide applications are made as liquid formulations sprayed over foliage during the growing season. The chemicals may be solely foliar absorbed (e.g., glyphosate), or they may have both foliar and soil absorption (e.g., imazapyr). There are a number of techniques for making foliar applications to forestry sites.

Backpack sprayers can be used for directed foliar application on woody competitors up to 2 meters tall, if not too dense. Flat fan or hollow cone nozzles are the most commonly used with a wand at recommended spraying pressures of 138 to 207 kPa (Miller and Mitchell, 1988). Vegetation is sprayed to wet but not drip, covering as much of the leaf surface area as possible. The principal advantages of backpack sprayers are low cost and control over which individual stems receive treatment. The disadvantage is a high labor requirement.

Ground machines are used for spray applications on forestry sites and may be used for broadcast, band or spot applications. Sprayers have been mounted on crawlers, farm tractors, skidders, and all-terrain vehicles (ATVs). Spraying systems are normally custom manufactured or home made and require a tank, pump, pressure regulating system, strainer, and spray nozzle. Nozzle setups may be of several types. However, most forestry spray systems use cluster or manifold nozzles, which are often mounted up to 5 meters off the ground to reach over brush during site preparation and release applications. Nozzles mounted on a horizontal boom are used only for certain herbaceous weed control applications where the crop trees are small and the terrain is relatively uniform and even. Tractor or ATV mounted boom sprayers can be effectively used for herbaceous weed control on agricultural land planted to trees.

The advantage of using ground machines is that they provide a spraying system which can be used on virtually any size tract. Additionally, ground applications are more easily controlled when compared to aerial applications in regard to drift, application windows (periods when application is feasible), and public risk perceptions. Site factors that limit mobility are the principal disadvantage to ground sprayers. Machines cannot operate safely on steep slopes and rugged terrain. In addition, they cannot be used where heavy logging slash or dense vegetation restricts machine movement.

The use of aircraft is a widely practiced and effective application technique in forest weed management. Most applications are made by helicopter, and rarely are fixed wing aircraft employed in forestry situations. The maneuverability and slow flying speed of helicopters are advantageous in forest management because of irregular stand boundaries, existing snags and live trees that may be found on a tract, and relatively small size of treated areas (Newton and Knight, 1981). Also, helicopters may be used in remote areas without the need of a nearby airport. Spray systems are much the same as for row crop applications although a microfoil drift control attachment is often used (Kidd, 1987). Drift control is important to keep chemicals within the tract boundary and maintain straight flight paths to reduce skips and overlaps. Unlike row crop applications, the characteristic uneven terrain, irregular boundaries, and variable canopy height make it difficult to maintain straight flight lines consistently. Aerial applications, in forestry, require considerable effort in developing and establishing flight guidance systems such as flagging, balloons, tall poles, and electronic navigation.

Soil Application

Hexazinone is a forestry herbicide which may be applied directly to the soil for site preparation and release operations. Although it can be absorbed through leaves, it is transported in the water conducting tissue and, similar to the other triazines, is most effective when absorbed into the roots. There are several granular formulations of hexazinone, which may be applied by hand, ground machine, or aerial spreaders. In addition, the liquid formulation of hexazinone may be applied directly to the soil in a grid pattern covering the entire tract or applied around individual plants and away from crop trees. This type of application is made with a special "spotgun" or "gunjet" with a backpack sprayer placing a calibrated amount of chemical (usually 1 to 3 ml) at each spot.

Stem Application

Several methods involve applying the herbicide directly to the stem of woody competitors. These include basal stem sprays, basal stem streamlining, stem injection, and stump spraying. Basal stem sprays and streamlining may be used to control hardwoods with diameters of 7.5 cm or less for the removal of unwanted hardwood competitors in conifer stands and undesirable individuals in hardwood stands. Both techniques require the application of a herbicide to the main stem of the tree between 15 and 60 cm above the ground. Chemicals may be applied with a regular backpack and spray nozzle or streamlined by shooting a thin stream of mixture from a backpack through a "gunjet" to the stem. In both cases, ester formulations of acid herbicides and/or special penetrants are used to carry the active ingredient through the bark into the stem. Most stems are treated on only one side, although larger trees may require applications to both sides. Each stem in a clump of sprouts must receive the chemical. Streamline applications using the "gunjet" have the advantage of being able to direct the chemical to the stem at a distance of about 1 to 3 meters. This saves time and energy. Sprays tend to be more judicious with the amount of chemical used per individual stem. The cost effectiveness of both techniques is very sensitive to the number of stems per hectare to be treated.

Individual trees over 5 cm DBH can be injected with herbicides by using one of several techniques. Special tubular injectors are available that are jabbed into the base of the tree, and a measured amount of chemical (usually 1 to 2 ml) is released into the cavity by pulling a lever on the injector. Although effective, they can be cumbersome to use. Another injection tool is the "hypo-hatchet," which uses a specially designed hatchet that administers a regulated amount of chemical each time the stem is struck. The same effect can be achieved when using a regular hatchet and immediately spraying a precalibrated amount of chemical into the wound from a plastic spray bottle. This is known as "hack and squirt." In both cases the wounds are made at a convenient height. For all injection techniques, the chemical must be placed in wounds strategically spaced around the trees. It is important that the sapwood be penetrated and that wounds be sufficiently angled to hold the chemical long enough for absorption. Heavy sap flow or rain can wash chemical from the wound and diminish its effectiveness.

Chemical application to freshly cut stumps is another technique for delivery to individual stems. Backpack sprayers with solid cone or flat fan nozzles are used to apply herbicides to stumps after tree harvest. Phloem transported chemicals (e.g., picloram, glyphosate, and triclopyr) are applied to wounded phloem tissue (inner bark) for absorption and transport to the root system. The freshness of the wound greatly influences the degree of control, and applications should be made minutes or at the most a few hours after cutting. The chemical should be placed completely around the outer edge of the stump, with care taken to wet the entire cambial and inner bark areas. Small stumps may be completely wetted. When the stump was cut many hours prior to treatment, an ester formulation in an oil carrier is more effective. Stump spraying may be done throughout the year, although the most effective time is late summer and the least effective time is during spring sap flow.

B. Prescribed Burning

Prescribed burning is an inexpensive technique commonly used in site preparation to improve mechanical or chemical effectiveness, facilitate planting, and reduce wildfire danger from logging slash. Prescribed burning is also an effective vegetation management tool in established stands of fire resistant conifer species. In the southeastern United States fire is routinely used to reduce hardwood competition in plantations and natural stands. Proper use of fire does not result in soil erosion (Van Lear and Danielovich, 1988) or decrease water quality (Douglas and Van Lear, 1983).

Even though longleaf, slash, loblolly, and shortleaf pines are highly fire resistant, they must have developed the thick bark necessary to insulate the cambium from fire damage as well as be tall enough to prevent crown scorch. Most prescribed burning begins in stands 8 to 10 years old or older with an average stand height of 5 to 7 meters (Wade and Lunsford, 1988). Burning has been done in younger stands, but it requires considerable skill in matching fuel and weather conditions with a suitable ignition strategy.

There can be numerous additional benefits to a prescribed burning program, including the reduction of hazardous fuel loads, improved wildlife habitat, improved access for timber marking or recreation, control of certain diseases, and promotion of fire dependent plant communities. If the primary goal is competition control, the stand should be burned during the summer on a regular basis every 1 to 3 years. Repeated burning will generally control hardwoods less than 5 cm in diameter. When a regular prescribed burning program is employed in conjunction with the judicious use of single stem applications of herbicides, the result can be a very cost effective system of vegetation management to maximize stand productivity. Furthermore, the benefits of this program will carry over into the next rotation through reduced woody plant competition.

V. VEGETATION MANAGEMENT PRESCRIPTIONS

A. Prescription Planning

Prescribing vegetation management practices requires a detailed description of how a specific weed control problem will be dealt with at a particular time. A vegetation management prescription is done on an individual stand level but must always be in concert with the overall management of the forest. Since each management decision may have future consequences, a particular prescription at one point in stand development is likely to affect those that will come afterward. In fact, a prescription should be developed in a manner that anticipates silvicultural practices that may be necessitated by the planned actions. Whenever possible, prescriptions should be integrated with other silvicultural decisions, such as the density of planting, the type of seedlings used, the type of regeneration method (natural versus artificial, hand versus machine planting), and the thinning schedule. Considering the vegetation problems that one encounters, the various alternatives in the management of a timber stand, the variety of site conditions and qualities encountered, and the social, economic, and climatic environments in which a stand may be located, the difficulty in providing a generic approach to prescription of vegetation management practices is obvious. Nevertheless, all thoughtful comprehensively developed prescriptions have a common basis. This common basis encompasses the types of considerations and approaches in the development of the specific recommendations.

In developing a prescription for solving a particular vegetation management problem, a number of decisions need to be made. The importance of each may depend on the management philosophy of the landowner (e.g., some managers believe that extensive forestry is better than intensive, others the reverse;

some stands are surrounded by the same owners on all sides, others are bordered by other landowners, some of whom may be hostile to the management practices applied to the site). However, every prescription should consider the following questions: 1) Where, on the list of management priorities, does this site fall? 2) What is the specific vegetation problem to be addressed? 3) What alternative approaches are available to address the problem? 4) Which species is managed? 5) What existing site conditions may influence selection of alternatives? 6) What weather or climatic conditions may influence implementation of the selected alternative? 7) Are sensitive areas present in the vicinity of the site? 8) What will be the anticipated costs and benefits of the treatment? and 9) What records of application and treatment efficacy will be made?

B. Stand Priority

Funds available for forest management are quite often limiting. Selection of what is done and where it is done is often constrained by available personnel and funding.Thus developing some strategy that allows for efficient distribution of resources available for vegetation management is an important first step in developing a prescription for the stand. Stand priority for a particular treatment will involve a host of considerations including the specific vegetation problems that exist on particular sites; however, certain inherent considerations that are specific to that site or locale need to be considered. The quality of the site is one of the first considerations. Sites with greater inherent productivity (high site index) will justifiably attract greater investment than those of lower productivity. Some locations command greater value for specific products. If a site is near a pulp mill, stumpage values for pulpwood will be inflated by the lower transportation costs and this site will command more value.

In addition to inherent site productivity and market conditions, sites may be prioritized by the problem that is present. For instance, a landowner may place as the highest priority treating all land that is infested with kudzu. Kudzu is a particularly noxious and aggressive plant that becomes progressively more difficult, and more expensive to treat, if left unattended. Some landowners believe that it is important to regenerate an area as soon after harvest as possible; thus all land harvested in the year preceding the planting season would be given the highest priority. Another landowner may wish to hold all areas to be chemically site prepared for at least a year to encourage sprouting of hardwoods and development of sufficient crown area in order to increase treatment efficacy. Sites needing timber stand improvement before harvesting to increase access to the site may be a priority if a landowner has experienced greater stumpage prices resulting from this type of treatment. Thus it is important to develop a sense of priority for particular problems as well as site conditions.

C. Problem Assessment and Site Condition

One of the most important aspects of problem assessment is to describe species composition of the stand in question. Particularly on large tracts, species composition may not be homogenous throughout the stand. Often the vegetation that can be seen from the road is not representative of that within the stand. Windshield surveys may not give the accuracy or precision needed in the vegetation survey. At the same time that the vegetation survey is done, the soil should be sampled. Soil colloids have a large surface to volume ratio and bind with many of the soil-applied herbicides, making soil characterization important. A thorough search for any protected species is required, along with notification of the proper authorities if an endangered species is located in the county where the site is located.

In addition to identifying vegetation and soils the site survey can accomplish several other functions. The topography and drainage characteristics of the site should be analyzed in ways that can assist with a prescription. For instance, if the watershed is steep and drainage patterns lead directly to an adjoining neighbor, the prescription needs to consider the presence of a homesite, garden, pond, or other important features. In addition, areas that can be used for mixing or heliports should be located.

After the site survey is completed, a reconnaissance of all adjacent areas should be done. Notes should be taken regarding the presence of sensitive crops (cotton and tomatoes are extremely sensitive to 2,4-D and similar herbicides), homesites, wells, gardens, and bodies of water (lakes, ponds, and streams).

After conditions of the site and nearby areas are adequately documented, the treatment selection and prescription development begin. The first step in selecting among alternatives is listing the variety of treatments that are both effective and legal for the particular problem on the site. This will vary with the type of problem. For example, the tools used for site preparation and pine release from hardwood competition may be quite different. Alternatives may be restricted by terrain. Steep terrain can preclude mechanical site preparation because of access and environmental considerations. Local ordinances can affect legal alternatives. For instance, one county in Alabama has banned aerial application of forest herbicides. Taking these factors into account, narrowing choices about chemical versus mechanical site preparation can often be accomplished. When site conditions or legal constraints do not favor one or the other, the choice is often based on the availability of equipment and expertise of those involved. Organizations or individuals who have made large capital outlays for mechanical site preparation and machine planting systems may find it economically unfeasible to switch to a chemical site preparation system.

Once the decision has been made to initiate a vegetation control program based on chemicals, then several steps need to be followed to select the most cost effective treatment. First, a product must be specifically labeled for the purpose (i.e., site preparation) and for that state. Rates, crop species, and warnings on the labels need to be adhered to carefully. All of these can eliminate selection of a particular herbicide. The next step in selection involves species sensitivity. A herbicide or mixture of herbicides should be selected to control the major competitors at rates that are within cost constraints. Lastly, selection is based on cost/benefits ratios and risks associated with a particular treatment. Risks can include potential for off site damage as well as variable performance of the product.

After an alternative is selected, the type of application and the specific actions need to be prescribed. Application methods can vary from hand crew to ground equipment to aerial application. Each type of application has its own idiosyncratic details that are important for the treatment to be successful. Several texts provide exceptional detail on the specifics of both ground application (Miller and Mitchell 1988) and aerial applications (Cantrell, 1985). However, it may be important to include some of these considerations into the prescription, particularly if a contractor is going to be hired for the application. Items such as acceptable weather conditions that will be tolerated during application should be detailed and adhered to. If a wind speed of 8 kph is the maximum acceptable and the job has been delayed for 3 days waiting for acceptable conditions, requirements on the fourth day should not be relaxed.

Finally, the prescription should detail the records that will be taken during the application as well as any posttreatment efficacy surveys. An example application record sheet taken from Nelson and Cantrell (1990) is provided (Figure 6). Postapplication surveys generally assess the degree of success, defined by either reduction in competition levels or crop condition (i.e., 350 trees in a "free to grow" class).

VI. FUTURE NEEDS AND CONSIDERATIONS

A. Search for Tools

In one of the first references to vegetation management in forests, Filber Roth, as early as 1902, suggested that forest "weeds" were important constraints to the survival and growth of "useful" trees. Furthermore, he goes on to state, "While we can hardly afford to grub them (weeds) out, yet we shall try to keep them down, but how?" Walstad and Kuch (1987) provide several observations. First, survival and growth may be important yet separate responses to competition. Historically this distinction has caused some confusion in both practice and research regarding the importance of competition on regeneration sites. One may infer, from his statement, the importance of economic thresholds in forest vegetation management. This subject is only recently dealt with in the for-

HERBICIDE APPLICATION RECORDS SHEET

GENERAL

Recorded by: ____________ Date: ____________ County: ________ State: _____
Objective: __________________________ Ownership: ____________________
Location: ____
Site Index: _____ Tract Acreage: _____ Proposed Treatment Acreage: _______
Description of Residual Stand: ______________________________________
Physiographic Region: ____________________________ Tract: _____________

APPLICATION EQUIPMENT

Method (helicopter, injection, etc.): Model/Type: __________
Contractor: ___________________________ Pilot: ____________________
Type of Boom: ___________ Nozzle Tips: _____ Number of Nozzles: ________
Application Height: ___ Application Speed (mph): ___ Boom Pressure (psi):
Direction Spraying Progressed (N-S, etc): ________ Swath Width: _______
Other (injection spacings, etc.): ___________________________________
____

HERBICIDE

Herbicide(s): ____________________________ Rate(s)/Acre: ____________
Total Spray Volume/Acre: ________ Carrier (water, diesel, etc.): ___ , and
Source (creek, pond, etc.): ____________________________
Additive(s) and Rate(s): _______________________________________
____

SOIL CONDITIONS

Series: _________________________ Texture: ______________________
Moisture (wet, moist, dry, very dry): ________Soil Temperature (°F): _____

VEGETATION

Weed Species: ____
Weed Species Condition (drought,stress, fall colors, insect or disease damage, etc.):
____
Crop Species: ____
____
Crop Species Condition (drought, stress, etc.): ________________________

WEATHER CONDITIONS

	Session 1		Session 2		Session 3	
	Start	Finish	Start	Finish	Start	Finish
Time of Day	____	____	____	____	____	____
Relative Humidity (%)	____	____	____	____	____	____
Air Temperature (°F)	____	____	____	____	____	____
Wind Speed (mph)	____	____	____	____	____	____
sind Direction	____	____	____	____	____	____
Cloud Cover (%)	____	____	____	____	____	____
Fog	____	____	____	____	____	____

Dew (none, light, moderate, heavy): ______________ Dry By: ___________ a.m.
Rainfall (date(s) and amount(s)):
--before application: ____
--after application: ____
Comments: ____
____

Figure 6 Sample application record sheet. (From Nelson and Cantrell, 1990.)

estry literature (Wagner et al., 1989). Lastly, Roth asks how noxious plants can be controlled. Certainly, since 1902 tremendous technological advancements have provided a number of alternatives for controlling (or at least inflicting injury upon) unwanted plants in forests, but far too little work has been directed toward how managers can use these tools to achieve objectives better from both an economic and an ecological perspective.

The search for tools has dominated the forestry literature. The underlying philosophy behind this work is that weeds are bad; thus better technology applied to their eradication will inherently improve forest management. With this in mind, Stewart et al. (1984) catalogue more than 260 studies that describe the results of competition in various experimental settings. Yet, with this large body of work it is surprising how little can be generalized in a manner that would assist a forest manager in making quantitative decisions regarding the silvicultural options that are available.

Although this literature has shown that competition from associated vegetation dramatically constrains forest productivity, predicting the degree that a particular stand will respond to various levels of competition control, the consequences of delaying the implementation of control measures, or the potential interactions with other silvicultural practices across a range in site productivity is well beyond the reach of a forest manager. Wagner (1989) suggests that predicting responses to various competitive situations requires that 1) competition is quantified in a manner that is both biologically meaningful (i.e., robust in its prediction of competition effects) and operationally feasible; 2) a quantitative link between the measures of competition and growth of important crop species needs to be defined across a broad range of site qualities and competition intensities; 3) these relationships need to be dynamic with time and stand development (for both the competitors and the crop trees); and 4) a quantitative understanding of how competition is influenced by specific silvicultural practices is attained. Simply improving our ability to predict competitive responses is not sufficient. Improving management and increasing our understanding of the domain in which these predictions are appropriate require a more fundamental and comprehensive view of the mechanism controlling plant-plant interactions.

B. Social and Political

Because of the social and political environment in a number of areas (the Pacific Northwest, Canada, Europe) an active search for herbicide alternatives in vegetation management programs is under way. Although the search for alternatives is important work, an argument should be made that the criteria for assessing alternatives need to be more rigorously and thoughtfully developed. For an alternative to be better there has to be some demonstrable improvement in performance; it is not enough that the alternative merely be different. Performance considerations range from those associated with the vegetation management objectives of the tools to attributes that influence the environment and human health and risks assessment. Improvement in any of these areas is welcome, yet it must be real improvement. Some sectors of society believe that any reduction in the use of chemicals inherently improves management from a human and environmental health perspective. This belief simply cannot be supported with data. The risks of mechanical alternatives should be compared with the risks due to specific herbicide treatments. Reducing risks associated with treatments as benign as some commonly used forestry herbicides such as glyphosate, hexazinone, and imazapyr may be difficult. Often the use of chemical alternatives preserves the physical properties of the soil and reduces the potential for erosion, which can be an important part of sustaining long term productivity. It is within these parameters that we should be searching for alternatives, not simply using perceptions that are without a substantive base.

APPENDIX A

List of Common and Scientific Names of Crops and Weeds Mentioned in Chapter

Common name	Scientific name	BAYER code
American beautyberry	*Callicarpa americana* L.	
Ash	*Fraxinus* spp.	
Beggarweed	*Desmodium tortuosum* DC.	DEDTO
Bermudagrass	*Cynodon dactylon* Pers.	CYNDA
Black cherry	*Prunus serotina* Ehra	
Black locust	*Robinia pseudoacacia* L.	ROBPS
Blackberry	*Rubus* spp.	
Blackgum	*Nyssa sylvatica* Marsh	
Blackjack oak	*Quercus marilandica* Muenchh.	
Broomsedge	*Andropogon virginicus* L.	ANOVI
Buckeye	*Aesculus* spp.	
Carpetweed	*Mollugo verticillata* L.	MOLVE
Clover	*Trifolium* spp.	
Cocklebur	*Xanthium* spp.	
Common purselane	*Portulaca oleracea* L.	POROL
Crabgrass	*Digitaria* spp.	
Crowfootgrass	*Dactyloctenium aegyptium* Willd.	DTTAE
Dogfennel	*Eupatorium capillifolium* Small	EUPCP
Dogwood	*Cornus* spp.	
Eastern red cedar	*Juniperus virginiana* L.	IUPVI
Elm	*Ulmus* spp.	
Fall panicum	*Panicum dichotomiflorum* Michx.	PANDI

Appendix A Continued

Common name	Scientific name	BAYER code
Flathead sedge	*Cyperus compressus* L.	CYPCP
Florida pusley	*Richardia scabra* L.	RCHSC
Goosegrass	*Eleusine indica* Gaertn.	ELEIN
Greenbriar	*Simlax* spp.	
Groundsel	*Senecio* spp.	
Hawthorn	*Crataegus* spp.	
Hickory	*Carya* spp.	
Holly	*Ilex* spp.	
Honeysuckle	*Lonicera japonica* Thunb.	LONJA
Horseweed	*Conyza canadensis* Cronq.	ERICA
Johnsongrass	*Sorghum halepense* Pers.	SORHA
Kudzu	*Pueraria lobata* Ohwi.	PUELO
Lespedeza	*Lespedeza* spp.	
Loblolly pine	*Pinus taeda* L.	
Longleaf pine	*Pinus palustris* Mill.	
Magnolia	*Magnolia grandiflora* L.	
Morningglory	*Ipomoea* spp.	
Multiflora rose	*Rosa multiflora* Thunb. ex Murr.	ROSMU
Nutsedge	*Cyperus* spp.	
Persimmon	*Diospyros virginiana* L.	DOSVI
Post oak	*Quercus stellata* Wangenh.	QUESL
Privet	*Ligustrum* spp.	
Prostrate spurge	*Chamaesyce humistrata* Small	EPHHT
Red oak	*Quercus falcata* Michx.	
Red maple	*Acer rubrum* L.	
Redbud	*Cercis canadensis* L.	
Redroot pigweed	*Amaranthus retroflexus* L.	AMARE
Sassafras	*Sassafras albidum* Nutt.	
Shortleaf pine	*Pinus echinata* Mill.	
Sicklepod	*Cassia obtusifolia* L.	CASOB
Slash pine	*Pinus elliottii* Engelm.	
Sumac	*Rhus* spp.	
Sweetgum	*Liquidambar styraciflua* L.	
Tropical croton	*Croton glandulosus* var. *septentrionalis* Muell.-Arg.	CVNGS
Trumpet creeper	*Campsis radicans* Seem. ex Bureau	CMRIA
Viginia creeper	*Parthenocissus quinquefolia* Planch	PRTQU
Waxmyrtle	*Myrica cerifera* L.	MYRCE
White oak	*Quercus alba* L.	
White snakeroot	*Eupatorium rugosum* Houttuyn.	
Willow	*Salix* spp.	
Yellow poplar	*Liriodendron tulipifera* L.	

APPENDIX B

List of Herbicides Mentioned in Chapter

Common name	Chemical name
Alachlor	2-Chloro-*N*-(2,6-diethylphenyl)-*N*-(methoxymethyl)acetamide
Atrazine	6-Chloro-*N*-ethyl-*N′*-isopropyl-1,3,5-triazine-2,4-diamine
Bentazone	3-(1-Methylethyl)-(1*H*)-2,1,3-benzothiadiazin-4(3*H*)-one-2,2-dioxide
Bifenox	Methyl 5-(2,4-dichlorophenoxy)-2-nitrobenzoate
Dazomet	Tetra-hydro-3,5-dimethyl-2*H*-1,3,5-thiadiazine-2-thione
Dicamba	3,6-Dichloro-2-methoxybenzoic acid
Dichlorprop	(±)-2-(2,4-dichlorophenoxy)propanoic acid
Diphenamid	*N*,*N*-Dimethyl-α-phenylbenzeneacetamide
EPTC	*S*-ethyl dipropyl carbamothioate
Fluazifop-P	(±)-2-[4-[[5-trifluoromethyl)-2-pyridinyl]oxy]phenoxy]propionate
Fomesafen	5-[2-chloro-4-(trifluoromethyl)phenoxy]-*N*-(methylsulfonyl)-2-nitrobenzamide
Fosamine	Ammonium ethyl(aminocarbonyl)phosphanate
Glyphosate	*N*-(Phosphonomethyl)glycine
Hexazinone	3-Cyclohexyl-6-(dimethylamino)-1-methyl-1,3,5-triazine-2,4(1*H*,3*H*)-dione
Imazapyr	(±)-2-[4,5-dihydro-4-methyl-4-(1-methylethyl)-5-oxo-1*H*-imidazol-2-yl]-3-pyridinecarboxylic acid
Lactofen	(±)-2-ethoxy-1-methyl-2-oxoethyl 5-[2-chloro-4-(trifluoromethyl) phenoxy]-2-nitrobenzoate
Methyl bromide	Bromomethane
Metsulfuron-methyl	2-[[[[(4-Methoxy-6-methyl-1,3,5-triazin-2-yl)amino]carbonyl]amino] sulfonyl]benzoic acid
Napropamide	*N*,*N*-Diethyl-2-(1-naphthalenyloxy)propanamide
Nitrofen	2,4-dichloro-1-(4-nitrophenoxy)benzene
Oryzalin	4-(dipropylamino)-3,5-dinitrobenzenesulfonamide
Oxadiazon	3-[2,4-dichloro-5-(1-methylethoxy)phenyl]-5-(1,1-dimethylethyl)-1,3,4-oxadiazol-2(3*H*)-one
Oxyfluorfen	2-Chloro-1-(3-ethoxy-4-nitrophenoxy)-4-(trifluoromethyl)benzene
Perfluidone	1,1,1-Trifluoro-*N*[2-methyl-4-(phenyl-sulfonyl)phenyl] methanesulfonamide
Picloram	4-Amino-3,5,6-trichloro-2-pyridinecarboxylic acid
Prometryn	*N*,*N′*-Bis(1-methylethyl)-6-(methylthio)-1,3,5-triazine-2,4-diamine
Sethoxydim	2-[1-(ethoxyimino)butyl]-5-[2-(ethylthio)propyl]-3-hydroxy-2-cyclohexen-1-one]
Sufometuron	2-[[[[(4,6-Dimethyl-2-pyrimidinyl)amino]carbonyl]amino]sulfonyl] benzoic acid
Toxaphene	Chlorinated camphene (content of combined chlorine, 67-69%)
Triclopyr	[(3,5,6-trichloro-2-pyridinyl)oxy]acetic acid
Trifluralin	2,6-Dinitro-*N*,*N*-dipropyl-4-(trifluoromethyl)benzenamine
2,4-D	(2,4-Dichlorophenoxy)acetic acid

REFERENCES

Abbott, H. G. and Eliason, E. J. (1968). Forestry tree nursery practices in the United States, *J. Forestry 66*: 704.

Abbott, H. G. and Fitch, S. D. (1977). Forest nursery practices in the United States, *J. Forestry 75*: 141.

Aldhous, J. R. (1972). *Nursery practice*, U.K. Forestry Comm. Bulletin No. 43, H.M.S.O., London.

Armson, K.A. and Sadreika, V. (1979). *Forest tree nursery soil management and related practices*, Ontario Ministry of Natural Resources, Toronto.

Barham, R. O. (1980). Handweeding times reduced in hardwood seedbeds by a modified rolling cultivator, *Tree Planters' Notes, 31*: 30.

Beasley, R. S., Granillo, A. B., and Zillmer, V. (1986). Sediment losses from forest management: Mechanical vs. chemical site preparation after clearcutting, *J. Environ. Qual. 15*: 413.

Belli, M. L., Straka, T. J., Dubois, M., and Watson, W. F. (1993). Costs and cost trends for forestry practices in the South, in *Forest Farmer 29th Manual Edition*, Forest Farmers Association, Atlanta, pp. 25–31.

Blake, J. I. and South, D. B. (1987). Weekly herbicide applications prove beneficial in forest nurseries, Agric. Exp. St. of Auburn University, *Highlights of Agricultural Research 34*: 12.

Boyd, R. J. (1971). Effects of soil fumigation on production of conifer nursery stock at two northern rocky mountain nurseries, USDA Forest Service Res. Pap. INT-91.

Boyer, J. N. and South, D. B. (1984). Forest nursery practices in the South, *S. J. Appl. Forestry 8(2)*: 67.

Britt, J. R., Zutter, B. R., Mitchell, R. J., Gjerstad, D. H., and Dickson, J. F. (1990). Influence of herbaceous interference on growth and biomass partitioning in planted loblolly pine (*Pinus taeda*), *Weed Sci. 38*: 497.

Burkhart, H. E., Glover, G. R., and Sprinz, P. T. (1987). Loblolly pine growth and yield response to vegetation management, in *Forest Management for Conifer Production* (J. D. Walstad and P. J. Kuch, eds.), John Wiley & Sons, New York, p. 243–271.

Burkhart, H. E. and Sprinz, P. T. (1984). A model for assessing hardwood competition effects on yields of loblolly pine plantations, School of Forestry and Wildlife Resources, Virginia Polytechnic Institute and State University, Blacksburg, Virginia, Publ. FWS-3-84.

Campbell, S. J. and Kelpasas, B. R. (1988). Comparison of three soil fumigants in a bareroot conifer nursery, *Tree Planters' Notes 39*: 16.

Cantrell, R. L. (1985). *A Guide to Silvicultural Herbicide Use in the Southern United States*, Auburn University School of Forestry, Alabama Agricultural Experiment Station, Auburn.

Chapman, W. A. (1962). *Annual Report of Tree Nursery*, Kimberly-Clark Corporation, Coosa, Alabama.

Cossitt, F. M. (1951). Weeding with oil, *Tree Planters' Notes 4*: 3.

Creighton, J. L., Zutter, B. R., Glover, G. R., and Gjerstad, D. H. (1987). Planted pine growth and survival responses to herbaceous vegetation control treatment duration, and herbicide application technique, *S. J. Appl. Forestry 11*: 223.

Douglas, J. E. and Van Lear, D. H. (1983). Prescribed burning and water quality of ephemeral streams in the Piedmont of South Carolina, *Forest Sci. 29*(1): 181.

Duthie, E. C. (1927). Nursery Weeding: a new method for weeding in the nursery, *Scottish Forestry J. 41*: 52.

Dutton, D. W. (1983). "Chinese weeder geese - do they or do they not weed in the nursery? Proceedings, 1982 Western Nurserymen's Conference," Bedford, Oregon, pp. 44–48.

Foster, A. A. (1961). Control of black root rot of pine seedlings by soil fumigation in the nursery, Georgia Forest Research Council Report No. 8.

Garrett, H. E., Stenberg, R. C., Jr., Cox., G. S., Mitchell, R. J., and Yoder, W. G. (1991). "A case for herbicidal weed control in forest nurseries," Proceedings, Chemical Vegetation Management, Plant Growth Regulator Society of America, San Antonio, Texas, pp. 83–101.

Gill, D. S. and Marks, P. L. (1991). Tree and shrub colonization of old fields in central New York, *Ecol. Monographs 61*: 183.

Glover, G. R., Creighton J., and Gjerstad, D. H. (1989). Herbaceous weeed control increases loblolly pine growth, *J. Forestry 87*: 47.

Glover, G. R. and D. R. Dickens (1985). Impact of competing vegetation on yield of the southern pines, Georgia For. Comm, Ga. For. Res. Pap. 59.

Goldberg, D. E. (1990). Components of resource competition in plant communities, in *Perspectives on Plant Competition*, (J. B. Grace and D. Tilman, eds.), Academic Press Inc., San Diego, California, p. 27.

Green, T. H., Mitchell, R. J., Paliwal, K. C., Pathre, U. V., Zutter, B. R., and Gjerstad, D. H. (1991). Effects of herbaceous weeds on fourth year water relations and gas exchange of loblolly pines, *Weed Techn.* 5: 753–758.

Grierson, D. G. (1989). "Methyl bromide fumigation at the Lone Peak State Forest Nursery, Utah," Proceedings, Intermountain Forest Nursery Association. USDA Forest Service Gen. Tech. Rep. RM-184., Bismarck, North Dakota, pp. 38–39.

Harstela, P. and Tervo, L. (1983). Technology of the production of bare-root seedlings, *Commun. Inst. For. Fenn. 110*: 1.

Hill, J. A. (1955). Methyl bromide gas controls weeds, nematodes, and root rots in seedbeds, *Tree Planters' Notes 21*: 11.

Hodges, C. S., Jr. (1962). Diseases in southeastern forest nurseries and their control, USDA Forest Service Southeastern For. Exp. Sta. Station Pap. 142.

Kennedy, H. E. (1984). Hardwood growth and foliar nutrient concentrations best in clean cultivation treatments, *For. Ecol. Manage. 8*: 117.

Kidd, F. A. (1987). Application and use of herbicides in forest and industrial right-of-way sites, in *Methods of Applying Herbicides*, Monograph No. 4 of the Weed Science Society of America, Champaign, Illinois, pp. 297–312.

Landis, T. D. and Campbell, S. J. (1989). "Soil fumigation in bareroot tree nurseries," Proceedings, Intermountain Forest Nursery Association, USDA Forest Service Gen. Tech Rep. RM-184, pp. 13–28.

Lowery, R. and Gjerstad, D. H. (1991). Chemical and mechanical site preparation, in *Forest Regeneration Manual,* (M. L. Duryea and P. M. Dougherty, eds.), Kluwer Academic Publishers.

Macdonald, J. (1953). Recent developments in nursery practice in Great Britain, *Quarterly J. Forestry 47*: 89.

Mangold, R. D., Moulton, R. J., and Snellgrove, J. D. (1991). Tree planting in the United States 1990, USDA Forest Service, State and Private Forestry, Cooperative Forestry, Washington, D.C.

McKee Jr., W. H. and Wilhite, L. P. (1988). Response of newly planted loblolly pines to herbicide and nitrogen, *S. J. Appl. Forestry* 12: 33.

McKellar, A. D. (1936). The weed problem at the Stuart Forest Nursery, Pollock, La., USDA Forest Service, South. For. Exp. Sta. Occasional Pap. No. 55.

Mikola, P. (1957). Nursery practices in Finland, *Commun. Inst. For. Fenn., 48(1)*: 1.

Miller, J. H. and Mitchell, R. J. (1988). *A Manual on Ground Applications of Forestry Herbicides*, USDA Forest Service, Southern Region, Manage. Bull. R8-MB 21.

Miller, J. H., Zutter, B. R., Zedaker, S. M., Edwards, M. B., Haywood, J. D., and Newbold, R. A. (1991). A regional study on the influence of woody and herbaceous competition on early loblolly pine growth, *S. J. Appl. Forestry 15*(4); 169.

Morris, L. A., Pritchett, W. L., and Swindel, B. F. (1983). Displacement of nutrients into windrows during site preparation of a flatwoods forest, *Soil Sci. Soc. Am. J. 47*: 591.

Morris, L. A. and Lowery, R. F. (1988). Influence of site preparation on soil conditions affecting stand establishment and tree growth, *S. J. Appl. Forestry 12*: 170.

Neary, D. G., Bush, P. B., and Michael, J. L. (1986). "Herbicides in southern forestry - improving water quality," Proceedings of the Southern Weed Science Society, January 1986, Nashville, Tennessee, pp. 335–341.

Nelson, L. R. and Cantrell, R. L. (1990). Herbicide prescription manual for southern pine management, Cooperative Extension Service, Clemson University Publ. EC 659.

Newton, M. and Knight, F. B. (1981). *Handbook of Weed and Insect Control Chemicals for Forest Resource Managers*, Timber Press, Beaverton, Oregon.

Owston, P. W. and Abrahamson, L. P. (1984). Weed management in forest nurseries, in *Forest nursery manual: Production of bareroot seedlings*, (M. L. Duryea and T. D. Landis, eds.), Martinus Nijhoff/Dr. W. Junk Publishers, The Hague/Boston/Lancaster, p. 193.

Scholtes, J. R. (1989). "Soil fumigation at J. Herbert Stone Nursery," Proceedings, Intermountain Forest Nursery Association, USDA Forest Service Gen. Tech. Rep. RM-184, pp. 35–37.

Schubert, G. H. and Adams, R. S. (1971). *Reforestation Practices for Conifers in California*, California Department of Conservation, Division of Forestry, Sacramento, California, p. 359.

Shoulders, E., Hollis, J. P., Merrifield, R. G., Turner, E. E., and Verrall, A. F. (1965). Test of soil fumigants in Louisiana, *Tree Planters' Notes 73*: 14.

Smethurst, P. J. and Nambiar, E. K. S. (1989). Role of weeds in the management of nitrogen in young *Pinus radiata* plantations, *New Forests 3*: 203.

South, D. B. (1982). "Relationship between amount of epicuticular wax and activity of oxyfluorfen on sweetgum leaves," Proceedings, 35th Annual Meeting of the Southern Weed Science Society, Atlanta, p. 245.

South, D. B. (1983). "Oxyfluorfen impregnation on ammonium nitrate for use in forest nurseries," Proceedings, 36th Annual Meeting of the Southern Weed Science Society, Atlanta, p. 259.

South, D. B. (1986). Herbicides for southern pine seedbeds. *S. J. Appl. Forestry 10*(3): 152.

South, D. B. (1987). "A look back at mechanical weed control," Proceedings, Southern Forest Nursery Association, Pensacola, Florida, pp. 97–106.

South, D. B. (1988). Mechanical weed control for the forest nursery, Georgia Forestry Comm. Res. Rep. No. 1.

South, D. B. (1991). "Nursery seed efficiency can affect gains from tree improvement," Proceedings, Southern Forest Nursery Association, Biloxi, Mississippi, pp. 46–53.

South, D. B., Crowley, R. H., and Gjerstad, D. H. (1976). "Herbicide weed control results in pine seedbeds," Proceedings, 1976 Southeastern Area Nurserymen's Conference, USDA Forest Service, pp. 95–106.

South, D. B. and Gjerstad, D. H. (1980). Oxyfluorfen: an effective herbicide for southern pine nurseries, *S. J. Appl. Forestry 4*(1): 36.

South, D. B., Gjerstad, D. H., and Crowley, R. H. (1978). Bifenox: a promising new herbicide for southern pine nurseries, *S. J. Appl. Forestry, 2*(1): 11.

Steven, H. M. (1932). Weeds in forest nurseries, *Forestry 6*: 175.

Stewarts, R. E., Gross, L. L., and Honkala, B. H. (1984). Effects of competing vegetation on forest trees: a bibliography with abstracts, USDA Forest Service Gen. Tech. Rep. WO-43.

Stoeckeler, J. H. (1951). Control of weeds in the nursery by chemicals, *Tree Planters' Notes 7*: 14.

Stringfield, D. (1987). "Weed management incorporated," Proceedings, Southern Forest Nursery Association, Pensacola, Florida, pp. 97–106.

Sutherland, J. R. and Adams, R. E. (1965). Stand, growth, and nitrogen content of red pine seedlings following chemical treatment of the soil to control diseases, *Tree Planters' Notes 73*: 7.

Sutherland, J. R. and van Eerden, E. (1980). Diseases and insect pests in British Columbia forest nurseries, British Columbia Ministry of Forests/Canadian Forestry Service, Joint Rep. No. 12.

Tanaka, Y., Russell, K. W., and Linderman, R. G. (1986). "Fumigation effect on soilborne pathogens, mycorrhizae, and growth of Douglas-fir seedlings," Proceedings, Combined Western Forest Nursery Council and Intermountain Nursery Association Meeting, USDA Forest Service Gen. Tech. Rep. RM-137, Tumwater, Washington, pp. 147–152.

Umland, C. B. (1946). Nursery weeding costs reduced by mechanical cultivation, *J. Forestry 44*: 379.

Van Lear, D. H., and Danielovich, S. J. (1988). Soil movement after broadcast burning in the southern Appalachians, *S. J. Appl. Forestry 12*: 49.

Wade, D. D. and Lunsford, J. D. (1988). *A Guide for Prescribed Fire in Southern Forests*, USDA Forest Service, Southern Region, Tech. Publ. R8-TP 11.

Wagner, R. G. (1989). Interspecific competition indices for vegetation management decisions in young Douglas-fir plantations of the Oregon coastal range, Ph.D. Thesis, Oregon State University, Corvallis, Oregon, 199 pp.

Wagner, R. G., Peterson, T. D., Ross, D. W., and Radosevich, S. R. (1989). Competition thresholds for survival and growth of ponderosa pine seedlings associated with woody and herbaceous vegetation, *New Forests 3*: 151.

Wakeley, P. C. (1935). *Artificial reforestation in the southern pine region*, Government Printing Office, Washington, D.C. USDA Tech. Bull. 429.

Walstad, J. D. and Kuch, P. J. (1987). Introduction to Forest Management, in *Forest Vegetation Management for Conifer Production*, (J. D. Walstad and P. J. Kuch, eds), John Wiley & Sons, New York, pp. 3–14.

Wood, C. W., Mitchell, R. J., Zutter, B. R., and Lin, C. L. (1992). Loblolly pine plant community effects on soil carbon and nitrogen, *Soil Sci. 154*(5): 410–419.

Woodrun, H. G. (1952). Comments on grazing geese, *Tree Planters' Notes 12*: 5.

Wycoff, H. B. (1954). An analysis of nursery stock production costs in relation to the use of mineral spirits, *Tree Planters' Notes 18*: 25.

Zutter, B. R., Nelson, L. R., Minogue, P. J., and Gjerstad, D. H. (1987). Hardwood plantation growth following weed control using herbicides and cultivation, *S. J. Applied Forestry 11*(3): 134.

Herbicide Index

Weed Index

Crop Index

Subject Index

Rs:- 9,6,995